L.C. Catalog Card No. 79-53978
ISBN 0-88318-153-3
DOE CONF- 790347

Meson-Nuclear Physics–197

(Houston)

AIP Conference Proceedings
Series Editor: Hugh C. Wolfe
Number 54

Meson-Nuclear Physics–1979
(Houston)

Editor
E.V. Hungerford III
University of Houston

American Institute of Physics
New York 1979

Proceedings of the

2nd International Topical Conference on

Meson-Nuclear Physics

University of Houston

Houston, Texas

March 5-9, 1979

EDITOR'S FOREWORD

After much discussion, several of my colleagues and I concluded that the field of Meson-Nuclear physics was maturing very rapidly. In fact, it was felt that there was a wealth of new, exciting data, theoretical analyses, and predictions which required exposition and discussion in a topical conference organized along the lines of the previous Carnegie-Mellon Conference. I believe that these conclusions were substantially correct as is evidenced by the quantity and quality of the contributions to the conference. Of course these papers themselves speak to the interest, excitement, and depth of the field better than a brief introduction.

The proceedings are organized into major areas with the contributed papers following the invited presentations. The initial paper in each major area was intended to be an overview of the subject - a paper that was to set the stage for the more detailed descriptions of the specific topics to follow. In general this worked quite well, though it is really for others not quite so close to the conference organization to draw these conclusions. The contributed papers contain excellent physics and it is hoped that by publication in these proceedings they may provide useful reference material.

It is with much appreciation that I thank the Organizing Committee and the International Advisory Committee for their help. The sponsorship by the U. S. Department of Energy, National Science Foundation, and Department of Physics at the University of Houston is also gratefully acknowledged. Finally I must say that without the help of Mrs. Lois Johnson, Mrs. Shirley Double, and Mrs. Mary Hungerford the conference could not have been undertaken. Certainly most of the insoluable problems of conference organization were handled well under their guidance. I express my admiration for their dedicated efforts.

"....Such knowledge is too wonderful for me, it is high, I cannot attain it." Psalms 139:6

SPONSORS

U. S. Department of Energy
U. S. National Science Foundation
Department of Physics, University of Houston

INTERNATIONAL ADVISORY COMMITTEE

A. Arima
E. Boschitz
T. Bressani
G. Brown
K. Crowe
T. De Forest
T. Ericson
H. Feshbach
A. Gal
H. Gerber
J. Hüfner
C. Joseph
T. Kamae
L. Kisslinger
H. Palevsky
G. Phillips
H. Pilkuhn
P. Radvanyi
L. Rosen
J. Sample
R. Siegel
N. Tanner
G. Tibell
A. Yavin

ORGANIZING COMMITTEE

E. Auld
P. Barnes
J. Domingo
E. Gross
E. Hungerford - Chairman
B. Preedom
G. Stephenson
M. Sternheim
C. Wilkin

TABLE OF CONTENTS

I - ELEMENTARY INTERACTIONS

II - MESON PRODUCTION AND ABSORPTION

III - NUCLEAR STRUCTURE WITH MESONS

IV - MESONIC DEGREES OF FREEDOM

V - ELECTROMAGNETIC INTERACTIONS

VI - PION-NUCLEUS SCATTERING I

VII - PION-NUCLEUS SCATTERING II

VIII - Kaon-Nuclear Reactions

IX - NEW FACILITIES

MESON-NUCLEAR PHYSICS 1979: A PERSPECTIVE

J. M. Eisenberg*
Department of Physics and Astronomy, Tel-Aviv University
Tel-Aviv, Israel

ABSTRACT

A survey is made of developments in meson-nuclear physics as they were anticipated at the International Topical Conference on that subject in 1976. The evolution of the intervening nearly three years is sketched. Open questions which may receive discussion, and at least partial resolution, at the present conference are noted.

INTRODUCTION

This work is not to be read as a review - or even as a preview of the conference which it opens - but rather as an effort to achieve a certain perspective on meson-nuclear physics as we enter the conference. It will attempt to search out main lines of new development in the subject, sketching the status of the various topics as of the previous conference in this area in 1976, surveying advances since, and trying to anticipate the questions and issues for which one should remain alert at the present conference. As such, the present effort is more one of the history of science than science itself. Methodologically, it takes as its point of departure various remarks made at the preceding conference[1] in this series which seemed to characterize the state of evolution of a main line of investigation and attempts to follow this line through to expected developments to be reported here. (This approach may at least have the virtue of helping to promote responsibility in public utterances at the present conference.) Due to obvious limitations of time and space, this look to the past and to the future must be restricted to only a few of the topics covered at these conferences and this selection has been made in a somewhat capricious way. The author extends regrets to those whose works or words have had to be curtailed here. He also apologizes profusely for any slights which may be caused by his over-hasty and over-simplified evaluations of the pros and cons of various approaches.

PION-NUCLEUS ELASTIC SCATTERING

The bewildering state produced at the Carnegie-Mellon Conference in 1976 with regard to competing theoretical treatments of pion-nucleus elastic scattering was well characterized there

* Work supported in part by the U.S.-Israel Binational Science Foundation.

ISSN:0094-243X/79/540001-14$1.50

by Tabakin (Ref. 1, p. 53):

> Unfortunately, we now have several "complete dynamical formulations" - each proponent disclaims any ambiguities. I wonder if they agree.

In the intervening years, these several formulations have achieved much fuller development, and we shall hear detailed reports on this here. Since the theoretical claims and counterclaims have not yet all been sorted out (will they be now in Houston?), it seems in order first to reassure the reader that truly impressive strides forward have been taken experimentally since 1976 in the area of pion-nucleus scattering, and these will be reported here by R. Eisenstein, Q. Ingram and others. Moreover, our understanding of the basic πN interaction, needed as input for the πA (pion-nucleus) problem, is also improving, as we shall hear from M. Banerjee, J. Miller and others, thus narrowing some of the range of uncertainties in calculating πA scattering.

The conventional optical potential approach. In one form or another, virtually all of the theoretical approaches make contact with the well-known Watson multiple-scattering theory[2], which is then generalized in one direction or another. The multiple-scattering theory is intended to supply the link between knowledge of the basic πN scattering amplitude and nucleon distribution on the one hand, and πA scattering on the other. ("Pion" here stands for "projectile" whatever it may be.) It is based on the nontrivial supposition that a potential v exists which describes πN interactions, and which may be summed over the nucleons to give the full πA interaction, which then appears in a Lippmann-Schwinger equation as

$$T = \sum_{i=1}^{A} v_i + \sum_{i=1}^{A} v_i \, G_o T, \tag{1}$$

where T is the πA scattering amplitude and G_o is the pion propagator in the presence of nucleon dynamics,

$$G_o = \frac{1}{E_A + \omega - H_A - H_\pi + i\varepsilon}, \qquad \varepsilon \to 0^+, \tag{2}$$

with E_A and ω the initial nucleus and pion energies, and H_A and H_π the corresponding hamiltonians. As is well known, the multiple-scattering series then rearranges Eq. (1) so that the successive scatterings on a given nucleon are presummed,

$$T = \sum_{i=1}^{A} t_i + \sum_{i=1}^{A} \sum_{j\neq i} t_i G_o t_j + \sum_{i=1}^{A} \sum_{j\neq i} \sum_{k\neq j} t_i G_o t_j G_o t_k + \ldots, \tag{3}$$

where propagation is always governed by Eq. (2), so that

$$t_i = v_i + v_i G_o t_i \tag{4}$$

is not the free πN amplitude but rather the amplitude for scattering on the nucleon within the nuclear medium.

When complete sets of nuclear states are introduced between the successive rescatterings in Eq. (3), and we develop a hierarchy of nuclear matrix elements of which the largest are expected to be those taken between nuclear ground states, we obtain an approximate expression for the elastic scattering amplitude

$$T_{oo} \cong (\sum_{i=1}^{A} t_i)_{oo} + (\sum_{i=1}^{A} t_i)_{oo} G_o(0) T_{oo}, \tag{5}$$

where the subscripts indicate elastic nuclear ground state matrix elements and the propagator is taken for H_A referring to the ground state energy. From Eq. (5) we can identify the optical potential V as that interaction which in a single-particle Lippmann-Schwinger equation for the pion produces the elastic πA amplitude T_{oo}, i.e.

$$V = (\sum_{i=1}^{A} t_i)_{oo}, \tag{6}$$

or, for a local amplitude t_i in configuration space,

$$V(\underset{\sim}{r}) = t\rho(\underset{\sim}{r}), \tag{7}$$

where $\rho(\underset{\sim}{r})$ is the nuclear density and we have suppressed distinctions between protons and neutrons. The corrections arising from nonelastic contributions lead to an expansion of this optical potential, with higher-order terms involving two-nucleon, three-nucleon,....correlations, as well as "reflection terms" in which various of the nucleons are revisited. The expression (7), with corrections for important nonlocalities in the πA case, say, has been the basis of most of the traditional work in medium-energy elastic scattering, where it is at least a plausible approximation to ignore the distinction between the t_i in the nuclear medium of Eq. (4) and the free amplitude and one can hope that higher-order terms involving several-nucleon correlations will be small.

In a condensed and oversimplified evaluation of the conventional optical potential approach, we note that in its favor is the fact that the Watson formalism provides a complete, consistent and systematic theory in which the lowest-order result is easily developed and higher-order corrections may be written down relatively straightforwardly and unambiguously. A major deficiency of the approach lies in its restriction to potential models, with no treatment of absorption or emission or any other relativistic effects.

In more modern times, theorists have been greatly concerned with improvements over the simple $V = t\rho$ form of Eq. (7), especially for πA scattering. The improvements in question have attempted to address relativistic features of πA scattering (at the 3,3 resonance the pion lab kinetic energy is 195 MeV, which is considerably greater than its rest mass energy), both of the kinematic variety and those pertaining to the absorption and emission of particles and field theory features. They have also tried to deal more fully with the kinematics and dynamics of the Δ resonance as formed in the nucleus during the πN scattering governed by Eq. (4), as well as taking into account other aspects of nuclear binding effects on t_i. We shall attempt now to sketch the main lines of these concerns.

Absorption and emission features. At the 1976 conference, a plaintive cry was heard from Shcherbakov (Ref. 1, p. 53), who asked

> Is there some hope to introduce absorption processes in the optical model calculations in the near future?

The difficulty is well known: As soon as an effort is made to introduce specifically absorptive features which the pion, as a boson, possesses, one risks double-counting due to the fact that important parts of πN and NN scattering are mediated by processes in which the pion is absorbed and then emitted (or vice versa) on nucleon lines. Thus, the nucleon pole term, uncrossed and crossed, in πN scattering and the one-pion exchange piece of NN scattering are iterations of absorption and emission vertices, as are processes in which a pion is absorbed on one nucleon which then transfers a heavy meson to a second nucleon, the latter then emitting a pion. To treat all of this consistently, so that absorptive processes enter, and when iterated yield the scattering events once and only once, requires a careful, punctilious formalism. This has been provided recently by Mizutani and Koltun[3], the first notices of it already having appeared at Pittsburgh, where Mizutani mentioned with regard to their approach (Ref. 1, p. 174), "We have studied systematically how to avoid the pion overcounting in π-nucleus scattering ... and the conclusion is that overcounting is associated with the πN P_{11} Born term." We can expect to hear more about this work - at least tangentially - in D. Koltun's talk here.

The approach of Mizutani and Koltun attempts to give a unified description, for the A = 2 system, of the channels

$$\pi + d \begin{cases} \to \pi + d \\ \to \pi + N + N \end{cases} , \quad \pi + d \to N + N , \quad N + N \to N + N ,$$

by means of the Hamiltonian

$$H = H_o + V_{NN} + V_{\pi N} + U + U^\dagger, \qquad (8)$$

in which H_o refers to free pion and nucleon propagation, V_{NN} is the nucleon-nucleon forces due to heavy meson exchange, without one-pion exchange, and $V_{\pi N}$ is the πN interaction due to ρ-exchange and the like, without one-nucleon pole terms. The missing parts of these interactions, as well as the absorptive processes, are supplied by U and $U^\dagger$, the absorption and emission vertices.

By using a projection operator formalism, Mizutani and Koltun are able to construct effective interactions, $\mathcal{V}_{NN}$ for the NN system and $\mathcal{V}_{\pi N}$ for the πN system, from which the corresponding amplitudes can be constructed for use in a Watson or Faddeev multiple-scattering series, as in Eq. (3). There appear, as well, specifically absorptive graphs, thus allowing consistent treatment of absorption or emission channels. (Certain πNN three-body graphs, in which, say, a heavy meson may be exchanged between nucleon lines in the midst of crossed-pion emission and reabsorption, also occur and are ignored; this jeopardizes the full treatment of crossing.) In a thumbnail evaluation of this approach, we note that its great advantage is the inclusion of absorption processes without double-counting, while its main drawback is its greater complexity as compared with Watson's approach.

Linearized Chew-Low equations. Amongst the efforts to deal with field theory aspects of πA interactions has been an approach based on the Chew-Low equation, but in linearized form so as to make it more readily applicable to multiple-scattering situations. This approach was still in rather primitive stages in 1976; indeed Tabakin (Ref. 1, p. 45) noted with regard to it

> Also field theory approaches ... have been advocated .... Hopefully we will learn about this problem at this meeting

a hope which we can perhaps echo at the present conference as well, where G. Miller will enlighten us on this subject. Miller's line of development[4] is to start from the Chew-Low equation

$$T_{fi} \cong \frac{V_{fi}}{z} + \sum_n \frac{T^\dagger_{nf} T_{ni}}{z - E^n_A} + (\text{crossing operator}) . \sum_n \frac{T^\dagger_{nf} T_{ni}}{z - E^n_A}, \qquad (9)$$

where $z = E_A + \omega$ is the total pion and nucleus energy (with a small positive imaginary part as required), V_{fi} is, approximately, a sum of single-particle πN interactions, the levels i,n,f refer to initial, intermediate and final nuclear states, and we have indicated the action of the inclusion of full crossing symbolically. (Crossing is here, too, not quite fully respected in the actual, detailed execution.)

The Chew-Low equation is then brought to linear form and the usual multiple-scattering manipulations carry it into the structure of Eq. (3), but for a modified πA amplitude

$$T \Rightarrow \frac{z}{h_o} T \frac{z}{h_o} , \tag{10}$$

where h_o is the free Hamiltonian for pion plus nucleon. The individual πN amplitudes are similarly modified,

$$t \Rightarrow \frac{z}{h_o} t \frac{z}{h_o} , \tag{11}$$

and the propagator goes over to a Klein-Gordon form

$$G_o \Rightarrow \frac{2h_o}{z^2 - h_o^2 + i\varepsilon} . \tag{12}$$

The appearance of the factors z/h_o in the πN amplitude of Eq. (11) act effectively as modifications of the πNN vertex cutoff behavior, and as such lead to a "softening" of the form factor, i.e. to a longer-range cutoff in configuration space. This naturally influences all those features of πA scattering which are sensitive to the cutoff, such as the Kisslinger catastrophe (anomalous behavior at high momenta arising from the $\rho(\underset{\sim}{k}'-\underset{\sim}{k})\underset{\sim}{k}'.\underset{\sim}{k}$ p-wave term in the Kisslinger potential), the Lorentz-Lorenz effect and the size of reflection terms in the multiple-scattering series. As a brief evaluation, we note that this approach has the advantage of including field theory effects and some crossing. It lacks a natural way to include full coupling to other channels, including the absorptive ones, and it is not fully relativistic.

Generalized Blankenbecler-Sugar approaches. A central difficulty in meson-nucleus physics is the need to wed the relativistic features of the meson kinematics and dynamics to the basically nonrelativistic theories of nuclear structure which have thus far been most fully evolved. Tabakin (Ref. 1, p. 54) captured the problem well:

> Even in covariant theories one comes back to non-relativistic wave functions at which stage the benefits of covariance are endangered.

Nonetheless, quite ambitious efforts[5] have been made in recent years to develop covariant theories of pion-nucleus elastic scattering in which the focus is primarily on the relativistic features of the πN and πA scattering amplitudes; these will be discussed at this conference by C. Shakin.

The basic approach here is to develop a covariant, but three-dimensional version of Eq. (5), by selecting a suitable propagator with the minimum requisite analyticity structure. Those working with this approach have applied it with the inclusion of off-shell effects relating to binding and Fermi motion (i.e., they have relaxed the fixed-scatterer assumption) and have also been able to include aspects of true absorption. The virtue of the method is clearly its relativistic character, while its less advantageous features are the by now well-known ones: it does not incorporate true absorption in a very automatic or natural way and it must find a way to mesh with nonrelativistic nuclear wave functions.

The ΔN^{-1} approach. No less than three or four groups have been developing a theory of pion-nucleus elastic scattering based on the dominance of the Δ-resonance in the relevant kinematic region. This feature is introduced into the formalism in a doorway sense, that is one develops the multiple-scattering concept through the creation of a Δ-particle/nucleon-hole (ΔN^{-1} for short) configuration which then undergoes further interaction with the nuclear system. These methods have reached full flower in the years since Pittsburgh, thereby fulfilling a comment of Kisslinger (Ref. 1, p. 162) concerning them:

> If one restricts oneself to Δ-hole states then one can easily [sic] diagonalize potentials given by meson exchange processes .... but it is stlll difficult to carry out accurate dynamic calculations.

The results of the easy, but still a little uncertain, ΔN^{-1} calculating will be discussed here by M. Dillig (work[6] of M. Huber, M. Dillig and others of the Erlangen group), E. Moniz (work[7] of F. Lenz, E. Moniz, and others at M.I.T. and SIN), and W. Weise (work[8] of W. Weise, E. Oset and others at Regensburg); some mention of this problem may also be made by L. Kisslinger (work[9] of the Carnegie-Mellon group).

The basic formalism of the ΔN^{-1} approach relies on the lowest-order optical potential of Eq. (6),

$$\langle \underset{\sim}{k}' | V^{(1)} | \underset{\sim}{k} \rangle = A \langle \underset{\sim}{k}';0 | t(E) | \underset{\sim}{k};0 \rangle$$

$$= A \langle \underset{\sim}{k}';0 | g^{\dagger}_{\Delta\pi N} \frac{1}{D(E-H_{\Delta})} g_{\Delta\pi N} | \underset{\sim}{k};0 \rangle \, , \qquad (13)$$

where $g_{\Delta\pi N}$ is the $\Delta\pi N$ absorption vertex and $D(E-H_{\Delta})$ is the resonance denominator, in which are incorporated isobar propagation and binding effects by means of the ΔN^{-1} hamiltonian

$$H_{\Delta} = T_{\Delta} + V_{\Delta} + H_{A-1} . \qquad (14)$$

This denominator involves, roughly,

$$D(E-H_\Delta) \Rightarrow E + \varepsilon - E_{res} - \frac{(\underset{\sim}{k} + \underset{\sim}{p}_N)^2}{2M^*} + \frac{i\Gamma}{2}, \qquad (15)$$

where E is the total energy, ε is the hole separation energy, E_{res} is the position of the Δ and Γ is its width, and $\underset{\sim}{k}$ and $\underset{\sim}{p}_N$ are pion and nucleon momenta. The Δ kinetic energy in the fourth term on the right-hand side of (15), together with the width or finite Δ lifetime, lead to a nonlocality in the optical potential due to the finite propagation of the Δ after its formation. This nonlocality is measured by

$$(M^*\Gamma)^{-\frac{1}{2}} \sim 0.6 \text{ fm}. \qquad (16)$$

In practice, the optical potential of Eq. (13) is evaluated by diagonalizing the reciprocal of the ΔN^{-1} propagator to produce admixed Δ-hole states which are then used with explicit expressions for the $\Delta\pi N$ vertex,

$$\langle \underset{\sim}{k}' | \mathcal{V}^{(1)}(E) | \underset{\sim}{k} \rangle = \sum_{\substack{NN' \\ \Delta\Delta'}} F^*_{\Delta' N'}(\underset{\sim}{k}') \langle \Delta' N'^{-1} | \frac{1}{D(E-H_\Delta)} | \Delta N^{-1} \rangle F_{\Delta N}(\underset{\sim}{k}), \qquad (17)$$

with

$$F_{\Delta N}(\underset{\sim}{k}) = \langle \Delta | g_{\Delta\pi N} | \underset{\sim}{k}; N \rangle$$

$$= \int \frac{d\underset{\sim}{k}'}{(2\pi)^3} \psi^*_\Delta(\underset{\sim}{k}+\underset{\sim}{k}') \frac{f_{\Delta\pi N}}{m_\pi} h(\kappa^2) \underset{\sim}{\kappa} \cdot \underset{\sim}{S} \, \psi_N(\underset{\sim}{k}'), \qquad (18)$$

where the Δ and nucleon wave functions appear, $f_{\Delta\pi N}$ is the relevant vertex coupling constant with form factor $h(\kappa^2)$, $\underset{\sim}{\kappa}$ is the πN relative momentum and $\underset{\sim}{S}$ is the $N \to \Delta$ transition spin.

The great step forward in this method is its inclusion of the dynamics of the ΔN^{-1}, i.e. of the interacting system of isobar and core nucleus (with A-1 nucleons). For this one pays the inevitable price of relatively time-consuming computations (shortened[7] somewhat by exploiting features of the schematic model which enter in diagonalizing the ΔN^{-1} matrix). The method also does not yet fully take into account relativistic features of the problem and microscopic treatment of true absorption, though the latter have been included more recently[6,8] by calculating the self-energy of the Δ arising from its coupling to N^2N^{-1} true-absorption, continuum channels.

Starting energy. Before leaving the formal issues surrounding the use of the multiple-scattering and optical potential formalisms, we note that Tabakin, in his remarks at Pittsburgh, stressed at some length the role of the choice of the energy parameter E, taken as $E_A+\omega$ in Eqs. (2), (4) and (6), noting that (Ref. 1, p. 41)

> Variations of 20 to 30 MeV due to binding and other effects can make a considerable difference since t(E) resonates.

The matter of the optimal choice for the starting energy has continued to be debated since 1976, with those[10] who favor the energy dictated by the three-body formalism and others[11] who attempt to choose this parameter so as to minimize the second-order binding correction.

Hadronic matter distribution in nuclei. With regard to the very basic goal of medium-energy physics having to do with probing the strongly-interacting matter distribution in nuclei, Koltun remarked at Pittsburgh (Ref. 1, p. 8)

> Questions to which we shall not have new answers given at this conference:
> "How can we measure the difference between the distribution of neutrons and that of protons in nuclei?"
> "How can we demonstrate the presence of short-range correlations between nucleons in nuclei?"

On both of these questions there has been some progress in the nearly three years since Pittsburgh. The developments have been more clear cut on the neutron distribution issue than on short-range correlations. On the former, careful analysis of proton-nucleus data has begun to pin down the neutron radii much better than hitherto, though the pion seems to remain an ambiguous tool for this purpose[12] with some hope held out in the anlaysis of π-atoms (talk of Friedman). As regards the latter, effort[13] has been invested in studying the effects of short-range correlations in πA - or more precisely π-^{4}He - scattering though the results are not yet fully persuasive.

In other aspects of πA elastic scattering, various open questions have lingered which may be addressed here, such as ambiguities in the low-energy πA scattering optical potential parameters (H. McManus), continued doubts about the cutoff in the very influential πNN vertex function, and Coulomb effects in πA scattering (C. Wilkin).

PION-NUCLEUS INELASTIC REACTIONS

True absorption. The inelastic pion-nucleus reaction which in many ways dominated our thinking between 1976 and 1979 was that of true absorption, especially with reference to the effects which the true absorption channel is likely to have on elastic scattering. The concern was voiced clearly at the Pittsburgh conference by Koltun (Ref. 1, p. 8)

> Absorption is a rather probable reaction, yet we have few measurements and no real theory .... yet it must have an effect on elastic scattering.

This concern was echoed by Lenz (Ref. 1, p. 415)

> To resolve this problem of the physical origin of these damping mechanisms [in the elastic channel] more experimental data specifying the major contributions to the reactive cross section are needed

and by others. The effects of the true absorption channel on elastic scattering will no doubt be touched upon by the speakers noted in the previous subsection. The experimental situation has been improved by measurements[14] of $\pi^{\pm}$ absorption cross sections for a wide range of energies all through the resonance region and for a large number of nuclei ranging from lithium to bismuth. These find that, for all the energies and nuclei, the shape of the summed, inclusive scattering cross section for angles greater than 70° follows that of the πN cross section averaged over protons and neutrons. Moreover, when the effective number of neutrons (for π^-) or protons (for π^+) is extracted by dividing these two quantities a dependence of roughly $(N \text{ or } Z)^{0.4}$ is found near resonance, while the true absorption cross section varies roughly as $A^{0.7}$, (and the elastic cross section - from other sources - as $\sim A^{0.7}$). The true absorption dominates for heavier nuclei, being some 70% of the reactive channel for bismuth, but for light nuclei contributes equally with the inelastic scatterings. The variation of the total cross section according to $\sim A^{0.65}$ thus seems to be the resultant of different A-dependence in the different components. More will be said on meson absorption by D. Koltun at this conference.

<u>Low-lying excitations</u>. A memorable moment in the Pittsburgh conference came when some skepticism was voiced as to the possibility of obtaining precise data with good statistics for pion inelastic scattering to low-lying nuclear states. The issue was joined when new and impressive excitation curves were shown (Ref. 1, p. 681):

<u>Arvieux</u>: I would like to announce that we have very recently taken data for $^{12}C(\pi,\pi')^{12}C^*$....

<u>Schiffer</u>: How long did it take to get that spectrum?

<u>Arvieux</u>: About one hour.

In the meanwhile, the method has proved itself and spectra are becoming readily available for many nuclei and many energies; we shall be hearing about this during the conference from F. Bertrand, G. Walker and others.

Pion charge exchange scattering. Pion double charge exchange scattering has for a long time been billed as one of the unique and interesting features of the pion probe. Since Pittsburgh, data[15] have become available on this reaction and have proved puzzling and challenging to theory. In fact, with regard to double charge exchange, there took place at Pittsburgh a dramatic exchange (Ref. 1, p. 692):

> Negele: I believe that following M. Goldhaber's example with the antiproton the real test of one's belief in a theory should be a willingness to make a public wager about the outcome of theoretical predictions.
>
> Miller: I'm ready.

Fortunately, to the best of the knowledge of the present author, stakes were never set. The data came in and almost all theoretical work was shown to be inadequate; we will hear more about this from some of our speakers.

Meanwhile, even the single charge exchange process has remained enigmatic[16], in common perhaps with[17] (π,γ) showing too much absorption arising from the optical potential. It was suggested at Pittsburgh that new data, obtained by methods other than activation, might help to clarify this difficulty (Koltun, Ref. 1, p. 6):

> There is the exciting possibility of future measurements of the π^o in charge exchange ... [using] the proposed π^o spectrometer .....

With luck, preliminary data from this device should at least form the subject of corridor rumors at our present conference.

Meson production and absorption. The years from 1976 to 1979 have seen a sizable increase in angular distributions for (p,π) reactions and remarkable and puzzling data[18] on pion asymmetries in polarized beam $(\vec{p},\pi)$ experiments. Various aspects of these reactions will be dealt with in the talks of G. Jones and others, while I. Duck will discuss the fundamental $NN \rightarrow NN\pi$ reaction. G. Bertsch will deal with pion production in heavy-ion collisions, where there are indications[19] recently of an analog of the Hanbury-Brown and Twiss effect, and photopion production will also be reviewed at the conference.

KAON-NUCLEUS INTERACTIONS

Striking new data on the (K^-,π^-) reaction have been obtained since Pittsburgh and will be reported on here by M. May. The implications of these for the still-unknown ΛN spin-orbit force will no doubt be explored in the talks of R. Dalitz and A. Gal. This was noted as an unresolved and difficult question in 1976 by Giai (Ref. 1, p. 517)

> The problems of the spin-orbit potential and effective mass of a Λ inside a nucleus are also open

and by Negele (Ref. 1, p. 693)

> [Regarding extracting the ΛN spin-orbit force,] how can you possibly expect to do better in hypernuclei than has been done in ordinary nuclei?

Indeed the problem has remained vexing, with even the new data[20] not completely resolving the issue, while the indications[21] are that theory may tend to favor a ΛN spin-orbit force about 1/3 the magnitude of the NN one and with the same sign.

The K-nucleus scattering problem will be discussed here by C. Dover and others. The kaon gives indications of being an interesting probe, which makes it gratifying that thought is being given to building kaon factories at some time in the future, as we shall hear from D. Nagle in the context of new experimental facilities. Lastly, B. Gibson will discuss problems of kaon interactions in light nuclei and S. Wycech will deal with kaonic and Σ-atoms. The possibility of ΣN and ΣA bound states will be touched upon in both these talks, as well as in the remarks of Dalitz and of May. Such states may show up in Σ^--atoms or in (K^-,π^+) reactions, and comparison between these will be in order; Wycech, in particular, has attempted to examine this situation[22].

PION STRONGLY-BOUND STATES, CONDENSATES, NON-NUCLEON DEGREES OF FREEDOM

At Pittsburgh, the topic of pion strongly-bound states[23] had not yet been born, while pion condensates received little treatment. The first of these refers to the intriguing possibility that the p-wave, momentum-dependent part of the πA optical potential may be sufficiently strong to produce hadronically-bound states of pions, extending over nuclear dimensions, which however would appear to be exceedingly broad in energy. This will discussed further here by E. Friedman. The speculation[24] on possible pion condensates - perhaps even in convential nuclei - receives at least some experimental probing in recent efforts[25] to detect pion exchange currents, and a possible excess of pions, in nuclear $(\pi,2\gamma)$ reactions, to be discussed here by J. Deutsch. Other aspects of non-nuclear degrees of freedom in nuclei will be dealt with by L. Kisslinger, who speaks on isobars in nuclear ground state wave functions, and by E. Lomon, who will address questions on mesonic exchange currents.

FUTURE EXPERIMENTAL FACILITIES

The truly dramatic improvement in meson-nucleus data between 1976 and 1979 sharpens our appetites for even more remarkable developments on the experimental front in the future -- to be previewed here by M. Craddock, J. Domingo, D. Nagle and R. Siegel. It seems singularly superfluous for me to try to preview these previews.

ACKNOWLEDGMENTS

It is a pleasure to thank Messrs. W. Gibbs, B. F. Gibson, E. V. Hungerford and G. J. Stephenson for valuable suggestions at the inception of the planning for this talk which helped to determine its form. Likewise, Mr. A. Gal gave very useful advice during its development, and the speakers at the conference were very generous in making available preliminary information on their anticipated remarks here.

REFERENCES

1. Meson-Nuclear Physics - 1976, P. D. Barnes, R. A. Eisenstein and L. S. Kisslinger, eds. (American Institute of Physics, New York, 1976).
2. Collision Theory, M. L. Goldberger and K. M. Watson (Wiley, New York, 1964). For a review of meson-nucleus applications see J. M. Eisenberg, lecture at the NATO Advanced Seminar Institute, Banff, August, 1978, and references therein.
3. T. Mizutani and D. S. Koltun, Ann. Phys. (N.Y.) 109, 1 (1977). See also A. S. Rinat, Nucl. Phys. A287, 399 (1977), A. W. Thomas, doctoral thesis, Flinders University, 1973. A. W. Thomas and A. S. Rinat, preprint, Weizmann Institute of Science, 1979, and A. S. Rinat, preprint, Weizmann Institute of Science, 1979. Note that no effort at exhaustive bibliography has been made in the present sketch since this is expected to be supplied by the talks at the conference.
4. G. A. Miller, Phys. Rev. C16, 2325 (1977). See also R. L. Ingraham, Ann. Phys. (N.Y.) 99, 231 (1976), J. B. Cammarata and M. K. Banerjee, Phys. Rev. C12, 1595 (1975) and C. B. Dover and R. H. Lemmer, Phys. Rev. C7, 2312 (1973).
5. See for example, L. S. Celenza, M. K. Liou, L. C. Liu and C. M. Shakin, Phys. Rev. C10, 435 (1974); L. S. Celenza, L. C. Liu and C. M. Shakin, Phys. Rev. C13, 2451 (1976), L. C. Liu and C. M. Shakin, Phys. Rev. C16, 1963 (1977), and references therein. Other viewpoints are given by L. Heller, Ref. 1, p. 93, and A. W. Thomas, Ref. 1, p. 375.

6. M. G. Huber and K. Klingenbeck, lecture at the Mainz-Darmstadt seminar on Nuclear Physics with Electron Accelerators, January, 1978; K. Klingenbeck, M. Dillig and M. G. Huber, Phys. Rev. Lett. 41, 387 (1978), and references therein.
7. F. Lenz, in Seventh International Conference on High Energy Physics and Nuclear Structure, Zurich, August, 1977, M. P. Locher, ed. (Birkhäuser, Basel, 1977) p. 175; E. J. Moniz, lecture at the NATO Advanced Summer Institute, Madison, Wisconsin, June, 1978; E. J. Moniz, Ref. 1, p. 105, and references therein.
8. E. Oset and W. Weise, Phys. Lett. 77B, 159 (1978) and preprint, University of Regensburg, 1978.
9. L. S. Kisslinger, Ref. 1, p. 159.
10. R. H. Landau and A. W. Thomas, Nucl. Phys. A, to be published. See also A. W. Thomas, lecture at the NATO Advanced Summer Institute, Madison, Wisconsin, June 1978, as well as Ref. 5.
11. S. A. Gurvitz, J.-P. Dedonder and R. D. Amado, Phys. Rev. C19, 142 (1979).
12. M. M. Sternheim and K.-B. Yoo, Phys. Rev. Lett. 41, 1781 (1978).
13. T.-S. H. Lee and S. Chakravarti, Phys. Rev. C16, 273 (1977).
14. I. Navon et al., preprint, Tel-Aviv University, 1978 and contributed paper to this conference; D. Ashery, G. Azuelos and I. Navon, private communications.
15. R. L. Burman et al., Phys. Rev. C17, 1774 (1978).
16. J. Warszawski, A. Gal and J. M. Eisenberg, Nucl. Phys. A294, 321 (1978).
17. I. Blomqvist et al., Phys. Rev. C15, 988 (1977).
18. E. G. Auld et al., Phys. Rev. Lett. 41, 462 (1978). See also H. J. Weber and J. M. Eisenberg, Nucl. Phys. A312, 201 (1978).
19. S. Y. Fung et al., Phys. Rev. Lett. 41, 1592 (1978).
20. W. Brückner et al., Phys. Lett. 79B, 157 (1978).
21. C. B. Dover and A. Gal private communication.
22. See also S. Wycech, W. Stepien-Rudzka and J. R. Rook, preprint, Institute for Nuclear Research, Warsaw, 1978.
23. T. E. O. Ericson and F. Myhrer, Phys. Lett. 74B, 163 (1978); E. Friedman, A. Gal and V. B. Mandelzweig, Phy. Rev. Lett. 41, 794 (1978); V. B. Mandelzweig, A. Gal and E. Friedman, preprint, Hebrew University, 1979. For the πNN system see T. Ueda, Phys. Lett. 74B, 123 (1978) and, for the possibility of a stable π^-nn bound state, G. Kalbermann and J. M. Eisenberg, J. Phys. G 5, 35 (1979), and W. A. Gale and I. M. Duck, Nucl. Phys. B8, 109 (1968).
24. See for example, the review of A. B. Migdal, Sov. Phys. Usp. 20(11), 879 (1977).
25. J. Deutsch et al., Phys. Lett. 80B, 347 (1979).

THE PRESENT STATUS OF THE PION NUCLEON INTERACTION AT LOW ENERGY

J. Miller
DPh-N/HE, CEN Saclay, BP 2, 91190 Gif-sur-Yvette, France

ABSTRACT

This overview covers the main features of the pion nucleon interaction below 300 MeV. It deals with experimental results and phase shift analyses, Coulomb corrections and the various theoretical approaches to the pion nucleon problem. The results on the values of low energy-parameters such as S- and P-wave scattering lengths, σ-commutator are also discussed. The connection between off-shell and on-shell amplitudes is mentioned.

INTRODUCTION

The subject of this conference, Meson Nuclear Physics, is strongly connected with the fundamental Pion Nucleon Interaction which is the basic ingredient for any description of the pion-nucleus interaction. Because the relatively low density of the nucleons inside the nucleus, compared to the range of the pion-nucleon interaction (< 0.5 fm), the π-nucleus amplitude $T_{\pi A}$ can be expressed in terms of the π-nucleon amplitudes ($t_{\pi N}$) plus corrections from the nuclear medium (nuclear Hamiltonian H_N) : $T_{\pi A} = T(t_{\pi N}, H_N)$, by a kind of multiple scattering theory. That involves some terms corresponding, very schematically, to diagrams like those in Fig. 1. The first order diagrams correspond, in the impulse approximation, to the π-N on-shell amplitude $< \vec{k}' | t(E) | \vec{k} >$, with $\vec{k}'^2 = \vec{k}^2$ $E = (M_N^2 + k^2)^{1/2} + (m_\pi^2 + k^2)^{1/2}$, directly available from measurements of $(\pi^\pm, p)$ scattering. In the second (and higher) order diagrams the amplitudes, with $\vec{k}'^2 \neq \vec{k}^2$, are off-shell. These are not available from the experiments without a fundamental theory of the pion-nucleon dynamics. Unfortunately we do not know the strong interaction dynamics. The only way to proceed is to take more or less plausible phenomenological models for extrapolating the amplitudes from the on-shell data to their half- or fully-off-shell values.

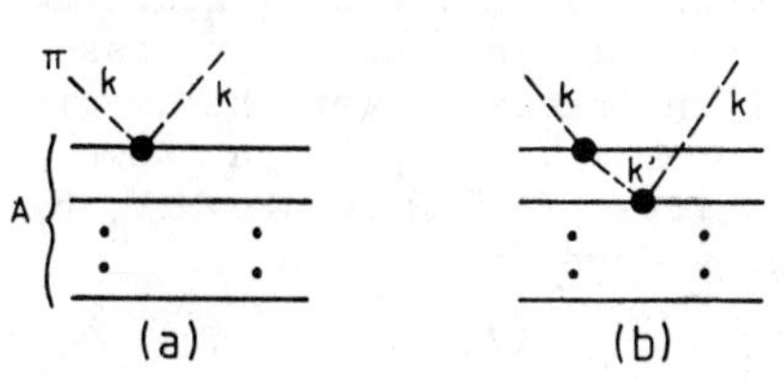

Fig. 1 - First (a) and second (b) order π-nucleus diagrams.

I will try to give you an overview of the present status of our knowledge of the pion nucleon interaction. I will limit this overview to the low energy region, below 300 MeV, where only S- and P-partial waves are involved. I will begin with the "experimental values" and their analysis, including Coulomb corrections. You will see why I put the word "experimental" in quotation marks. I will summarize very briefly the different theoretical approaches. At the end I will com-

ISSN:0094-243X/79/540015-15$1.50

pare the various results obtained for some of the fundamental parameters of the pion-nucleon interaction. I will mention ways that were used to derive the off-shell amplitudes

EXPERIMENTAL DATA AND THEIR ANALYSES

a) Experimental data and phase-shift analysis

Any discussion of the pion-nucleon interaction must start with the experimental values of the on-shell scattering amplitudes. Many measurements of (π^+p) and (π^-p) scattering cross-sections have been performed over the years, near the Δ_{33} resonance ($\sim$ 197 MeV). The major experiments in the resonance region have been reviewed by Kelly[1], and a list of the most important low energy data can be found in the paper of E. Pietarinen[2] at the last Zurich Conference (1977). The standard way to discuss the scattering amplitudes is in terms of a partial wave expansion. The partial wave amplitude, for each value of the total angular momentum $J = \ell \pm 1/2$ and (due to charge independence) for each value of the isospin I (1/2 or 3/2), is given in terms of a phase-shift δ_{2I2J} :

$$f_{2I,2J} \approx \frac{1}{k} e^{i\delta_{2I2J}} \sin\delta_{2I2J}.$$

The phase-shifts are extracted from the cross-section data by two general types of analysis. One is an energy independent fit where data are analysed separately at each energy e.g. CBC[3]). The other is an energy dependent fit where a certain, reasonably smooth dependence of the phase-shift is assumed e.g. CERN[4]). The five most frequently used, phase-shift sets are given in the table I in the review of Pietarinen[2]. All of these analyses used dispersion relations which express the real part of the amplitude in term of an integral over the imaginary part. In a different way, by using an empirical approach, Rowe and Salomon[5] fit the experimental phase shifts, from 0 to 400 MeV, by a very simple parametrization of the energy dependence :

$$\frac{\operatorname{tg} \delta_p}{q^{2\ell+1}} = a + bq^2 + cq^4 + \frac{x\,\Gamma_o\,\omega_o\,q_o^{-(2\ell+1)}}{\omega_o^2 - \omega^2}$$

where q and ω are the c.m. momentum and energy : x, Γ_o, q_o and ω_o are the resonance parameters for the particular partial wave as given in tables[6] ; a,b,c are free parameters to be determined by fitting to the phase-shifts. Less ambitious than the dispersion relation fits, such a parametrization has the advantage of being nearly free of theoretical inputs.

At this point I would like to emphazise the importance of the choice of the available π-N phase-shift fits in the π-nuclear problem. Fig. 2 shows an analysis by Landau and Thomas[7] of a ($\pi^\pm$, ^{4}He) elastic scattering experiment at 51 MeV. The two upper curves are obtained in the impulse approximation by using the on-shell π-N amplitudes given by two different phase-shift fits : the 1968 CERN tabulation which

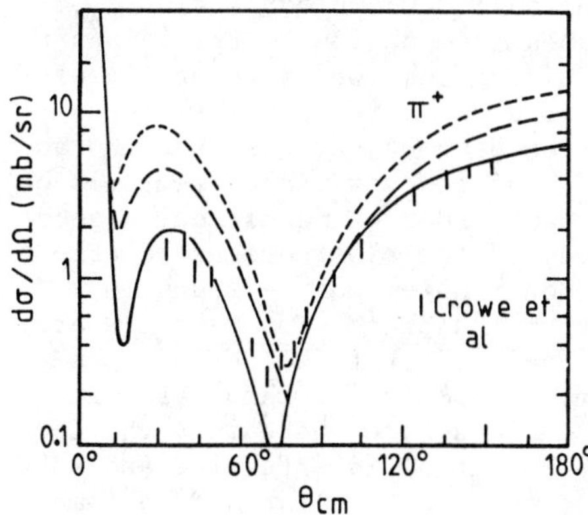

Fig. 2 - π^+, ^4He scattering at 51 MeV, analysed by Landau[7]. Dashed curves are first order approximation with two different π-nucleon phase shift fits : CERN 68 (short dashes), Salomon (long dashes). Solid curve is the final Landau result.

gives an extrapolation at low energy from data available only at higher energy (short dashes) ; the 1974 Salomon parametrization which includes the low energy data of Bussey[8] (long dashes). By using new and better low energy data, still in the first order approximation, one finds that the predicted cross-sections, at some angles, change by as much as a factor of two, coming nearer the experimental values. Assuming the correctness of these last on-shell amplitudes, quite good agreement with the data is obtained by second order calculations (solid curve). One could raise the question : is it relevant to make such second order calculations if the first order might change again with new and better very low energy π-nucleon data? In fact the nuclear theoreticians ask for these new data.

The most recent and accurate low energy measurements are, beyond those of Bussey[8] (1973) (which cover the range 90 to 300 MeV in π^+ and in π^-), the 1976 data of Saclay[9] in π^+ (i.e. only for the 3/2 isospin) from 20 to 100 MeV. Portions of these results have been confirmed at 47.9 MeV[10] and 49.9[11]. Some measurements have just been completed at LAMPF (π^+,π^-) from 30 to 90 MeV[12] and at Saclay from 20 to 70. Their analyses are in progress : the expected accuracy would be better than 5 % in the cross-sections. One of the greatest problems in making measurements of such accuracy is the beam monitoring.

b) Electromagnetic corrections

What I call "experimental" values for the amplitudes or the phase shifts are not really measured values. To extract the true nuclear phase shift from these data, it is necessary to take account of the Coulomb interaction between the pion and the proton. This is not a simple problem. The treatment of the Coulomb corrections to the scattering of strong interacting particles is not yet an entirely resolved matter. A relatively trivial correction comes from the long distance part of the Coulomb potential. It introduces a Coulomb phase shift σ_ℓ which is fully calculable. The ℓ-wave amplitude, when the pure Coulomb part is removed, becomes

$$f_\ell = \frac{1}{k}\, e^{2i\sigma_\ell}\, e^{i\delta'_\ell} \sin\delta'_\ell$$

where δ'_ℓ, the pseudo-nuclear measured phase-shift, is equal to the pure nuclear phase-shift δ_ℓ plus a residual Coulomb correction c_ℓ, i.e. $\delta_\ell = \delta'_\ell - c_\ell$. The quantity c_ℓ comes from photon contributions to the pion-nucleon diagrams (Fig. 3). This correction is generally of the order of a few percent. The situation becomes serious when the accuracy of the new measurements is of the same order of magnitude. A good survey of the electromagnetic correction problem can be found in a paper by Rasche[13]. Two technics are employed to try to solve this problem. The potential theories[14] consider regions, one where only the Coulomb potential is effective and some where the two, nuclear and Coulomb, potentials are more or less involved. They need some models for the potentials, point or extended charge distributions. The other method tries to put the theory of the electromagnetic effects in a more satisfactory basis by using a dispersion approach. Other difficulties remain the $n\gamma$ channel contribution and the mass difference of the neutral and charged pion, neutron and proton, etc. However using this technic Tromborg and al.[15] recently published tables giving the c_ℓ term for S- and P-waves from 20 to 500 MeV, in π^+ and π^-.

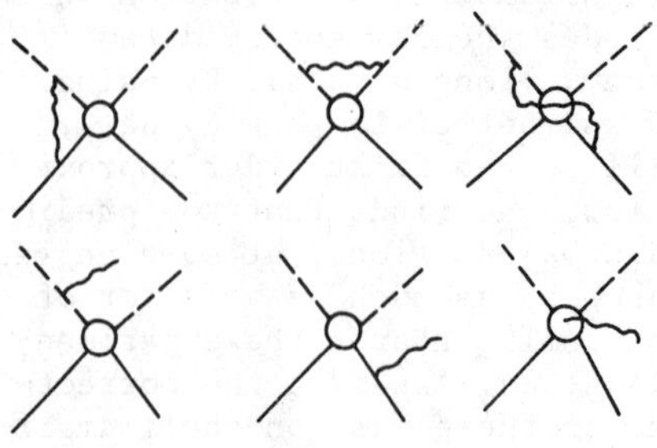

Fig. 3 - Diagrams contributing to the electromagnetic corrections.

At this point one has to be very careful with the so-called "experimental values" for the phase-shifts and for the extrapolated scattering lengths etc. With accurate measurements it will be necessary to have very accurate estimates of the Coulomb corrections, or to increase the systematic error bars in order to take account of these unknown corrections.

The above phase-shift fits, corrected as well as possible for the electromagnetic effects, are the experimental material that the theoretician will utilize to try to understand, or less ambitiously to describe, the pion-nucleon interaction.

c) Invariant amplitudes and low energy parameters

Let me remind you the quantities generally used in calculations and comparisons with the experiments. In a Lorentz invariant formalism the canonical form of the transition T-matrix for the process $\pi(q_1) + N(p_1) \to \pi(q_2) + N(p_2)$ (p_i, q_i are four momenta) is $T = - A + i/2\ \gamma_\mu(q_1 + q_2)^\mu B$. A and B are scalar functions of two of the Mandelstam (Lorentz invariant) kinematical variables[31] : $s = -(q_1 + p_1)^2$, $t = -(q_1 - q_2)^2$, $u = -(q_1 - p_2)^2$. Most often one chooses t (the negative of the square of the four momentum transfer) and $\nu = -(s-u)/4M$ (which is $-(p_1 + p_2)(q_1 + q_2)/4M$). In the interchange of π^+ and π^- (crossing symmetry, $q_i \to -q_i$, $s \leftrightarrow u$) $\nu \to -\nu$. One generally considers the amplitudes $T^\pm = \pm 1/2\,(T_{\pi^-} \pm T_{\pi^+})$ and the corresponding quantities $A^\pm$, $B^\pm$ or combinations of them. By crossing symmetry $T^\pm(\nu,t) = \pm T^\pm(-\nu,t)$. In the ν,t plane (Fig. 4) the physical

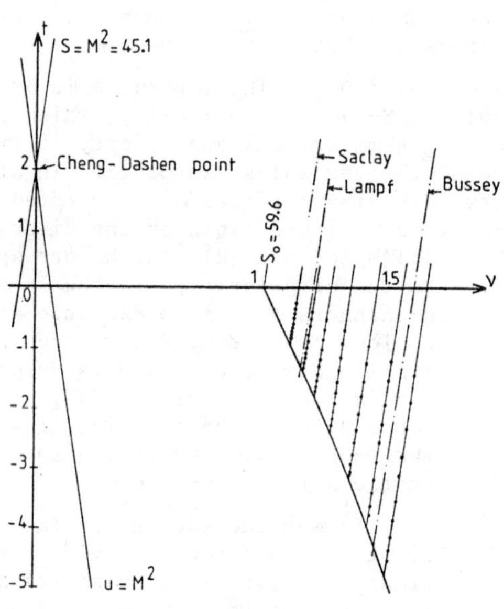

Fig. 4 - ν,t plane (units = ν/m_π t/m_π^2 s/m_π^2). The points correspond to the Saclay measurements.

region (where we are able to measure something) corresponds to $t \leq 0$, $|\cos\theta| < 1$. The experimental amplitudes at a given energy correspond for each angle to a couple of ν and t on the parallel lines s (total c.m. energy). One aim of the theory will be to determine the analytical form of the amplitude in the physical region and to continue it anywhere in the ν,t plane.

Some points of this plane have a particular interest. By extrapolation of the amplitudes at the physical threshold ($\nu = m_\pi$, $t = 0$) you can obtain the various partial waves scattering lengths $a_{2I,2J}$ ($\lim_{q\to 0} \delta_{2I,2J}/q^{2\ell+1}$). The continuation of the $A^+(\nu,t)$ amplitude to the so-called Cheng-Dashen[16] point ($\nu = 0$, $t = 2\,m_\pi^2$) gives the σ-commutator, which measures the breaking of the chiral SU2 ⊗ SU2 symmetry (σ would be zero for the perfect symmetry which includes the conservation of the vector- and axial vector currents) : $A^+(0,2m_\pi^2) = G^2/M_N + \sigma_{NN}/f_\pi^2 + O(m_\pi^4)$ (f_π is the pion decay constant, and G is the pion-nucleon coupling constant). The values of the $A^\pm$ or $B^\pm$ amplitudes around the unphysical crossing symmetric point ($\nu = 0$, $t = 0$) were expanded in powers of ν and t by various authors[17]. Then the coefficients that they obtained (so-called "crossing symmetric expansion coefficients") were considered as parameters to test or to fit the various models.

It is clear that lower the energy of the known, measured amplitudes, the easier and more accurate will be their extrapolations. Fig. 4 shows the limit reached by the more recent low energy measurements.

PION-NUCLEON THEORIES

Various theoretical approaches are used to describe the pion-nucleon interaction or, at least to extract information from the experimental data. I will consider : i) S-matrix theory and dispersion relation techniques ; ii) current algebra models and iii) Lagrangian theories. This classification is a little bit arbitrary, all of these

methods being overlapping. All of them have field theory bases. All use underlying Lagrangian : pseudoscalar ($\mathcal{L}_{ps} = i\,G\,\overline{N}\gamma_5\,N\pi$ where π is the pion field) or pseudovector ($\mathcal{L}_{pv} = i\,f/m_\pi\,\overline{N}\gamma_5\gamma_\mu N\partial^\mu\pi$). In the lowest order of the perturbation expansion and in the non relativistic limit they are equivalent provided that $G/\sqrt{4\pi} = 2M_N\,f/m_\pi$.

i) S-matrix theory and dispersion relations.- This approach is based on the fundamental properties of the S-matrix : unitarity, which gives $T^*T = -(i/2)(T^* - T)$, crossing symmetry and analyticity of the amplitudes in the complex plane of the variables except for singularities required by the unitarity. The singularities are connected to the diagrams of Fig. 5. The P-wave is well described by the diagrams (a)(b)(c) and (d), which correspond to the singularities in the s- and u-channels : nucleon exchange and nucleon pole term, Δ exchange and Δ pole term. The S-wave is described by the diagrams (e)(f)(g) corresponding to the singularities in the t-channel. They give also corrections to the p-wave.

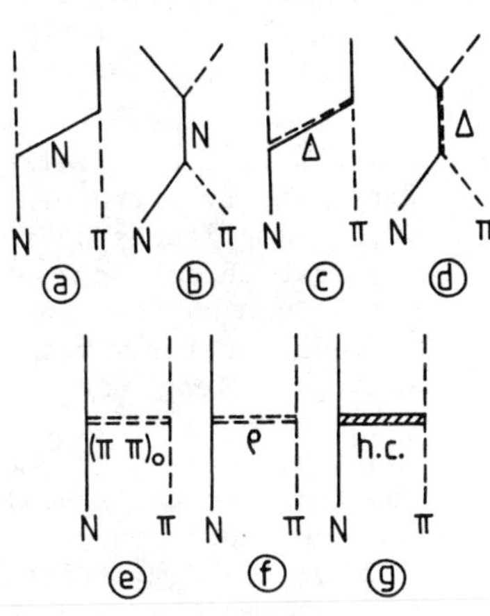

Fig. 5 - The dominant diagrams for P-wave (a),(b),(c),(d) and S-wave (e),(f),(g), low energy pion-nucleon interaction.

It was the success of the first Chew and Low theory[19] to estimate the contribution of (c) and (d) diagrams in term of the rescattering of (a) and (b) (Born term). They gave the first qualitative description of the four P-waves in the elastic region, especially for the P_{33} resonance.

Assuming the analyticity of the amplitudes, except the above singularities, the dispersion relations express the real part of the amplitude in terms of pole contributions plus a principal value integral on the imaginary part. A typical dispersion relation has the form (here, at fixed-t) :

$$\mathcal{R}e\,X^{\pm}(\nu,t) = \left[\text{Pole terms}\right] + \frac{1}{\pi}\,\mathcal{P}\int_{\nu_1}^{\infty} \mathcal{I}m\,X(\nu',t)\left[\frac{1}{\nu'-\nu} \pm \frac{1}{\nu'+\nu}\right]d\nu'$$

with, if necessary, some subtraction constants or discrepancy functions to take account of the unknown quantities. The problem is then the solution of an integral equation. The calculations are often tedious, with some ambiguities on the subtraction constant, the convergence of the integrals, the stability of the discrepancy functions... The problem is also with the inputs, taking account of the inelasticities, the unknown high energy region, the propagation of the experimental errors... Considerable sophistication has been achieved by using various forward[20], fixed-t[21], fixed-angle[22], hyperbolic[23] and weighted[24] dispersion relations. This list is not exhaustive :

references can be found in the compilations of Pilkuhn 1973[25], Nagels 1978[26].

The dispersion relations were used successfully to extract the coupling constant G and the low-energy parameters (scattering lengths, crossing symmetric expansion coefficients, σ-commutator term) or, more simply, to fit the phase-shifts.

The following methods pretend to have more fundamental bases with dynamical content and to predict the behaviour of the π-N scattering.

ii) Current algebra models.- By using the current commutation relations of Gell-Man[27] and P.C.A.C, or the equivalent statement of the approximate chiral SU2 ⊗ SU2 symmetry, Weinberg[28] succeeded in 1966 to predict the S-wave scattering lengths relations : $a_1 - a_3 = 0{,}3\ m_\pi^{-1}$ and $a_1 + 2a_3 = 0$, in satisfactory agreement with the experimental values. It was the starting point of the current algebra models for π-N interaction. The current algebra, by the equal time commutators (E.T.C.) between vector and axial vector currents (and their divergences) imposes some constraints to the invariant on-shell amplitudes namely that they are written on the form :

$$X(\nu,t) = \overline{X}(\nu,t) + \frac{\delta X}{f_\pi^2}$$

where δX is directly given by the E.T.C. (f_π^2 is the pion decay constant) : for A^+ amplitude δX is nothing else than the σ-commutator, for A^- and B^-, δX contains the nucleon isovector electromagnetic form factors. The remainder of the amplitude $\overline{X}(\nu,t)$ contains nucleon, nucleon resonances, ρ exchange, etc. Typical of these models is the work of Olsson and Osypovsky[29] where the contributions to $\overline{X}(\nu,t)$ are estimated with an underlying pseudovector Lagrangian including Δ and ρ, by using seven free parameters to fit simultaneously the "experimental data" like the crossing symmetric expansion coefficients. The agreement between the fit and the experimental S- and P-phase shifts at low energy is satisfactory, except for δ_{11} , which desagrees above 200 MeV (Fig. 6). By considering the contribution of other terms, namely N*(1470) and N*(1520) the agreement is made better for the P_{11} wave but results in unacceptable changes for S_1. Among the seven parameters, one, called Z, which takes account of the off-shell effects of the Δ, takes the value Z = - 0.47, in contradiction with field theory considerations which suggest Z = + 1/2. In a more recent calculation of the same type Haider[30] includes the N*, but no the ρ-exchange, and he fixes Z = + 1/2. The rest of the parameters are fitted to the crossing symmetric expansion coefficients. They deduce scattering lengths which agree with the values obtained by dispersion relations. However the authors do not show the comparison with the various phase-shifts. Remember that Olsson too found good values for the scattering lengths, but disagreed with δ_{11} at higher energy.

iii) Lagrangian theories.- These calculations are based on a fundamental Lagrangian density and use the Feynman diagram formalism. In principle they do not need parameters if we are able to take account

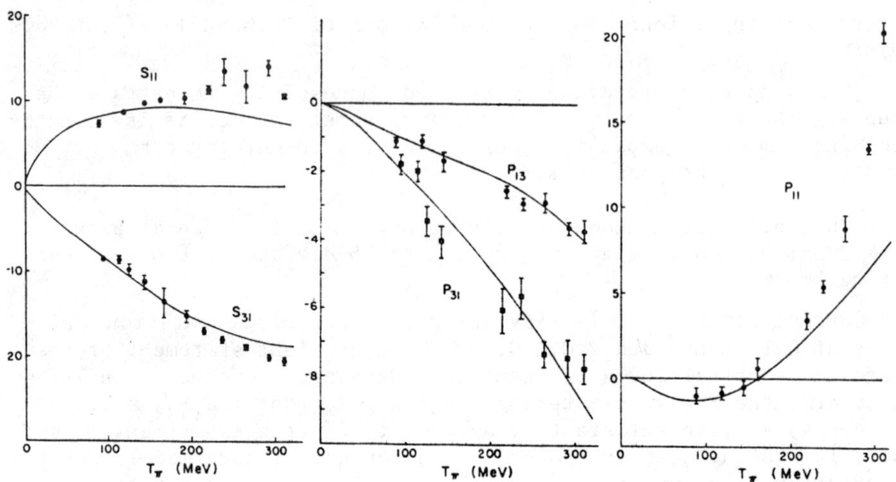

Fig. 6 - Phase-shift fits of the current algebra model of Olsson[29].

of all diagrams and if we know, in other way, the coupling constants which appear at the vertices. The g^2 power series for the amplitudes

$$T = g^2 T_1 + g^4 T_2 + \dots$$

is purely formal in view of the magnitude of the strong interaction coupling constant g. In many calculations the difficulty has been tackled with Padé approximant techniques[31] which take the information contained in the first few perturbative terms with the aim of reconstructing the true solution.

The Lagrangian formalism is the convenient way to make explicit the chiral symmetry by writing $L(x) = L_0(x) + \varepsilon\, L_{SB}(x)$, where L_0 is a SU2 ⊗ SU2 invariant and L_{SB} is the symmetry breaking part, the quantity ε reminding us that this breaking is small. This form allows, in principle, a perturbation treatment of the π-N interaction around the perfect chiral symmetry. In the famous linear σ-model[32] L_0 describes the interaction of an isodoublet of nucleons, an isotriplet of pions and an isoscalar σ meson. The standard form for the breaking term is $\varepsilon\, L_{SB}(x) = \varepsilon\, \sigma(x)$ (σ is the equal time commutator $[A(x), \partial_\mu A^\mu(x)]$ of the axial current and its divergence (the pion field) $\varepsilon = m_\pi^2 f_\pi$). Although demonstrating many qualitative features, especially for the S-wave part, and reproducing directly the results of the current algebra, the σ-model was inadequate for the detailed description of the π-nucleon system. The main failures are : 1) the underestimate of the attractive P_{33} partial wave ; 2) it does not consider the I = 1 ρ exchange, although we all know that the ρ-meson exists (that is more than we can say about the σ meson!) and should couple strongly. More sophisticated chiral models try to resolve these weaknesses : non linear σ-model[33], involving only pions and nucleons ; the chiral gauge model[33] which shows that the isospin odd S-wave comes from ρ exchange. Finally there are phenomenological chiral models[34] which include N, π, σ, ρ and Δ exchanges but with some free parameters.

As an illustration I will consider a recent work of Lin and Willey[35] which uses a non-linear σ-model. They calculated S- and P-wave amplitudes by taking account of second order diagrams (one loop) by a Padé approximant, with only one parameter. Fig. 7 shows their results, the most spectacular disagreement being for the P_{11} wave.

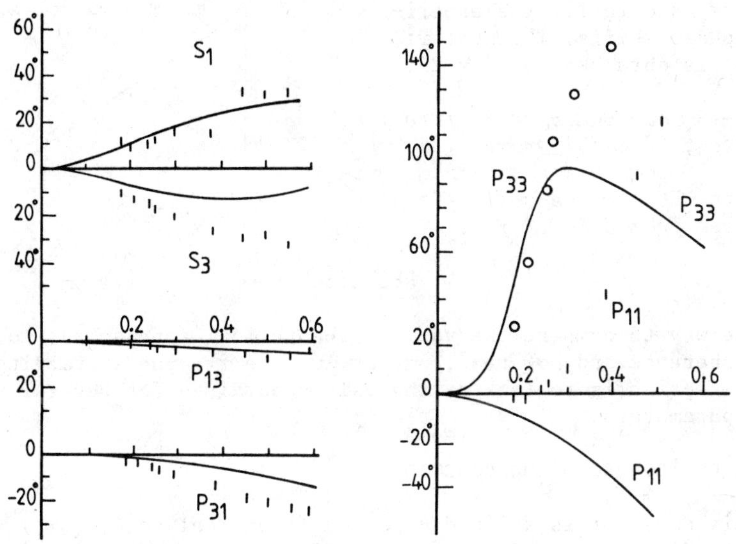

Fig. 7 - Phase-shift versus momentum (q/m_π) of the phenomenological chiral Lagrangian of Lin[35].

In an other way, with a pseudoscalar Lagrangian, Brunet[36] uses a so-called democratic approximant and gives phase-shifts δ as a function of the energy by

$$f_1(x)\ \cot\delta(x) = B_0 + B_1\ f_1(x)/f_2(x)$$

f_1 is the Born term corresponding to the first order in g^2 and f_2 is the second order term. This approximation is based on the fact that for the higher terms the ratio $f_{i>2}/f_1$ would be constant. The B_0 and B_1 parameters are fitted to the experimental phase-shifts, with some different values for each phase. The scattering lengths are determined by extrapolating the fits to the threshold.

I will mention here the work of Banerjee and Camerata[37] although it does not appear, strictly speaking, like a Lagrangian theory. It is an extension of the original work of Chew and Low, with some improvements based on the information given by Lagrangian models and current algebra. By using explicitly PCAC and soft pion limit, they resolve the Low equation, without static approximation and with dynamical inputs including nucleon and antinucleon exchanges. They obtain the S-wave amplitude with one pion off-shell. The σ-commutator appears as one of their six free parameters which are used to fit the experimental phase shifts. The best fit (Fig. 8) is obtained for σ = 25.

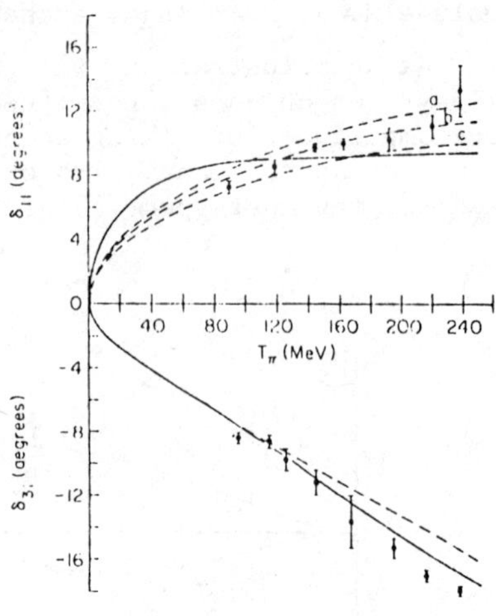

Fig. 8 - S-wave Banerjee [37] fits for σ = 25 (a), 25.5 (b) and 26 (c) (dashes). Solid curves are from equivalent to CERN[4] phase-shift fit points are from C.B.C.[3] fit.

RESULTS

One way to compare the various theoretical approaches, to see their coherence and how near, or far, we are from understanding the π-N interaction is to look at the values obtained for several fundamental parameters.

a) Pion nucleon coupling constant

This constant is defined by $G^2/4\pi$ (identical to $4(M_N/m_\pi)^2f^2$ in terms of effective pseudo-vector coupling constant f). Fig. 9 shows

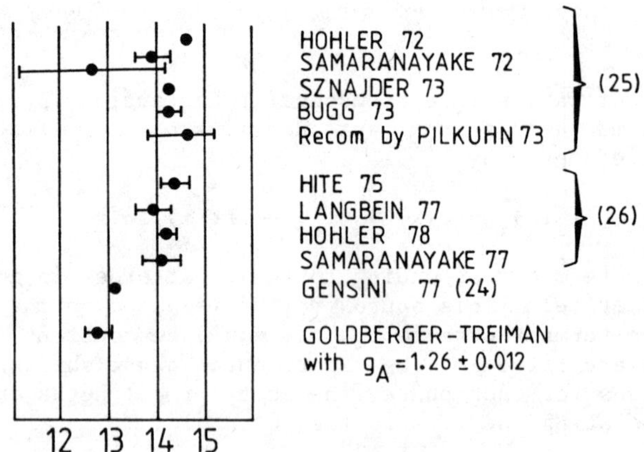

Fig. 9 - π-nucleon coupling constant $G^2/4\pi$.

the various estimates, all obtained by dispersion relations. The average is around 14.3. Gensini[24] by using weighted dispersion relations and taking account of the SU2 symmetry breaking in the mass spectrum of the hadrons ($M_n \neq M_p$), found a smaller value $\sim$ 13.16. Recall that the Goldberger-Treiman relation, which connects G and the weak axial vector coupling constant g_A via the PCAC hypothesis, gives $G^2/4\pi = 12.8 \pm 0.24$, if we believe $g_A = 1.26 \pm 0.012$[26].

b) S-wave scattering lengths

They are defined by $a^- = 1/3\ (a_1-a_3)$ and $a^+ = 1/3\ (a_1+2a_3)$ which are the extrapolations at the threshold ($\nu = m_\pi$, $t = 0$) of the S-part of the $T^\pm(\nu,t)$ invariant amplitudes. Fig. 10 shows the values obtained by different methods. Untill 1973 the dispersion relation analyses gave a_1-a_3 around 0.29, the value recommended in the Pilkuhn[25] compilation. With the availability of the new data by Bussey[8], the dispersion relation analyses give somewhat lower results, around the Moir[22] value 0.254. The current algebra and Lagrangian theory estimates give a small value of a^-, except for the phenomenological chiral

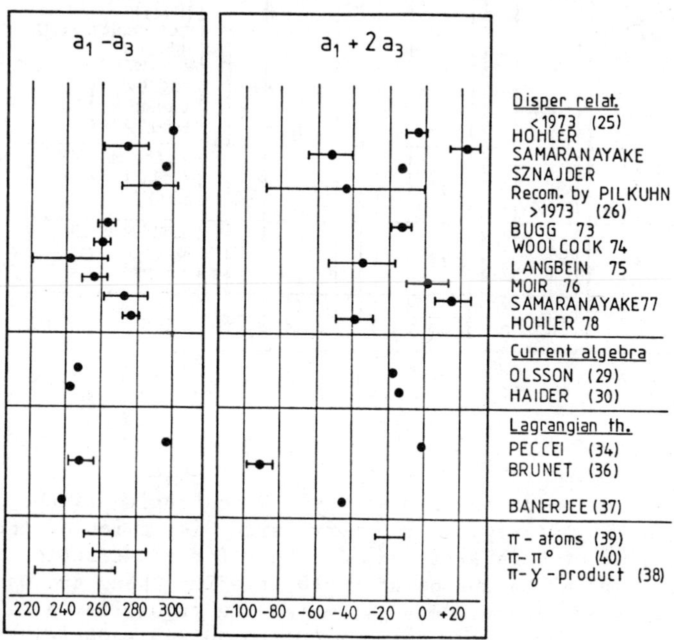

Fig. 10 - S-wave scattering lengths ($10^{-3}\ m_\pi^{-1}$).

Lagrangian model of Peccei[34] which, in 1968, fitted the dispersion relation data available at that time. I add to this table the estimate of a^- from pion-photoproduction measurements and the Panovsky ratio[38] a determination from pionic atom data[39] and a more or less direct measurement by pion charge exchange at 180°[40]. For a^+, the more recent estimates of the dispersion relations come to 0, the predicted value of Weinberg[28] in 1966 by the current algebra. The Langrangian predictions are a little bit further from zero, except for Peccei's but he adjusted his model to obtain $a^+ = 0$!

c) P-Wave scattering lengths

If we compare the P-wave scattering lengths (Fig. 11) the dispersion of the results is much less than for the S-wave, except for the P_{11} -wave like it is reflected too in the δ_{11} phase-shift itself[29,38]. That is not surprising : the P_{11} amplitude has a number of curious features[18], the most notable being an unusually large (for a P-wave) short-range contribution that we do not know to really identify.

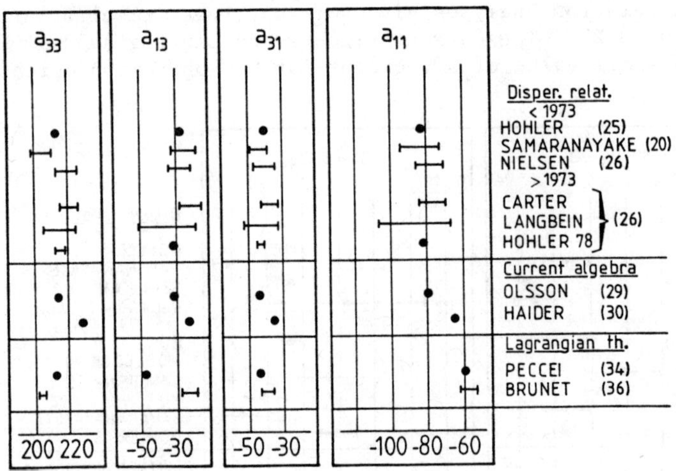

Fig. 11 - P-wave scattering lengths ($10^{-3}\ m_\pi^{-3}$).

d) Σ-commutator

An other datum of the low energy π-N scattering is the σ-commutator. By looking at Fig. 12 the very large dispersion of the "experimental" value of σ is striking. In their first estimation of σ by extrapolation of A^+ to the point $\nu = 0$, $t = 2m_\pi^2$ Cheng and Dashen[16] found σ ∿ 100 MeV. The various estimates lie around three values : ∿ 25 to 30 from pionic atom data, ∿ 40 from dispersion relation with data before 1973, ∿ 60 with new data.Gensini[24] by using new data and dispersion relations, taking account of the mass difference of the hadrons, finds $\sigma = 40 \pm 9$. Banerjee[37], using it like a free parameter, needs $\sigma \simeq 25$. He claims that the higher value obtained by dispersion relations is due to extrapolation errors[41].

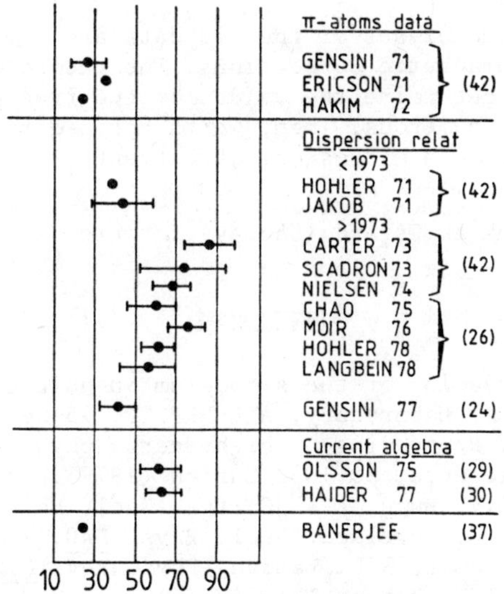

Fig. 12 - σ_{NN}-commutator (MeV).

e) Off-shell amplitudes

I mentioned, at the beginning, that one aim of a π-N theory would be to give the off-shell amplitudes. In principle the Lagrangian or current algebra models give directly those amplitudes, and the on-shell amplitudes are calculated only to check the model or to help to choose the parameters of the model. However many more or less empirical approaches try to relate the off-shell to the measured on-shell amplitudes by using separable potentials[43] or pion-nucleon form-factors[44]. I think that D.J. Ernst will go in the details of this problem in this talk, in the same session of this conference.

CONCLUSION

This review does not pretend to he exhaustive. I did not mention many important works. You will find a complete list of these works in the articles given in references. My aim was simply to give an idea of our present knowledge of the pion-nucleon interaction.

It is clear that we are faced a lot of difficulties. The experimental material used by theoriticians, namely the phase-shift fits, is not actually independent of theoretical inputs or assumptions about unknown quantities. The coulomb corrections, most often, are not properly included. The dispersion relation techniques always need some approximations and present some ambiguities. Although the models of the current algebra or Lagrangian theories give nearly the "experimental" scattering lengths they do not correctly reproduce the phase-shifts, except may be at lower energy. The approach of these models is still phenomenological and they need many parameters : it is the fate of any strong interaction problem! However new and accurate data are coming in at very low energy. New phase-shift fits will be necessary, including a reanalyses of the old data and a more careful treatment of the electromagnetic corrections. The theoretician will have better material to determine the values of the free parameters, hopefully less and less numerous. Then, whith refined models, it may become possible to begin a microscopic treatment of the pion interaction with the nucleus.

I want to thank A. Gerard (Saclay) for his help with the preparation of this review.

REFERENCES

1. R. Kelly, Proceeding of the symposium on hadron spectroscopy, Argonne National Laboratory, ANL-HEP.CP. 75-58 (1975).
2. E. Pietarinen, Proceeding of high energy physics and nuclear structure conference, p. 103, Zurich (1977).
3. J.R. Carter, D.V. Bugg, A.A. Carter, Nucl. Phys. B58, 378 (1973).
4. S. Almehed and C. Lovelace, Nucl. Phys. B40, 157 (1972).
5. G. Rowe, M. Salomon, R.H. Landau, Phys. Rev. C18, 584 (1978). M. Salomon, TRIUMF Report TRI 74-2 (1974).
6. Particle Data Group, Rev. of Mod. Phys. 48, 1 (1976).
7. M.H. Landau, A.W. Thomas, TRIUMF Report TRI 77-4 (1977).
8. P.J. Bussey and al., Nucl. Phys. B58, 363 (1973).
9. P.Y. Bertin and al., Nucl. Phys. B106, 341 (1976).
10. E.G. Auld and al., Can. J. of Phys. 57, 73 (1979).
11. M.A. Moinester and al., Phys. Rev. C18, 2678 (1978).
12. J.S. Frank and al., Contributed paper to this conference.
13. G. Rasche, W.S. Woolcock, Helv. Phys. Acta 49, 435 (1976).
14. e.g. P.R. Auvil, Phys. Rev. D4, 240 (1971).
15. B. Tromborg and al., Phys. Rev. D15, 725 (1977).
16. T.P. Cheng, R. Dashen, Phys. Rev. Lett. 26, 594 (1971).
17. G. Höhler, H.P. Jakob, R. Strauss, Nucl. Phys. B39, 237 (1972) and Karlsrühe preprint TKP 78-11 (1978) ; P. Lichard, CERN Preprint TH (1977) ; H. Nielsen, G.C. Oades, Nucl. Phys. B72, 310 (1972).
18. e.g. J. Hamilton, Pure and applied physics 25 (I). High energy physics, Ed. by E.K.S. Burhop (1967).
19. G.F. Chew and F.E. Low, Phys. Rev. 101, 1570 (1976).
20. e.g. V.K. Samaranayake and W.S. Woolcock, Nucl. Phys. B49, 128 (1972).
21. e.g. W. Langbein, Nucl. Phys. B94, 519 (1975).
22. D.C. Moir, R.J. Jacob, Nucl. Phys. B103, 477 (1976).
23. Y.A. Chao and al., Phys. Lett. 57B, 150 (1975).
24. P.M. Gensini, SLAC-PUB 1967 (1977).
25. H. Pilkuhn and al., Nucl. Phys. B65, 460 (1973).
26. M.M. Nagels and al., Nucl. Phys. B147, 189 (1979).
27. M. Gell-Man, Phys. Rev. 125, 1067 (1962).
28. S. Weinberg, Phys. Rev. Lett. 616 (1966).
29. M.G. Olsson and E.T. Osypowski, Nucl. Phys. B101, 136 (1975).
30. Q. Haider and al., Phys. Rev. D15, 766 (1977).
31. see B.H. Bransden and R.G. Moorhouse, The pion nucleon system, Princeton Univ. Press (1973).

32. J. Schwinger, Ann. Phys. 2, 407 (1957) ;
D.K. Campbell, Field theory, chiral symmetry and pion nucleon interaction, LA-UR-78-394, Los Alamos (1977) ;
B.W. Lee, Chiral dynamics (Gordon and Breach 1972).
33. S. Weinberg, Phys. Rev. 166, 1568 (1968).
34. R.D. Peccei, Phys. Rev. 176, 1812 (1968).
35. W.C. Lin and R.S. Willey, Phys. Rev. D14, 196 (1976).
36. R.C. Brunet, Phys. Rev. D16, 85 (1977).
37. M.K. Banerjee, J.B. Camerata, Phys. Rev. C17, 1123 (1978).
38. A. Donnachie and G. Shaw, Nucl. Phys. 87, 556 (1966).
39. L. Tauscher, W. Schneider, Zeit. Phys. 271, 409 (1974).
40. J. Duclos and al., Nucl. Phys. 43B, 245 (1973).
41. M.K. Banerjee and J.B. Camerata, Phys. Rev. C17, 1125 (1978).
42. E. Reya, Rev. of Mod. Phys. 46, 545 (1974).
43. e.g. J.T. Londergan and al., Ann. Phys. 86, 147 (1974).
44. D.J. Ernst and M.B. Johnson, Phys. Rev. C17, 247 (1978).

PION NUCLEON INTERACTION AT LOW ENERGY†

Manoj K. Banerjee
Department of Physics and Astronomy
University of Maryland, College Park, Maryland 20742

The period from the mid 1950's to the late 1960's saw intensive developments in the field of πN interaction. The classic work of Chew and Low[1] was followed by developments in dispersion theories,[2] current algebras, PCAC, Chiral symmetry,[3] etc. During the last six or seven years there has been a resurgence of interest in πN interactions at low energies inspired entirely by the growth of various pion experimental facilities which have provided increasingly precise data on pion nucleus scattering and reactions. To understand the interaction of a pion with a nucleus we require the knowledge of πN scattering amplitudes for π and N 4-momenta which are off the mass shell, but real. These amplitudes can be determined only with the help of a dynamical theory. It should be noted that the usual dispersion theories deal with amplitudes where the 4-momenta are on the mass shell and, in general, complex. Since the work of Chew and Low there has not been a great deal of work in particle physics attempting to explain the low energy πN scattering data in terms of a dynamical theory. So when the need arose the intermediate energy theorists had to take up the task.

In the literature there are several papers which deal with the S and P wave πN scattering amplitudes in separable form. Some of these use separable potentials;[4] others are extensions of the static theory of Chew and Low.[5]

In this talk I will describe a theory of πN interaction at low energy developed by Barry Cammarata and myself. An early version of the theory was presented at the First Meson Nucleus Conference at Carnegie-Mellon University.[6]

I will try to convey the spirit of our approach by analogy with an unusual approach to potential scattering problem. Let t and v be the kinetic and potential energy operators, respectively. Let $|\vec{p}\rangle$, $|\vec{q}\rangle$ be plane wave states and $|\psi^{(+)}(n)\rangle$ be a scattering eigenfunction of $H = t + v$ with eigenvalue E_n. The half-off-shell scattering amplitude $\langle n|T(E_n)|\vec{p}\rangle = \langle\psi^{(+)}(n)|v|\vec{p}\rangle$ satisfies the Low equation

$$\langle\vec{k}|T(E_k)|\vec{p}\rangle = \langle\vec{k}|v|\vec{p}\rangle + \sum_n \frac{\langle n|T(E_n)|\vec{k}\rangle^* \langle n|T(E_n)|\vec{p}\rangle}{E_k - E_n + i\eta} .$$

If one knew all the potential matrix elements $\langle\vec{k}|v|\vec{p}\rangle$ one could consider solving the nonlinear integral equation. In a field theoretic Low expansion of the scattering amplitude the so-called seagull terms play the role of potential matrix elements. If we assume a Lagrangian we can write down, in principle, the form of the seagull terms with a few unknown parameters, such as coupling constants, form factor masses, etc. But there are serious practical problems. For

†Work supported by the U. S. Department of Energy.

ISSN:0094-243X/79/54030-10$1.50

example, consider a Lagrangian involving nucleons, a scalar-isoscalar meson (σ meson) and pions with pseudoscalar coupling. We are all aware of the pair suppression mechanism in the isoscalar amplitude where the huge repulsive contribution of $-\frac{g_\pi^2}{M}$ from the Z-graphs is almost completely cancelled by $\frac{g_{\sigma NN}\, g_{\sigma\pi\pi}}{m_\sigma^2}$, the attractive contribution from the σ meson exchange. The remainder, presumably of the same order as the physical amplitude, is about $\frac{1}{100}$th of $\frac{g_\pi^2}{M}$. It will be very difficult for any practical calculation to reproduce this kind of result. We avoid these dangers by taking a different path. Suppose we have a way of knowing all elements of a fully off-shell T matrix, $\langle\vec{k}|T(E_o)|\vec{p}\rangle$ for some value of E_o which may be quite unphysical. Then not only can we write down an integral equation

$$\langle\vec{k}|T(E_k)|\vec{p}\rangle = \langle k|T(E_o)|\vec{p}\rangle + \sum_n \langle\vec{k}|T^\dagger(E_n)|n\rangle\langle n|T(E_n)|\vec{p}\rangle \left\{\frac{1}{E_k - E_n + i\eta} - \frac{1}{E_o - E_n + i\eta}\right\} \qquad (1)$$

without any reference to the potential matrix elements but we gain the additional advantage that the equation is once-subtracted, thus more rapidly convergent.

In our theory[7] a soft pion limit plays the role of $T(E_o)$. This requires two conjectures.

Conjecture 1: Strong PCAC.

$$\phi_\alpha = \frac{\sqrt{2}}{f_\pi}\frac{\partial A_\mu^\alpha}{\partial x_\mu},$$

with $f_\pi = 0.939\ m_\pi^3$, is the canonical pion field.

Conjecture 2: Notion of σ commutator.

$$i\int d^4y\ \delta(x_o - y_o)[A_o^\alpha(x), \partial^\mu A_\mu^\beta(y)] = \delta_{\alpha\beta}\sigma(x).$$

The locality of $\sigma(x)$ follows from microcausality. The statement that it is an isoscalar field is the conjecture.

It is very important to note that the foregoing conjecture implies that

$$\int d^4y\ \delta(x_o - y_o)[A_o^\alpha(x), A_\mu^\beta(y)]$$

is a conserved vector isovector current. We may assume, if we wish, that this is proportional to the vector current, i.e.,

$$\int d^4y\ \delta(x_o - y_o)[A_o^\alpha(x), A_\mu^\beta(y)] = \zeta\{i\varepsilon_{\alpha\beta\lambda}V_\mu^\lambda(x)\}.$$

But our conjecture says nothing about the constant ζ, which in current algebra equals one. Thus the notion of σ commutator is compatible with current algebra but does not imply it completely. We do not assume current algebra and, therefore, we can use it as an external test of our theory.

When the notion of PCAC was first introduced a popular accompanying assumption was that the matrix elements of the axial vector current or of its divergence are slowly varying functions of the relevant momentum variables. We should emphasize that we make no such assumption.

The soft pion limit can be used to eliminate not all the seagull terms, but only the isosclar part. So a conjecture about the isovector part must be made.

Conjecture 3: The seagull terms are purely isocalar.

The third conjecture does not allow a term in the underlying Lagrangian involving the coupling of a vector-isovector field [ρ meson] to the hadronic isospin current. Thus the ρ meson must appear as a two pion resonance and not as a canonical field. In other words, we cannot draw a Feynman diagram with a ρ meson exchange.

While in our approach it is not necessary to spell out what the underlying Lagrangian is we may note that the Gell-Mann - Levy σ model[8] is completely compatible with our conjectures, and, as we shall see shortly, with some of our results. Several years back Basdevant and Lee[9] had shown how the ρ and the f resonances appear from the Gell-Mann - Levy σ model.

Finally, our theory gives an integral equation which is shown below pictorially.[10]

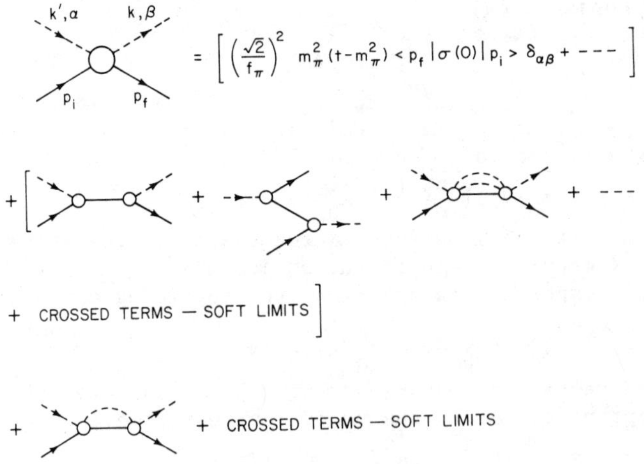

k', α $\quad$ k, β $\quad$ p_i $\quad$ p_f

$$= \left[\left(\frac{\sqrt{2}}{f_\pi}\right)^2 m_\pi^2 (t - m_\pi^2) < p_f | \sigma(0) | p_i > \delta_{\alpha\beta} + --- \right]$$

+ [+ + + ---

+ CROSSED TERMS — SOFT LIMITS]

+ + CROSSED TERMS — SOFT LIMITS

Fig. 1: Once-subtracted Low expansion.

The analogy between this integral equation and Eq. (1) is transparent. The first group of terms on the right-hand side make up the soft pion limit. Since pions can be absorbed and emitted and can dissociate into $N\bar{N}$ pair we get the middle group of terms. The last group involve integrals over the πN scattering amplitudes and their complex conjugates. The first two groups are the driving terms for the integral equation. In the work on low energy S wave phase shifts that we did two years back we did not include S wave inelasticity.[7,10] The nucleon pole terms have only one unknown ingredient, namely, the πN form factor for $t < 0$. Note that our assumption of PCAC fixes $g_\pi(0) = \frac{\sqrt{2}Mm_\pi^2}{f_\pi} g_A(0)$ which is the Goldberger-Treiman[11] relation. The nucleon pole terms play a relatively minor role and so we do not determine the form factor for $t < 0$ very well.

Our main driving terms are the σ commutator term, the main isoscalar mechanism, and the Z-graphs, the main isovector mechanism. The subtraction of the soft limits kills the huge isocalar part of the Z-graphs, thus simulating pair suppression.

Evaluation of the Z-graphs require the quantity $g_\pi(t \geq 4M^2)$. At present it is not practical to calculate its value reliably. However, the various studies[12] of πN form factor assure us that $g_\pi(t \geq 4M^2)$ is not likely to be small compared to $g_\pi(0)$. Through a slight approximation we arrange matters such that we need to know only $|g_\pi(t \geq 4M^2)|$ which we parametrize as

$$|g_\pi(t \geq 4M^2)| = \frac{\bar{g}_\pi = |g_\pi(4M^2)|}{1 + \frac{t - 4M^2}{4\, m_o^2}} .$$

We parametrized the σ commutator matrix element as

$$\langle p_f | \sigma(0) | p_i \rangle = \frac{\sigma(\pi N)}{(1 - \frac{t}{m_1^2})(1 - \frac{t}{m_2^2})^2} .$$

Commonly the quantity $\sigma(\pi N)$ is referred to as the σ commutator.

Our S wave equations have another type of driving term which is not manifest in the pictorial representation. Our choice of the CM frame guarantees that the πN system in the right-hand cut is at rest. But the πN system is not at rest in the crossed term nor in the soft pion limits. Because of this the P wave (and, in principle, all higher waves) amplitudes do provide a small but non-negligible effect. The resulting driving term is parametrized in terms of the P wave cross sections and form factors by making a separability ansatz. The form factor is assumed to be of the form[7,10] $k/(1 + k^2/\mu^2)^{5/2}$.

Naturally, our low energy S wave work cannot fix the various form factors very precisely. In Refs. 7 and 10 we quoted the values $m_0 = 8.6\, m_\pi$, $m_1 = 8.24\, m_\pi$, $m_2 = 7.5\, m_\pi$ and $\mu = 8\, m_\pi$. But for any given set of values of the form factor masses the sensitivity of the S wave phase shifts on the main parameters $\sigma(\pi N)$ and $\bar{g}_\pi$ is very marked. In Fig. 2 we exhibit experimental and theoretical phase shifts. The solid lines are the energy dependent fits of

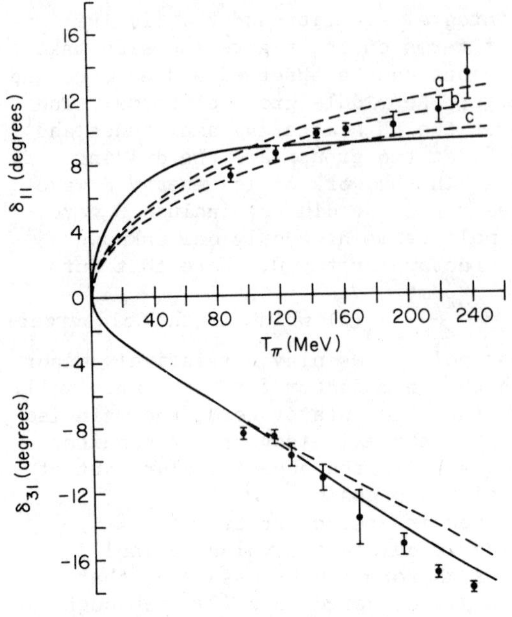

Fig. 2: Experimental and theoretical S wave phase shifts.

Zidell, Roper and Arndt;[13] the flagged points are the energy independent fit phase shifts of Carter, Bugg and Carter.[14] The dashed lines are theoretical calculations for three sets of $\sigma(\pi N)$ and $\bar{g}_\pi$, listed below.

Curve	$\sigma(\pi N)$	$\bar{g}_\pi$
a	25.0 MeV	11.85
b	25.5 MeV	11.69
c	26.0 MeV	11.54

The values were chosen such that they gave the same δ_{31}.

We draw two conclusions. First, the energy dependent fits are unacceptable at low energies. It is nearly impossible to reproduce theoretically the sharp change of slope of δ_{11}. Second, as long as the true phase shifts are close to those of Carter, Bugg and Carter, the parameters $\sigma(\pi N)$ and $\bar{g}_\pi$ will have values close to the ones shown here. We obtained the following scattering lengths, $a^{(+)} = \frac{1}{2}[a(\pi^- p)+a(\pi^+ p)] = -0.016\ m_\pi^{-1}$ and $a^{(-)} = \frac{1}{2}[a(\pi^- p)-a(\pi^+ p)] = 0.079\ m_\pi^{-1}$, which agrees exactly with the current algebra prediction of Weinberg.[15]

Current algebra provides another check of our theoretical work in the form of the following sum rule,[7]

$$\left(\frac{\sqrt{2}}{f_\pi}\right)^2 \frac{m_\pi^4}{2} = \frac{1}{2M^2}|g_\pi(4M^2)|^2 + \frac{M}{6\pi^2}\int \frac{q^2 dq}{\omega(q)E(q)} \frac{|f_{11}(q,0)|^2 - |f_{31}(q,0)|^2}{[E(q)+\omega(q)-M]^2},$$

where $f_{2I,2J}(q,0)$ is a half-off-mass shell S wave scattering amplitude. We satisfy the sum rule within the numerical accuracy of our work. The sum rule may also be interpreted as providing a check on our value for $\bar{g}_\pi$.

Our value of $\sigma(\pi N) \sim 25$ MeV is in sharp disagreement with the value of ~ 65 MeV obtained in recent years by various authors using dispersion relations method.[16,17,18] But significantly our value is close to those obtained in the late 1960's[19] using the Fubini-Furlan theory.[20] The dispersion relations method is based on an idea proposed by Osypowski[21] and by Cheng and Dashen.[22] It deals with the amplitude

$$\bar{C}^{(+)}(\nu,t) = A^{(+)}(\nu,t) + \frac{\nu}{1-\frac{t}{4M^2}} B^{(+)}(\nu,t) - \frac{G_\pi^2}{M(1-\frac{t}{4M^2})} \frac{\nu_B^2}{\nu_B^2-\nu^2}$$

where A and B are the usual CGLN[2] amplitudes, with pions and nucleons in their mass shell. $\nu = (s-u)/4M$, $\nu_B = (t-2m_\pi^2)/4M$ and $G_\pi = g_\pi(m_\pi^2)$ is the πN coupling constant. The authors observed that

$$\bar{C}^{(+)}(0,2m_\pi^2) = \left(\frac{\sqrt{2}m_\pi^2}{f_\pi}\right)^2 \sigma(\pi N) - \frac{m_\pi^2 G_\pi}{2M^3} + O(m_\pi^4)\ldots. \qquad (2)$$

The left-hand side is an on-mass-shell amplitude for unphysical energy and momentum transfer and, therefore, can be determined from dispersion relations. The authors conjectured that the terms $\sim O(m_\pi^4)$ are negligible and so the above relation provides a method of determining $\sigma(\pi N)$. In practice there are three steps involved. First, one must determine a set of amplitudes $\bar{C}^{(\pm)}(\nu > m_\pi, t \leq 0)$ and $B^{(\pm)}(\nu > m_\pi, t \leq 0)$ for physical values of ν and t using fixed t dispersion relations in the variable ν.[23] It is in the course of this that one also determines the coupling constant $f^2 = \frac{G_\pi^2}{4\pi}\frac{m_\pi^2}{4M^2}$. In the second step one determines the values of $\bar{C}^{(+)}(0,t<0)$. Finally $\bar{C}^{(+)}(0,2m_\pi^2)$ is evaluated with some extrapolation technique.

The amplitudes $\bar{C}^{(+)}(0,t<0)$ involve pure imaginary space momentum $\sqrt{(\frac{t^2}{16} - M^2m_\pi^2)/(M^2+m_\pi^2-\frac{t}{2})}$ for small t. But we noted that as long as the magnitude of the space momentum is small it is in the domain of analyticity of the scattering amplitude as a function of the momenta. Thus we could use our theory to determine $\bar{C}^{(+)}(0,t<0)$ for small t.

The quantity $O(m_\pi^4)$ in Eq. (2) can be determined only with a dynamical theory. We evaluated it and found it to contribute $\sim$ 4 MeV to $\sigma(\pi N)$. While the error is small compared to the discrepancy between 25 MeV and 65 MeV it is totally unacceptable for a phase shift calculation.

In Fig. 3 we present $\bar{C}^{(+)}(0,t<0)$s determined by Langbein[17] as illustrative of typical dispersion relation method calculations. The solid curves are our theoretical calculations.[24] The curve a is based on parameters quoted in our paper on S wave phase shifts. For a theoretical calculation, with all parameters fixed a priori, the agreement for small t is most encouraging. We then found that we can noticeably improve the fit by changing the P wave form factor mass μ from 8 m_π to 10 m_π. To get the same low energy S wave phase shifts it was necessary to increase $\bar{g}_\pi$ to 11.9 and decrease $\sigma(\pi N)$ to 24.9 MeV. The curve b is due to this set of parameters.

In view of the nice agreement for small t we are inevitably led to the conclusion that there is no real discrepancy between our value and the results of the dispersion relation methods. While the art of determining the best set of amplitudes from dispersion relation is at a very high level the art of determining the error bars on these amplitudes is far behind. The errors on $\sigma(\pi N)$

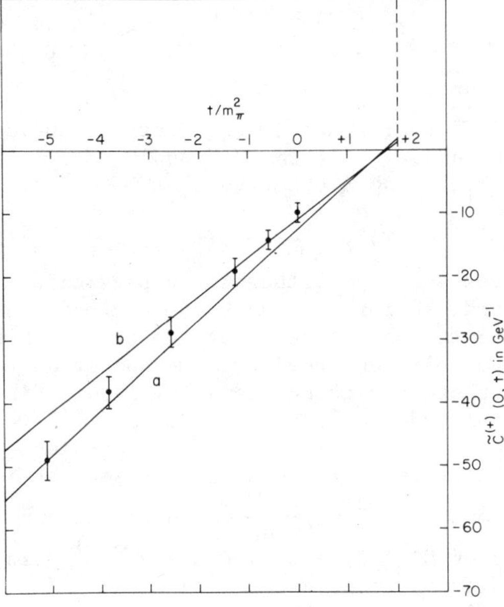

Fig. 3: $\bar{C}^{(+)}(0,t)$ amplitudes.

quoted in many papers are simply not based on careful analysis of the problem. For the evaluation of $\sigma(\pi N)$ considerable care must be exercised as the amplitude $\bar{C}^{(+)}(0,t)$ has a node very close to $t = 2m_\pi^2$.

We should also point out that any theory of πN interaction which claims to be reasonable should be capable of yielding values of $\bar{C}^{(+)}(0,t)$ for small t. In fact these amplitudes are of the "off-energy shell" category a la Blancenbecler and Sugar.[25] Thus dispersion relations provide a check on models of off-energy shell πN amplitudes. A parallel situation does not exist for NN amplitudes.

Before moving off this subject we may note that the σ model coupled with the notion of pair suppression gives $\sigma(\pi N) \sim 26$ MeV. According to the σ model the σ commutator $\sigma = \sigma_{field} \frac{f_\pi}{\sqrt{2}}$. Thus $\sigma(\pi N) = \frac{f_\pi}{\sqrt{2}} \frac{g_{\sigma NN}}{m_\sigma^2}$. But pair suppression requires $g_{\sigma NN}/m_\sigma^2 = g_\pi/M^2$. Combining the two and using Goldberger-Treiman relation we have $\sigma(\pi N) = \frac{m_\pi^2 g_A(0)}{M} = 26\,\text{MeV}$.

Up until now we have tested our theoretical amplitudes for small values of momenta only. It is possible that our successes have been spurious. Another source of worry is the value of $f^2 = \frac{G_\pi^2}{4} \frac{m_\pi^2}{4M^2}$ which according to dispersion theorists is 0.079 ± 0.001.[26] As PCAC is the central assumption of our theory we are locked into a value ~ 0.0725 which includes a 3% enhancement due to form factor. However, there is a class of tests, to which we can subject our theory, which touches on both these issues.

Consider the unsubtracted forward dispersion relation

$$\frac{M+m_\pi}{Mm_\pi} a^{(-)} - \frac{2}{m_\pi^2} \frac{f^2}{1-\frac{m_\pi^2}{4M^2}} = J = \frac{1}{2\pi^2}\int \frac{dk}{\omega} \sigma^{(-)}(\omega)\ldots \tag{3}$$

and the once-subtracted relations

$$\frac{4\pi(M+m_\pi)}{Mm_\pi}\, a^{(-)} - \bar{C}^{(-)}(0,0) - \frac{8\pi f^2}{M(1-\frac{m_\pi^2}{4M^2})} = I^{(-)} = \frac{2m_\pi^2}{\pi}\int \frac{dk}{\omega^3}\,\sigma^{(-)}(\omega)$$

$$\frac{4\pi(M+m_\pi)}{M}\, a^{(+)} - \bar{C}^{(+)}(0,0) + \frac{4\pi f^2}{M(1-\frac{m_\pi^2}{4M^2})} = m_\pi\, I^{(+)} = \frac{2m_\pi^2}{\pi}\int \frac{dk}{\omega^2}\,\sigma^{(+)}(\omega).$$

It is useful to remind ourselves that f^2 is determined in the course of evaluation of the amplitudes $\bar{C}^{(\pm)}(\nu,t)$ and $B^{(\pm)}(\nu,t)$ using fixed t dispersion relations of which the above are special cases. Also, we may note that if $a^{(-)}$ was directly measurable one could use Eq. (3) to determine f^2. Unfortunately, $a^{(-)}$ is not directly measurable.

From our low energy theory we can calculate[27] $a^{(\pm)}$ and $\bar{C}^{(\pm)}(0,0)$. As explained before $f^2 = 0.0725$ is practically a fixed parameter of the theory. Thus we have predictions for the cross section integrals via dispersion relations. We can then test our theory by comparing our results with those obtained by the dispersion relation methods. In Table I we present our theoretical results for parameter set b, described earlier and the results obtained by Höhler, Koch and Pietarinen.[18]

TABLE I

Integral	Ours (m_π^{-2})	Höhler, et al. (m_π^{-2})
J	-0.056	-0.053
$I^{(-)}$	-0.202	-0.206
$I^{(+)}$	-1.39	1.47

We consider the agreement eminently pleasing.

Work on the P wave is still in progress.[28] One of the things we have learnt is that the dynamical picture which emerges from a static theory is not the whole story. Details will be presented in a contributed paper.

REFERENCES

1. G. F. Chew and F. E. Low, Phys. Rev. 101, 1570 (1956).
2. For example, G. F. Chew, M. L. Goldberger, F. E. Low and Y. Nambu, Phys. Rev. 106, 1337 (1957).
3. For a review see Current Algebras by S. L. Adler and R. F. Dashen (Benjamin, N. Y., 1968).
4. For example, R. Landau and F. Tabakin, Phys. Rev. D5, 2746 (1972).
5. For example, D. J. Ernst and M. Johnson in Meson-Nuclear Physics-1976, proceedings of the International Topical Conference.
6. J. B. Cammarata and M. K. Banerjee, Meson-Nuclear Physics-1976, proceedings of the International Topical Conference.
7. M. K. Banerjee and J. B. Cammarata, Phys. Rev. D16, 1334 (1977).

8. M. Gell-Mann and M. Levy, Nuovo Cimento 16, 705 (1960).
9. J. L. Basdevant and B. W. Lee, Phys. Rev. D2, 1680 (1970).
10. For technical details see M. K. Banerjee and J. B. Cammarata, Phys. Rev. C17, 1125 (1978).
11. M. L. Goldberger and S. B. Treiman, Phys. Rev. 110, 1178 (1958).
12. H. F. Jones and M. D. Scadron, Phys. Rev. D11, 174 (1875); H. Pagels and A. Zepeda, Phys. Rev. D5, 3262 (1972); J. B. Cammarata and M. K. Banerjee, Bull. Am. Phys. Soc. 20, 1192 (1975); W. Nutt and B. Loiseau, Nucl. Phys. B104, 98 (1976).
13. V. S. Zidell, L. D. Roper and R. A. Arndt, private communication.
14. J. R. Carter, D. V. Bugg and A. A. Carter, Nucl. Phys. B58, 378 (1973).
15. S. Weinberg, Phys. Rev. Lett. 17, 616 (1966).
16. H. Nielson and G. C. Oades, Nucl. Phys. B72, 310 (1974).
17. W. Langbein, Nucl. Phys. B94, 519 (1975).
18. G. Höhler, R. Koch and E. Pietarinen, Karlsruhe Report No. TKP77-11 (unpublished).
19. F. Von Hippel and J. K. Kim, Phys. Rev. D1, 151 (1970); 3, 2923(E) (1970); W. T. Huang, C. A. Levinson and M. K. Banerjee, Phys. Rev. C5, 651 (1972).
20. S. Fubini and G. Furlan, Ann. Phys. (N.Y.) 48, 322 (1968).
21. E. T. Osypowski, Nucl. Phys. B21, 615 (1970).
22. T. P. Cheng and R. Dashen, Phys. Rev. Lett. 26, 594 (1971).
23. For example, E. Pietarinen, Nucl. Phys. B107, 21 (1976).
24. M. K. Banerjee and J. B. Cammarata, Phys. Rev. D18, 4078 (1978).
25. R. Blankenbecler and R. Sugar, Phys. Rev. 142, 1051 (1966).
26. G. E. Hite, R. J. Jacob and D. C. Joir, Phys. Rev. D12, 2677 (1975).
27. M. K. Banerjee and J. B. Cammarata, Phys. Rev. D19, 145 (1979).
28. In collaboration with Nien-Chih Wei.

Question: L. Kisslinger, Carnegie-Mellon University
1) A technical question: How does replacing your integral equation by another form take care of pair suppression? 2) I do not understand why you do not include a vector-isovector coupling equation to a ρ-field.
Response: Banerjee
First a Lagrangian with ρ is not renormalizable. Second, Z graphs gave adequate isovector strength and so, for simplicity, we let the Gell-Mann-Levy σ model be our guide. It is not possible for us to eliminate every other possibility. We made a particular choice, fitted the phase shifts and passed the various tests mentioned. The soft pion limit is the limit of the whole amplitude (Z graphs plus σ exchange) and thus includes pair suppression. The Z graphs terms of the low expansion are suppressed by the soft limits of these terms.

OFF-SHELL BEHAVIOR OF THE PION-NUCLEON ELASTIC AMPLITUDE

David J. Ernst
Texas A&M University, College Station, Tx. 77843

ABSTRACT

Simple models of the off-shell behavior of pion-nucleon amplitudes are reviewed. The separable potential model which includes the coupling of inelastic channels to elastic channels is derived. It is shown that the dominant, low-momentum behavior of the separable potential is determined by kinematic considerations. Recent results in the S, D, and F-wave channels of Ernst and Johnson are presented. It is argued that in P-wave channels, the presence of the nucleon pole in the amplitude requires a model that is similar to the Chew-Low model. The off-shell behavior of this amplitude is shown to differ from the separable potential model. Results of the model of Ernst and Johnson for P-wave amplitudes are presented. Off-shell effects in elastic pion-nucleus elastic scattering are examined. First, it is shown that different off-shell extrapolations are required for different choices of the basic propagator. Secondly, the dependence of elastic pion-nucleus scattering on the off-shell extrapolation and the choice of propagators is examined.

I. OFF-SHELL BEHAVIOR OF PION-NUCLEON AMPLITUDE

In order to calculate the interaction of a pion with a nucleus using any of the standard nuclear physics techniques (such as multiple scattering approaches, distorted wave impulse approximations, etc.) one requires the knowledge of the pion-nucleon scattering amplitude off the energy shell. We mean here "off the energy shell" in the usual nonrelativistic quantum mechanical sense that one continues the two-body scattering amplitude, $\langle \vec{k}'|T(\omega)|\vec{k}\rangle$, away from the line $\omega=\omega_k=\omega_{k'}$ which is determined by the scattering data. In this section, we shall review some of the important features of present models. These models are designed to be input into the many-body problem and are not intended to be a theory of pion-nucleon scattering. They should, however, contain as much of the physics of pion-nucleon theories[1] as is possible. In the next section, we shall describe the use of these models in the impulse approximation for elastic pion nucleus scattering.

One of the most popular models for the off-shell T-matrix is the separable potential model. The first of these potentials was due to Landau and Tabakin.[2] A generalization of this model has been given in Ref. (3) which provides a rather general justification for the popular LMM potential.[4] In Ref. (3) the interaction between a pion and nucleon is assumed to be of the form

ISSN:0094-243X/79/54040-14$1.50

$$(k'|V_\alpha|k) = \begin{bmatrix} \lambda_{11}\,\bar{V}_\alpha(k')\bar{V}_\alpha(k) & \lambda_{12}\,\bar{V}_\alpha(k')\,w^2_\alpha(k) & \lambda_{13}\bar{V}_\alpha(k')\,w^3_\alpha(k)\ldots \\ \lambda^*_{12}\,w^2_\alpha(k')\bar{V}_\alpha(k) & V^{22}_\alpha(k',k) & V^{23}_\alpha(k',k) \quad \ldots \\ \lambda^*_{13}\,w^3_\alpha(k')\bar{V}_\alpha(k) & V^{23}_\alpha(k',k)^\dagger & V^{33}_\alpha(k',k) \quad \ldots \\ \vdots & \vdots & \vdots \end{bmatrix}, \quad (1)$$

where i stands for a channel number with i = 1 being the elastic channel and i ≠ 1 inelastic channels. The subscript α labels a particular angular momentum and isospin state. The form of Eq. (1) was chosen because in the elastic channel, the effective interaction will be of the form

$$(k'|V_{eff}|k) = \tilde{\lambda}_\alpha(\omega)\,\bar{V}_\alpha(k')\bar{V}_\alpha(k)\,. \quad (2)$$

Straightforward partial wave N/D dispersion theory will then allow[5] a derivation of the full scattering amplitude. The approach of Ref. (6) will be followed closely here. First, the scattering amplitude is divided according to

$$\langle k'|T_\alpha(\omega)|k\rangle = \frac{N^\alpha_{k'k}}{D_\alpha(\omega)}\,. \quad (3)$$

The denominator function $D_\alpha(\omega)$ will contain the elastic scattering cut <u>and</u> the inelastic scattering cuts. The numerator function $N^\alpha_{k'k}$ will then contain the other singularities (called "potential singularities"), and for a separable potential is just

$$N^\alpha_{k'k} = -\lambda_\alpha\,\bar{V}_\alpha(k')\,\bar{V}_\alpha(k)\,. \quad (4)$$

The known analytic structure of $D_\alpha(\omega)$ and unitarity allow one to solve for $D_\alpha(\omega)$ explicitly to give

$$D_\alpha(\omega) = 1 - \lambda_\alpha \int k^2\,dk\, \frac{\bar{V}^2_\alpha(k)}{\hat{\eta}_\alpha(k)}\, \frac{1}{\omega - \omega(k) + i\eta}\,, \quad (5)$$

where $\hat{\eta}_\alpha(k)$ is the ratio of the elastic to inelastic cross section in channel α. Eqs. (3), (4), and (5) produce a simple model of the T-matrix whose off-shell behavior is

$$\langle k'|T_\alpha(\omega)|k\rangle = \frac{\bar{V}_\alpha(k')}{\bar{V}_\alpha(k_0)}\,\langle k_0|T(\omega(k_0))|k_0\rangle\,\frac{\bar{V}_\alpha(k)}{\bar{V}_\alpha(k_0)} \quad (6)$$

If one takes the inelasticity parameter $\hat{\eta}_\alpha(k)$ from the experimentally determined phase shifts, the function $\bar{v}_\alpha(k)$ can then be determined by adjusting it until the T-matrix given by Eqs. (3-5) fit the data. An inverse problem for the model has been derived[3]; it can produce $\bar{v}_\alpha(k)$ directly from the data. However, if certain kinematic factors are included in $\bar{v}_\alpha(k)$ explicitly, then the remaining part of $\bar{v}_\alpha(k)$ is quite smooth and can easily be parameterized by simple functions.

First, as k tends to zero, $\bar{v}_\alpha(k)$ must tend toward k^ℓ. Secondly, the T-matrix which is customarily used is not an invariant amplitude but is related to an invariant amplitude by

$$\langle \vec{k}'|T(\omega)|\vec{k}\rangle = \left[\frac{M^2}{\omega_\pi(k)\,\omega_\pi(k')\,\omega_N(k)\,\omega_N(k')}\right]^{1/2} \langle \vec{k}'|\mathcal{M}|\vec{k}\rangle \qquad (7)$$

At low momentum, the T-matrix behaves like $[\omega_\pi(k)\,\omega_\pi(k')]^{-1/2}$. We must then have $\bar{v}_\alpha(k)$ behave like $[\omega_\pi(k)]^{1/2}$ near k=0. In Ref. (6), $\bar{v}_\alpha(k)$ was taken to be of the form

$$\bar{V}_\alpha^2(k) = k^{2\ell}\,\frac{V_\alpha^2(k)}{\omega_\pi(k)} \qquad (8)$$

The importance of this parameterization is that the dominant low momentum dependence of the potential is contained in the kinematic factors. Simple functions of $v_\alpha^2(k)$ were then adjusted to fit the elastic scattering pion-nucleon phase shifts in the S, D and F channels for pion laboratory kinetic energies up to 1.2 GeV.

For example, in the S_{31}-channel $v_\alpha^2(k)$ was taken as

$$V_\alpha^2(k) = 8.082\ \exp\left(-\frac{k^2}{10.54}\right)(1 + .4053\,k^2) \qquad (9)$$

where all numbers are in units with $m_\pi = 1$. This form of $v_\alpha^2(k)$ produces the curve in Fig. (1). The data is from (7). For D and F-waves a simple Gaussian (or any other smooth function) was found to be sufficient to produce phase shifts in agreement with the data. For example, in the D_{15} channel, $v_\alpha^2(k)$ given by

$$V_\alpha^2(k) = -1.465\times 10^{-5}\ \exp\left(-\frac{k^2}{102.4}\right) \qquad (10)$$

produces the results depicted in Fig. (2).

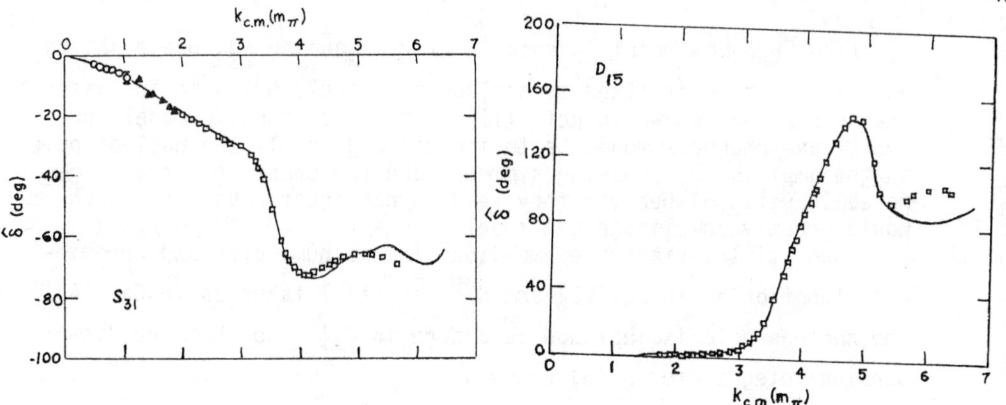

Fig. 1. The S_{31} phase shifts versus center of mass momentum.

Fig. 2. The D_{15} phase shifts versus center of mass momentum.

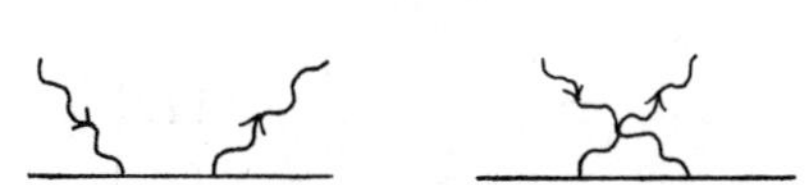

Fig. 3. Absorption and emission of pions.

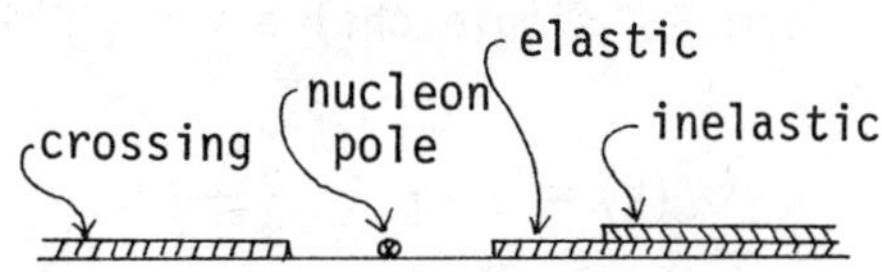

Fig. 4. Analytic structure of pion-nucleon amplitude.

Clearly simple separable potentials are capable of reproducing pion-nucleon elastic scattering quite well. One must still question the validity of such a model. The study[8] of approximating interactions by separable interactions leads to the conclusion that any energy independent, short range interaction can be well approximated by a separable potential of low rank. In the S, D, and F-channels the underlying interaction is thought to be predominantly exchanges of heavy mesons or intermediate anti-nucleons. These interactions are quite short range and can lead to energy indepent[9] potentials.

In the P-wave channels, however, the dominant interaction is thought to be the absorption and emission of pions as is depicted in Fig. (3). This does not lead to an energy independent interaction. Rather, the analytic structure of the T-matrix, in the complex variable z, is pictured in Fig. (4). The nucleon-pole and crossing cut are analytic structures which cannot be reproduced by a potential model. One is thus lead naturally to a Chew-Low[10] model which is simple but still incorporates this basic physics.

The Chew-Low model is capable of reproducing P_{33} phase shifts only if it is generalized to include the coupling to the inelastic channels as was shown in Ref. (11). Thus a reasonable model for the P-wave channels would be to include explicitly the nucleon pole in the amplitude. Crossing symmetry and the crossing cut are ignored in Ref. (6) because they lead to nonlinear equations which would greatly complicate the model.

One still divides the amplitude into a numerator and denominator function as in Eq. (3) and $N^{\alpha}_{k\,k}$ is still taken as in Eq. (4). The nucleon pole is included as a zero in $D_\alpha(\omega)$ so that the dispersion integral for $D_\alpha(\omega)$ becomes

$$D_\alpha(\omega(k)) = \omega_\pi(k)\left[1 - \lambda_\alpha \int k'^2 dk' \frac{v_\alpha^2(k')\,k'^2}{\omega_\pi^2(k')\,\hat{\eta}_\alpha(k')} \frac{1}{\omega(k)-\omega(k')+i\eta}\right] . \quad (11)$$

This is precisely the Chew-Low model with some corrections for the finite mass included, although it is written in a somewhat different form. A simple choice for $v_\alpha^2(k)$ in the P_{33} channel of

$$v_\alpha^2(k) = \exp\left(\frac{k^2}{24.56}\right) \quad (12)$$

produces the results pictured in Fig. (5). The range of $v_\alpha(k)$ is 978 MeV/c.

The difference between the form factor in the Chew-Low model and a separable potential form factor which would produce exactly the same data can easily be seen by comparing Eq. (5) and Eq. (11). If they were to be equal on shell the separable potential $v^{SEP}(k)$ in Eq. (8) would be related to the Chew-low form factor $v^{CL}(k)$ by

$$V^{SEP}(k) = \frac{V^{CL}(k)}{\omega_\pi(k)^{1/2}} \quad (13)$$

Combining Eq. (13) with Eq. (8) one finds that the P-wave separable potential models behave like

$$\bar{V}^{SEP}(k) = \frac{k\,V^{SEP}(k)}{\omega_\pi(k)^{1/2}} = \frac{k\,V^{CL}(k)}{\omega_\pi(k)} \quad (14)$$

They thus have an additional $\omega_\pi(k)^{-1/2}$ dependence which incorporates erroneously the presence of the nucleon pole into the momentum dependence of the amplitude.

The P_{11}-channel presents its own set of peculiar problems. First, the Chew-Low model in the no-crossing approximation only has a solution[12] if one includes a CDD pole in the denominator. Secondly, the phase shifts change sign at low energies; the interaction changes from repulsive to attractive as the energy increases. Although the separable potential model is able to produce amplitudes with zeros, it is not able to reproduce the particular behavior exhibited by the P_{11} phase shifts. The model thus must be generalized to a form where there is a term which dominates at low energies and contains the repulsive nucleon-pole plus a term which is attractive and dominates at high energy. To do this the numerator function is chosen as

$$N_{kk} = \left(v_1(k) \;\; v_2(k) \right) \begin{pmatrix} D_{22}(\omega) & -D_{21}(\omega) \\ -D_{12}(\omega) & D_{11}(\omega) \end{pmatrix} \begin{pmatrix} v_1(k) \\ v_2(k) \end{pmatrix} \frac{k^2}{\omega_\pi(k)} \tag{15}$$

and the denominator is given by the determinant of the matrix D_{ij}, which is

$$\mathcal{D}(\omega) = D_{11}(\omega)\, D_{22}(\omega) - D_{12}(\omega)\, D_{21}(\omega). \tag{16}$$

Unitarity and the assumption that $D_{11}(\omega)$ contains the zero which produces the nucleon-pole term while $D_{22}(\omega)$ will dominate at high energies give

$$D_{11}(\omega(k)) = \frac{\omega_\pi(k)}{\lambda_{11}} \left[1 - \lambda_{11} \int k'^2 dk' \, \frac{v_1^2(k')\, k'^2}{\omega_\pi^2(k')} \, \frac{1}{\omega(k) - \omega(k') + i\eta} \right] , \tag{17}$$

$$D_{22}(\omega(k)) = \frac{1}{\lambda_{22}} \left[1 - \lambda_{22} \int k'^2 dk' \, \frac{v_2^2(k')\, k'^2}{\omega_\pi(k')\, \bar{\eta}_{22}} \, \frac{1}{\omega(k) - \omega(k') + i\eta} \right] , \tag{18}$$

$$D_{12}(\omega(k)) = D_{21}(\omega(k)) = -\,\omega_\pi(k) \int k'^2 dk' \, \frac{v_1(k')\, v_2(k')\, k'^2}{\omega_\pi^2(k')} \, \frac{1}{\omega(k) - \omega(k') + i\eta} \tag{19}$$

The form factor $v_1(k)$ and $v_2(k)$ were chosen to be simple Gaussians, and they and $\bar{\eta}_{22}$ in Eq. (18) were adjusted to fit the elastic phase shifts and inelasticity parameters. The results are shown in Fig. (6). This model has not yet been incorporated into the many-body problem so one cannot yet make any statements about the details of its implications there.

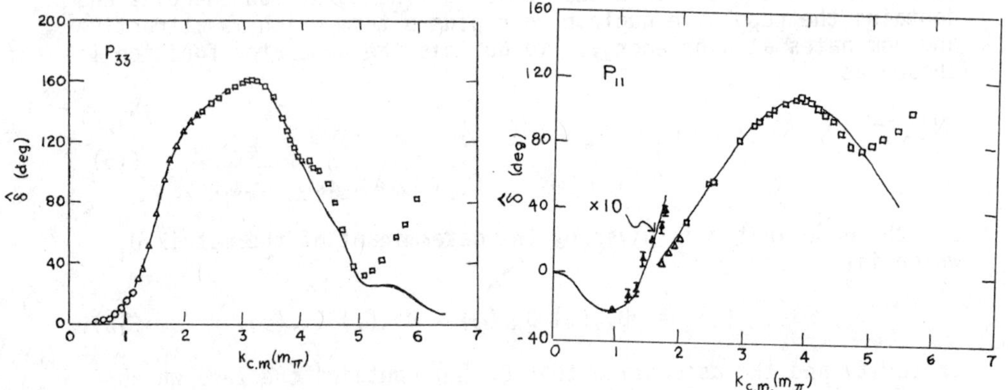

Fig. 5. The P_{33} phase shifts versus center of mass momentum.

Fig. 6. The P_{11} phase shifts versus center of mass momentum. Notice that the lower part of the curve and the data have been multiplied by ten.

II. ELASTIC PION-NUCLEUS SCATTERING

In this section we shall examine the effects that off-shell two-body amplitudes have on elastic pion-nucleus scattering. We follow the approach of Ref. (13). The first point that must be made is that the appropriate off-shell behavior of the two-body amplitude depends upon the propagator one is using. The Klein-Gordon equation for scattering from an optical potential is

$$\psi^{KG}_{\vec{k}_0}(\vec{k}) = \phi_{\vec{k}_0}(\vec{k}) + \int d\vec{k}' \frac{1}{k_0^2 - k^2 + i\eta} \langle \vec{k} | U_{KG} | \vec{k}' \rangle \psi^{KG}_{\vec{k}_0}(k') . \quad (20)$$

We now break the propagator into its forward going and backward going piece according to

$$\frac{1}{k_o^2 - k^2 + i\eta} = \frac{1}{2\omega_\pi(k)} \frac{1}{\omega_\pi(k_o) - \omega_\pi(k) + i\eta} - \frac{1}{2\omega_\pi(k)} \frac{1}{\omega_\pi(k_o) + \omega_\pi(k)}$$

If we now neglect the backward going piece, we find

$$\tilde{\Psi}_{\vec{k}_o}^{KG}(\vec{k}) = \phi_{\vec{k}_o}(\vec{k}) + \int d\vec{k}' \frac{1}{\omega_\pi(k_o) - \omega_\pi(k) + i\eta} \frac{1}{2\omega_\pi(k)} \langle \vec{k} | U_{KG} | \vec{k}' \rangle \Psi_{\vec{k}_o}^{KG}(\vec{k}') . \quad (21)$$

The relativistic Schroedinger equation may be obtained from Eq. (21) if we make[14] the identifications

$$\tilde{\Psi}_{\vec{k}_o}^{KG}(\vec{k}) \equiv \left(\frac{\omega_\pi(k_o)}{\omega_\pi(k)} \right)^{1/2} \Psi_{\vec{k}_o}^{RS}(\vec{k}) , \quad (22)$$

and

$$\langle \vec{k} | U_{RS} | \vec{k}' \rangle = \frac{1}{(\omega_\pi(k))^{1/2}} \langle \vec{k} | U_{KG} | \vec{k}' \rangle \frac{1}{(\omega_\pi(k'))^{1/2}} \quad (23)$$

This makes it clear that ψ^{KG} and ψ^{RS} will differ[15] substantially as is shown explicitly by Eq. (22).

The potentials which one uses in the two equations will also differ according to Eq. (23). The models in Section I were derived using the relativistic Schroedinger propagator. One sees from Eq. (23), that the appropriate Klein-Gordon potential will <u>not</u> contain the factors of $\omega_\pi(k)^{-1/2}$ which were present. Thus the range in momentum-space for a Klein Gordon potential is the large momentum cutoff found in $v_\alpha(k)$. This is only true in the P-wave channels if one uses the Chew-Low like model; a potential model would still leave a factor of $\omega_\pi(k)^{-1/2}$ in the Klein Gordon potential in P-waves.

In order to examine the differences between these two equations for elastic pion-nucleus scattering, the optical potential was approximated by the first order impulse approximation. In the impulse approximation, the factors relating the optical potentials can be incorporated directly into the two-body amplitudes contained in U. The off-shell factor $v_\alpha(k)$ was taken as $\exp(-k^2/\beta^2)$ for all channels. The model of Ref. (6) gives an average β of about 900 MeV/c. Using this value of β , one can compare ψ^{kG} with ψ^{kG}. The results[13] are pictured in Fig. (7) for $u \equiv \psi/r$. One sees that at 40 MeV there are substantial differences between the two equations.

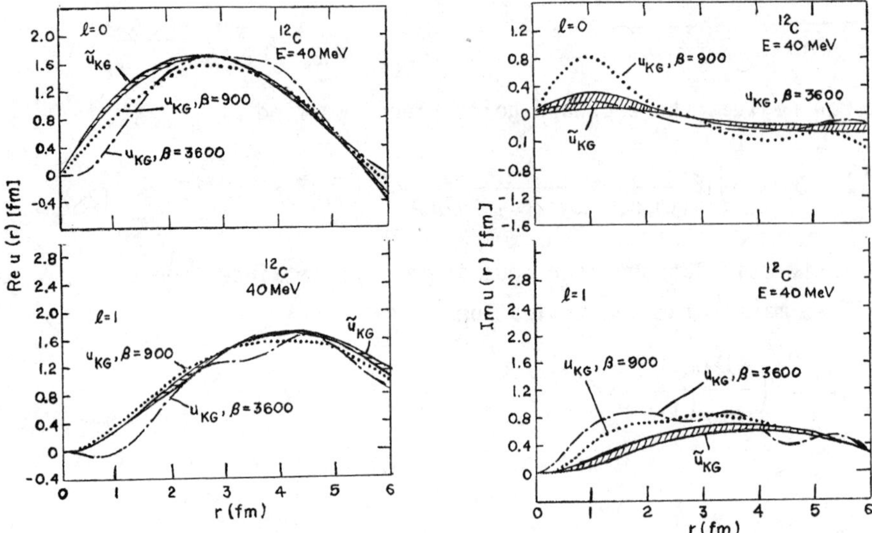

Fig. 7. The real and imaginary part of the wave functions for elastic pion ^{12}C scattering at 40 MeV. Curves are from Ref. (13).

One can also examine the dependence of pion-nucleus scattering on the range parameter β. One should not be free to vary β at will; it should be consistently fixed according to one's two-body model. Much work has been done using models of varying range and thus we will allow β to vary in order to determine the general sensitivity of elastic pion-nucleus scattering to this range. Also in Fig. (7), is shown the dependence of the wave function on β. For the relativistic Schroedinger equation, the wave function u^{kG} lies within the shaded area for all values of β. It shows only a small dependence on the range because the potential contains the functions $\omega_\pi(k)^{-1/2}$ which provides a natural cutoff. The Klein-Gordon equation, however, shows a dramatic dependence on the value of β. As β is increased the wave functions clearly acquire substantial large momentum components. The Kisslinger potential arises from the limit $\beta\to\infty$ and produces wave functions which are quite different than those for β =900 MeV/c. We only pictured wave functions here, but the dependence of differential cross sections on the equation being used or the value of β follows very much[13] the trends seen in the wave functions.

This behavior can easily be understood[13] in terms of the Kisslinger anamoly and the propagation of pions in nuclear matter. For the Klein Gordon equation the momentum of the pion in nuclear matter K is given by

$$K^2 = \frac{k^2}{1-\alpha} \tag{24}$$

where k is the momentum associated withthe pion's energy, ρ is the density of nuclear matter, and α is given by

$$\alpha = 4\pi\ f(k)/k^2 \tag{25}$$

with f(k) the forward pion-nucleon scattering amplitude. The value of K is very large for physical values of f(k) and ρ because α for nuclear matter and low energy pions is very close to one. This gives rise to wave-functions with high momentum components. Any form factor will reduce these high momentum compnents and change the character of the interaction as we have seen.

The dispersion relation for the relativistic Schroedinger equation is in the limit $\beta \to \infty$,

$$(K^2 + m_\pi^2)^{1/2} = \frac{\omega_\pi(k) \pm \sqrt{k^2 - (1-\alpha)m_\pi^2}}{2-\alpha} \tag{26}$$

where there are two values of the momentum for each energy. The important point is that K is well behaved in the physical region of small k and α near one. The introduction of a form factor would introduce a smooth and not very dramatic change in K.

So far we have discussed only the case of small values of k. As one increases k and approaches the resonance, α acquires a large imaginary part. In this case, both equations for K are smooth and well behaved. In Ref. (13), it was found that near resonance, both equations give quite similar cross sections and wave functions. Both show a geometric dependence on the size of the two-body interaction. If one uses the total cross sections, total reaction cross sections, or positions of the minima in the differential cross sections to determine the effective size of the optical potential, R_U, one finds

$$R_U^2 = R_{Nuc}^2 + R_{v^2}^2 \tag{27}$$

where R_{Nuc} is the r.m.s. radius of the target nucleus and R_{v^2} is the r.m.s. radius of the Fourier transform of $v^2(k)$.

An alternative approach has been proposed by G. Miller.[16] Because the two-body amplitude and the optical potential contain a pole it may be incorrect to interate these pole terms in either the Klein Gordon equation or the relativistic Schroedinger equation. It is certainly true that such an iteration in the two-body problem would not reproduce the Chew-Low amplitudes. He developes a linear theory which requires the use of what we have called U_{RS} (the nucleon pole is in U_{RS} as we have defined it) in an equation with a modified propagator. His wave function would satisfy

$$\psi^{GM}_{\vec{k}_0}(\vec{k}) = \phi_{\vec{k}_0}(k) + \int d\vec{k}' \frac{1}{\omega_\pi(k_0) - \omega_\pi(k) + i\eta} \left(\frac{\omega_\pi(k_0)}{\omega_\pi(k)}\right)^2 \langle \vec{k}|U_{RS}|\vec{k}'\rangle \psi^{GM}_{\vec{k}_0}(\vec{k}') . \tag{28}$$

If we define

$$\tilde{\psi}^{GM}_{\vec{k}_0}(\vec{k}) = \frac{\omega_\pi(k)}{\omega_\pi(k_0)} \psi^{GM}_{\vec{k}_0}(\vec{k}) , \tag{29}$$

with

$$\langle \vec{k}'|U_{GM}|\vec{k}\rangle = \frac{\omega_\pi(k_0)}{\omega_\pi(k')} \langle \vec{k}'|U_{RS}|\vec{k}\rangle \frac{\omega_\pi(k_0)}{\omega_\pi(k)} \tag{30}$$

then Eq. (28) is equivalent on-shell to a relativistic Schroedinger equation for $\tilde{\psi}^{GM}_{\vec{k}}(\vec{k})$ in terms of U_{GM}. Thus, this approach provides additional factors of $\omega_\pi(k)$ which would even further surpress high momentum components.

CONCLUSIONS

We have seen that simple pion-nucleon models are able to fit the pion-nucleon data over an extended range. The off-shell behavior of these models is dominated at low momentum by simple kinematic factors. If one properly includes the effect of the nucleon pole, the form factor will differ from the separable potential model by an additional factor of $\omega_\pi(k)^{1/2}$. The behavior of the form factor when the kinematical factors are removed is smooth and, within the context of the models, well determined.

The proper off-shell extrapolation is seen to depend on the propagator which one uses. At low energies, the choice of propagators has a drastic effect on pion-nucleus elastic scattering. Moreover, the Klein-Gordon equation is seen to be dramatically dependent on the range of the two-body interaction at low energies. The differences between the two equations becomes small as the energy is increased and near the resonance, the dependence on the off-shell behavior of the two-body amplitude becomes geometrical in nature.

This work was supported, in part, by the National Science Foundation.

REFERENCES

1. M. G. Olsson and E. J. Osypowski, Nucl. Phys. B101, 136 (1975); R. Hackman, Phys. Rev. D 8, 3920 (1973);
M. D. Scadron, Few Body Dynamics (ed. A. N. Mitra, I. Slaus, V. S. Bhasin, and V. K. Gupta, North Holland, 1976).
M. K. Banerjee and J. B. Cammarata, Phys. Rev. D 16, 1334 (1977); Phys. Rev. C 17, 1125 (1978).
2. R. Landau and F. Tabakin, Phys. Rev. D 5, 2746 (1972).
3. D. J. Ernst, J. T. Landergan, E. J. Moniz, and R. M. Thaler, Phys. Rev. C 10, 1708 (1974).
4. J. T. Landergan, K. W. McVoy, and E. J. Moniz, Ann. Phys. (N.Y.) 86, 147 (1974).
5. D. J. Ernst, "Interaction of Pions with Nucleons at Low Energies", C.I.E.A. del I.P.N., Mexico (1976).
6. D. J. Ernst and M. B. Johnson, in preparation.
7. D. J. Herdon, A. Barbaro-Galtieri, and A. H. Rosenfeld, Lawrence Radiation Laboratory Report (unpublished) UCLR-20030 (1970);
J. R. Carter, D. V. Bugg, A. A. Carter, Nucl. Phys. B58, 378 (1973);
H. Zimmerman, Helv. Phis. Acta 48, 191 (1975);
P. Y. Bertin, B. Coupat, A. Hivernat, D. B. Isabelle, J. Duclos, A. Gerard, J. Miller, J. Morgenstern, J. Picard, P. Vernin, R. Powers, Nucl. Phys. B106, 341 (1976).
8. See R. J. McLeod and D. J. Ernst, Phys. Rev. C 18, 1060 (1978) and references contained therein.
9. M. B. Johnson, Ann. Phys. (N. Y.) 97, 400 (1976).
10. G. F. Chew and F. E. Low, Phys. Rev. 101, 1570 (1956).
11. D. J. Ernst and M. B. Johnson, Phys. Rev. C 17, 247 (1978).
12. D. J. Ernst, R. A. Friedenberg, and M. B. Johnson, Z. Physik A287, 363 (1978).
13. M. B. Johnson and D. J. Ernst, submitted to Phys. Rev. C.
14. J. B. Cammarata and M. Banerjee, Phys. Rev. Lett. 31, 610 (1973); Phys. Rev. C 12, 1595 (1975).
15. E. R. Siciliano and R. M. Thaler, Phys. Rev. Lett. 41, 927 (1978).
16. G. A. Miller, Phys. Rev. C 16, 2325 (1977).

Question: M. Banerjee, University of Maryland

You stated that nobody understands the difference between "off mass shell" and "off energy shell" amplitudes. Obviously you were exaggerating. Blankenbecker and Sugar gave very clear definitions. I want to bring to everyone's attention the curious fact that a model of off energy shell amplitude can be tested by "πN data. The amplitude $\bar{C}^{(+)}(0,t,<0)$ is an amplitude of the "off energy shell" category and thus can be used to test any model. Of course, we realize that $\bar{C}^{(+)}(0,t,<0)$ does have dynamics in it. The second point I would like to draw your attention to is the fact that several years ago Cammarata and I discussed the whole question of Schrodinger and Klein-Gordon equations.

Response: Ernst

I overstated the problem of the "nonrelativistic" reduction somewhat. The points to make are: first, there are ambiguities in Blankenbecker-Sugar reductions as shown by A. Jackson some years ago, and secondly, for the pion (unlike the nucleon) one has a particle which can be absorbed out of the problem. Therefore it does not satisfy a linear equation, and thus the reduction techniques developed for the nucleon-nucleon problem cannot be directly taken over to the pion problem. The relation between the wave functions and optical potentials used here are taken from the work of Banerjee and Cammarata.

Question: M. Bawin, University of Liege

The relativistic Schrodinger equation, on one hand, corresponds to summing ladder graphs only in the context of the Bethe-Salpeter equation. On the other hand, the Klein-Gordon equation sums up all crossed ladder graphs. Why would you then favor the relativistic Schrodinger equation?

Response: Ernst

I tried here not to favor either equation but rather point out that different groups have used each and thus wanted to clarify the relationship between the two and point out the behavior of these differences. This brings up the further point made by Cammarata and Banerjee that if you sum additional terms into the relativistic Schrodinger optical potential, it will be equivalent to the Klein Gordon optical potential. Since one uses both to lowest order in the impulse approximation however, it is pertinent to compare the predictions of both to lowest order.

Question: A. Thomas, TRIUMF

You described a model for the off-shell πN t-matrix in P-wave, with a pole in the energy variable (at w=0) in the D-function. This is correct in the P_{11} channel, but as shown by F. Myhrer and myself (NORDITA preprint, Sept. 1978, to be published) if this is to be done in P_{33}, P_{13}, or P_{31} it will lead to violations of unitarity in systems of more than one nucleon! Thus, the only proper place in an off-shell theory for the Chew-Low nucleon pole in other than P_{11} is as left-hand structure - i.e. for separable models (if these can be used), in $g_\alpha(p)$. Much confusion exists in the literature because, although the Chew-Low model is basically an on-shell theory, the t-matrix is usually written as a function of k,k' and w as if these were independent.

Response: Ernst

To derive the formulas for the two-body model one needs only the numerator on-shell and thus does not have to commit oneself to the off-shell extension. In the many-body problem one must be careful to treat the crossed pole as a u-channel pole (not an s-channel pole) or it will indeed, as pointed out by Thomas, produce a spurious contribution to the absorption. This does not mean that one puts the pole back into the separable potential factors, although this is the simplest way of maintaining a model which is consistent with the two-body data and has reasonable unitarity properties in the many-body problem.

ACCURATE MEASUREMENTS OF $\pi^{\pm}$ + PROTON ELASTIC SCATTERING AT LAMPF*

M. E. Sadler, M. Arman**, F. Borcherding, W. J. Briscoe,
D. H. Fitzgerald, P. F. Glodis, R. P. Haddock, N. Matz,
B.M.K. Nefkens, C. Seftor, and B. Silverman
University of California, Los Angeles, CA 90024

Differential cross section measurements of $\pi^{\pm}$ + p elastic scattering at 378-625 MeV/c are now being completed. Scattered pions and recoil protons are detected in coincidence using scintillation counter hodoscopes. Counter pulse heights, proton scattering angle (resolution <5°), coplanarity information and time-of-flight differences of the scattered and recoil particle are recorded for each event and used as cuts in off-line analysis to minimize backgrounds and determine inelastic contributions. Preliminary results have been obtained using a liquid hydrogen target over an angular interval of $-0.8 \leqslant \cos \leqslant 0.4$. Measurements up to $\cos \approx 0.8$ are being completed using a solid CH_2 target. Statistical uncertainties are ∿1% and systematic uncertainties of ∿2% should be attained when all corrections are taken into account. The current estimated systematic uncertainty is 4%.

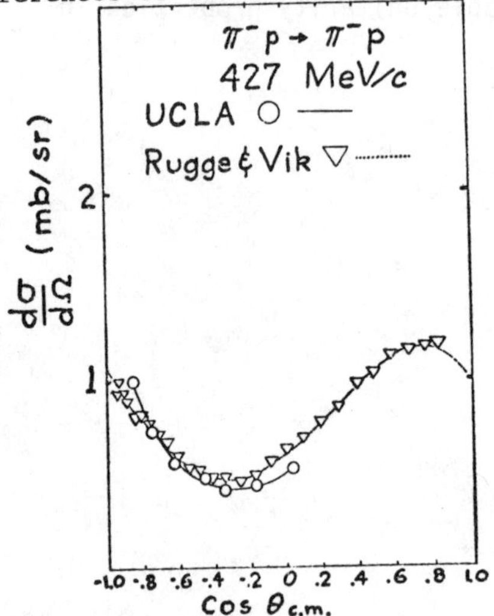

Fig. 1
Comparison at 427 MeV/c of present results to those of Ref. 5.

Our results for 408 MeV π^+ agree with references 1 and 2 to the 2% level. For 408 MeV/c π^- we are systematically 3% lower than 1) and 3% higher than 2). At 427 MeV/c π^+ we are ∿7%, 10% and 13% lower than references 2, 3 and 4, respectively.

* Supported by U.S.D.O.E.

**On sabbatical leave from Pahlavi University, Shiraz, Iran.

REFERENCES

1. P. J. Bussey et al., Nucl. Phys. B58, 363 (1973).
2. V. A. Gordeev et al., Sov. J. Nucl. Phys. 24, 599 (1976).
3. P. M. Ogden et al., Phys. Rev. B4, 1115 (1965).
4. E. H. Rogers et al., Rev. Mod. Phys. 33, 356 (1961).
5. H. R. Rugge and O. T. Vik, Phys. Rev. 129, 2300 (1963).

ISSN:0094-243X/79/540054-01$1.50

LOW ENERGY π^+p AND π^-p ELASTIC SCATTERING*

J. S. Frank, A. A. Browman, P. A. M. Gram, R. E. Mischke,
D. E. Nagle, J. M. Potter, R. P. Redwine, and M. A. Yates-Williams
Los Alamos Scientific Laboratory, Los Alamos, New Mexico 87545

We have measured the differential cross sections for both π^+p and π^-p elastic scattering at pion kinetic energies of 29.1, 49.3, 69.4, and 89.4 MeV. The data was obtained on the Low Energy Pion channel at the Clinton P. Anderson Meson Physics Facility (LAMPF). The beam of pions was incident upon a 1.4 cm-thick liquid H_2 target which was rotated at 45° with respect to the beam line. The amount of H_2 which was in the beam line was determined by optical measurements of the target thickness and by thermometry and pressure measurements of the surrounding H_2 gas volume. This thickness was checked by differential range techniques. The beam flux of 10^4 to 10^5 π/sec was monitored by ion chambers, by scaling a beam counter and by counter telescopes which were sensitive to $\pi \to \mu\nu$ decay.

Coordinates of the scattered particles were observed with a series of four multiwire proportional chambers. The reconstructed trajectories were extrapolated back to the beam line and cuts were made to demand that the scattered particles came from the liquid H_2 region. The pions were identified from other backgrounds from both energy loss and range information within stacks of scintillation counters. Residual backgrounds which were typically < 10% of the πp signal were subtracted by taking data with the liquid H_2 replaced by H_2 gas. Approximately 40,000 πp elastic events were obtained for each energy-polarity combination between the laboratory scattering angles of 45° to 145°.

Figure 1a - d show preliminary π^-p elastic scattering cross sections for T_π = 29.1, 49.3, 69.4, and 89.4 MeV. The normalization of the data, which is presently uncertain to $\sim$ 15%, has been floated to agree with an energy dependent phase shift fit of all low energy pion-nucleon measurements below 300 MeV.[1] The relative cross sections have been allowed to effect the energy dependent fit to low energy pion-nucleon data. The error bars shown are due only to the statistics of the πp elastic signal level folded in with the Monte Carlo statistics. Data obtained with differing beam conditions and detector rotations which were taken to check any possible systematics have not yet been combined.

At the current level of analysis, the angular distributions for π^+p are of similar quality to the measurements of Bertin et al.[2] The π^-p measurements tie in well to Bussy et al.[3] 88.5 MeV differential cross sections and are of much higher precision than other published π^-p angular distributions below 100 MeV.

*Work supported by the U.S. Department of Energy.

ISSN:0094-243X/79/540055-02$1.50

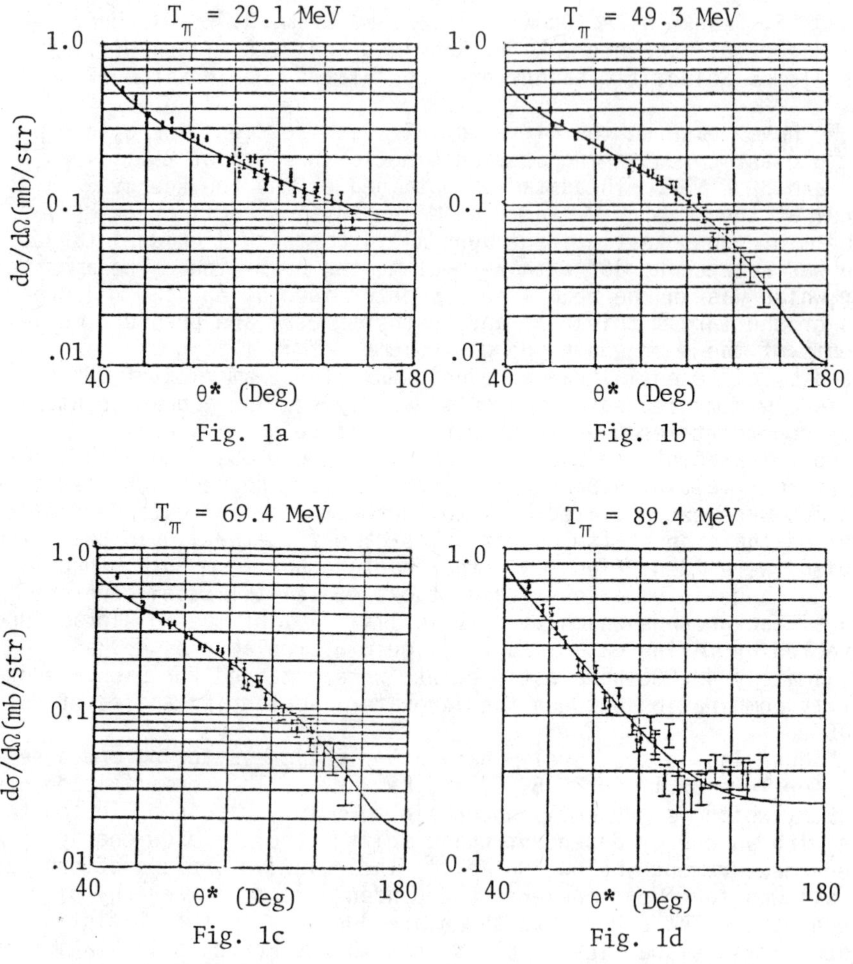

Fig. 1. π^-p elastic differential cross sections in the center of mass system at T_π = a) 29.1 MeV, b) 49.3 MeV, c) 69.4 MeV, and d) 89.4 MeV (preliminary).

---

[1] D. Dodder, private communication.

[2] P. Y. Bertin, B. Coupat, A. Hivernat, D. B. Isabelle, J. Duclos, A. Gerard, J. Miller, J. Morgenstern, J. Picard, P. Vernin, and R. Powers, Nucl. Phys. B106, 341 (1976).

[3] P. J. Bussey, J. R. Carter, D. R. Dance, D. V. Bugg, A. A. Carter, and A. M. Smith, Nucl. Phys B58, 363 (1973).

A THEORY OF THE HALF-SHELL πN T-MATRIX

M. J. Reiner*
University of Rochester, Rochester, N.Y. 14627

ABSTRACT

The theory of the half-shell πN t-matrix (the final pion being off its mass shell) as described in Ref. 1 has been generalized to all partial waves in such a way as to exactly maintain the crossing symmetry. As previously the theory incorporates the on-shell information, including the effects of the inelasticities, and derives in a natural and consistent way the additional information required to continue off-shell from the underlying field theory.

The off-shell kinematics in the CM system are defined as shown in Fig. 1, and, in the spirit of CGLN[2], it is assumed that the corresponding invariant half-shell amplitudes satisfy a crossing symmetric dispersion representation in the W_k-plane ($W_k=\omega(k)+E(k)-M$). A dispersion representation for the partial-wave half-shell amplitudes is then derived by making a partial wave projection. The absorptive parts are deduced using a further partial wave expansion and half-shell unitarity. This leads to an infinite set of coupled, singular, linear integral equations for the half-shell partial-wave t-matrix, which is manifestly crossing symmetric. This representation is then subtracted at the on-shell point and finally reduced to Fredholm form. The final set of coupled Fredholm integral equations can be represented in the form,

$$F_\alpha(p,W_k)=F_\alpha(W_p)\Omega_\alpha(p,W_k)+B_\alpha(p,W_k)$$
$$+\frac{W_k-W_p}{\pi}\Omega_\alpha(p,W_k)\int_\mu^\infty dW'\,\frac{\Omega(p,W')^{-1}e^{i\phi_\alpha(W')}\sin\phi_\alpha(W')B_\alpha(p,W')}{(W'-W_p)(W'-W_k-i\varepsilon)}$$
$$+\frac{1}{\pi}\sum_\beta\int_\mu^\infty dW'K_{\alpha\beta}(p,W_k,W')F_\beta(p,W') \qquad (1)$$

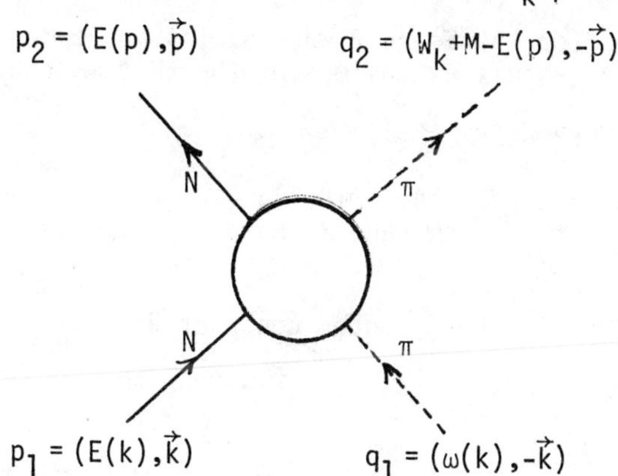

Fig. 1. The off-shell kinematics in the CM system. Here $\omega(k) = \sqrt{k^2+\mu^2}$; $E(k) = \sqrt{k^2+M^2}$, $\mu(M)$ being the pion (nucleon) mass.

ISSN:0094-243X/79/540057-02$1.50

where $F_\alpha(p,W_k)$ is the half-shell t-matrix in the state α, $F_\alpha(W_p)$ is its on shell value, and $\phi_\alpha(W)$ is a phase angle which is related to the πN phase shift, $\delta_\alpha(W)$, and inelasticity parameter, $\eta_\alpha(W)$, by

$$\phi_\alpha(W) = \tan^{-1}\left(\frac{1 - \eta_\alpha(W)\cos 2\delta_\alpha(W)}{\eta_\alpha(W)\sin 2\delta_\alpha(W)}\right) \tag{2}$$

Also in (1), $\Omega_\alpha(p,W)$ is the usual Omnes function given in terms of $\phi_\alpha(W)$, $B_\alpha(p,W_k)$ is the Born term which is calculated from the off-shell extension of the covariant Feynman amplitudes corresponding to the exchange of the nucleon in the s- and u-channels and the ρ and σ mesons in the t-channel, and finally the kernel, $K_{\alpha\beta}(p,W_k,W')$, is determined from the πN (nonstatic) crossing matrix.

I attempt to incorporate the PCAC constraints by using coupling constants in the Born term which are consistent with the chiral symmetry and by making a further subtraction at the Adler soft pion point.

Results will be presented for the S and P wave states and compared to other commonly used off-shell extensions of the πN t-matrix.

REFERENCES

1. M. J. Reiner, Phys. Rev. Lett. 38, 1467 (1977).
2. G. F. Chew, M. L. Goldberger, F. Low, and Y. Nambu, Phys. Rev. 106, 1337 (1957).

*Supported by the U.S.Department of Energy under Contract No. EY-76-S-02-2171.

FIELD THEORETIC MODEL FOR LOW ENERGY PI-NUCLEON SCATTERING

R. J. McLeod and D. J. Ernst
Texas A&M University, College Station, Tx. 77843

ABSTRACT

A relativistic model for pion-nucleon scattering is derived by assuming pseudoscalar coupling and suppressing the contribution of the intermediate anti-nucleon states. The leading term in the model is isospin and angular momentum decomposed and partial wave N/D dispersion theory is used to generate a unitary amplitude. The leading term in an expansion in inverse powers of the nucleon mass is similar to but differs from the Chew Low model. In particular, the degeneracy between the P_{31} and P_{13} channels present in the Chew Low model is removed. The phase shifts in the P_{33} channel are reproduced well by the model for pion laboratory energies below 1 GeV.

DERIVATION

The lowest order ("Born" term) approximation for the invariant pion-nucleon amplitude M is given by, assuming pseudoscalar coupling,

$$M^{\pm} = - g_o^2 \left[\frac{-\not{p}_1 - \not{q}_1 + M}{M^2 - (p_1+q_1)^2} \pm \frac{-\not{p}_1 + \not{q}_2 + M}{M^2 - (p_1-q_2)^2}\right] , \qquad (1)$$

with $p_1(q_1)$ the incident nucleon (pion) momentum. The success of the Chew Low model[1] which contains no antinucleon states and suppression of the antinucleon states in the sigma model motivates us to remove the contribution to (1) which arises from intermediate anti-nucleon states. We do this by decomposing the propagator in (1) into its nucleon and anti-nucleon parts and then discarding the contribution of the intermediate anti-nucleon part of (1). We then expand the resulting amplitude in inverse powers of the nucleon mass and keep the lowest two nonvanishing terms. These amplitudes are then angular momentum decomposed.

The lowest order (in powers of M^{-1}) approximation in the dominant P_{33} channel is

$$f_{33}^B = \frac{g_o^2 \, (E_N+M)}{12M^2 \, u} \, \frac{p^2}{M} \, v^2(p) + O(M^{-4}) , \qquad (2)$$

with E_N the nucleon energy in the center-of-mass frame and $u = E_N - \omega_\pi - M$. This amplitude differs from the Chew Low amplitude in a significant way. The Chew Low amplitude contains a pole in ω_π while our amplitude contains a pole in u. Since this amplitude

ISSN:0094-243X/79/540059-02$1.50

arises solely from the crossed Born term, one should expect this u-channel pole rather than the Chew Low s-channel pole.

The final amplitude is generated by applying partial wave N/D dispersion theory to the amplitude in (2). The contribution of the inelastic, pion production cut is included using the technique of Ref. (1). The resulting amplitude fits all the elastic scattering data in the P_{33} channel for pion laboratory energies below 1300 MeV. The model also removes the degeneracy of the P_{31} and P_{13} channels present in the Chew Low model. We have not yet examined the P_{11} channel.

CONCLUSION

We have developed a model of the low energy pion-nucleon scattering amplitudes by beginning with a relativistic amplitude, discarding the anti-nucleon states, expanding in inverse powers of M, and utilizing partial wave N/D techniques. The resulting amplitude is not complicated, and this will hopefully be useful in pion-nucleus studies.

REFERENCES

1. D. J. Ernst and M. B. Johnson, Phys. Rev. C 17, 247 (1978).

CHARGE-SYMMETRY VIOLATION IN THE REACTION $np \rightarrow d\pi^o$*

C. Y. Cheung, E. M. Henley and G. A. Miller
Physics Department, University of Washington, Seattle, WA 98195

In the reaction $np \rightarrow d\pi^o$, an accurate measurement of the pion-production differential cross section could give a clear-cut demonstration of charge-symmetry violation. In this reaction, the presence of a charge-symmetry breaking nuclear interaction would render the pion-production differential cross section slightly asymmetric about 90° in the center of mass frame. At present, we are concerned with only the energy region of the (3,3) resonance. Several charge-symmetry breaking mechanisms have been calculated or considered: (a) direct electromagnetic interactions, (b) ρ^o-ω mixing, (c) η-π^o mixing, (d) neutron-proton mass difference, and (e) Δ^+-Δ^o mass difference. The above five mechanisms are depicted diagrammatically in Fig. 1, where the blob represents the (3,3) resonance.

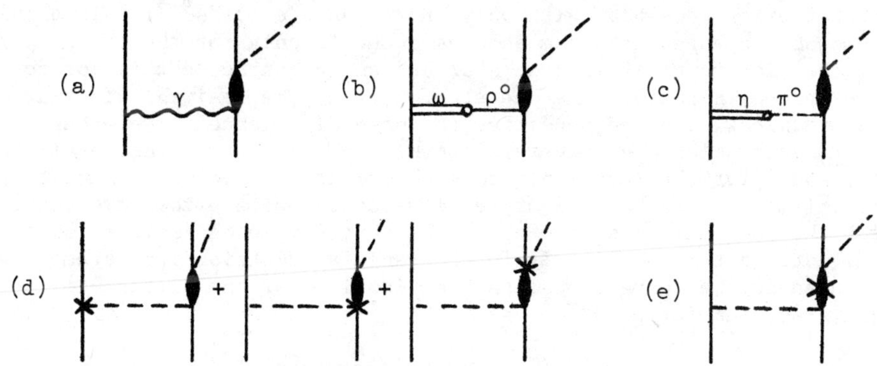

Fig. 1. Charge-Symmetry Breaking Mechanisms

In order to calculate the effects of various charge-symmetry breaking mechanisms, we need to have a model for computing the differential cross section of the reaction via strong interactions, and the model by Alberg et al.[1] is adopted. It is found that the η-π^o mixing mechanism gives the greatest asymmetry, which is about three parts in a thousand. The terms of Fig. 1(a) and 1(b) give effects which are approximately seven and nine times respectively smaller than that caused by η-π^o mixing. The calculation of mechanisms (d) and (e) is still in progress. Our preliminary result is that the asymmetry due to neutron-proton mass difference is about half of that due to η-π^o mixing, while Δ^+-Δ^o mass difference is not expected to give a large effect.

* Supported in part by US DOE and GSRF(Univ. of Washington).

1)M. A. Alberg, E. M. Henley, G. A. Miller and J. F. Walker, Nucl. Phys. A306 (1978), p.447.

ISSN:0094-243X/79/540061-01$1.50

ON THE ANGULAR DISTRIBUTIONS FOR THE REACTION $pp\to d\pi^+$

J. Chai and D.O. Riska
Department of Physics, Michigan State University
East Lansing, Michigan 48824

ABSTRACT

It is shown that the angular distribution parameters γ_j for the reaction $pp\to d\pi^+$ are as sensitive to the nuclear wavefunctions as to the model for the production mechanism.

The total cross section for the reaction $\pi^+d\to pp$ can be reasonably well explained in terms of (a) a two-nucleon pion rescattering model for the absorption mechanism if the pion exchange operator is cut off at relatively long range[1,2] or (b) a pion and ρ-meson rescattering mechanism with only short range cutoffs[3,4]. Since the ρ-meson exchange mechanism acts as a cutoff only for the tensor part of the pion rescattering operator but as an enhancement factor for the spin-spin part of the operator, it would be desirable to find observables which are sensitive to these differences. Unfortunately most observables are either relatively insensitive to the details of the model for the reaction mechanism or alternatively, as sensitive to the model for the nuclear force used to generate the wavefunctions for the two-nucleon system as to the reaction mechanism. We shall demonstrate this explicitly for the angular distribution parameters γ_j used in the conventional parametrization for the differential cross section for $pp\to d\pi^+$:

$$\frac{d\sigma}{d\Omega} = \frac{1}{32\pi}\,(\gamma_0 + \gamma_2\cos^2\theta + \gamma_4\cos^4\theta) \tag{1}$$

In Figs. 1-3 we show the parameters γ as calculated using a model involving single pion rescattering, with monopole vertex factors with a mass scale of 700 MeV/c^2 and a model involving π- and ρ-meson rescattering using monopole vertex factors with mass scales 1.2 GeV/c^2 for the pion nucleon vertices and 2.0 GeV/c^2 for the ρ-nucleon vertices. These results were obtained with the formalism of Ref. 4 amended to include ^3F scattering states for all parts of the reaction mechanism[5]. Results are given both as obtained using the Reid Soft Core[6] (RSC) and the de Tourreil-Sprung super soft core[7] (SSC) models for the nuclear wavefunctions. The meson-nucleon coupling constants were those used in Ref. 4 but the energy of the exchanged meson is here taken to be half of the produced pion energy. The data points in the figures are taken from Ref. 8.

The parameter γ_2 shows far more sensitivity to the nuclear potential than to the reaction model, whereas γ_4 is more sensitive to the reaction model. The parameter γ_0 is roughly as sensitive to

ISSN:0094-243X/79/540062-02$1.50

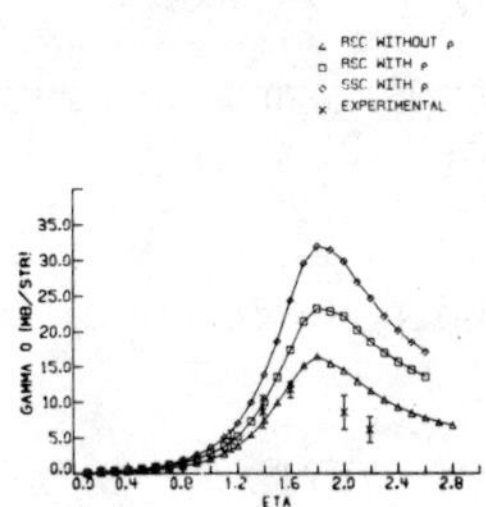

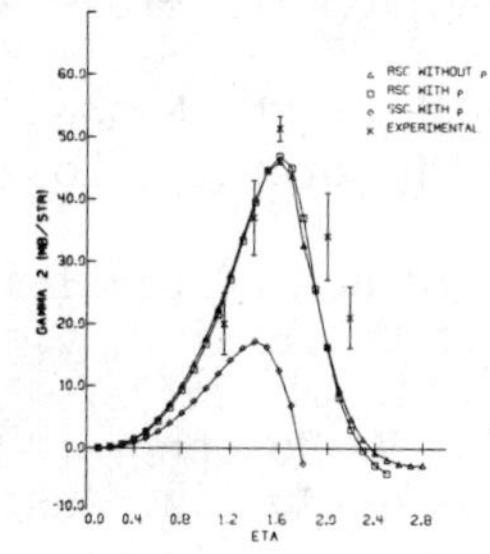

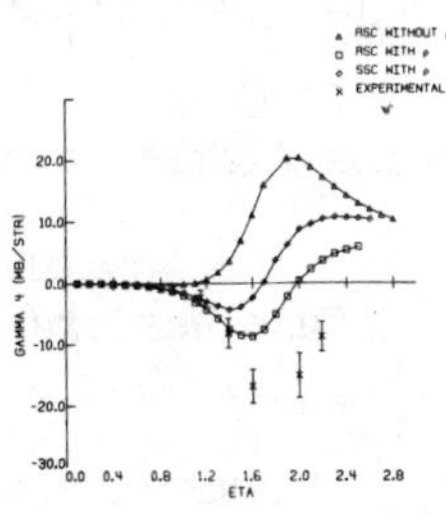

the wavefunctions as to the reaction model. While the model with no ρ-meson is somewhat better for γ_0 it is to the same extent worse for γ_4 allowing no strong conclusions.

Paradoxically, the main evidence against the single pion exchange production model and in favor of the π+ρ exchange model is therefore that provided by the total cross section near threshold; it is very difficult to obtain a large enough cross section at low energies with single pion exchange cutoff with a long range form factor. In this region the dominant production mechanism is S-wave πN rescattering with no ρ-meson modification. Use of the ρ-meson to cut down the P-wave rescattering mechanism allows shorter range πNN vertex factors and thus a larger cross section at low energies which agrees well with experiment may be obtained.

REFERENCES

1. G. Goplen, W. Gibbs and E. Lomon, Phys. Rev. Lett 32, 1012 (1974).
2. W. Gibbs, B.F. Gibson and G.J. Stephenson, Jr. In "Meson-Nuclear Physics--1976", P.D. Barnes et al., eds. AIP Conf. Proc. Vol. 33 464 (1976).
3. D.O. Riska, M. Brack and W. Weise, Phys. Lett. 61B, 41 (1976).
4. M. Brack, D.O. Riska and W. Weise, Nucl. Phys. A287, 425 (1977).
5. Some errors in the formulae in the Appendix of Ref. 4 have been corrected.
6. R.V. Reid, Ann. Phys. 50 411 (1968).
7. R. de Tourreil and D.W. Sprung, Nucl. Phys. A201, 193 (1973).
8. M.G. Albrow, et al., Phys. Lett. 34B, 332 (1971).

ANALYZING POWER FOR THE REACTION $\uparrow PP \rightarrow d\pi^+$ AT 0.8 GeV PROTON ENERGY

F. Cverna, M. McNaughton, C. Hwang, B. Bonner, and N. King
Los Alamos Scientific Laboratory, Los Alamos, NM 87545

H. Willard and E. Winkleman
Case Western Reserve University, Cleveland, OH 44106

R. Joseph
University of Texas at Austin, Austin, TX 78712

A 0.8 GeV polarized proton beam was used to measure the analizing power for the reaction $\uparrow pp \rightarrow d\pi^+$. Eight multiwire proportional counters were used in a two-arm system to detect both the deuteron and pion for left and right scattering. Reversal of the beam polarization every three minutes and use of the two-arm system removed systematic errors from the analyzing power measurement. A fit of the analyzing power to a phenomenological model indicates the importance of d-wave pions. Comparison is made with other data in the 0.6 to 1.0 GeV range.

*Work supported by the U. S. Department of Energy

ISSN:0094-243X/79/540064-01$1.50

A Generalized Low Equation Approach to $\pi D \to pp$

Roger H. Hackman*
Hollins College
Hollins, Va.
and
Virginia Polytechnic Institute
Blacksburg, Va. 24061

ABSTRACT

Using non-perturbative, field-theoretic techniques based on a generalization of the Low equation, an inhomogeneous, coupled, linear, integral equation is developed for the reaction matrix for $\pi D \to pp$. The driving term for this equation, which consists of a truncated, multiplescattering expansion of the breakup process, is evaluated at the reaction threshold. P-wave πN rescattering is found to make a non-negligible contribution to the s-wave absorption process at threshold.

DISCUSSION

Problems involving the absorption and emission of mesons are best suited to the techniques of relativistic quantum field theory. Since perturbation theory is almost guar nteed not to converge for strong interactions, a non-perturbative approach such as that embodied in the Low equation seems more suitable. The Low equation has been applied with considerable success to the study of pion-nucleon and pion-nucleus elastic scattering[1]. However, there is an important qualitative distinction between these elastic scattering applications and the $\pi D \to pp$ reaction considered in the present work that diminishes the attractiveness of the Low equation. In pion elastic scattering, the Bose statistics of the pion are the most important symmetry consideration and the Low equation provides a natural vehicle for the expression of this symmetry, since it is explicitly crossing symmetric. In the breakup reaction considered here, the Fermi statistics of the two final state nucleons is similarly an important consideration. The difficulty with the Low equation is that this antisymmetry is not manifest, that is, the two nucleons seem to play quite different roles even when both are on the mass-shell. Furthermore, the matrix element of the Seagull term, which is an essential part of this approach, is not well known. Banerjee, Levinson, Shuster and Zollman[2] have evaluated this term in a soft-pion approach for $NN \to NN\pi$, but their results are valid only for low energy and they do not consider the rescattering mechanism in any detail.

*Work supported in part by the National Science Foundation under Grant PHY 77-04408.

ISSN:0094-243X/79/540065-02$1.50

In the present work, we generalize the Low equation to a form more suitable to the two-nucleon system. The resulting expression is manifestly antisymmetric in the final state nucleon variables and satisfies the requirements generated by crossing the external pion leg separately with each external nucleon leg. This last feature is important since these crossing terms lead to both the direct absorption and the backward rescattering graphs, and represents an improvement on the earlier work of Lazard, Ballot and Becker[3]. The Seagull term mentioned above plays an important role in both the forward and the backward rescattering contributions. We disagree with Alberg, Henley, Miller and Walker's[4] interpretation of this matrix element as a simple absorption vertex.

In our approach, there is a clear separation of the final state interaction, the initial state interaction, and the rescattering mechanism, to the order of approximation considered. We find no need to modify either the NN interaction or the πN rescattering vertex to avoid double-counting of meson exchanges. Our final result defines a linear, coupled, inhomogeneous integral equation for the reaction matrix. The driving term for this equation consists of a direct absorption contribution, and both a forward and a backward rescattering contribution. The integral terms involve the couplings of this reaction to the nucleon-nucleon and the pion-deuteron elastic scattering channels.

The potential has been numerically evaluated at the reaction threshold, using relativistic forms for both the absorption and the rescattering vertices. Probably our most important result is that p-wave πN rescattering makes a non-negligible contribution to threshold absorption. Thus, the success of models such as that of Koltun and Reitan[5] in the description of threshold absorption with only an s-wave rescattering mechanism, is probably fortuitous.

REFERENCES

[1]Some examples are G. F. Chew and F. E. Low, Phys. Rev. 101, 1570 (1956), Roger H. Hackman, Phys. Rev. D8, 3920(1973), M. K. Banerjee and J. B. Cammarata, Phys. Rev. D16, 1334(1977), and J. B. Cammarata and M. K. Banerjee, Phys. Rev. C12, 1575(1975).

[2]M. K. Banerjee, C. A. Levinson, M. D. Shuster and D. A. Zollman, Phys. Rev. C3, 509(1971).

[3]C. Lazard, J. L. Ballot and F. Becker, Nuovo Cimento 65B, 117(1970).

[4]M. A. Alberg, E. M. Henley, G. A. Miller and J. F. Walker, Nucl. Phys. A306, 447(1978).

[5]D. S. Koltun and A. Reitan, Phys. Rev. 141, 1413(1966).

A POSSIBLE POLE IN THE πNN VERTEX FUNCTION AND THE LOW ENERGY πN SCATTERING

T. Mizutani and P. Rochus

Physique Nucléaire Théorique, Institut de Physique, Université de Liège, Sart Tilman, B-4000 Liège 1, Belgium.

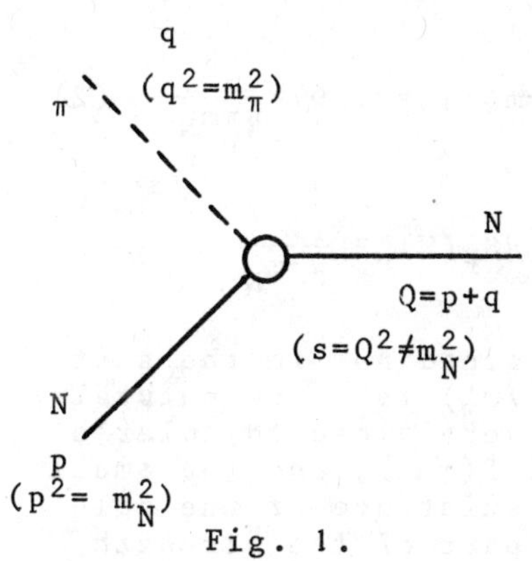

Fig. 1.

Using the πN phase shifts and inelasticities in the P_{11} and S_{11} partial waves, we have calculated the pion nucleon vertex function with one nucleon off-mass-shell through a dispersion approach[1] (the kinematical situation is shown in Fig. 1). We can summarize the result as follows. (i) As a function of W ($\equiv \pm\sqrt{s}$) the vertex function Γ(W) has two poles in the unphysical region $|W| < m_N + m_\pi$: at $W_+ \sim 1030$ MeV and $W_- \sim -125$ MeV respectively, whereas the dressed nucleon propagator $S_F'(W)$ has zeros at the same positions. (ii) The pole position W_+ is very much sensitive to input variations whereas W_- is pretty stable. Considering the nature of approximations we have made, the existence of the pole at W_- (and the corresponding zero of $S_F'(W)$ at W_-) will be definite. (iii) The physical amplitudes in the P_{11} and S_{11} may be written as

$$f_\pm^{tot}(W) = -R_\pm(W)\ \Gamma(\pm W)\ S_F'(\pm W)\ \Gamma(\pm W) + f_\pm^{irr}(W)$$
$$(+ \ldots P_{11},\ - \ldots S_{11}) \qquad (W > 0) \qquad (1)$$

In eq. (1) the first term which is the dressed (direct) nucleon pole term ($R_\pm$ (W) is some kinematical factor) has turned out to have pole(s) at $W_-(W_+)$ with negative residue(s) (ghost(s) !). However, any such kind of pole has been shown to be cancelled exactly by the second term, viz. $f_\pm^{irr}(W)$ [2]. Thus even if the vertex function does have a pole, we may not be able to conceive of its existence through physical scattering data.

Fortunately there appears one limiting situation where the contribution comes only from the dressed

SSN:0094-243X/79/540067-02$1.50

(direct plus crossed) nucleon pole term so that we may observe some signal from the possible vertex pole(s). This is the well known soft pion limit : although not exactly physical, it can be close to physical situations. In such a limit Adler[3] showed using PCAC that the iso-symmetric πN invariant amplitude $A^{(+)}(s,u,t)$ at $s = u = o$ gets contributions from nucleon pole term and is equal to $G^2/4\pi m_N$ with G the strong πNN coupling constant.

We have calculated $A^{(+)}$ ($s = u = o, t$) from the dressed nucleon pole term (direct and crossed) using our $\Gamma(W)$ and $S'_F(W)$ with the result :

$$A^{(+)}(s=u=o,t) = \frac{G^2}{4\pi m_N}\,[1-\Gamma^2(-m_N)J(-m_N)] = 1.07\times\frac{G^2}{4\pi m_N} \quad (2)$$

with $$J(W) \equiv S'_F(W)/S_F(W) \quad , \quad S_F(W) \equiv \frac{1}{W-m_N}$$

(note that the above result is obtained not in the soft pion limit, so a deviation of $O(m_\pi^2/m_N^2)$ is to be naturally expected). The fact that it is very close to Adler's value is due to the small value of $\Gamma(-m_N)$, and the small $\Gamma(-m_N)$ has resulted from the very existence of the pole at W_- which has absorbed a good part of the strength in $\Gamma(W)$. Furthermore, the above result indicates that the effective πNN coupling to be used in low energy πN processes prefers pseudo-vector type.

As for the possible existence of the pole at W_+ we can say nothing definite at present.

REFERENCES

1. T. Mizutani and P. Rochus, to appear in Phys. Rev. C (1979).
2. C.J. Goebel and B. Sakita, Phys. Rev. Lett. 11, 293 (1963) ; Y.S. Jin and S.W. MacDowell, Phys. Rev. 137B, 688 (1965).
3. S.L. Adler, Phys. Rev. 137B, 1022 (1965).

Δ^+ PRODUCTION IN THE REACTION $pp \rightarrow pp\pi^\circ$ AT 800 MeV*

G. P. Pepin, G.S. Mutchler, M. Furić,† I. M. Duck,
R. D. Felder,†† J. Hudomalj-Gabitzsch,‡ W. H. Dragoset,
and G. C. Phillips

Rice University, Houston, TX 77001

J. C. Allred, E. V. Hungerford, B. W. Mayes, L. S. Pinsky, and T. M. Williams

University of Houston, Houston, TX 77004

Δ^+ production in the reaction $pp \rightarrow pp\pi^\circ$ has been studied in a kinematically-complete experiment using the LAMPF 800 MeV External Proton beam. Details of the experimental setup and data analysis are given more extensive discussion in reference 1. The data were taken in six angular settings, covering a range of $|t_{\Delta^+}|$ from 0.20 to 0.47 $(GeV/c)^2$.

Preliminary cross sections $d^5\sigma/dp_1 d\Omega_1 d\Omega_2$ are shown in the figure as a function of P_1, the measured proton momentum. The peak in the cross section at about 900 MeV/c corresponds to the Δ^+. This peak is most pronounced at low momentum transfer, decreasing in magnitude as the momentum transfer increases. The peaks at the high and low ends of the spectra are phase space effects due to tangency at the ends of the kinematically-allowed locus.

The cross sections were compared to the peripheral model for the reaction $pp \rightarrow p\pi^+ n$ of reference 1, modified for the $pp \rightarrow pp\pi^\circ$ reaction. The calculations involved π and ρ exchanges for the Δ^+ resonance, nonresonant s and p wave one pion exchange, and the N*(1470) resonance. The parameters used for the calculations are the same as those used in reference 1 except for the values of Λ_π. Good qualitative agreement with the data was obtained by using Λ_π=820 MeV/c for the π-N range parameter. Quantitively, the position of

*Work supported by U.S. D.O.E. Contract EY-76-C-05-1316.
†On leave from Institute "Ruder Boskovic," Zagreb, Yugoslavia.
††Present address: Exxon Production Research Co., Houston, TX.
‡Present address: Teledyne Exploration Co., Houston, TX.

ISSN:0094-243X/79/540069-02$1.50

the calculated peak for the Θ_1, Θ_2=15°,41° spectrum is displaced with respect to the data. Also, the curve falls below the data for the low momentum portion of the Θ_1, Θ_2=14°,30° spectrum. This may be improved with additional diagrams not included in the model or by varying parameters other than Λ_π, such as M_Δ. From the preliminary data we conclude that the theory constructed for the $pp\to p\pi^+ n$ channel is also adequate for the $pp\pi^0$ channel. This goes beyond the assumption of charge dependence since we experimentally explored different regions of phase space for the two channels. We expect to find that a number of spectra for the two reaction channels can be fitted by a unique choice of the parameter Λ_π.

REFERENCE

[1]J. Hudomalj-Gabitzsch *et al.*, Phys.Rev. *C18* (1978) 2666.

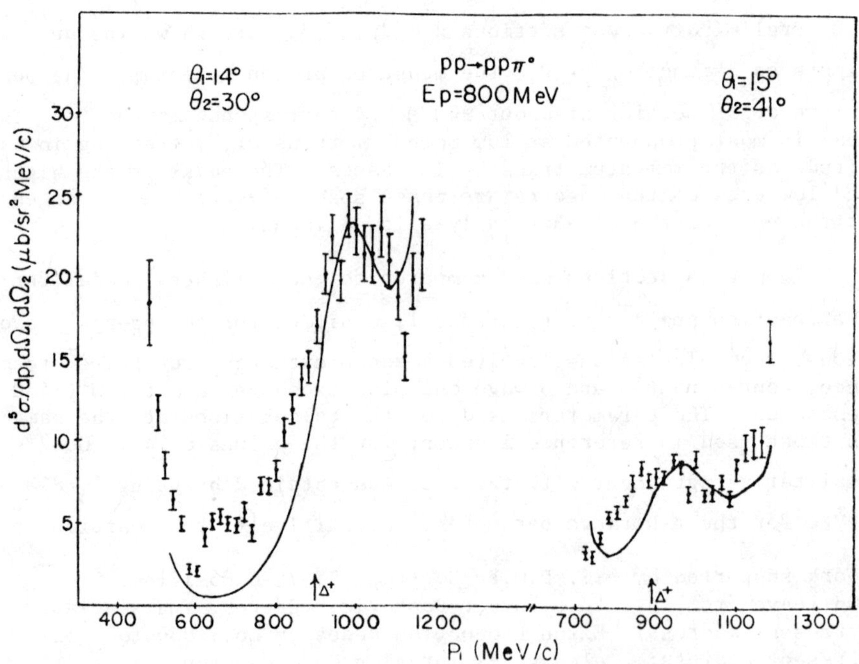

Cross section versus measured proton momentum.
Solid lines are the predictions using the peripheral model.

SEPARABLE APPROXIMATIONS TO SHORT RANGE INTERACTIONS

D. J. Ernst and R. J. McLeod
Texas A&M University, College Station, Tx. 77843

ABSTRACT

A technique for approximating short range interactions by separable interactions of simple analytic form is presented. Approximations to exponential potentials and the Reid soft core potential have been investigated. The unique feature of the approximation is that it can produce separable potentials whose scattering amplitudes are good approximations near a predetermined energy. The range of validity of the approximation is found to expand rapidly as one increases the rank of the potential. We conclude that any energy-independent, short range interaction can be well approximated by a separable potential of low rank.

INTRODUCTION

Because low rank separable potentials are quite convenient to use, it has been quite common to describe any two-body interaction (nucleon-nucleon, meson-nucleon, or meson-meson interaction) by a separable potential. There is often no justification for the separable form of the interaction other than its usefulness. We find, however, that short range, energy independent interactions can be well approximated over an extented energy region by a separable interaction. Thus the justification that a low rank, separable potential is a reasonable model requires only a knowledge that the basic interaction is energy independent.

APPLICATIONS

The general form of the separable approximation[1] to the potential V of rank N is

$$\mathcal{V}_{SEP} = \sum_{i,j=1}^{N} V|f_i\rangle M_{ij}\langle f_j|V \ , \tag{1}$$

with

$$\sum_j M_{ij} \langle f_j|V|f_k\rangle = \delta_{ik} \ . \tag{2}$$

We use $\langle r|f_i\rangle = j_\ell(k_i r)$ for scattering in the orbital angular momentum state ℓ.[1] This choice yields an analytic expression for $\mathcal{V}_{SEP}$ in both coordinate space and momentum space if V is a local exponential or Yukawa interaction. One may choose the parameter k_i to yield an approximation which is valid in a predetermined energy range. A judiscious choice of N values of k_i can yield a rank N separable potential which is valid over a large energy region centered about a predetermined energy. A detailed study of this approximation to

ISSN:0094-243X/79/540071-02$1.50

an expotential potential was given in Ref. (2).

We have applied this approximation to the Reid soft core nucleon-nucleon potential. Since this potential has regions of strong attraction which are cancelled by regions of strong repulsion, we feel this is a severe test of the approximation. We find that a rank three potential works well up to energies of E_{LAB} = 325 MeV in all partial waves. The results for the $^1D_2$ phase shifts are presented in the figure.

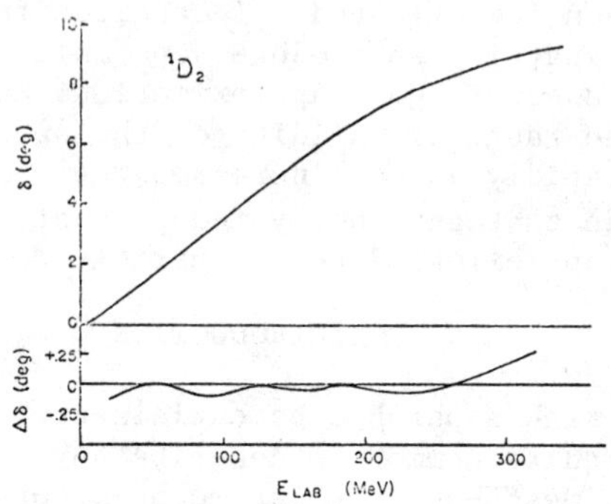

The upper curve shows the phase shifts produced by the Reid potential; the lower curve shows the difference, Δδ, between the Reid phase shifts and those produced by our separable approximation. The largest errors are at E = 0 and E = 320 MeV because we chose to fit over this range. These can be reduced by either increasing the rank of the potential or reducing the range of energies over which one requires the fit.

It has long been known that near a resonance or bound state pole an interaction can be well approximated by a separable form. We have found that one need not be near such a singularity for a separable approximation of low rank to be reasonable even for a rather violently behaved potential such as the nucleon-nucleon potential.

The success of the separable approximation studied here may be used as justification of a separable potential model for any energy independent interaction, for no matter what the form of the real physical interaction it most probably could be approximated well by a separable potential. This, however, does not imply that the pion-nucleon interaction can be represented by a separable potential. This is because the nucleon pole introduces an inherent energy dependence that must be treated explicitly.

REFERENCES

1. D. J. Ernst, C. M. Shakin, and R. M. Thaler, Phys. Rev. C10, 2176 (1974).
2. R. J. McLeod and D. J. Ernst, Phys. Rev. C18, 1060 (1978).

STUDY OF THE FINITE RANGE STRUCTURE OF THE πNN AND πNΔ VERTICES

M. Brack
University of Regensburg, Regensburg, W-Germany

M. Dillig
University of Erlangen-Nürnberg, Erlangen, W-Germany

Off shell effects due to the finite range structure of the πNN and the πNΔ vertex play a very important role in the calculation of various processes, to mention only their influence on the two-meson-exchange contribution to the NN interaction[1] or on the dynamics of medium energy processes, which are in general dominated by Δ excitation induced by π exchange[2]. As the extraction of finite range corrections from the experiment is presently rather ambiguous, we study the structure of the πNN, the πNΔ and the πΔΔ vertices in a microscopic model. In detail the following steps are performed:

- the πNN and the πNΔ vertices are build up as a sum of vertex corrections (triangle diagrams with virtual πσ, πρ, ρ, ω, N and Δ components) and renormalization terms to the π-propagator (with similar intermediate components included);
- the contribution of these various diagrams are calculated in 1. order perturbation theory;
- the leading pieces, i. e. the πρ triangle contribution and the virtual $N\bar{N}$ excitation in the π propagator are summed by solving a set of coupled integral equations.

We find that the πNN and πNΔ vertex functions can be approximated for virtual pions with $\omega_\pi = 0$ and $q_\pi \lesssim 800$ MeV/c by a monopole form factor with a cut off mass $\Lambda_{\pi NN} = 8\ m_\pi$ and $\Lambda_{\pi N\Delta} = 6\ m_\pi$; they scale approximately as $\Lambda^2 \sim f^{-1}_{\rho NN}$ with the ρNN coupling constant as the most sensitive parameter (the values above were derived from $f^2_{\rho NN}/4\pi = 5$)[3].

ISSN:0094-243X/79/540073-02$1.50

Further applications of the model, in particular for the study of the energy dependence of Λ as a function of ω_π and for the derivation of finite range corrections of the vector meson-nucleon vertices are under investigation.

1. K. Holinde and R. Machleidt, Nucl. Phys. A 256, 479 (1976)
2. M. Brack, D. O. Riska and W. Weise, Nucl. Phys. A 287, 425 (1977)
3. M. Dillig and M. Brack, Jour. Nucl. Phys. 5 (1979)

NN SCATTERING ABOVE THE PION THRESHOLD

A.S. Rinat
Weizmann Institute of Science, Rehovot, Israel

A.W. Thomas
TRIUMF, University of British Columbia,
Vancouver V6T 1W5

ABSTRACT

We studied the NNπ system in a non-relativistic field theory with given NN and πN interactions (V_{HBE}, i.e. heavy boson exchanges) which cannot be generated by the also given NNπ vertex [1]. Using graph summations or reduction techniques, we obtain rigorous solutions in the one-pion approximation [2]. Coupled equations connect as follows amplitudes for the NN, NΔ (N,πN isobars) and πd (generally 'a') channels

$$T_{ab} = B_{ab} + \sum_c B_{ac} G_c T_{cb} \quad (1)$$

B_{ab} is a single π or N exchange graph except for a=b=NN, when B is the sum of OPE + HBE, i.e. the total one-boson-exchange potential; G_c are propagators for an interacting pair c and spectator. Explicit solution for $T_{NN,NN}$ yields a V_{NN} which under neglect of the πd channel reads ($V_{OBE}=V_{HBE}+V_{OPE}$)

$$V_{NN} = V_{OBE} + B_{NN,\pi\Delta}\left(G_{\Delta}^{-1} - B_{N\Delta,N\Delta}\right)^{-1} B_{\pi\Delta,NN} \quad (2)$$

The model has the following features:
1) The two N's are treated symmetrically and the NN channel is not part of the input. These were deficiencies in previous treatments [3, 4].
2) The approach given in [5] can theoretically be based.
3) One demonstrates that (1) is a version of a coupled channel approach [6] which in the one pion approximation is unitary. Further information can be found in [2].

REFERENCES

1. T. Mizutani and D.M. Koltum, Ann. Phys. NY 109, 1 (1977).
2. A.W. Thomas and A.S. Rinat, preprint WIS-78/53-Ph.
3. I. Afnan and A.W. Thomas, Phys. Rev. C10, 109 (1974).
4. A.S. Rinat, Nucl. Phys. A287, 399 (1977).
5. H.J. Weber, J. Eisenberg and M. Schuster, Nucl. Phys. A278, 491 (1977).
6. See for instance A. Green, Progress in Physics 39, 1109 (1976).

ISSN:0094-243X/79/540075-01$1.50

ANGULAR DEPENDENCE OF NN→NNπ CROSS SECTIONS AND THE PION FORM FACTOR

B. J. VerWest
Department of Physics, Texas A&M University
College Station, Texas 77843

ABSTRACT

The effect of the π-baryon form factors on the angular dependence of inclusive spectra for NN→NNπ is studied in the context of a Feynman diagram based peripheral model calculation.

By combining the ideas of the peripheral model and Feynman diagram calculations it is possible to get a unified description of all the present data for the reaction NN→NNπ in the energy region 700-800 MeV[1]. In this calculation the πNN and πNΔ form factors play an important role in fitting the data, and the scale of these form factors can be well determined from the present data. The calculation extends the basic idea of the one pion exchange peripheral model to include ρ meson exchange as well and considers both NN and NΔ intermediate states. At the πNN and πNΔ vertices where the pion is off-shell, the form factors

$$\Gamma_{\pi NN} = \frac{\Lambda^2 - \mu^2}{\Lambda^2 - q^2}\,, \quad \Gamma_{\pi N\Delta} = \left[\frac{2\Lambda^2 - \mu^2}{2\Lambda^2 - q^2}\right]^2 \tag{1}$$

have been used where q is the pion four-momentum. The form of $\Gamma_{\pi N\Delta}$ used was chosen since it falls off a factor of q^{-2} faster asymptotically than $\Gamma_{\pi NN}$ and yet for $q^2 \leq 0$ the two form factors are essentially the same.

The reaction np→pX at forward angles[2] is very sensitive to this form factor and using the angular dependence of this reaction the value of Λ along with an effective πNΔ coupling can be well determined. Shown in fig. 1 are two fits to the $\theta=0^0$ spectra using Λ = 1150 MeV (solid curve) and Λ = 750 MeV (dashed curve). The Λ = 1150 MeV calculation gives the proper angular dependence for this reaction and also fits the spectra for pp→π+X[3] well as is shown in fig. 2. The calculation with Λ = 750 MeV shows too rapid a fall off with increasing angle in the reaction np→pX. In the reaction

ISSN:0094-243X/79/540076-02$1.50

pp→π+X the results for Λ = 750 MeV are about 30% smaller than for Λ = 1150 MeV. At forward angles Λ = 750 MeV gives very poor results compared with the data while at 60^0 it seems to agree fairly well. Recently, however, new preliminary data on this reaction was released[4] which confirms the 20^0 data in fig. 2 but that data at 40^0 and near 60^0 is considerably higher than the data in fig. 2. This indicates that these spectra may be larger than shown in fig. 2 and close to the Λ = 1150 MeV curves. Overall, only values near Λ = 1150 MeV give good agreement with the available inclusive spectra.

1. B. J. VerWest, Field Theoretical Calculation of NN Pion Production, TAMU preprint ORO-5223-13.

2. G. Glass, et al., Nucleon-Nucleon Interactions - 1977 (Vancouver) (American Institute of Physics, N.Y., 1978) 544.

3. Case Western Reserve University/LASL group (F. Cverna private communication).

4. T. M. Williams, et al., Bull. Amer. Phys. Soc. 23, 951 (1978).

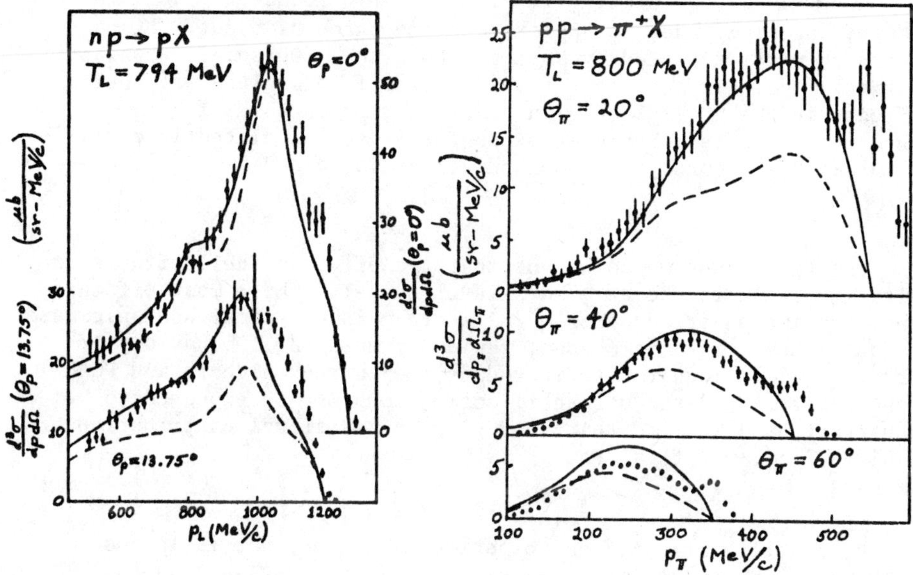

Fig. 1. Data from ref. 2, Curves for Λ=1150 MeV (solid) and Λ=750 Mev (dashed).

Fig. 2. Data from ref. 3, Curves as in fig. 1.

THE πNN VERTEX FUNCTION AND OFF SHELL πN SCATTERING

Anna Cass and Bruce H.J. McKellar
University of Melbourne, Parkville, Vic. 3052. Australia.

It is well known that the πNN vertex function exhibits structure as a function of t, the square of the pion mass. Since the vertex function is known to be the πNN coupling constant G at $t=\mu^2$, it is convenient to describe the structure by introducing a form factor F(t) and writing the vertex function as $\bar{u}\gamma_5 uGF(t)$. One can then write a dispersion relation for F(t) [1]

$$F(t) = 1 + \frac{t-\mu^2}{\pi} \int_{(3\mu)^2}^{\infty} dt' \frac{\text{Im } F(t')}{(t'-t)\,(t'-\mu^2)}$$

Physically F(t) may be probed in a variety of processes involving pion exchange, but it is often difficult to separate F(t) from other effects. Some success in this program has been achieved in the analysis of $pn \to np$ and $\bar{p}p \to \bar{n}n$ reactions [2]. However the most readily accessible test of F(t) is the Golberger-Treiman discrepancy [1].

$$\Delta_{GTR} = 1 - F(t=0)$$

The present experimental value of Δ_{GTR} is [3] 0.06±0.01.

One way of calculating Im F(t) is to introduce the $\pi N \to \pi N$ scattering amplitude, to approximate the three pion cut. This approach was introduced by Nutt and Loiseau [5]. Durso, Jackson and VerWest [6] pointed out that it was essential to include off shell effects in the πN amplitude in order to reproduce the field theoretic results for the ρπ exchange terms, calculated in detail by Braathen [7]. Durso, Jackson and VerWest obtained

$$\Delta_{GTR}^{DJV} = 0.02$$

In this paper we point out that the off mass shell effects are also important in the $\pi\pi \to NN$ s-wave channel. While most off mass shell variation if of order q^2/m^2 where m is a typical hadronic mass (about 1 GeV), in this channel rapid off mass shell behaviour of order q^2/μ^2 is required to reproduce the current algebra and PCAC constraints. Thus the off shell corrections may be expected to be important, and we find that they give an additional contribution of 0.01 to Δ_{GTR}. Our final result is

$$\Delta_{GTR}^{theo.} = 0.03$$

which is still only half the experimental value, but is in good agreement with the field theoretic value of Jones and Scadron [4].

ISSN:0094-243X/79/540078-02$1.50

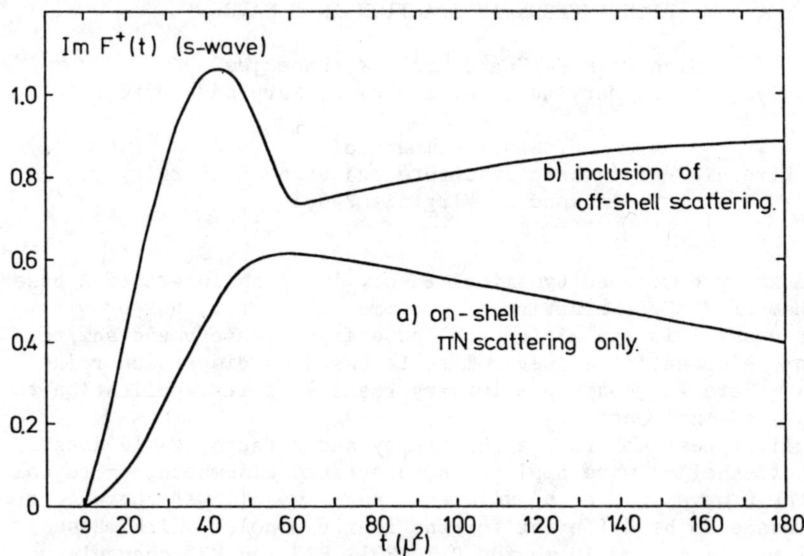

Fig.1. Plot of the s-wave part of Im F(t) calculated using the Höhler and Pietarienen helicity amplitudes [11] for

a) the on shell case and

b) with off shell πN scattering included

REFERENCES

1. M.L. Goldberger and S.Treiman, Phys. Rev. 111, 354 (1958).
2. A. Cass and B.H.J. McKellar, Phys. Rev. D18, 3269 (1978).
3. M.M. Nagels, J.J. deSwart, H. Nielsen, G.C. Oades, J.L. Petersen, B. Tromborg, G. Gustafson, A.C. Irving, C.Jarlskog, W. Pfeil, H. Pilkuhn, F. Steiner, L. Tauscher, Nucl. Phys. B109, 1 (1976).
4. H.F. Jones and M.D. Scadron, Phys. Rev. D11, 174 (1975).
5. W.T. Nutt and B. Loiseau, Nucl. Phys. B104, 98 (1976). W.T. Nutt, Phys. Rev. C16, 1124 (1977).
6. J.W. Durso, A.D. Jackson and B.J. VerWest, Nucl. Phys. A282, 404 (1977).
7. H.J. Braathen, Nucl. Phys. B44, 93 (1972).
8. G.C. Chew, M.L. Goldberger, F.E. Low and Y. Nambu, Phys. Rev. 106, 1337 (1957).
9. S.A. Coon, M.D. Scadron, P.C. McNamee, B.R. Barrett, D.W.E. Blatt and B.H.J. McKellar, Nucl. Phys. (to be published).
10. T.P. Cheng and R. Dashen, Phys. Rev. Letts. 26, 594 (1971).
11. G. Höhler, F. Kaiser, R. Koch and E. Pietarienen, Physics Data 12-1 (1978).

PION-NUCLEON INTERACTION IN P WAVE

Nien-Chih Wei* and Manoj K. Banerjee*
University of Maryland, College Park, Maryland 20742

J. Barry Cammarata†
Virginia Polytechnic Institute and State University
Blacksburg, Virginia 24061

A recently developed dynamical theory[1,2] of πN interaction based on the use of PCAC, the notion of σ commutator, etc., has proved quite successful in explaining the low energy S wave phase shifts. The theory also satisfies several tests based on dispersion relations.[3,4] Here we report preliminary results of its application to the p wave πN interaction.

The first test was to use the theory and a factorizable ansatz for the off-shell P wave amplitudes, described elsewhere,[1,2] to calculate the P wave scattering volumes. Here, we allowed the $\pi N\Delta$ form factor masses to be different for the four channels. With masses 8 m_π, 11 m_π, 11 m_π and 10 m_π for P11, P31, P13 and P33 channels we obtained values listed in the first line of Table I. The second line gives the experimental values.[5]

Table I. Theoretical and experimental scattering volumes in units of m_π^{-3}

	P11	P31	P13	P33
Theory	-0.122	-0.029	-0.024	0.214
Experiment	-0.089 ± 0.02	-0.043 ± 0.011	-0.037 ± 0.016	0.214 ± 0.009

Noting that all parameters relevant to the P33 channel were determined by earlier works[1-4] the good fit for the scattering volume indicates the basic soundness of the theory. The agreement is not bad for the other three channels, where the interaction is weaker.

At other energies the theory can be similarly tested by substituting the off-shell amplitudes of the factorable form into the integral equation and checking if the resulting amplitudes agree with the input set. The comparison worsens with increasing energy.

As yet a resolution of this problem has not been achieved. However, our studies have revealed several new and important features of the P wave dynamics. Most notably we find that the dynamical

*Supported in part by the U. S. Department of Energy.
†Supported in part by the U. S. National Science Foundation.

ISSN:0094-243X/79/540080-02$1.50

picture which emerges from a static approximation to the problem (Chew-Low) must be revised. Mere inclusion of nucleon recoil and relativistic kinematics reduces the size of the integrals in the equations for the amplitudes. This reduction is sufficient to make it impossible for the nucleon pole terms to produce the resonance.

These kinematic factors plus the soft pion subtraction make the left-hand cuts (crossed terms) and the inelastic channels much less important in the P33 channel compared to their roles in a static calculation.

We have also found that the Z graph contributions are very important. They enhance the attraction in the P33 channel.

REFERENCES

1. M. K. Banerjee and J. B. Cammarata, Phys. Rev. D16, 1334 (1977).
2. M. K. Banerjee and J. B. Cammarata, Phys. Rev. C17, 1125 (1978).
3. M. K. Banerjee and J. B. Cammarata, Phys. Rev. D18, 4078 (1978).
4. M. K. Banerjee and J. B. Cammarata, to appear in the Jan. 1, 1979 issue of Phys. Rev. D.
5. W. Langbein, Nucl. Phys. B94, 519 (1975).

A NEW SET OF EQUATIONS FOR THE COUPLED πNN–NN SYSTEM

Y. Avishai

D.Ph.N/HE, CEN Saclay, F-91190 Gif-sur-Yvette, France

T. Mizutani

Physique Nucléaire Théorique, Institut de Physique, Université de Liège, Sart Tilman, B-4000 Liège 1, Belgium.

The importance of correctly describing a set of coupled systems, πNN (or πd) and NN cannot be overstated: (i) on one hand there is a definite need for including the effect of true pion absorption in theories of pion-nucleus scattering[1], into which the study in the simplest π-d scattering should give some insight, (ii) on the other hand there has been some interest in calculating inelastic NN scattering quantities such as complex phase shifts, $\sigma_{inel.}$ etc.[2,3] in connection with the isobar coupled-equation model, e.g.

Starting from a coupled-channel approach described in ref. 4, we have found two sets of coupled equations for the πNN (πd) and NN systems. Combined with a separable ansatz to the two-body inputs (to be used in the πNN three-body sectors) these coupled integral equations read symbolically :

$$\begin{aligned} X^{(A)}_{\alpha\beta} &= Z^{(A)}_{\alpha\beta} + \sum_{\gamma} Z^{(A)}_{\alpha\gamma}\tau_{\gamma}X^{(A)}_{\gamma\beta} + \frac{1}{2} Z^{(A)}_{\alpha N}\tau_{N}X^{(A)}_{N\beta} \\ X^{(A)}_{N\beta} &= Z^{(A)}_{N\beta} + \sum_{\gamma} Z^{(A)}_{N\gamma}\tau_{\gamma}X^{(A)}_{\gamma\beta} + \frac{1}{2} (v^{(A)}_{o} + Z^{(A)}_{NN})\tau_{N}X^{(A)}_{N\beta} \end{aligned} \tag{1}$$

$$\begin{aligned} X^{(A)}_{\alpha N} &= Z^{(A)}_{\alpha N} + \sum_{\gamma} Z^{(A)}_{\alpha\gamma}\tau_{\gamma}X^{(A)}_{\gamma N} + \frac{1}{2} Z^{(A)}_{\alpha N}\tau_{N}X^{(A)}_{NN} \\ X^{(A)}_{NN} &= (v^{(A)}_{o}+Z^{(A)}_{NN}) + \sum_{\gamma} Z^{(A)}_{N\gamma}\tau_{\gamma}X^{(A)}_{\gamma N} + \frac{1}{2}(v^{(A)}_{o}+Z^{(A)}_{NN})\tau_{N}X^{(A)}_{NN} \end{aligned} \tag{2}$$

In the above expressions Greek subscripts (α, β, etc) denote channels with a correlated pair (NN or πN) plus a spectator, and "N" represents the NN state. X's, Z's and τ's are transition amplitudes, driving terms and propagators for a correlated pair plus a spectator, respectively, as commonly used in three-body theories. Furthermore, quantities with superscript (A) signify that they are antisymmetrized with respect to two nucleons. Finally v_o is some part of the NN interaction to be explained below.

The equations look like those found in standard

ISSN;0094-243X/79/540082-02$1.50

three-body theories with separable two-body inputs. However there are some differences : (i) the pion exchange driving term $Z_{NN}^{\dagger}$ consists of two terms reflecting the fact that both nucleons can equally emit or absorb the pion, (ii) similarly, in the two-nucleon propagator each nucleon has a self energy bubble, (iii) there is a contribution v_o to the N-N interaction which corresponds to some higher mass exchanges. More precisely one may establish a relation between v_o and the total N-N potential v, as follows (in the separable approximation)

$$v^{(A)} = v_o^{(A)} + Z_{NN}^{(A)} + \sum_{\alpha,\beta} Z_{N\alpha}^{(A)} \tau_\alpha [\delta_{\alpha\beta} + \tilde{X}_{\alpha\beta}^{(A)} \tau_\beta] Z_{\beta N}^{(A)} \qquad (3)$$

where $\tilde{X}_{\alpha\beta}^{(A)}$ is obtained from the equation for $X_{\alpha\beta}^{(A)}$ (eq. (1)) by omitting the last term responsible for the coupling to NN states.

It can be shown that our eqs. (1) and (2) are equivalent to those found in refs. 4 and 5 [eqs. (2.27), (2.28) of ref. 4, e.g.] but are far more practical in that, for example, they can automatically incorporate the non-static one-pion-exchange N-N interactions, which is crucial to fulfil exact two- and three-body unitarity. Also it may be worth pointing out that our equations are free of difficulties (such as violation of Pauli principle, a difficulty in including the non-nucleon pole part of the $P_{11}(\pi N)$ scattering, etc.) encountered in the so-called "bound state picture" where one of the two nucleons in the N-N sector is identified as a $P_{11}(\pi N)$ bound state with the binding energy equal to the pion mass.[6]

An actual calculation using eqs. (1) and (2) together with the relativistic kinematics is in progress.

REFERENCES

1. See, for example, F. Lenz, Proc. of Int. Conf. on High Energy Physics and Nuclear Structure (Zürich) 175 (1977).
2. W.M. Kloet, R.R. Silbar, R. Aaron and R.D. Amado, Phys. Rev. Lett. 39, 1643 (1977).
3. A.M. Green, J.A. Niskanen and M.E. Sainio, J. of Phys. G4, 1055 (1978).
4. T. Mizutani and D.S. Koltun, Ann. Phys. 109, 1 (1977).
5. A.S. Rinat, Nucl. Phys. A287, 399 (1977).
6. I.R. Afnan and A.W. Thomas, Phys. Rev. C10, 109 (1974).

† also $Z_{\alpha N}$ and $Z_{N\beta}$

FIELD-THEORETIC ASPECTS OF πN INTERACTION

R. F. Alvarez-Estrada
Departamento Fisica Teorica, Universidad Complutense, Madrid-3, Spain

ABSTRACT

For certain low-energy field-theoretic πN hamiltonians, rigorous bounds on the probability for finding n π's in the dressed N-state are given. They justify the neglection of states with many π's, which gives rise to the Tamm-Dancoff approximation. One of the resulting disconnectedness problems is studied.

We use a generalization of the static Chew-Low theory which includes N-recoil[1]. The total hamiltonian is

$$H=H_0+H_1 \ , \ H_0=\frac{\bar{p}^2}{2M_0}+\sum_{j=1}^{3}\int d^3\bar{k}\,\omega(k)a^+(\bar{k}j)a(\bar{k}j) \ , \quad k=|\bar{k}| \tag{1}$$

$$H_1=\frac{f_0}{\mu}\sum_{j=1}^{3}\int d^3\bar{k}\frac{V(k)}{[(2\pi)^3 2\omega(k)]^{1/2}}\left\{i[\bar{k}(1+\lambda_1)+\lambda_2\bar{p}]\bar{\sigma}\tau_j\,a(\bar{k}j)\exp i\bar{k}\bar{x}+h.c.\right\} \tag{2}$$

$M_0, \bar{p}, \bar{x}$ are the nucleon bare mass, threemomentum and position operators other notations are standard[2], and λ_1, λ_2 are dimensionless coupling constants, which either vanish or are of order μ/M_0. Let Ψ_+ be the normalized dressed N-state with energy $E_{\bar{q}}$ $((H-E_{\bar{q}})\Psi_+=0)$ and small total threemomentum $\bar{q}$. Spin and isospin dependences in wave functions will not be made explicit. The results below are also valid for the static case if $M_0=\infty$, $\lambda_1=\lambda_2=0$ and the $\bar{q}$-dependence is omitted. Let ρ_n be the total probability for finding n π's in Ψ_+ $(\sum_0^\infty \rho_n=1)$. By generalizing previous work[3,4], one finds the rigorous bound:

$$\rho_n\leqslant\frac{4}{n!}\cdot\left[\frac{3}{(2\pi)^2}\left(\frac{f_0}{\mu}\right)^2\right]^n\cdot\prod_{s=1}^{n}\int_0^{+\infty}dk_s\,\frac{k_s^2\,V(k_s)^2}{\omega(k_s)}\cdot X_n^2 \ \equiv \rho_n' \tag{3}$$

$$X_n=\prod_{h=1}^{n}\left\{\frac{k_h(1+\lambda_1)}{\sum_{r=h}^{n}\omega(k_r)+E_{\bar{q}=0}-E_{\bar{q}}}+\frac{\lambda_2\cdot Y\left(\sum_{r=h}^{n}\omega(k_r)\right)}{\left[\sum_{r=h}^{n}\omega(k_r)+E_{\bar{q}=0}-E_{\bar{q}}\right]^{1/2}}\right\}+\Lambda \tag{4}$$

where Λ denotes the sum of all $n!-1$ terms obtained from the first term on the right-hand-side of Eq.(4) by permutation of $(k_1,k_2,\dots k_n)$ and $Y(\omega)$ is the norm of $(\omega+H-E_{\bar{q}})^{-1/2}\cdot\bar{\sigma}\bar{p}$ in the subspace of states with total threemomentum $\bar{q}$ (a divergent factor $\delta^{(3)}(\bar{0})$ being factored out). One shows that, at least for small f_0, $Y(\omega)<\infty$ and $\rho_n'<\infty$, $\rho_n'\to 0$ as $n\to\infty$. For $\lambda_1=\lambda_2=0$, one finds a stronger result:

$$\rho_n'=4\frac{a^n}{n!} \ , \quad a=\frac{3}{(2\pi)^2}\cdot\left(\frac{f_0}{\mu}\right)^2\cdot\int_0^{+\infty}dk\,\frac{k^4\,V(k)^2}{\omega(k)[\omega(k)+E_{\bar{q}=0}-E_{\bar{q}}]^2} \tag{5}$$

so that $\rho_n'\to 0$ as $n\to\infty$ for any f_0, and a numerical estimate of ρ_n' for the physical values[2] of $f_0, V(k)$ indicates that the contribution of n

ISSN:0094-243X/79/540084-03$1.50

π's to Ψ_+ can be neglected for $n \geq 6$. This neglection, upon which the Tamm-Dancoff approximation[5] is based, is, probably, also reliable for smaller n, both with and without recoil. In the static case, we have seen that the N self-energy, when $n \geq 3$ π's are neglected, is in qualitative numerical agreement with estimates based upon sum rules and low-energy data[2]. We shall concentrate on one of the dynamical equations appearing in the Tamm-Dancoff approach[5] to dressing, scattering and production(thus, including π creation and absorption and finite renormalization effects).The momentum space wave function $c_3 = c_3(\bar{k}_1 \bar{k}_2 \bar{k}_3)$ for three π's with threemomenta $\bar{k}_1 \bar{k}_2 \bar{k}_3$, the N one being $\bar{q} - \sum_1^3 \bar{k}_i$, satisfies the linear integral equation:

$$c_3 = c^{(0)} + \left[\sum_{i=1}^{3} T_i^{(1)} + \sum_{i,j=1,\, i<j}^{3} T_{ij}^{(2)} + T^{(3)} \right] . c_3 = c^{(1)} + \left[\sum_{i=1}^{3} T_i^{(1)} \right] . c_3 \qquad (6)$$

$c^{(0)}$ is a given inhomogeneous term, the T's are given kernels depending on $\bar{k}_1 \bar{k}_2 \bar{k}_3, \bar{k}'_1 \bar{k}'_2 \bar{k}'_3$ and an integration over $\bar{k}'_1 \bar{k}'_2 \bar{k}'_3$ is understood.In general, $T_i^{(1)} (T_{ij}^{(2)})$ contains $\delta^{(3)}(\bar{k}_r - \bar{k}'_r) . \delta^{(3)}(\bar{k}_h - \bar{k}'_h), r \neq h \neq i, r \neq i$ $(\delta^{(3)}(\bar{k}_s - \bar{k}'_s), i \neq s \neq j)$ and $T^{(3)}$ is δ-function free.Eq. (6), which gives rise to disconnectedness problems, is obtained from the Schrodinger equation for H via the Tamm-Dancoff approach[5], when:i) either the contributions from n π's $(n \geq 5)$ are neglected,ii) or one neglects states with $n \geq m$ π's $(m \geq 6)$ and is able to express all m-pion wave functions in terms of c_3. We shall reduce Eq.(6) to new equations with connected kernels.The disconnectedness problem associated with $T_i^{(1)}$,which corresponds to one π interacting with the N, while the other two π's behave as spectators, is solved by applying suitably Yakubovski's theory[6]:

$$c_3 = c^{(1)} + \sum_{i=1}^{3} \left[G_i c^{(1)} + \sum_{j=1,\, j \neq i}^{3} M_{ij} \right] , \quad G_i = T_i^{(1)} + T_i^{(1)} G_i \qquad (7)$$

$$M = (I_6 - W_1)^{-1} M^{(0)} + (I_6 - W_1)^{-1} W_2 M \qquad (8)$$

$$M = \begin{pmatrix} M_{23} \\ M_{32} \\ M_{13} \\ M_{31} \\ M_{12} \\ M_{21} \end{pmatrix} , \quad M^{(0)} = \begin{pmatrix} G_2 G_3 \\ G_3 G_2 \\ G_1 G_3 \\ G_3 G_1 \\ G_1 G_2 \\ G_2 G_1 \end{pmatrix} . c^{(1)} , \quad W_1 + W_2 = \begin{pmatrix} 0 & G_2 & 0 & G_2 & 0 & 0 \\ G_3 & 0 & 0 & 0 & 0 & G_3 \\ 0 & G_1 & 0 & G_1 & 0 & 0 \\ 0 & 0 & G_3 & 0 & G_3 & 0 \\ G_1 & 0 & 0 & 0 & 0 & G_1 \\ 0 & 0 & G_2 & 0 & G_2 & 0 \end{pmatrix} \qquad (9)$$

I_6 is the 6×6 unit matrix times the unit operator and W_1 is the block matrix formed by the three 2×2 submatrices inside the dotted boxes along the main diagonal on the right-hand-side of the last Eq. (9), and W_2 is formed by the remainder. One shows that any term in $[(I_6 - W_1)^{-1} . W_2]^3$ is a connected kernel(δ-function free).Then, by iterating Eq.(8) twice, one gets an equation for M with connected kernels.By solving the latter and using the definition of $c^{(1)}$, one gets $c_3 = c^{(2)} + (\sum_1^3 \sigma_i) c_3$, where $c^{(2)}$ is a new known inhomogeneous term and σ_i is a known kernel, which contains $\delta^{(3)}(\bar{k}_i - \bar{k}'_i), i = 1,2,3.$ This last disconnectedness problem is solved through a Faddeev-like decomposition:

$$C_3 = C^{(2)} + \sum_{j=1}^{3} U_j \quad , \qquad F_j = \sigma_j + \sigma_j F_j \quad , \; j=1,2,3 \tag{10}$$

$$\begin{pmatrix} U_1 \\ U_2 \\ U_3 \end{pmatrix} = \begin{pmatrix} F_1 \\ F_2 \\ F_3 \end{pmatrix} \cdot C^{(2)} + \begin{pmatrix} 0 & F_1 & F_1 \\ F_2 & 0 & F_2 \\ F_3 & F_3 & 0 \end{pmatrix} \cdot \begin{pmatrix} U_1 \\ U_2 \\ U_3 \end{pmatrix} \tag{11}$$

One sees that the square of the 3×3 matrix on the right-hand-side of Eq.(11) is formed by connected kernels. Thus, since $\tau^{(3)}$ is a connected kernel, we have solved all disconnectedness problems posed by Eq. (6).

REFERENCES

1. H. W. Ho, M. Alberg and E. M. Henley, Phys. Rev. C12, 217(1975). M. V. Barnhill III, Nucl. Phys. A131, 106(1969).
2. E. M. Henley and W. Thirring "Elementary Quantum Field Theory" Part III, McGraw-Hill, New York, 1962.
   S. S. Schweber "An Introduction to Relativistic Quantum Field Theory", Chapter 12, Evanston and Row, Illinois, 1961.
3. J. Frohlich, Ann. Inst. Henri Poincare, 19, 1(1973).
4. R. F. Alvarez-Estrada in "Few Body Systems and Nuclear Forces I" Proceedings Graz 1978, Lecture Notes in Physics 82, Springer, Berlin, 1978.
5. H. A. Bethe and F. de Hoffmann "Mesons and Fields.VolumeII: Mesons", section 40, Row, Evanston and Co.,Illinois, 1955.
6. O. A. Yakubovski, J. Nucl. Phys.(USSR) 5, 1312(1967).

AN OVERVIEW OF ABSORPTION AND PRODUCTION OF MESONS

Daniel S. Koltun*
Department of Physics and Astronomy
University of Rochester, Rochester, New York 14627

INTRODUCTION

This talk will present a brief look at some of the recent developments in experiment and theory of meson absorption and production. This is not a review, but more an attempt to show that a number of advances have been made in understanding the connection of absorption, inelastic scattering, and the optical model. Some other recent material on the general subject can be found in Refs. 1, 2, and in proceedings of recent conferences.

1. NEW TRENDS IN EXPERIMENT

Let us begin with experiments on pion absorption. Ten years ago there was considerable interest in correlation studies: $(\pi^-,2n)$, $(\pi^+,2p)$, and such, with particular interest in the mechanism for these particular reaction modes, and on the nuclear states which could be reached. After a burst of this activity at the older accelerators (which still continues at a slower rate), almost the only kind of experiments done were one particle out: (π,p) or the inverse (p,π) production reactions, to low excited states of the final nuclear targets, and "no particles out," that is, the $(\pi,\gamma_{nuclear})$ studies, in which only the final particle-stable products of absorption are identified. The first of these investigates an exclusive process, a branch of absorption with small probability, whose angular distributions are typical of direct reactions. The second gives rather inclusive data on the total rates for absorption, without telling as much about how the initial reaction takes place.

Recently we have begun to see results of a newer class of experiments, which I think will prove to be quite interesting for the future. These are inclusive reactions, which look at large branches of absorption, but in such a way that we can correlate the information with data on scattering reactions. The first of these is the simple experiments at LAMPF which have already been published[3]

$$\pi + A \to p + \text{anything} \tag{1}$$

and the second, the new studies at SIN, to be reported here[4] on counter studies which decompose the total cross section for π scattering into its major branches: elastic, inelastic, charge exchange, and absorption,

$$\sigma = \sigma_{el} + \sigma_{inel} + \sigma_{ex} + \sigma_{abs} \; . \tag{2}$$

*Work supported in part by the U. S. Department of Energy.

ISSN:0094-243X/79/540087-08$1.50

These data are presented at fixed energy, as log σ_m vs log A, which suggest power-law behavior of the cross sections for each mode of (2) (see Figure of Ref. 4):

$$\sigma_m \propto A^{P_m} . \tag{3}$$

The interest in these newer experiments is in the information they provide for understanding both scattering and absorption reactions of pions, as I hope to indicate in this talk.

There will be talks on π production following this in the conference. I do not know if we shall see inclusive data, that is, for protons or heavy ions:

$$p + A \rightarrow \pi + \text{anything} \tag{4a}$$

$$A + B \rightarrow \pi + \text{anything} \tag{4b}$$

but I shall say a little later on how this information fits in with approaches to scattering theory.

2. NEW TRENDS IN THEORY

In the last several years it has become clear to many theorists that it is necessary to understand absorption and production of mesons in order to understand scattering as well. This has several aspects: first to put the theory of multiple scattering on a firm footing for meson projectiles, where the field aspects of the mesons, including absorption, are treated consistently. There has been some success in this direction, although the practical applications have been restricted almost entirely to the πd case.[5,6,7]

A second aspect, which will be discussed below in greater detail, has to do with the connections between elastic and nonelastic channels, or put another way, with the roles of multiple scattering and meson absorption in the theory of the optical potential.

A third aspect of the interaction of scattering and absorption emerges from the many complicated modes of excitation of a nucleus by an energetic meson. New methods of treating the statistical aspects of such reactions are just emerging.[8,9] Previous approaches to this have been based on cascade calculations, following in a simplified way the many branches available to the system. The new approach is that of the transport equation, which has been used with success recently in discussing some aspects of heavy ion collisions. One might expect that this will also have applications to meson production in ion-ion scattering, Eq. (4b). We return to this at the end.

Another trend in the theory of absorption is the renewed interest in dealing with relativistic aspects of the problem. These arise both in dealing with the relativistic motion of particles, e.g. in (p,π),[10] or even in the question of the correct form of the absorption vertex for absorption of stopped pions.[11] I shall not

not discuss this further.

3. OPTICAL POTENTIAL AND MESON ABSORPTION

A problem which has puzzled many of us for some years is how properly to include meson absorption (or production) in the optical model for elastic scattering, U. There are several related questions here:

a) In the low-density or uncorrelated target approximation to multiple scattering theory, the optical potential for π-nucleus scattering is given by

$$U^{(1)}(r) \simeq t_{\pi N}\rho(r) = -\frac{2\pi}{\varepsilon}\rho(r)f(0) \qquad (5)$$

in terms of the t-matrix $t_{\pi N}$ or amplitude f(0) for forward πN scattering. How should (5) be modified to take account of absorption of the projectile?

b) A given optical potential predicts the differential cross section $\sigma_{el}(\theta)$, and also the total cross section for nonelastic processes σ_r (reactions). The sum of the integrated cross sections gives the total cross section, $\sigma = \sigma_{el} + \sigma_r$. There is a useful relation between σ_r and U, which has been much used in some recent papers:

$$\sigma_r = -\frac{2}{v}\langle\phi^{(+)}|\mathrm{Im}U|\phi^{(+)}\rangle, \qquad (6)$$

which expresses the distribution of the probability flux lost from the elastic wave, through the imaginary part of the optical potential, to all other channels. (Here $\phi^{(+)}$ is the optically distorted meson wave, with initial velocity v.) The question here is how to decompose σ_r into the specific channels involved, e.g. inelastic, charge exchange, absorption, as in (2), and in particular, how to do so in a particular theory of the optical potential; this has been denoted the reactive content of the optical potential.

c) There are different kinds of absorptive processes, which should affect the optical potential (5) and the reactive content in different ways. We shall consider two cases: absorption in one step and in two. This refers to the elementary processes; for the first, we have $\bar{K}$ absorption of the type

$$\bar{K}p \rightarrow Y\pi \qquad (7)$$

which may take place on a single free nucleon, as well as in a nuclear target. An analogous production process is $pp \rightarrow pp\pi$, which also can happen on single free nucleons, bombarded by protons. On the other hand, π-absorption cannot take place on free nucleons; we often think of it as a two-nucleon process in nuclei, not excluding the participation of more nucleons as well. It is sometimes thought that therefore only correlated targets can absorb pions; this is not so, as we shall see.

Absorption or production processes like (7) which may take place on single nucleons fit in a natural way into the theory of the optical potential for targets of uncorrelated nucleons particularly in the high energy limit. We use the form (5), with the forward amplitude f(0) approximated by its value for free nucleons, including the inelastic process (7) along with the elastic: $\bar{K}p \rightarrow \bar{K}p$. We obtain the imaginary part of (5) from the optical theorem: $\mathrm{Im}\, f(0) = k\sigma/4\pi$:

$$\mathrm{Im}\, U^{(1)}(r) = -\frac{k\sigma}{2\varepsilon}\rho(r) = -\frac{v}{2}\sigma\rho(r), \tag{8}$$

where σ is the total cross section (on a nucleon) in this case for $\bar{K}p$ scattering. The reaction cross section on the nuclear target may be obtained from (6):

$$\sigma_r = \sigma \int d^3 r \rho(\vec{r}) |\phi^{(+)}(\vec{r})|^2 \tag{9a}$$

$$\simeq \int d^2 b\ (1-e^{-L(b)/\lambda}) \tag{9b}$$

where the second line is a well-known result in the high energy limit, for a uniform density nucleus, in terms of the impact parameter b and the straight-line path length L(b); the mean free path is given by $\lambda = (\rho\sigma)^{-1}$. Now the cross section σ can be decomposed into σ_s for scattering ($\bar{K}p \rightarrow \bar{K}p$) and σ_a for absorption (7): $\sigma = \sigma_s + \sigma_a$, and ImU (8) reflects this separation. But the separation of the reaction cross section into $\sigma_{inel} + \sigma_{abs}$ (ignoring charge exchange) does not have such a simple form, and we cannot use (9) directly. The well-known results (e.g. Ref. 12) in the high energy approximation, are given by

$$\sigma_{abs} = \int db^2\ (1-e^{-L(b)\lambda_a}) \tag{10a}$$

$$\sigma_{inel} = \sigma_r - \sigma_{abs} = \int d^2 b\ e^{-L(b)/\lambda_a}(1 - e^{-L(b)\lambda_s}) \tag{10b}$$

with $\lambda_a = (\rho\sigma_a)^{-1}$, $\lambda_s = (\rho\sigma_s)^{-1}$. We note that although σ_s and σ_a enter symmetrically in the optical potential (8), they contribute quite differently in (10) to the inelastic and absorptive cross sections on a nucleus. Equation (10a) says that the probability of absorption is independent of scattering, while (10b) says that the probability of scattering is reduced by absorption, so that for large nuclei, for which most $L(b) >> \lambda_a$, σ_{inel} becomes much smaller than σ_{abs}. (Remember that Eqs. (9) and (10) also apply to production, from $pp \rightarrow pp\pi$.)

Now how are we to understand nuclear absorption of pions? The first-order optical potential given by (5) or (8) contains only elastic scattering $\pi N \rightarrow \pi N$; and $\sigma = \tilde{\sigma}_s$ in this case, since there is no one-nucleon absorption. It has been common to add a term to (5) for true absorption, depending on $\rho(r^2)$, representing correlations, as for π-atoms.[13] However, it is possible for pions to be absorbed by scattering one or more times on uncorrelated nucleons, then to be absorbed off-energy-shell, on another nucleon. Let us therefore consider a simple model of absorption on an uncorrelated target, as follows: The pion cannot be absorbed by the first nucleon it encounters; it must scatter at least once (to put it off-energy-shell), after which it may be absorbed by any nucleon, with an effective cross section $\tilde{\sigma}_a$. The optical potential in this case is simply (5), and

$$\text{Im } U^{(1)}(r) = -\frac{v}{2}\tilde{\sigma}_s\rho(r) \tag{8'}$$

as in (8), but with $\sigma = \tilde{\sigma}_s$. The reaction cross section in the high energy approximation takes the form

$$\tilde{\sigma}_r = \int d^2b\,(1 - e^{-L(b)/\lambda_s}) \tag{9'}$$

and looks like (9), but without σ_a; this follows since every reaction is initiated by a πN scattering in this model. The scattering and absorption cross sections can also be calculated in the high energy approximation (these are new results[14])

$$\tilde{\sigma}_{inel} = \left(\frac{\lambda'}{\lambda_s}\right)\int d^2b\, e^{-L(b)/\lambda_a}\,[1-e^{-L(b)/\lambda'}] \tag{11a}$$

$$\tilde{\sigma}_{abs} = \int d^2b\,\{(1 - \frac{\lambda'}{\lambda_s}e^{-L(b)/\lambda_a}) - \frac{\lambda_a}{\lambda_s}e^{-L(b)/\lambda_s}\} \tag{11b}$$

where $(\lambda')^{-1} = \lambda_s^{-1} - \lambda_a^{-1} = \rho(\tilde{\sigma}_s - \tilde{\sigma}_a)$. These results are distinctly different from those of the one-nucleon process, given in Eq. (10). If we take $\tilde{\sigma}_s$ for this model such that $\tilde{\sigma}_s = \sigma = \sigma_s + \sigma_a$ for the one-nucleon model, and $\tilde{\sigma}_a = \sigma_a$, then for a given target, comparing (9) and (9'), we find $\tilde{\sigma}_r = \sigma_r$. However, from (10) and (11) we find that $\tilde{\sigma}_{inel} > \sigma_{inel}$, and therefore $\tilde{\sigma}_{abs} < \sigma_{abs}$. This is quite reasonable: if the projectile must scatter before absorption, it travels a mean free path λ_s before absorption may begin, unlike in the one-step case. Note that the absorption does not appear at all in the optical potential in this case; this is because we have used a high energy approximation in which off-shell effects are not apparent. This would be modified both by correlations, and at lower energies, e.g. near a resonance. The point is that it is possible to have absorption, with no apparent change in the optical potential, which seems to neglect only scattering processes.

This last point is related to another recent development in the theory of multiple scattering. It has long been assumed that the inelastic process implied by the first-order optical potential (5) is knockout of uncorrelated nucleons. This has been studied is some detail by Tandy, Bolle, and Redish, who derived an interesting expression for the reaction cross section in a theory where there is no true absorption of the projectile. Their result may be written in terms of one-nucleon knockout probabilities:

$$\sigma_r = \frac{2\pi}{v}\sum_{i=1}^{A}\int |\langle\chi^{(-)}|t_i|\phi^{(+)}\rangle|^2 d\rho_f \tag{12}$$

where the sum on i is over the target nucleons, the integral is over the phase space for a knocked-out nucleon, and the meson in the final (knockout) state $\chi^{(-)}$ is a plane-wave, undistorted by the optical potential. We have also been able to show that this formula is actually related to a sum rule[16] for inelastic scattering which includes knockout of any number of nucleons, $1 \leq n \leq A$,

$$\sigma_r = \sigma_{inel} = \sum_{n=1}^{A} \sigma(n) \tag{13}$$

where $\sigma(n)$ is an exclusive cross section for n-body knockout to be calculated by DWIA. The sum (13) or (12) then gives the inclusive reaction cross section, for no absorption. In the high energy limit the expression (12) goes over to (9) as we would expect.

It turns out that Eq. (12) also gives the reactive cross section, when there is true absorption of the projectile, for a model, like our second one above, for which at least one scattering must precede the absorption. In fact one can prove that under suitable conditions,[14] for this model, that another sum rule obtains, yielding the inclusive inelastic cross section

$$\sigma_{inel} = \frac{2\pi}{v} \sum_{i=1}^{A} \int |\langle\psi^{(-)}|t_i|\phi^{(+)}\rangle|^2 d\rho_f. \tag{14}$$

This, like (12), is an integral over one-nucleon knockout probabilities, where, however, the final state $\psi^{(-)}$ now is distorted by a potential representing true meson absorption (after the initial scattering). In the high energy limit, this potential takes the form $\mathrm{Im}U(r) = -v/2\ \tilde{\sigma}_a \rho(r)$, and leads to Eq. (11a), which is equivalent, in this limit, to (14). (Details will be published later.[14])

These sum rules and their extensions can actually teach us a lot about what to expect from different models of absorption in terms of multiple scattering, and the decomposition of the total cross section, as in Eq. (2). However, the simple expressions only hold for oversimplified models: they are not a substitute for better, more realistic calculations of cross sections. David Schneider is working on models with multiple scattering and absorption, of the sort I have alluded to.[14] The delta-hole scattering model of Lenz and collaborators[17] is an example of a two-step model of absorption, with the difference that the first step makes a Δ, and the second step is Δ-absorption: $\Delta N \to NN$. Models of this sort, but with correlations, have not developed yet (to my knowledge).

The statistical approach to calculating absorption is another way of dealing with the complexities of multiple scattering and absorption in a nucleus. Two recent proposals[8,9] to use transport equations in this context take different detailed approaches, but have the same general principle in mind: to follow the projectile through the target, allowing energy transfer through scattering, statistically treated. Hufner and Thies follow the pion through a kind of high-energy, semiclassical picture; Agassi and I have taken the Δ as our particle which exchanges energy, and may eventually be absorbed-- an adaptation of the model of Lenz and collaborators.[17]

I would like to acknowledge the collaboration of D. Agassi and D. Schneider on several of the subjects mentioned in this report.

REFERENCES

1. D. S. Koltun, Pion Absorption by Nuclei, lectures at 1978 Nato Advanced Institute in Nuclear Theory, Banff, Alberta, to be published.
2. J. M. Eisenberg and D. S. Koltun, Theory of Meson Interactions with Nuclei, (Wiley Interscience, New York, to be published: 1979-80).
3. H. E. Jackson, et al., Phys. Rev. Lett. 39, 1628 (1977).
4. I. Navon, et al., contribution to this conference.
5. A. W. Thomas, Ph.D. dissertation, Flinders University, Australia (1973).
6. T. Mizutani, Ph.D. dissertation, University of Rochester (1975), Technical Report UR-53 (1975), T. Mizutani and D. S. Koltun, Ann. Phys. 109, 1 (1977).
7. Y. Avishai and T. Mizutani, contribution to this conference.
8. D. Agassi and D. S. Koltun, contribution to this conference.
9. J. Hufner and M. Thies, preprint.
10. L. D. Miller and H. J. Weber, Phys. Rev. C17, 219 (1978).
11. J. L. Friar, Phys. Rev. C10, 955 (1974).
12. R. J. Glauber, in Lectures in Theoretical Physics, Vol. 1, W. E. Brittin et al., eds. (Interscience, New York, 1959) p. 315.
13. M. Ericson and T. E. O. Ericson, Ann. Phys. 36, 323 (1966).
14. D. M. Schneider, Ph.D. dissertation, Rochester (1979?), D. M. Schneider and D. S. Koltun, to be published.
15. P. C. Tandy, E. F. Redish and D. Bolle, Phys. Rev. C16, 1924 (1977).
16. D. S. Koltun and D. M. Schneider, Phys. Rev. Lett. 42, 211 (1979).
17. M. Hirata, J. H. Koch, F. Lenz, and E. J. Moniz, MIT preprint CTP #748 (1978).

Question: W. Wharton, Carnegie-Mellon University

Using your simple high energy models with L(b) and λ can you explain the A dependence of σ_{abs} and σ_{inel} shown at the beginning of your talk.

Response: Koltun

The results presented are quite new, and have not been applied to data. However, the high energy approximation here is not so useful for the experimental energies involved, i.e. below resonance. We are working on models more appropriate to lower energy.

Question: A. Rinat, Weismann Institute of Science

You mentioned that the division of $\mathrm{Im} f_{\pi A}$ into inelastic and true absorption cross sections may be model dependent. It appears that in the single pion approximation model applied to πA scattering the σ_{abs} can be associated only to part of $\mathrm{Im} f_{\pi A}$, but apparently not to all of it.

Response: Koltun

Yes.

Question: R. McAdams, Utah State University

I would just like to point out with respect to the pion-gamma-ray work mentioned at the beginning, that the USU group recently conducted a coincidence experiment at LAMPF with 190 MeV $\pi^{\pm}$ in ^{27}Al where the outgoing particles ($\pi^{\pm}$, p and d) were detected in coincidence with the gama rays. This gives considerably more information on the reaction that detecting just the particles or gamma rays. For example it shows that at this energy absorption is probably less than 50%. This work was published in Phys. Rev. Letters 41,1023(Oct 9, 1978).

Response: Koltun

It is good to have the additional information from measuring the outgoing particles.

Question: L. Kisslinger, Carnegie-Mellon University

The reason the scattering and absorption must be considered together is because the scattering provides the off-mass-shell pions (and other kinematic features) which enhances the absorbtion. But there are integrals over the variables which do not factorize except in special cases. My question is: can you neglect these extra variables and use the factorized on-shell eikonal description?

Response: Koltun

Off-shell propagation is treated only approximately in the high energy limit I have discussed here; the dependence on the kinematic variables is suppressed, and the average effect appears in the effective absorption parameter $\tilde{\sigma}_a$.

PION PRODUCTION IN PP COLLISIONS*

Ian Duck

T. W. Bonner Nuclear Laboratories
Rice University, Houston, TX 77001

Since this problem was reviewed by Kloet at the Graz Conference[1] some progress has been made. In this talk I will bring to your attention our own work[2] which has been published; some recent analysis by our group which is nearly ready for publication;[3] some modifications to our work resulting from suggestions by Kisslinger;[4] a fine contribution by VerWest which has been mentioned at previous conferences and is now ready for publication.[5] I will close my report by emphasizing the importance of the polarization-asymmetry experiments[6] to motivate progress on the very difficult question of the effect of unitarity on these production amplitudes.[7,8]

In our work[2] we seek to understand the exclusive pion production experiments in terms of the OPE and VME amplitudes of Fig.1. As is well known, the OPE amplitude has to be sharply reduced by form factors (which take the pion off-shell) at the $NN\pi$ and $N\Delta\pi$ vertices, effectively reducing the pion wave function within the baryon. The vertices are evaluated by using static model interaction Lagrangians of Brack <u>et al</u>.[9] and Goplen <u>et al</u>.[10] made Galilean invariant following Nishanen[11] and Bhasin.[12] In our initial fits to the data we used a single dipole form factor[9] for the $NN\pi$ and $N\Delta\pi$ vertices with a range parameter Λ_π and similarly for the $NN\rho$ and $N\Delta\rho$ vertices with Λ_ρ=1.8 GeV.[9] The total burden of fitting the data is put on the single parameter Λ_π. We require a value of Λ_π=.78 GeV for a "global" fit to the exclusive $pp\rightarrow p\pi^+n$ data of the Rice-UH group, shown in Fig.2. With this global fit we are able to examine various effects:

1) the dominance of the Δ^{++} resonance
2) the dominance of the OPE amplitudes
3) the importance of πN S- and P-wave non-resonant terms

and we extract a $pp\rightarrow n\Delta^{++}$ angular distribution and cross section shown in Fig.3. The dominance of the Δ^{++} is almost complete as documented in Table I and Fig.4. The π-ρ interference makes an important reduction in the cross section and significantly reduces the burden on the pion form factors. The correction for Galilean invariance is important especially at larger values of the momentum transfer. The importance of the form factors is shown also in Fig.3. The derivative couplings at the vertices give rise to delta functions in the configuration space potentials. Kisslinger[4] maintains that it is unsafe to simply use the artifice of dropping these delta functions but form factors should be used instead.

*Work supported by U.S. D.O.E. Contract No. EY-76-C-05-1316.

ISSN:0094-243X/79/540095-12$1.50

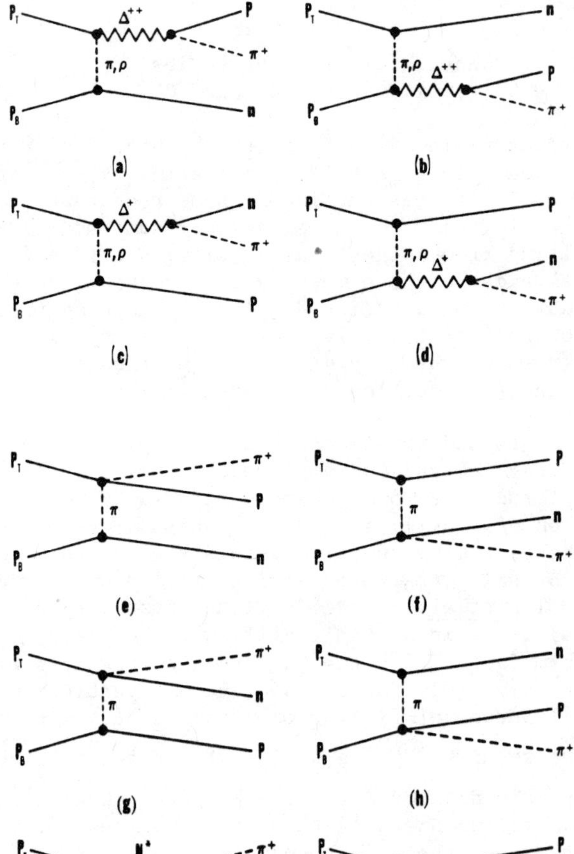

Fig.1. Feynmann Diagrams for OPE and VME model. (a) - (d) OPE and VME leading to the Δ^{++} and Δ^{+} resonance (e)-(h) nonresonant S- and P-wave OPE; (i) and (j) OPE leading to the N*(1470) resonance.

Kisslinger[4] makes further criticisms which we have incorporated into later calculations. He uses Wolf's[13] $\Delta N\pi$ form factor which has a different functional form from the $NN\pi$ form factor but is equivalent to a dipole form factor with range $\Lambda_{\pi N\Delta}$=.683 GeV for momentum transfers encountered here. So we fix $\Lambda_{\pi N\Delta}$ and just use $\Lambda_{\pi NN}$ as a free parameter. Kisslinger also points out that the static quark model prediction for $f_{\rho N\Delta}$ used by Brack <u>et al</u>.[9] exceeds the value obtained from an extrapolation of the Regge fit[14] to the $\pi N \rightarrow \pi\pi N$ data by thirty percent, and he advocates reducing the rho exchange

TABLE I: $d^5\sigma/dp_1 d\Omega_1 d\Omega_2$ [μb/sr² MeV/c]:(a) $\theta_p = 15°$ $\theta_\pi = 20.8°$; (b) $\theta_p = 25°$, $\theta_\pi = 40°$. Cross sections are calculated for three values of proton momentum p_1 (MeV/c) assuming contributions of the following graphs from Fig. 1 OPE-π exchange in (a)–(b); VME-ρ exchange in (a)–(b); Δ^{++} graphs (a), (b); Δ^+ graphs (c), (d); Δ graphs (a)–(d); TOT graphs (a)–(j); SING *NN* spin singlet part of TOT; TRIP *NN* spin triplet part of TOT. Range parameter values are $\Lambda_\pi = 0.78$ GeV/c, $\Lambda_\rho = 1.8$ GeV/c. For values in parentheses, the range parameters are $\Lambda_\pi = 0.72$ GeV/c, $\Lambda_\rho = 1.5$ GeV/c.

p_1	OPE	VME	Δ^{++}	Δ^+	Δ	TOT $\Delta + (S\&P)$	SING *FS*	TRIP *FS*	EXP
					(a)				
620	19.4	4.9	8.0	0.12	9.2	9.8	2.8	7.0	10
	(16.5)	(2.5)	(7.8)	(0.13)	(9.0)	(9.6)	(2.8)	(6.8)	
820	47.4	11.1	24.0	0.65	24.3	24.0	7.5	16.5	28
	(43.0)	(5.0)	(24.8)	(0.72)	(25.0)	(25.4)	(7.8)	(17.6)	
980	14.5	3.2	5.6	1.3	6.9	6.0	1.7	4.3	8
	(13.1)	(1.4)	(6.0)	(1.4)	(7.5)	(6.5)	(2.1)	(4.4)	
					(b)				
600	3.3	1.6	0.32	0.09	0.52	0.48	0.09	0.39	2
	(2.5)	(0.86)	(0.37)	(0.07)	(0.56)	(0.52)	(0.09)	(0.43)	
800	6.7	2.8	2.2	0.09	2.3	2.4	0.8	1.6	3
	(5.5)	(1.4)	(2.1)	(0.10)	(2.2)	(2.3)	(0.68)	(1.6)	
960	13.5	6.6	5.0	0.02	5.0	5.0	1.6	3.4	4
	(10.2)	(2.9)	(4.3)	(0.02)	(4.4)	(4.3)	(1.2)	(3.2)	

contribution accordingly. (He has further criticism of using rho exchange potentials in configuration space which do not affect our calculations.)

We have rerun our calculations using Kisslinger's parameters and readjusting only $\Lambda_{\pi NN}$ to .70 GeV. In fact we get an improved fit to the data of Fig.2., although we still have several problems with the data set at $\Theta_p=20°$, $\Theta_\pi=22°$. This data set has been criticized, but a careful re-examination gives us no reason to change the experimental results. The model fit would be much easier if this data set were renormalized upwards by 25%. Our earlier conclusions are still valid, except now the OPE is even more dominant. The cross section is >90% saturated by the Δ^{++}.

The dominance of the $\Delta^{++}>>\Delta^{+}>>$NR makes it difficult to really test the parametrization of the non-resonant contribution. Since the $pp\to pn\pi^+$ data was published, the $pp\to pp\pi°$ exclusive results are being completed.[3] Our preliminary results are very encouraging, although there is the suggestion of an energy calibration problem, as seen in Fig.5.

In an already publicized and soon to be published paper, VerWest[5] reports a very fine calculation in which he does a fully

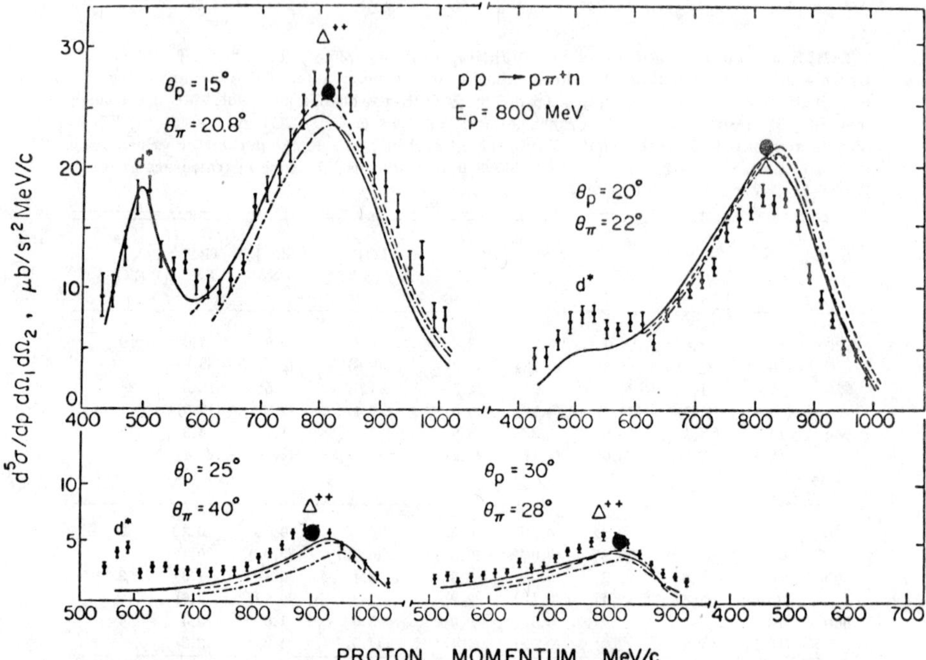

Figure 2: Differential cross section $d^5\sigma/dpd\Omega_1 d\Omega_2$ as a function of proton momentum and angle. The parameters for the curves are in units of GeV/c: (a) solid curve Λ_π=0.78, Λ_ρ =1.8, (b) dashed curve Λ_π=0.62, Λ_ρ=1.5, and (c) dot-dashed curve Λ_π =0.6, Λ_ρ =1.2, M_Δ=1220 MeV for all curves; Galilean invariance was used. Solid circles are results using Kisslinger's parameters [see text].

relativistic OPE+VME calculation of $NN \to NN\pi$. He includes the $\pi N \to \pi N$ amplitude at one vertex with nucleon and Δ intermediate states and their cross diagrams, and ρN-πN amplitudes are treated similarly. The amplitudes corresponding to the Feynmann diagrams are evaluated using the covariant extension of the interaction Lagrangians of Brack _et al._[9] to describe the vertices. The coupling constants of Brack _et al._[9] are used, with one exception. All the Dirac algebra has been done on the computer. Form factors are basically in accord with Brack _et al._[9] with Λ_ρ=1.8 GeV and Λ_π adjusted. VerWest evaluates the parameter Λ_π and the coupling constant $f_{\pi N\Delta}$ by fitting the Δ peaks of two inclusive experiments that correspond to the minimum momentum transfer t. These are the most peripheral amplitudes and correspond to impact parameters >1 fm. VerWest fits the forward and near forward inclusive spectra at their peaks by setting $f_{\pi N\Delta}$=1.882, Λ_π=1.150 GeV and M_Δ= 1.220 GeV. ($f_{\pi N\Delta}$=2.0 gives the Δ width). The fit to the inclusive spectrum for $np \to pX$ is shown in Fig.6. The

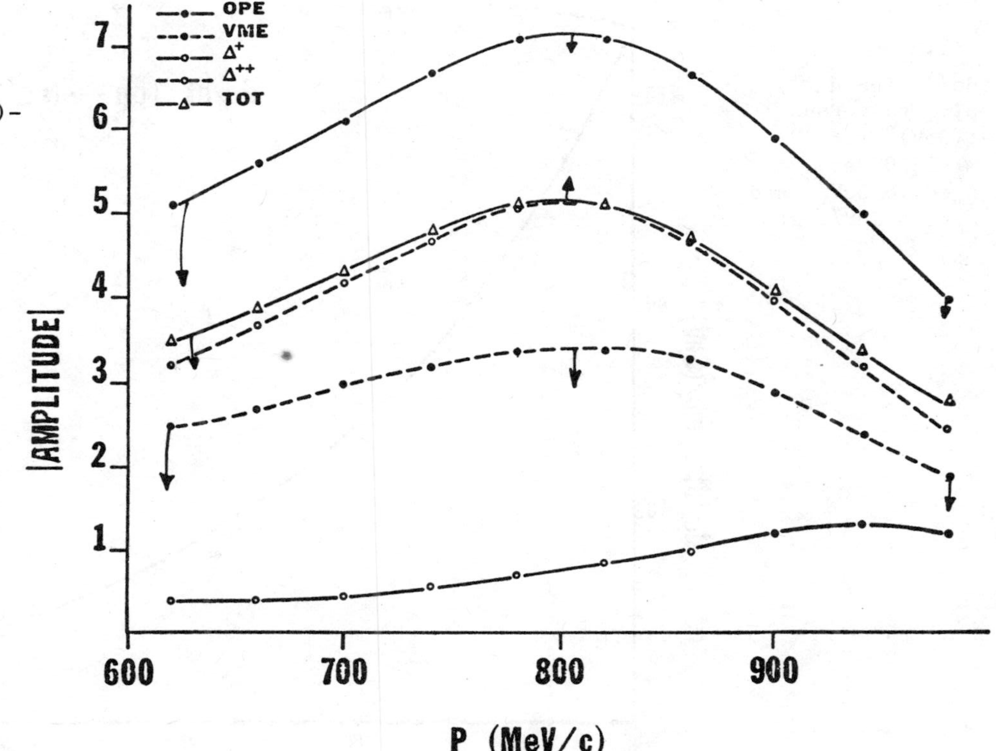

Fig.4. Relative amplitudes for $pp \rightarrow p\pi^+n$ at $\Theta_p=15°$ and $\Theta_\pi=20.8°$: (a) OPE, contribution from diagrams 4(a)-4(d) for pion exchange; (b) VME, contribution from diagrams 4(a)-4(d) for rho exchange; (c) Δ^{++} contribution from diagrams 4(a) and 4(b) for pion and rho exchange combined; (d) Δ^+, contribution of diagrams 4(c) and 4(d) for OPE and VME; (e) TOT, total amplitube (coherent sum of all diagrams shown in Fig.1. Arrows indicate results with Kisslinger's parameters and Λ_π=.7 GeV.

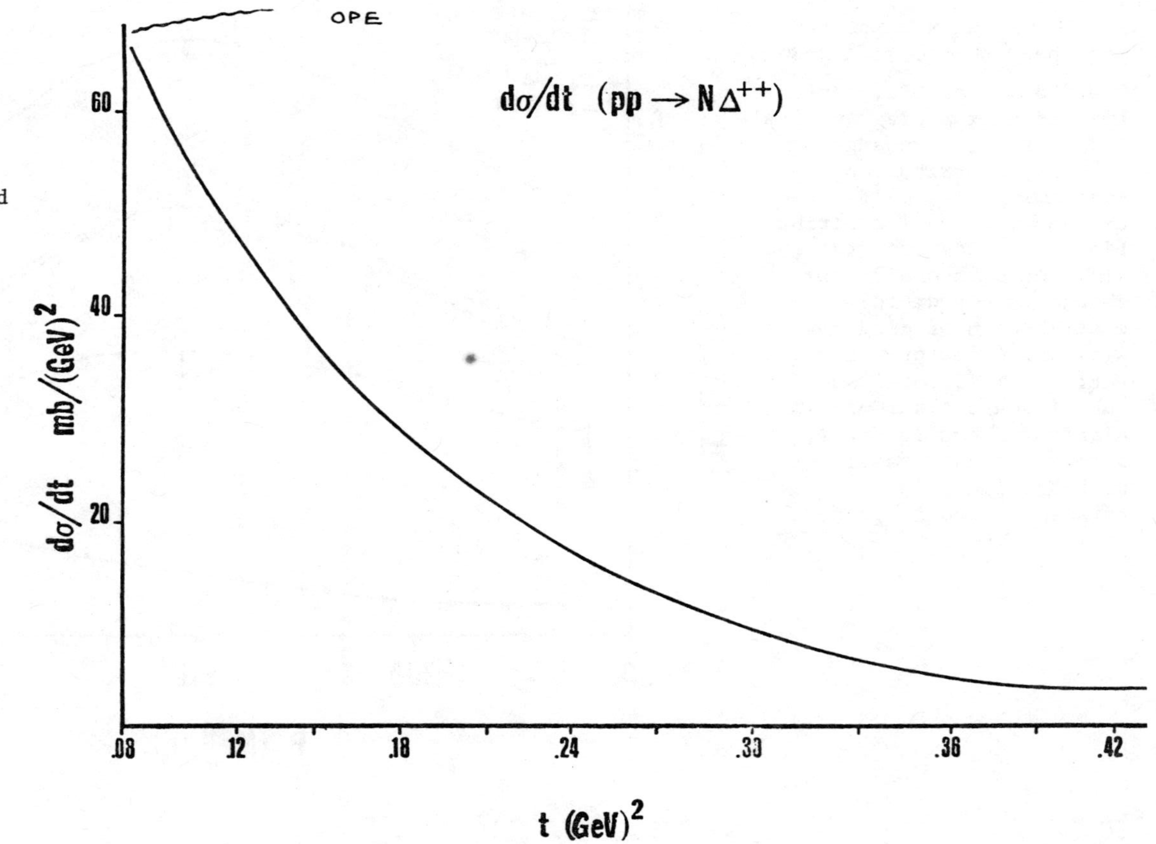

Fig.3.

$d\sigma/dt$ ($pp \to n\Delta^{++}$) plotted versus $t(GeV)^2$ for M_Δ=1220 MeV. Λ_π=0.78 GeV/c and Λ_ρ=1.8 GeV/c at E_p=800 MeV.

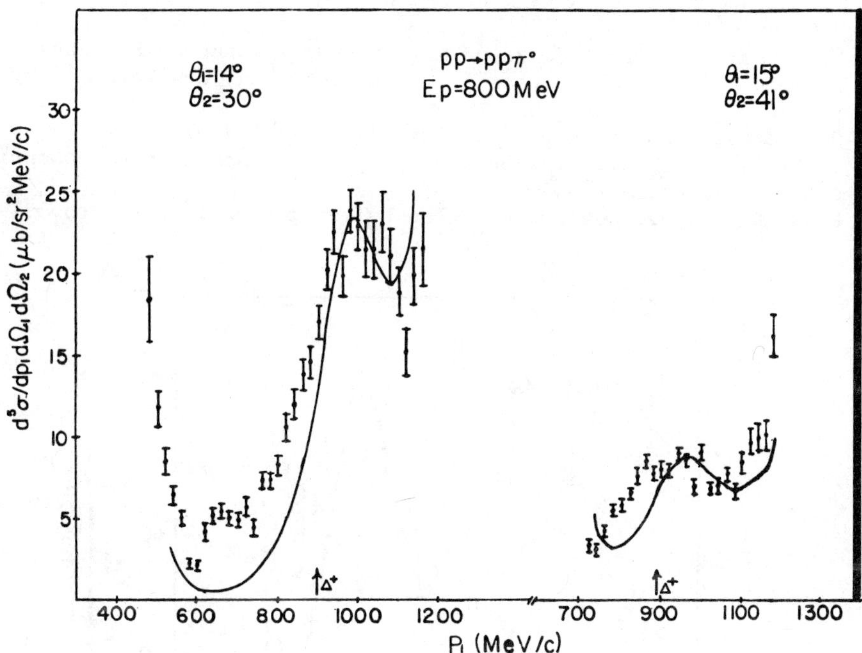

Fig.5. pp→ppπ° cross section versus measured proton momentum. Solid lines are the predictions using the peripheral model

predictive power of his model is then tested against other near forward, low t inclusive and exclusive data with great success in Fig. 7 and Fig.8. His model overshoots the larger angle cross sections, which he leaves outside his peripheral model. We look forward to the detailed report of his work on this problem and especially the tests of his parametrization at higher energies.

Whether one chooses, as VerWest does, to fit the peripheral model parameters to the small t cross section or, as we do, to the full range of exclusive data, the peripheral model is doomed to failure at an early stage when unitarity corrections upset the balance of individual partial waves. The full consequences of unitarity on the inelastic amplitudes will no doubt be required in the next year or so when polarization-asymmetry experiments are completed. It is a formidable task to impose unitarity, whether by Lomon's coupled channel isobar model, Kloet's three body model, or VerWest's proposed DWBA. The experiments promise a tough test and therefore a powerful direct motivation for the theory to rigorously incorporate the inelastic channels into the NN scattering problem.

REFERENCES

1. W. M. Kloet in Proceedings of the Graz Conference on Few Body Systems and Nuclear Forces II (Springer-Verlag, New York, 1978) p. 164.
2. J. Hudomalj-Gabitzsch et al., Phys.Rev. C18 (1978) 2666.
3. G. P. Pepin et al., Rice University preprint and private communication, February 1979.
4. L. S. Kisslinger, Carnegie-Mellon University preprints, Feb.1979.

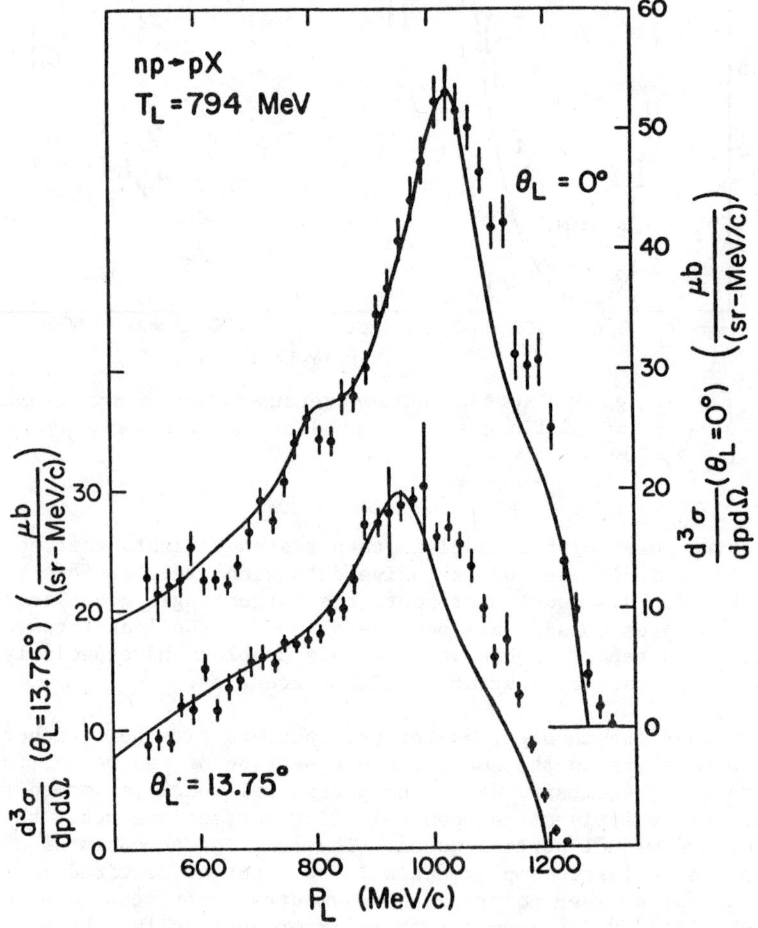

Fig.6. Differential cross section $d^3\sigma/dp_p d\Omega_p$ in the lab for n+p→p+X as a function of proton momentum for Θ_p=0° and Θ_p=13.75° from reference 5.

5. B. VerWest, Texas A&M University preprint ORO-5223-13, Feb.1979.
6. G. S. Mutchler et al., private communication, Feb.1979.
7. W. M. Kloet et al., Phys.Rev.Lett. 39 (1977) 1643.
8. E. Lomon in Proceedings of the Graz Conference on Few Body Systems and Nuclear Forces II (Springer-Verlag, New York, 1978)p.48.
9. M. Brack, D. O. Riska, W. Weise, Nucl.Phys. A287 (1977) 425.
10. B. Goplen, W. R. Gibbs, and E. L. Lomon, Phys.Rev.Lett.32(1974) 1012.
11. J. A. Niskanen, Phys.Lett. 71B (1977) 40.
12. V. S. Bhasin (unpublished).
13. G. Wolf, Phys.Rev. 187 (1966) 1538.
14. G. Renninger and K. Sarma, Phys.Rev. 178 (1969) 2201.

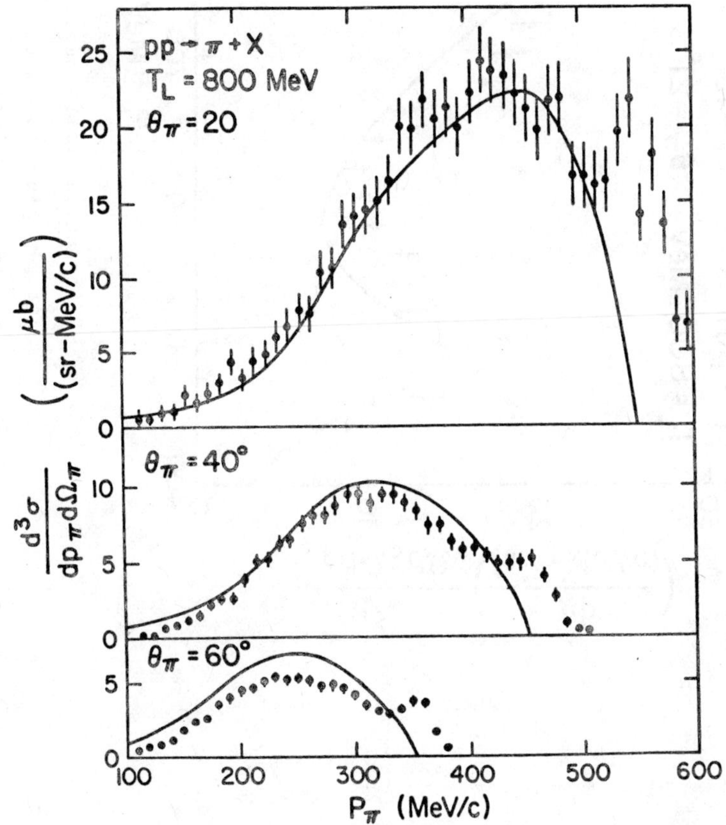

Fig.7. Differential cross section $d^3\sigma/dp_\pi d\Omega_\pi$ in the lab for $p+p\rightarrow\pi^++X$ as a function of pion momentum for Θ_π=20°, 40°, and 60° from reference 5.

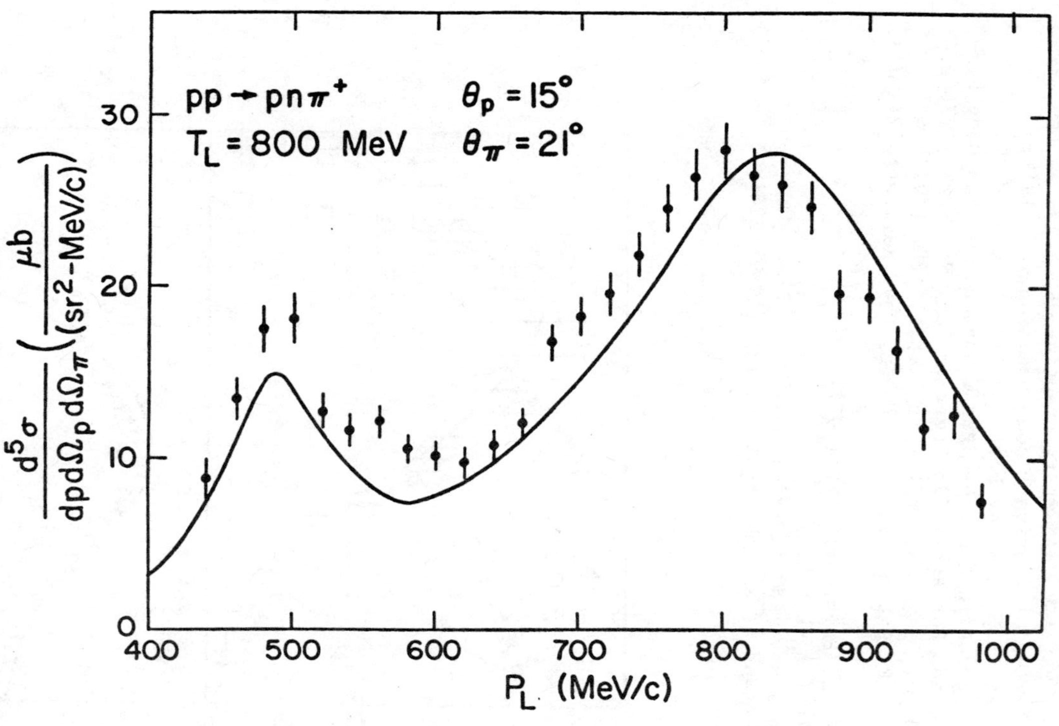

Fig.8. Differential cross section $d^5\sigma/dp_p d_{\pi p} d\Omega_\pi$ in the lab for $p+p \rightarrow p+n+\pi^+$ as a function of proton momentum for $\Theta_p = 15°$ and $\Theta_\pi = 21°$ from reference 5.

Question: A. Thomas, TRIUMF

In the conclusion to your talk you mentioned the words polarization and unitarity. When will the polarization data be available and why did you connect these concepts? It is very hard for me to see how the theoretical approaches described today can be made unitary. In this respect I would like to mention contributions to this conference by Avishai and Mizutami and also by Rinat and myself where a set of integral equations are developed for the coupled channels $NN \rightarrow NN$, $NN \rightarrow N\Delta$ etc. Such a formalism will allow us to guarantee both two and three body unitarity, and to incorporate multiple pion rescattering to all orders--thus avoiding assumptions like "peripheral-ity"

Response: Duck

Polarization-asymmetry calculations inevitably involve expressions of the form Im(F*G). The only significant source of such terms in our calculation is the phase difference between Δ^{++} and Δ^{+} amplitudes. I just don't believe the peripheral model when it comes to predicting the phase of amplitudes and feel that a fully unitary dynamical calculation will be required for the explanation of polarization-assymmetry effects in the inelastic channel. Manageable unitary models involve compromises. The neglect of relativity for the baryons and the isobar approximation are not so bad but models which butcher the fundamental interactions should, perhaps, finally be scrapped. In this regard, I would like to remind you of a N/D unitary isobar model described by Bruno and Duck [Nucl.Phys. B20 (1970) 354] which is relativistic, easily incorporates various meson exchanges and coupling to elastic channels, and involves one dimensional integral equations for partial wave amplitudes. The analyticity properties of the model are somewhat complicated, though.

Question: A. Cass, Melbourne University

Why should the $\pi N\Delta$ vertex and a πNN vertex have different asymptotic forms when a naive quark model would lead one to expect that these should have the same behaviour?

Response: Duck

Let me amplify Professor Kisslinger's remark to the effect that the naive quark model has nothing to say about the $\pi N\Delta$ form factor. He has told me, and I don't know any details, that he and Miller have made a Chew-Low calculation in which the $\pi N\Delta$ form factor is dominated by low energy processes and that the N-Δ mass splitting is crucial. I share your concern that these old fashioned dispersion type calculations are producing a dramatic difference between the $\pi N\Delta$ and πNN form factors, and also that they are producing such a hard πNN vertex. It would be interesting to see if and how the quark model is failing here when it can be made to succeed for γNN form factors.

Comment: M. Banerjee, University of Maryland

There is a remarkable fact about πN amplitude which is not well known: it is that a model for off-shell πN amplitude can be tested against results of dispersion relations. The amplitude $\bar{C}^{(+)}(0,t<0)$ is an "off energy shell" amplitude and its evaluation depends on the

$\pi N\Delta$ form factor. In fact, from the current values of $\bar{C}^{(+)}(0,t<0)$ one can set a range of 26 m_π^2 to 40 m_π^2 for the square of the monopole form factor mass.

Comment: B. VerWest, Texas A & M University

The wide angle $NN\rightarrow N\Delta$ cross sections predicted by my calculation are in fact not as high as have been indicated. My calculations are about as high of the data at wide angles as yours are low. Actually, the calculations are very similar and the differences really come from what data is fitted. You have fit the exclusive data while I fit very forward nucleon inclusive data where the reaction is peripheral and then extend the calculation to all other regions and find very good overall agreement with a great deal of data. I agree with you completely though in that the real information and real tests of these models lies in the polarization experiments. These models are really just tools to probe the data and separate what we can explain simply from the more complicated parts of the interaction which require more in-depth study both experimentally and theoretically.

INCLUSIVE PION PRODUCTION IN PROTON AND ION COLLISIONS WITH NUCLEI

Rubin H. Landau*
Oregon State University, Corvallis, Oregon 97331

ABSTRACT

Inclusive pion production in nuclear collisions is examined with models for hard scattering from single nucleons or clusters within the nucleus. The occurrence of scaling and the shapes of the forward and backward pion spectrums are examined.

INTRODUCTION

Much of "intermediate energy nuclear physics" deals with either pion and proton interactions with nuclei, or relativistic heavy ion reactions. In many ways the former is an extension of lower energy techniques, whereas the latter is an application of particle physics and cosmic ray tools. In this short talk I will try to act somewhat as a mediator between the "exclusive" meson factory π production researchers and the more "inclusive" heavy ion workers. To clarify the reasons for this class distinction it is necessary to describe the meaning of "π production" to each group.

To meson factory physicists (to wit $\gtrsim$ 80% of the talks this afternoon) π production usually denotes the (p,π) reaction

$$p + A \rightarrow \pi + (A + 1)^* \quad , \tag{1}$$

in which there is a two body final state with a well defined nuclear state $(A + 1)^*$. This is an "exclusive" reaction in that many (and probably most) of the possible final states are excluded from observation. Although the final state may be "simple," I think it still fair to say that the (p,π) problem, though extensively studied, still contains many ambiguities and is without a solid theoretical framework. To relativistic heavy ion physicists (to wit Dr. Gyulassy's 16%), pion production frequently denotes the much less exclusive

$$B + T \rightarrow \pi + X$$

In this inclusive reaction, which I discuss today, the beam B can be a wide variety of projectiles (p, d, α...Ar) with energies from $\sim$500 MeV to $\sim$10 GeV per projectile nucleon, the target T can span the periodic table, but the spectrum of only the pion in the final state is observed; thus the X. (Examples of these spectra are in Figs. 1 and 2.) Of course the single particle inclusive spectrum of other particles (p, d, t, α...) are also observed--but we shall concentrate on the more direct pions. Because of the mixed degree of coherence present in the $T(B,\pi)X$ reaction, this is an even more complicated mixture of many body physics, relativistic field theory, and nuclear physics than the simpler $A(p,\pi)B^*$ reaction and therefore

*Supported in part by the National Science Foundation under Grant PHY76-82659.

ISSN:0094-243X/79/540107-09$1.50

less open to a full theoretical treatment. Consequently, the interpretation of these inclusive spectra requires a collection of models, each of which presumably contains some of the same physics as occur in experiment.

THE PROJECTILE FRAGMENTATION REGION

If the incident beam energy is high enough, a high energy pion produced in the forward direction ($\theta_\pi \simeq 0°$) is quite likely a "fragment" of the incident projectile. Conversely, a high energy pion in the backward direction ($\theta_\pi \simeq 180°$) is considered a "target fragment." The modern Bevelac π production experiments began with the projectile fragmentation ($\theta_\pi = 2.5°$) studies of Papp, et al.[1] As we see in Fig. 1 for p, d, and α beams with the same energy per nucleon, the doubly differential cross section depends sensitively on the number of nucleons in the projectile (this is the projectile fragmentation region) with heavier projectiles naturally producing more pions. In addition, for proton beams (Fig. 2), the production increases smoothly with beam energy (Dr. Gyulassy will discuss pion multiplicities later today).

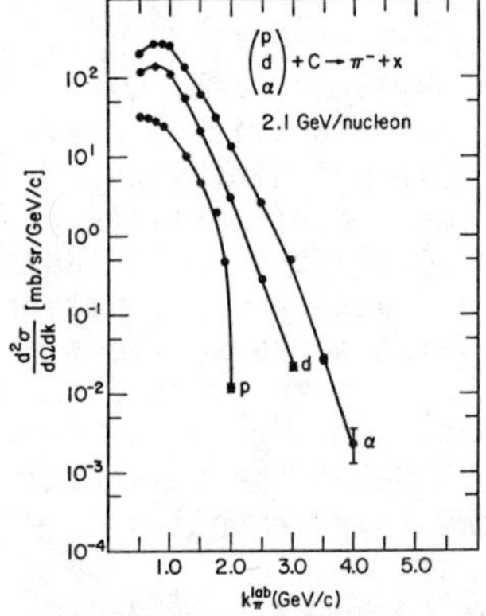

Fig. 1. Momentum spectrum[1] for pions produced at 2.5° for different beams.

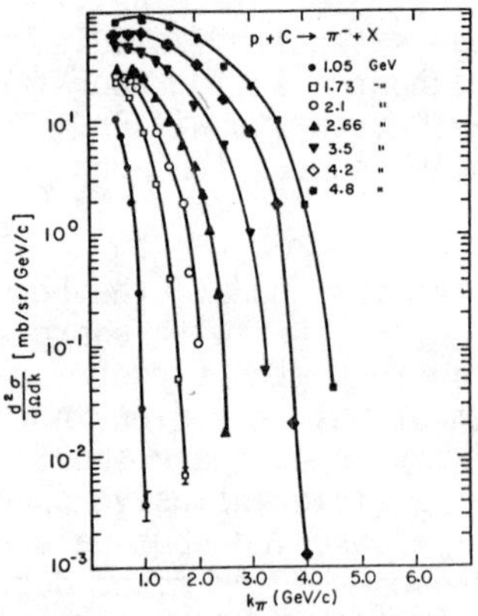

Fig. 2. Momentum spectrum for pions produced at 2.5° for 1-5 GeV protons[1] incident on C.

A very interesting aspect of the projectile, but especially the target, fragmentation measurements is that the pion fragments have momenta larger than those obtainable in free, nucleon-nucleon reactions at these same energies. Such fragments can arise either as the result of the coherent interaction of more than one target nucleon, or from the interaction with one target nucleon having a very high internal momentum (e.g. p $\simeq$ 1 GeV/C); in either case via nuclear binding effects. In fact, by counting the number of initially frozen nucleons which--on kinematic grounds alone--must recoil together to produce the highest energy pion observed, we obtain Baldin, et al.'s[2] "degree of cumulativity" for a reaction.

Another technique reported to isolate the contribution of the high momentum tail of the nuclear wavefunction, is to plot the Lorentz invariant cross section, $Ed\sigma/d^3p$, vs. the scaling variable x. For example, Papp, et al.,[1] found that their data for B + T → π(2.5°) + X fall on a universal curve for each beam (see Fig. 3) when plotted as a function of the scaling variable

$$x = k^*_\pi(BT)/k^*_\pi(BT)_{max}. \quad (3)$$

Here k^*_π is the observed pion momentum transformed to the beam-target c.m., and $(k^*_\pi)_{max}$ is the maximum momentum possible. Although we do not discuss the field theoretic origin of this variable, it is manifestly useful in that the one point x = 1 is always the maximum allowed π momentum, and is achieved only if the beam and target particles recoil off together--e.g. the (p,π) reaction. For the lower momentum pions shown in Figs. 2 and 3, the number of coherent target nucleons must be less than T. In analogy with QED, the very steep drop off as x → 1 is assumed to indicate the presence of hard, point-like, collisions (the (p,π) reaction should enter as a delta function at x = 1).

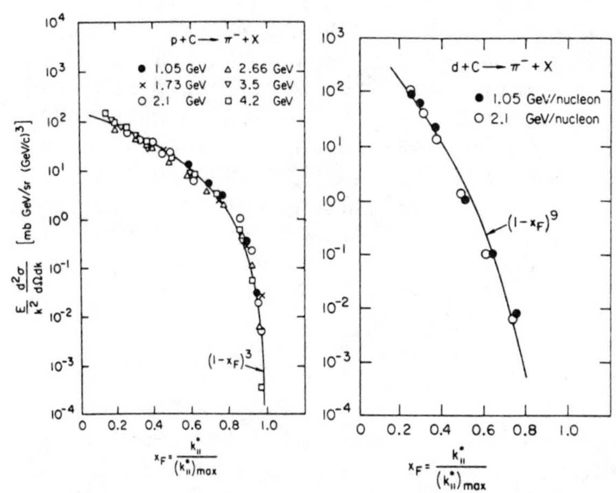

Fig. 3. The data of Figs. 1 and 2 plotted[3] vs. scaling variable.

HARD SCATTERING MODELS

The curves in Fig. 3 were predicted by Schmidt and Blankenbecler[3] with a hard scattering parton model treating the nucleons as partons. Actually this is one of many calculations which should calculate the diagram 4a, but instead ignore final state interactions and calculate the diagram 4b. Consequently, these models concentrate only on the shapes of the spectra. In these PWIA models, the π production cross section is written as

$$Ed\sigma/d^3p\ [BT \rightarrow \pi X] = \int d^3p\ \rho_c(p)\ r\ (Bc \rightarrow \pi x)\ P \quad , \quad (4)$$

where $\rho_c(p)$ is a pseudo-probability distribution for finding a cluster of c nucleons with momentum $\vec{p}$ in the nucleus, "r" is the elementary projectile-cluster, π production rate, and P is the phase space constraints. Since final state interactions and off-shell

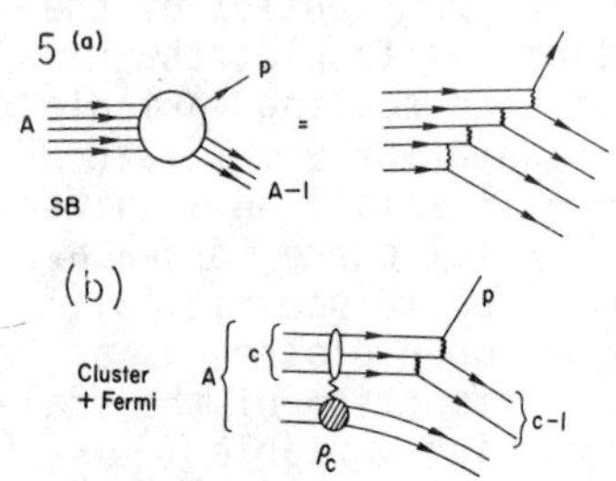

effects are ignored in fitting eq. (4) to data, we call $\rho(p)$ a "pseudo" distribution.

The most unique aspect of the parton model[3] is its assumption that the nucleon distribution function (Diag. 5b) is valid for the discussed data. A $p_B \to \infty$, $x_\pi \to 1$ collision can only arise from a very hard projectile-nucleon collision in which the nucleon has an asymptotically large momentum. For large enough momentum, the remaining A-1 nucleons must recoil coherently and share an equally large and opposite momentum amongst themselves (Fig. 5a). This coherence leads to the prediction

$$E d\sigma/d^3p\ [BT \to \pi(0°)X]\ \alpha\ (1-x)^{6B-3}\ , \qquad (5)$$

which appears to agree with the data of Fig. 3.

Target Fragmentation Region

If the 0.73-7.5 GeV (proton beam energy) data[4] for pions produced in the target fragmentation region, $\theta_\pi \simeq 180°$, are plotted as a function of the scaling variable x, they do not fall on a universal curve. As we see from the data in Fig. 6a, the power describing the fall off $[(1-x)^s]$ is much less than the $\sim$379 predicted by the parton model[3], eq. (5) with B$\leftrightarrow$T. In addition, the power s increases with energy from $\sim$13-90.[5] Although it might not be surprising that the infinite-momentum, $x \to 1$ parton theory is not yet valid, the striking behavior in Fig. 6a still represents a challenge to explain, particularly since scaling appeared to occur in the forward hemisphere.

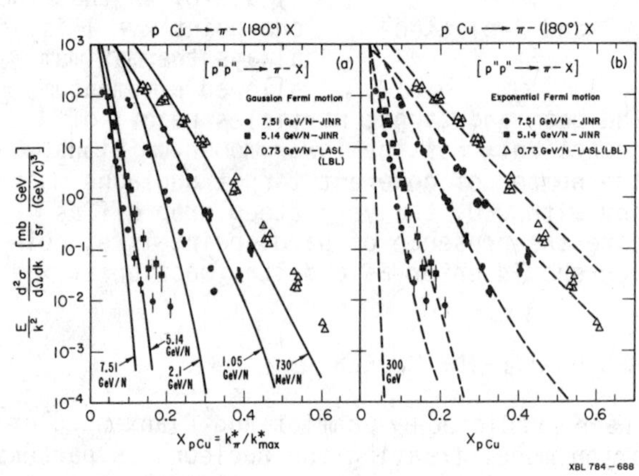

Fig. 6. 180° π production cross section from Ca[4] with 1N cluster production.

Landau and Gyulassy[5] explained the magnitudes and energy dependences of these slopes by employing a hard scattering, cluster model with Fermi motion, eq. (4) and Fig. 5b. One unique aspect of this model is that it employs a phenomenological, invariant parameterization of the elementary beam and cluster π production rate "r", and several models for the pseudo-probability density $\rho(p)$. Explicitly, rather than assume all T target nucleons act coherently to produce a backward (inclusive pion), we assume only a "cluster" of $c < T$ nucleons have a hard collision. The parton counting rules are then used to determine the x-dependence of this small c production rate:

$$r\ (Bc \to \pi(180°)x)\ \alpha (1-x_{Bc})^{S_{Bc}}\ ,\ S_{Bc} = 6c - 3 - 2\ \delta_{B,1}\ . \qquad (6)$$

This quite naturally produces smaller powers for the fall off (after x_{Bc} in the B - c c.m. is transformed to the beam-target c.m.). For a proton "cluster," the elementary data are available and approximately scale, so we employ an empirical, x-dependent fit.[5] The phase space factor P in eq. (4) contains the energy-momentum conservation constraint which we apply to the beam-target and beam-cluster systems via

$$P\ \alpha\ \ \theta(1 - x_{BT})\ \theta(1 - x_{Bc})\quad . \qquad (7)$$

The elementary rate of eq. 6 describes the absorption of a very large momentum by a cluster of c nucleons. We account for the overall c.m. motion of the c nucleons by averaging "r" over a distribution of internal cluster momentum with a phenomenological weighting factor $\rho_c(p)$ in eq. (4). For $\rho_c(p)$ we consider two possible forms which appear in the literature and are independent of cluster size. The first is a Gaussian, of the type observed in the projectile fragmentation studies of Greiner, et al.,[6] ($\sigma_z \equiv 130$ MeV/c), and the second is an exponential suggested by Amado and Woloshyn,[7] ($k_0 \simeq$ 100 MeV/c):

$$\rho_{Gauss}(p)\ \alpha\ \exp[-p^2/2\sigma^2]\ ,\quad \rho_{exp}(p)\ \alpha\ \exp[-p/k_0]\quad . \qquad (8a,b)$$

As we see in Figs. 6a and b, the simple folding of $pp \to \pi^- x$ rates account well for both magnitude and energy dependences of the slopes of the data. The Gaussian form for eq. (8a) leads to slightly steeper slopes, whereas the exponential form (8b) leads to flatter slopes. For the p, d, α, and C projectiles examined, the calculated π spectrums generally lie within the band produced by these two distributions.

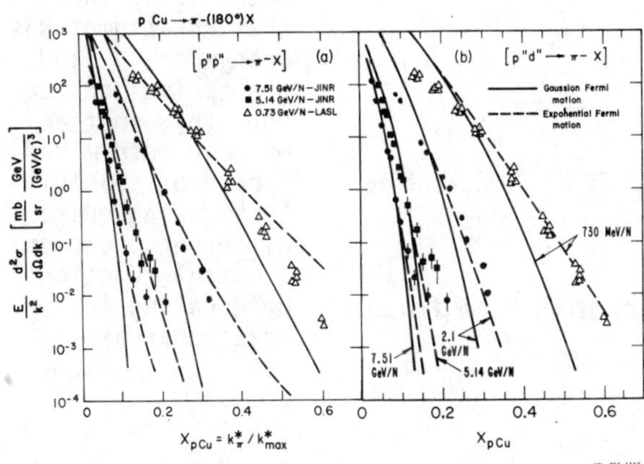

Fig. 7. Same as Fig. 6, for 1N and 2N clusters.

In Fig. 7 we compare the agreement between theory and experiment if a 1 ("p") or 2 ("d") nucleon cluster model is used. We see that the deuteron cluster model with an exponential internal momentum distribution fits the p Cu $\to \pi(180°)$X data equally well as the single nucleon cluster model. The 4N ("α") cluster model, however, does not fit as well for low energies.[5] Possibly a different $\rho_\alpha(p)$ is needed, or the assumed form for the (unmeasured) pα cross section, $\sigma\alpha(1 - x)^{19}$, is inaccurate. In this regard it is important to note (Fig. 8) that the assumed parton model rate for pd $\rightarrow$ $\pi^-(180°)X$ agrees with the few data available, as does our PWIA model based on pp rates. Apparently diagram 5a is more reliable for two nucleons than 64 and/or the energy dependence which enters the theory in the Lorentz transformation of x_{Bp} to x_{BA}, is much weaker for A = 2 than 64.

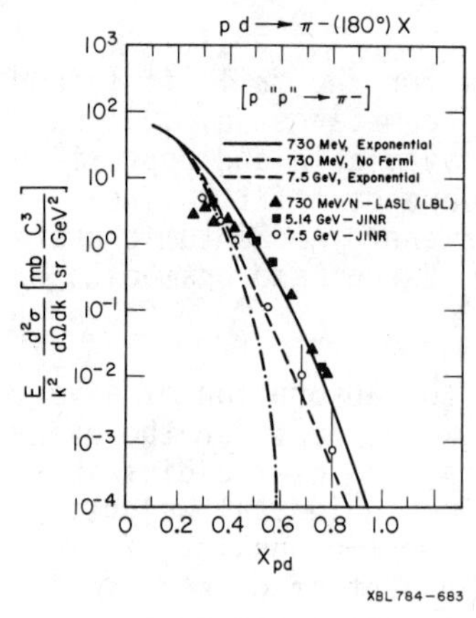

Fig. 8. Backward π production from deuterium.

A critical test of the above model is pion production in the beam fragmentation region where scaling occurs. We find here that the shape of the pC $\rightarrow$ $\pi^-(2.5°)X$ cross sections in Figs. 2 and 3 are not sensitive to the reaction mechanism ($\rho(p)$ and r) but instead arise mainly from a careful treatment of kinematic effects. For example, we show in Fig. 9 that our model leads to the apparent scaling found in the dC $\rightarrow$ $\pi^-(2.5°)X$ and αC $\rightarrow$ $\pi^-(2.5°)X$ data, and that the αC data can be explained (9b) via either a "d"C or "pC" mechanism.[8] In examining these results, the viewer should keep in mind that no attempt has been made to find the optimal choice for $\rho(p)$; instead we have determined a band of feasibility by using quite different phenomenological functions originating in related areas.

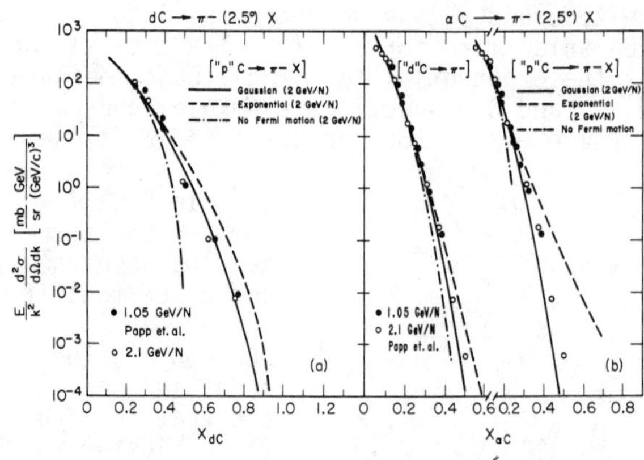

Fig. 9. Comparison of theory with beam fragmentation data.

Other Models

Burov, et al.,[9] (Fig. 10) have analyzed the 7.5 GeV pC $\rightarrow$ $\pi^-(180°)X$ data in terms of an incoherent sum over p-"fluctuon" (correlated cluster) rates with the high energy parts of the π spectrum

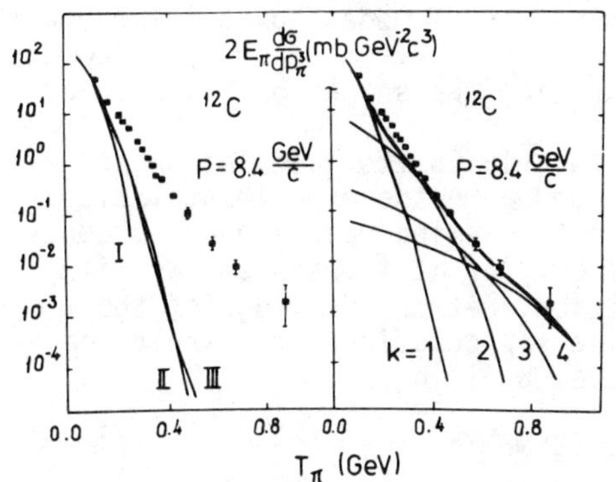

Fig. 10. 7.5 GeV backward π production[9] compared to model using (a) I-frozen nucleons, II,III-nucleon Fermi motion, and (b) fluctuons with k nucleon masses.

arising from progressively heavier clusters ($M_C = kM_p$). Although this model appears successful at this energy for pions, and lower energies for proton production,[10] it contains vastly different assumptions than the model previously discussed. In particular, the p-cluster π production rates are assumed to be independent of cluster size. Furthermore, the clusters occur with probabilities given by a binomial distribution and with c.m. momentum distributions of Gaussian form with increasing width for heavier clusters, $\rho(r) \propto \exp(-p^2/2k\sigma^2)$. In addition, the important phase space constraints and Lorentz transformation of momentums appear to be handled differently.

The quasi-two-body scaling phenomenology of Amado, Woloshyn, and Frankel[7] has also been applied to π production with an apparently high degree of success.[11] Aside from kinematic differences, the original form of QTBS derives from a PWIA model of the form eq. (4), in which the exponential pseudo probability function $\rho_{exp}(p)$, eq. (8b), dominates the integral. This yields a cross section proportional to $\rho(k_{min})$, where k_{min} is the minimum cluster momentum necessary to produce a pion with the observed momentum. In the limited range of applicability of this assumption ($k_{min} \gtrsim k_0$), $d\sigma \propto \exp(-k_{min}/k_0)$ does indeed produce a reasonable fit to the slopes of the second half of the pion spectrum [x > (.28, .17, .08) at (.73, 2.1, 7.5) GeV]. For the other half, the momentum dependence of the elementary subprocess is influential.

Conclusion

For both proton and heavy ion beams it seems possible to account for the shapes of the inclusive π production spectrum with a fairly simple hard scattering model employing either single nucleon or multi nucleon "cluster" production. This is true in both the target ($\theta_\pi \simeq 180°$) and projectile ($\theta_\pi \simeq 0°$) fragmentation regions. Much of the shape of these spectrums arises from Lorentz transformations and kinematic constraints, as we have seen in a model making optimal use of invariance and scaling. The proper internal momentum distribution to use for high momentum nucleons, or varying mass clusters, and the relation of $\rho(p)$ to the true Fermi motion, we[5] consider open questions. The presently employed functions are at best pseudo distributions since the theory does not include the important final state interactions [absorption, dispersion, and cascade], off-shell effects, wave function orthogonality, or explicit NN correlations. Yet this type of model, particularly with the exponential pseudo distribution, does have the purely phenomenological value of succinctly summarizing the shapes of a wealth of experimental data.

More theoretical work, particularly in the calculation of reliable normalizations, is sorely needed. On the experimental side, it would be a fascinating exercise to examine back angle π production for $|x| > 0.6$ and for $T_p \lesssim 1$ GeV (or at forward angles with heavy, heavy ion beams). In addition, better knowledge of the elementary pp, pd, and pα inclusive π production rates would remove some of the uncertainty in our model building.

ACKNOWLEDGEMENTS

I would like to thank Drs. M. Gyulassy and R. Woloshyn for helpful discussions, and Dr. L. Schroeder for both helpful interactions and use of his preliminary 180° data. In addition I would like to acknowledge the help of Ms. J. Westfall for her able assistance in the preparation of the manuscript.

REFERENCES

1. J. Papp, J. Jaros, L. Schroeder, J. Stapes, H. Steiner, A. Wagner, and J. Wiss, Phys. Rev. Lett. 34, 601 (1975).
2. A. M. Baldin, N. Giordenescu, V. N. Zubarev, L. K. Ivanova, N. S. Moroz, A. A. Poutorpiko, U. B. Radomanov, and U. S. Stavinskii, Yad. Fiz. 20, 1201 (1974).
3. I. A. Schmidt and R. Blankenbecler, Phys. Rev. D15, 3321 (1977).
4. S. Chessin, J. Geaga, J. E. Grossiord, D. L. Hendrie, L. S. Schroeder, B. Treuhaft, K. van Bibber, and E. Wuest, Bull. Am. Phys. Soc. 23, 48 (1978), plus private communication of preliminary data;
   D. R. F. Cochran, et al., Phys. Rev. D6, 2085 (1972) as reanalyzed by R. H. Landau, Phys. Rev. C17, 2144 (1978); Ref. 2.
5. R. H. Landau and M. Gyulassy, Phys. Rev. C19, 149 (1979).
6. D. E. Greiner, P. J. Lindstrom, H. H. Heckman, B. Cork, and F. S. Bieser, Phys. Rev. Lett. 35, 152 (1975).
7. R. D. Amado, R. M. Woloshyn, Phys. Rev. Lett. 36, 1435 (1970); Phys. Lett. 69B, 400 (1977), Phys. Rev. C16, 1255 (1977).
8. See also analysis of Ref. 1 and Nakai, et al., Zurich 1977.
9. V. V. Burov, V. K. Lukyanov, and A. I. Titov, Phys. Lett. 67B, 46 (1977).
10. T. Fujita, Phys. Rev. Lett. 39, 174 (1977).
11. C. F. Perdrisat, S. Frankel, and W. Frati, Phys. Rev. C18, 1764 (1978).

Question: S. Wallace, University of Maryland

Is it clear that the concept of cluster recoil is essential in your model? Is it equally acceptable to view the production as arising from a multiple scattering process?

Response: Landau

The cluster arises to account for kinematic thresholds. At present the model is not capable of distinguishing production from a cluster or from production in a multiple scattering process that is also kinematically allowed. I think our model indicates that a two nucleon mechanism with appropriate elementary rates can explain the shapes and energy dependences of the shapes. Whether you should view the 2N mechanism as indicating cluster production from two correlated nucleons is not answered by our model-if in fact there is any real difference. We have seen that the kinematics favor a multi-nucleon mechanism. A more conclusive understanding, unfortunately, requires a full study of normalization and relative weightings of each component. This same problem is true for all inclusive production and future work should help clarify how to handle absorptions in non-coherent environments.

Question: M. Furic, Rice University

We recently calculated the kinematically complete $pd \to d\pi^+ n$ differential cross sections. We found that the multiple scattering is important for large momentum transfers. The remarkable success of your theory in inclusive spectra arises most probably from the fact that each point is still heavily dominated by the smallest allowed momentum transfer, while the integrated multiple scattering contribution does not reach that level.

Response: Landau

As I indicated at the end of my talk, the Frankel-Amado-Woloshyn analysis essentially confirm your statement. The point being that for $k_\pi^{min} > k_0$ in the exponential pseudo-distribution, $\rho(p)$, is so rapidly varying the $\rho(p_{min})$ dominates and $\sigma \sim e^{-kmin/k_0}$. For our one nucleon model this is true for approximately one-half the spectra (the higher k_π parts); for 2N cluster production, Fermi motion is less important and there is less dominance of the pseudo-probability distribution.

$\vec{p}\pi$ REACTION ON LIGHT NUCLEI

Garth Jones
University of British Columbia, Vancouver, B.C., Canada, V6T 1W5

INTRODUCTION

"Pion Production resembles stripping--Protons from the Uppsala University synchrocyclotron have produced positive pions with angular distributions for (p,π^+) that are quantitatively similar for distributions in the well-known deuteron stripping (d,p) reaction." This announcement in the 'Search & Discovery' section of Physics Today (February 1972) was in some sense the "Birth Announcement" of a research area which has developed extensively and been the subject of much discussion at meetings and conferences since that time. A principle motivation for initiating such a research program was the hope of gaining information concerning the high-momentum components of the nuclear single-particle wave functions. However, in the years that have elapsed since the initial observations of the reaction, the research groups have been more involved with questions concerning the detailed nature of the reaction mechanism itself, a question that obviously requires understanding before meaningful nuclear structure information can be extracted from the experimental measurements.

An indication of the shape of the differential cross-section for reactions of the type: $p + A \rightarrow B^* + \pi^{\pm}$ is given in Figs. 1, 2 which show the experimental results that have been published by the Uppsala group.[1)]

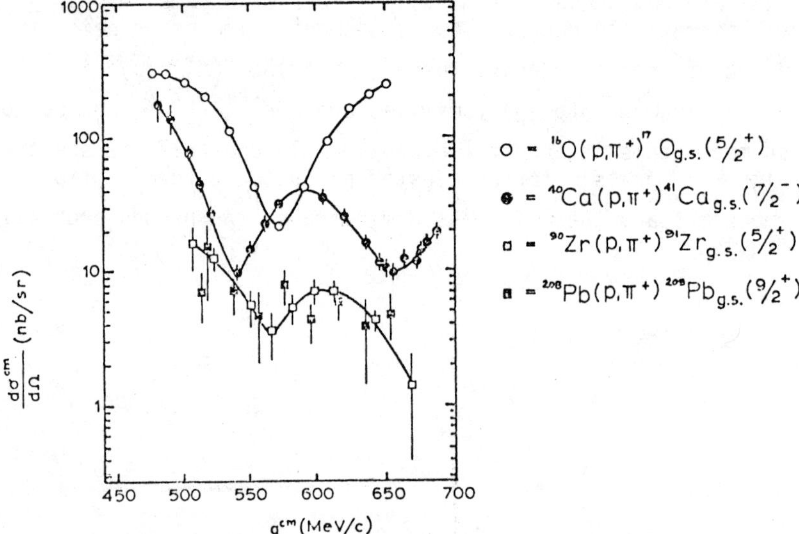

Fig. 1. Angular distributions of (p,π^+) reactions from different targets leading to the ground states of the residual nuclei. The solid lines are guides for the eye.

ISSN:0094-243X/79/540116-15$1.50 Copyright 1979 American Institute of Physics

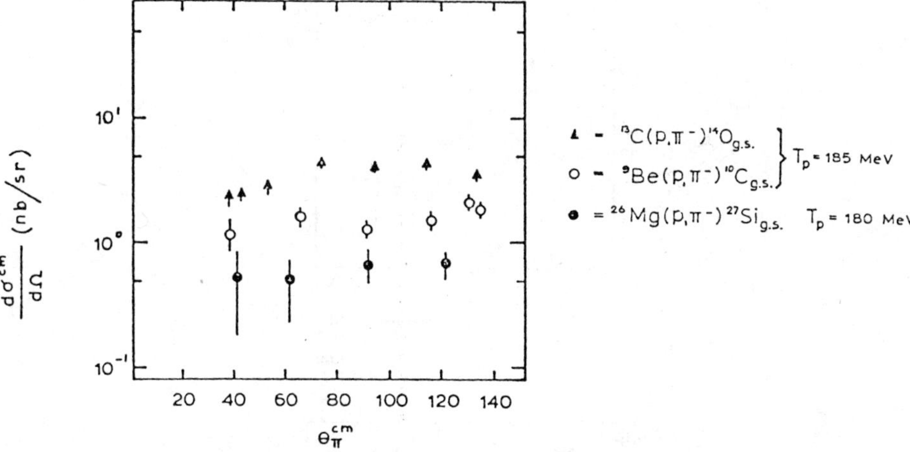

Fig. 2. Angular distributions of (p,π^-) reactions from several targets leading to the ground states of the residual nuclei.

Additional measurement of such reactions are now being reported from Saclay,[2] Indiana,[3] LAMPF,[4] and TRIUMF.[5] Much of the concern has centred around the nature of the basic reaction mechanism itself, whether a "one-nucleon model" (ONM) approach which is basically the stripping model referred to earlier, is adequate, or whether a more complex mechanism such as the "two-nucleon model" (TNM) is required. These mechanisms are illustrated diagrammatically in Fig. 3. Examples of the degree of fit attainable by modern practitioners of these techniques are illustrated[1] in Fig. 4.

The ability to achieve fits of this quality does require, however, some "fine-tuning" of the basic models, the inclusion for example, of appropriate 'distortion'

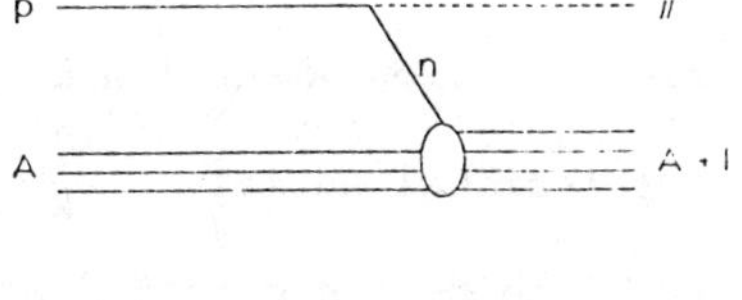

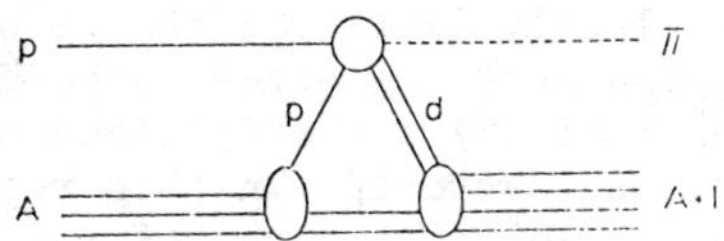

Fig. 3. Typical diagrams illustrating pion production.
a) One-nucleon model (ONM).
b) Two-nucleon model (TNM).

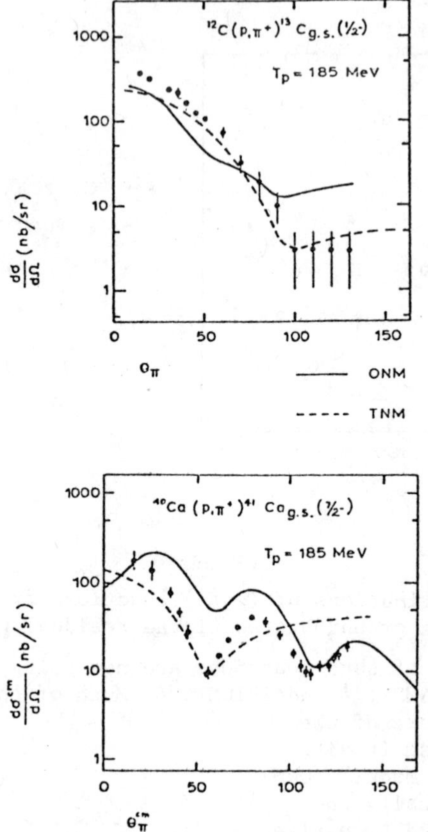

Fig. 4. Experimental and theoretical cross sections for $^{12}C(p,\pi^+)^{13}C$ and $^{40}Ca(p,\pi^+)^{41}Ca$ leading to the ground states of the residual nuclei.

of the ingoing and outgoing proton and pion waves, respectively. On the basis simply of the quality of the fit of the differential cross-section for pion production at a fixed incoming proton energy, there is clearly little to choose between the different theoretical approaches. As stressed by Höistad,[1] systematic experimental (and theoretical) investigations of the energy dependence of such reactions is long overdue. In this regard, the initiation of measurements in the 300-650 MeV range of incident proton energies by the Saclay group[6] is most timely. One of the problems in this area, however, arises from the fact that each of the different experimental facilities is only capable of measuring a portion of the energy region of interest, and there are serious problems concerning the consistency of the results from the different laboratories. Fig. 5 illustrates the extent to which such discrepancies exist, with results shown from Indiana, Orsay and Uppsala. The resolution of such problems will only be achieved when measurements over a large energy range

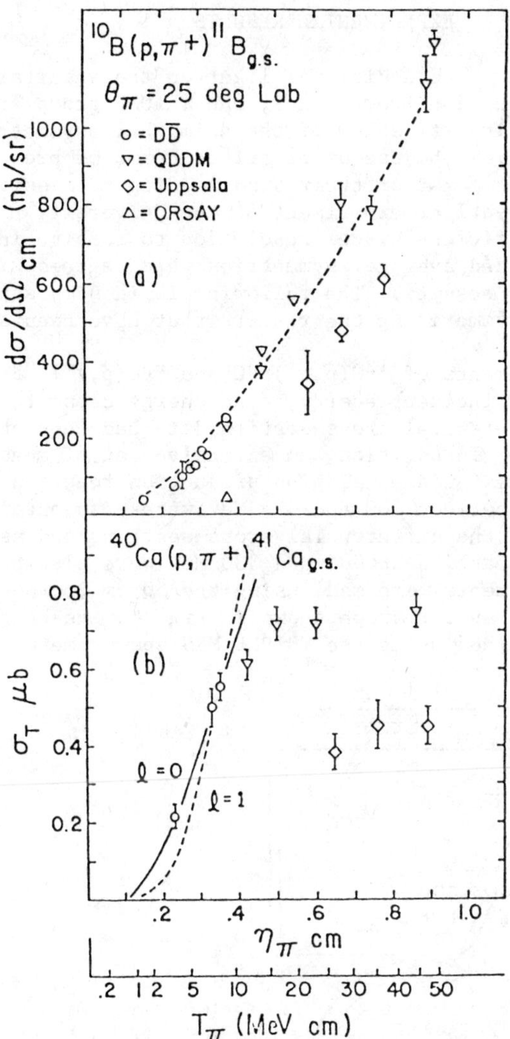

Fig. 5. Energy Dependence of (p,π^+) cross sections.[7]
a) Differential cross-section for $^{10}B(p,\pi^+)^{11}B_{g.s.}$ at 25°
b) Total cross-section for $^{40}Ca(p,\pi^+)^{41}Ca_{g.s.}$

are performed at a single facility. The value of such a program has been underlined recently by the beautiful sets of cross-section data from threshold to 200 MeV incident energy coming out of Indiana.[7] The extension of such a comprehensive program to the energy range above 200 MeV is very much in order.

EXPERIMENTAL RESULTS

In an attempt to shed additional light on the uncertainties concerning pion production from nuclei, the TRIUMF group embarked a few years ago on an investigation of the azimuthal asymmetry of pion production arising from the use of polarised beams of protons. These measurements were the first of their kind leading to resolved final nuclear states. An earlier experiment at the University of Chicago,[8] performed with insufficient system resolution to resolve individual nuclear states, yielded average asymmetries which agreed with the more detailed TRIUMF results. The following is in some sense, a 'progress report', summarizing the results that have been attained to date at TRIUMF.

Initial measurements of $^{12}C(\vec{p},\pi^+)^{13}C$ and $^{9}Be(\vec{p},\pi^+)^{10}Be$ were conducted at 200 MeV incident energy,[5] an energy close to that where extensive differential cross-section data had been obtained by the Uppsala group.[1] In addition, an extensive set of measurements of asymmetries in the fundamental pion production reaction $\vec{p}p \to d\pi^+$ for proton energies between 300 and 425 MeV were also obtained.[9] Some measurements of the differential cross-sections and asymmetries for the $\vec{p}\,^2H \to {}^3H\pi^+$ reaction at 305 and 330 MeV were also performed. All of these measurements were made using the 50 cm Browne-Buechner broad-range magnetic spectrograph shown in Fig. 6, except the 22.5° data which were obtained using the TRIUMF MRS spectrometer.

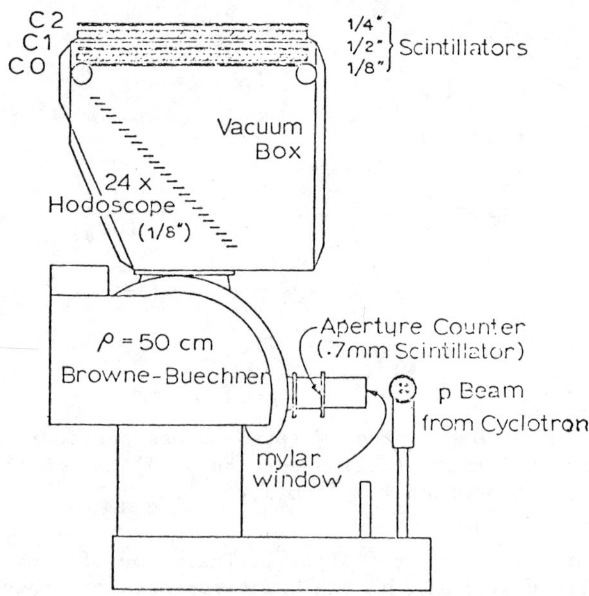

Fig. 6. Broad-range magnetic spectrograph used for pion detection at TRIUMF.

The asymmetry results for the pions leading to the ground and first few excited states of ^{13}C and ^{10}Be are shown in Fig. 7, 8.

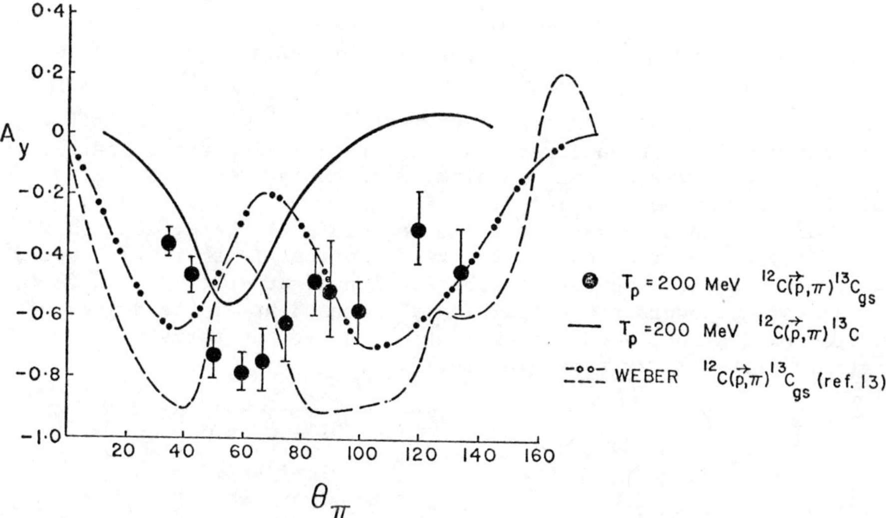

Fig. 7. Analysing powers for the reaction $^{12}C(\vec{p},\pi^+)^{13}C_{g.s.}$. Dashed lines are results of theoretical calculations,[13)] whereas the solid line represents the results for the $\vec{p}p \rightarrow d\pi^+$ reaction (kinematically transformed).

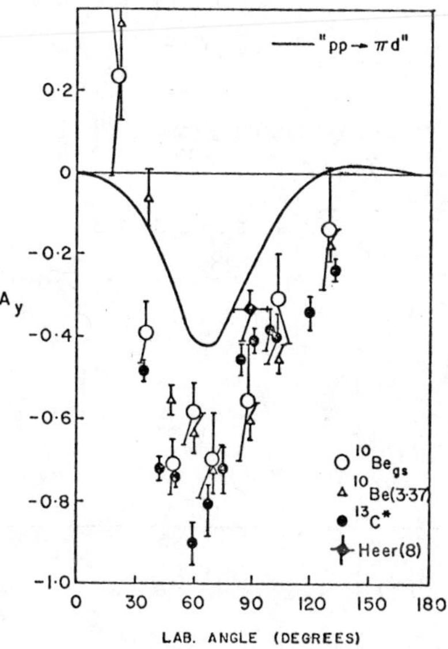

Fig. 8. Analysing powers for the reactions: $^9Be(\vec{p},\pi^+)^{10}Be_{g.s.}$, $^9Be(\vec{p},\pi^+)^{10}Be$(3.4 state), and the $^{12}C(\vec{p},\pi^+)^{13}C^*$ (average of 3 excited states).

The quantity displayed is the Analysing Power,

$$A_y = \frac{\frac{d\sigma}{d\Omega}(+) - \frac{d\sigma}{d\Omega}(-)}{P^- \frac{d\sigma}{d\Omega}(+) + P^+ \frac{d\sigma}{d\Omega}(-)}$$

where the (+) sign refers to measurements with $\vec{k}_p \times \vec{k}_\pi$ parallel to the proton polarisation direction, and the (-) sign the opposite. The P^- and P^+ terms are the magnitudes of the proton polarisation for each of these cases.[10)]

Quite unexpected were both the magnitude of the effect and the similarity of the results for the two reactions (with little dependence on the quantum numbers of the final nuclear states). In fact, the analysing powers are surprisingly independent of the degree of nuclear excitation, even for excitations above the threshold for particle emission, as shown in fig. 9.

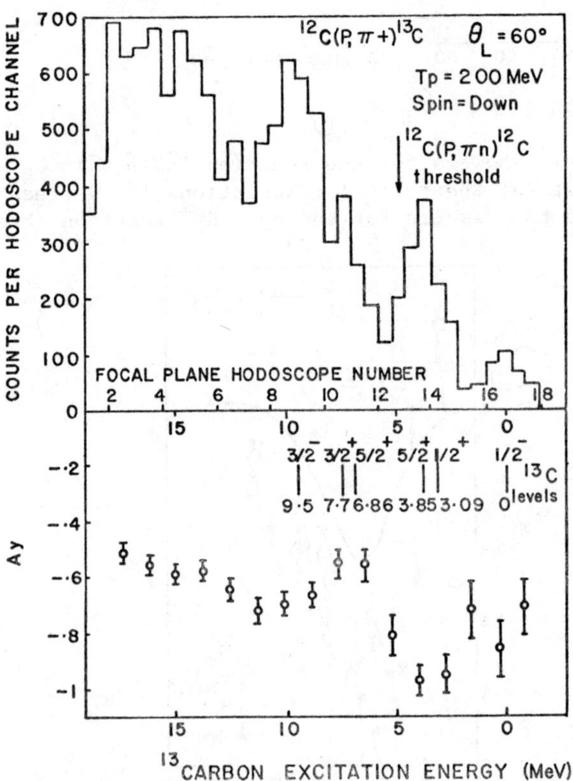

Fig. 9. $^{12}C(\vec{p},\pi^+)^{13}C$ results for a range of ^{13}C excitation energy. The top curve shows the counting rate as a function of ^{13}C excitation for incident proton spin "down", and the lower curve shows the analysing power for the reaction as a function of excitation energy.

Except for the magnitude and the position of the peak, the shapes of these results are very reminiscent of the asymmetries observed in the $\vec{p}p \rightarrow d\pi^+$ reaction at a similar energy above threshold[9] (Fig. 10).

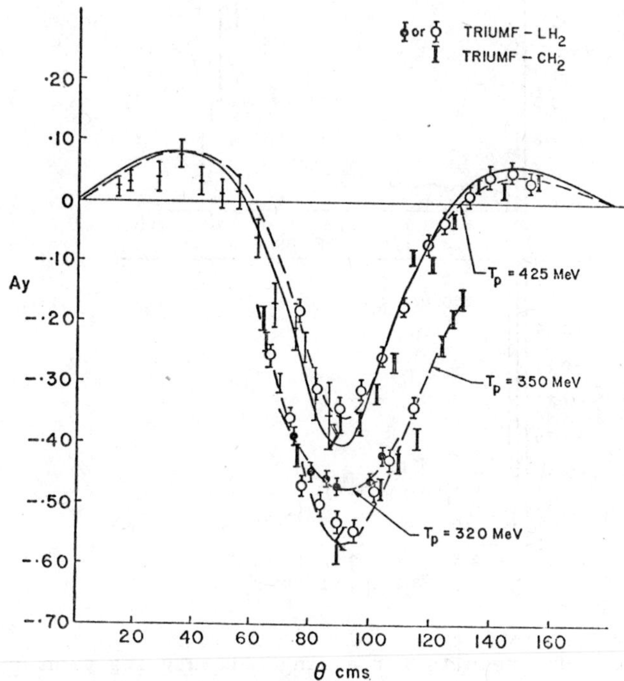

Fig. 10. Analysing powers for the $\vec{p}p \rightarrow d\pi^+$ reaction as a function of pion cms angle for proton energies of 320, 350 and 425 MeV. The dashed curves are fits to the data.

In this regard, it is worth noting that the analysing powers for those energetic pions associated with the break-up reaction: $\vec{p} + p \rightarrow p + n + \pi^+$ are also very similar to those of the $\vec{p}p \rightarrow d\pi^+$ reaction, as shown in Fig. 11.

DISCUSSION

Theoretically, all predictions of analysing powers for nuclear production performed to date[11,12,13] have involved the ONM with the pion asymmetries arising from the effect of distortions in either the incoming proton wave[11] or outgoing pion wave.[12] The most recent of these calculations[13] includes distortions in both the incoming and outgoing waves, finding, however, that the effect is dominated by the distortions characterizing the incoming wave. Conventional proton distortions were employed, with both absorptive and spin-orbit parts. A comparison of these latter theoretical expectations with the experimental results are displayed in Fig. 7. Although there is qualitative agreement at least in sign, the

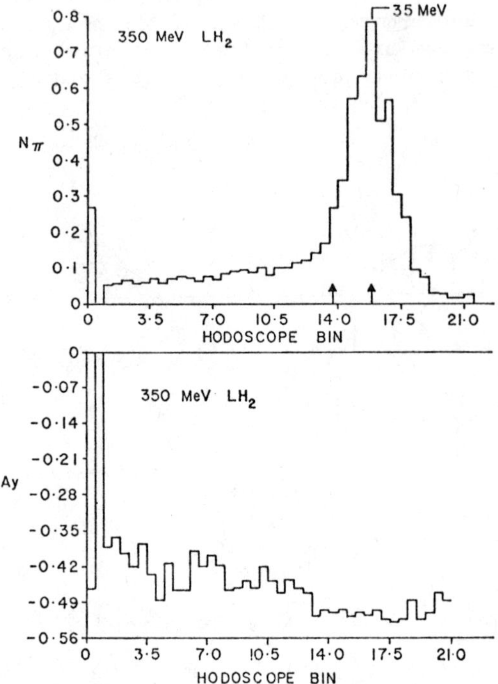

Fig. 11. $\vec{p}p \rightarrow d\pi^+$ results for a range of outgoing pion energy. The top curve shows the counting rate per hodoscope bin and the lower shows the analysing power as a function of hodoscope bin. The width of the $pp \rightarrow d\pi$ peak is about 3 MeV (FWHM).

theoretical expectations still anticipate a far greater dependence on the detailed characteristics of the bound neutron wave-function than we observe in the experimental results.

As shown in Fig. 12 the experimental analysing powers for the $\vec{p} + {}^2H \rightarrow {}^3H + \pi^+$ reaction at proton energies of 305 and 330 MeV also bear a striking similarity to the results observed both for the heavier nuclei as well as for the elementary $\vec{p}p \rightarrow d\pi^+$ reaction. In this case, like the $\vec{p}p \rightarrow d\pi^+$ reaction, the results display a single prominent maximum (negative) peak at 90° (cms). Again, no significant variation in analysing power is observed for the most energetic pions arising from the break-up reaction: $\vec{p} + {}^2H \rightarrow \pi^+$ + anything. Unfortunately, no theoretical predictions exist for the asymmetry effects characterizing pion production from this most simple of nuclei. Instead, the theoretical analyses of the $p + {}^2H \rightarrow {}^3H + \pi^+$ reaction have been restricted to considerations of the dependence of the unpolarised differential cross-section on angle and incident proton energy.[14,15)]

The striking features of all these experimental analysing powers are:

a) the similarities in shape (and, indeed, similarity in shape to the results characterizing the elementary $\vec{p}p \rightarrow d\pi^+$ reaction),
b) independence of nuclear structure details, and
c) independence on the degree of excitation (or even of binding) of the residual nucleus.

In fact, the maximum analysing powers occur at a common lab angle (55-60°) for all the reactions investigated, even including the $\vec{p}p \rightarrow d\pi^+$ reaction. Thus, it is difficult to avoid interpreting the nuclear analysing powers in terms of the basic $\vec{p}p$ reaction (which in turn points to the role of the TNM for such processes). Interpreted this way, the nuclear pion production at 200 MeV would occur through the interaction of the incident proton with a target proton near the top of the Fermi sea (a momentum of about 195 MeV/c is required in order for the pion production to occur via the $\vec{p}p \rightarrow d\pi^+$ reaction). Such a model has previously been applied by a number of authors to pion production by unpolarised protons.[14,15,16] Disregarding all nuclear structure effects, we can compare the $\vec{p}$ + nucleus pion production analysing powers with the corresponding results for the $\vec{p}p \rightarrow d\pi^+$ reaction by simply performing the appropriate kinematical transformation. The prescription employed involves selecting the $\vec{p}p \rightarrow d\pi^+$ values of analysing power appropriate to the same value of (cms) outgoing pion momentum and angle as those characterizing the microscopic $\vec{p}p \rightarrow d\pi^+$ reaction assumed in the nucleus. This rather unconventional choice of dynamical variables was chosen purely for convenience, as the analysing power data for the $\vec{p}p \rightarrow d\pi^+$ reaction is traditionally parameterized in terms of η and θ^* (the outgoing pion momentum in units of μc and angle in the cms).[9] The results of such transformations are shown by the curves so indicated in Figs. 7, 8, 12. The transformation for heavier nuclei is little different from that for carbon. The qualitative agreement in shape with the experimental measurements, and the quantitative agreement of the positions of the maxima are indeed striking, although it is clear that the magnitude of the maxima in the analysing power results for the nuclear cases are substantially larger than would be

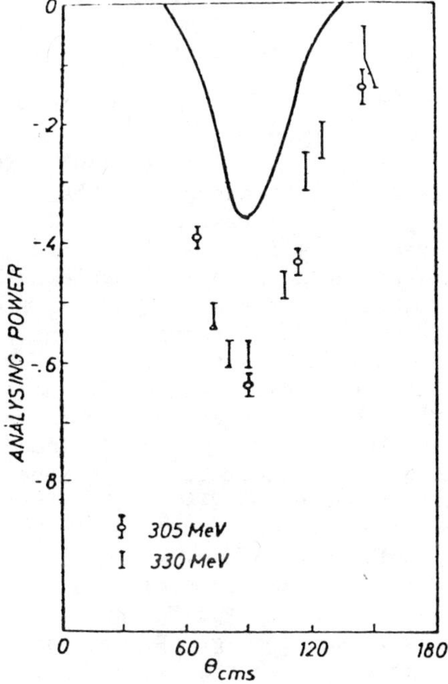

Fig. 12. $^2H(\vec{p},\pi^+)^3H$ analysing powers at 305 and 330 MeV incident energy.

expected from the elementary $\vec{p}p \rightarrow d\pi^+$ reaction. For the nuclear case, it is also true that the shape of the analysing powers appears to be deviating somewhat from the simple $\vec{p}p \rightarrow d\pi^+$ expectations at angles far removed from the maximum.

CONCLUSIONS

What does all this mean? Given the almost complete lack of theoretical guidance that characterizes this subject, one cannot help but be impressed by the degree to which the experimental observations in this field parallel those of elastic and inelastic scattering of polarised nucleons by nuclei. In their pioneering treatment of the scattering of fast nucleons from nuclei,[17] Kerman, McManus and Thaler noted in their introduction, that as a result of a number of investigations "there is a strong reason to believe that the elastic scattering and polarization of the nucleons on nuclei can be understood in terms of the two nucleon interaction."

They pointed out that since the polarisation in such reactions is given to good accuracy by the Born approximation it is thereby determined primarily by the scattering coefficients and little affected by the form factor. Since the condition for this theorem to hold is that the destructive interference between the contribution of various partial waves to the scattering should not be too marked, the conclusion breaks down near diffraction minima of the elastic scattering. Beyond this point, the geometrical effects associated with the nuclear form factor dominate, so that the polarisation at small forward angles (and the integrated potentials) are the quantities most directly connected with the two nucleon scattering coefficients. Very similar conclusions were found for inelastic scattering as well. Experimental polarisation (or analysing power) results characterizing inelastic proton scattering in light nuclei are reproduced[18] in Fig. 13. The similarities between the different sets of results is most suggestive of the situation we have encountered in pion production.

Our observation that the analysing powers characterizing pion production from nuclei is similar in shape but larger than that for the elementary $\vec{p}p \rightarrow d\pi^+$ reaction also has its precedent in the field of nucleon elastic and inelastic scattering. In 1955, Tamor published a paper[19] in which he attempted "to reconcile the large polarizations produced by complex nuclei with the smaller effects found in nucleon-nucleon scattering." He further concluded that for a particular class of nuclei (the deuteron and the alpha-particle nuclei) the polarisation is independent of the nuclear wave function.

Clearly, the interaction underlying the production of pions is quite different from that associated with the scattering of nucleons. The results referred to above for nucleon scattering are not applicable as they stand for pion production. However, the resemblances are striking--with very similar characteristics observed. It would seem timely for some of our theoretical colleagues to investigate pion production in nuclei in the same

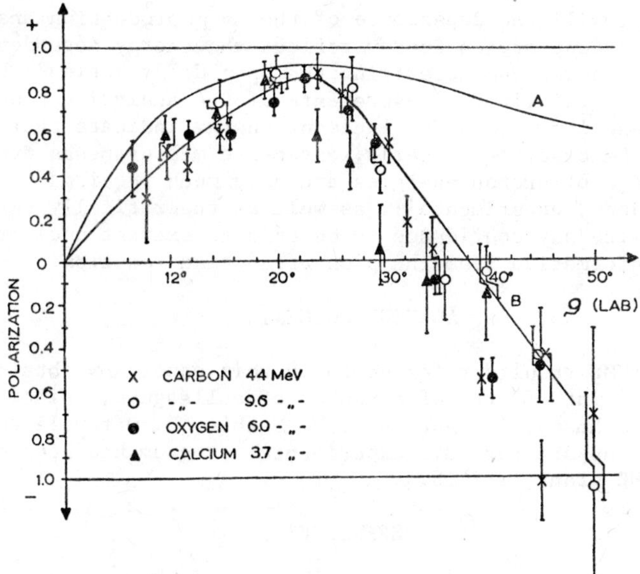

Fig. 13. Polarisation (or analysing power) results for the inelastic scattering of 173 MeV protons from carbon, oxygen and calcium.

spirit (impulse approximation employing the free nucleon-nucleon interaction) that proved so effective for the case of nucleon scattering. Certainly, as experimentalists, we are very much interested in extending our analysing power measurements out to larger angles, for heavier nuclei and for significantly different incident proton energies. Will we, for pion production, also find that these "free nucleon-nucleon like" asymmetries hold only "at small angles". Will specific nuclear structure effects become more apparent past the minimum of the pion production differential cross-section? Will the lack of nuclear structure effects so apparent in our $^{12}C(\vec{p},\pi)^{13}C$ and $^{9}Be(\vec{p},\pi)^{10}Be$ results extend also to heavier nuclei like $^{40}Ca(\vec{p},\pi)^{41}Ca$ (as was observed for the nucleon scattering shown in Fig. 13)? Recent analysing power results for the $^{12}C(\vec{p},\pi^+)^{13}C$ and $^{40}Ca(\vec{p},\pi^+)^{41}Ca$ reactions from Indiana,[20] made at about 10 MeV above threshold, show very little energy dependence for the ^{12}C case (other than a shift of the maximum toward larger angles). Their analysing power results for $^{40}Ca(\vec{p},\pi^+)^{41}Ca_{g.s.}$ (the first measured for this element) also yield negative values. The maximum for this case, however, was at much larger angles (about 130°). Clearly, more extensive investigation of the analysing powers as a function of proton energy is called for.

Finally, will the dependence of the π^+ production analysing powers on incident proton energy reflect the energy dependence of the analysing power characterizing the $\vec{p}p \rightarrow d\pi^+$ reaction? Recent evidence from preliminary measurements of the analysing power in the $^2H(\vec{p},\pi)^3H$ reaction at 800 MeV incident energy indicate that such may not be the case.[21] Clearly, extensive measurements over an extended range of proton energies are very much required. There is much to be done, experimentally as well as theoretically before we can expect with any confidence to be able to extract nuclear structure information from the pion production reaction.

ACKNOWLEDGEMENTS

The TRIUMF results referred to in this paper were obtained only through the joint efforts of a number of colleagues, principally E. Auld, R. Johnson, T. Masterson, E. Mathie, D. Ottewell and P. Walden. Support for this experimental program has been derived from an NSERC grant, IEP 18.

REFERENCES

1. B. Höistad, Proceedings of the Seventh International Conference on High Energy Physics and Nuclear Structure; Zurich (1977).
2. P. Couvert et al, Phys. Rev. Lett. 41 (1978) 530.
3. R. D. Bent et al, Phys. Rev. Lett. 40 (1978) 495.
4. J. Källne et al, Phys. Rev. Lett. 40 (1978) 378; J. Bolger, Ph.D. thesis, University of Texas at Austin (1977).
5. E. Auld et al, Phys. Rev. Lett. 41 (1978) 462.
6. "Experimental Study of the (p,π^+) Reaction at Intermediate Energies", (1978) DPh-N/ME,CEN SACLAY.
7. P. H. Pile et al, to be published in Phys. Rev. Lett. R. Marrs, private communication.
8. E. Heer et al, Phys. Rev. 111 (1958) 640.
9. P. Walden et al, Phys. Lett. B, (to be published).
10. Madison convention, in "Proceedings of the Third International Symposium on Polarization Phenomena in Nuclear Reactions, Madison, Wisconsin, 1970".
11. J. V. Noble, Nucl. Phys. A244 (1975) 256.
12. S. K. Young and W. R. Gibbs, Phys. Rev. C17 (1978) 837.
13. H. J. Weber and J. M. Eisenberg, private communication.
14. M. P. Locher and H. J. Weber, Nucl. Phys. B76 (1974) 400.
15. H. W. Fearing, Phys. Rev. C16 (1977) 313.
16. C. H. Q. Ingram et al, Nucl. Phys. B31 (1971) 331. R. H. Landau and M. Gyulassy, to be published in Phys. Rev. C.
17. A. K. Kerman, M. McManus and R. M. Thaler, Annals of Phys 8 (1959) 551.
18. H. S. Köhler, Nucl. Phys. 9 (1958/59) 49.
19. S. Tamor, Phys. Rev. 97 (1955) 1077.
20. P. H. Pile, private communication.
21. K. K. Seth, private communication.

Question: W. Wharton, Carnegie-Mellon University

In the elementary $\vec{p}+p\to d+\pi^+$ reaction the asymmetry is understood in terms of the interference of the s-wave and p-wave pion amplitudes. Has there been any estimate of how these amplitudes change in complex nuclei? Could an increase in the s-wave amplitude explain the increased asymmetry?

Response: Jones

I have discussed this point with Thomas at TRIUMF and would prefer turning this question over to him to answer.

Response: Thomas

You have raised a very interesting question. That is, it is quite possible that the different off-shell behavior of the s and p wave elementary πN amplitudes may imply that the s and p wave $\pi d\to pp$ amplitudes vary differently off-shell. As you suggest, such off-shell variation (which is ignored in all quasi-deuteron models published to date) could significantly alter the $(\vec{p},\pi^+)$ predictions. In particular it could easily increase $|A_y(\theta)|$.

Question: G. Phillips, Rice University

At TRIUMF the main reaction mechanism of π^+ production is $pp\to d\pi^+$, is it not? At higher energies, LAMPF for example, the mechanism is $pp\to n\Delta^{++}$. There the asymmetry should have very different angular dependence. The polarization effects are therefore energy and angle dependent because of the changing role of different reaction mechanisms and indeed should provide sensitive tests and comparisons of reaction mechanisms.

Response: Jones

I agree with these general remarks. The analyzing power basically samples the spin structure of the reaction mechanism, and so depends on the spin state of the outgoing nucleon. For those pions near maximum energy (i.e. pions from $pp\to d\pi$ and those from $pp\to pn\pi$ where the outgoing p and n have low relative energy) both "inclusive" pion data from TRIUMF at proton energies less than 425 MeV and the Rice "complete kinematic" $pp\to pn\pi$ measurements at 800 MeV indicate that the pion is associated predominantly with a pn triplet spin state. Thus, I would expect the $\vec{p}p\to d\pi^+$ analyzing powers to be applicable to a significant range of pion energies near this maximum. Since the type of nuclear pion production which I have been discussing in this talk is concerned with the maximum energy pions possible then I would expect the $\vec{p}p\to d\pi$ analyzing powers to still be relevant elementary process, even at LAMPF energies.

Question: K. Seth, Northwestern University

Let me add a little more information on our $\vec{p}+p\to d+\pi^+$ and $\vec{p}+d\to t+\pi^+$ experiments at 800 MeV done at HRS at LAMPF. We have measured differential cross sections with an unpolarized beam and analyzing power with a polarized beam. The A_y measurements are only preliminary and incomplete so far, but we find that at 800 MeV the minimum in A_y for

$\vec{p}+p\rightarrow d+\pi^+$ is at $\theta_{cm}\simeq 78°$ but for $\vec{p}+d\rightarrow t+\pi^+$ it is at $\theta_{cm}\simeq 53°$. I realize that one doesn't quite know how to transform the p+p minimum to correspond to the p+p collision in the deuteron, but it does look like that we do not find the identity that is seen in the near threshold data. In our case a better correspondence is seen between the differential cross sections and A_y for $\vec{p}+d\rightarrow t+\pi^+$. The tather sharp minimum in A_y is found to be approximately at the same angle as the minimum, or rather the inflection in the differential cross section. This is similar to the connection one sees in elastic scattering differential cross sections and A_y. It will be interesting to see if we find more structure in A_y since $\sigma(\theta)$ has more than one "inflection" in it.

<u>Response: Jones</u>

It is certainly possible that the mechanism responsible for the analyzing power changes as one goes from 150 MeV above threshold to 600 MeV. Clearly more experimental data within this energy range is required before we can answer such questions.

MULTIPION PRODUCTION IN NUCLEAR COLLISIONS*

M. Gyulassy
Lawrence Berkeley Laboratory
University of California
Berkeley, California 94720

ABSTRACT

We discuss two aspects of multipion production in nuclear collisions: (1) the negative pion multiplicity distribution, and (2) the $\pi^-\pi^-$ correlation function. The emphasis is on how these observables could be used to search for signals of collective phenomena in nuclear collisions.

INTRODUCTION

One of the prominent features of relativistic ($E_{lab} \sim 1$ GeV/nucleon) nuclear collisions is multipion production. In a typical reaction[1] such as Ar + Pb_3O_4 at 1.8 GeV/nucleon, up to 14 negative pions (π^-) alone are observed in certain events. The topic of my talk is what such multipion final states can teach us about the dynamics in nuclear collisions. Some aspects of what single pion inclusive cross sections can teach us are discussed elsewhere in these proceedings by R. Landau.[2]

The primary motivation for colliding relativistic nuclei has been the expectation that through such collisions the properties of dense, highly excited nuclear systems can be studied. Of special interest has been the search for collective phenomena. Several theoretical studies[3] have suggested in particular that pionic instabilities could possibly occur when nuclear matter is compressed to $\rho \sim (2\text{-}4)\rho_0$, $\rho_0 = 0.17$ fm^{-3}, at temperatures $T \sim 50$ MeV. The elusive problem, however, has been to identify specific experimental signatures of such novel phenomena.

One natural direction to search for evidence of collective phenomena has been multi-pion production. In this talk, I will review our work on two aspects of multipion production. The first is the nature of the π^- multiplicity distribution, $P(n_\pi)$, and the second is the dynamical and geometrical information contained in the $\pi^-\pi^-$ correlation function[5,6] $R(\tilde{k}_1,\tilde{k}_2)$. For a more complete review of other aspects of relativistic nuclear collisions see, e.g., Ref. 7.

Multiplicity Distribution[4] (with S. K. Kauffmann)

Our first expectation is that the nuclear dynamics in this energy range is governed simply by intranuclear cascading,[8] i.e., multiple nucleon-nucleon collisions. In that case pions are produced mostly one at a time [via $\Delta_{33}(1232)$ decay] in separate

ISSN:0094-243X/79/540131-11$1.50

inelastic nucleon-nucleon collisions. To compute the multiplicity distribution in such a cascade picture, let $\gamma_{\pm}(t)$ denote the time-dependent π^- production and absorbtion rates during the nuclear collision. Then the pion multiplicity distribution $P(n,t)$ can be shown[4] to be a Poisson,

$$P(n,t) = e^{-\langle n_\pi(t)\rangle} \frac{\langle n_\pi(t)\rangle^n}{n!} \tag{1}$$

with a time-dependent mean satisfying

$$\frac{d}{dt}\langle n_\pi(t)\rangle = \gamma_+(t) - \gamma_-(t)\langle n_\pi(t)\rangle \quad . \tag{2}$$

If the interaction time were long enough for the nuclear system to come to chemical and thermal equilibrium with the pions, then $\langle n_\pi(t)\rangle$ would approach an equilibrium value $\langle n_\pi\rangle_{eq}$ which is independent of the detailed pion production and absorbtion history, $\gamma_{\pm}(t)$, and given by[4]

$$\langle n_\pi\rangle_{eq} = \frac{\gamma_+(\infty)}{\gamma_-(\infty)} = \int \frac{Vd^3k}{(2\pi)^3} \frac{1}{e^{\sqrt{k^2+m_\pi^2}/T} - 1} \quad . \tag{3}$$

In the fireball model,[9] V and T follow from geometrical, kinematical, and equilibrium considerations.[4] The crucial parameter in this model is the freezeout nuclear density $\rho_c \approx 1/3\ \rho_0$, below which no further scattering is assumed to take place.

The results of this thermodynamic calculation[4] for Ar + Pb_3O_4 are shown in Fig. 1. While the features of the impact parameter averaged multiplicity distribution data[1] are indeed well accounted for, we stress that this agreement does not prove that chemical or thermal equilibrium have been reached. The only necessary condition is that the final average pion multiplicities, $\langle n_\pi(b)\rangle$, are well approximated. An example of a completely different dynamical model that also reproduces the impact-parameter averaged multiplicity data is given in Ref. 10.

What makes $P(n)$ insensitive to dynamical assumptions about $\gamma_{\pm}(t)$ is that $\langle n_\pi(b,t_f)\rangle$ at the final (freezeout) time, t_f, can be the same for very many dynamical paths, $\gamma_{\pm}(t)$, in Eq. (2). However, it should be recalled that there is one major (though reasonable) assumption leading to Eq. (2). That assumption, which we call the cascade assumption, is that pion production and absorbtion involve only <u>one</u> pion at a time.

Consider now what happens if the cascade assumption were relaxed. If there were unusual collective phenomena, then pions could possibly be produced also in correlated groups or clusters.. For example, there may be a rate $\gamma_{+k}(t)$ for producing correlated groups of k-negative pions at a time. The effect of such unusual dynamics has been invesitgated in Ref. 4. The result is that if some $\gamma_{+k}(t) \neq 0$ for $k \geq 2$, then the resulting multiplicity distribution for a given impact parameter becomes a convoluted multiple Poisson[4]

$$\frac{P(n,t)}{P(0,t)} = \sum_{n_1=0}^{n} \cdots \sum_{n_p=0}^{[n/p]} \cdots \sum_{n_n=0}^{1} \prod_{k=1}^{n} \left\{ e^{-C(k,t)} \frac{(C(k,t))^{n_k}}{n_k!} \right\} \delta(n - \sum_{r=1}^{n} r n_r) \tag{4}$$

with combinants[4] $C(k,t)$ satisfying the system of equations

$$\frac{d}{dt} C(k,t) = \gamma_{+k}(t) - \gamma_{-}(t)\left(k\, C(k,t) - (k+1)\, C(k+1,t)\right) \quad . \tag{5}$$

In the case $C(1,t_f) \gg C(k \geq 2,t_f)$, then Eq. (4) of course reduces to Eq. (1), with $C(1,t_f) \approx \langle n_\pi(t_f) \rangle$.

The signature of unusual pion production dynamics would thus be a non-Poisson form of the multiplicity distribution in central collisions. Preliminary streamer chamber data[11] taken in a central trigger mode (with high associated nucleon multiplicity) are as yet inconclusive as to whether the high ($n_\pi \gtrsim 10$) tail of the π^- multiplicity distribution is Poisson or not. (We note[4] that the low ($n_\pi \lesssim 5$) portion of $P(n_\pi)$ depends sensitively on the trigger bias.) Should any significant deviation be observed from Poisson behavior for large n_π, then these deviations would reflect on the inadequacy of the cascade assumption and thus provide us with a hint that more exotic pion production mechanisms are at work in nuclear collisions.

We also want to call attention to the results of Ref. 10 which show that even the impact parameter averaged P(n) can be sensitive to certain non-linear phenomena. The non-linearity studied in Ref. 10 involved the assumption that $\langle n_\pi(b) \rangle \propto [N(b)]^\alpha$ with $\alpha \geq 1$ and N(b) being the number of nucleons in the interaction region for a given b. A simple multiple scattering model would give $\alpha = 1$. However, non-linearity ($\alpha > 1$) could arise if pions are produced coherently. Analysis of the data of Ref. 1 suggests[10] that, in fact, $\alpha \approx 1$. A nonlinear growth of $\langle n_\pi(b) \rangle$ as a function of N(b) seems to be inconsistent with those data.

In summary, existing multiplicity distribution data can be well accounted for by simple multiple collision or thermal models. Thus far, no evidence for unusual pion dynamics is suggested by these data. However, whether the form of the high n_π tail is Poisson or not still needs future experimental study.

Pion Interferometry[5,6] (with S. K. Kauffmann and Lance W. Wilson)

While the multiplicity distribution P(n) offers some insight into the dynamics of multipion production, it is clear that more detailed information is contained in multipion correlation data. We consider here specifically the $\pi^-\pi^-$ correlation function defined as

$$R(\tilde{k}_1,\tilde{k}_2) = \frac{\langle n_\pi\rangle^2}{\langle n_\pi(n_\pi-1)\rangle}\,\sigma_\pi\,\frac{d^6\sigma(\pi^-\pi^-)/d^3k_1d^3k_2}{(d^3\sigma(\pi^-)/d^3k_1)(d^3\sigma(\pi^-)/d^3k_2)}, \quad (6)$$

where $\langle n_\pi\rangle$ and $\langle n_\pi(n_\pi-1)\rangle$ are the first and second binomial moments of P(n), $d^3\sigma(\pi^-)$ and $d^6\sigma(\pi^-\pi^-)$ are the single and double pion inclusive cross sections, and σ_π is the total negative pion cross section.

The information that $R(\tilde{k}_1,\tilde{k}_2)$ is expected[12,13] ideally to provide is twofold: (1) R can reveal the space-time structure of the pion production source, and (2) R can determine the degree of coherence of the pion field. The first property is the result of the well known[12] Bose-Einstein interference between two identical pions (the Hanbury-Brown and Twiss effect) also known as the Goldhaber effect. The second property of R has only recently been emphasized in Ref. 13.

To understand the content of $R(\tilde{k}_1,\tilde{k}_2)$, we derived[5] the form of R for partially coherent fields in a simple dynamical model. The basis of the calculation was the approximate solution of the pion field equation

$$(\partial_\mu\partial^\mu + m_\pi^2)\,\phi(x) = J(x) \quad . \quad (7)$$

To solve Eq. (7), the pion source current operator, J(x), was first approximated by its expectation value, thereby decoupling the pion and nuclear field equations. The physical motivation for this approximation is again the expectation that the nuclear dynamics is dominated by multiple nucleon-nucleon interactions. In effect we assume that the number of produced pions is small enough that pion-nucleon rescattering would not significantly affect the source current J(x). This physical picture of pion production is thus equivalent to that of bremsstrahlung radiation.[14]

The pion final state produced by a classical source current J in Eq. (7) is well known[14] to be a coherent state $|J\rangle$ given by

$$|J\rangle = e^{-\bar{n}/2}\exp\{i\int d^3k\, J(\tilde{k})\, a^+(k)\}\,|0\rangle \quad (8)$$

where $J(\tilde{k})$ is the on-shell Fourier transform[5] of $J(\tilde{x},t)$ and $\bar{n} = \int d^3k\,|J(\tilde{k})|^2$.

The m-pion inclusive distribution is then given by[14]

$$P_m(\tilde{k}_1, \ldots, \tilde{k}_m) = \frac{1}{\sigma_\pi} \frac{d^{3m}\sigma(m \times \pi^-)}{d^3\tilde{k}_1 \ldots d^3\tilde{k}_m}$$

$$= \langle J|a^+(\tilde{k}_1) \ldots a^+(\tilde{k}_m)\, a(\tilde{k}_m) \ldots a(\tilde{k}_1)|J\rangle$$

$$= |J(\tilde{k}_1)|^2 \ldots |J(\tilde{k}_m)|^2 \quad . \tag{9}$$

Also, the multiplicity distribution is rigorously Poisson.[14] Thus, the correlation function in Eq. (6) is just

$$R(\tilde{k}_1, \tilde{k}_2) = \frac{P_2(\tilde{k}_1, \tilde{k}_2)}{P_1(\tilde{k}_1)\, P_1(\tilde{k}_2)} \quad . \tag{10}$$

From Eqs. (9,10) we see that for a coherent pion field, $R \equiv 1$, and there are no momentum space correlations of pions produced via this bremsstrahlung mechanism.

However, it is clear physically that no fixed form of $J(\tilde{x},t)$ can represent the pion source in all nuclear collisions. For example, the number of inelastic scatterings, N, will differ from event to event as will the impact parameter b. Therefore, the source current $J_\alpha(x)$ must depend on some set of parameters $\alpha = \{N, \tilde{b}, \ldots\}$ that can vary from event to event. The observed inclusive distribution thus samples an <u>ensemble</u> of final states $|J_\alpha\rangle$. We can include such an ensemble average in Eq. (9) by

$$P_m(\tilde{k}_1, \ldots, \tilde{k}_m) = \sum_\alpha p(\alpha)\, |J_\alpha(\tilde{k}_1)|^2 \ldots |J_\alpha(\tilde{k}_m)|^2 \quad , \tag{11}$$

where $p(\alpha)$ is the ensemble distribution of the parameters α. As a result of the ensemble average, it is clear that $R(\tilde{k}_1, \tilde{k}_2) \neq 1$ in general.

As an example of Eq. (11), we first consider a space-time parametrization of the pion production dynamics in the spirit of the intranuclear cascade model.[8] In this space-time picture, pions are produced in N separate inelastic nucleon-nucleon collisions at different space-time points x_i. If $J_\pi(x)$ is taken to parametrize the source current for an inelastic nucleon-nucleon collision at the origin, then the total source current can be written as

$$J_\alpha(x) = \sum_{i=1}^{N} J_\pi(x - x_i) \quad . \tag{12}$$

If we now assume that the "inelastic scattering centers," x_i, are distributed in a space-time reaction volume specified by a density distribution $\rho(x)$, and the number of inelastic collisions, N, is distributed according to $P_s(N)$, then from Eq. (11),

$$P_m(\tilde{k}_1,\ldots,\tilde{k}_m) = \sum_N P_s(N) \int d^4x_1\rho(x_1)\ldots d^4x_N\rho(x_N)\ |J_\alpha(\tilde{k}_1)\ldots J_\alpha(\tilde{k}_m)|^2 , \quad (13)$$

with $J_\alpha(\tilde{k})$ given by the on-shell Fourier transform of Eq. (12), $\alpha = \{N, x_i\}$. An impact parameter average can also be easily included in Eq. (13). Evaluating Eqs. (10,13), we obtain finally[5]

$$R(\tilde{k}_1,\tilde{k}_2) \approx 1 + \left\{ 1 - \frac{\langle N \rangle}{\langle N^2 \rangle} \right\} |\rho(\tilde{k}_1 - \tilde{k}_2,\ \omega_1 - \omega_2)|^2 , \quad (14)$$

in the limit when the interaction time $T = [\int d^4x\ x_o^2 \rho(x)]^{\frac{1}{2}} >> m_\pi^{-1}$. In particular, we have neglected terms of order

$$O[\langle N \rangle |\rho(\tilde{q}=0,\ q_o=m_\pi)|^2] << 1 .$$

This limit is satisfied in all but the most peripheral nuclear collisions.

If the average number of inelastic collisions is large, then Eq. (14) reduces to the well known form of the correlation function in Ref. 12. Furthermore, we have also shown[5] that in the limit $\langle N \rangle \to \infty$, the multiplicity distribution for a given mode $\tilde{k}$ becomes a Bose-Einstein distribution with a mean $\langle n_\pi(\tilde{k}) \rangle = P_1(\tilde{k})\,((2\pi)^3/V)$. Therefore, we are justified in calling the $\langle N \rangle \to \infty$ limit as the <u>chaotic field limit</u>.

For chaotic pion fields, Eq. (14) shows then that R-1 measures the space-time Fourier transform of the pion source region, $\rho(\tilde{x},t)$. Therefore, $R(\tilde{k}_1,\tilde{k}_2)$ provides geometrical information about nuclear collisions.

Next we consider the question of partially coherent fields. We suppose that in addition to pion production from separate inelastic nucleon-nucleon collisions, pions can also be produced coherently as a result of the collective action of some group of nucleons. Let $J_o(x)$ denote that coherent source current. The total pion source current is then

$$J_\alpha(\tilde{k}) = J_o(\tilde{k}) + J_\pi(\tilde{k}) \left(\sum_{i=1}^{N} e^{i(\omega_k t_i - \tilde{k}\cdot\tilde{x}_i)} \right) . \quad (15)$$

Observe that the chaotic component involves random phases distributed according to $\rho(\tilde{x}_i, t_i)$.

Evaluating[5] the ensemble average, the single pion inclusive distribution is now found to be a sum of a coherent and chaotic part,

$$P_1(\tilde{k}) = |J_o(\tilde{k})|^2 + \langle N \rangle |J_\pi(\tilde{k})|^2 \equiv n_o(\tilde{k}) + n_{ch}(\tilde{k}) , \quad (16)$$

where $n_o(\tilde{k})$ and $n_{ch}(\tilde{k})$ are the average number of pions in mode $\tilde{k}$ in the coherent and chaotic component respectively. From Eq. (16) we obtain therefore a natural definition of the <u>degree of coherence</u> of the pion field in mode $\tilde{k}$ as

$$D(\tilde{k}) = n_o(\tilde{k})/P_1(\tilde{k}) \tag{17}$$

As we now show, $D(\tilde{k})$ can be measured from $R(\tilde{k},\tilde{k})$. Evaluating the double pion inclusive distribution, the correlation function is found to be[5]

$$R(\tilde{k}_1,\tilde{k}_2) \approx 1 + (1-D(\tilde{k}_1))(1-D(\tilde{k}_2))\ |\rho(k_1-k_2)|^2 \tag{18}$$

$$+\ 2(D(\tilde{k}_1)(1-D(\tilde{k}_1))\ D(\tilde{k}_2)(1-D(\tilde{k}_2)))^{1/2}\ \mathrm{Re}\rho(k_1-k_2) \quad .$$

In deriving Eq. (18), we have taken $\langle N\rangle >> 1$ in the chaotic component but have again used the long interaction time condition, $\langle N\rangle|\rho(\tilde{0},m_\pi)|^2 << 1$.

Equation (18) is our main result showing that the degree of coherence of the pion field can be measured <u>mode by mode</u> from $R(\tilde{k}_1,\tilde{k}_2)$ via the relation

$$R(\tilde{k},\tilde{k}) = 2 - D^2(\tilde{k}) \quad . \tag{19}$$

Therefore, pion interferometry (the analysis of $R(\tilde{k}_1,\tilde{k}_2)$) provides both geometrical and dynamical information about the pion source in nuclear collisions.

The first data[15] on pion interferometry in nuclear collisions are now available and are shown in Fig. 2. At the present, only impact parameter-averaged and momentum-averaged correlation data are available. Therefore, detailed study of $D(\tilde{k})$ is not yet possible. Nevertheless. these data seem to indicate that at least the average (over $\tilde{k}_1+\tilde{k}_2$ and b) degree of coherence is small. Planned correlation experiments at the Bevalac with much higher statistics will be able to determine whether there is coherence in at least some modes for central collisions.

While the ideal form of the $\pi^-\pi^-$ correlation function in Eq. (18) demonstrates how both dynamical and geometrical information can be extracted from $R(\tilde{k}_1,\tilde{k}_2)$, it is important to recognize that effects of final state interactions must also be taken into account.[6] In nuclear collisions, the nuclear remnant has, for example, a residual nuclear charge $Z_{eff} \sim 100$ which can distort the π^- wavefunctions and lead to distortions of $R(\tilde{k}_1,\tilde{k}_2)$ for $|\tilde{k}_1-\tilde{k}_2| \sim (Z\alpha m_\pi/R)^{1/2} \sim 50$ MeV/c. To incorporate distortion due to some effective optical potential, $V(\tilde{x})$, into the calculation of $P_m(\tilde{k}_1...\tilde{k}_m)$, we need to know the form of the incoming Møller scattering wavefunction, $\psi_{\tilde{k}}^{(-)}(x)$, for that potential. The final result[5] for partially coherent fields has the same structure as Eq. (18) except that the Fourier transform, $\rho(k_1-k_2)$, is replaced by the <u>distorted</u> transform

$$\rho(k_1-k_2) \to \bar{\rho}_V(\tilde{k}_1,\tilde{k}_2) = \rho_V(\tilde{k}_1,\tilde{k}_2)(\rho_V(\tilde{k}_1,\tilde{k}_1)\ \rho_V(\tilde{k}_2,\tilde{k}_2))^{-\frac{1}{2}} \tag{20}$$

where

$$\rho_v(\tilde{k}_1,\tilde{k}_2) = \int dtd^3x\ \rho(\tilde{x},t)\ e^{i(\omega_1-\omega_2)t}\ \psi_{\tilde{k}_1}^{(-)*}(x)\ \psi_{\tilde{k}_2}^{(-)}(x) \qquad (21)$$

Also, the degree of coherence is distorted[5] ($D \to D_V$) via $\psi_k^{(-)}(k')$. In Ref. 6 we have evaluated Eqs. (20) and (21) using Coulomb wavefunctions. We find significant distortions of the shape of $R(\tilde{k}_1,\tilde{k}_2)$ in the region $20 \lesssim |\tilde{k}_1-\tilde{k}_2| \lesssim 100$ MeV/c. These distortions are also sensitive to the orientation of the relative momentum. However, we find[6] that distortions are minimized (~20%) for large pion momenta ($\gtrsim m_\pi$) especially for that orientation of $\tilde{k}_1-\tilde{k}_2$ corresponding to $|\tilde{k}_1| = |\tilde{k}_2|$. Experimentally, it will be important to concentrate on this equal energy configuration at high momenta to get the cleanest determination of the pion source geometry.

Finally, we note a potential problem in determinind $D(\tilde{k})$ due to relative $\pi^-\pi^-$ Coulomb interaction. An apparent non-zero degree of coherence can result from a Coulomb hole[6] in $R(\tilde{k}_1,\tilde{k}_2)$ for $|k_1-k_2| \lesssim \sqrt{\alpha}m_\pi \lesssim 10$ MeV/c. If there were no strong external fields, then such a distortion could be easily taken into account[5,6] and unfolded from the data via a Gamow penetration factor,

$$|\psi^{(-)}_{(\tilde{k}_1-\tilde{k}_2)/2}(\tilde{x} = 0)|^2$$

In the presence of an external potential though, this Coulomb hole tends to be filled up since the typical momentum transfer provided by the external optical potential $V(x)$ is much larger than that provided by the relative $\pi^-\pi^-$ potential, $U(\tilde{x}-\tilde{y})$. The structure of the correlation function in this case is the same as that given in Eqs. (18, 20, 21), except that the correlation function is multiplied by a penetration factor $\mathcal{P}(\tilde{k}_1,\tilde{k}_2)$. In the limit $V \to 0$, $\mathcal{P}$ reduces to the familiar Gamow factor; for $U \to 0$, $\mathcal{P} \to 1$. However, for $V \neq 0$, $\mathcal{P}$ is much more difficult to evaluate. The formula for $\mathcal{P}$ given in Ref. 5 involves a 12-dimensional intergral over products of Coulomb wavefunctions. Since the effect of is mainly to distort $D(\tilde{k})$, we feel at this time that only if there were data indicating a significant degree of coherence would the effort to evaluate $\mathcal{P}$ be worth it. Here we only emphasize that $R(\tilde{k},\tilde{k}) \neq 2$ could in principal arise from relative $\pi^-\pi^-$ interactions. We note that for chaotic fields $R(\tilde{k},\tilde{k}) = 2$ if $U = 0$, regardless of the form of V.

In summary, two-pion correlation data can provide both geometrical and dynamical information that can be used in the search for collective phenomena in nuclear collisions. However, final state interactions must be first unfolded from the raw data. In the case of chaotic fields ($D(k) \equiv 0$) this unfolding will be much easier than if partially coherent fields are produced. Nevertheless, we are still excited to have found at least one observable, $R(\tilde{k}_1,\tilde{k}_2)$, that is sensitive to possible pion field coherence.

References

1. S. Y. Fung, et al.; Phys. Rev. Letters 40, 292 (1978).
2. R. H. Landau, M. Gyulassy; Phys. Rev. C 19, 149 (1979).
3. V. Ruck, M. Gyulassy, W. Greiner; Z. Phys. A 277, 391 (1976); M. Gyulassy, W. Greiner; Ann. Phys. 109, 485 (1977); and references therein.
4. M. Gyulassy, S. K. Kauffmann; Phys. Rev. Lett. 40, 298 (1978); S. K. Kauffmann, M. Gyulassy; J. Phys. A 11, 1715 (1978).
5. M. Gyulassy, S. K. Kauffmann, Lance W. Wilson; Lawrence Berkeley Laboratory preprint LBL-8759 (1979).
6. Lance W. Wilson, M. Gyulassy, S. K. Kauffmann, Lawrence Berkeley Laboratory preprint LBL-8760 (1979).
7. A. S. Goldhaber, H. H. Heckman; Ann. Rev. Nucl. Part. Sci. 28, 161 (1978).
8. R. K. Smith, M. Danos, Proceedings of Topical Conf. on Heavy Ion Collisions, Fall Creek Falls, Tennessee; CONF-770602, p. 363 (1977); Z. Fraenkel, Y. Yariv, to be published.
9. J. Gosset et al.; Phys. Rev. C 18, 844 (1978).
10. J. Knoll, et al.; Nucl. Phys. A 308, 500 (1978); Y. Afek et al.; Phys. Rev. Lett 41, 849 (1978).
11. R. T. Poe, A. Sandoval, private communication.
12. G. I. Kopylov, M. I. Podgoretskii; Sov. J. Nucl. Phys. 18, 336 (1974); C. Enzell, et al.; Phys. Rev. Lett. 38, 873 (1977); F. B. Yano, S. E. Koonin; Phys. Lett. 78B, 556 (1978).
13. G. N. Fowler, R. M. Weiner; Phys. Letts. 70B, 201 (1977).
14. J. D. Bjorken, S. D. Drell; Relativistic Quantum Fields (McGraw-Hill, New York, 1965), p. 202.
15. S. Y. Fung, et al.; Phys. Rev. Lett. 41, 1592 (1978).

* This work has been performed under the auspices of the Division of Nuclear Physics of the U. S. Department of Energy under Contract W-7405-ENG-48.

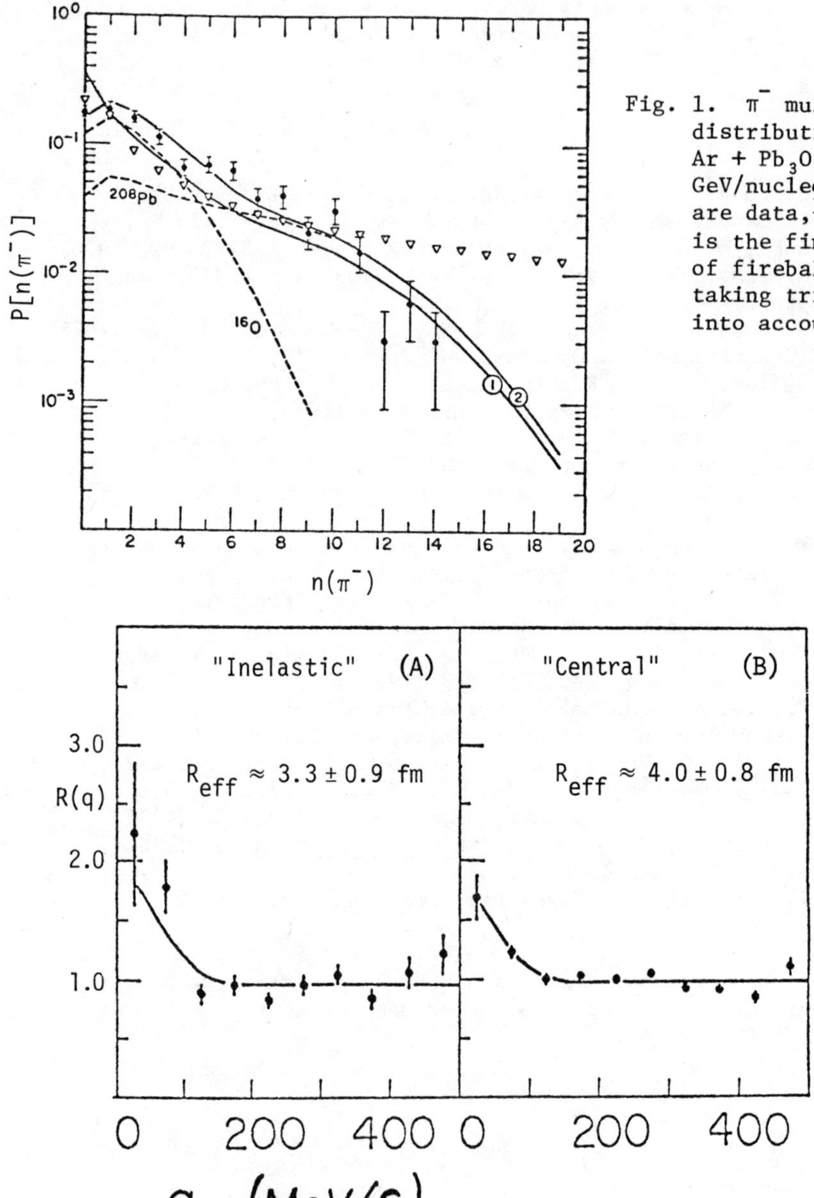

Fig. 1. π^- multiplicity distribution for $Ar + Pb_3O_4$ at 1.8 GeV/nucleon : dots are data,[1] curve 2 is the final result[4] of fireball model taking trigger bias into account.

Fig. 2. First data[15] on the averaged $\pi^-\pi^-$ correlation function $R(q) \equiv \langle R(\tilde{k}, \tilde{k}+\tilde{q})\rangle_k$ for the same reaction as for Fig. 1. Solid curve is fit[15] of data based on pion interferometry[12] assuming a chaotic field in Eq. (18). The effective pion source radius is indicated in two trigger modes.

Question: C. Shakin, Brooklyn College
How does the Coulomb interaction between the various charged particles modify the correlation functions?
Response: Gyulassy
Coulomb final state interactions make major, non-trivial modifications of the correlation function. Our calculations using Coulomb wave functions indicate that the shape of $R(k_1,k_2)$ is distorted by 20-100% for low CM energies (k;<100 MeV/c). However, for high CM energy, distortions are reduced to <20%. Therefore, clean extraction of geometrical information can only be made with high energy pion pairs. Secondly, the relative $\pi^-\pi^-$ Coulomb interaction by itself produces a Coulomb hole given by the usual Gamov factor. However, the coupled three-body Coulomb problem is yet unsolved. This difficult problem will have to be addressed if a finite degree of coherence is actually observed in future data.

THE (p,π^+) REACTION NEAR THRESHOLD*

Robert D. Bent
Indiana University Cyclotron Facility, Bloomington, Ind. 47405

ABSTRACT

New data on the energy-dependence of the $^{10}B(p,\pi^+)^{11}B_{g.s.}$ and $^{40}Ca(p,\pi^+)^{41}Ca_{g.s.}$ reactions havebeen obtained at IUCF. The measurements extend from 60 MeV down to only a few MeV above threshold. Some of the main features of the data are reproduced by DWBA calculations.

The $A(p,\pi^+)A+1$ reaction is of interest because of its sensitivity to the high momentum components of nuclear wavefunctions and the possibility that it may provide new information about the production and propagation of pions in strong nuclear fields. The near threshold region is of particular interest from a nuclear structure point of view, because at $T_p(\text{lab}) \lesssim 290$ MeV there is not enough kinetic energy in the center-of-mass system to make pions in free nucleon-nucleon collisions, and it is only through nuclear effects that pions can be produced at all. In addition, the (p,π) reaction near threshold produces low-energy pions that can be used to study the pion-nucleus interaction in an energy region (0 to 20 MeV) that is inaccessible in pionic atom and pion-nucleus scattering experiments. The theoretical interpretation of experimental results, however, has turned out to be quite complex because of interrelated questions regarding the reaction mechanism, pion rescattering and nuclear structure. The present work was motivated by the expectation that data on the (p,π^+) reaction covering a broad range of outgoing pion energies and momentum transfers would provide useful constraints on reaction models and aid in the unraveling of the different parts of the theory.

Pions were first detected at the Indiana University Cyclotron Facility (IUCF) in the summer of 1976, nine months after the first extraction of beam from the new accelerator. Since that time several different kinds of (p,π^+) experiments have been carried out using a variety of experimental techniques. I will discuss here some of the new results on positive pion production in the near-threshold region extending from 1.7 to 61.4 MeV outgoing center-of-mass pion energy. The experimental techniques employed are described elsewhere.[1,2]

Experimental information on the energy dependence of the (p,π^+) reaction on nuclei is limited, and there are discrepancies in the absolute normalization of the data from different laboratories.[3] The present measurements were made with two different magnetic spectrographs, one for pion energies above 9 MeV and the other for pion energies below 13 MeV. Pions of energies between 9 and 13 MeV were

ISSN:0094-243X/79/540142-12$1.50

measured in both instruments and the overlapping data are in good agreement. This argues strongly for a correct absolute scale for data from either instrument.[2]

The variable-energy proton beam at IUCF is singularly well-suited to a detailed investigation of the energy dependence of the (p,π^+) reaction. The data we present for ^{10}B and ^{40}Ca targets overlap several earlier measurements but extend much closer to threshold, changing significantly the experimental picture of the energy dependence of these reactions in the near threshold region.

Differential cross sections for the reactions $^{10}B(p,\pi^+)^{11}B_{g.s.}$ and $^{40}Ca(p,\pi^+)^{41}Ca_{g.s.}$ plotted against momentum transfer q for several proton energies in the 140 to 200 MeV range are shown in Figs. 1 and 2. Excitation functions are shown in Fig. 3. These data were collected over a two-year period, starting at 154 MeV, as beams at the various energies were developed at IUCF.

The contrast in the energy-dependence of these two reactions is striking. The $^{10}B(p,\pi^+)$ cross sections are large (the largest observed for any nucleus in the forward direction), and the yield at forward angles increases markedly with bombarding energy. To a first approximation the shape of the angular distributions is exponential with q below 540 MeV/c, and the slope of the exponential changes only slightly over this substantial energy range. The 25 degree excitation function shown in Fig. 3a appears to be dominated by Coulomb and angular momentum barrier factors over the entire energy range,[2] and it approaches zero more gradually than the earlier data from Uppsala[3,4] and Orsay,[5] that extrapolates linearly to zero about 7 MeV above threshold. The $^{40}Ca(p,\pi^+)$ cross sections are much smaller, and the angular distributions exhibit a well-defined and systematic variation in shape with changing bombarding energy, which is in sharp contrast to the behavior of the $^{10}B(p,\pi^+)$ reaction. The deep minimum seen at forward angles becomes deeper (ratio $\sigma_{90°}/\sigma_{min}$ approaches 100/1 at T_π = 8.1 MeV) and moves toward smaller angles as the energy is lowered toward threshold. The yield at the backward maximum stays nearly constant until T_π (CM) is below 10 MeV. Since the motion of the minimum in these data can confuse the interpretation of the energy dependence of forward angle cross sections for this nucleus, the total cross sections obtained by integration of the differential cross sections are shown plotted against pion momentum in Fig. 3b. In contrast with Fig. 3a, Coulomb penetrability arguments can explain the energy variation only for the lowest energy points. There is strong suppression of yield at pion energies greater than 10 MeV.

Points from Uppsala[3,4] and Orsay[5] are plotted in Fig. 3 to show the discrepancies that exist in absolute normalization of the data from the different laboratories. Although the cross sections measured at IUCF with the two different spectrographs ($D\bar{D}$ and QDDM) using entirely different detectors agree well within the systematic measurements errors, they are larger than those measured at Uppsala and Orsay by factors of about 1.7 and 5, respectively.

The experimental results presented here establish for the first time the energy-dependence of the $^{10}B(p,\pi^+)^{11}B_{g.s.}$ and $^{40}Ca(p,\pi^+)^{41}Ca_{g.s.}$ reactions in the near threshold region. The measurements cover a sufficiently wide energy range to offer a stringent test for any theoretical explanation of the (p,π^+) process. Although the reaction may not be dominated by a single mechanism over this entire energy range, it is nevertheless instructive to compare the data with calculations based on simple models. We have examined the extent to which the data can be described by the DWBA pionic-stripping model using a code originally written by Keating and Wills[6] and later modified by Tsangarides and Wills.[7] In this model, the proton emits a pion and the resulting neutron is captured directly into an available single-nucleon state. In the plane-wave approximation the differential cross section is proportional to the square of the single-particle wavefunction in momentum space, so that the angular distributions are a sensitive and direct measure of the momentum distribution of the captured neutron. Because the momentum transfer is large in (p,π) reactions even near threshold, the reaction measures the tail end of the neutron wavefunction in momentum space. Realistic calculations must, of course, include distortions of the incoming proton and outgoing pion waves; however, the data should bear some resemblance to the plane-wave calculations if the DWBA model is realistic. If the pion distortions required to fit the data are so large as to completely destroy the main features of the plane-wave calculation, then it probably makes more sense to incorporate the most important pion rescattering effects explicitly into the calculation. It is therefore of interest from the standpoint of the DWBA model to compare the data with the plane-wave calculations. This is done for the case of the $^{40}Ca(p,\pi^+)^{41}Ca_{g.s.}$ reaction in Fig. 4, where the differential cross sections are plotted against pion center-of-mass angle rather than momentum transfer. Calculations with proton distortions included are also shown. While the latter give a fair representation of the 172.8 MeV data, the energy dependence of the calculations is completely wrong. The forward angle minimum in the calculated angular distributions occurs at approximately constant momentum transfer and, hence, at larger angles as the bombarding energy is decreased, whereas the forward minimum in the data behaves in the opposite way.

The sensitivity of the DWBA predictions to distortions of the incoming and outgoing waves is illustrated by the calculations shown in Fig. 5, which were made with a standard proton optical potential and a pion potential of the Laplacian form with parameters adjusted from those that gave the best fit to 30 MeV π^+ - ^{40}Ca elastic scattering data in order to improve the fit to the (p,π^+) data. The neutron bound-state wavefunction was calculated using a Woods-Saxon potential with r_o=1.1 fm and a=0.55 fm. The calculations including pion distortions agree qualitatively with the data shown in Fig. 4 insofar as the two minima occur at about the right places and curve in opposite directions; however, the absolute magnitude of the calculations is too large by about an order of magnitude (presumably due to unrealistic off-shell effects), and the pion potential used lacks theoretical justification and does not give a good fit to pion elastic scattering data. Improved fits to the data achieved by Tsangarides et al.[8] using a Laplacian-like pion-nucleus potential with off-shell damping corrections and parameters obtained by fitting 50, 40, and 30 MeV π^+ - ^{40}Ca elastic scattering data are shown in Fig. 6. The behavior of the first minimum in the angular distributions is reproduced reasonably well, and the discrepancy in absolute magnitude is reduced to only a factor of two; however, the second minimum seen at the highest energies was not reproduced by these calculations, and the fits to pion-elastic scattering data were still poor at large angles. One, of course, does not expect to obtain a good fit to the data at high momentum transfers with a simple Woods-Saxon neutron wavefunction.

The deep minimum at $\theta_\pi=0^o$ and $T_p=142$ MeV in the plane-wave calculations shown in Fig. 5 is caused by the factor $|\vec{k}_\pi-\lambda(m_\pi/m_p)\vec{k}_p|^2$, which comes from the non-relativistic reduction of the pion production operator, but the other minima come from the momentum distribution of the captured neutron. Although they are filled in and shifted by distortion effects, the distortions are not so large as to completely mask the sensitivity of the reaction to nuclear structure effects. Information about the neutron wavefunction in principle could be extracted from the data if the present uncertainties and ambiguities in the pion-nucleus optical potential could be resolved. It may be fruitful to attack the problem the other way around. If the high momentum components of the nuclear wavefunction could be determined by other means, for example by (d,p) reaction studies at the same momentum transfer, then the (p,π^+) data might yield new information about the low-energy pion-nucleus interaction.

There is an additional complication in the DWBA calculations that comes from the ambiguity in the non-relativistic reduction of the pion production operator. We used the standard Galilean invariant form of this operator (λ=1) for calculations shown in Fig. 6 and λ=0.5 for those shown in Fig. 5. It seems premature to attempt to draw conclusions about this part of the problem from existing data in view of the present uncertainties in the nuclear wavefunctions and pion rescattering effects.

The data presented here on the energy-dependence of the $^{10}B(p,\pi^+)^{11}B_{g.s.}$ reaction near threshold, and also similar but less extensive data on the energy dependence of the $^{90}Zr(p,\pi^+)^{91}Zr_{g.s.}$ and $^{208}Pb(p,\pi^+)^{209}Pb_{g.s.}$ reactions, are compared with DWBA calculations of Tsangarides et al. in a contribution to this conference. These calculations, together with those presented here, covering a wide range of pion energies (T_π=1.7 to 61.4 MeV), momentum transfers ($q\approx$450 to 760 MeV/c), target masses (A=10 to 208) and target configurations, reproduce some of the main features of the data and suggest that the DWBA model, in which the pion rescattering effects are included in an average way through the use of an optical potential, may be able to provide a useful description of the (p,π^+) process. At the same time, it would be quite interesting to confront the various proposed two-nucleon models, in which the most important pion rescattering terms are calculated explicitly, with the new data.

It is a pleasure to acknowledge the many contributions of my colleagues in this work. The experiments were carried out with P.H. Pile, R.E. Pollock, P.T. Debevec, R.E. Marrs, M.C. Green, T. Sjoreen, and F. Soga, and the DWBA calculations with M.C. Tsangarides and J.G. Wills.

REFERENCES

*Work supported in part by the National Science Foundation.

1. R.D. Bent, P.T. Debevec, P.H. Pile, R.E. Pollock, R.E. Marrs, and M.C. Green, Phys. Rev. Lett. 40, 495 (1978).
2. P.H. Pile, Ph.D. Thesis, Indiana University (1978) (unpublished); P.H. Pile, R.D. Bent, R.E. Pollock, P.T. Debevec, R.E. Marrs, M.C. Green, T.P. Sjoreen, and F. Soga, Phys. Rev. Lett. 42, 1461 (1979)
3. B. Höistad, High Energy Physics and Nuclear Structure Conference Proceedings, Zürich, Switzerland, edited by M.P. Locher, 215 (1977).
4. S. Dahlgren, P. Grafström, B. Höistad, and A. Åsberg, Nucl. Phys. A227, 245 (1974), and references therein.
5. Y. Le Bornec, B. Tatischeff, L. Bimbot, I. Brissaud, J.P. Garron, H.D. Holmgren, F. Reide, and N. Willis, Phys. Lett. 49B, 434 (1974); Y. Le Bornec, B. Tatischeff, L. Bimbot, I. Brissaud, H.D. Holmgren, J. Källne, F. Reide, and N. Willis, Phys. Lett. 61B, 47 (1976).
6. M.P. Keating and J.G. Wills, Phys. Rev. C7, 1336 (1973).
7. M. Tsangarides and J.G. Wills, private communication (1977).
8. M. Tsangarides, J.G. Wills, and R.D. Bent, Bull. Am. Phys. Soc. 23, 952 (1978).

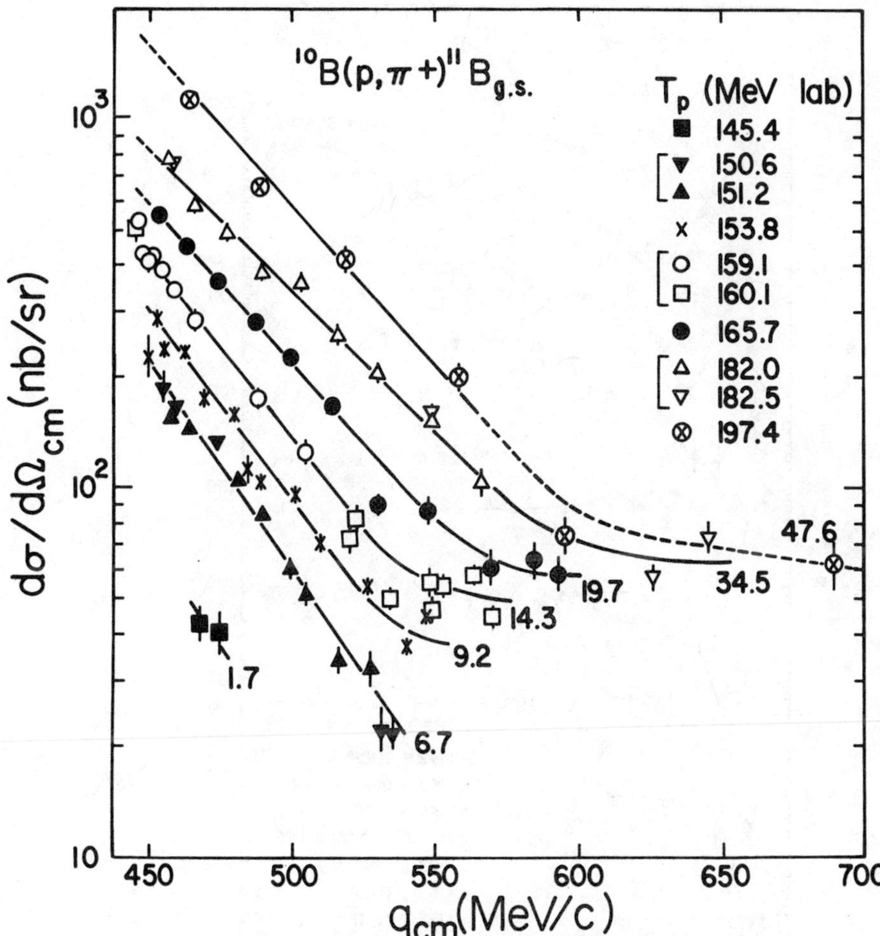

Fig. 1. The $^{10}B(p,\pi^+)^{11}B_{g.s.}$ differential cross sections at seven energies plotted as a function of momentum transfer. The curves are guides for the eye.

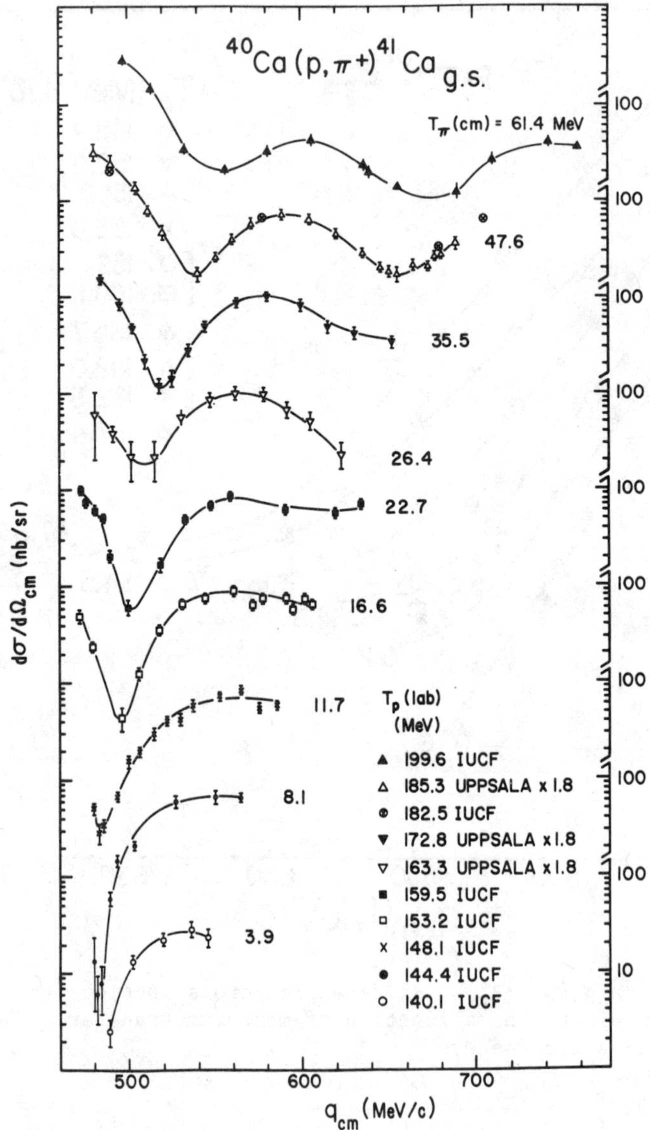

Fig. 2. The $^{40}Ca(p,\pi^+)^{41}Ca_{g.s.}$ differential cross sections at nine energies plotted as a function of momentum transfer. The Uppsala cross sections have been multiplied by a normalization factor of 1.8, which is based on the overlapping data taken at approximately the same energies (182.5 and 185.3 MeV). The curves are guides for the eye.

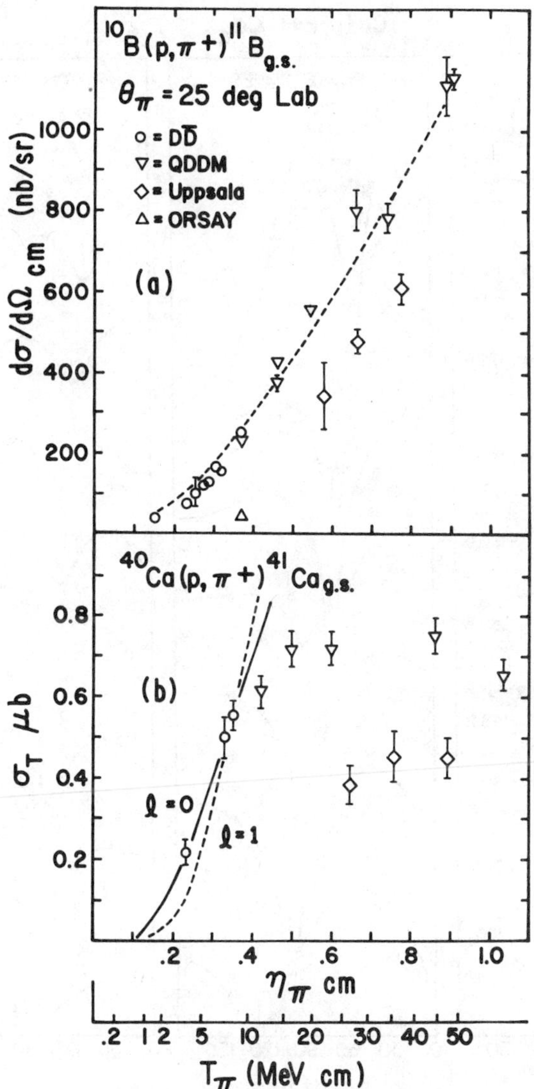

Fig. 3. (a) The dependence of the $^{10}B(p,\pi^+)^{11}B_{g.s.}$ differential cross section at $\theta_\pi(\text{lab})=25^0$ on pion center-of-mass momentum (in units of $m_\pi c$). Where not shown, the error bars are smaller than the data points. The dotted line is the combined s and p wave Coulomb penetrability function for the outgoing pions.

(b) The dependence of the $^{40}Ca(p,\pi^+)^{41}Ca_{g.s.}$ total cross section on pion center-of-mass momentum. The Coulomb penetrability functions plotted here show the ℓ=0 and ℓ=1 components separately (normalized to the data at η_π=0.35).

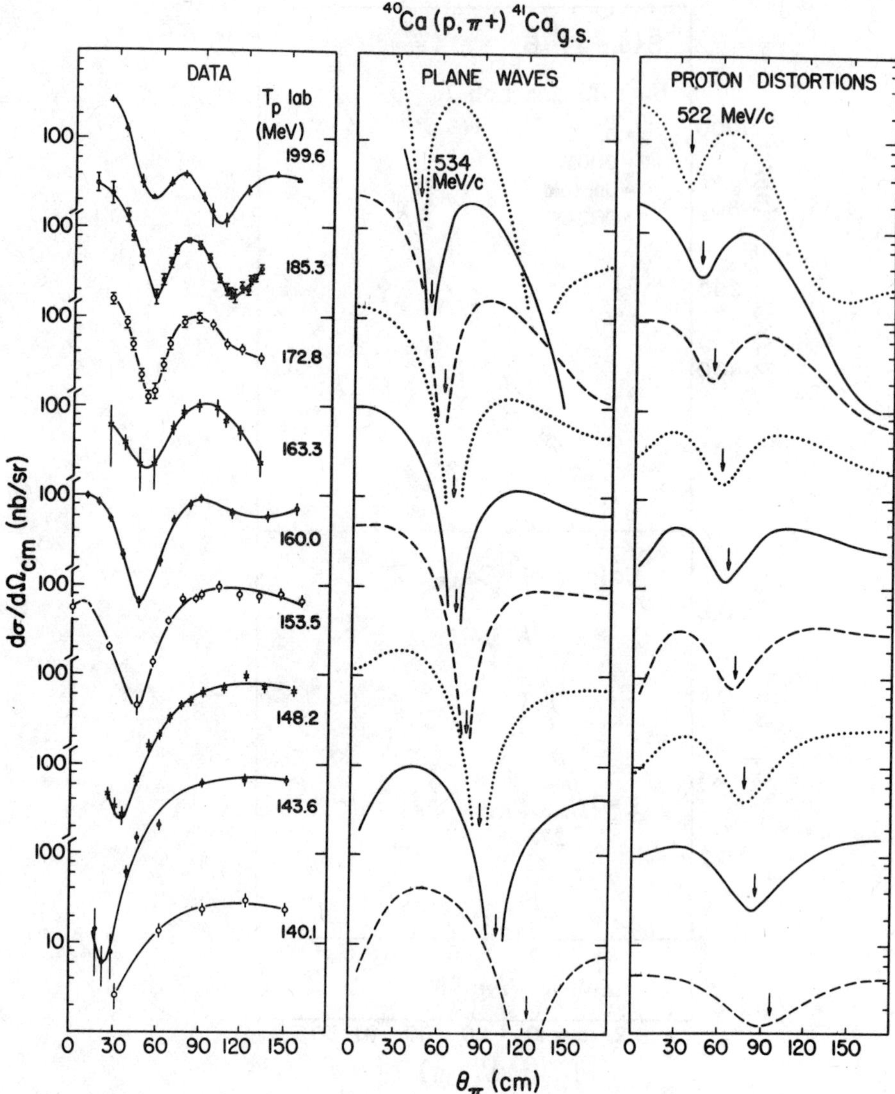

Fig. 4. The $^{40}Ca(p,\pi^+)^{41}Ca_{g.s.}$ differential cross sections at nine energies plotted against pion center-of-mass angle, and calculations based on the pionic stripping model in the plane-wave approximation and with proton distortions added.

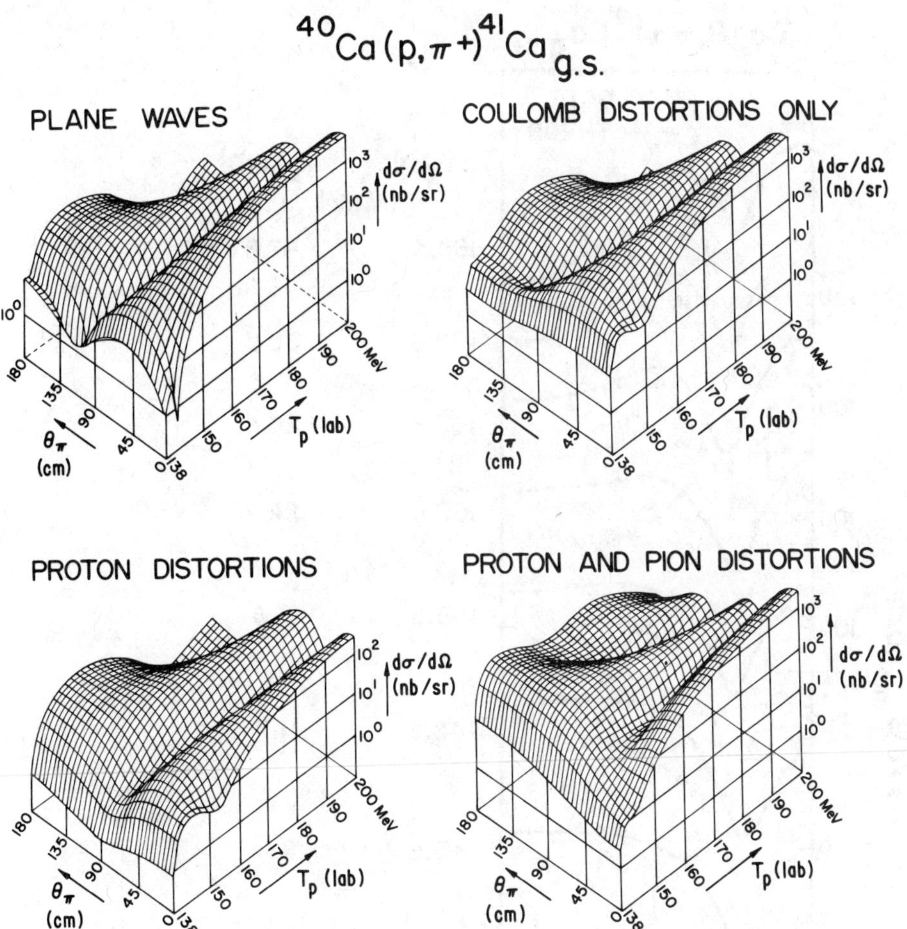

Fig. 5. Calculated differential cross sections for the $^{40}Ca(p,\pi^+)^{41}Ca_{g.s.}$ reaction as a function of incident proton energy and outgoing pion angle to show the sensitivity of the pionic stripping model to distortion effects.

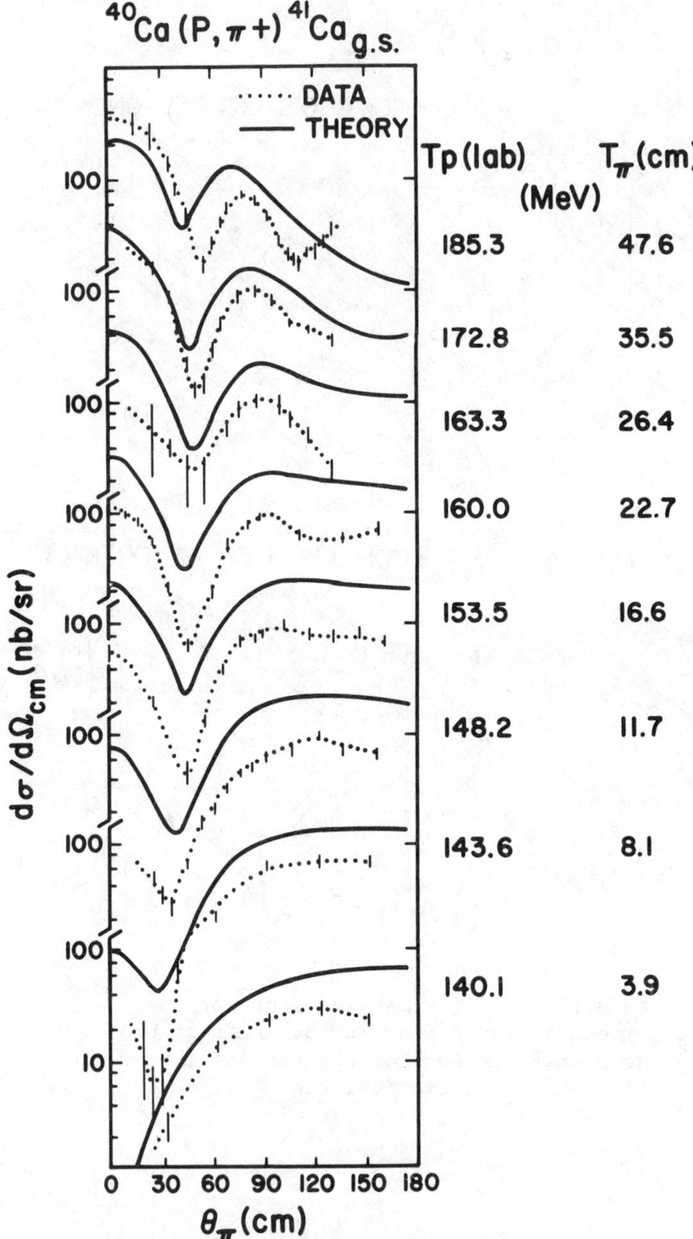

Fig. 6. DWBA fits to the $^{40}Ca(p,\pi^+)^{41}Ca_{g.s.}$ data using a Laplacian-like pion-nucleus optical potential with off-shell damping corrections and parameters obtained by fitting π^+ - ^{40}Ca elastic scattering data.

Question: J. Eisenberg, Tel-Aviv University

Without wanting to be unduly pessimistic, I think one must note that at the very lowest energies discussed here (say $T_\pi \lesssim 10$ MeV) the usual DWIA calculations of (p,π) do not respect orthogonality and so must be treated with some considerable caution.

Response: Bent

If the effects of using nonorthogonal wave functions are large, they presumably should be seen in the data presented here. This would add to the confusion and the interpretation of the data, however, perhaps the near-threshold behavior of the (p,π) reaction can be used to place limits on the size of these effects.

STUDY OF THE PION ABSORPTION VIA THE OBSERVATION OF THE SUBSEQUENT PARTICLE EMISSION

H. Ullrich
Kernforschungszentrum Karlsruhe, Institut für Kernphysik and University Karlsruhe, Institut für Experimentelle Kernphysik, Postfach 3640, 7500 Karlsruhe, Federal Republic of Germany

I. INTRODUCTION

Despite many theoretical and experimental efforts, the mechanism of pion absorption in nuclei is still not very clear[1]. In the microscopic description on the basis of the 2N-model a variety of effects appear: initial state NN-correlations, rescattering, final state interactions and distortions due to scattering of the emitted particles by the residual nucleus. The relative importance of these effects is very difficult to disentangle experimentally - and sometimes even conceptually.

The aim of our experiments is to gain better understanding of these effects and the absorption process in general, by careful investigations of as many observables as possible, in order to produce multiparameter distributions which allow extensive comparisons with respective theoretical calculations. Therefore, precision measurements of the particle emission following pion absorption have been performed, preferentially with two particles in coincidence. In order to minimize complications caused by secondary interactions with the residual nucleus, as well as to facilitate a proper description of the nuclear wave function of the initial and final state, light target nuclei have been chosen. All results reported here have been obtained with stopped negative pions.

A) At the CERN SC a (π^-2n)-experiment on ^6Li and ^7Li [2], ^9Be, ^{10}B and ^{12}C [3], ^{14}N [4], ^{16}O [5] and ^{40}Ca [6] as well as an experiment on single neutron emission[7] have been performed.

B) At SIN the π^- absorption on ^3He [8] and the Lithium isotopes is being studied in a coincidence experiment where also the $2P \rightarrow 1S$ pionic X-ray is detected.

II. INVESTIGATION OF THE 2N-PROCESS

Assuming a quasifree reaction mechanism, the absorption on a nucleon pair inside the nucleus can be studied in a kinematically complete way by the coincident measure-

ISSN:0094-243X/79/540154-13$1.50

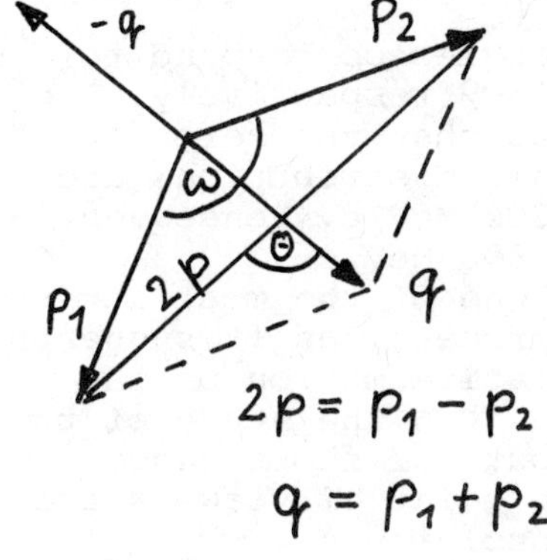

Fig.1. Kinematical variables for the 2N-process.

ment of the momenta of the two nucleons which leave the nucleus after absorption. The kinematical variables that have to be considered are shown in Fig. 1. The momenta of the emitted nucleons, their vector difference and sum are denoted by p_1, p_2, 2p and q, respectively. The excitation energy E_x of the residual nucleus is essentially given by:

$$E_x = Q - \frac{p_1^2}{2m} - \frac{p_2^2}{2m} - \frac{q^2}{2m(A-2)} \quad (1)$$

where Q is the Q-value of the reaction. In the case of the absorption of stopped pions, there are three independent variables. A complete independent set of variables to describe the data could be e.g. E_x, q and ω, the angle between the emitted nucleons.

It should be noted that q - which equals the recoil momentum - is also the momentum of the nucleon pair before the absorption, regardless of the momentum exchanged between the nucleons. With the relations given in Fig. 1. we find immediately from (1):

$$4\,p^2 + \left(1 + \frac{2}{A-2}\right) q^2 = 4m\,(Q - E_x) \quad (2)$$

As q, the momentum of the pair before the absorption is limited, the numerical value of p must be fairly large in order to satisfy (2). An experimental proof for this fact

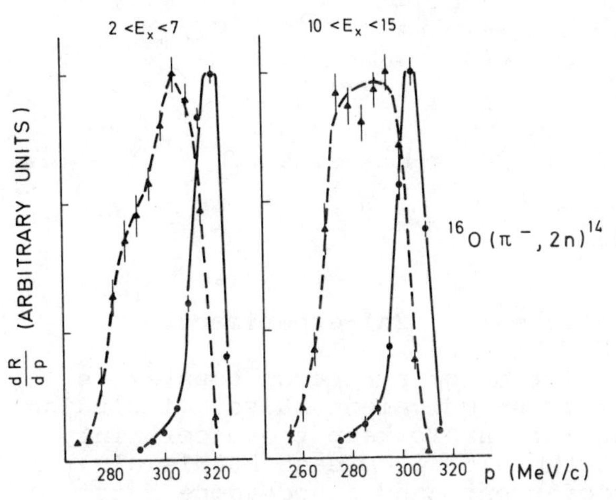

Fig.2. p-distribution for the reaction $^{16}O(\pi^-, 2n)^{14}N$

is given in Fig.2. The dots and triangles correspond to ω-ranges of 155° - 180° and 125° - 160° respectively. The lines are just to guide the eye. Note that the zeros on the p-axis are far at the left and all distributions are restricted to narrow ranges around 300 MeV/c. Consequently, the value of $p_1 - p_2 = 2p$ is around 600 MeV/c.

This high relative momentum is one of the most attractive features of the 2N-absorption process, as it suggests that the short range part of the relative motion of the two nucleons plays an important role. Thus, the study of the absorption can yield information about the short range NN-correlation[9]. Another interesting point of view is the information about 2-hole states in nuclei[10].

Fig.3 shows the experimental set-up of our (π^-, 2n)-experiment at the CERN SC. Negative pions from the LEP are degraded and stopped in targets of 5 g/cm^2. The neutrons are detected in two large area (2m x 48cm x 9cm) time-of-flight counters. In order to determine the direction of the neutrons, the counters are position sensitive. The position information along the counters is taken from the time difference between the multiplier signals at their ends. To localize the neutron impact point vertically, the counters are subdivided into bundles of scintillators, each of 6 cm height.

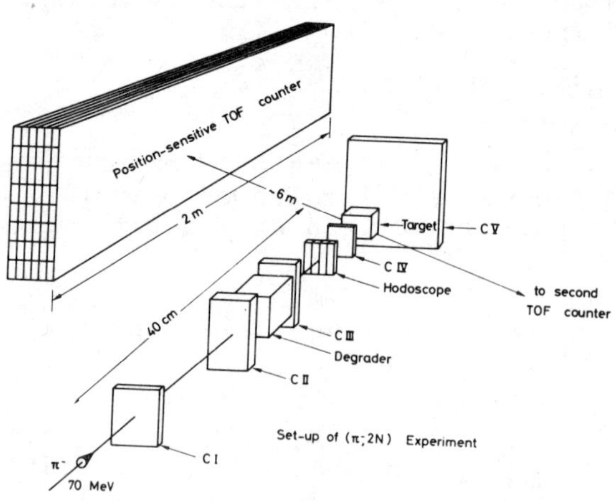

Fig.3. Set-up of the (π^-, 2n)-experiment.

A typical time resolution of a neutron counter is 800 psec for 20 MeV electron-equivalent pulses, including start-counter (CII) resolution. To keep the uncertainty in the neutron path length below 2 cm, each bundle of scintillators consists of 6 bars, and a hodoscope determines the stop position of the pion in the target. In total an excitation energy resolution of 3 - 6 MeV FWHM has been achieved, depending on flight path length, excitation energy and pulse-height threshold. This resolution is at least a factor of three better than in previous experiments of this type.

Fig. 4 shows an example of a two-dimensional presentation of measured data. Here the energies of the two neutrons from the pion absorption on ^{6}Li are shown. Transitions to the ground state of the residual nucleus ^{4}He cluster around a line which is clearly visible at the upper right part of the figure. As the residual nucleus has four times the nucleon mass, the line is only slightly bent. The length of the line indicates the large range of energy sharing among the nucleons. Transitions to excited states of the residual nucleus are also visible at smaller neutron energies.

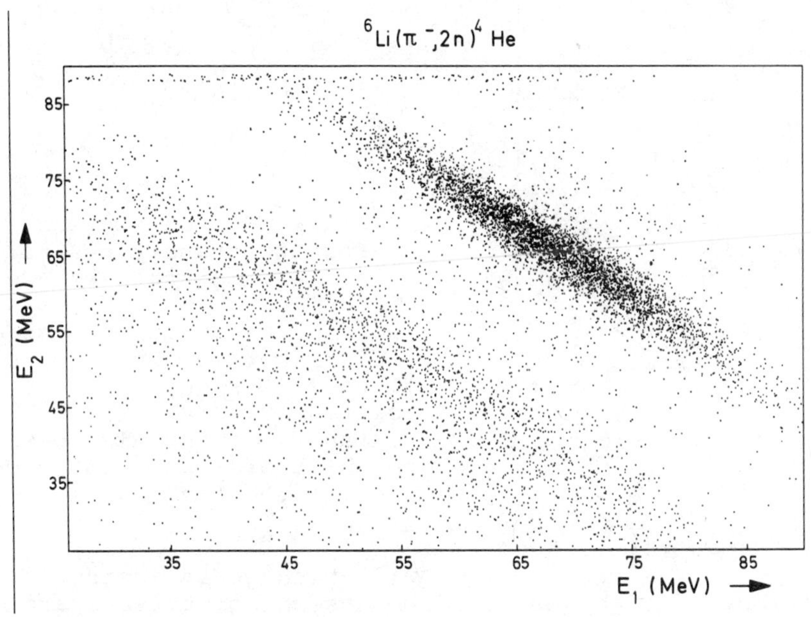

Fig. 4. Density distribution as a function of the two neutron energies.

Fig. 5 (lower part) shows the ω-distribution for the ground state transitions seen in Fig.4. The upper part includes also the transitions to the excited states. Both distributions exhibit strong peaking at 180°,proving the dominance of the quasifree reaction mechanism. The curves shown are theoretical predictions [11,12] based on cluster models. These predictions are primarily determined by the intercluster motion. The pion-specific effects are not taken fully into account.

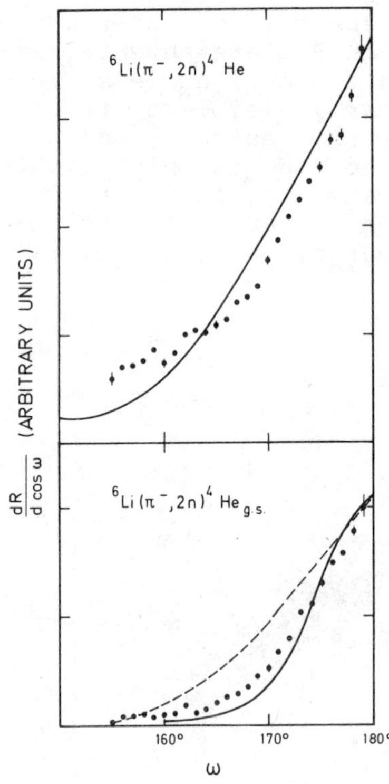

Fig.5. ω-Distributions

Fig.6. θ-Distribution for the transitions to the ground state of ^4He

Fig.6 shows an example of a θ-distribution. The importance of this variable was pointed out first by Koltun[13]. It has also been shown[2] that a pion interaction of the form

$$\mathcal{O} = \nabla\pi - a\nabla N \qquad (3)$$

will produce an isotropic θ-distribution if either the two absorbing nucleons are in a relative s-state, or if only a single value appears for their center of mass orbital angular momentum Λ. At least one of these conditions seems to be fulfilled, as the data points exhibit isotropy.

For a pair of p-shell nucleons with relative angular momentum $l = 0$, the possible Λ-values are 0 and 2.

In the $(\pi^-, 2n)$-reaction the π^- is absorbed by an np-pair which can have T = 0 or T = 1.

Rigorous treatment of the pion absorption is rather complicated because of the various effects that have to be considered. Detailed theoretical investigations taking into account the quantum number of the absorbing pair and calculating momentum and angular distributions for specific residual states are necessary and are being attempted.[14,15,16,17]. A comparison of our results with such a calculation could be made in the case of ^{16}O.

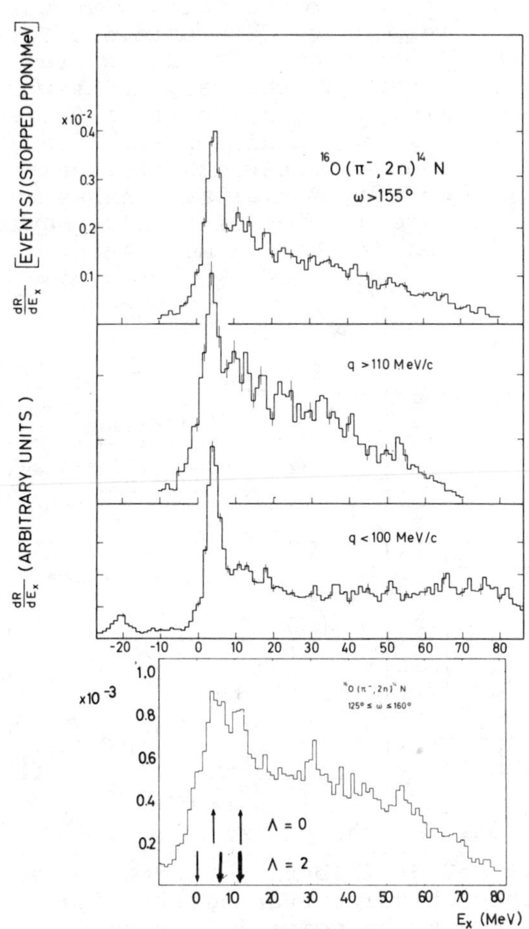

Fig.7. Excitation energy spectra

Fig.7. shows E_x-spectra of ^{14}N. The upper part contains data for $155^o < \omega < 180^o$ a.) without additional restriction in q, b.) for q>110 MeV/c and c.) q<100 MeV/c (top to bottom). The lower part of the figure contains data for $125^o < \omega < 160^o$. T=0 two-hole states are indicated in the lower part of the figure. In the 1p-shell the following states are known[18].

1) $p_{1/2} \cdot p_{1/2}$: 1^+, GS, $\Lambda = 2$

2) $p_{1/2} \cdot p_{3/2}$: 1^+ 3.9 MeV, $\Lambda = 0$

3) $p_{1/2} \cdot p_{3/2}$: 2^+ 7.0 MeV, $\Lambda = 2$

4) $p_{3/2} \cdot p_{3/2}$: 3^+ 11.05 MeV, $\Lambda = 2$

5) $p_{3/2} \cdot p_{3/2}$: 1^+ 11.5 MeV, $\Lambda = 0,2$

All $\Lambda = 0$ states should be peaked at 180°, whereas $\Lambda = 2$ states should be visible at smaller ω only. This is in good agreement with Fig. 7. Structures at higher excitation energies indicate two-hole states involving 1S-nucleons.

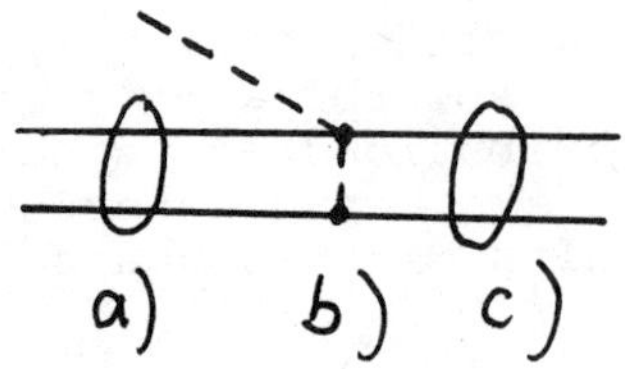

Fig.8. Effects considered in ref 17.
a) Initial state correlation
b) Rescattering
c) Final state correlation

A detailed calculation of the 2P-pion absorption in ^{16}O, taking into account the effects indicated in Fig.8 plus final nucleon-nucleus interaction has been performed recently by Shimizu and Faessler [17]. In Fig.9 their calculated q-distributions for different final states are compared with our respective data obtained for $155^{o}<\omega<180^{o}$. The rates are given in arbitrary units which are, however, the same for all 4 measured histograms and the respective theoretical curves. Normalization has been made at one point only, namely at 50 MeV/c in the distribution of the 3.9 MeV state.

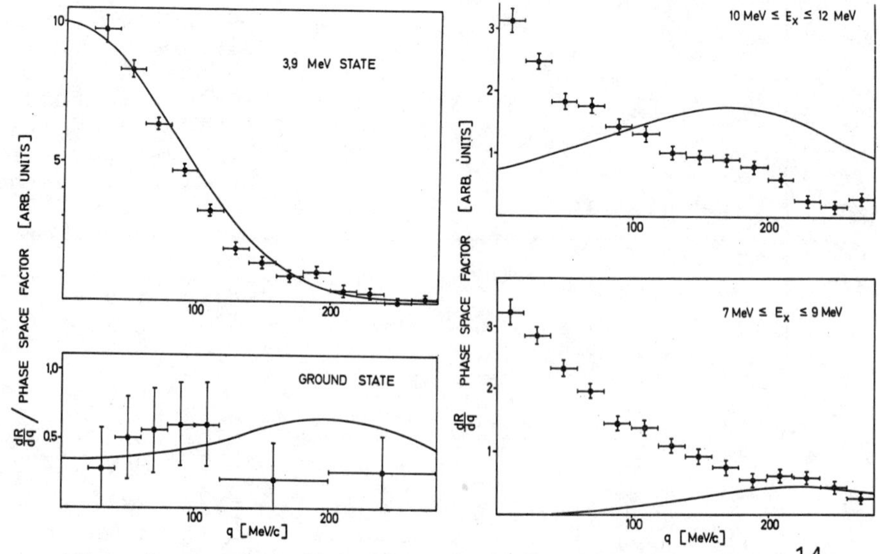

Fig. 9. q-Distributions for the states 1)-5) in ^{14}N

The shape of the 3.9 MeV distribution (state 2) is well reproduced by the theoretical curve and also the ground state (state 1) seems to be compatible with the calculation within the experimental error. State 3 is in the 7-9 MeV region, but the experimental distribution shows a rather large $\Lambda = 0$ contribution, possibly from the 3.9 MeV tail. The 10-12 MeV region covers states 4 and 5, but - for unknown reasons - shows again predominant $\Lambda = 0$ character. The inclusion of different types of rescattering processes increases the absolute height of the theoretical curves up to a factor of 5, but because of the normalization to the 3.9 MeV distribution, the situation does not change drastically.

Because of space limitations, results from other nuclei cannot be presented here. In summary the following can be said however: For a detailed comparison with theory, the energy and quantum numbers of two-hole states are of importance. The population of these states compared with the known cfp-factors yields essential information about the pion absorption process. The q- and ω-distributions are mainly determined by Λ and the separation energy (see figs. 5 + 10) but also contain reaction specific information.

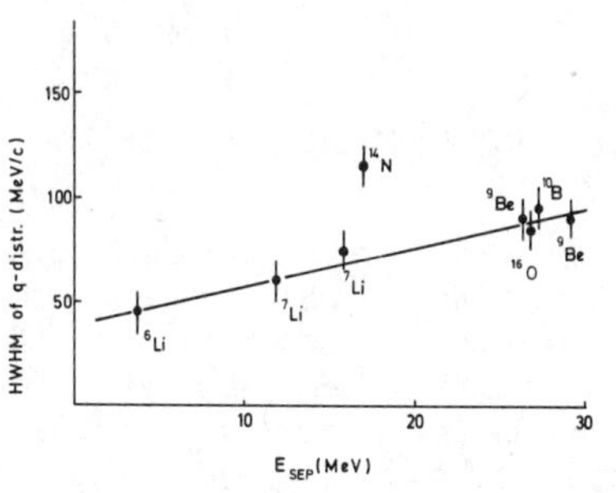

Fig.10. Width of q-distributions for Λ = 0 transitions as function of the pair separation energy.

One important result from the study of the 1p-shell nuclei should be quoted here: π^--Absorption on np-pairs with T = 1 is strongly suppressed. This fact can be understood by the reasonable assumptions: a) the pion cannot change Λ and b) it is mainly captured from the 2P atomic orbit. The selection rules (spin, isospin and parity) then forbid absorption on a T = 1, l = 0 np-pair.

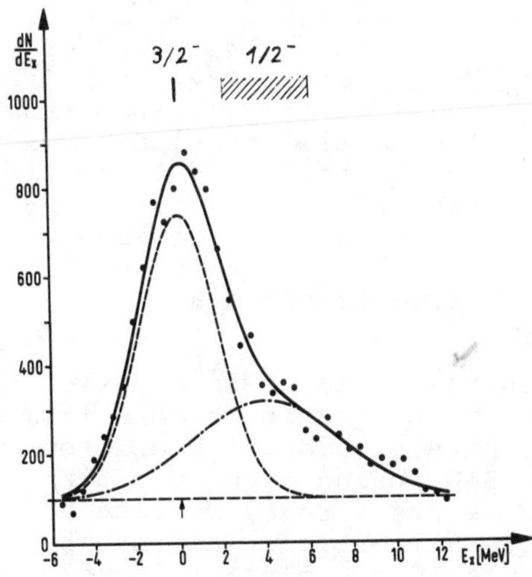

Fig.11. Transitions to ground ($3/2^-$) and first excited state ($1/2^-$) of ^{5}He.

In this context the lithium isotopes are exceptional as there already 40% of the pions are captured from the 1S-orbit[13] and the T = 0 and T = 1 absorption rates R_0 and R_1 can be compared. In the cluster model ^{7}Li can be considered as an α-cluster and a triton. Removing the T = 0 (S=1) np-pair from the triton leads to the broad first excited $1/2^-$ state in ^{5}He. The population of this state is proportional to R_0.

Removing on the other hand the T = 1(S = 0) np-pair

leads to the 3/2 ground state, the population of which is proportional to $R_0 + 0.4 \times R_1$. From the observed ratio of the population of these two states the ratio of the absorption rates could be determined. The result is:

$$R_1/R_0 = 0.8 \pm 0.2$$

III. SEARCH FOR OTHER ABSORPTION MECHANISMS.

Though it is generally believed that in pion absorption the 2N-process plays the most important role, there is no a priori justification against the existence of other hadronic processes. We have searched for evidences of such mechanisms A) in the single neutron emission, B)in the absorption on the three nucleon system and C) in the emission of composite particles.

A) A search for the (π^-,n)-reaction has been made via a careful investigation of the high energy part of the neutron spectra following pion absorption on 6 1p-shell nuclei from ^6Li to ^{14}N. Strong evidence for single neutron emission is obtained in the cases of ^6Li (see Fig.12), ^7Li, ^{10}B and ^{12}C. The spectrum from ^9Be is also consistent with this reaction. A lower limit of 2×10^{-3} per stopped pion is established for the effect on the Lithium isotopes and ^{12}C.

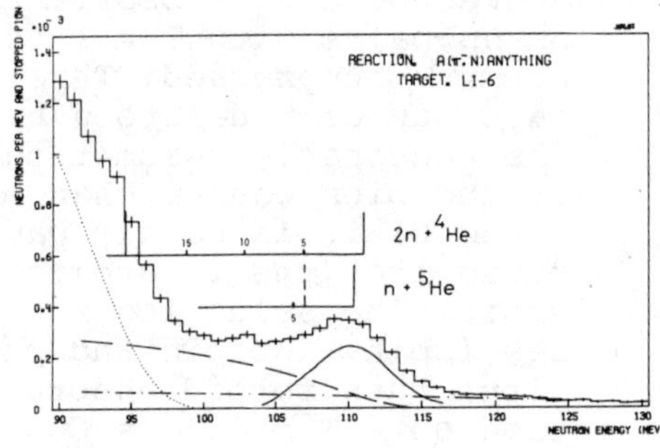

Fig.12. High energy part of the neutron spectrum from ^6Li.

B) The pion absorption in the three nucleon system is of principal importance in order to find deviations from the 2N-absorption process. We have studied pion absorption on ^3He at the π E1-channel at SIN, using a cooled (5^oK) gas target of 80 mg/cm^2. Besides the already mentioned neutron time-of-flight counters, a plastic scintillator hodoscope with large solid angle ($\Omega = 0.5$ sr) in connection with a system of multiwire proportional chambers is used for the charged particles. Energy- and angular resolutions are 5% and $\pm 0.5^o$, respectively. Mesonic X-rays from the $(2P \rightarrow 1S)$-transition are also detected in coincidence by an array of 6 thin NaI-detectors. In about 37% of the observed events the pion has been captured from the atomic S-state.

Fig.13 shows preliminary results from a np-coincidence measurement. As energy and emission angle of both nucleons are measured, the data contain still one constraint which will be used later to suppress background from the target surroundings. Like in a Dalitz plot the phase space density is constant in such a presentation and the density of events reflects directly the matrix element of the reaction. At the region with $E_n = E_p$ we

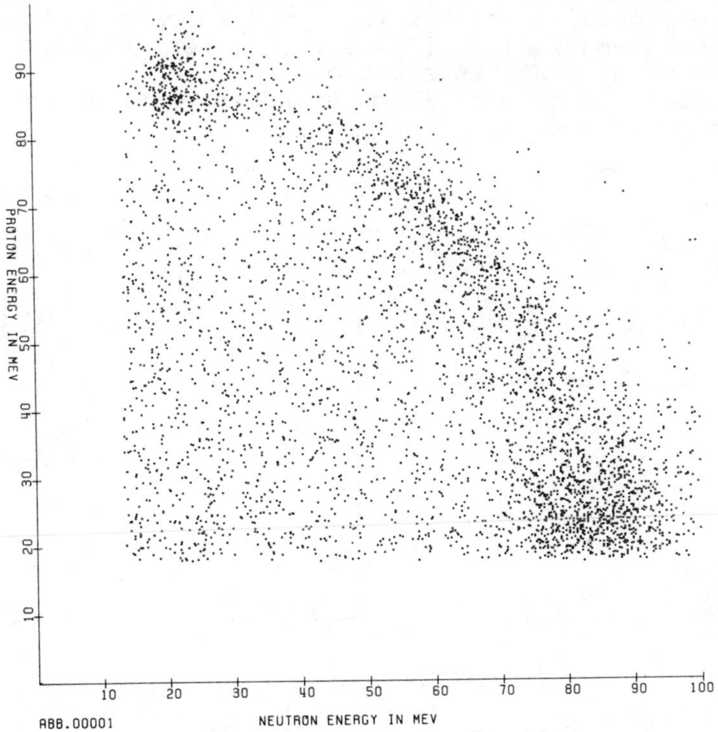

Fig. 13. np-Coincidences from ^{3}He

find a density maximum, which is attributed to the absorption by a pp-pair, the undetected neutron being a spectator with low energy. This maximum is expected e.g. from a comparison with Fig. 4. It is however surprising to find two more even stronger maxima in the regions where one of the two detected nucleons has maximum energy. This means that the two other particles have the same - parallel - momenta, i.e. these are just the regions of final state interaction. This classification is consistent with the facts that a) the density peak is stronger for np- (T=0.1) than for nn- (T=1) final state interaction and b) a large number of nd-coincidences (30% not shown in Fig. 13) is observed.

C) The coincident detection of composite particles offers access to some exotic reaction channels. Using the same apparatus and pion beam as described in B), we have measured (n,X)- and (X,X)-reactions (X=P,d,t) on the Lithium isotopes. As an example, Fig.14 shows results from (t,t)-coincidences on natural Lithium. The circular spot with high density at the diagonal line stems from the 7.5% admixture of ^{6}Li in the natural isotope admixture (2-body decay), whereas the two stretched maxima inside the kinematically allowed region for ^{7}Li indicate strong n-t final state interaction.

It seems as if pion absorption has provided us with some surprises.

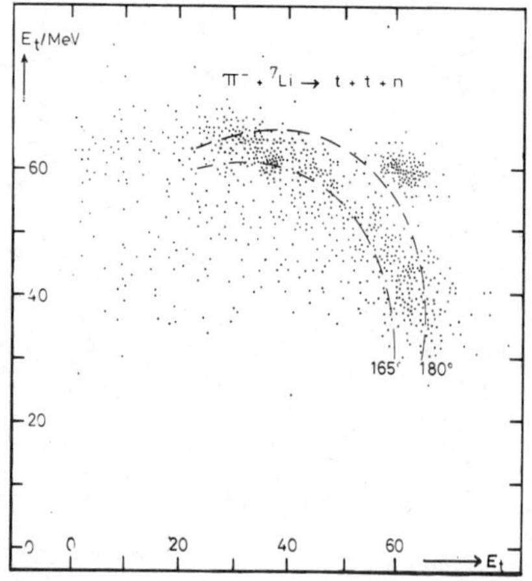

Fig.14. tt-Coincidences from natural Lithium

REFERENCES

1. J.Hüfner, Phys.Reports 21, 1 (1975)
2. B.Bassalleck, E.L.Haase, W.-D.Klotz, F.Takeutchi, H.Ullrich, M.Furic and Y.Sakamoto, submitted to Phys. Rev.C and CERN preprint Oct.1978
3. B.Bassalleck, W.-D.Klotz, F.Takeutchi,H.Ullrich,M.Furić, Phys.Rev.C16, 1526 (1977) and CERN preprint, March 1977
4. B.Bassalleck et al. Phys.Letters 65B, 128 (1976)
5. B.Bassalleck et al. in preparation
6. B.Bassalleck et al. Z.Physik A 286, 401 (1978)
7. B.Bassalleck et al. accepted by Nucl.Phys.A, and CERN preprint 1978
8. B.Backenstoss, P.Blüm, M.Dörr, W.Fetscher, D.Gotta, W.Klotz, H.Koch, W.Kowald, U.Raich, G.Schmidt, I. Schwanner, L.Simons, L.Tauscher, H.Ullrich and H.J. Weyer, 8.Intern.Conf. Few Body Systems and Nuclear

Forces Graz, Austria Aug. 1978
9. M.G.Huber, Annales de Physique 5, 239 (1970)
10. T.Ericson, Phys.Letters 2, 278 (1962)
11. B.K.Jain and B.Banerjee, Nuovo Cimento 69A, 419 (1970)
G.Alberi and L.Taffara, Nuovo Cimento 58A, 441 (1968)
12. T.I.Kopaleishvili and I.Z.Machaberi, Sov.J.Nuclear Phys.4, 27 (1967)
13. D.S.Koltun, Phys.Rev.162, 963 (1967)
14. H.Garcilazo and J.M.Eisenberg, Nucl.Phys.A220,13 (1974)
15. R.S.Blalerao and Y.R.Waghmare, Nucl.Phys.A298, 367 (1978)
16. W.Schneider, Ph.D. Thesis, University of Erlangen-Nürnberg, 1977
17. K.Shimizu and A.Faessler, Phys.Rev.C17, 1891 (1978)
K.Shimizu and A.Faessler, KFA Jülich preprint 1978 and to be published in Nucl.Phys.A, private communication
18. A.Ajzenberg-Selove, Nucl.Phys.A152, 42 (1970)
19. W.W.Sapp, M.Eckhause, G.H.Miller and R.Welsh, Phys. Rev. C5, 690 (1972)

Question: C. Cernigoi, Universita di Trieste

Regarding the (π^-,2n) absorption process at rest, I have preliminary data on ^{4}He which shows that this reaction can account for only 10% of the total channels of the reaction. This shows also that correlated events around 180° are not so important as was emphasized in the past.

Question: Schlepuetz, University of Zurich

The t-n final state reaction you showed in the last slide could be looked at as one ^{4}H resonance. Do you know the ground state energy and the width of that resonance?

Response: Ullrich

Not yet, but we hope to find this out analyzing the data.

Question: S. Hoath, Rutherford Laboratory

Why was ^{14}N "Special" on the plot of the width of the Q-distribution versus separation energy?

Response: Ullrich

Probably this state is not a pure 2S-state as the others but contains 1S (i.e. a different principal quantum number).

Question: R. Bhalerad, Brooklyn College

If there are no events at w = 180° theories seem to be in disagreement with the data. Any comment?

Response: Ullrich

There is no such principal disagreement. At strictly 180° the counting rate is zero just because of the vanishing solid angle dΩ. The excitation energy spectra of Fig. 7 (upper part) were therefore obtained taking a larger angular range around 180°. For the q and w distributions dΩ/dw has of course been taken into account together with phase space and geometrical corrections.

Question: E. Auld, UBC

In your scatter plot of T_p vs T_n in the π^-+He→nnp there appears to be two peaks in the T_p spectrum near the nn final state interaction region. Is this significant?

Response: Ullrich

I do not think that is statistically significant.

DOUBLY COHERENT PRODUCTION OF π^- BY ^{3}He IONS OF 910 MeV

E. Aslanides,[††] T. Bressani,[†] E. Chiavassa,[‡]
G. Dellacasa,[‡] P. Fassnacht,[††] M. Gallio,[‡]
F. Hibou,[††] A. Musso,[‡] and G. Puddu[†]

We report here the first results on the search of the "doubly coherent" reaction:

$$^3_2He + ^6_3Li \rightarrow ^9_6C + \pi^-$$

induced by ^{3}He ions of 910 MeV. The experiment was carried out at the CERN Synchro-Cyclotron where we have been able to observe the reaction in spite of its very low cross-section (∿1 pb/sr MeV/c) taking advantage of the high intensity (∿2x10^{12} p.p.s.) of the extracted ^{3}He beam.

Inclusive pion spectra from the reaction:

$$^3He + ^6Li \rightarrow \pi^- + X$$

were measured at 0° with moderate momentum resolution using the standard magnetic elements of a secondary beam transport line as an achromatic ($\Delta p/p \simeq 3.2\%$) spectrometer for pion momenta up to 800 MeV/c. The overall absolute normalization uncertainty is ±18% and includes uncertainties in the target thickness, the beam intensity, the spectrometer solid angle and momentum acceptance.

Experiments of pion production were already performed in other laboratories with different incident ions at various energies (see ref.1), but they extended down to cross section sensitivities of about 100 nb/sr.MeV/c. In the present experiment we were able to increase the above sensitivity level by more than five order of magnitude. Fig.1. shows the measured inclusive π^- spectrum, from 200 to 750 MeV/c; the insert shows with more detail the region from 680 to 750 MeV/c.

Typical processes responsible for the π^- production range from the "incoherent" pion production in nucleon-nucleon collisions to the "doubly coherent" production. More precisely, we could try to establish the following naive classification:

(a) "incoherent" production in neutron (projectile)-neutron (target) collisions. The maximum pion momentum of the corresponding free reaction $n+n \rightarrow d+\pi^-$ at 303 MeV would be 115 MeV/c, with a laboratory cross section at 0° of ∿700 μb/sr.

CERN Collaboration:
† Università di Cagliari, Cagliari, Italy and INFN, Sez.di Torino.
†† Centre de Recherches Nucléaires et Université Louis Pasteur, Strasbourg, France.
‡ Università di Torino, Torino, Italy and INFN, Sez.di Torino.

ISSN:0094-243X/79/540167-03$1.50

(b) coherent production on ^{6}Li by a single nucleon of the ^{3}He projectile. The corresponding maximum pion momentum would be 250 MeV/c and the differential cross section at 0° for the free reaction $n+{}^6_3Li \rightarrow {}^7_4Be + \pi^-$ of the order of 1 μb/sr.

(c) coherent production of π^- by ^{3}He on single nucleons of the ^{6}Li target. The maximum pion momentum would be 325 MeV/c and the differential cross section at 0° for the free reaction ${}^3_2He+n \rightarrow {}^4_3Li+\pi^-$ of the order of 80 μb/sr.

(d) "doubly coherent" production of pions through the reaction: ${}^3He+{}^6Li\rightarrow{}^9C+\pi^-$. The maximum pion momentum, corresponding to the final nucleus ${}^9_6C$ left in the ground state, would be 725 MeV/c.

Pion production in processes at intermediate coherence levels, i.e. in interactions between substructures in the projectile and in the target, are also possible. It must be noticed that, whereas for the first three processes the Fermi momentum of the nucleons is expected to smear out completely the structures in the pion spectra arising from the elementary reactions, for the doubly coherent process we expect pion peaks at well defined energies if the ^{9}C nucleus is left in well defined states.

The measured pion spectrum has a smooth behavior up to 650 MeV/c and its fall-off looks similar to that obtained in previous experiments.[1] Beyond 650 MeV/c we observe:

i) a break around 675 MeV/c, which is the pion momentum expected for the reaction: ${}^3He+\alpha\rightarrow {}^7B+\pi^-$ at 910 MeV.

ii) A deviation of the experimental points, beyond 700 MeV/c, from a smooth, exponentially falling curve. More precisely, the cross sections are nearly constant from 700 to 725 MeV/c, whereas in a similar momentum range, at a different momentum, we observe a decrease of a factor 5. This is just the region where we expect the contribution from the doubly coherent reaction. Due to our poor momentum resolution we cannot establish which final nuclear states are produced. The integrated differential cross section in this region, assuming we are looking at genuine "doubly coherent" events, is ∿1 pb/sr. This order of magnitude is in good agreement with previous results of the (^{3}He,π°) reaction at 180 and 200 MeV).[2]

REFERENCES

1. H. Steiner, in Proc. VIIth Int.Conf. on High-Energy Physics and Nuclear Structure, Zurich, 29 Aug.-2 Sept.1977 (Birkhauser, Basel, 1977), p. 261.
2. N. S. Wall, J. N. Craig, and D. Ezrow, Nucl.Phys. *A268* (1976) 459.

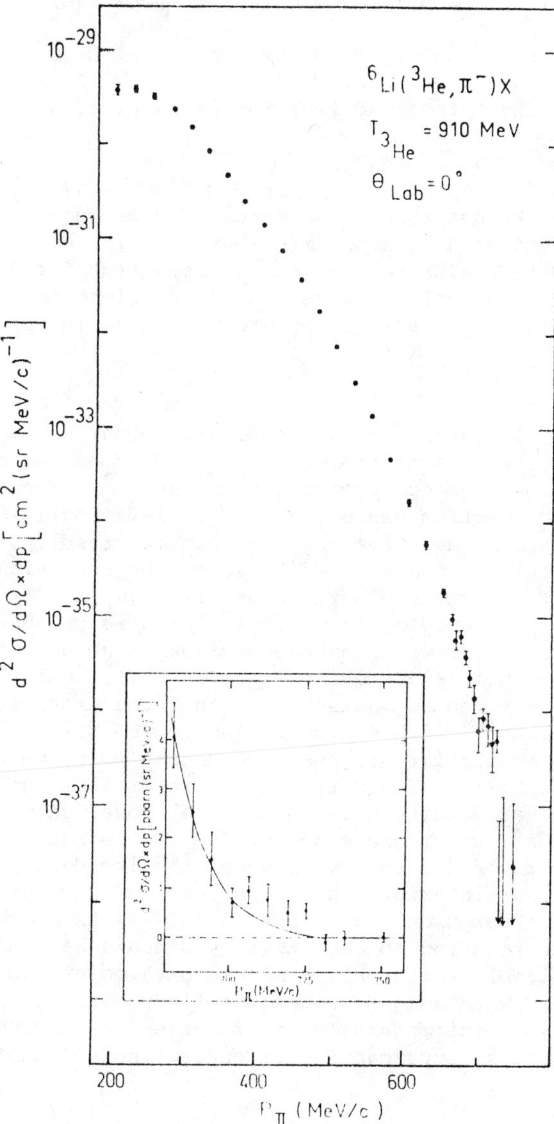

Fig.1. Inclusive π^- Spectrum

ABSORPTION IN FLIGHT OF POSITIVE PIONS IN CARBON*

D. I. Sober, W. J. Briscoe‡ , P. E. Burt, Hall Crannell,
J. T. O'Brien and W. J. Stapor
The Catholic University of America, Washington, DC 20064

ABSTRACT

We describe an experiment to measure the absorption in flight of π^+ mesons in carbon for pion energies between 80 and 140 MeV. The measurement is important for discriminating between competing theoretical predictions for the energy dependence of the absorption cross section in this region.

INTRODUCTION

The absorption of energetic pions by nuclei is an important process both in its own right and as an ingredient in the calculation of meson production processes[1]. Nonetheless the absorption cross section has never been satisfactorily measured[2], and calculations have led to contradictory results. The π-nucleus reaction cross section σ_{reac} can be subdivided into the cross sections for inelastic scattering (σ_{inel}) and for absorption (σ_{abs}) according to whether a pion is present in the final state. σ_{reac} is known to have a maximum at about the energy of the (3,3) resonance (~150 MeV in light nuclei). Calculations of σ_{abs}, however, do not agree on whether the cross section is rising[1,2] or falling[3,4] with energy in the region below the resonance. Since the absorption process cannot take place on a single nucleon, the probability of absorption is proportional to the square of the nuclear density. In the optical model calculation of Werntz and Lucas[4], the increasing opacity of the nucleus in the vicinity of the resonance energy prevents the incident pion from reaching the nuclear interior, where the density is greatest, and consequently the absorption cross section falls with energy above 100 MeV. The data required to test this hypothesis have not hitherto been available, with only a single data point (on carbon, at 130 MeV)[5] cited with some degree of confidence in the literature. Thus, a systematic measurement of the energy dependence of σ_{abs} will fill a serious gap in our understanding of pion-nucleus interactions.

METHOD

We have measured the absorption cross section on carbon by the simultaneous measurement of pulse heights in an array of plastic scintillation counters (NE102), which served as both

*Supported by National Science Foundation Grant #PHY76-81565, and by a C.U.A. Biomedical Sciences Support Grant.

‡Present address: University of California, Los Angeles.

ISSN;0094-243X/79/540170-02$1.50

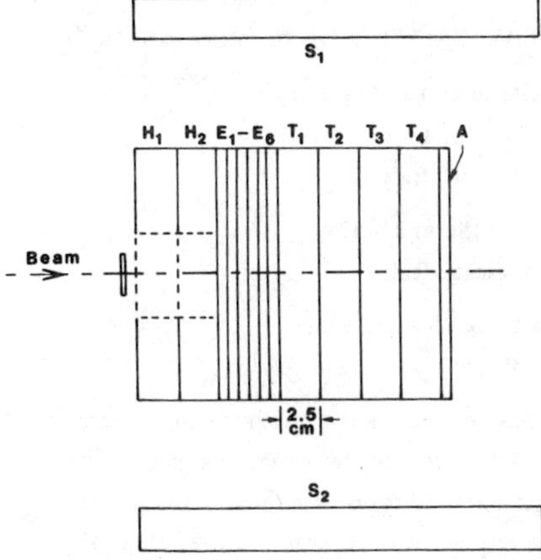

Figure 1. Scintillation counter array (light pipes omitted for clarity).

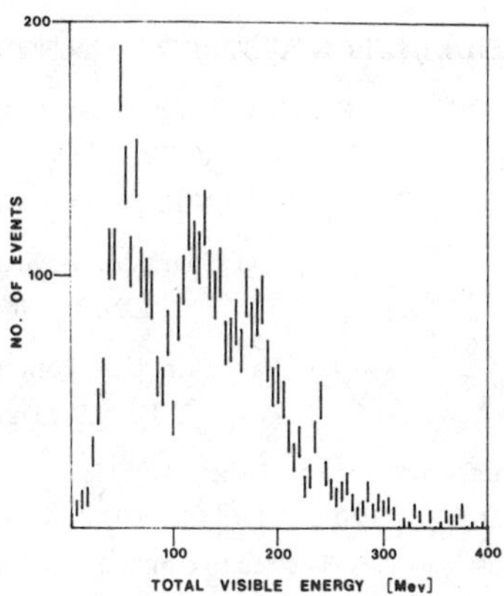

Figure 2. Total visible energy spectrum at 140 MeV, after side-counter correction.

target and detector (Fig. 1). Well-collimated pion beams of energy 80, 100, 120 and 140 MeV were produced at the SREL synchrocyclotron. The event trigger was a beam coincidence in anticoincidence with the downstream veto counter A which helped to reject non-interacting particles and small-angle scatters. The "side counters," S_1 and S_2, which covered 30% of the transverse solid angle, served to identify elastic and near-elastic scattering events which could then be subtracted, with proper weighting, from the observed spectra. A total-energy spectrum, after such subtraction, is shown in Fig. 2. The low-energy peak is believed to consist primarily of forward scattering events which escaped the veto counter, with a contribution from charge exchange and other inelastic scattering events, while the peak with visible energy above 100 MeV contains the absorption events. Final analysis of the data is now in progress.

REFERENCES

1. R. Silbar and M. Sternheim, Phys. Rev. C 8, 492 (1973); D. Sparrow et al., Phys. Rev. C 10, 2215 (1974).
2. R. R. Silbar, Phys. Rev. C 11, 1610 (1975).
3. D. Beder and P. Bendix, Nucl. Phys. B26, 597 (1971). D. Beder, Can. J. Phys. 49, 1211 (1971).
4. C. Werntz and C. Lucas, in "Nuclear Cross Sections and Technology," ed. R. Schrack and C. Bowman, NBS Special Publication 425 (1975), pp. 472-475; and Report to Los Alamos Scientific Laboratory for Contract NPG-12321-1, Catholic University preprint, unpublished (1976).
5. E. Bellotti et al., Nuovo Cimento 18A, 75 (1973).

QUASIDEUTERON ABSORPTION IN NON RADIATIVE CAPTURE OF π^+ MESONS ON ^{4}He

F. Balestra, M.P. Bussa, L. Busso, R. Garfagnini and G. Piragino
Istituto di Fisica dell'Universitá di Torino, Torino, Italy
I.N.F.N., Sezione di Torino, Torino, Italy

C. Guaraldo, A. Maggiora and R. Scrimaglio
L.N.F. dell'I.N.F.N., Frascati, Italy

I.V. Falomkin, G.B. Pontecorvo and Yu. A. Shcherbakov
J.I.N.R., Dubna, U.S.S.R.

The total cross section of the $\pi^{+4}\text{He} \rightarrow 3pn$ reaction has been measured[1] with the pion beam of the LNF at $E_\pi = (120 \pm 15), (145 \pm 10), (165 \pm 10)$ MeV. The events of this reaction appear as stars of three fast particles. The relative high energy of all the particles indicates that the number of reactions with change exchange $^4\text{He}(\pi^+, \pi^0 3pn)$ is negligible respect to the number of the $^4\text{He}(\pi^+, 3pn)$ events. We selected the two faster particles of each event and they resulted mostly emitted with large opening angle, in agreement with the hypothesis of two fast protons emission from π^+ absorption by a quasideuteron: $\pi^+ + ^4\text{He} \rightarrow 2p_f + d_s$ or $(np)_s$. Fig. 1 shows the distribution of the events as a function of the opening angle between the two faster particles, for $E_\pi = 145$ MeV. A strong correlation between the two faster protons is clearly shown at large angle. In the figure the distribution of fast proton pairs produced in π^+ absorption in nuclear emulsion versus their opening angle is also reported[2], for $E_\pi = 45$ MeV. Notwithstanding the different pion energy the agreement is very good and confirms the role of this absorption mechanism. The third charged particle should be as a spectator. Fig. 2 shows the angular distribution of the "spectator", compared with the phase space previsions[3]. As one can see, it is impossible to deduce if the spectator is a deuteron or a proton.

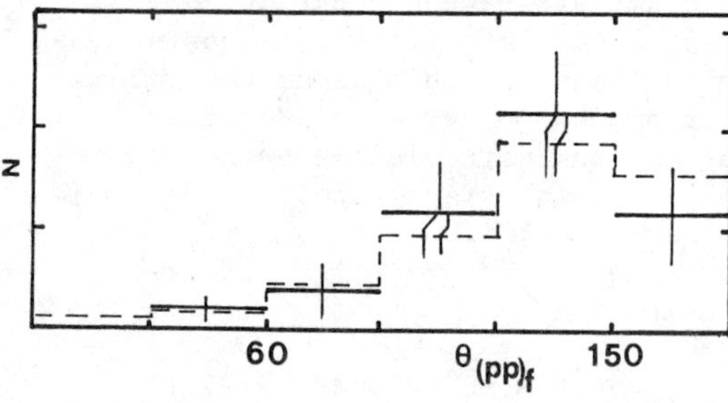

Fig. 1. Distribution of ^{4}He (π^+, 3pn) events versus the opening angle between the two faster protons compared with the distribution of fast proton pairs produced in absorption of π^+ in nuclear emulsion[2] (dashed line).

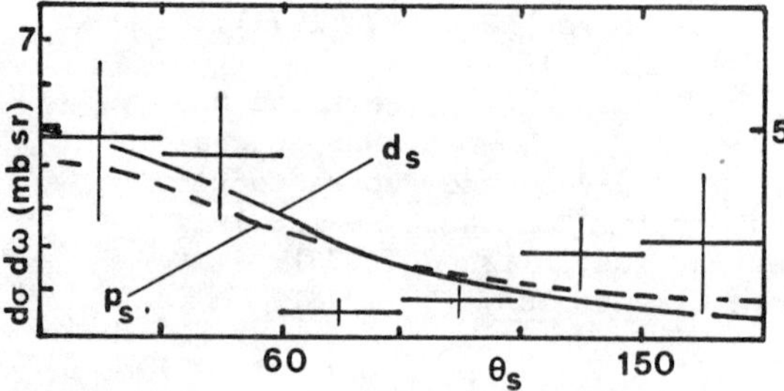

Fig. 2. Angular distribution of the spectator particle (crosses), compared with the phase space distribution with the hipothesis that the spectator is a deuteron (full line) or a proton (dashed line) from a n-p pairs. $E_\pi = 145$ MeV.

ISSN:0094-243X/79/540172-02$1.50

In this experiment, we have also deduced that the cross section of the $\pi^+ {}^4He \rightarrow p {}^3He$ reaction is less than the $10^o/_o$ of the cross section of the $\pi^+ {}^4He \rightarrow \pi^o p {}^3He$ reaction. In order to distinguish between the two above reactions we tested the coplanarity of the p and ${}^3He$ tracks and deduced all the kinematic parameters in the phase space for the two and three body reactions[3]. The values of the cross section of the ${}^4He(\pi^+, 3pn)$ reaction are close to the values of the $\pi^+ {}^2H \rightarrow 2p$ reaction[4], as is shown in fig. 3. It must be noted that, in the case of quasideuteron absorption of photons on ${}^4He$, the (γ, np) cross section values are also close to the deuteron photodisintegration values[5].

It seems possible to conclude that the absorption by a quasideuteron is the principal mechanism in the non radiative capture of π^+ on ${}^4He$. Theoretical calculations of absorption cross section, taking into account the quasideuteron mechanism, have not yet been performed. The only calculation by Hirata et al.[6], considering the excitation of Δ-nucleon states, gives values very higher than the experimental ones, as shown in fig. 3.

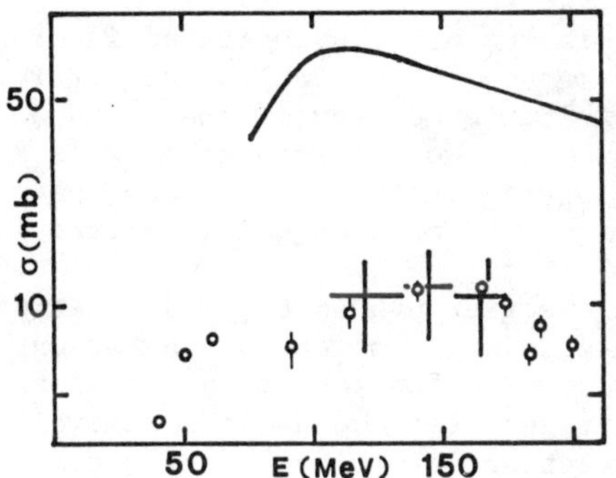

Fig. 3. ${}^4He(\pi^+, 3pn)$ cross section (crosses), compared with $\pi^+ {}^2H \rightarrow 2p$ cross section values, (Open circles). The full line is a calculation of the absorption cross section which takes into account only Δ-nucleon states[6].

REFERENCES

1. L. Busso et al., Nucl. Instr. and Meth., 102, 1 (1972).
2. S.C. Chakravartty and J. Hebert, Phys. Rev., C16, 1097 (1977).
3. F. James, report CERN 68-15.
4. C. Richard-Serre et al., Nucl. Phys., B20, 413 (1970) and literature thereby quoted and B.M. Preedom et al , Phys. Rev. C17, 1402 (1978).
5. F. Balestra et al., Nuovo Cim., (in press).
6. M. Hirata et al., Ann. Phys., 108, 116 (1977).

RESULTS FROM AN EXTENSIVE STUDY OF THE (π^-,2n) REACTION*

B. Bassalleck**, D. Engelhardt, M. Furić***, W. D. Klotz†, F. Takeutchi†† and H. Ullrich
Institut für Kernphysik, University and Kernforschungszentrum Karlsruhe, Germany and CERN, Geneva, Switzerland

ABSTRACT

An extensive study of the (π^-,2n) reaction with stopped π^- was completed at the CERN-SC. Target nuclei were from the 1p-shell ranging from ^{6}Li to ^{16}O and also ^{40}Ca.[1)] Large-area neutron counters with subnanosecond time resolution were used in the kinematically complete experiment.

Distributions in different variables were investigated in order to extract information on reaction mechanism and nuclear structure. Among the most important are the excitation energy of the residual nucleus showing the location of observed two-hole states as well as angular and momentum distributions giving information on the quantum state and dynamical characteristics of the nucleon pair involved in the absorption.

Transitions to low excitation energies proved to be dominated by the direct quasifree reaction mechanism. In general the strongest observed two-hole strength agreed with theoretical predictions on coefficients of fractional parentage[2)] for two-nucleon removal.

Clear evidence for a direct reaction involving s-shell nucleons was found in the two Li isotopes, see Fig. 1. In the heavier nuclei of the 1p-shell some indications of direct absorption by inner shell nucleons have been seen.

In general our excitation spectra are very similar to the corresponding (π^+,2p) results,[3,4)] within the experimental uncertainties. This indicates a very similar reaction mechanism. Higher resolution measurements are, however, needed for a more detailed comparison.

The measurement on ^{16}O was extended to angles between the two neutrons away from the quasi-deuteron angle of 180°, see Fig. 2. This gave very important information on absorption by nucleon pairs with total angular momentum $L \neq 0$.

* Work supported in part by the Bundesministerium für Forschung und Technologie of the Federal Republic of Germany

** Present address: University of British Columbia, Vancouver, British Columbia, Canada

*** Present address: Rice University, Houston, Texas 77001

† Present address: PTB, Berlin, Germany

†† Present address: Carnegie-Mellon University, Pittsburgh, Pennsylvania 15213.

ISSN:0094-243X/79/540174-02$1.50

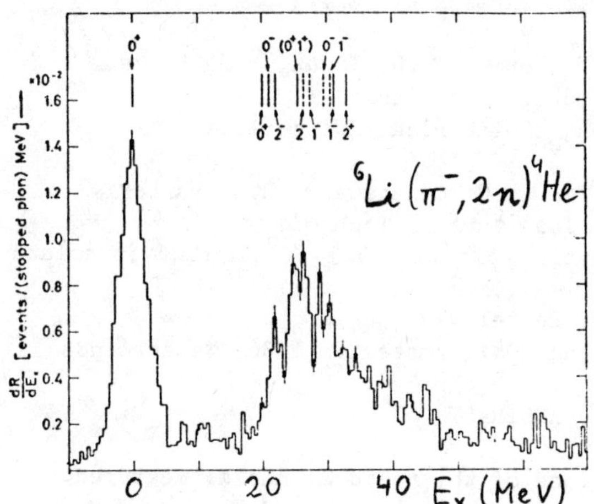

Fig. 1. Excitation energy spectrum for the reaction on ^{6}Li.

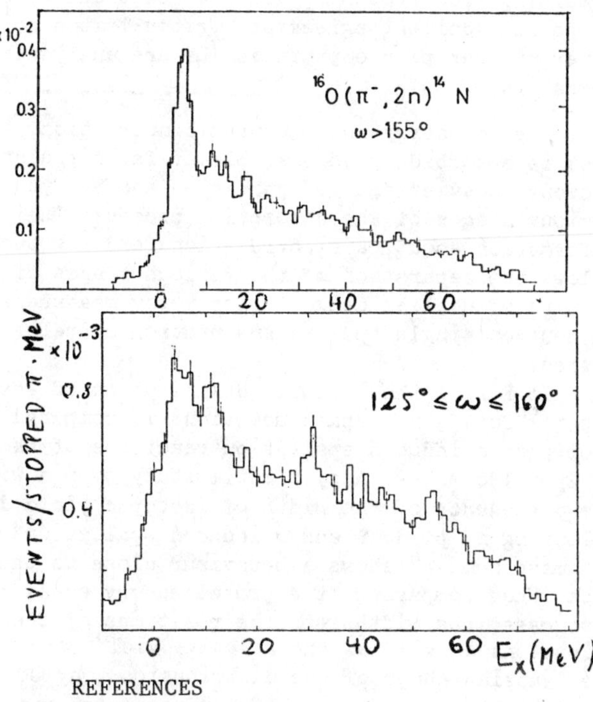

Fig. 2. Excitation energy spectrum from the reaction $^{16}O(\pi^-,2n)^{14}N$ for different regions in the angle between the two neutrons.

REFERENCES

1. B. Bassalleck et al., Phys. Lett. 65B, 128 (1976), Phys. Rev. C16, 1526 (1977), Z. f. Physik A286, 401 (1978) and submitted to Phys. Rev.
2. S. Cohen and D. Kurath, Nucl. Phys. A141, 145 (1970).
3. T. Favier et al., Nucl. Phys. A169, 540 (1971).
4. E. D. Arthur et al., Phys. Rev. C11, 332 (1975).

NUCLEAR PION CAPTURE AT REST

R. Engfer, R. Hartmann†, E. Hermes, H.P. Isaak, H.S. Pruys,
F.W. Schlepütz, A. Zglinski*
University of Zürich, 8001 Zürich, Switzerland

T. Kozlowski, H.J. Pfeiffer††, U. Sennhauser, H.K. Walter
Swiss Federal Institute of Technology,
Laboratory of High Energy Physics, c/o SIN, 5234 Villigen, Switzerland

W. Hesselink*
Vrije Universiteit de Boelelaan 1081, Amsterdam 1006, Netherlands

ABSTRACT

Isotopic yield distributions of the residual nuclei measured following negative pion capture at rest are compared to distributions resulting from proton spallation reactions at Ep = 136 MeV and found to be in excellent agreement. Triton-Triton coincidences observed after nuclear pion capture in ^{7}Li are analysed with respect to the formation of ^{4}H.

Due to energy and momentum conservation, a negative pion at rest is absorbed by an ensemble of few correlated nucleons. Neutrons, protons, heavier charged particles and γ-rays are emitted during the various stages of the absorption process. Their measurement provides information about pair correlations and few nucleon clusters and allows the separation of the various stages of pion absorption. The results of nuclear deexcitation γ-ray measurements[1] as well as data on neutron single spectra and neutron correlations[2] have been published.

In Fig. 1 the isotopic distribution of the residual nuclei after $^{59}Co(\pi^-, n + ypxn)$ reactions is compared to the one resulting from proton induced spallation reactions of the type $^{62}Ni(p,p+ypxn)$ at Ep = 136 MeV [3]. A systematic study over a wide energy range of the dependence of the width of isotopic yield distributions obtained following n, p, d, t and α induced spallation reactions on the target mass number[4] shows a behaviour close to that of the present pion data. When comparing at a proton energy equal to that of the pion rest mass, the widths and the positions of the isotopic yield distributions as well as the relative yields are in excellent agreement. The Gaussian shape of the distributions may be understood from the purely statistical nature of the secondary reaction stages, setting out at an energy high compared to the depth of the valley of stability.

* and SIN, 5234 Villigen, Switzerland

† Present address: Physics Department, College of William and Mary, Williamsburg, Va 23 185

†† Present address: Abteilung für die Sicherheit von Kernkraftwerken, c/o EIR, 5303 Würenlingen, Switzerland

ISSN:0094-243X/79/540176-02$1.50

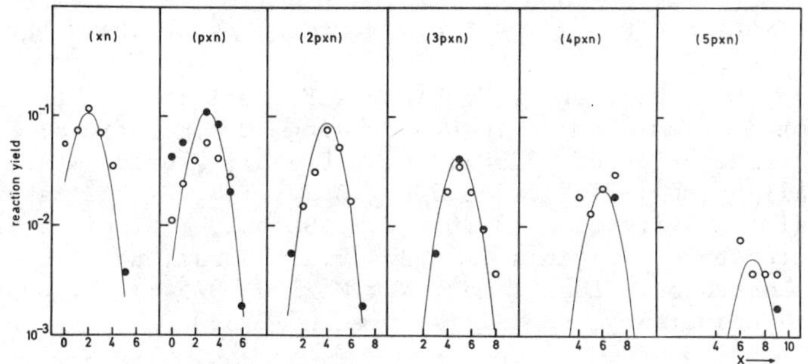

Fig. 1: Comparison of isotopic yield distributions from $^{59}Co(\pi^-, n + ypxn)$[1] (Gaussian curves) and $^{62}Ni(p, p + ypxn)$[3] (open circles from in beam measurements and black circles from activation measurements).

Whereas in heavy nuclei nucleon emission predominantly takes place during the evaporation stage, pion capture in light nuclei is well suited for the observation of particles emitted during early reaction stages. Charged particle correlations were measured for several targets, with special attention to the measurement of t-t coincidences following nuclear pion capture in ^7Li. Such coincidences may stem from the pure 3-body process

$$\pi^- + {}^7Li \rightarrow t + t + n \tag{1}$$

or may be the result of a two-body breakup of the ^7Li nucleus into a triton and a ^4H, followed by the decay of the ^4H.

$$\pi^- + {}^7Li \rightarrow t + {}^4H, \quad {}^4H \rightarrow t + n \tag{2}$$

Since the kinematics of the two processes are distinctly different, it is possible to decompose this distribution into the contributions from (1) and (2) and thus obtain information on the position and width of the ^4H ground state - and possibly excited states, too. At present, calculations are being carried out which simulate the energy and angular correlation distributions of (1) and (2) arising from pure phase space, folded with the acceptance of the experimental apparatus, and fit these to the measured distributions.

REFERENCES

1 H.S. Pruys et al., SIN Preprint PR-78-007 and Nucl. Phys., in press
2 R. Hartmann et al., Nucl. Phys. A300 (1978) 345
3 M. Sadler et al., Phys. Rev. Lett. 38 (1977) 950
4 G. Rudstam, Zeitschrift für Naturforschung 21a (1966) 1027

EVIDENCE THAT FOR NON ABSORPTION REACTIONS OF 190-MeV π^+ π^- on ^{27}AL MULTINUCLEAR REMOVAL OFTEN OCCURS AS SEPARATE PARTICLES

R.E. McAdams, O.H. Otteson, V.G. Lind, W.F. Denig; Utah State University, Logan, Utah 84322: C.A. Goulding, M. Greenfield; Florida A&M, Tallahassee, Florida 32307: H.S. Plendl; Florida State University, Tallahassee, Florida 32306: B.J. Lieb; George Mason University, Fairfax, Virginia 22030: C.E. Stronach; Virginia State College, Petersburg, Virginia 23803: P.A.M. Gram; Los Alamos Scientific Laboratory, Los Alamos, New Mexico 87544: T. Sharma; Oregon State University, Corvallis, Oregon 97330

ABSTRACT

Gamma-ray spectra in coincidence with scattered pions at 35° were obtained for π^+ and π^- incident on ^{27}Al at LAMPF. The gamma-ray spectra in coincidence with the scattered π^+ or π^- were separated into two energy bins each: $E_\pi'<80$ MeV and $E_\pi'>80$ MeV. The average ratio of the differential cross section for the high-energy bin to the low-energy bin is 2.0±0.2 for single particle removal and 1.3±0.02 for multiparticle removal indicating that, in addition to possible cluster removal, the particles are removed as separate particles for a significant portion of the time.

INTRODUCTION

In late fall 1977 a 2 gm/cm^2 ^{27}Al target was bombarded with 190-MeV π^+ and π^- on the P^3 channel at LAMPF. Gamma-ray spectra were accumulated in coincidence with emerging particles using a Ge(Li) gamma-ray detector set at 90° to the beam and a E-ΔE scintillator particle telescope set at 35° to the beam. Preliminary results of the gamma spectra in coincidence with scattered pions for the π^- case has been published.[1] In sorting the spectra, which was recorded event by event using a PDP 11/34 computer, events corresponding to particles passing through the E detector were easily separated from ones which were stopped in the E detector, since a scintillator placed immediately behind the E detector detected particles passing completely through. This allows easy particle separation into 2 energy bins: E<80 MeV and E>80 MeV.

DATA

Figure 1.a. shows the gamma-ray spectrum for π^+ on ^{27}Al with scattered pions of energy less than 80 MeV and figure 1.b. the spectrum with scattered pions of energy greater than 80 MeV. One obtains similar spectra for π^- on ^{27}Al.

Table I gives the differential cross sections for scattered π^+ of energy less than and greater than 80 MeV. Table II is similar but for π^- and also includes an average set of ratios for the two cases combined. Note the average ratio of the higher energy bin to the lower bin is 2.0±0.2 for inelastic scattering and single nu-

ISSN:0094-243X/79/540178-02$1.50

cleon removal. For multinucleon removal the ratio is 1.3±0.2, which indicates the pion often loses considerably more energy during the process which leads to multinucleon removal. This infers that frequently the particles come out separately rather than as a cluster, since this requires more energy, even though cluster removal may also occur part of the time.

Table I. The differential cross sections for scattered π^+ of energy less than 80 MeV and greater than 80 MeV.

Residual Nucleus and Transition	E_γ(keV)	$\frac{d\sigma}{d\Omega}$ $(35^o\pi^+)$μb/str $E_\pi'<80$ MeV	$E_\pi'>80$ MeV	$\frac{E_\pi'>80 \text{ MeV}}{E_\pi'<80 \text{ MeV}}$
^{27}Al, II→0	1014.5	130 ± 45	280 ± 50	2.2 ± 0.8
^{26}Al, II→0 (100%)	416.8	38 ± 20	84 ± 20	2.2 ± 1.3
^{26}Mg, I→0	1808.9	390 ± 110	1400 ± 140	3.6 ± 1.1
^{26}Mg, II→I (90%)	1129.6	190 ± 50	310 ± 60	1.6 ± 0.5
^{25}Mg, I→0	585.1	63 ± 25	130 ± 30	2.1 ± 0.4
^{24}Mg, I→0	1368.6	300 ± 70	350 ± 80	1.2 ± 0.3
^{23}Na, I→0	439.9	110 ± 24	130 ± 20	1.2 ± 0.3
^{21}Ne, I→0	350.5	47 ± 14[a]	47 ± 15[a]	1.0 ± 0.4

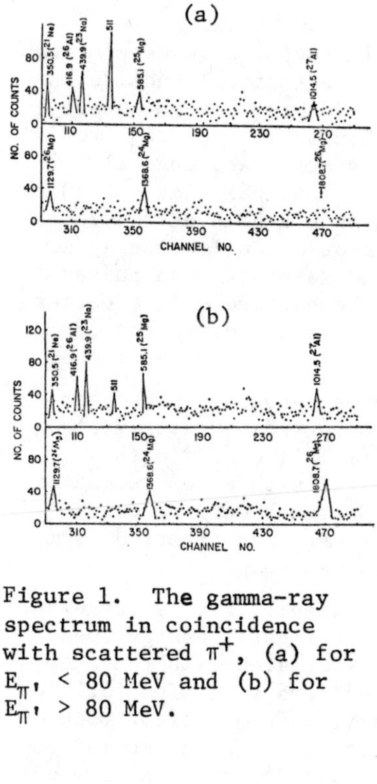

Figure 1. The gamma-ray spectrum in coincidence with scattered π^+, (a) for $E_{\pi'}$ < 80 MeV and (b) for $E_{\pi'}$ > 80 MeV.

Table II. The differential cross sections for scattered π^- of energy less than 80 MeV and greater than 80 MeV.

Residual Nucleus and Transition	E_γ(keV)	$\frac{d\sigma}{dr}$ $(35^o\ \pi^-)$ μb/str E_π' 80<MeV	$E_\pi'>80$ MeV	$\frac{E\pi'>80 \text{ MeV}}{E_\pi'<80 \text{ MeV}}$	Ave of π^+ and π^- Ratio
^{27}Al, II→0	1014.5	140 ± 70	300 ± 80	2.1 ± 1.2	2.2 ± 0.7
^{26}Al, II→0 (100%)	416.8	48 ± 29	190 ± 30	4.0 ± 2.5	3.1 ± 1.2
^{26}Mg, I→0	1808.9	650 ± 160	1300 ± 200	2.0 ± 0.6	2.8 ± 0.5
^{26}Mg, II→I (90%)	1129.6	250 ± 100	270 ± 110	1.1 ± 0.6	1.4 ± 0.4
^{25}Mg, I→0	585.1	68 ± 42	180 ± 50	2.6 ± 1.8	2.4 ± 0.4
^{24}Mg, I→0	1368.6	430 ± 120	770 ± 140	1.8 ± 0.6	1.5 ± 0.3
^{23}Na, I→0	439.9	95 ± 32	180 ± 34	1.9 ± 0.7	1.6 ± 0.3
^{21}Ne, I→0	350.5	75 ± 20[a]	42 ± 22[a]	0.6 ± 0.4	0.8 ± 0.3

[a] These values may be low because this line is on the edge of the electronic discriminator cut-off.

REFERENCES

1. V.G. Lind *et al*. Phys. Rev. Lett. 41, 1023, (1978).

NEUTRON SPECTRA FROM (π,nγ) COINCIDENCE EXPERIMENTS WITH PIONS STOPPED ON ^{28}Si

R. Legrain, Y. Cassagnou, J. Julien, L. Roussel,
DPh-N, CEN Saclay, BP 2, 91190 Gif-sur-Yvette, France

R. Fonte, A. Palmeri
Institut Physique Nucléaire, Catane, Italy

E. Gadioli, E. Gadioli Erba
Institut Physique Nucléaire, Milano, Italy

Recently, protons from π^- absorption in nuclei have been measured in coincidence with γ rays assigned to the residual nuclide in the various exit channels[1,2]. In comparison to previous inclusive experiments, a better knowledge of the mechanism is expected from more selective data such as partial energy spectra and channel multiplicities as a challenge of present theoretical calculations. More specifically, partial energy spectra would show a high energy component from the initial absorption of the pion and a low energy component from the subsequent equilibration and de-excitation phases. It is hoped that the low energy part would be derived with a better accuracy from coincidence experiments involving the detection of gamma rays and neutrons (no coulomb barrier).

Such an experiment was performed using a secondary pion beam from the 600 MeV Saclay electron linear accelerator. Measurements were made for stopped pions on a sample of metallic ^{28}Si with a thickness of 3 g/cm^2. Gamma rays were detected with a resolution of 3 keV at 1 MeV in a 80 cm^3 GeLi detector. Neutrons were detected with eight liquid scintillators and their energy was inferred from time of flight measurement. The flight path was equal to 150 cm.

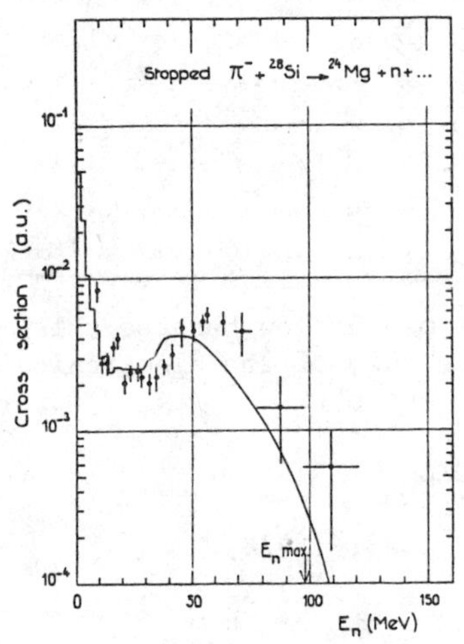

As an example, Fig. 1 represents the experimental energy spectrum associated with the residual nucleus ^{24}Mg. The full curve results from exciton model predictions[3]. The agreement is rather satisfactory in view of the low counting rates and the efficiency and background corrections which were made.

Fig. 1 - Neutron spectrum in coincidence with γ rays identifying the residual nucleus ^{24}Mg in the reaction $\pi^- + {}^{28}Si \rightarrow {}^{24}Mg^* + ...$ The full curve results from exciton model predictions.

ISSN:0094-243X/79/540180-02$1.50

Yields and neutron multiplicities in the various exit channels are also well reproduced by the calculations. This experiment shows the possibility to investigate the different stages of the reaction mechanism and the agreement with the exciton model.

REFERENCES

1. R. Legrain et al., Phys. Lett. 74B, 207 (1978).
2. R.E. Mc Adams et al., Clustering aspects of nuclear structure and nuclear reactions, Winwipeg 1978.
3. E. Gadioli et al., Proceedings of the International Conference on nuclear reaction mechanism, Varenna 1977.

A STUDY OF NON-RELATIVISTIC APPROXIMATIONS TO THE PION PRODUCTION OPERATOR IN $^4He(p,n\pi^+)^4He$

J.M. Greben
TRIUMF and Univ. of British Columbia, Vancouver, B.C., Canada V6T 1W5

R.M. Woloshyn
TRIUMF, Vancouver, B.C., Canada V6T 1W5

Although it is possible in principle to construct an effective non-relativistic pion production operator[1] from a relativistic theory, it is difficult in practice to avoid uncertainties.[2] It has been suggested that a practical way of determining an effective non-relativistic pion-nucleon vertex could be by measuring the $(p,n\pi^+)$-reaction on spinless, isoscalar nuclei.[3] The expectation is that the dominant contribution to this process is external pion emission, which will be quite sensitive to the pion-nucleon vertex but not very sensitive to the details of the nucleon-nucleus scattering amplitude. We have studied the adequacy of a non-relativistic treatment of this process by calculating the production cross section at fixed proton and pion energy with fully relativistic and approximate non-relativistic external emission amplitudes. The non-relativistic approximation to the relativistic expressions contains two ingredients. First, one deletes the negative frequency part of the relativistic propagator. Second, one keeps only the lowest powers in the p/m expansion of the pion-nucleon vertex. For the pseudoscalar vertex we find that the first approximation reduces the cross section drastically. The second approximation leads to a further, smaller, reduction. In the pseudovector case we find a—somewhat smaller—overall reduction in the cross section. Even if we allow the coefficient of the nucleon recoil term to vary at will we are unable to reproduce the relativistic cross section satisfactorily. We conclude that the usual non-relativistic expressions for the pion-nucleon vertex cannot simulate the effects of the negative frequency intermediate states.

The apparent ambiguity in the relativistic calculations between pseudovector and pseudoscalar coupling can be resolved to lowest order in the pion momentum by using the soft-pion theorem.[4] Application of this theorem dictates the use of pseudovector coupling up to this order. Comparison of the relativistic pseudovector calculations with experiment will then tell us whether calculations up to lowest order in the soft-pion theorem are sufficient, or whether one has to take into account higher-order terms. The latter terms will depend on the off-shell properties of the nuclear scattering amplitude.

1. J.L. Friar, Phys. Rev. C10, 955 (1974).
2. T.S.H. Lee and S. Pittel, Nucl. Phys. A256, 509 (1976).
3. H.S. Sherif, S.W. Leung, A.W. Thomas, and G. Brookfield, preprint TRI-UAE-5013 (1978).
4. S.L. Adler and Y. Dothan, Phys. Rev. 151, 1267 (1966).

ISSN:0094-243X/79/540182-01$1.50

MICROSCOPIC CALCULATION OF PION SCATTERING AND PION ABSORPTION IN ^{4}He

H. M. Hofmann
University of Erlangen-Nürnberg, Erlangen, W-Germany

In the regime of the free Δ-resonance, Δ-dominance is the starting point for a microscopic description of pion-nucleus scattering. The restriction to 1 Δ-particle 1 nucleon-hole intermediate states, allows to reproduce the form of the observed angular distributions. However, the magnitude of integrated cross-sections is over-estimated by 30 to 300 %, see fig. 1. The necessary additional damping can be included phenomenologically by a spreading potential[1] or microscopically by considering the effects of true absorption[2,3]. With the inclusion of true absorption also the magnitude of integrated (fig.1) and total absorption (fig. 2) cross-sections are well reproduced. The energy dependence of integrated cross-sections is modestly dependent on dispersive shifts induced by the coupling to various open and closed channels. For the absorptive parts and the dispersive corrections off-shell pion and ρ-exchange and nucleon-nucleon short-range-correlations may contribute. The off-shell behaviour of these exchanges, parametrized by cut-off masses, and the form of the N-N-correlations is not well known. We consider various choices of these para-meters, in agreement with the $\pi^+ d \rightarrow pp$ reaction, and study the effects on scattering and absorption of pions on helium.

1. H. Hirata et al., Ann. Phys. 108, 116 (1977)
2. H. M. Hofmann, Z. Phys. 289, 273 (1978)
3. E. Oset and W. Weise, Phys. Lett. 77 B, 158 (1978) and preprint

ISSN:0094-243X/79/540183-02$1.50

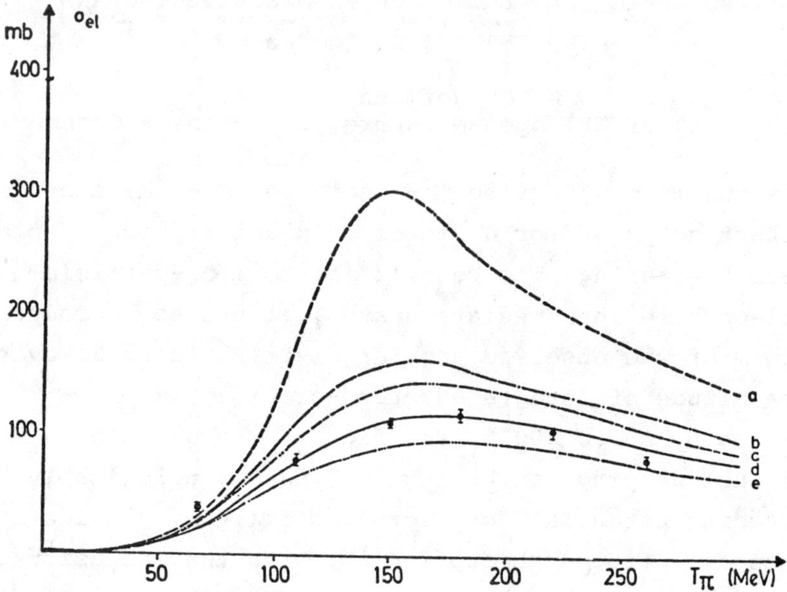

Fig. 1. Integrated elastic $\pi\ ^4He$ cross section. Curve (a) without true absorption effects, (b) and (c) with direct absorption contributions, curves (d) and (e) show the full absorption calculation. For the curves (b) and (d) a weak ρ -coupling and for (c) and (e) a strong ρ-coupling is used.

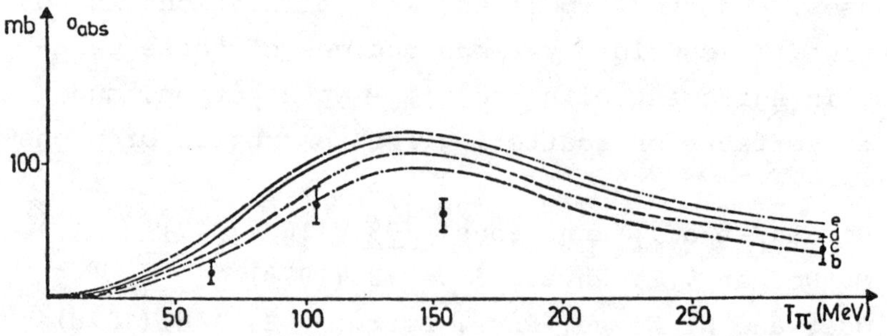

Fig. 2. Total pion absorption cross-section in ^4He. The curves are labeled as in fig. 1.

ANALYSIS OF THE (p,π^+) AND THE (d,p) STRIPPING REACTION

A. Boudard, P. Couvert
CEN Saclay, Gif-sur-Yvette, France

M. Dillig
University of Erlangen-Nürnberg, Erlangen, W-Germany

To test the various theoretical models for simple stripping processes at medium energies the differential cross section for the (p,π^+) and the (d,p) reaction at scattering energies above the 33-resonance as well as the excitation functions in proton-induced π-production on ^{10}B and ^{208}Pb were investigated. In order to get in particular an idea about the importance of the stripping term relative to the rescattering term and higher order multiple scattering contributions we investigate the excitation function in a microscopic model including thereby

- the stripping term;
- the rescattering term from π,ρ exchange with virtual N or Δ excitation (p-wave rescattering) as well as effects from the s-wave πN interaction (within the ansatz of Koltun-Reitan[1]);
- distortion corrections from the Coulomb distortion as well as for the coupling to inelastic channels.

As presently the absolute magnitude of the various contributions cannot be calculated reliably, we normalize the cross section without the stripping term, to the experimental cross section at the resonance (where Δ-dominance is expected to hold rather well) and keep the ratio to the stripping term as a free parameter.
Applying our model for the excitation function of $^{10}B(p,\pi^+)\,^{11}B(g.s.)$ where data from $T_p \cong 150$ MeV up to $T_p \cong 600$ MeV[2] are available, we find that

(i) rescattering contributions dominate the reaction mechanism at practically all energies;

ISSN:0094-243X/79/540185-02$1.50

(ii) the influence of the stripping term is negligable expect very close to the production threshold, however, its detailed extraction is prohibited due to the delicate

(iii) s and p-wave interference in the πN-t-matrix far off the energy shell;

(iv) there is no indication for an anomalous behaviour of the excitation function of ^{10}B or ^{208}Pb near the π-production threshold.

In a simplied form - the effective NN → NNπ operator is treated more phenomenologically - the same model was applied for the calculation of the (p,π^+) and (d,p) differential cross section for light nuclei at scattering energies far above the π-production threshold. It is found that the model reproduces the characteristic trends in the experimental data reasonably, in addition it accounts in a rather natural way for the strong similarities between the (p,π) and (d,p) reaction seen in the experiment at large scattering energies[3].

1. D. Koltun and A. Reitan, Phys. Rev. 139 B, 1372 (1965)
2. B. Hoistad, 7. Intern. Conf. on High Energy Physics and Nuclear Structure (1977) 215 (Ed. M. P. Locher, Birkhäuser Verlag, Basel)
3. M. Dillig et al., preprint (Saclay, 1979)

INCLUSIVE (p,π^+) CROSS SECTIONS NEAR THRESHOLD

R. E. Marrs* and R. E. Pollock
Indiana University Cyclotron Facility, Bloomington, Indiana 47401

ABSTRACT

Angle-integrated (p,π^+) cross sections on several 1p-shell targets have been measured over the energy range $0.5 \lesssim T_p - T_{threshold} \lesssim 10$ MeV by detecting μ^+ decays. For heavier targets, (p,π^+) inclusive cross sections at energies well above threshold are surprisingly different from previous (p,π^0) measurements.

ENERGY DEPENDENCE

Inclusive (p,π^+) cross sections have been measured near threshold at the Indiana University Cyclotron Facility by stopping the pions in or near the target and counting energetic positrons from $\pi^+ \to \mu^+ \to e^+$ decay. The positrons were detected in plastic Cherenkov counters between 1 μs long beam bursts. Figure 1 shows the measured $^{16}O(p,\pi^+)^{17}O$ excitation function for energies extending downward to the π^+ Coulomb barrier, where only the ^{17}O ground state contributes. Similar measurements were obtained for ^{9}Be, ^{11}B, ^{12}C, ^{13}C and ^{14}N. Previous $^{16}O(p,\pi^+)^{17}O(g.s.)$ angular distribution measurements at $T_p = 154$ MeV have been integrated to obtain a total cross section,[1] which is included in Fig.1 to extend the ground-state excitation function over a larger energy range.

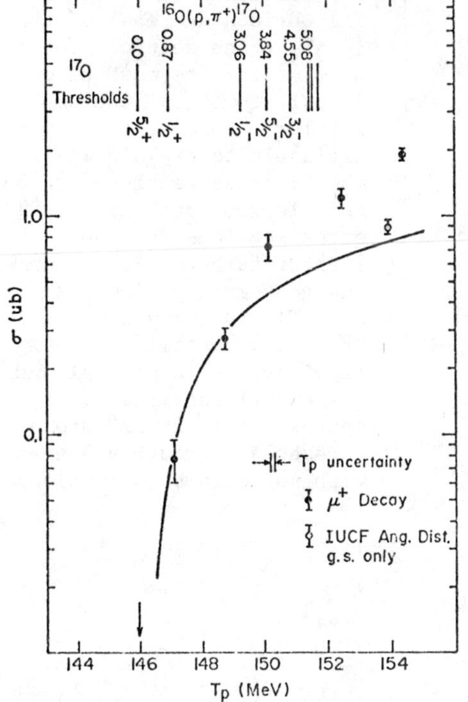

Fig. 1. Inclusive $^{16}O(p,\pi^+)^{17}O$ cross section as a function of proton energy in the lab. Solid curve is a Coulomb factor.

The energy dependence of the angle-integrated (p,π^+) cross section to a specific final state near threshold should be dominated by the phase space and barrier penetration of the outgoing pion. The solid curve in Fig.1 is the (arbitrarily normalized) expression $\sigma(p,\pi^+) \alpha (k_\pi/k_p) S(k_\pi)$ where $\hbar k_\pi$ and $\hbar k_p$ are the barycentric

*Present Address: TRIUMF, University of British Columbia, Vancouver, B.C., V6T 1W5, CANADA

ISSN:0094-243X/79/540187-02$1.50

momenta; and $S(k_\pi) = 2\pi\gamma/(\exp(2\pi\gamma)-1)$, with $\gamma = Z(e^2/\hbar^2c^2)\,\frac{E_\pi}{k_\pi}$, is the usual Coulomb factor which approximately accounts for the pion penetrability. The measured ground-state cross sections for all of the light targets have an energy dependence consistent with this expression near threshold.

A DEPENDENCE

In order to study the A dependence of inclusive (p,π^+) cross sections additional targets were run at two fixed beam energies. The results are grouped in Fig. 2 according to excess energy above threshold. In addition to the errors shown there is a normalization uncertainty of about 10%. The broken lines in Fig. 2 represent the measured A dependence of inclusive (p,π^0) cross sections on many of the same targets at two different beam energies.[2) The obvious decrease of the (p,π^+) cross sections for heavier targets must be due in part to the π^+ Coulomb barrier. However it is more difficult to explain why the (p,π^+) cross sections are so much larger than the (p,π^0) cross sections for the lighter targets. For example, charge symmetry requires $\sigma(p,\pi^+)/\sigma(p,\pi^0) = 2$ for the $^{40}Ca(p,\pi)$ reaction to $T = \frac{1}{2}$ final states in mass 41, but the experimental ratio at approximately 16 MeV above threshold is roughly 5 even with no Coulomb corrections.

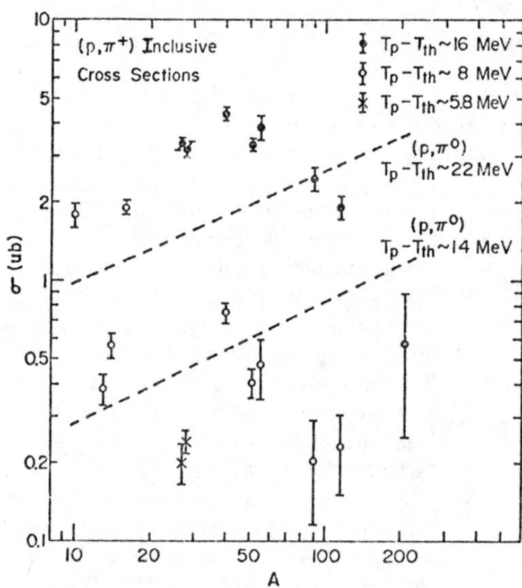

Fig. 2. Inclusive (p,π^+) cross sections at nearly constant (± 1 MeV) energy-above-threshold.

REFERENCES

1. P. H. Pile, Ph.D. Thesis, Indiana University, 1978 (unpublished).
2. A. D. Backer, P. T. Debevec, G. T. Emery, M. A. Pickar, K. Gotow, D. A. Jenkins and P. L. Roberson, Indiana University Cyclotron Facility Technical and Scientific Report, November 1975 - January 1977 (unpublished), p. 31.

THE $(p,\pi^{\pm})$ REACTIONS ON ^{9}Be AT 800 MEV

Bo Höistad
Los Alamos Scientific Laboratory, Los Alamos, NM 87545

G. S. Adams, M. Gazzaly, G. Igo, and F. Iram
University of California, Los Angeles, CA 90024

H. Nann
Northwestern University, Evanston, IL 60201

One striking feature of the (p,π) reaction is the large difference in the π^+ and π^- production cross sections. This is demonstrated by previous data obtained close to the threshold energy[1] as well as at 600 MeV.[2] Furthermore, these data show almost isotropic (p,π^-) angular distributions, while the (p,π^+) distributions exhibit a great variety of shapes and thereby indicate a large sensitivity to the nuclear wave functions involved. Although the (p,π^-) distributions are probably not very sensitive to nuclear structure, important information can be obtained about the reaction mechanism, since each reaction model predicts a characteristic π^+/π^- ratio.

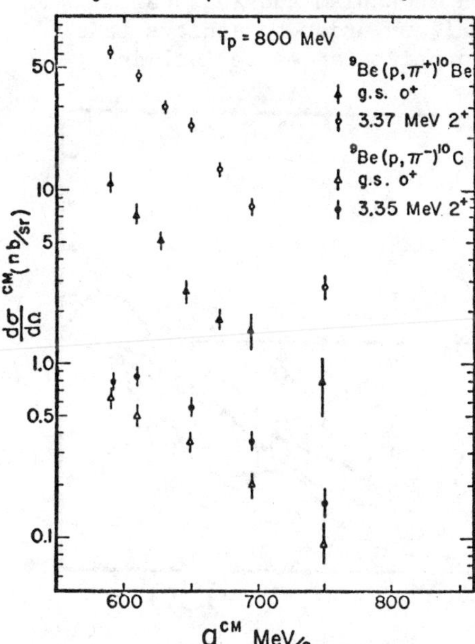

Fig. 1. Angular distributions from the $^{9}Be(p,\pi^{\pm})$ reactions leading to isobaric analog states in ^{10}Be and ^{10}C.

In this paper we present (see fig. 1) angular distributions from the $^{9}Be(p,\pi^{\pm})$ reactions at 800 MeV obtained by the High Resolution Spectrometer (HRS) at LAMPF. These data are recorded at an energy and in a momentum transfer interval where no data is reported before. From fig. 1 we immediately observe that even at 800 MeV the π^+ cross section dominate over the π^-. In a comparison with the low-energy data we find the surprising result that at 800 MeV the difference in the π^+ and π^- cross section is even larger than at low energy. Another observation from fig. 1 is that the π^- distributions are not isotropic but show a slope very similar to the one of (p,π^+). This result was neither obtained at 185 MeV[1] nor indicated by the 600 MeV data.[2] Whether this is an effect of the energy dependence of the reaction mechanism and/or a consequence of the very large momentum transfer remains to be seen from future data at intermediate energies.

Ref. 1. S. Dahlgren et.al., Nucl. Phys. A204 (1973) 53.
Ref. 2. P. Couvert et.al., Phys. Rev. Lett. 48 (1978) 530.

ISSN:0094-243X/79/540189-01$1.50

PROTON DISTORTION AND ASYMMETRIES IN THE $(\vec{p},\pi)$ REACTION†

H. J. Weber

Department of Physics, University of Virginia
Charlottesville, VA 22901

and

J. M. Eisenberg

Department of Physics and Astronomy
Tel-Aviv University, Tel-Aviv, Israel

Recent experiments on the $(\vec{p},\pi^+)$ reaction in light nuclei at 200 MeV proton kinetic energy show a striking uniformity in the asymmetry for varying targets and final nuclear states.[1] We find that this qualitative feature can be explained by conventional proton distortion using a single-nucleon mechanism and reasonable neutron bound states. It is essential, in order to achieve this effect, that both spin-orbit and absorptive forces be included in the continuum proton potential.

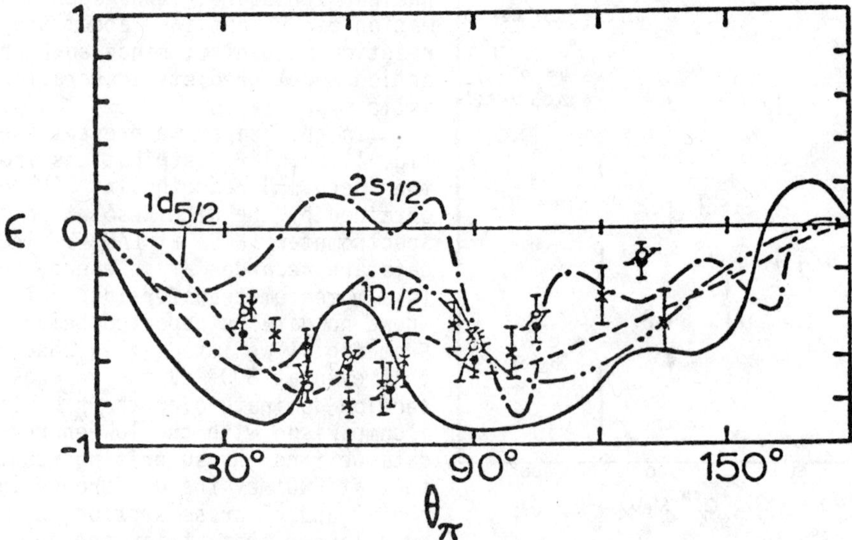

Fig. 1. The $(\vec{p},\pi)$ asymmetry ε for protons distorted in a Woods-Saxon potential with V = 19.4 MeV, W = 15.6 MeV, V_s = 4.12 MeV, W_s = -0.1 MeV, c = 2.3 fm, a = 0.45 fm. The $1p_{1/2}$ neutron (solid line) is bound in a well with V = 42 MeV, V_s = 4.8 MeV, c = 2.78 MeV and a = 0.65 fm, while the $2s_{1/2}$ (dot-dash) and $1d_{5/2}$ (dash) use V = 52.5 MeV, V_s = 9 MeV, c = 3.02 fm and a = 0.55 fm. The case for a $1p_{1/2}$ neutron bound also in this latter well is shown by a dot-dot-dash line. The data[1)] are for a ^{9}Be target (closed dots: final ground state, open dots: first excited state) and for ^{12}C (crossed), ground state.

ISSN:0094-243X/79/540190-02$1.50

The results shown in Fig. 1 are fairly sensitive to the distorting and binding potentials, but are quite stable over a reasonable range of parameter values for these potentials.

We incorporate the fairly weak pion distortion only crudely, using the Sopkovich method, in terms of the small phase shifts found in low-energy pion scattering from p-shell nuclei. Lastly, we study the consequences of orthogonality for the (p,π^+) process concluding that, in the context of a practical calculation in our kinematic regime, it must yield in importance to the crucial incorporation of the more physical distortion for the proton wave.

1. E. Auld et al., Bull. Am. Phys. Soc. 22 (1977) 590, Phys. Rev. Lett. 41 (1978) 462.

†Supported in part by the U. S. National Science Foundation and the U. S. - Israel Binational Science Foundation.

PROTON-INDUCED PION PRODUCTION AND THE ONE-NUCLEON MODEL*

M. C. Tsangarides, J. G. Wills and R. D. Bent
Indiana University Cyclotron Facility, Bloomington, Indiana 47401

In this work, still in progress, we've calculated differential cross sections for the reaction $A(p,\pi^+)A+1_{g.s.}$ with outgoing pion energies from near threshold to ∿50 MeV and compared with data for the nuclei ^{10}Ba, ^{40}Ca, ^{90}Zr and ^{208}Pb[1]. A DWBA code[2] was used with standard proton and coulomb distortions, a Galilean invariant production operator, and different forms of the pion-nucleus optical potential. The best fits so far (figures 1 to 4) were obtained with the Laplacian-like potential, which includes corrections for off-shell damping, Pauli-blocking, and angle transformation of the form

$$2M_\pi V = \{-b_o k^2 + \frac{k^4}{4\pi}(b_o{}^2 + 2b_1{}^2)\langle\frac{1}{r}\rangle\}\rho + b_1 k^2(\rho_n - \rho_\rho) - 4_\pi B_o \rho^2$$

$$- \{k^2 c_o + \frac{c_o}{2}\nabla^2 + \frac{E_\pi}{2M} c_o \nabla^2\}\rho \qquad (1)$$

For numerical expediency the off-shell damping corrections to the πN t-matrix were of a Modified Gaussian form and were incorporated with the nuclear form factor, S(q), in the expresion

$$S(q) \times \{a + bq^2\}e^{-cq^2}, \quad q = |\vec{k} - \vec{k}'| \qquad (2)$$

the convolution of which in r-space gives the effective densities ρ, ρ_η and ρ_ρ used in (1). The parameters b_o, c_o and b_1 were computed from πN phaseshifts[3]. These calculations, covering a wide range of energies and target masses, indicate that the ONM contributes significantly to the (p,π^+) reaction.

REFERENCES

1. R.D. Bent, P.T.Debevec, P.H. Pile, R.E. Pollock, R.E. Marrs, and M.C. Green, Phys. Rev. Letters 40, 495 (1978); P.H. Pile, R.D. Bent, R.E. Pollock, P.T. Debevec, R.E. Marrs, and M.C. Green, Submitted to Phys. Rev. Letters (January, 1979).
   S. Dahlgren, P. Grfstrȯm, B. Höistad, and A. Asberg, Nucl. Phys. A227, 245 (1974) and references there in.
2. M.P. Keating, and J.G. Wills, Phys. Rev. C 7, 1336 (1973).
3. G. Rowe, M. Salomon, and R.H. Landau, Phys. Rev. C 18, 584 (1978).

*Work supported by the NSF.

ISSN:0094-243X/79/540192-02$1.50

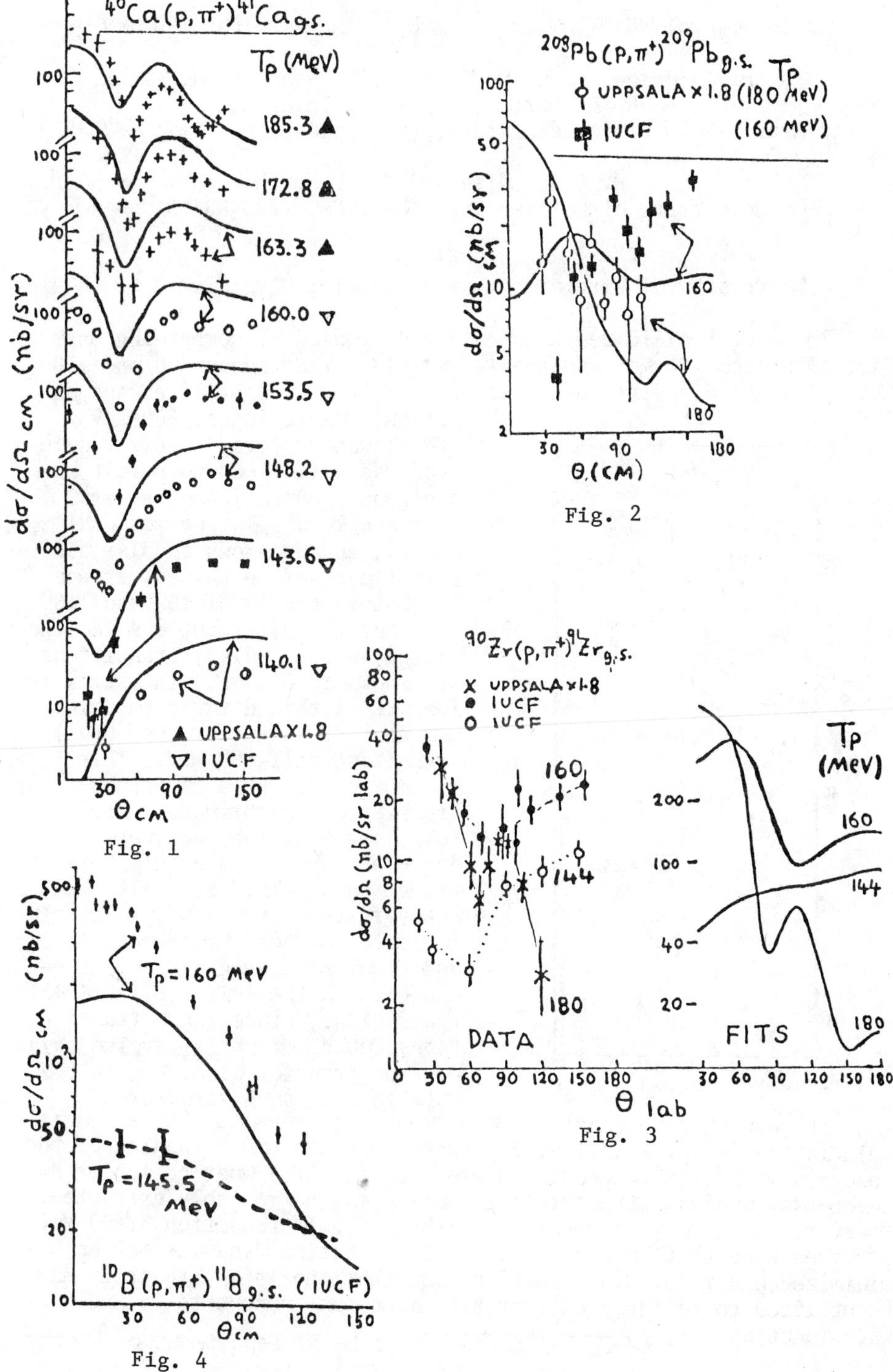

Fig. 1

Fig. 2

Fig. 3

Fig. 4

THE (π^+,p) REACTION ON ^{12}C AND ^{13}C NEAR 90 AND 180 MeV

R. E. Anderson, E. W. Hoffman, B. Höistad, R. J. Macek,
C. L. Morris, and H. A. Thiessen
Los Alamos Scientific Laboratory, Los Alamos, New Mexico 87545

R. L. Boudrie and G. R. Smith
University of Colorado, Boulder, Colorado 80309

J. Källne
University of Virginia, Charlottsville, Virginia

The $^{12}C(\pi^+,p)^{11}C$ reaction has been studied at bombarding energies of 90 and 180 MeV and the $^{13}C(\pi^+,p)^{12}C$ reaction at 90 and 170 MeV. The EPICS spectrometer facility at LAMPF was utilized and an optimal resolution of 500 keV (FWHM) was achieved. This resolution was sufficient to permit unambiguous separation of several discrete levels in the residual nuclei, and numerous angular distributions were obtained between laboratory angles of 10° and 115°.

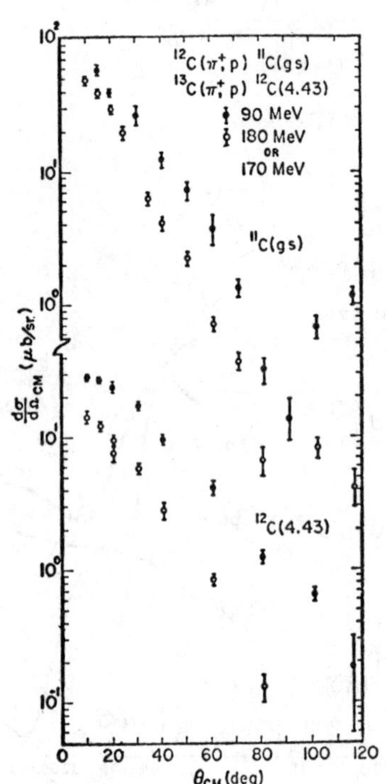

Angular distributions for the $^{11}C(gs)$ and $^{12}C(4.43)$ transitions are shown in Fig. 1. The striking feature of this data is the minimum near 85° seen for the $^{11}C(gs)$ transition at T_π=90 MeV. This minimum occurs at a momentum transfer of approximately 650 MeV/c and would be seen near an angle of 55° in the angular distribution obtained at T_π=180 MeV. However, the data shown in Fig. 1 indicate no such structure. In addition, similar minima might be expected in the $^{13}C(\pi^+,p)^{12}C(4.43)$ transitions, since those transitions would nominally involve $p_{3/2}$ neutron transfer also. Again the data show no such structure.

It must thus be concluded that reaction dynamics play a crucial role in the (π^+,p) reaction, and that a precise understanding of the reaction mechanism (including distortion effects) must be achieved before any nuclear structure information may be reliably extracted. However, the structure observed in the $^{11}C(gs)$ transition at 90 MeV provides hope that uncertainties in the reaction dynamics may be minimized under certain conditions and that the (π^+,p) reaction may be utilized to reliably extract high momentum components in nuclear wave functions.

*Work supported by the U. S. Department of Energy

ISSN:0094-243X/79/540194-01$1.50

THE (π,p) REACTION IN $^{6,7}Li$ AT 75 AND 175 MeV

Jan Källne, J. Davis, P. C. Gugelot and J. S. McCarthy
University of Virginia, Charlottesville, VA 22901

G. Smith
University of Colorado, Boulder, Colorado 80309

R. Boudrie, B. Höistad and C. L. Morris
Los Alamos Scientific Laboratory, N. M. 87545

The (π^+,p) reaction in $^{6,7}Li$ has been measured using the EPICS pion channel and spectrometer system at LAMPF. Angular distributions ($\theta \approx 20$-$100°$) were measured for (π^+,p) transitions to discrete final states and for ^6Li at $\theta = 20°$ both (π^+,p) and (π^-,p) were studied over an extended part of the continuous spectrum (See Fig.1.).

The angular distributions recorded were featureless (except for $\theta \gtrsim 80°$ at 75 MeV) showing a general exponential dependence in terms of momentum transfer (q), $e^{-q/\lambda}$. We find the fall-off to be λ = 40-50 MeV/e which is quite similar to the result[1] for (π,N) in $^{3,4}He$. A large difference, however, is observed in cross section magnitude where those for $^{3,4}He$ exceed those of $^{6,7}Li$ 5 times or more.

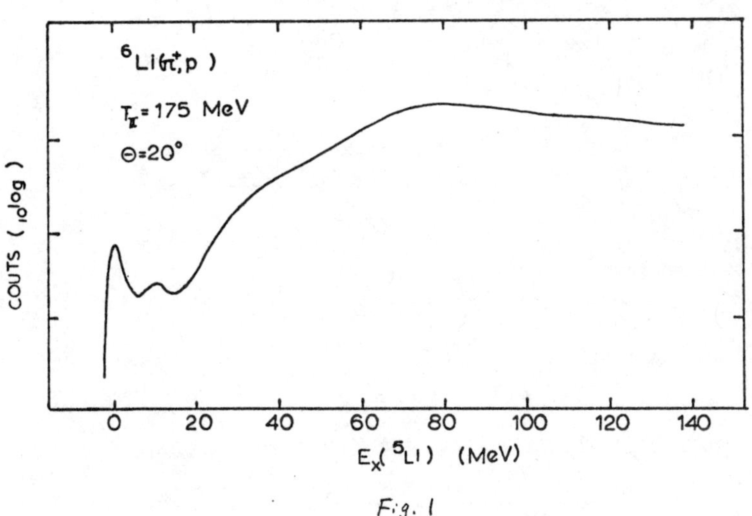

Fig. 1

ISSN:0094-243X/79/540195-02$1.50

The broad minimum appears at an energy close to that defined by the kinematics of the reaction $\pi d \rightarrow pp$ from which one can realize the importance of pion rescattering in the (π,p) reaction. A comparison between (π^+,p) and (π^-,p) in this energy region shows a striking difference between the two reactions with respect to πN rescattering since $d^2\sigma(\pi^+,p) \gg 10\ d^2\sigma(\pi^-,p)$.

REFERENCE

1. J. Källne *et al*. Phys. Rev. Lett. 40, 378 (1978).

NEGATIVE PION ABSORBTION AND ISOBAR FORMATION AS OBSERVED IN STREAMER CHAMBER PARTICLE NUCLEI INTERACTIONS*

R. Allen[+], A. Kanofsky[+†], and A. Hasan[+]
Dept. of Physics, Lehigh University, Bethlehem, Pa. 18015
and
Brookhaven National Laboratory[†]
Upton, L.I., NY 11973

We discuss here negative pion interactions with nuclei as observed in a He and Ne filled streamer chamber. We have observed isobar production, and evidence is presented here for this.

If we look at the angular distributions of negative produced pions in the laboratory, we find that they are almost isotropic. This is to be contrasted with the angular distributions observed for positive pions and protons, which are peaked in the forward direction.

If we go into the c.m. system of the incoming beam particle and a fixed nucleus nucleon, we then see two distinct peaks for the c.m. momenta of the negative pions which are at the values one would expect for pions coming from the N*(1420) and $\Delta(1230)$ resonances produced at rest in this system. The negative pion momenta distribution is shown in figure 2.

If, as well, we plot up the angular distribution of the negative pions in the incoming pion nucleon c.m., we obtain a non isotropic angular distribution. In fact, this distribution obeys one expected from an Adair analysis for decay of an intermediate 1/2, 3/2, or even 5/2 resonance. The angular distribution is shown in figure 3, and we have fit both a curve of the form A $(1 + 3\cos^2\theta)$ for a J = 1/2 or J = 3/2 resonance to it and a curve of the form $A + B\cos^2\theta + C\cos^4\theta$ for a J = 5/2 resonance to it. The 2 values we obtain for these fits are 47.6 for the 3/2 resonance fit and 36 for the 5/2 resonance, for 39 degrees of freedom.

We may investigate negative pion production just from incident positive pions and protons. We have separated out the data just from these two particle types and we show the results in fig. 4. Fig. 5 gives a diagram for production.

*Supported in part by DOE under contract EY-76-5-02-2894-A002.

ISSN:0094-243X/79/540197-02$1.50 Copyright 1979 American Institute of Physics

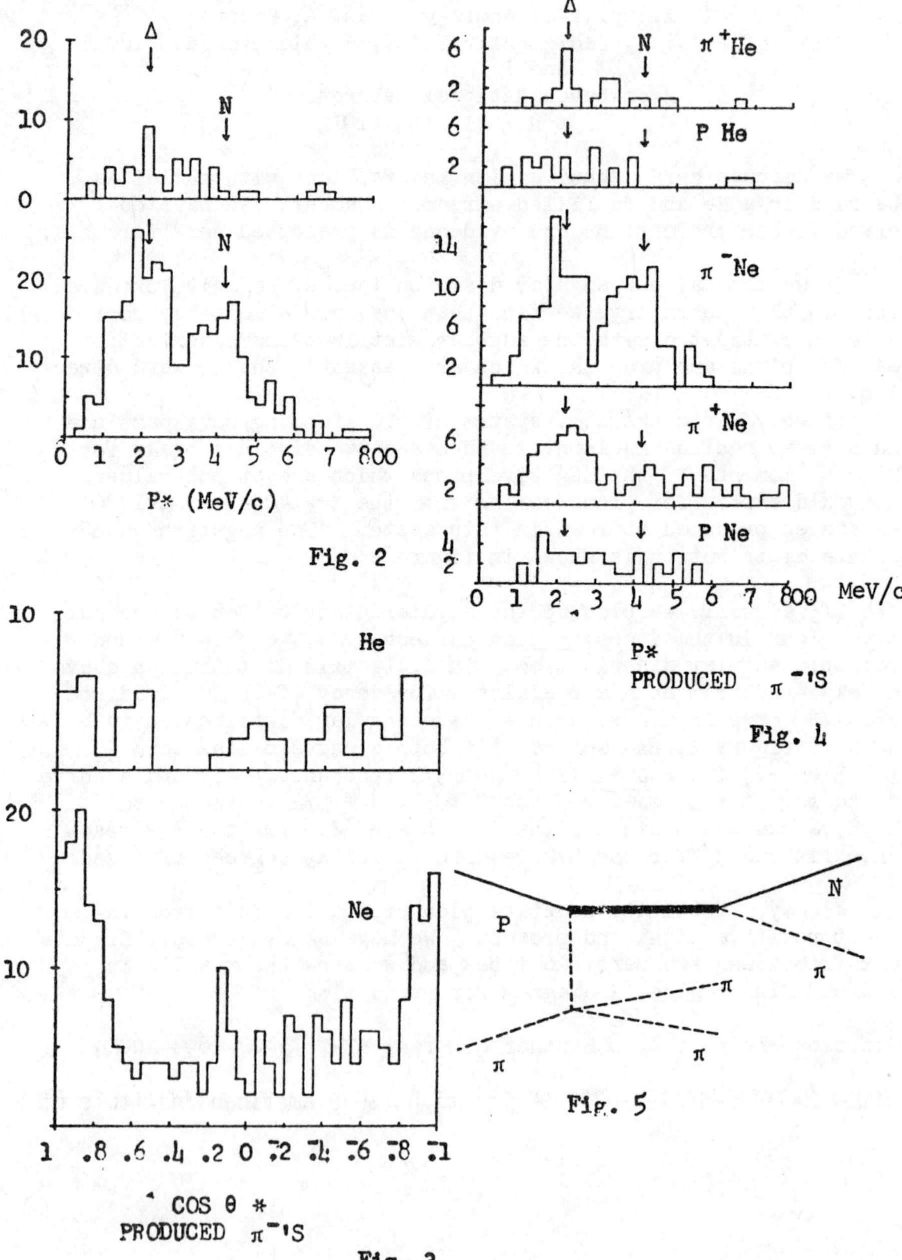

Fig. 2

Fig. 4

Fig. 3

Fig. 5

OBSERVATION OF FRANKEL EFFECT EVENTS IN PARTICLE NUCLEI COLLISIONS IN A STREAMER CHAMBER*

R. Allen[+], A. Kanofsky[+†], and A. Hasan[+]
Dept. of Physics, Lehigh University, Bethlehem, Pa. 18015
and
Brookhaven National Laboratory[†]
Upton, L.I., NY 11973

Frankel has hypothesized that the high momentum backward particles observed in medium energy nuclear reactions are the result of internal high momentum nucleons which are excited onto the mass shell by interaction of the incoming beam particle with the nucleons.[(1)] The momentum distribution of the observed backward particle is then an indication of the internal nucleon distribution.

In our case, we define a "Frankel event" as one where the leading particle momentum, as measured by the track curvature, is not more than 300 MeV/c below the incoming beam momentum. In this way, we are assured that the inelasticity of the event is not greater than about 300 MeV. In table 1 we show the fraction and multiplicity of Frankel events for the various beam and target conditions.

The magnitude of the beam momentum and leading particle momentum can only be measured with an accuracy of about 200 MeV/c because of the large radii of curvature of these tracks. The value of k, the internal nucleon momentum, is given by the momentum conservation equation for a two body collision: $p + k = p' + q$ where p is the incoming beam momentum, k is the internal nucleon momentum, p' is the inelastically scattered forward leading particle momentum, and q is the momentum of the highest momentum nuclear fragment. The vector k is very sensitive to the momenta magnitudes, and therefore we required a better determination of the magnitudes than could be provided from direct measurements of track curvature.

The value of the magnitude of the beam momentum is taken as that determined by the beam magnet settings. The magnitude of the leading particle is calculated using kinetic energy conservation: $T_p = T_q + T_{p'}$ where T_p, T_q, and $T_{p'}$ are the kinetic energies of the beam, leading particle, and highest momentum fragment respectively. This equation assumes the excitation energy given to the nucleus is small, an assumption which is supported by the small number of fragments observed in our selection of "Frankel events".

The distribution for He and Ne obtained in polar angle and momentum magnitude for the k vectors is shown in figure 1. The distribution of the momentum magnitudes is given in figure 2. This satisfies in Ne the distribution postulated by Frankel for the low momentum nucleons: e^{-k/k_o}, with k_o about 80 MeV, and a $1/k^4$ distribution, Frankel's tail, for He.

ISSN:0094-243X/79/540199-02$1.50

Our values of k for Ne do not extend up into the high momentum tail hypothesized by Frankel where his data has been obtained, but the events we observe satisfy his hypothesis that the nucleus recoils coherently and/or with very little excitation, as well as giving the hypothesized low momentum distribution.

NO. OF ADDITIONAL TRACKS IN A FRANKEL EVENT

Incoming beam type, target, momentum (GeV/c)	0	1	2	3	4	5	6	Total Frankel	Total Events	Fraction
PHe 2	22	10						32	96	33%
PHe $2\frac{1}{2}$	11	14						25	104	24%
π + He 2	19	7						26	64	41%
π + Ne 2	4	2	1				1	8	101	8%
π + Ne $2\frac{1}{2}$	4	4		1	1			10	100	10%
PNe 2	14	5						19	93	20%
PNe $2\frac{1}{2}$	32	4	2	1				39	150	26%
πNe $1\frac{1}{2}$	12	2	1					15	101	15%
πNe 2	12	2						14	87	16%
πNe $2\frac{1}{2}$	13	1	2					16	96	17%

Table 1

(1) S. Frankel and W. Frati, Phys. Rev C 16, Nr. 4 pg 1499 (1977) S. Frankel, Phys. Rev. C 17, pg 694 (1978); S. Frankel, W. Frati, R. M. Wolyshyn, and D. Yang Phys. Rev. C 18, pg 1379 (1978)

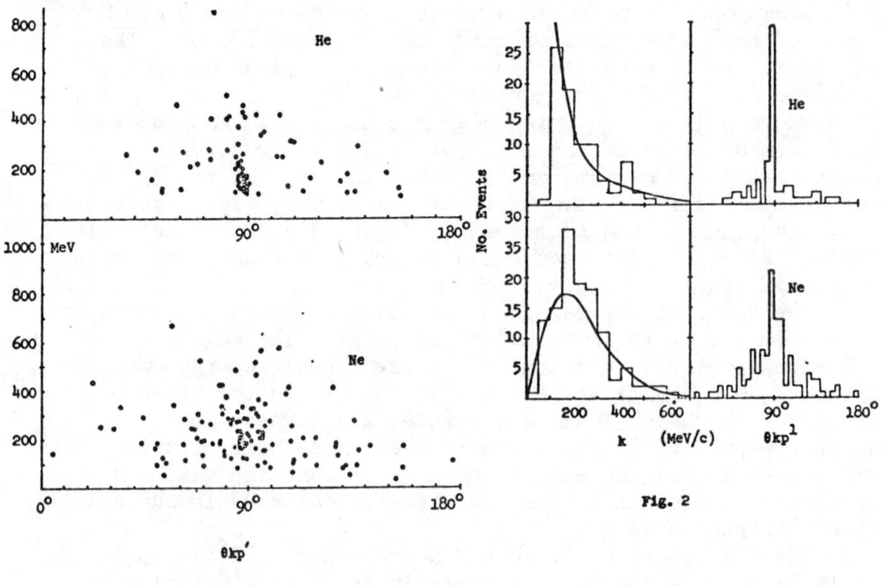

Fig. 1

Fig. 2

*Supported in part by DOE under contract EY-76-5-02-2894-A002

PARTICLE MULTIPLICITY CORRELATIONS OBSERVED IN PARTICLE NUCLEI COLLISIONS*

R. Allen[+], A. Kanofsky[+†], and A. Hasan[+]
Dept. of Physics, Lehigh University, Bethlehem, Pa. 18015
and
Brookhaven National Laboratory[†]

We present here results on particle total multiplicities and multiplicity correlations on an event by event basis as observed in a streamer chamber particle-nuclei experiment performed at Argonne National Laboratory. The experimental apparatus has been described in previous articles.[(1)] Suffice it to say here that it was possible to see all minimum ionizing tracks and fragments resulting from collisions of pions and protons with He and Ne nuclei in a streamer chamber. The trigger requirement was that a high momentum particle of approximately 1500 - 2500 MeV/c be detected in counters at 8° to the beam direction.

It was then possible to count the number of positive minimum ionizing tracks, negative minimum ionizing tracks, and heavily ionizing fragments for each event and then bin events according to the numbers of positive and negative minimum ionizing tracks and fragments.

The results are presented in table 1. We may then calculate the correlations between the various event types to, for example, determine whether the π^- multiplicity is increasing along with the π^+(minimum ionizing) and p tracks multiplicity.

We may, as well, just obtain total multiplicity distributions without correlating the various track types for each event. Some of these results are given in table 2.

We have compared these results with the theoretical predictions of Sandel and Vary which use an isobar model. We have found that the agreement is reasonable where theoretical results exist.

For the data of table 1, we may calculate correlation coefficients ρ to give the correlation between the various particle multiplicities. This linear coefficient ρ is defined as:
$\rho = \sigma(\pi^+ \pi^-)/(\sigma(\pi^+)\ \sigma(\pi^-)), \sigma(x,y) = \langle xy\rangle - \langle x\rangle \langle y\rangle,\ \sigma(x) = \sqrt{\langle x^2\rangle - \langle x\rangle^2}$
and ρ is ±1 if the correlation is exact and 0 with no correlation. The correlation coefficient for the various cases has been calculated and no strong correlations have been found.

*Supported in part by DOE under contract EY-76-5-02-2894-A002.

ISSN:0094-243X/79/540201-02$1.50

Number of MI(-) tracks in an event:

(NEON GAS)

Number of MI(+) Tracks in an EVENT:		pi+ BEAM 0	1	2	3	P BEAM 0	1	2	3	pi- BEAM 0	1	2	3
2½ GeV Beam	0	3					1			5	141	4	1
	1	135	28	3		232	18	3		5	121	42	
	2	64	102	11		123	53	3		5	35	38	2
	3	17	46	14		33	17			3	21	13	2
	4	16	21	7		8	7	1		1	4	5	1
	5	3	6	6		1	2	1			2	1	
	6	2	1	1		1					2	3	1
	7				486				504				458
2 GeV Beam	0	2	3								260	9	
	1	155	30			302	10			7	102	46	
	2	92	86	5		103	23	1		5	35	10	
	3	32	43	8		22	9			1	13	5	1
	4	8	20	2		4	1				3		1
	5	3	5	2		1							
	6	2	1	1									
	7				500				476				498
1½ GeV Beam	0									1	323	4	
	1									2	64	34	
	2									1	26	15	
	3										8	5	
	4										3		
	5												
	6												
	7												486

(HELIUM GAS)

Number of MI(+) Tracks in an Event:		pi+ BEAM 0	1	2	3	P BEAM 0	1	2	3	pi- BEAM 0	1	2	3
2½ GeV Beam	0												
	1					160	8	1					
	2					91	30						
	3					13	17	1					
	4						2	2					
	5												
	6												
	7								325				
2 GeV Beam	0	3	2			4							
	1	224	14			292	7						
	2	42	79	2		85	12	1					
	3	6	17	3		4	8						
	4	1	1	1			1						
	5												
	6												
	7				395				414				

Table 1

NOTE: These numbers are not corrected for the out-going beam particle which should appear as an MI track.

TOTAL MULTIPLICITY DISTRIBUTIONS°

Number of out going tracks	Helium p 2	Helium p 2½	Helium pi+ 2	Neon pi+ 2	Neon pi+ 2½	Neon p 2	Neon p 2½	Neon pi- 1½	Neon pi- 2	Neon pi- 2½
1 :	10	7	13	10	32	68	32	180	68	33
2 :	168	106	144	92	66	143	135	121	154	113
3 :	201	136	126	94	94	118	106	60	111	88
4 :	18	37	46	77	56	54	62	36	49	61
5 :	12	32	57	63	57	35	60	30	35	45
6 :	4	5	4	49	49	28	31	23	29	23
7 :		2	4	33	42	12	27	16	17	31
8 :				36	23	8	26	7	17	22
9 :				17	26	7	15	8	9	21
10 :				17	21	1	7	5	7	12
11 :				8	10	2	2		2	5
12 :				5	2		1			2
13 :				1	6					2
14 :					2					
TOTAL :	413	325	393	500	486	476	504	486	498	458

Table 2

FITS TO PARTICLE MOMENTA SPECTRA IN STREAMER CHAMBER EXPERIMENT WITH TWO TEMPERATURE BOLTZMANN MODEL*

R. Allen[+], A. Kanofsky[+†], and A. Hasan[+]
Dept. of Physics, Lehigh University[+], Bethlehem, Pa. 18015
and
Brookhaven National Laboratory[†]
Upton, L.I., NY 11973

We present here results of Boltzmann two temperature fits to the momenta spectra of low energy particles resulting from collisions of beam particles with the nuclei of atoms of a streamer chamber gas. The experiment utilized the streamer chamber gas both as the detecting medium and as the target, and provided 4π solid angle detection as well as detection of very low energy nuclear fragments. The apparatus has been described previously.(1)

The momentum distributions of non minimum ionizing tracks (thus excluding π's and high momenta protons) were obtained and fitted to a two temperature Boltzmann distribution of the form:

$$A\, p^2 e^{-p^2/2MT_A} + B\, p^2 e^{-p^2/2MT_B}$$

We were able to obtain these distributions for various beam particle and target nuclei conditions. The values of the parameters obtained by doing a three moment fit to the data are given in table 1 for the various configurations.

We have also obtained fits to the transverse momenta distributions:

$$A\, p_T e^{-p_T^2/2MT_A} + B\, p_T e^{-p_T^2/2MT_B}$$

These results are given in table 2. A typical fit to one of these distributions is shown in fig. 2. As well, we may obtain fits to the longitudinal momenta, p_L, distributions. The results for the p_L distribution for the various configurations is given in table 3.

The ambiguity in particle identification could result in different effective temperatures. Table 4 presents the effective temperatures resulting from the misidentification of various fragments as protons. The effect of misidentifying particles is expected to be small since most of the tracks, as determined by ionization measurements on our data, are protons.

The results are consistent with subtracting an isotropic angular distribution, which varies as $\sin\theta$, and of magnitude equal to the fraction of the low energy isotropic component, as given by the fitted values of A_1 and B_1, from the total distribution. The remaining angular distribution looks very similar to the high temperature distribution.

ISSN:0094-243X/79/540203-02$1.50

GeV/c	A	T_A	B	T_B
	P			
pHe2	.78	7.9	.22	70.9
pHe2-1/2	.81	8.2	.19	79.2
π^+He2	.67	13.3	.33	78.9
π^+Ne2	.78	8.2	.22	103.3
π^+Ne2-1/2	.79	9.6	.21	160.8
PNe2	.67	5.8	.33	42.0
PNe2-1/2	.77	9.8	.23	118.9
π^-Nel-1/2	.58	5.4	.42	68.4
π^-Ne2	.59	6.3	.41	51.0
π^-Ne2-1/2	.71	6.5	.29	71.2
He	.77	9.9	.23	77.2
Ne	.74	7.8	.26	94.0

Two Temperature Hypothesis

$F(p) = Af(p,TA) + Bf(p,T_B)$

$A+B = 1$

Table 1

GeV/c	A	T_A	B	T_B
	P_t			
pHe2	.44	3.9	.56	28.7
pHe2-1/2	.35	2.1	.65	24.5
π^+He2	.61	14.7	.39	68.7
π^+Ne2	.31	1.98	.69	17.2
π^+Ne2-1/2	.39	2.45	.61	41.6
PNe2	.15	.83	.85	12.8
PNe2-1/2	.42	4.1	.58	50.5
π^-Nel-1/2	.53	2.73	.47	34.9
π^-Ne2	.75	7.06	.25	22.2
π^-Ne2-1/2	.62	4.73	.38	55.7
He	.47	5.96	.53	37.6
Ne	.47	3.46	.53	35.5

Two Temperature Hypothesis

$F(p) = Af(p,TA) + Bf(p,T_B)$

$A+B = 1$

Table 2

GeV/c	A	T_B	B	T_B
	P_L			
pHe2	.99	11.2	.01	75.3
pHe2-1/2	*	*	*	*
π^+He2	.53	12.8	.47	17.7
π^+Ne2	.19	2.53	.81	16.3
π^+Ne2-1/2	.53	3.98	.47	39.6
PNe2	.31	3.34	.69	20.1
PNe2-1/2	.39	2.41	.61	39.9
π^-Nel-1/2	.39	2.01	.61	30.5
π^-Ne2	.41	5.55	.59	21.9
π^-Ne2-1/2	.35	3.18	.65	25.9
He	*	*	*	*
Ne	.40	3.17	.60	28.3

Two Temperature Hypothesis

$F(p) = Af(p,TA) + Bf(p,T_B)$

$A+B = 1$

Table 3

TABLE OF ERRORS MADE WHEN USING THE IDENTIFICATION OF ALL TRACKS AS PROTONS:

$m = \alpha m_p, \; q = \beta e$

	α	β	$\frac{\alpha}{\beta^2}$	T_{true}
p	1	1	1	T_p
d	2	1	2	$2\, T_p$
t	3	1	3	$3\, T_p$
He^3	3	2	3/4	$3/4\, T_p$
He^4	4	2	1	T_p

IT SHOULD BE KEPT IN MIND, THAT MOST TRACKS SHOULD BE PROTONS FROM BARRIER AND ENERGY CONSIDERATIONS.

Table 4

(1) A. Kanofsky, N.I.M. 140,433 (1977)
R. Allen, A. Kanofsky, A. Hasan APS Bulletin 23, pg. 626 (1978), 3rd International Conference on Clustering in Nuclei, AIP conf. Proc. No. 47 pg. D17, 19 (1978)

A STUDY OF THE TWO NUCLEON MECHANISM FOR PION PRODUCTION *

D. R. Giebink
Los Alamos Scientific Laboratory
and
University of Texas, Austin, TX

W. R. Gibbs and E. R. Siciliano
Los Alamos Scientific Laboratory, Los Alamos, NM 87545

ABSTRACT

The one pion exchange approximation in the reaction $p + p \to p + n + \pi^+$ provides an ideal testing ground for many of the approximations which are currently employed in the many-body environment. The reason for this is that the intermediate state pion is far off-shell and there exists a kinematic region where initial and final state NN interactions are expected to be small. We study this process using the relativistic three body helicity formalism of Wick,[1] which allows for a correct relativistic treatment while admitting the freedom to systematically address the following points: (1) The relation between the off shell πN scattering amplitude in the two body center of mass and that amplitude in the many body center of mass. (2) The effect of using different off shell prescriptions for the πN vertex. (3) The importance of s-wave and small p-wave contributions. (4) The effect of removing the nucleon pole terms from the πN scattering amplitude.

The first three points were investigated by considering their contribution to the pion asymmetry and differential cross section in the reaction $\vec{p} + p \to p + n + \pi^+$. Reasonable agreement with current cross section data was obtained in a kinematic region where the initial and final state NN interactions could be neglected. The fourth point addresses the problem of over counting exchange diagrams when correlated NN wave functions are used. A detailed theoretical investigation of this problem within a fully relativistic framework has recently been completed by Kowalski, Siciliano, and Thaler.[2] They suggest that the πN scattering vertex must <u>not</u> include either the direct or the crossed nucleon pole terms if one uses correlated NN wave functions. Without actually including NN wave functions in our calculation, this suggestion was investigated by comparing our previous calculations for the reaction $p + p \to p + n + \pi^+$ to a calculation with the nucleon pole terms explicitly removed from the scattering amplitude.

*Work performed under the auspices of the US DOE and Associated Western Universities.

1. G. C. Wick, Annals of Physics <u>18</u>, 65 (1962).
2. K. L. Kowalski, E. R. Siciliano, and R. M. Thaler, preprint.

ISSN:0094-243X/79/540205-01$1.50

NUCLEAR STRUCTURE WITH MESONS

G. E. Walker*
Indiana University, Bloomington, IN 47405

ABSTRACT

The general possibilities for studying nuclei with kaon and pion projectiles is discussed. Typical information that one might expect to obtain includes nuclear density distributions in the tail region, energy and strength distributions of giant resonances for various strangeness and isospin modes, details of the nuclear structure of states prominent in the excitation spectrum for special values of momentum transfer (such as high spin states), the masses of exotic nuclei obtained via double charge or strangeness exchange. The DWIA, despite present uncertainties associated with its validity for mesons, has been used to make theoretical predictions for inelastic and charge exchange reactions. These results are compared with experimental data. This approach currently appears to have both successes and difficulties and therefore further tests are suggested. The utility and often necessity of incorporating results obtained with other probes is stressed.

I. INTRODUCTION

There does not exist as yet a satisfactory theory of meson-nucleus interactions. It is not obvious that a multiple scattering theory, based upon a potential model, is applicable. In fact, since mesons are absorbed and emitted it is clear that the usual non-relativistic potential model is not applicable. Whether an "effective" potential can be defined and incorporated in an appropriate (perhaps field theoretic) multiple scattering theory in such a way as to handle relativistic, absorption and emission, crossing, and resonance effects is not known. Considerable theoretical effort should be directed toward investigation of this fundamentally important question.

In such a situation it may appear premature to discuss the elucidation of nuclear structure with mesons, especially if the basis for the discussion are theoretical predictions based upon the distorted wave impulse approximation which depends on a weakly motivated multiple scattering theory. However it is also unknown how important the various complications associated with meson-nucleus interactions are - how sensitive the predicted results are to features not incorporated in current theories. Many of the properties that we discuss today are associated with general features of the meson-nucleon interaction and depend primarily on nuclear spin and isospin selection rules and the gross properties of the nuclear density known from other nuclear probes. Our view

*Work supported in part by the National Science Foundation.

ISSN: 0094-243X/79/540206-016$1.50

is that a balanced approach to meson-nuclear physics should, at this stage, include application of microscopic, parameter free, theories based on the DWIA to compare with the flood of recent high quality experimental data.

In the next section we review the type of results expected for kaon-nucleus and pion-nucleus inelastic scattering based upon the properties of the elementary meson-nucleon interaction and the DWIA. In section III selected results of actual calculations for pions are presented and compared with inelastic scattering data. Appropriate related results for electron and proton-nucleus scattering is presented and found to be extremely important in interpreting the pion results. In section IV we discuss more generally the kinds of information one might expect to obtain from pion-nucleus inelastic scattering and single and double charge exchange. It is hoped that this section will serve as an introduction to the other talks in this session on giant resonances and pion single and double-charge exchange. Finally, in section V we summarize some possibilities for studying nuclear structure using kaons based on recent DWIA predictions.

II. THE DWIA AND MESON-NUCLEUS SCATTERING

The procedure used to obtain the differential cross section for meson-nucleus inelastic scattering from an initial state J_o to a final nuclear excited state J is discussed in detail in references 1-3. The resulting expression is given by

$$\frac{d\sigma_{JJ_o}}{d\Omega} = \frac{E(p_f)p_f}{E(p_i)p_i} \left| F_{JJ_o}(p_f, p_i) \right|^2 \qquad (1)$$

where $E(p_{i,f}) = (M^2 + P^2_{i,f})^{\frac{1}{2}}$ and p_i and p_f are the initial and final meson momenta, respectively. Using the DWIA, assuming the transition operator can be written as a sum of single meson-nucleon amplitudes f_{E_i}, normalized so that $(d\sigma/d\Omega)_{\text{two-body}} = |f_{E_i}|^2$, treating the nuclear wavefunctions in the TDA approximation, assuming a separable form for the elementary transition operator, and making liberal use of standard techniques for angular momentum recoupling allows one to write $|F_{JJ_o}(p_f, p_i)|^2$ as follows;

$$|F_{JJ_o}(p_f, p_i)|^2 = (2\pi)^{-4}\ 8/9 \sum_{J_z} \Big| \sum_{\ell_p \ell_h j_p j_h} \alpha^{\ell_p \ell_h}_{j_p j_h} M^{JJ_z}_{\ell_p \ell_h j_p j_h} \Big|^2 \qquad (2)$$

where

$$M^{JJ_z}_{\ell_p \ell_h J_p J_h} = \sum_i a_{n\ell_a} a_{n'\ell_b} \alpha\beta\gamma(N+\hat{N}) \qquad (3a)$$

$$\alpha = \begin{pmatrix} \ell & \ell_3 & \ell_b \\ o & o & o \end{pmatrix}\begin{pmatrix} \ell & \ell_4 & \ell_a \\ o & o & o \end{pmatrix} \hat{\ell}_3 \hat{\ell}_4 (\hat{\ell}_a \hat{\ell}_b \hat{\ell}_p \hat{\ell}_h \hat{J})^{\frac{1}{2}}\, i^{\ell_3-\ell_4}(-1)^{\ell} \qquad (3b)$$

$$\beta = (-1)^{m_b}\begin{pmatrix} J & \ell_a & \ell_b \\ -J_z & -m_a & m_b \end{pmatrix} Y^{+}_{\ell_b m_b}(r_{p_f})\, Y_{\ell_a m_a}(r_{p_i}) \qquad (3c)$$

$$\gamma = \int_o^{\infty} r^2 dr\, R_{\ell_p}(r)\, R_{\ell_h}(r)\, j_{\ell_3}(k_{n'} r)\, j_{\ell_4}(k_n r) \qquad (3d)$$

$$N = (-1)^{j_h+\frac{1}{2}}\begin{pmatrix} J & \ell_3 & \ell_4 \\ o & o & o \end{pmatrix}\begin{pmatrix} \ell_p & \ell_h & J \\ o & o & o \end{pmatrix}\begin{Bmatrix} \ell_p & j_p & \frac{1}{2} \\ j_h & \ell_h & J \end{Bmatrix}$$

$$\times \left(A^{\Delta S=0}_{\Delta T=0} + B^{\Delta S=0}_{\Delta T=1}\right) \qquad (3e)$$

$$\hat{N} = \sqrt{6}\sum_{\bar{J}}\begin{pmatrix} \bar{J} & \ell_3 & \ell_4 \\ o & o & o \end{pmatrix}\begin{pmatrix} \ell_p & \ell_h & \bar{J} \\ o & o & o \end{pmatrix}\begin{Bmatrix} \ell_p & \frac{1}{2} & j_p \\ \ell_h & \frac{1}{2} & j_h \\ J & 1 & J \end{Bmatrix}\begin{Bmatrix} \ell & \ell_4 & \ell_a \\ \ell & \ell_3 & \ell_b \\ 1 & \bar{J} & J \end{Bmatrix}$$

$$\times \hat{\bar{J}}\hat{\ell}^{\frac{1}{2}}(-1)^{\ell_p}\left(C^{\Delta S=1}_{\Delta T=0} + D^{\Delta S=1}_{\Delta T=1}\right) \qquad (3f)$$

where $i = \ell, \ell_3, \ell_4, \ell_a, \ell_b, n, n', m_a, m_b$ and $\hat{\ell} \equiv 2\ell + 1$. In eq. (3a-f), the factor β contains the angular dependence and γ contains the radial overlap integrals. The single nucleon bound state wavefunctions, $R(r)$, are normalized by $\int_o^{\infty} r^2 R^2(r)\, dr = 1$. The factors $a_{n\ell_a}$, $a_{n'\ell_b}$ arise because the configuration space meson distorted waves, $X_E^{\ell}(p_i r_k)$, have been expanded in terms of spherical Bessel functions $j_o(k_n r_k)$ where the k_n are chosen so that $j_\ell(k_N r_k)\big|_{r_k = R} = 0$.

After taking the Fourier Transform, $X_E^{\ell}(p_i,k)$ can be written

$$X_E^{\ell}(p_i,k) = \sum_n a_{n\ell}(p_i) \frac{\pi}{2k^2} \delta(k_n-k) \quad (4)$$

The $\alpha_{J_p J_h}^{\ell_p \ell_h}$ are the admixture amplitudes of the pure p-h states obtained for the configuration mixed p-h state.

For pions the transition amplitudes A, B, C, D are given by

$$A_{\Delta T=0}^{\Delta S=0} = \ell\left(f_{\ell-}^{1/2} + 2f_{\ell-}^{3/2}\right) + (\ell + 1)\left(f_{\ell+}^{1/2} + 2f_{\ell+}^{3/2}\right) \quad (5a)$$

$$B_{\Delta T=1}^{\Delta S=0} = \ell\left(f_{\ell-}^{1/2} - f_{\ell-}^{3/2}\right) + (\ell + 1)\left(f_{\ell+}^{1/2} - f_{\ell+}^{3/2}\right) \quad (5b)$$

$$C_{\Delta T=0}^{\Delta S=1} = \sqrt{\ell(\ell+1)}\left(f_{\ell+}^{1/2} - f_{\ell-}^{1/2} + 2f_{\ell+}^{3/2} - 2f_{\ell-}^{3/2}\right) \quad (5c)$$

$$D_{\Delta T=1}^{\Delta S=1} = \sqrt{\ell(\ell+1)}\left(f_{\ell+}^{1/2} - f_{\ell-}^{1/2} + f_{\ell-}^{3/2} - f_{\ell+}^{3/2}\right) \quad (5d)$$

The quantities $f_{\ell\pm}^{I}$ are the appropriate off-shell meson-nucleon amplitudes for isospin I and $J = \ell \pm 1/2$. The fully off-shell t-matrix required for calculating inelastic scattering is related to the on-shell t-matrix (assuming a separable form) by

$$T_\alpha(k_1,k_2;E(k)) = \frac{g_\alpha(k_1)\, g_\alpha(k_2)}{g_\alpha^2(k)} T_\alpha(k,k;E(k)) \,. \quad (6)$$

In order to use eqs. (2), (3a-f) for meson-charge exchange one simply sets A = C = 0, since $\Delta T = 1$ only, and for kaons (pions) multiplies the remaining cross section by a factor of two (one). The constants appearing in the equations above are correct for pions. The minor changes required for treating kaons will be given in a later section.

The ΔS and ΔT labels on A-D (eqs. 5a-d) indicate the final spin and isospin (spin-flip, iso-spin flip character) of the final nuclear excited state reached via that part of the transition operator if the ground state has S = T = 0. The terms A-D will be found to provide a simple and accurate means for predicting the spin and isospin structure of the dominant states excited by meson probes. For example, the J = 3/2, t = 3/2, p wave (ℓ = 1) pion-nucleon amplitude is the dominant component in the pion nucleon interaction in the region below 250 MeV, thus, for the purposes of obtaining qualitative estimates of the most strongly excited states we ignore the other amplitudes. This means we retain only the $f_{\ell+}^{3/2}$ term in eqs. 5a-d. We immediately obtain the result for the amplitudes $A = 4f_{\ell+}^{3/2}$, $B = -2f_{\ell+}^{3/2}$, $C = 2\sqrt{2}f_{\ell+}^{3/2}$, $D = \sqrt{2}f_{\ell+}^{3/2}$. This means that (as usual) normal parity, $T = 0$, states are predicted to be most strongly excited, but the next most strongly excited states would come from $C(\Delta T = 0, \Delta S = 1)$ and would be responsible for the excitation of high spin, non-normal parity $\Delta T = 0$ states at large momentum transfer (in contrast to the high spin $T = 1$ states excited via electrons). A knowledge of the location of these "stretched states" for both $T = 0$ and $T = 1$ as a function of the number of nucleons, as we shall discuss later, would be valuable information for the study of nuclear structure. For kaons it turns out that non-spin-isospin flip term, A, is completely dominate so essentially normal parity, $\Delta T = 0$, excited states will be seen for $K^{\pm}$ mesons.

Before examining the results of DWIA (π,π') calculations we note one further characteristic of pion and kaon induced excitations that can be seen from examining eqs. (3a-f) - these mesons cannot excite 0^- p-h states via a parity conserving transition from a 0^+ target (since the π and K have zero intrinsic spin). Thus, if one could detect characteristic γ rays from the decay of low lying 0^- states reached via (π,π') or (K,K') this should yield an estimate of the relative importance of parity non-conserving processes or parity mixing in the nuclear wavefunctions.

III. PION-NUCLEUS INELASTIC SCATTERING

One of the first considerations in discussing inelastic pion scattering is to study the sensitivity of the results to the various calculational inputs. In figure 1 we show representative results[1)] for the sensitivity of the predictions to the optical potential used to generate the distorted waves. The results are compared with experimental data for (π,π') leading to the lowest $T = 0$, 2^+ and 3^- states of ^{12}C. The solid curve was generated using distorted waves derived from the momentum space optical potential obtained by Landau and Tabakin.[4)] The dashed curve was derived using distorted waves from a Laplacian optical potential. From studies of this type and those of Gupta and Walker[2)] we conclude that the variations due to different input are not large enough to qualitatively change the shape of the angular distributions of states predicted to be relatively strongly excited. It is necessary to use distorted waves in general because of the important absorptive part of the optical potential.

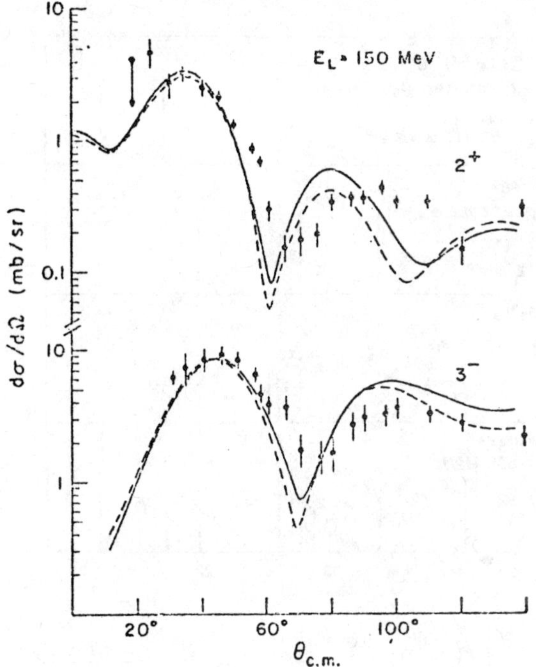

Fig. 1 - See text, figure taken from ref. 1

In figure 2 we show the predictions[2)] for $^{16}O(\pi,\pi')^{16}O$ for T_π(lab) = 180 MeV. In all figures shown subsequently in this talk T = 1 (T = 0) states are represented by solid (dashed) lines. The giant dipole T = 1, J = 1^- state is predicted to dominate the spectrum at forward angles (low momentum transfer). At large momentum transfers high spin states, as expected, dominate the spectrum. The 4:1 ratio predicted above for the 4^-, T = 0 to 4^-, T = 1 cross section is essentially reproduced in the detailed calculations.

The stretched high spin states such as the 4^- states in ^{16}O are attractive candidates for study because they are pure particle-hole states in a simple model and are predicted to dominate the large momentum spectrum for a variety of probes. Another example of such a stretched state is a 6^- state in ^{28}Si. In the TDA this state might be described as a pure $1f_{7/2}\,(1d_{5/2})^{-1}$ configuration. The 6^- T = 1 state was seen earlier in inelastic electron scattering where a reduction factor of the theory between two and three is required to give agreement between theory and experiment.[5)] This is presumably due to oversimplified nuclear

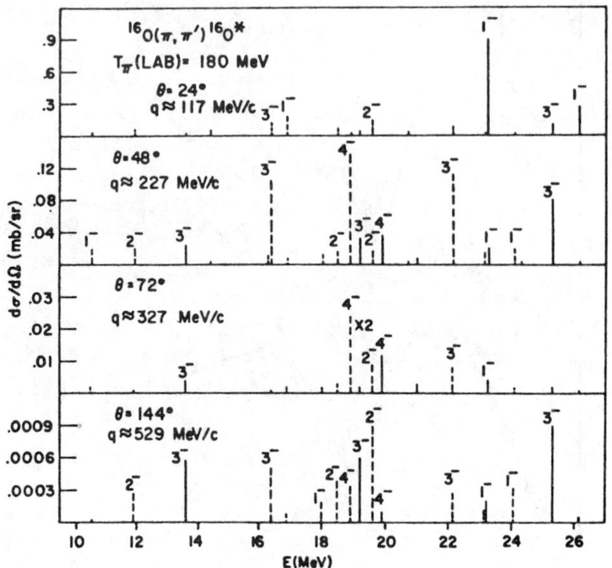

Fig. 2 - $^{16}O\ (\pi,\pi')\ ^{16}O^*$ for T_π(lab) of 180 MeV

wavefunctions. The 6^-, T = 0 state is not seen in electron scattering because of the dominance of the isovector magnetic moment.

Recently the 6^- T = 1 and T = 0 states have been studied using the (p,p′) reaction at IUCF.[6] In fig. 3 we show a comparison between theory[7] and experiment. The theoretical results use the DWIA and a nucleon-nucleon transition operator obtained by Picklesimer and Walker.[7] Note that a reduction factor of ~ 2.72 is required of the theory for agreement with experiment. In this case both the T = 1 and T = 0 states are strongly excited; the T = 0 state has a strong contribution from the exchange term.

At this preliminary stage we conclude that there is reasonable consistency between the electron and proton results and that one can use the DWIA and (p,p′) to study states - such as the 6^-, T = 0 state, not easily seen using inelastic electron scattering. One interesting problem is that one predicts that the 6^-, T = 1 state should be strongly excited in (p,p′) on ^{24}Mg, ^{28}Si, ^{32}S, and ^{40}Ca. It is apparently only seen in ^{24}Mg and ^{28}Si.[8] We know of no satisfactory explanation for its absence in the ^{32}S and ^{40}Ca spectrum. Experimental results using electron and pion probes for these nuclei would be, potentially, very useful.

In the past few weeks experimental results[9] and theoretical predictions[10,11] for the 6^-, T = 0 and T = 1 states excited in ^{28}Si (π,π') ^{28}Si have become available. The comparison between theory and experiment is shown in fig. 4. The main difference

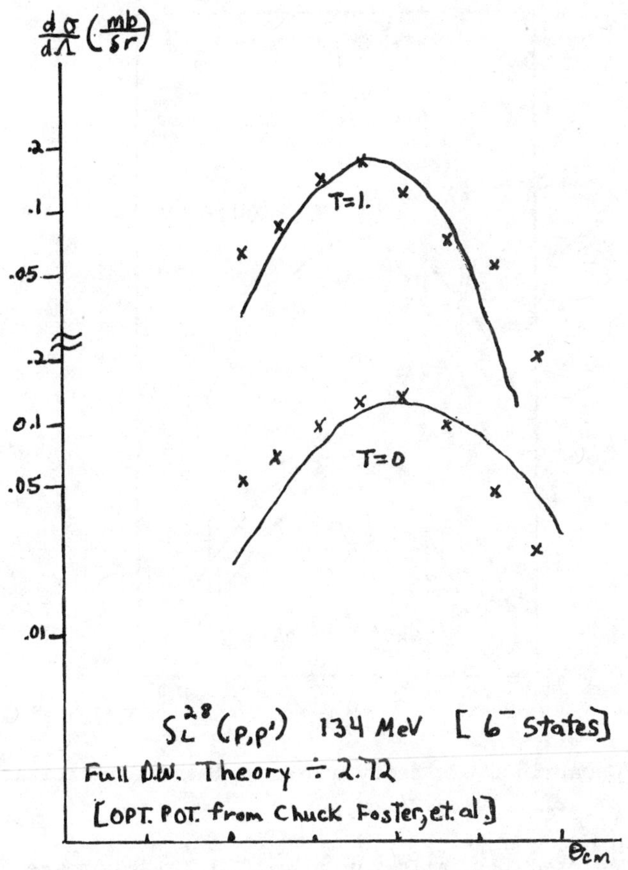

Fig. 3 - Comparison between theory and experiment for the 6^- T = 0, 1 states excited in ^{28}Si (p,p′) ^{28}Si*

between the two theoretical curves results from the use of a different oscillator parameter for the harmonic oscillator single particle wavefunctions (Siciliano adopted b = 1.8fm, Lee 1.64fm). The larger oscillator parameter is motivated by electron scattering results and coulomb energy differences. It is interesting to see the sensitivity of the results to this variable. More realistic calculations using Saxon-Woods wavefunctions are underway. At this early stage however one can see that the pion DWIA results are in rough qualitative agreement with experiment when one compares the renormalization factors with those required using electrons and protons. There are some discrepancies between the probes and it is premature to state definitely that these differences are primarily due to the inadequacy of the DWIA for calculating (π,π').

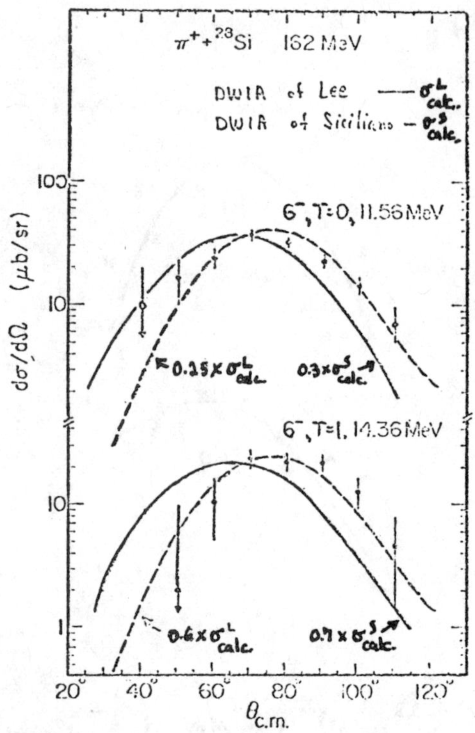

Fig. 4 - Comaprison of theory and experiment for excitation of 6^- states in $^{28}Si\ (\pi,\pi')\ ^{28}Si^*$

The predicted 4:1 ratio for spin-flip T = 0 compared to spin-flip T = 1 nuclear excitations is based on the $P_{3/2}$ dominance and assumes the same spin and configuration space wavefunctions for the states being compared (i.e. only the isospin wavefunctions differ). Boudrie et al[11] have determined the ratio of the 1^+ T = 0 (12.71 MeV) state to the 1^+ T = 1 (15.1 MeV) state excited in (π,π') on ^{12}C. For example the ratio is 3 at T_n = 116 MeV and 1.4 ± .5 at 180 MeV. The fact that the ratio is not 4:1 could be attributable to the fact that the configuration space-spin wavefunctions of the two states are not identical. However the fact that the ratio changes so much in a region where the $p_{3/2}$ wave is still dominant implies that other complications such as multistep processes, pion absorption in the tail region, etc. may be non-negligible. Clearly the ability to vary the pion kinetic energy is vital in clarifying whether nuclear structure or reaction mechanism complications are present.

The pion can be used to elucidate isospin mixing in selected excited states of self conjugate nuclei. Because the J = 3/2, t = 3/2 partial wave is dominant in the pion-nucleon interactions, the π^+-proton (π^--neutron) interaction is much stronger than the

π^+-neutron (π^--proton) interaction. This result can be exploited to study isospin mixing in self-conjugate nuclei by subtracting the π^+ results from the π^- results for a given set of states. A specific example is shown in fig. 5.[13] If isospin were a good quantum number there would be no significant difference for the $\pi^\pm$ results for a given state. The fact that the π^+ cross section is significantly larger for the 4^- (19.25 MeV) state in ^{12}C and the π^- cross section is dominant for the 4^- (19.65 MeV) state implies there is significant isospin mixing between these two states. We are currently in the process of using these results to make quantitative estimates of the isospin mixing. It is important to consider several effects of coulomb mixing including the different tails in the proton and neutron single particle wavefunctions. This technique shows promise for studying isospin mixing in other light self conjugate nuclei.

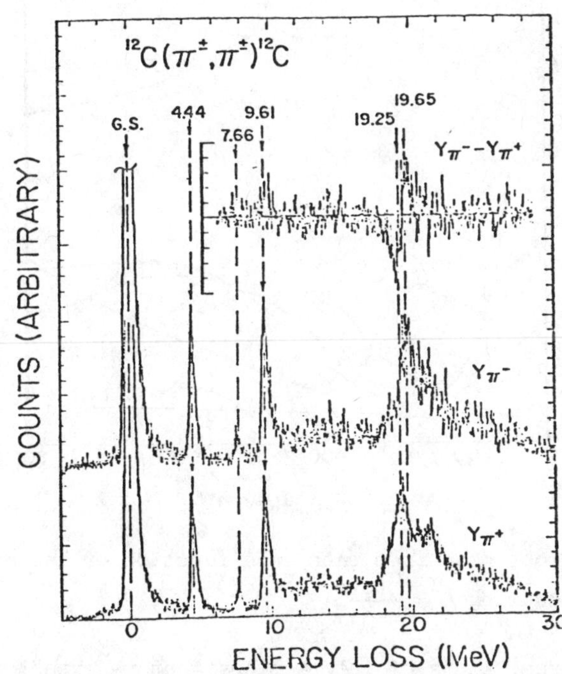

Fig. 5 - Difference of π^+ and π^- cross sections in ^{12}C conjectured to be due to isospin mixing in the two 4^- states.

IV. ADDITIONAL INFORMATION USING PIONS

In this section we briefly discuss some selected examples of the kinds of information one might expect to obtain using pions. One purpose of this section is to introduce the topics to be discussed by the subsequent speakers in this session. One of the obvious main points to be kept in mind is that the pion results will be most useful when compared with results obtained using other probes. In fig. 6 we show the mean free path of various probes in nuclear matter as a function of laboratory momentum. In addition to each probe's characteristic spin and isospin dominance rules the strength of the elementary interaction as a function of lab. momentum points out the different ranges with which the various probes penetrate into the nucleus and the different energy dependence of such complications as multistep processes.

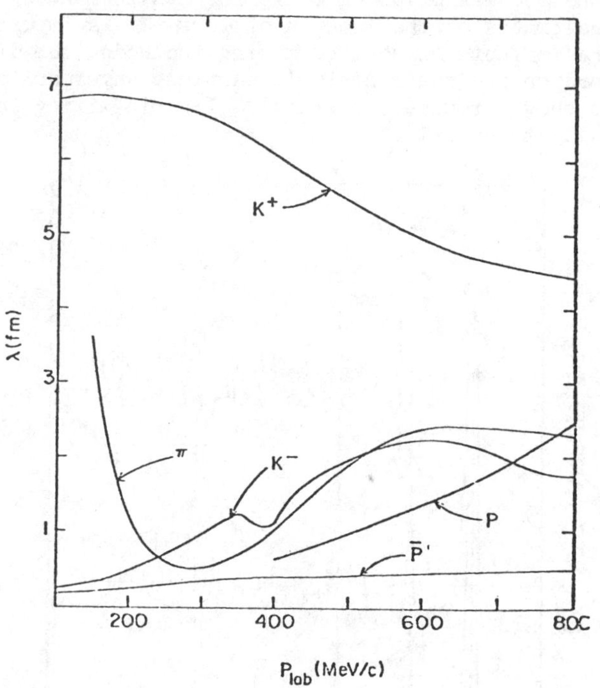

Fig. 6 - Probe mean free path vs a function of lab momentum assuming $\lambda = \frac{\sigma_{tot}}{\rho_0}$

The different kinds of interactions that we expect to consider and compare in intermediate energy physics is quite large. Some selected examples are shown below. Those for which spin flip may be unusually important are so designated. These processes can, of course, be used to study individual states or giant resonances. Dr. Bertrand will discuss the general situation with respect to giant resonances and Dr. Cooper will discuss the charge exchange reaction. For charge exchange it is useful to keep in mind that new information from IUCF on the (p, n) reaction will provide useful complementary data. As will be discussed in the next section kaon charge exchange data should be available and extremely useful as we look toward the future.

(N-1,Z+1)	(N,Z)	N+1,Z-1
(π^+,π^o) S.F. high q only	(e,e') S.F. high q	(π^-,π^o) S.F. high q only
(p,n) S.F.	(p,p')	(n,p) S.F.
(γ,π^-) S.F.	$(K^\pm,K^\pm{}')$	(π^-,γ) S.F.
(π^+,γ) S.F.	(π,π')	(μ^-,ν) S.F.
(K^+,K^o)		(γ,π^+) S.F.
		$(K^-,\overline{K}^o)$

Charge Exchange ⇒ (Isospin Identifier)

Finally, in this session, Dr. Seth will discuss pion double charge exchange. The exciting new data he will present can be used in a variety of ways. There has always been interest in this reaction since, because of its two step nature there is the possibility of studying the two nucleon correlation function. Because of the many uncertainties associated with this reaction it may not yield "clean" information on correlations in the foreseeable future. However, starting from a T = 0 target just locating the energy difference between the T = 2 ground state and low lying T = 2 excited states is of considerable interest in nuclear structure physics. In addition, Seth may discuss how the (π^+,π^-) reaction may be used to measure the masses of exotic nuclei such as ^{48}Ar, ^{26}Ne and perhaps ^{7}H and ^{7}He.

V. KAONS AS PROBES OF NUCLEAR STRUCTURE

Because kaons possess non-zero strangeness, S (S = ±1 for $K^\pm$), which is conserved in strong interactions, the K^+ and K^- differ significantly in their strong interaction with the nucleus. This situation is in strong contrast to that for the π^+ and π^-. The basic difference in the kaons arises because the only "light" strange baryons stable with respect to strong interactions have S = -1 (Λ and Σ) and therefore the two-body channel, K^+ + nucleon, has no open inelastic channels below pion production threshold. The K^- + nucleon channel does have open channels (even at zero energy!)

As in the case of the (π,π') predictions presented, the $K^\pm$ calculations reported[3)] utilize the DWIA, TDA, and assume a separable form for the off-shell kaon-nucleon transition t-matrix. Once again, s, p, and d waves are included in the elementary transition amplitude. We find that the kaon-nucleon interaction is dominated by s waves in the region of interest (the interaction has, apparently a short effective range) and is relatively weak and energy independent. Thus, one may hope that the K^+-nucleus interaction should be relatively easily understood and <u>quantitatively</u> reliable calculations should be possible (if the appropriate two-body data is available).

Only minor modifications of eqs. 2-5 are necessary for applying them to kaon-nucleus interactions. All the modifications result because the isospin Clebsch-Gordon sums have been evaluated in these equations and the kaon has isospin 1/2 while the pion had unit isospin. The changes to be made are

a) replace the constant 8/9 by 1/2 in eq. (2)
b) redefine the terms A-D given by eqs. (5a-d) so that for kaons we have:

$$A^{\Delta S=0}_{\Delta T=0} \equiv \ell(f^0_{\ell-} + 3f^1_{\ell-}) + (\ell + 1)(f^0_{\ell+} + ef^1_{\ell+}) \tag{5a'}$$

$$B^{\Delta S=0}_{\Delta T=1} \equiv \ell(f^0_{\ell-} - f^1_{\ell-}) + (\ell + 1)(f^0_{\ell+} - f^1_{\ell+}) \tag{5b'}$$

$$C^{\Delta S=1}_{\Delta T=0} = \sqrt{\ell(\ell+1)}\,(f^0_{\ell+} - f^0_{\ell-} + 3f^1_{\ell+} - 3f^1_{\ell-}) \tag{5c'}$$

$$D^{\Delta S=1}_{\Delta T=1} = \sqrt{\ell(\ell+1)}\,(f^0_{\ell+} - f^0_{\ell-} - f^1_{\ell-} - f^1_{\ell+}) \tag{5d'}$$

where now the quantities $f^I_{\ell+}$ are the "lab" off-shell kaon-nucleon amplitudes for isospin I = 0,1 and J = $\ell \pm 1/2$.

The first experiments, which are currently being performed at Brookhaven, will utilize an 800 MeV/c $K^{\pm}$-beam and so we now turn our attention to this energy region. An examination of the terms $|A|^2$ --- $|D|^2$ suggests that T = 0, normal parity states will dominate the nuclear response at all momentum transfers however spin-flip transitions should not be completely negligible. The results of a DWIA calculation for $^{12}C(K^+,K^{+\prime})^{12}C^*$ is shown in fig. 7. The giant dipole T = 1 state dominates the spectrum at low q but except for this collective state the predictions obtained by looking at $|A|^2$ --- $|D|^2$ are reenforced by the more detailed calculations.

We conclude our brief discussion of K^+ inelastic scattering by making some suggestions on how to use the K^+ in connection with some other probes. Since the K^+ predictions should be reliable one should compare the K^+ results with those obtained using the other "reliable" probe, the electron, for the low excitation, 2^+ and 3^-, T = 0, states in ^{12}C. The same wavefunctions which give agreement with the electroexcitation experiments (remember the RPA was required) should also yield satisfactory agreement for K^+ inelastic scattering. If they don't, something may be wrong with our understanding of the reaction mechanism for one of these two probes. Note, that, especially at larger momentum transfers, exchange currents may play a role in the electromagnetic experiments.

At higher excitation energies the characteristic T = 0, normal parity states excited by the K^+ should be compared with the $\Delta T = 0$, $\Delta S = 0$ states excited by (α,α') where it is not clear whether one is dealing with single nucleon or cluster excitations.

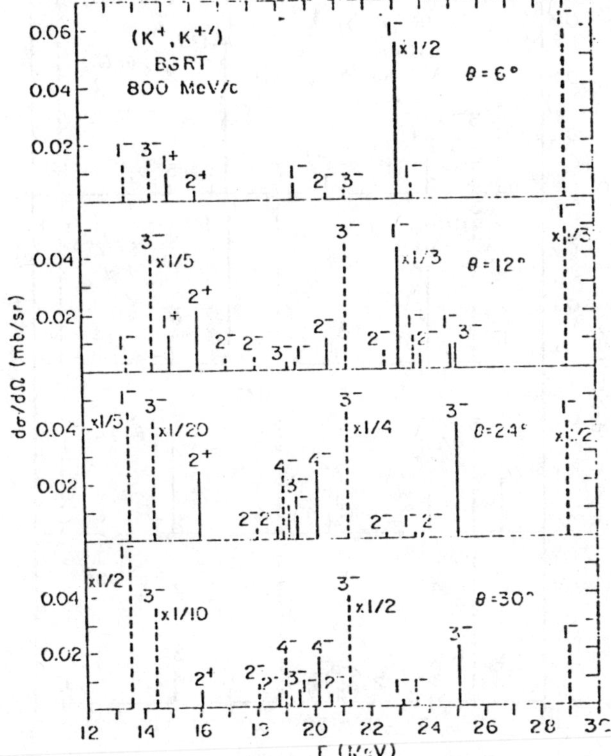

Fig. 7 - $^{12}C(K^+,K^{+\prime})^{12}C^*$ at P_{lab} = 800 MeV/c

Another example of the utility of the K^+ may be its use in investigating the concentration of strength associated with the isoscalar monopole and quadrupole resonances (T = 0, normal parity). Again by doing charge exchange, (K^+,K^0) on T = 0 nuclei all isoscalar excitations should disappear from the spectrum.

Typical results of the 800 MeV/c $^{12}C(K^-,K^{-\prime})^{12}C$ calculations are shown in fig. 8.[3] The excitation spectrum is very similar to the results obtained earlier, also using the DWIA, for $(K^+,K^{+\prime})$ with normal parity 1^- and 3^- states being dominant. This is important because, as discussed earlier, the spectroscopy of such states may be expected to be well understood from electron and K^+ inelastic scattering and thus one will be learning about the K^- reaction mechanism by studying these states via $(K^-,K^{-\prime})$. Once one has a better understanding of the K^- reaction mechanism one can use K^- charge exchange (as discussed below) to excite and study states not easily seen with other probes.

We conclude by briefly discussing the utility of the process $(K^-,\bar{K}^0)$. Starting from a $T = -T_z$ neutron excess ground state and using this change exchange reaction, then the final states reached would have $T_f \geq T_z + 1$. Two reasons for studying these isospin

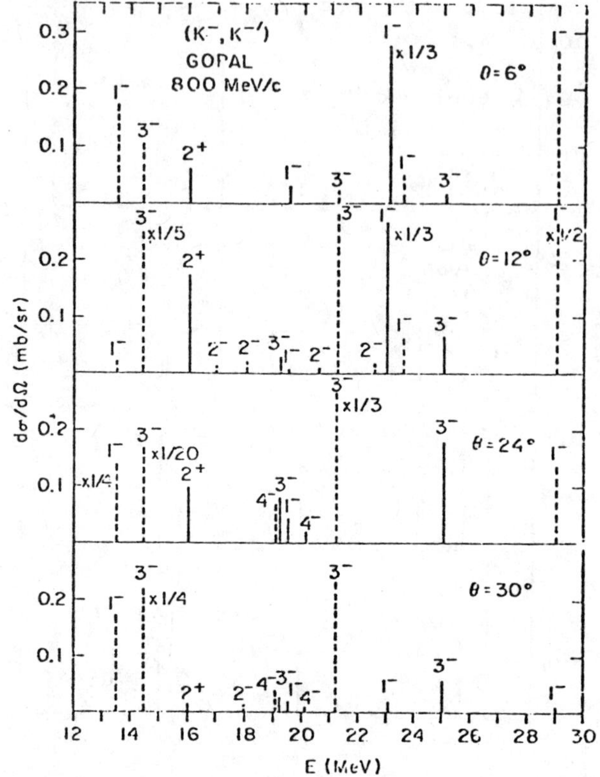

Fig. 8 - $^{12}C(K^-,K^{-\prime})^{12}C$ for P_{lab} = 800 MeV/c

stretched configurations are to investigate the energy and transition strength splitting of the $T_>$ and $T_<$ giant dipole resonances and also to obtain the same information regarding stretched angular momentum configurations. For those interested, more detailed calculations and results are presented in ref. 3.

REFERENCES

[†]Work supported in part by the NSF

1. T. S. H. Lee and F. Tabakin, Nucl. Phys. A226, 253 (1974).
2. M. K. Gupta and G. E. Walker, Nucl. Phys. A256, 444 (1976).
3. C. B. Dover and G. E. Walker, accepted for publication in Phys. Rev. C.
4. R. H. Landau and F. Tabakin, Phys. Rev. D5, 2746 (1972).
5. T. W. Donnelly, J. D. Walecka, G. E. Walker, and I. Sick, Phys. Lett. 32B, 545 (1970).
6. G. S. Adams et al., Phys. Rev. Lett. 38, 1387 (1977).
7. A. Picklesimer and G. E. Walker, Phys. Rev. C17, 237 (1978).
8. A. D. Bacher and G. T. Emery (private communication).

9. Argonne - Los Alamos collaboration, private communication from D. Geesaman and B. Zeidman.
10. E. Siciliano (private communication).
11. T. S. H. Lee (private communication).
12. R. L. Boudrie et al. (private communication).
13. Private communication by W. Braithwaite, W. B. Cottingame, and C. L. Morris (EPICS data).

PION SINGLE CHARGE EXCHANGE

Martin D. Cooper
University of California
Los Alamos Scientific Laboratory, Los Alamos, NM 87545

ABSTRACT

The LAMPF $\pi°$ spectrometer is described, and the first results of pion single charge exchange to discrete nuclear states are discussed. Angular distribution information has been obtained in ^3He and 0° excitation functions for the isobaric analog state have been observed in targets throughout the periodic table.

INTRODUCTION

The direct detection of neutral pions for nuclear spectroscopy has, for the first time, been accomplished in the past year. Their detection has been made possible by the development of the LAMPF $\pi°$ spectrometer, a joint project of the Los Alamos Scientific Laboratory, Case Western Reserve University and Tel Aviv University.[1] The resolution of the spectrometer is not comparable to charge particle spectrometers, although it is possible to conceive of a device which would compare favorably in both resolution and solid angle with magnetic spectrometers. The present spectrometer has now achieved a resolution of 2.5 MeV with a solid angle of about 1 msr. After its upcoming overhaul, it will obtain better than 2-MeV resolution with a 3 msr solid angle.

This talk will cover the operation of the $\pi°$ spectrometer and the results for analog charge exchange transitions in a variety of nuclear targets across the periodic table. The data exist at several energies for 0°, and there are angular distributions in the light elements. Unfortunately, the data analysis is not complete and only the qualitative features of the data can be presented. The targets include ^3He, ^7Li, ^{13}C, ^{27}Al, ^{58}Ni, ^{90}Zr, ^{120}Sn, and ^{208}Pb.

THE $\pi°$ SPECTROMETER

The principles of good resolution $\pi°$ detection are a good measurement of the angle between the decay γ-rays, a modest measurement of the energies of each γ-ray, and a restriction to nearly equal energy γ-rays. These ideas may be illustrated from the $\pi°$ decay kinematics in Fig. 1. The expressions for the total $\pi°$ energy W are

$$W = m_{\pi°} \sqrt{\frac{2}{(1-\cos\eta)(1-X^2)}} \quad \text{and} \quad X = \frac{E_2-E_1}{E_2+E_1} \quad , \qquad (1)$$

where η is the opening angle between the γ-rays and X is the energy-

ISSN: 0094-243X/79/540222-08$1.50

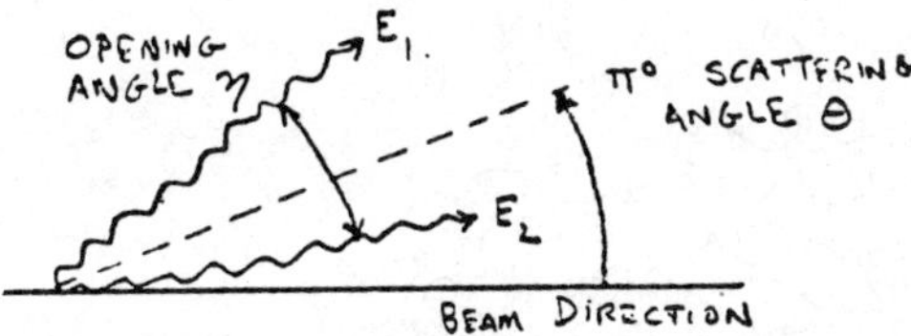

Fig. 1. The kinematics of π° decay.

sharing fraction. The expression for W has a weak dependence on X which is minimal for X=0. Thus, the best resolution is obtained for values of X near zero. The design of the spectrometer which allows better than 2-MeV resolution requires an accuracy for η of 3 mrad (FWHM) and for E_i of 25% (FWHM); the measured value of $|X|$ must be restricted to be less than 0.1. To date, the best η resolution obtained has been 5 mrad and the best γ-ray energy resolution obtained has been 30%. The resulting good resolution π° spectrum from CH_2 is shown in Fig. 2. The π° scattering angle is given by

$$\cos\theta_{\pi^\circ} = [(1+X)\cos\Theta_2 + (1-X)\cos\Theta_1]/2\beta_{\pi^\circ} \quad , \qquad (2)$$

where the Θ_i are the measured polar angles of the γ-rays and β_{π° is the π° velocity. Since the dependence on X is linear in Eq. 2, the scattering angle is only measured to about a degree (depending on the scattering angle).

The γ-ray detector is illustrated in Fig. 3. The active elements include a veto counter to eliminate charged particles, three position sensitive modules, and a 14-radiation-length total absorption lead glass counter. The position sensitive modules consist of 0.68 radiation lengths of lead glass, followed by 3 multiwire proportional chambers (MWPC) and a trigger scintillator. The thickness of lead glass is a compromise between conversion efficiency and shower complexity. Two of the three MWPCs are used to measure position in the plane containing the opening angle. Extrapolating the most forward-going track 1/6 the way back into the converter, a resolution of 2 mm (FWHM) may be obtained. The total absorption counters are segmented into 15 pieces arranged in a 5 x 3 array. Each piece has a 15 cm x 15 cm cross section. Each lead glass element is viewed by an EMI 9618R phototube; the converters have light pipe couplings. Both of the photon detecters are mounted on portable carts which allow the position of the MWPC's to be known relative to the target to 1 mm.

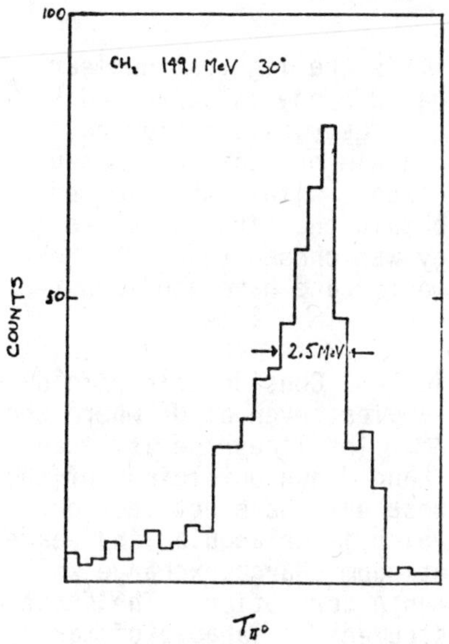

Fig. 2. A good resolution spectrum from CH_2.

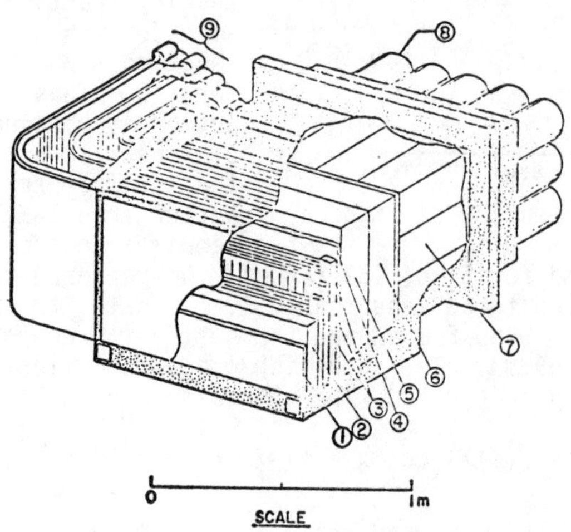

1. VETO SCINTILLATOR
2. CONVERTER #1, 0.68 R.L. GLASS
3. MWPC, X,Y,X PLANES
4. TRIGGER SCINTILLATOR
5. DETECTOR MODULE #2, REPEAT OF ITEMS 2,3,4
6. DETECTOR MODULE #3
7. LEAD GLASS (LF 5), 14 R.L.
8. 12.7 cm PM TUBE
9. PM TUBES FOR CONVERTERS AND SCINTILLATORS

Fig. 3. The active elements of one arm of the $\pi°$ spectrometer.

$^3He(\pi^-,\pi°)^3H$

^{3}He was chosen as a target because it is the lightest nuclear analog transition and the continuum charge exchange is separated by 5 MeV from tritium. Some back angle data already exists from recoil experiments,[2,3] but there were no forward angle or continuum measurements which contain most of the cross section. Within the limited life of the ^{3}He target, the goal was to obtain the first complete $\pi°$ angular distribution. The incident energy was chosen to be 200 MeV in order to join onto the recoil measurements[3] and have good counting rate.

The spectra obtained are shown in Fig. 4. Considerable unfolding of the continuum is necessary at all angles, even at 0° where the data actually includes everything up to 7°. The lineshape is taken from a CH_2 target at 0°. The curves are hand-drawn unfoldings of the analog and continuum charge exchange. These data have not been corrected for the spectrometer acceptance, which is triangular and peaked near 200 MeV. Thus, the shape of the continuum charge exchange will be considerably modified after the acceptance correction. The spectra point out the type of data which this instrument is capable of taking, i.e., covering a wide range of energies and angles. To obtain count rate, the resolution was compromised.

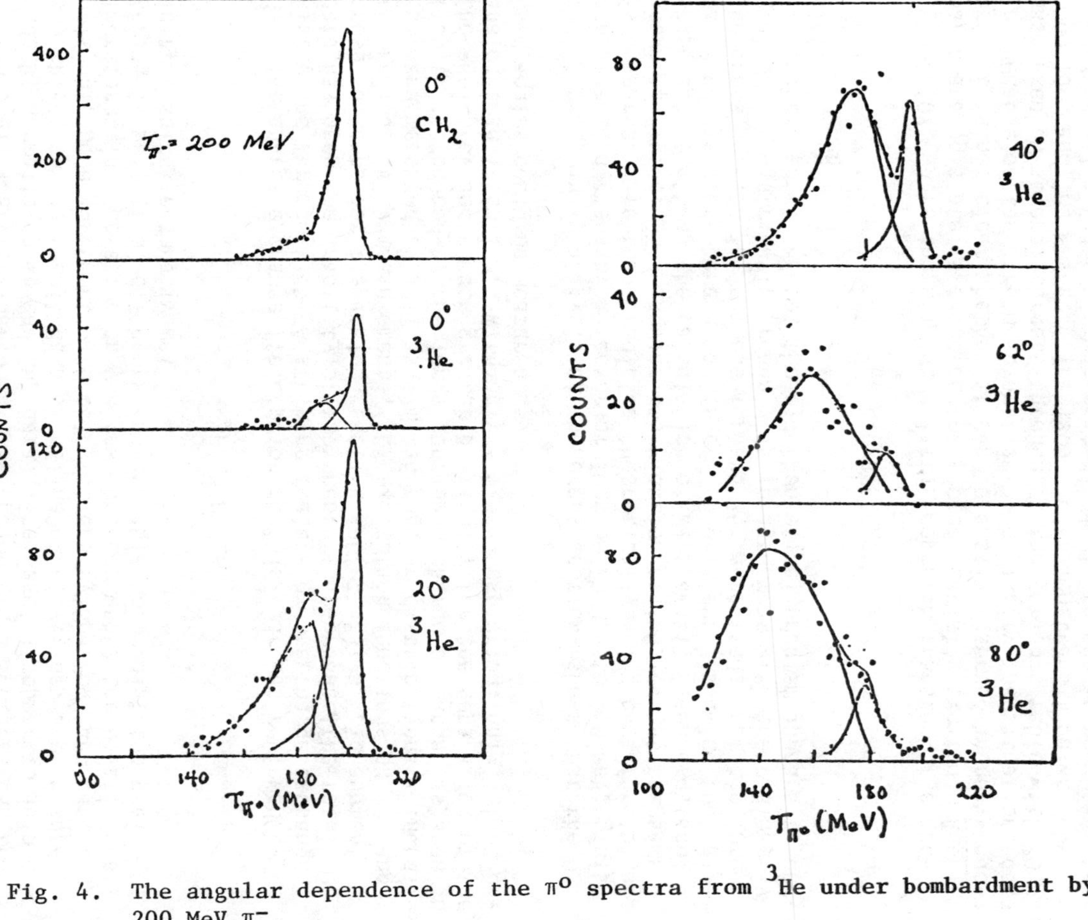

Fig. 4. The angular dependence of the π^0 spectra from ^{3}He under bombardment by 200 MeV π^-.

A preliminary evaluation of the analog cross sections can be made by normalizing to the CH_2 data. The results appear to join smoothly onto the recoil data of Källne et al.[3] They are in surprisingly good agreement with calculations by Sparrow[4] and Landau.[5] Therefore, in a case for only a few nucleons where the elastic scattering is well studied, theory appears able to predict the preliminary results reasonably well.

P-SHELL NUCLEI

The charge exchange field prior to this year was reviewed by Alster and Warszawski.[6] A striking feature of the charge exchange data has been the inability of the theory to reproduce the total cross sections for the analog reaction $^{13}C(\pi^+,\pi^\circ)^{13}N$. To augment the radioactivity measurements,[7] 0° excitation functions on ^{13}C and ^{7}Li were measured as well as limited angular distributions. The difficulty introduced by the large continuum charge exchange (as seen in the ^{3}He) provided a much greater difficulty for the separation of the analog transition ^{13}N. With only 2.5-MeV resolution, only a few forward angles could be measured.

A preliminary analysis of the cross sections shows that the 0° cross section rises by a factor of 10 from 70 MeV to 150 MeV. This is a sharper rise than predicted theoretically. The data for ^{7}Li are shown in Fig. 5. The angular distribution at 150 MeV has the same shape out to 45° as the calculation of Warszawski, Eisenberg, and Gal,[8] but is everywhere a factor of 2.5 greater. Assuming that all the cross section is in these forward 45°, one finds that the integral of the distribution roughly is in agreement with the radioactivity measurements. Hence, the old theoretical problem appears to persist.

A-DEPENDENCE

At 0°, where there is very little momentum transfer, the isobaric analog state is expected to be preferentially excited. *A priori*, it is not clear whether it can be easily observed in heavy nuclei because of the strong absorption. To clarify the situation, data was taken on several targets throughout the periodic table with 70, 100, 150, and 180 MeV π^+. The isobaric analog transition was observed in all targets; it was most difficult to see in the heavy targets as shown in Fig. 6. These data prove that when pions preferentially excite one nuclear state, π° detection can teach us about the charge exchange physics even in heavy nuclei.

The relative strength of the isobaric analog excitation is shown in Fig. 7. The shape of the A-dependence can be qualitatively understood as follows. While $N \approx Z$, the cross section drops because the increased absorption takes flux from all channels. As N becomes substantially greater than Z, the excess neutrons provide a new source of reaction channels for the charge exchange, and the cross section rises again.

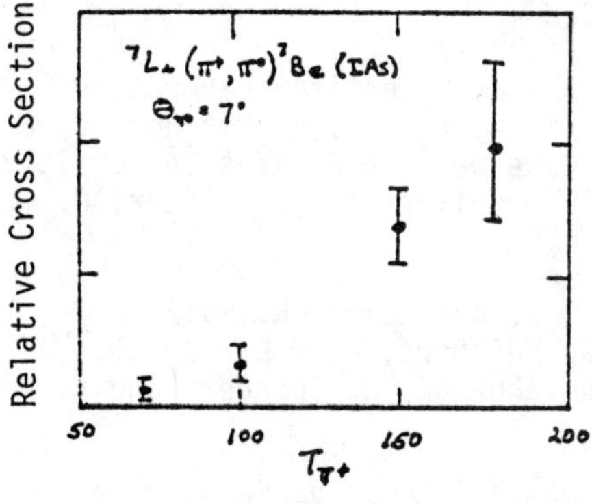

Fig. 5. The 7° excitation function for $^{7}Li(\pi^{+},\pi^{\circ})$ ^{7}Be (IAS).

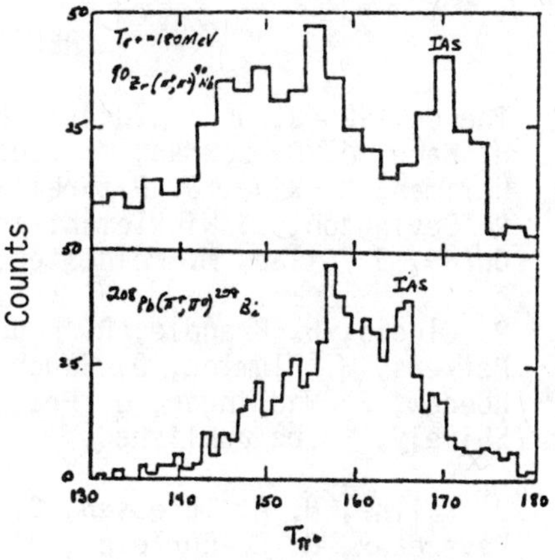

Fig. 6. The isobaric analog state as seen in the 0° spectra from the reactions a) $^{90}Zr(\pi^{+},\pi^{\circ})^{90}Nb$ and b) $^{208}Pb(\pi^{+},\pi^{\circ})^{208}Bi$ at $T\pi$ = 180 MeV.

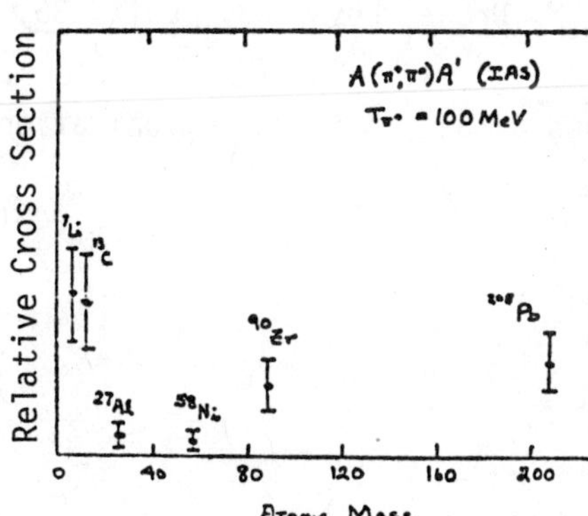

Fig. 7. The relative excitation strength of the isobaric analog state at T_{π} = 100 MeV as a function of mass, for $\theta_{\pi^{\circ}}$ = 7°.

CONCLUSIONS

The success of the π° spectrometer has provided the first pion charge exchange data to discrete nuclear states. Soon, some of the analysis problems will be solved, and these data will be converted to cross sections for detailed comparisons to theory. More experiments on a variety of other targets, energies and angles are in the offing. The resulting angular distributions should highlight the reaction mechanism and the mass-energy dependence should tell us about the strength of the isovector pion operator in the nuclear medium. The prospectus is very exciting that a new field of pion-nuclear physics is opening before us.

REFERENCES

1. The collaboration includes: Los Alamos Scientific Laboratory; H. Baer, J. D. Bowman, M. Cooper, F. Cverna, R. H. Heffner, C. M. Hoffman, N. King, J. Pifarelli: Case Western Reserve University; P. Bevington, E. Winkleman: Tel Aviv University; J. Alster, A. Doran, S. Gilad, M. Moinester.

2. P. Glodis, H. Brandle, R. Haddock, I. Kostoulas, N. Matz, B. Nefkens, W. Plumlee, O. Sander, R. Meinhart, J. McCarthy, D. Roeder, E. Wadlinger, J. Pratt, R. Sherman, J. Spencer, and F. Shively, to be published.

3. J. Kallne, H. A. Thiessen, C. L. Morris, S. L. Verbeck, M. J. Devereaux, G. R. Burleson, J. S. McCarthy, R. R. Whitney, J. E. Bolger, C. F. Moore and C. A. Goulding, to be published.

4. D. A. Sparrow, Phys. Lett. B58, 309 (1975).

5. R. Landau, Workshop on Pion Single Charge Exchange, Los Alamos (1979).

6. J. Alster and J. Warszawski, to be published in Physics Reports.

7. Y. Shamai, J. Alster, D. Ashery, S. Cochavi, M. A. Moinester A. I. Yavin, E. D. Arthur, and D. M. Drake, Phys. Rev. Lett. 36, 82 (1976).

8. J. Warszawski, J. M. Eisenberg, and A. Gal, private communication (1979).

Question: C. Wilkin, University College London

How does the forward cross section on ^7Li compare with the lower bound obtained, via, the optical theorem, from the difference between the π^+ and π^- total cross sections of ^7Li?

Response: Cooper

We have not yet done this comparison, but we plan to do so soon. There are two points in this regard to keep in mind. 1) The Coulomb effects must be handled carefully. 2) Near resonance, the real part of the scattering amplitude is small, and in some cases measured by Coulomb nuclear interference measurements, so that the bound may become a direct comparison.

Question: E. Oset, University of Barcelona

Several speakers have emphasized the importance of the isobar-hole model for π-A interactions because of the strong non-localities of the optical potential in the resonance region and the strong coupling of the intermediate Δ-h states to the medium. The effects due to the Δ-h self energies, already important in π-A elastic scattering, become dramatic in the single charge exchange reaction. According to one contributed paper to this conference the total charge exchange cross section oscillates within a factor of 3 when the different medium corrections are switched on gradually. This clearly indicates that the results obtained by using standard optical potentials should only be taken as qualitative results. On the other hand, when all the different medium corrections are taken into account, the total exchange cross section becomes flat in the resonance region around the value of 0.4 mb for the $^{15}N(\pi^+,\pi^0)^{15}O$. This seems to indicate that the solution to the ^{13}C puzzle is lying in the isobar-hole model and in any case that the exchange reaction are very promising to clarify the π-A interaction mechanism.

Response: Cooper

To say that you have solved the ^{13}C puzzle, you should calculate that case. In $^{15}N(\pi^+,\pi^0)^{15}O$, the data will be forthcoming in about a year for comparison with your predictions.

GIANT MULTIPOLE RESONANCES AND INELASTIC PION SCATTERING

Fred E. Bertrand
Physics Division, Oak Ridge National Laboratory*
Oak Ridge, Tennessee 37380

Invited talk presented at the 2nd International Conference on Meson-Nuclear Physics, Houston, Texas, March 5-9, 1979

During the past few years considerable interest has been generated by the experimental observation of non-dipole giant resonances. These so-called "new" giant resonances, or giant multipole resonances were first observed by and have been studied mostly by inelastic scattering of medium-energy hadrons, notably protons and alpha particles, and electrons. Through these measurements the isoscalar giant quadrupole resonance (GQR) has now been firmly established, there is good evidence for the existence of an isoscalar giant monopole resonance (GMR) and some observations of isovector giant quadrupole and isoscalar giant octopole resonances have been reported. Reference 1 provides some recent reviews of this subject.

This presentation will provide a short review of what has been learned about the non-dipole giant resonances during the last few years, describe what advantages might be gained through use of inelastic pion scattering and finally show what the current experimental situation is with regard to pion excitation of these resonances.

Giant resonances are often described as highly collective modes of excitation in which an appreciable fraction of the nucleons in a nucleus move together - so much so that it is appropriate to describe these modes in hydrodynamic terms like the oscillation of a liquid drop. From the experimental prospective of this presentation we might call a state a giant resonance when a large fraction of the theoretical limit for the strength of that particular class of transition is observed in the state. Such a theoretical limit that is particularly useful is the linearly energy-weighted sum rule (EWSR), computation of which is nearly model independent. It has long been known, for example, that in general the giant dipole resonance depletes all of its EWSR. However, the first excited 2^+ state in nuclei seldom contain more than 10% of the T=0, L=2, EWSR limit.

Schematically represented in figure 1 are single-particle transitions between shell-model states of a hypothetical nucleus. Collective transitions result from coherent superpositions of such single-particle transitions. Major shells are denoted as N, N + 1

*Operated by Union Carbide Corporation for U.S. Department of Energy under contract W-7405-eng-26.

ISSN:0094-243X/79/540230-22$1.50

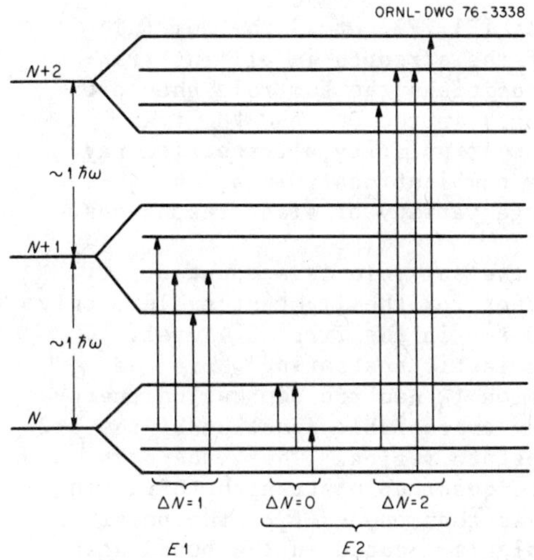

Fig. 1. Schematic representation of E1 and E2 single-particle transitions between shell-model states of a hypothetical nucleus. Major shells are denoted as N, N+1 and N+2 and lie ∿ 1ħω or 41 x $A^{1/3}$ MeV apart.

and N + 2 and within each major shell a few subshells are shown. The major shells are separated by 1ħω or ∿ 41/$A^{-1/3}$ MeV. Within a shell-model framework, giant resonances can in general be considered to result from transitions of nucleons from one major shell to another, under the influence of an interaction that orders these transitions into a coherent motion.

The giant dipole resonance (GDR) is pictured as built up of transitions spanning 1ħω (ΔN=1). The GDR might then be expected to be located at an excitation energy of ∿ 41 $A^{-1/3}$ MeV; however, it is located at ∿ 77 $A^{-1/3}$ MeV. This difference arises from the fact that the interaction between the nucleons in the nucleus is repulsive for the isovector mode, so that the excitation energy is pushed up from that expected. Conversely, the interaction is positive for isoscalar modes, thus pushing the excitation energy down from what might be expected.

Two different sets of E2 transitions are allowed. The first of these, with lowest energy, is comprised of transitions within a major shell, the so-called 0ħω (ΔN=0) transitions. A second set is comprised of transitions between shells N and N+2 (ΔN=2) and would have energy of 2ħω, pushed down or up for isoscalar or isovector modes respectively. While the first class (0ħω) of E2 excitations is identified with the familar low-lying 2^+ levels, the 2ħω class of E2 transitions had not been identified until a few years ago. By similar arguments E3 excitations of 1ħω and 3ħω and E4 excitations of 0ħω, 2ħω and 4ħω are expected. It is to be noted that for E3 excitation two distinct energy groups of 1hω excitation have been identified; an octopole state (states) at rather low excitation (2-4 MeV) and a clustering of L=3 strength near the energy 32 x $A^{-1/3}$ MeV, the so called low energy octopole resonance.[2]

For each class of transitions (E1, E2, etc.) the sum rule should be exhausted by the sum of the strength in all the transitions. For example, for E2 transitions the sum rule should be exhausted by the sum of the strength in the $0\hbar\omega$ and $2\hbar\omega$ transitions. Additionally, for each multipolarity a transition may have one of the four spin-isospin combinations. Thus, there is every reason to expect a very large variety of giant resonances.

For the GDR we know 100% of the sum rule is accounted for. What about other multipoles? Except for the lightest nuclei, only $\sim$ 10-15% of the EWSR is accounted for in the first 2^+ level. Table 1 shows a few cases from inelastic scattering where the entire bound state excitation region ($\lesssim$ neutron separation energy) was studied. Only for ^{24}Mg is any appreciable fraction of the L=2 EWSR strength found in the bound-state region. Thus, there is good reason to expect considerable quadrupole strength to lie in the $\Delta N=2$ excitations. Indeed, less than one half of the possible sum-rule strength for any multipole is located in the bound states for the nuclei studied.

How are these high excitation energy, non-dipole giant resonances to be studied? Since the photonuclear reaction proceeds so dominately by dipole absorption, this reaction is not an especially good choice for observing higher order multipoles. (Excitation of the GDR is 10-100 times stronger than E2 excitation via the photonuclear reaction.) On the other hand, it has long been known that the inelastic scattering reaction provides strong excitation of collective T=S=0 states. In addition, as shown on Table 2, variation of the particle type used for inelastic scattering can provide different strengths to isoscalar and isovector states. For example, the (e,e') reaction excites isoscalar and isovector states with equal probability (for same EWSR, J^{π} and E_x) while the (p,p') and (^3He,^3He') reactions preferentially excite isoscalar over isovector states by the ratios of $\sim$1/9 and $\sim$ 1/30 respectivley. The (α,α') and (d,d') reactions do not excite isovector states by an observable amount while the charge exchange reaction preferentially excite the isovector states. Comparison of the various reactions leading to giant resonance states can be of great help in unraveling the isospin makeup of the resonances. Indeed, inelastic pion scattering may be especially useful for excitation of T=0, S=1 resonances as has been pointed out in the presentation by G. E. Walker at this conference.[3]

In order to place the location of the giant resonances in the proper perspective I show on Figure 2 a plot of a complete inelastic proton spectrum.[4] The spectrum is complete in that nearly all protons emitted at 27 degrees by the ^{54}Fe target were observed (some protons are undetected for $E_p \lesssim 2$ MeV). The region of this spectrum above $\approx$ 10 MeV, the region which has been so often studied, contains only $\approx$ 5% of the integrated inelastic cross section. At the highest excitation energy end of the spectrum (lowest emitted proton energy) a large peak is observed which is

Table I. Percentage of isoscalar EWSR multipole strength depleted in bound states of ^{24}Mg, ^{40}Ca, and ^{208}Pb (Reference 1).

Nucleus	Multipole 0	1	2	3	4	5	6	7	8
^{24}Mg			40	10	3				
^{40}Ca	0	0	14	38	7	11	1	0.2	0
^{208}Pb	0	0	20	47	14	3	3	2	1

Table II. Relative cross sections[a] of isoscalar and isovector excitations for various reactions.

	Isoscalar	Isovector
(e,e')	1	1
(p,p')	1	∿ 1/9
(^{3}He, ^{3}He')	1	∿ 1/30
(α,α')(d,d')	1	∿ 0
(n,p)(t,^{3}He)	0	1
(p,n)(^{3}He,t)N = Z nuclei	0	1

[a]Relative cross section normalized to 1 for the stronger excitation.

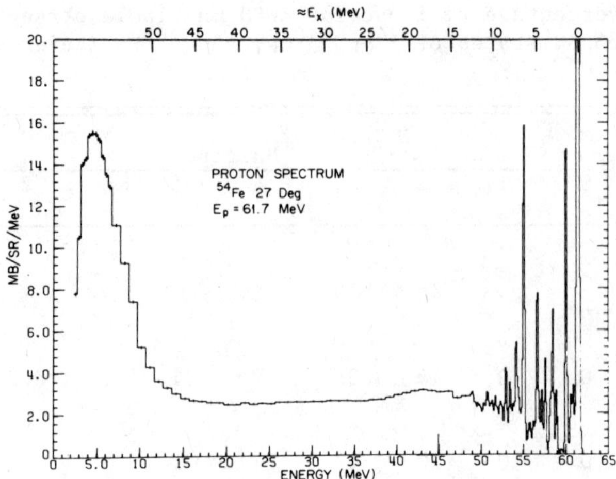

Fig. 2. Proton spectrum at 27° from 62-MeV protons in ^{54}Fe. The energy of the outgoing proton is plotted at the bottom of the figure while the approximate excitation energy is plotted at the top. Data have been plotted in ∿ 1-MeV-wide bins up to ∿ 49 MeV, then plotted in 50-keV-wide bins. Protons below ∿ 1.5 MeV were not detected in the experiment. The small, broad peak near E_x ∿ 16 MeV is identified as arising from excitation of the giant quadrupole and dipole resonances (ref. 4).

produced by the nuclear evaporation process. The rather flat region between the evaporation peak and the neutron separation energy is called the nuclear continuum. This region is often described as arising from a pre-equilibrium particle emission process. The topic of this talk and for that matter of all the giant multipole resonance work during the past years is the small bump located at ∿ 16 MeV of excitation. Systematic observation of this peak in many nuclei showed that it was not produced by kinematical effects or by GDR excitation. Rather the peak has been shown to arise from excitation of a giant quadrupole resonance (GQR).

Figure 2 serves to illustrate what is perhaps the most serious problem in inelastic scattering studies of giant resonances - the giant resonance cross section is only a small part of the total continuum cross section. Some assumption must be made about both the magnitude and shape of the continuum which lies under the resonance peak. It is assumed that the peak cross section does not mix with the underlying continuum and the peak is "stripped off" of the continuum by extrapolating the continuum magnitude and shape from higher excitation energies. Due to the uncertainty in the magnitude of the underlying continuum, I feel that cross sections for the giant resonances cannot be obtained from inelastic scattering with less than a 15-20% absolute uncertainty.

Figure 3 shows that proper selection of the reaction and incident projectile energy can enhance the resonance peak. This spectrum of ^{120}Sn was obtained using 152 MeV alpha particles[5] and shows a resonance cross section nearly equal to the cross section of the underlying continuum.

Over the past several years such giant resonance peaks have been observed in a large number of nuclei using inelastic scattering of a wide variety of incident hadrons and electrons. From these measurements the giant quadrupole resonance has now been well established.

In general, in nuclei having A $\gtrsim$ 40 it is found[1] that the GQR is a 2-4 MeV wide unstructured peak containing 50-100% of the T=0, E2 EWSR. However, in lighter nuclei the GQR strength is fragmented into a large number of discrete and resolvable states.[6]

Figure 4 shows the energy of the T=0, GQR as a function of nuclear mass number. The solid curve is the energy, 63 x $A^{-1/3}$ Mev. In general, the data agree with this energy rather well. However, there is a definite trend at higher mass for the energy to be above the 63 x $A^{-1/3}$ MeV systematic and for the energy to be below 63 x $A^{-1/3}$ MeV for A $\lesssim$ 40. You will recall that the GDR follows the systematic energy of $\sim$ 77 x $A^{-1/3}$ MeV for A $\gtrsim$ 50.

The percentage of the T=0, E2 EWSR strength depleted in the GQR is plotted on figure 5. These values do not include contributions to the L=2 EWSR strength depleted in bound state excitations. A trend to larger sum-rule depletion with increasing mass is clearly evident. For nuclei having A $\gtrsim$ 100 nearly 100% of the EWSR is found in the GQR peak (the $2\hbar\omega$ transitions), while for sd-shell nuclei 30-40% EWSR is typically found in the GQR states. However considerable EWSR strength is located in the low-lying 2^+ states of light nuclei, so that the sum of the GQR strength and low-lying quadrupole strength exhausts $\sim$ 100% of the T=0, L=2 EWSR in light as well as heavy nuclei.

Although the most thoroughly studied of the new giant resonances is the GQR, the resonance which has generated the most interest is the monopole. Observation of the monopole, "breathing," or compressional mode of nuclear excitation is of special significance because knowledge of its excitation energy could yield the value of the nuclear compressibility. During the past few years several candidates for the E0 resonance have appeared but most have not withstood the test of further measurements. These early measurements are discussed in reference 1. It is important to note that some early indirect evidence for an E0 resonance[17,18] placed the resonance at an excitation energy of

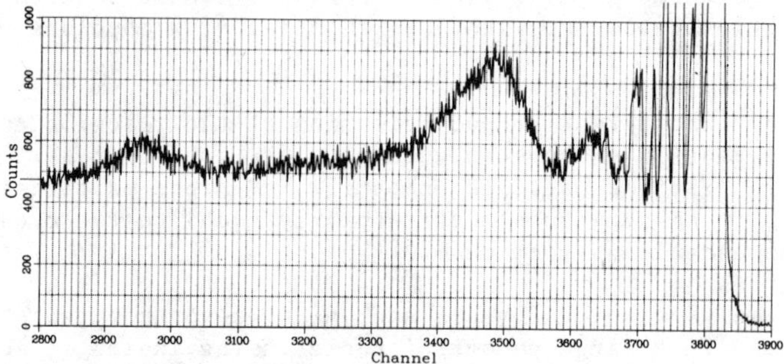

Fig. 3. Spectrum at 12 degrees from the $^{120}Sn(\alpha,\alpha')$ reaction using 152-MeV incident alpha particles (ref. 5). The broad peak observed near channel 3485 arises from excitation of giant resonances, mostly the GQR located at 13.3 MeV of excitation. The elastic scattering peak is in channel 3805.

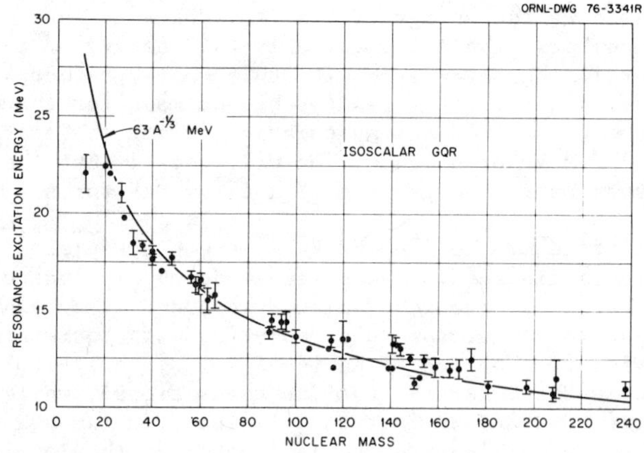

Fig. 4. Measured excitation energy of the isoscalar GQR plotted against nuclear mass number. The solid curve represents the energy 63 x $A^{-1/3}$ MeV.

∿ 80 x $A^{-1/3}$ MeV, a value in agreement with recent more direct observations.

Figure 6 shows spectra[7] at two angles from the $^{208}Pb(\alpha,\alpha')$ reaction for 120 MeV incident alphas. The 14-degree spectrum shows the presence of what appears to be two broad peaks (indicated by the dashed lines) in the giant resonance region of the spectrum. The larger peak located at 11 MeV (∿63 x $A^{-1/3}$ MeV)

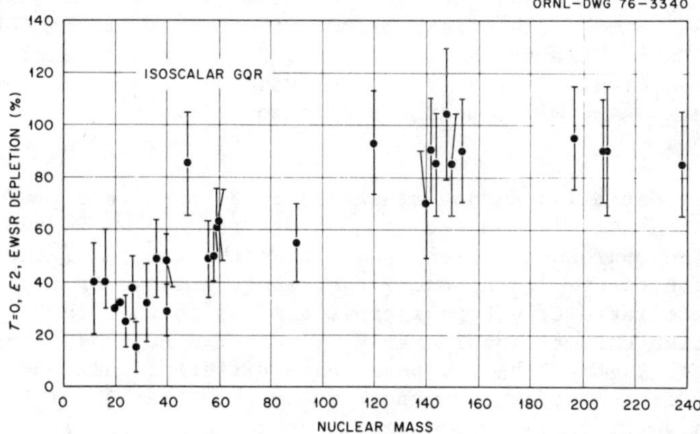

Fig. 5. Percentage of the T=0, E2 EWSR strength in the GQR plotted against nuclear mass number. The values do not include the EWSR depletion in low-lying 2^+ states.

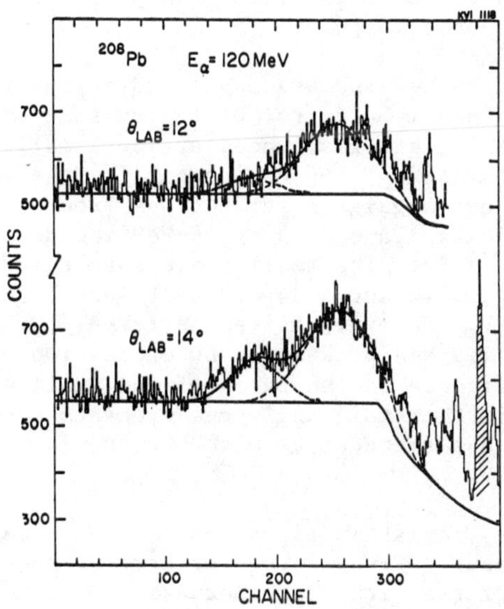

Fig. 6. Giant resonance spectra at 14 deg and 12 deg from the ^{208}Pb(α,α') reaction for 120-MeV incident alpha particles (ref. 7). The broad peak observed at both angles in channel 240 occurs at $\sim$ 11-MeV and is the GQR. In the 14 degree spectrum a second broad peak is observed at channel 180, or $\sim$ 13.9 MeV. This second peak is now identified as the E0 giant resonance.

is the now familiar GQR peak. The smaller peak is located at 13.9 MeV or $\sim$ 80 x $A^{-1/3}$ MeV. This peak is near the energy of the GDR in ^{208}Pb (13.6 MeV), however the isovector GDR will not be excited by the (α,α') reaction with nearly enough cross section to account for this new peak. A similar peak was also found[7] in ^{206}Pb, ^{209}Bi, and ^{197}Au.

It was thus established that a here-to-fore not directly observed resonance peak is located at $\sim$ 14 MeV for nuclei in the lead region. The obvious question was what is the nature of the peak? Comparison of the 12 deg and 14 deg spectra shown on figure 15 shows that the ratio of GQR peak cross section to the "new" peak cross section changes considerably in only two degrees. This observation suggests that the two peaks have different multipolarities. In addition, if the new peak is 2^+ then the L=2, T=0, EWSR would be significantly overdepleted; an unhappy situation theoretically. Unfortunately, in the angular region studied in reference 7 the L=0 and L=2 angular distributions are in phase. The only difference is in the larger peak to valley ratio for the L=0 calculation than for the higher multipoles. In fact, the L=0 calculation fits the data points better than the L=2 calculation, but the differences depend on only a few points and therefore the evidence is not very strong. If the peak is in fact E0 then comparison with the L=0 calculation indicates that $\sim$ 90% of the T=0, L=0, EWSR would be accounted for.[7]

At smaller angles the L=0 and L=2 angular distributions are considerably different (nearly a factor of ten at some angles). Thus, measurements of the giant resonances at very small angles could provide a more convincing L=0 assignment for the 14-MeV resonance. Figure 7 shows angular distributions from such measurements[8] made at Texas A and M using 96-MeV incident alpha particles on ^{208}Pb and ^{144}Sm. The small angle spectra of ^{208}Pb show the same two peaks, 11.0 and $\sim$ 14 MeV that were seen in the previous work at 120 MeV. However, as demonstrated in figure 7 the assignment of the 13.9-MeV peak as an E0 excitation can be made with far more confidence on the basis of the small angle measurements. For ^{144}Sm the small angle measurements showed the GQR peak at 12.4 Mev and a second peak with L=0 angular distribution located at 15.1 MeV.

For lighter nuclei, inelastic alpha scattering measurements for incident energies near 100-MeV encounter serious spectral contamination[9] which interferes with the observation of the monopole and quadrupole resonances. This contamination arises from the fact that the alpha particle may pick-up a neutron, (similar arguments hold for proton pickup) form ^{5}He and subsequently decay back into a neutron and alpha particle. The alphas from the ^{5}He decay form a broad spectral distribution which may obscure or interfere with (depending on the incident alpha energy) observation of the giant resonance peaks. Since no similar problem occurs for incident

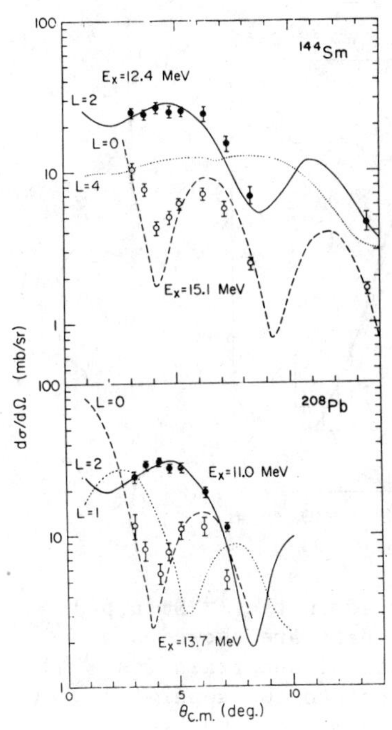

Fig. 7. Angular distribution for two peaks observed in the giant resonance spectra of ^{208}Pb and ^{144}Sm. The results are from the (α,α') reaction for E_α = 96 MeV. For both nuclei the lower excitation peak is the GQR while the angular distribution of the higher excitation peak is well described by the L=0 calculations and is thus, identified as the E0 giant resonance (ref. 8)

protons, (p,p') results may be useful for observations of the monopole resonance in lighter nuclei.

A spectrum at 20 degrees from the reaction ^{144}Sm(p,p') is shown on figure 8. The broad resonance structure appears to be composed[10] of two peaks as indicated on the figure. The lowest excitation peak falls at $\sim$ 12.8 MeV and is the GQR while the second peak is located at $\sim$ 15.5 MeV, the excitation energy of the GDR which is expected to be excited by inelastic proton scattering. Cross sections[10] for the two peaks are shown in figure 9 along with DWBA calculations for L=2, 1 and 0 excitations. The 12.8-MeV peak is well described by the L=2 calculation normalized to 80% T=0, L=2, EWSR depletion in agreement with previous measurement.[1] For the 15.5 MeV peak, the cross sections at 15 and 20 degrees are much larger than can be accounted for by the L=1 (T=1) calculation normalized to 100% of the EWSR. At those angles the L=0 calculation agrees very well with the data. However, the cross sections at 25 and 30 degrees are too large to be explained by the L=0 calculations alone, but a sum of the L=1 and L=0 calculations provides good agreement with the measurement. These results indicate that the 15.5 MeV peak is composed of both the GMR, which dominates the smaller angles and the GDR which is dominate beyond 25 degrees.

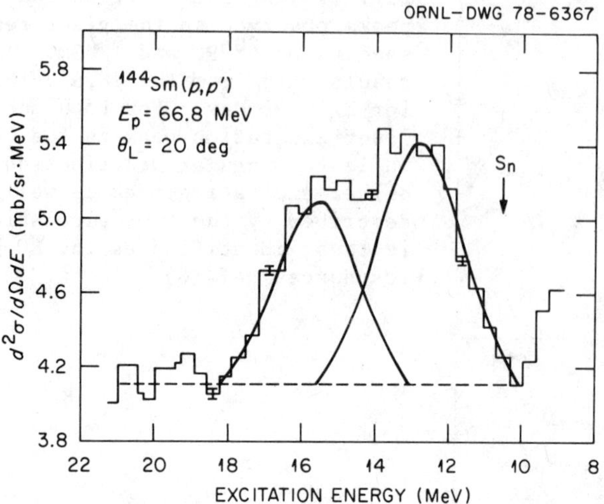

Fig. 8. Giant resonance spectrum observed in the ^{144}Sm(p,p') reaction at 20 deg; E_p = 66.8 MeV. The data are shown as a histogram. Error bars represent statistical uncertainties only. The resonance has been decomposed into two separate peaks as indicated (ref. 10).

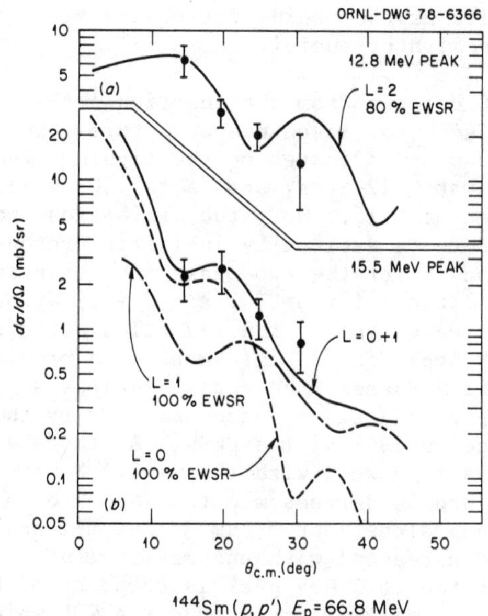

Fig. 9. Cross sections for the two peaks in ^{144}Sm compared with DWBA calculations (ref. 10).

We have applied[11] the same analysis to (p,p') data on ^{208}Pb, ^{197}Au, ^{120}Sn, ^{90}Zr, ^{58}Ni and ^{40}Ca in order to establish systematics for the GMR.

The results of our analysis provide us with an unexpected conclusion.[11] For nuclei with atomic number greater than $\sim$ 90 we find a rather compact E0 giant resonance located at $\sim 80 \times A^{-1/3}$ MeV and accounting for nearly 100% of the T=0, L=0, FWSR. However, the situation is quite different for the two lighter nuclei we have studied. In ^{58}Ni although we find evidence for a peak at $\sim 80 \times A^{-1/3}$ MeV the peak cross section contains only 30 ± 10% of the GMR, EWSR. For ^{40}Ca, the magnitude of the peak observed at $\sim 80 \times A^{1/3}$ MeV can be completely accounted for by excitation of only the GDR.

What happens to the E0 strength? Assuming the DWBA model for the breathing mode is correct for light as well as heavy nuclei, the most likely explanation is that the E0 strength becomes highly fragmented and spread out in light nuclei. If this is the case then observation of this strength would be very difficult in the inelastic scattering measurements because of the large nuclear continuum. It remains to be seen whether calculations support this possible explanation.

Figure 10 provides a summary of the GMR observations available at this time. The peak energies agree quite well with the systematic energy $80 \times A^{-1/3}$ MeV. Table 3 lists the measured parameters of the observed monopole resonances.

As I pointed out earlier, an important question is what value of the nuclear compressibility do the GMR energies yield? For an excitation energy of $80 \times A^{-1/3}$ MeV the liquid drop model yields a value of $\sim$ 200 MeV for the compressibility. However, no allowance for Coulomb, surface and neutron-excess effects have been made. Although theoretical estimates of these effects are very uncertain, at least one approach[12] yields compressibilities as small as 150 MeV, based on a GMR energy of $80 \times A^{-1/3}$ MeV.

Let me summarize the existing experimental situation, as I see it. Certainly, the GQR is now firmly established. I believe there is now good evidence for the existence of a GMR. The present status of the GMR is similar to that of the GQR in 1973 - a few measurements have been made on a few nuclei. What is needed now are systematics generated by a variety of inelastic reactions. There is some evidence for the existence of the isovector GQR at a systematic energy of $\sim 120 \times A^{-1/3}$ MeV. Very little evidence exists, all from inelastic electron scattering, for a giant octopole resonance,[1] and only "calculational evidence" and speculation exist for an L=4 resonance.[1]

Where are the rest of the giant resonances? Where are the isovector modes, the spin-flip modes, the high-L resonances? Can we expect help from inelastic pion scattering?

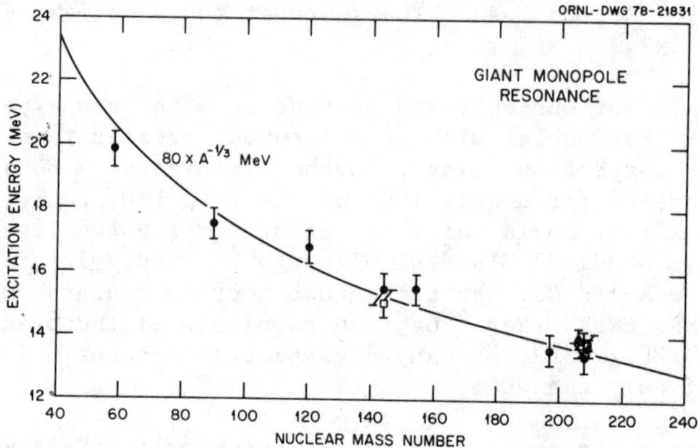

Fig. 10. Excitation energy (MeV) of the GMR for nuclei in which observations have been made. Circles - 60 MeV (p,p'), ref. 11; squares - 96 MeV (α,α'), ref. 8; triangles - 120 MeV (α,α'), ref. 7. The solid curve represents the energy 80 x $A^{-1/3}$ MeV.

Table III. GMR parameters.

Nucleus	Excitation Energy (MeV)	Width (MeV)	% EWSR	Reaction	Energy	Reference
^{209}Bi	13.7±0.3	2.5±0.4	100±25	(α,α')	120 MeV	19
^{208}Pb	13.9±0.3	2.5±0.6	110±25	(α,α')	120 MeV	19
	13.7±0.4	3.0±0.5	105±20	(α,α')	96 MeV	20
	13.4±0.5	3.0±0.5	90±20	(p,p')	61 MeV	24
^{206}Pb	14.0±0.3	2.5±0.4	100±25	(α,α')	120 MeV	19
^{197}Au	13.6±0.5	3.0±0.5	100±25	(p,p')	61 MeV	24
^{154}Sm	15.5±0.5	2.5±0.5	100±25	(p,p')	67 MeV	5
^{144}Sm	15.1±0.5	2.9±0.5	100±20	(α,α')	96 MeV	20
	15.5±0.5	2.5±0.5	100±25	(p,p')	67 MeV	5
^{120}Sn	16.8±0.5	3.5±0.5	100±25	(p,p')	61 MeV	24
^{90}Zr	17.5±0.5	3.0±0.5	60±25	(p,p')	61 MeV	24
^{58}Ni	19.8±0.5	3.5±0.5	30±10	(p,p')	61 MeV	24

It is clear from the activity at the major meson factories that there is keen interest in the observation of giant resonances by pion inelastic scattering. While the interest exists I think I can fairly and objectively say that the results are at present best described as conflicting. Before reviewing these results it seems reasonable to remind ourselves of some of the reasons giant resonance excitation by inelastic pion scattering might be expected to yield new or improved information. I describe three reasons here that directly affect the experimental observations; there are surely others.

1) It is anticipated that the unstructured nuclear continuum produced in pion inelastic scattering will be considerably lower in magnitude than that measured using other hadronic probes. Such a reduction in the continuum could lead to observation of resonances too weak to be seen observed above the continuum from other hadronic probes. Of course, if the reduction of the continuum is accompanied by an equal reduction in the giant resonance cross section then the advantage is lost.

   Calculations such as the intranuclear cascade model[13] have long predicted a low pion continuum cross section. There is now a little continuum data at lower incident pion energies. Figure 11 shows continuum spectra[14] from 50-MeV positive pions inelastically scattered from ^{40}Ca at 40- and 80-degrees. This is to be compared with the ^{40}Ca(p,p') spectrum for 60-MeV protons shown in figure 12. There is a factor of five to ten reduction of the high excitation energy continuum in the pion case. It is also to be noted that the GQR and GDR clearly stand out in the (p,p') reaction but not in the pion data.

2) Calculations indicate that a considerable difference is to be expected between L=0, and L=2 angular distributions at reasonable angles for excitation using $\sim$ 200-MeV pions. Figure 13 shows DWIA calculation performed by Arvieux _et al._[15] for the excitation of the GQR and GMR in ^{40}Ca by 241-MeV positive pions. There is a large difference in the expected angular distributions for the two multipolarities at $\gtrsim$ 20-degrees. This certainly presents a more favorable situation than that for 120-MeV alphas where such large differences are found only for angles less than $\sim$ 3 degrees.

3) Perhaps the most promising aspect to pion excitation of giant resonances is the prospect that resonances not easily excited in proton or alpha scattering may be more strongly excited in pion inelastic scattering. As has been discussed at this conference by G. E. Walker, T=0, S=1 states may be much stronger relative to T=0, S=0 states when excited by pions than other hadronic probes.

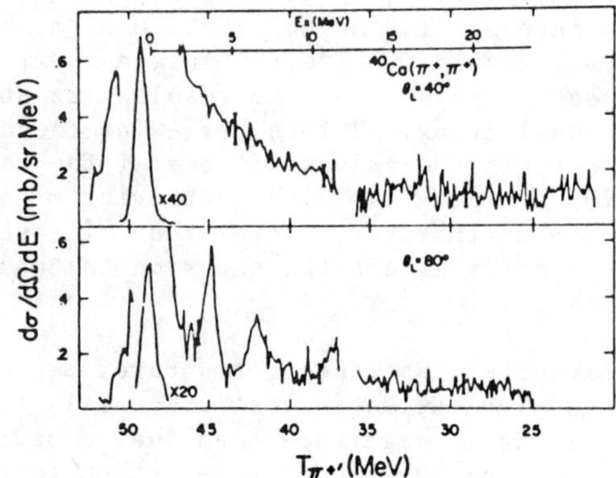

Fig. 11. Energy spectra of inelastically scattered π^+ mesons from ^{40}Ca at E_{π^+} = 50 MeV.

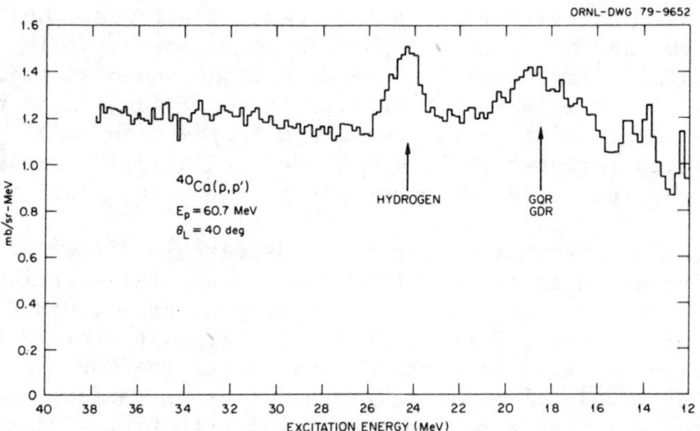

Fig. 12. Excitation energy of scattered protons from the reaction ^{40}Ca(p,p') for 60.7-MeV incident protons. The broad peak centered at ∿ 19-MeV contain the giant quadrupole (GQR) and giant dipole (GDR) resonances.

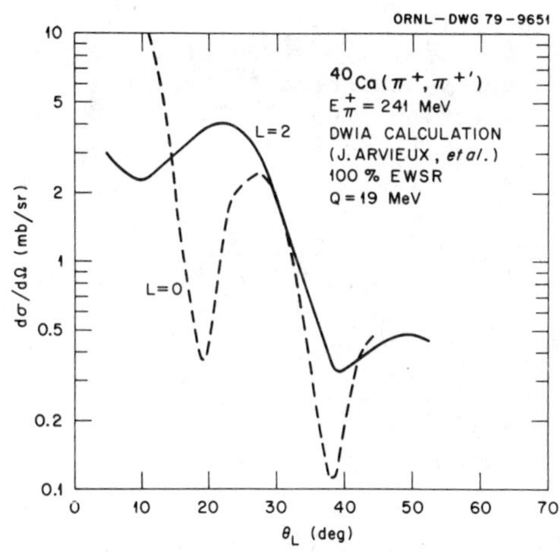

Fig. 13. DWIA calculations for the reaction $^{40}Ca(\pi^+,\pi^{+'})$ for E_{π^+} = 241 MeV (ref. 15).

However, before such "exotic" resonances can be found and understood, it is necessary to demonstrate that the T=0, S=0 giant resonances such as the GQR are observed in pion scattering.

Now, what about the experimental situation? There are some experimental results. I think for each "positive" result I will show there is an independent measurement that observes nothing using the same reaction with the same or nearly the same incident pion energy.

Perhaps the most convincing evidence presented so far for pion excitation of giant resonances is from the measurements by Ariveux <u>et al</u>.[15] In these measurements made at SIN, 241- and 163-MeV positive pions were inelastically scattered from ^{40}Ca. Figures 14, 15 and 16 show spectra at 16.5 and 19.5 degrees for 241 MeV incident pions and at 26 degrees for 163 MeV pions, respectively. In all of the spectra a distinct, broad, peak is located near channel 100 or at ∿ 18.2 MeV of excitation energy. Figure 17 shows a 12-degree inelastic proton spectrum for 60-MeV protons on ^{40}Ca. The large peak at ∿ 18-MeV has been well established as the GQR.[1] Excitation of the GDR provides the asymmetric tail on the proton spectrum.[11] Comparison of the pion and proton spectra indicate that the structure in the pion data near 18-MeV is the same as that observed with protons (and alphas also) and is thus probably the GQR.

Ariveux and co-workers have chosen to decompose this broad resonance peak into two components, a larger peak at 18 MeV and a smaller one at ∿ 20.6 MeV. Such an analysis is consistent with the analysis we performed for the proton spectra[11] although the statistical accuracy of the proton data was much better. Figure 18 shows angular distributions for the two proposed resonance peaks and for elastic scattering and excitation of the 3.74 MeV 3^-

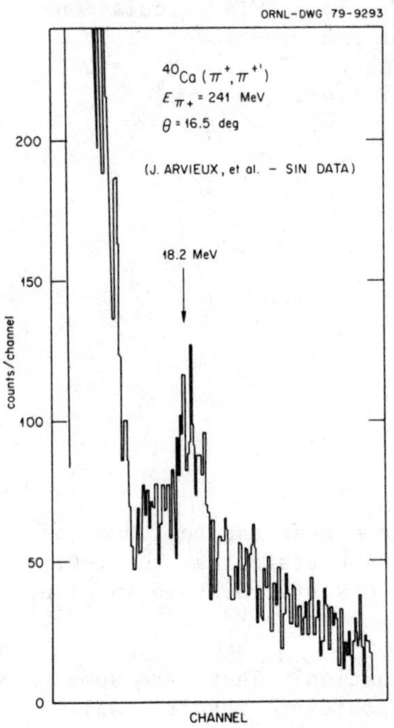

Fig. 14. Spectrum of inelastically scattered π^+ from the reaction $^{40}Ca(\pi^+,\pi^{+'})$ for E_{π^+} = 241 MeV and θ_L = 16.5 degrees (ref. 15).

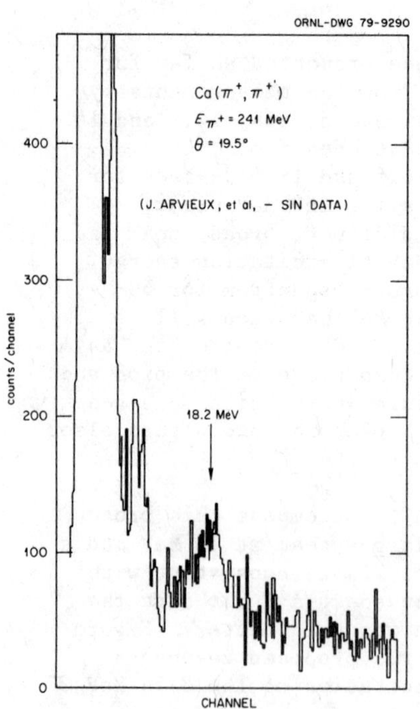

Fig. 15. Spectrum of inelastically scattered π^+ from the reaction $^{40}Ca(\pi^+,\pi^{+'})$ for E_{π^+} = 241 MeV and θ_L = 19.5 degrees (ref. 15).

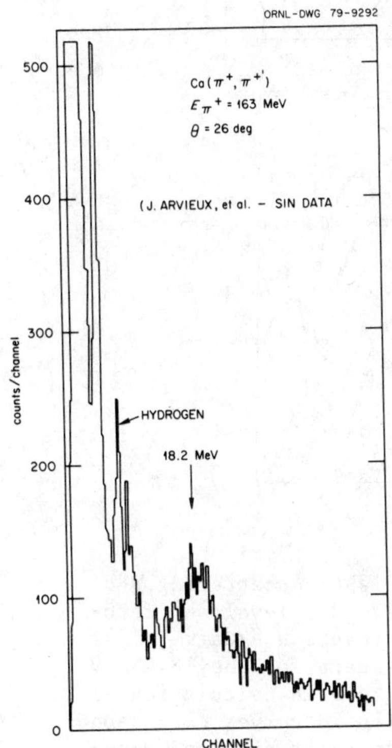

Fig. 16. Spectrum of inelastically scattered π^+ from the reaction $^{40}Ca(\pi^+,\pi^{+\prime})$ for E_{π^+} = 163 MeV and θ_L = 26 degrees (ref. 15).

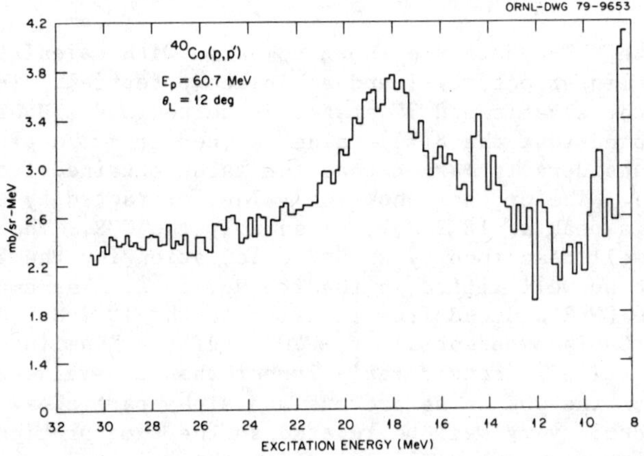

Fig. 17. Inelastic proton spectrum for the reaction $^{40}Ca(p,p')$. E_p = 60.7 MeV, θ_L = 12 degrees.

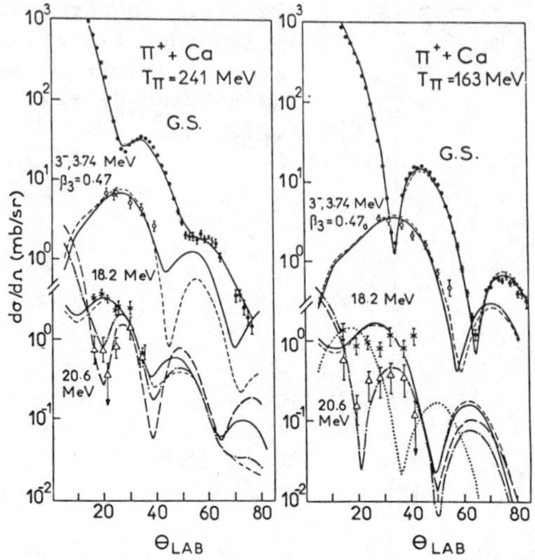

Fig. 18. Angular distributions for elastic scattering and inelastic scattering from the 3.47 MeV, 3^-, level and resonances at 18.2 and 20.6 MeV from the reaction $^{40}Ca(\pi^+,\pi^{+\prime})$ for 241 and 163 MeV incident π^+. The data for the 18.2 MeV and 20.6 MeV resonances are compared to DWIA calculations for L=2 and L=0, respectively, where the solid curves correspond to a Laplace optical potential and the short dash and dash-dot curves to a Kisslinger potential. The dotted curve shown on the 163-MeV results is the result of an L=1 calculation using a Goldhaber-Teller form factor assuming 100% of the L=1 EWSR is depleted (ref. 15).

excited state. The data are shown compared with calculations using a Kisslinger potential and a Laplace potential. The comparison to the elastic and 3^- states is quite good. However, it should be noted that the B(E3) value deduced from the pion data[15] is considerably larger than the value obtained from (α,α') measurements. The crosses show the values extracted by the authors for a peak at 18.2 MeV, presumably the GQR. The data points are well described by an L=2 calculation for the 241-MeV data but not so well fitted in the 163 MeV case. The amount of the L=2, T=0 EWSR deduced from the fits to the 18.2-MeV data is ∿ 70% for the Laplace potential or ∿ 200% for the Kisslinger potential. This value is considerably larger than the value of 40-50% obtained[1] for the GQR using protons and alpha particles. This difference could very well be related to the same problem I mentioned earlier with the B(E3) value for the 3.74, 3^-, state. I feel the most speculative part of this data analysis is that concerning the 20.6-MeV resonance extracted by the authors. The cross sections for the resonance are shown as triangles and the data are compared with an L=0 DWIA calculation. As is seen in figure 15, the 20.6 MeV data are well described by the L=0 calculation. The dotted curve compared with the 20.6 MeV results at 163 MeV is an L=1 calculation which does not yield good agreement with the data. The authors interpret these results as evidence for an L=0 resonance at 20.6 MeV in ^{40}Ca and deduce a T=0, L=0 EWSR strength of 77% or 17% depending on whether the Kisslinger or Laplace potential respectively, is used. Considering the low statistical accuracy of the data the uncertainty on these values must certainly be 30-50%. In addition the calculation itself is uncertain.

The interpretation of the 20.6 MeV peak as an E0 state depleting $\sim$ 70% of the EWSR would clearly be in disagreement with (p,p') results[11] and recent (^{3}He,^{3}He') results.[16] On the other hand, 10-20% of the GMR could be accommodated within the proton and ^{3}He results. It is more of a concern to ask why the GDR which should be located at $\sim$ 21 MeV is not more strongly excited. The proton results are quite consistent with the observation of 100% of the GDR EWSR strength at $\sim$ 21 MeV.

A very active search for giant multipole resonances using lower energy pions has been carried out at LAMPF over the past 2-3 years by a University of Washington, Carnegie-Mellon collaboration.[17] Figure 19 shows 80 degree spectra from the ^{40}Ca(π^+,$\pi^{+'}$) for 60-, 67- and 71-MeV incident pions. For 67-MeV pions a distinct peak is seen near channel 475 or $\sim$ 18 MeV of excitation. Such an observation would be in agreement with previous observations of the GQR in ^{40}Ca. However, by changing the beam energy only a few MeV the bump disappears - a very disconcerting result. The authors have repeated these measurements often enough to be confident that this effect is nuclear and not instrumental.

A peak such as the one seen in Ca from the 67-Mev pion scattering was also observed in Ca and Ni, very weakly in Zr and not al all in Sn and Pb. Again no peak is seen in Ca or Ni with 60 or 71 MeV pions.

It would be something of an understatement to say that the interpretation of these data is unclear. The disappearance of the resonance, if that is the origin of the bump, with only small changes in incident beam energy would be quite contrary to observations made with other hadronic probes. Could it be that the underlying continuum is really what is changing character so rapidly with incident energy? If so such changes might mask the appearance of giant resonance peaks. This and other possible interpretations of the data are being actively studied by the collaborators in these measurements.

As I suggested earlier in this presentation, the experimental situation regarding giant resonance excitation by pion inelastic

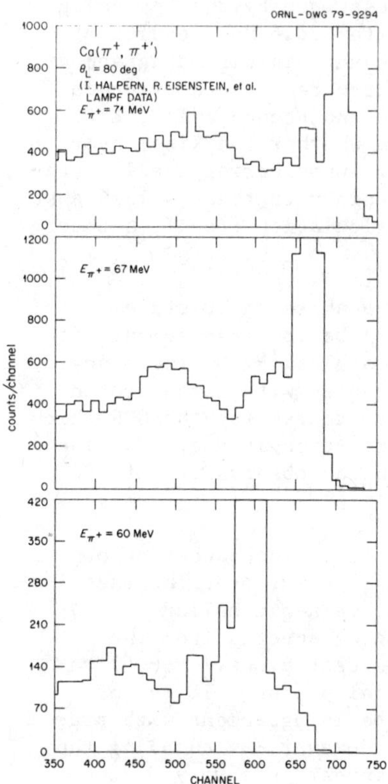

Fig. 19. Spectra of pions inelastically scattered from ^{40}Ca at 80 degrees for 71, 67, and 60 MeV incident pions (ref. 17).

scattering is indeed unclear. I feel it is still too soon to tell if pion inelastic scattering to giant resonances states will be the important tool many of us had hoped for - but we will only know through more experimental efforts.

REFERENCES

1. F. E. Bertrand, Ann. Rev. Nucl. Sci. 26, 457 (1976). F. E. Bertrand, "The New Giant Resonances - An Experimental Review," Proceedings of the NATO Advanced Study Institute on Theoretical Methods in Medium-Energy Physics, Banff, Canada, Aug. 21, 1979, to be published.
2. J. M. Moss et al., Phys. Rev. Lett. 37, 816 (1976).
3. M. K. Gupta and G. E. Walker, Nucl. Phys. A256, 444 (1976); also see G. E. Walker, invited presentation at this conference.
4. F. E. Bertrand and R. W. Peelle, Phys. Rev. C 8, 1045 (1973).

5. Unpublished results obtained at the Indiana University Cyclotron by F. E. Bertrand, D. J. Horen, G. R. Satchler, G. T. Emery, D. E. Muller, W. Jones A. D. Bacher and A. van der Woude.
6. K. van der Borg, M. N. Harakeh, S. Y. van der Werf, A. van der Woude and F. E. Bertrand, Phys. Lett. 67B, 405 (1977). F. E. Bertrand, K. van der Borg, A. G. Drentje, M. N. Harakeh, J. van der Plicht and A. van der Woude, Phys. Rev. Lett. 40, 635 (1978).
7. M. N. Harakeh, K. van der Borg, T. Ishumatsu, H. P. Morsch, A. van der Woude and F. E. Bertrand, Phys. Rev. Lett. 38, 676 (1977).
8. D. H. Youngblood, C. M. Rozsa, J. M. Moss, D. R. Brown and J. D. Bronson, Phys. Rev. Lett. 39, 1188 (1977).
9. A. Kiss, C. Mayer-Boricke, M. Rogge, P. Turek and S. Wiktar, Phys. Rev. Lett. 37, 1188 (1976).
10. F. E. Bertrand, G. R. Satchler, D. J. Horen and A. van der Woude, Phys. Rev. C 18, 2788 (1978) .
11. F. E. Bertrand, G. R. Satchler, D. J. Horen and A. van der Woude, Phys. Lett. 80B, 198 (1979).
12. C. Y. Wong, Phys. Rev. C 17, 1832 (1978).
13. H. W. Bertini, Phys. Rev. 131, 1801 (1963); 138, AB2 (1965).
14. F. Milder, M. Blecher, K. Gotow, D. Jenkins, P. Roberson, R. L. Burman, M. A. Moinester, R. P. Redwine, F. E. Bertrand, T. P. Cleary, E. E. Gross, C. A. Ludemann, C. W. Darden, R. E. Edge, D. Malbrough, T. Marks and B. M. Preedom, Phys. Lett. 72B, 159 (1977).
15. J. Arvieux, J. P. Albanese, M. Buenerd, D. Lebrun, E. Boschety, C.H.Q. Ingram and J. Jansen, Phys. Rev. Lett. 42, 753 (1979).
16. T. Yamagata, K. Iwamoto, S. Kishimoto, B. Saeki, K. Yuasa, M. Tamaka, J. Fakuda, O. Okada, I. Miura, M. Inoue and H. Ogata, Phys. Rev. Lett. 40, 1628 (1978).
17. The author would like to thank R. A. Eisenstein and W. R. Wharton for supplying their data prior to publication.

PION INDUCED DEUTERON BREAKUP AT 350 MeV/c*

J. H. Hoftiezer, G. S. Mutchler, W.v.Witsch†, W. H. Dragoset, M. Furić,†† D. M. Judd, G. P. Pepin, I. M. Duck, J. M. Clement, G. C. Phillips

Rice University, Houston, TX 77001

J. C. Allred, E. V. Hungerford, B. W. Mayes, L. S. Pinsky, T. M. Williams

University of Houston, Houston, TX 77004

The pion induced deuteron breakup reaction $d(\pi^{\pm},p\pi^{\pm})n$ was studied in a kinematically-complete counter experiment at 350 MeV/c using the LAMPF P^3 channel. Scattered reaction products were detected in a magnetic spectrometer arm in coincidence with a time of flight arm. Scintillation detectors and multiwire proportional counters were used to measure the velocity and scattering angles of particles in each arm and the momentum of protons in the magnet arm. Data were collected at eleven spectrometer settings in the region near the peak of the pion-proton subsystem Δ(1236) resonance. The experiment covers a range of spectator neutron momentum from 0 to approximately 200 MeV/c.

A typical spectrum is shown plotted versus the proton momentum in Fig.2 and versus the spectator neutron momentum in Fig.3. Also shown is a theoretical prediction for the differential cross section based on the one nucleon exchange model of Fig.1.

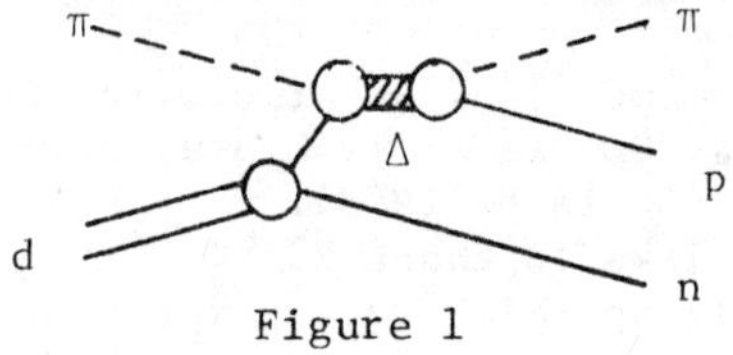

Figure 1

The calculation was done for the π^+ channel only, using no free parameters. Good agreement between the theory and data is observed. The spectral shapes are dominated by the Fourier transform of the deuteron wave function, while the presence of the Δ resonance mainly determines the absolute value of the cross section.

In the region of small momentum transfer where the quasifree scattering mechanism dominates, the ratio of π^+ to π^- data agrees with isospin predictions.

The preliminary results reported here are only a portion of the data recorded. Analysis of the complete assemblage is expected soon.

*Work supported by U.S. D.O.E. Contract EY-76-C-05-1316.
†Present address: Universität Bonn, West Germany
††On leave from Institute "Ruder Bosković," Zagreb, Yugoslavia.

ISSN: 0094-243X/79/540252-02$1.50

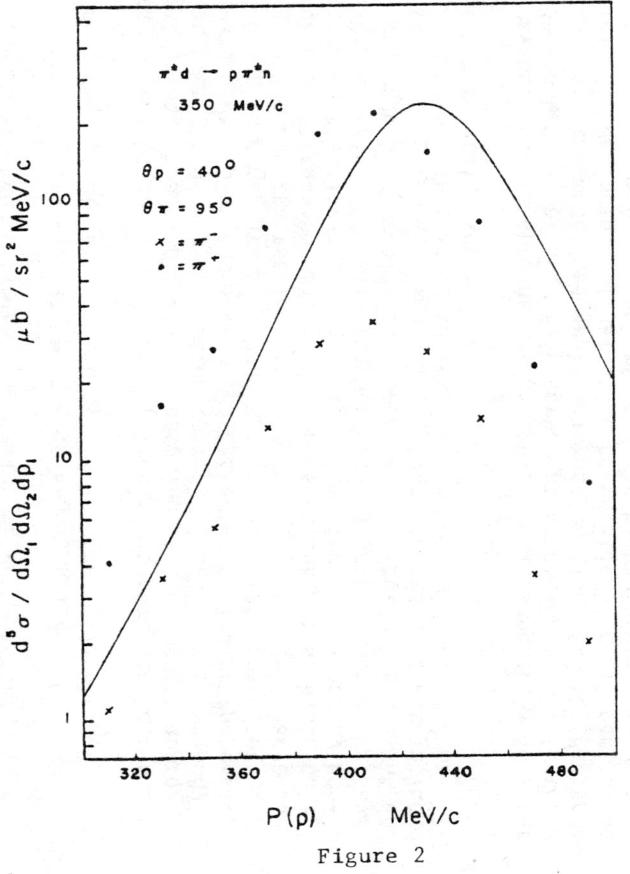

Figure 2

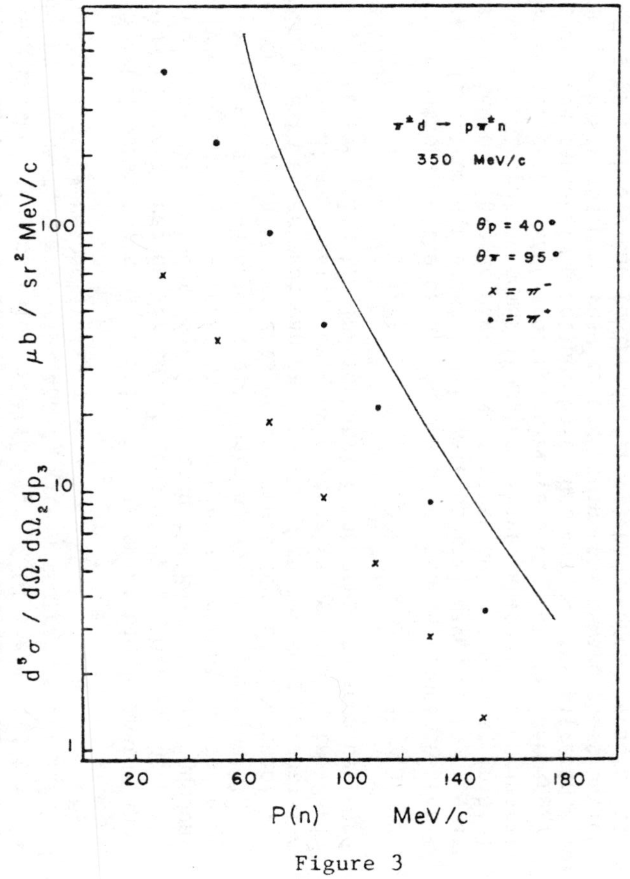

Figure 3

PIONIC DEUTERON DISINTEGRATION WITH THE PARIS POTENTIAL

O. V. Maxwell, M. Brack, and W. Weise
Inst. of Theoretical Phys., Univ. of Regensburg, W. Germany

We have obtained the total cross section for pionic deuteron disintegration within a simple model consisting of the impulse approximation plus a two-body absorption term. Included in the two-body term are p-wave pion and ρ-meson rescattering following the excitation of an intermediate Δ_{33} resonance and an s-wave rescattering contribution constructed from the phenomenological, zero range Lagrangain of Koltun and Reitan. To represent the ρ mass distribution, we have employed the f^1_- helicity amplitude obtained by Höhler and Pietarinen [1] with the Born contribution removed. We have also included phenomenological monopole form factors in the πNN and $\pi N\Delta$ vertices (with $\Lambda_{\pi N\Delta}=\Lambda_{\pi NN}=\Lambda_\pi$). ρ form factors were not included since the cross section is insensitive to Λ_ρ for Λ_π near 1.2 GeV [2]. For further details see ref. 2.

Previous results obtained within this model using the Reid interaction and $\Lambda_\pi \sim 1.2$ GeV show fair agreement with experiment. In the present work the model was applied in conjunction with the Paris potential [3]. We included final state correlations in all partial waves up to the d-wave and approximated the higher partial waves by plane waves (inclusion of correlations in higher partial waves alters the results by less than 1 %). For the $\pi N\Delta$ coupling constant, we adopted the Chew-Low value, $\alpha_\pi = f^*_\pi / f_\pi = 2.0$, and varied the $\rho N\Delta$ coupling constant, $\alpha_\rho = f^*_\rho / f_\rho$, and Λ_π within the ranges 1.7 - 2.0 and 1.2 - 1.5 GeV respectively.

Figure 1 shows the cross section vs. incoming pion momentum in the Δ_{33} resonance region for two choices of α_ρ and Λ_π. At the peak the Paris result is about 35 % lower than the Reid result, independently of α_ρ and Λ_π. We found that reducing the ρ contribution by decreasing α_ρ and enhancing the π contribution by increasing Λ_π have roughly equivalent effects on the cross section. Reasonable agreement with experiment can be obtained with the Reid potential using either a small value for α_ρ or a large value for Λ_π; the Paris results, however, lie too low for any of the parameter choices considered.

Contributions to the cross section near the Δ_{33} resonance peak ($q/m_\pi = 1.6$) from different final state partial waves up to $\ell = 4$ are shown in table 1 (higher partial wave contributions are negligible). The major contribution comes from the d-wave, which arises primarily from the deuteron s-wave and the relative s-wave of the incoming pion via the tensor interaction. Note that most of the difference between the Paris and Reid results occurs in this d-wave contribution and thus is probably due to the different short-range characteristics of the corresponding deuteron wave-functions. This suggests that with an independent determination of the other parameters, $\pi^+ d \rightarrow p+p$ ex-

ISSN: 0094-243X/79/540254-02$1.50

perimental data may ultimately provide a constraint on the short-range part of the deuteron wave function. Note also that the f-wave contribution is quite large and will probably yield interesting structural features in the differential cross section.

[1] G. Höhler and E. Pietarinen, Nucl. Phys. B 95 (1975) 210
[2] M. Brack, D.O. Riska, and W. Weise, Nucl. Phys. A 287 (1977) 425
[3] R. Vinh Mau, in "Mesons in Nuclei", ed. M. Rho and D. Wilkinson (preprint IPNO/TH 77-14) and private communication
[4] C. Richard-Serre et al., Nucl. Phys. B 20 (1970) 413

Table 1: Cross section contributions at $q/m_\pi = 1.6$ from different final state partial waves.

partial wave	σ(REID) (mb)			σ(PARIS) (mb)		
	$\alpha_\rho=2.0$ $\Lambda_\pi=1.2$	$\alpha_\rho=1.7$ $\Lambda_\pi=1.2$	$\alpha_\rho=2.0$ $\Lambda_\pi=1.5$	$\alpha_\rho=2.0$ $\Lambda_\pi=1.2$	$\alpha_\rho=1.7$ $\Lambda_\pi=1.2$	$\alpha_\rho=2.0$ $\Lambda_\pi=1.5$
s-wave	0.665	0.171	0.148	0.590	0.195	0.202
p-wave	1.585	1.779	2.376	1.403	1.526	1.983
d-wave	5.092	6.439	6.672	2.315	3.669	3.670
f-wave	2.037	2.122	2.174	1.809	1.906	1.946
g-wave	0.085	0.086	0.086	0.080	0.081	0.081
total	9.463	10.597	11.456	6.197	7.377	7.881

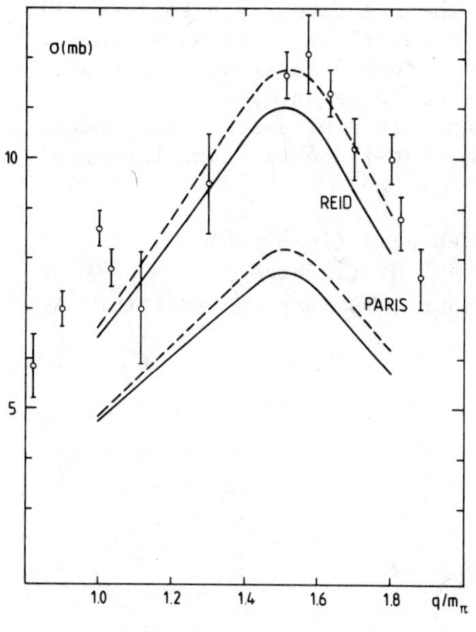

Figure 1:
Cross sections vs. pion momentum for $\alpha_\rho = 1.7$ and $\Lambda_\pi = 1.2$ (solid curves) and for $\alpha_\rho = 2.0$ and $\Lambda_\pi = 1.5$ (dotted curves). The experimental points are from ref. 4.

A MICROSCOPIC MODEL FOR THE pd → tπ+ REACTION USING REALISTIC NUCLEAR WAVEFUNCTIONS

A.M. Green and M.E. Sainio
Research Institute for Theoretical Physics, University of Helsinki
SF-00170 Helsinki 17, Finland

In a recent article[1] a microscopic model for the $pd \to t\pi^+$ reaction, in the proton laboratory energy range $300 < E_p < 600$ MeV, was constructed using two-body isobar wavefunctions containing the $\Delta(1236)$. The present work is a continuation of this development in which some of the approximations made in the earlier article are now removed. These improvements include:

a) the replacement of a simple Gaussian wavefunction for the deuteron by the more realistic Hulthén wavefunction

b) the introduction of short-range correlations into the triton wavefunction

c) the introduction of a D-state component into the triton wavefunction

d) the effect of pion production from states other than the $^5S_2$ isobar configuration. These addition states include purely nucleonic configurations and also the $^5D_2$ and $^5D_0$ isobar configurations.

e) the correlations in the initial state are taken directly from the solutions of the corresponding two-body problem[2] and are not approximated by simple expressions as in ref.[1].

In ref.[1] the cross section for the $pd \to t\pi^+$ reaction could be expressed as a double integral, which could then be integrated analytically because of the approximations made. However, with the above improvements the problem now reduces to a four dimensional integral, which is easily evaluated using Monte Carlo techniques.

The result of the above refinements is a parameter free model giving a fit to the experimental data that is better than the model described in ref.[1].

1. A.M. Green and E. Maqueda, Nucl. Phys. A (in press)
2. A.M. Green and M.E. Sainio, J. Phys. G: (in press)

ISSN: 0094-243X/79/540256-01$1.50

LOCAL ENERGY SHIFT EFFECTS IN THE π^+ AND π^- INELASTIC SCATTERING FROM ^{12}C

J. Piffaretti, R. Corfu, J.-P. Egger, P. Gretillat, C. Lunke, E. Schwarz, Univ. of Neuchâtel, Switzerland*; C. Perrin, Univ. of Grenoble, France; B. Preedom, Univ. of South Carolina, Columbia, SC†

The π^+ and π^- cross sections for the excitation of the 2^+ (4.44 MeV) and 3^-(9.64 MeV) states in ^{12}C measured at SIN exhibit large relative differences at 162 MeV (see fig.) in contrast with what is observed at 148 MeV[1] and 226 MeV. This is in apparent contradiction with expectations, given the charge symmetry of the π-N interaction and the identical distribution of protons and neutrons in ^{12}C.

If we assume that the only effect of the Coulomb field is to shift in opposite ways the energy of the incoming π^+ and π^-, the π^+/π^- relative differences at equal beam energy, E, and momentum, k, and the relative energy variations of the cross sections are simply related:

$$\frac{\sigma_+(E,\theta) - \sigma_-(E,\theta)}{\sigma_-(E,\theta)} \cong (2V_c(r)/\Delta E)\ \frac{\sigma_-(E-\Delta E,\theta)/k^2(E-\Delta E) - \sigma_-(E,\theta)/k^2(E)}{\sigma_-(E,\theta)/k^2(E)}$$

The similarities shown by the related graphs in the figure seem to justify this assumption. A π^+/π^- energy shift $2V_c(r)$ = 5.5 ± .5 MeV is deduced which corresponds to a radius r=3.15 ± .25 fm in agreement with the black disc radius of ^{12}C, R = 3.24 fm.

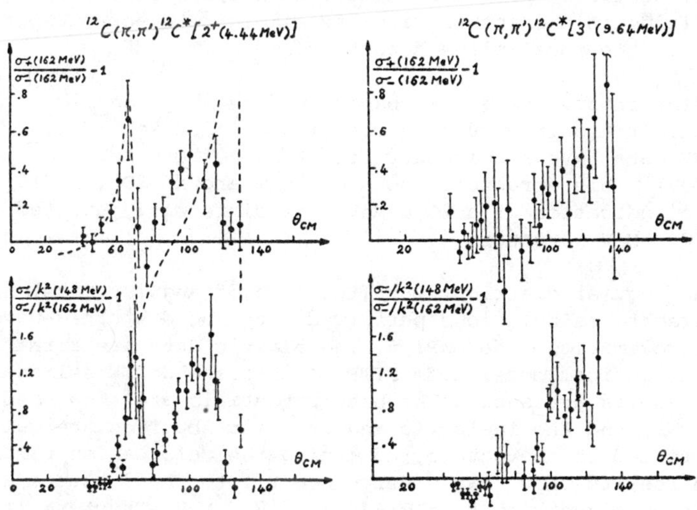

[1] J. Piffaretti et al., Phys. Lett. 67B, 289 (1977).

* Supported in part by the Swiss National Science Foundation
† Supported in part by the National Science Foundation of the U.S.A

ISSN: 0094-243X/79/540257$1.50

GIANT RESONANCE STUDIES IN CALCIUM VIA INELASTIC PION SCATTERING

J.Arvieux, J.P.Albanese, M.Buenerd and D.Lebrun
Institut des Sciences Nucléaires, IN2P3, 38026 Grenoble, France

E.Boschizt and C.H.Q.Ingram
IEKP, Postfach 3640, D7500, Karlsruhe, BRD

J. Jansen
SIN, 5234 Villigen, Switzerland

Inelastic scattering of pions has been performed with the SUSI spectrometer at SIN in the Giant Resonance region (GR) of ^{nat}Ca and ^{90}Zr. Sample spectra are shown in fig.1 for π^+ on Ca at 241 MeV and 163 MeV. They exhibit strongly excited bumps around 18 MeV for Ca and 14.5 MeV for ^{90}Zr not shown.

The G.R. structure in Ca has been extensively studied both at 163 MeV and 241 MeV. This study extended the data we have obtained so far on elastic scattering on this nucleus [1].

A special care has been taken to eliminate on line elastic events and muons coming from decay of the direct pion beam or from elastically scattered pions. Data have been taken at different magnetic fields in the spectrometer to eliminate eventual spurious structures in the focal plane detectors.

After removal of a 50%, background as shown on fig.1, all spectra exhibit a main structure centered at 18.2 MeV ($\Gamma \sim 3.-3.5$ MeV). The 241 MeV data have a secondary structure at 20.6 MeV ($\Gamma \sim 1.7-2$ MeV). This structure has a minimum around 20°. At 163 MeV there is no indication of a 20.6 MeV peak differently excited from the main 18.2 MeV structure.

The angular distributions (taken in 5° angular bins) have been compared to calculations performed with the distorted-wave impulse approximation code DWPI [2]. The elastic data are first fitted with the optical model code FITPI [3]. Laplace and Kisslinger potential results are shown. The best potentials are then used as input for DWPI and the inelastic angular distributions are calculated and fitted to the data by adjusting the deformation parameter β. The results show that a very good fit to the 3^- state at 3.74 MeV is obtained with $\beta_3 = 0.47$. The 18.2 MeV structure is compared in fig. 1b) and c) with an L = 2 angular distribution. A good fit (better at 241 MeV than at 163 MeV) is obtained with $\beta_2 = 0.22$ (or 77% of the EWSR), a value somewhat larger than va-

ISSN: 0094-243X/79/540258-03$1.50

lues obtained with other hadronic probes. This structure is then unambiguously identified with the GQR.

The 20.6 MeV structure has been compared with E0 and E1 angular distributions. The GDR angular distribution with β_1=0.1 corresponding to a Goldhaber-Teller form factor exhausting 100% of the EWSR [4] is shown in fig. 1c. It clearly does not agree with the 20.6 MeV data which width is also much smaller than the known width of the GDR.

We have compared E0 calculations to the angular distribution of the 20.6 MeV bump, using breathing mode form factors $F(r) = -3\rho - r\frac{\partial\rho}{\partial r}$. They give good fits with β_o = 0.15 (70% EWSR) at 241 MeV for a Laplace potential (β_o = .24 for Kisslinger). If one tries to divide the bump around 18 MeV into two structures at 18.2 MeV ($\Gamma \sim 3.5$ MeV) and 20.6 MeV ($\Gamma \sim 1.8$ MeV) one finds an angular distribution again in agreement with an E0 assignment although the best fit is obtained for a deformation β_o = 0.069 (17% EWSR) for Laplace, a value of the order of magnitude of what has been observed with other hadronic probes. Further studies with better statistics are needed to prove that it is not merely an infortunate statistical effect.

REFERENCES

1. J.P. Albanese et al., Phys. Lett. 76B, 173 (1978).
2. R.A. Eisenstein and G.A. Miller, Compt. Phys. Com. 11,95 (1976).
3. M.D. Cooper and R.A. Eisenstein, Report LA-5929 Ms, LASL (1975).
4. A.S. Rosenthal, Thesis (unpublished), University of Colorado, Boulder (1978).

FIGURE CAPTION

Fig.1. a) Energy spectra for π^+ + Ca.
b) angular distribution at 241 MeV for the ground-state, first inelastic and Giant-Resonance structures at 18.2 MeV (crosses) and 20.6 MeV (triangles) excitation energy. Full and long-dash curves correspond to DWIA calculations with the Laplace optical potential parameters, short-dash and dot-dash curves with the Kisslinger parameters.
c) same as b) but at 163 MeV. The dotted curve represents 100% of the EWSR for a giant dipole resonance with a Goldhaber-Teller form factor.

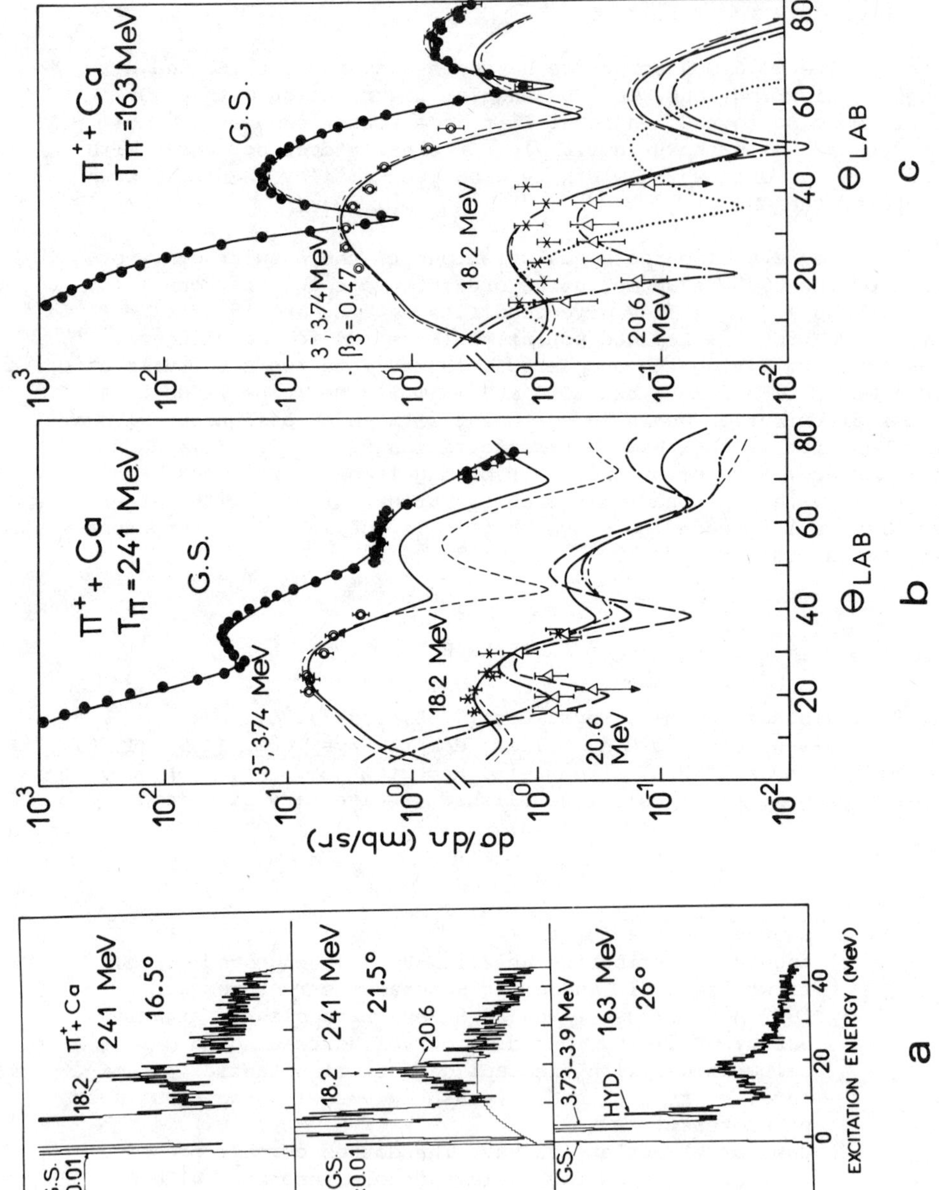
G.S. x0.01
18.2
π+ + Ca
241 MeV
16.5°
GS. x0.01
18.2
20.6
241 MeV
21.5°
GS.
3.73-3.9 MeV
HYD.
163 MeV
26°
EXCITATION ENERGY (MeV)
a
dσ/dΩ (mb/sr)
π+ + Ca
Tπ = 241 MeV
G.S.
3⁻, 3.74 MeV
18.2 MeV
20.6 MeV
θLAB
b
π+ + Ca
Tπ =163 MeV
G.S.
3⁻, 3.74MeV
β3 = 0.47
18.2 MeV
20.6 MeV
θLAB
c

CLUSTERING EFFECTS IN INELASTIC INTERACTION OF π^+ MESONS ON ^{4}He

F. Balestra, M.P. Bussa, L. Busso, R. Garfagnini and G. Piragino
Istituto di Fisica dell'Universitá di Torino, Torino, Italy
I.N.F.N., Sezione di Torino, Torino, Italy

C. Guaraldo, A. Maggiora and R. Scrimaglio
L.N.F. dell'I.N.F.N., Frascati, Italy

I.V. Falomkin, G.B. Pontecorvo and Yu. A. Shcherbakov
J.I.N.R., Dubna, U.S.S.R.

Clustering effects in (π^+, ^{4}He) inelastic interactions can be investigated searching for correlated emissions of nucleon-recoil nucleus systems in reactions as ^{4}He (π^+, π^0p ^{3}He), ^{4}He (π^+, π^+ p ^{3}H), ^{4}He (π^+, π^+n ^{3}He). Such processes have been studied using the pion beam of the LNF at the energies E_π = (120 ± 15), (145 ± 10), (165 ± 10) MeV. Fig. 1 shows the distribution of the events of the three above reactions as a function of the opening angle between the nucleon and the recoil nucleus at the energy E_π=145 MeV, normalized to the respective cross sections. The distribution of the (π^0p ^{3}He) events and of the (π^+p ^{3}H) events have symmetrical behaviours with a large maximum around 90°. On the contrary, the distribution of the (π^+n ^{3}He) events is clearly peaked around $\theta_{n^3He} \sim 150°$, wich indicates a strong correlation between the neutron and the ^{3}He recoil nucleus . Similar correlated n-^{2}H and n-^{3}H pair emission have been observed[2] in π^-capture on ^{12}C.

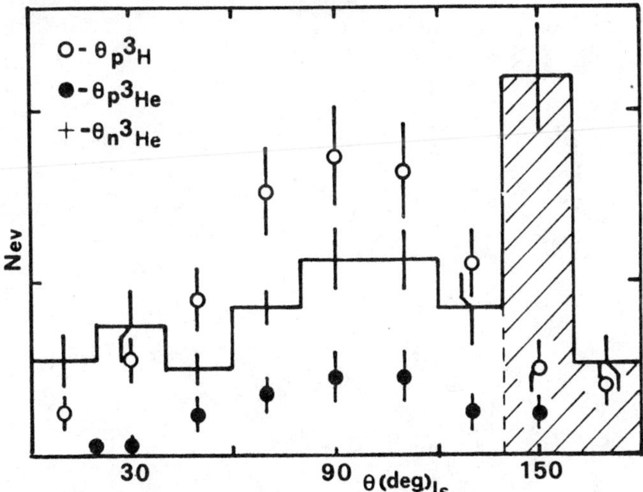

Fig. 1. Distributions of the events of the reactions ^{4}He (π^+, π^0p ^{3}He), ^{4}He (π^+, π^+ p ^{3}H) and ^{4}He (π^+, π^+ n ^{3}He) versus the opening angle of the pairs p-^{3}He, p-^{3}H, n-^{3}He respectively. E_π=145 MeV.

Further informations can be deduced from the analysis of the energy distributions of the particles emitted in the "clustering" reaction by selecting the events with $\theta_{n^3He} < 140°$ from those $\theta_{n^3He} > 140°$. In the first class ($\theta < 140°$) of events, π^+, n and ^{3}He have, respectively the average energies: $\bar{E}_\pi$=102.8 MeV, $\bar{E}_n$ =16.6 MeV, $\bar{E}_{^3He}$=6.9 MeV. In the class ($\theta > 140°$), the average energies are, respectively: $\bar{E}_\pi$ =89.6 MeV, $\bar{E}_n$ =27.4 MeV, $\bar{E}_{^3He}$= 7.3 MeV. In other words, the operating of the correlation (for $\theta_{n^3He} > 140°$) is characterized by a stronger relative energy of the n-^{3}He pair (and a lower scattered pion energy), compared to the energy of the incorrelated events ($\theta < 140°$).

ISSN:0094-243X/79/540261-02$1.50

Moreover, if we analize the average effective mass of the n-^{3}He system, selecting again between the $\theta_{n^3He} \lessgtr 140^o$ classes of events, we find $\bar{M}_{\theta<140^o}$=3760 MeV and $\bar{M}_{\theta>140^o}$=3780 MeV. The mass difference of 20 MeV is just close to the neutron binding energy in the ^{4}He nucleus.

REFERENCES

1. L. Busso et al., Nucl. Instr. and Meth. 102, 1 (1972).
2. D.M. Lee et al., Nucl. Phys. A197, 106 (1972).

SCATTERING OF A PION BY THE NUCLEAR PION CLOUD

William S. Truex* and Manoj K. Banerjee*
University of Maryland, College Park, Maryland 20742

J. Barry Cammarata†
Virginia Polytechnic Institute and State University
Blacksburg, Virginia 24061

The nuclear pion cloud can interact with any external probe including an external pion. The question is whether the effect can be significant in any situation. We find that in the case of pion nucleus scattering lengths (extracted from pionic atoms) the exchange current effect is about 10% of the total. Bearing in mind that the experimental accuracy is smaller than 5% one must regard the effect as significant.

We calculate the effect by considering the diagram where the external pion scatters off a pion in flight between two nucleons 1 and 2. For the pion-pion scattering amplitude we use a soft pion limit, regarding the exchanged pion as being soft. The resulting two nucleon operator in the coordinate space is

$$-\left[F_{\pi\pi}\frac{2m_\pi^4}{f_\pi^2}\right]\left[\frac{g_\pi^2}{4\pi}\frac{m_\pi^2}{4M^2}\right]\frac{1}{6}\tau_1\cdot\tau_2\left[\vec{\sigma}_1\cdot\vec{\sigma}_2\left(m_\pi-\frac{2}{r}\right)+S_{12}\left(m_\pi+\frac{1}{r}\right)\right]e^{-m_\pi r},$$

where $S_{12} = 3\,\dfrac{\vec{\sigma}_1\cdot\vec{r}\;\vec{\sigma}_2\cdot\vec{r}}{r^2} - \vec{\sigma}_1\cdot\vec{\sigma}_2$. The quantity $F_{\pi\pi}2m_\pi^4/f_\pi^2$ is the value of the invariant $\pi\pi$ scattering amplitude when one pion is soft and the other is physical and at rest. $f_\pi = 0.939\ m_\pi^3$ is the pion decay constant.

The value of $F_{\pi\pi}$ depends on the form of the chiral symmetry breaking term used. The predicted values range from m_π^2, due to Weinberg,[1] to $-4\ m_\pi^2$ due to Schwinger.[2]

We calculate the contribution of scattering by the nuclear pion cloud to $\pi-{}^{16}O$ scattering length whose value is $(-387\pm14)\times10^{-3}\ m_\pi^{-1}$.[3] The calculation was carried out using Weinberg's value of m_π^2 for $F_{\pi\pi}$. Because of the tensor term the tensor correlation in the nuclear wave function must be taken account of. This was done with the Reid soft core B potential[4] and including Pauli effect. The result is $+37.3\times10^{-3}\ m_\pi^{-1}$. (Without correlation one obtains $-36.1\times10^{-3}\ m_\pi^{-1}$.) Thus the pion cloud contribution is ∿10% of the total. If the Schwinger theory is correct the contribution will be $-149\times10^{-3}\ m_\pi^{-1}$.

This result makes a very strong case for developing a careful theory of scattering lengths including the effects of pion absorption, double scattering, etc. If the accuracy of the theoretical

---

*Supported in part by the U. S. Department of Energy.
†Supported in part by the U. S. National Science Foundation.

ISSN:0094-243X/79/540263-02$1.50

calculation is as low as that of the experimental data we should be able not only to detect the effect of scattering by the pion cloud but also to distinguish between the various theories of $\pi\pi$ scattering.

REFERENCES

1. J. Hufner, L. Tauscher and C. Wilkin, Nucl. Phys. A231, 455 (1974).
2. S. Weinberg, Phys. Rev. Lett. 17, 616 (1966).
3. J. Schwinger, Univ. of Hawaii Lectures (unpublished) (1977).
4. R. V. Reid, Jr., Ann. Phys. 50, 411 (1968).

CORRELATION EFFECTS IN THE ELASTIC SCATTERING OF $\pi^{\pm}$ FROM ^{18}O

I. Morrison and K. Amos
University of Melbourne, Victoria, Australia.

Angular distributions from the elastic scattering of 163 MeV pions (π^+ and π^-) from both ^{16}O and ^{18}O have been recently measured [1] and a marked shift in the position of the first minimum for π^- scattering from ^{18}O vis-à-vis from ^{16}O observed in contrast to that shift in the π^+ elastic scattering data. This distinction not only was cited [1] as evidence for the sensitivity of pion scattering to neutron distributions in the nucleus but also should reflect nuclear ground state correlations. In the cases of ^{16}O and ^{18}O, any effects of ground state correlations should be most apparent since a number of studies including, and referenced in, a recent large basis shell model study [2], have ascertained that the ^{18}O ground states has as much as 28% 4 particle-2 hole components (with respect to ^{16}O being closed).

To investigate the effects of correlations, and as a prelude to larger basis calculations and to analyses of double charge exchange (π^+, π^-) data, we have used Glauber theory to specify the elastic scattering amplitudes. The method of calculation and the π-nucleon profile functions we have used are those specified by Guardiola [3] while the intrinsic nuclear wave functions required in the evaluation of the many body overlap functions were determined from p-sd shell model calculations. The f-psd matrix elements of McGrory and Wildenthal [4] were used where core breaking in this study is limited to exatations from the $0p_{1/2}$ orbit.

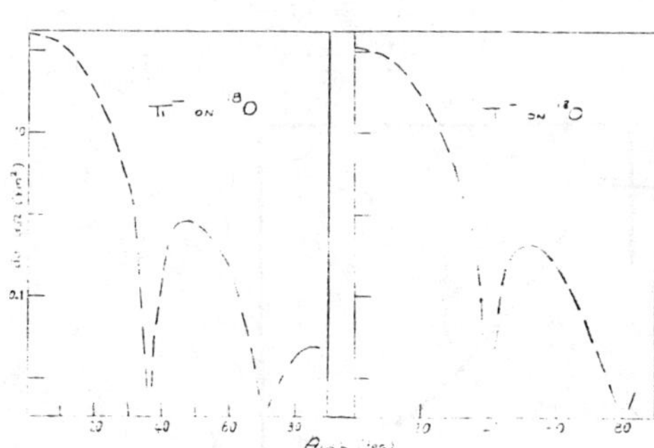

As shown in the accompanying figure, a shift between predicted minima in π^+ and π^- elastic scattering occurs but no appreciable change in either cross-section was observed with inclusion of ground state correlations in this model.

1) J. Jansen et. al, Phys. Letts. 77B (1978) 359.
2) I. Morrison et. al., Phys. Rev. C17 (1978) 1485.
3) R. Guardiola, in 'The Structure of Nuclei' IAEA Vienna 1972
4) J.B. McGrory and B.H. Wildenthal, Phys. Rev. C7 (1973) 974.

ISSN:0094-243X/79/540265-01$1.50

ISOBARIC RESONANCE EXCITATION IN INELASTIC INTERACTION OF π^+ MESONS ON ^4He

F. Balestra, M.P. Bussa, L. Busso, R. Garfagnini and G. Piragino
Istituto di Fisica dell'Universitá di Torino, Torino, Italy
I.N.F.N., Sezione di Torino, Torino, Italy

C. Guaraldo, A. Maggiora and R. Scrimaglio
L.N.F. dell'I.N.F.N., Frascati, Italy

I.V. Falomkin, G.B. Pontecorvo and Yu. A. Shcherbakov
J.I.N.R., Dubna, U.S.S.R.

All the inelastic channels in (π^+, ^4He) interaction have been measured with the pion beam of the L.N.F. at E_π=(120 ± 15), (145 ± 10) and (165 ± 10) MeV. The obtained total inelastic cross section, together with the existing data of Binon et al[2], of Fowler et al[3] and of Kozodaev et al[4], present resonant behaviour with the maximum at $E_\pi \simeq$165 MeV. However, among the different reaction channels wich contribute to the total inelastic cross section, only the processes ^4He (π^+,π°p ^3He), ^4He (π^+,π^+p ^3H) and ^4He (π^+,π^+n ^3He) exhibit a marked resonant behaviour. In fig. 1, as an example, the cross section values of the (π^+p ^3He) reaction are reported, compared with the experimental result for the charge exchange reaction π^-p→π°n on free nucleon[5]. The reaction mechanism appears clearly as the isobaric resonance excitation. This conclusion is significantly emphasized by the effective mass diagrams, in which the number of events for the three above reactions are plotted as a functions of the masses of the (π°p) (π^+ p) and (π^+n)systems at the energy E_π=165 MeV. In fig. 2a,b and c the experimental effective mass distributions are compared with the phase space behaviours calculated with the FOWL program[6] taking into account the Δ(1236) resonance excitation. Both experimental and calculated distributions show a significant maximum value at an energy around 1200 MeV. It is possible to conclude that the resonance behaviour of the total inelastic (π^+,^4He) cross section can be essentially ascribed to the isobaric resonance excitation mechanism of the (π°p ^3He), (π^+p ^3H) and (π^+n ^3He) reaction channels.

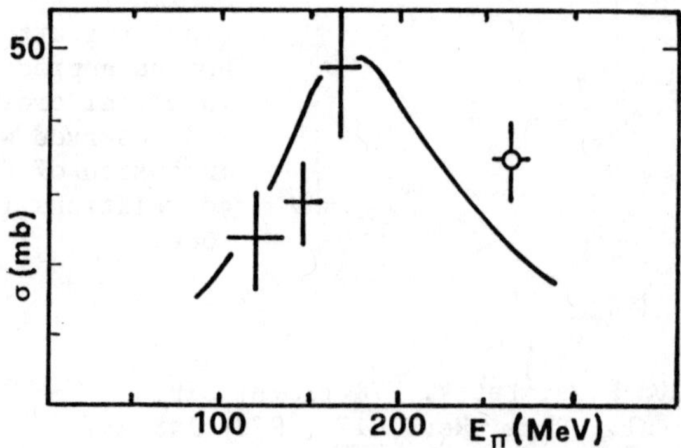

Fig. 1. Total cross section of the ^4He (π^+, π^op ^3He) reaction. Crosses: present data; open circle: ref. 4; full line: π^+p π^on reaction cross section from ref. 5.

ISSN:0094-243X/79/540266-02$1.50

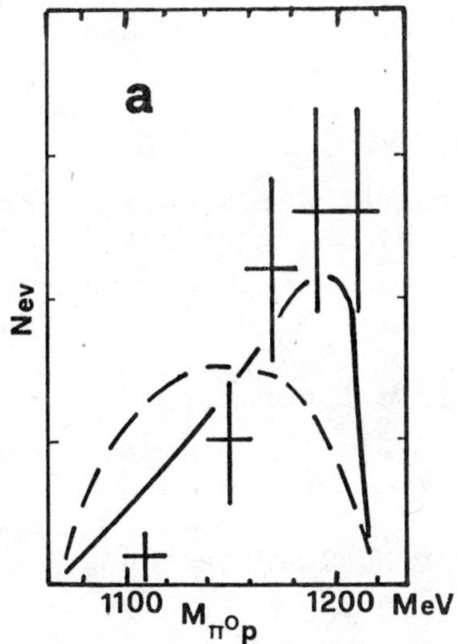

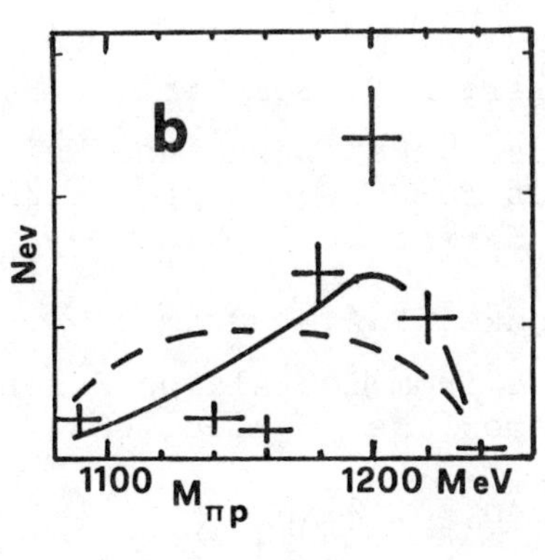

Fig. 2a,b,c. Effective mass distributions at $E_\pi = 165$ MeV for the $(\pi^o p)$, $(\pi^+ p)$, $(\pi^+ n)$ systems Crosses: present data;full and dotted lines: phase space calculations taking and without taking into account the $\Delta(1236)$ resonance excitation.

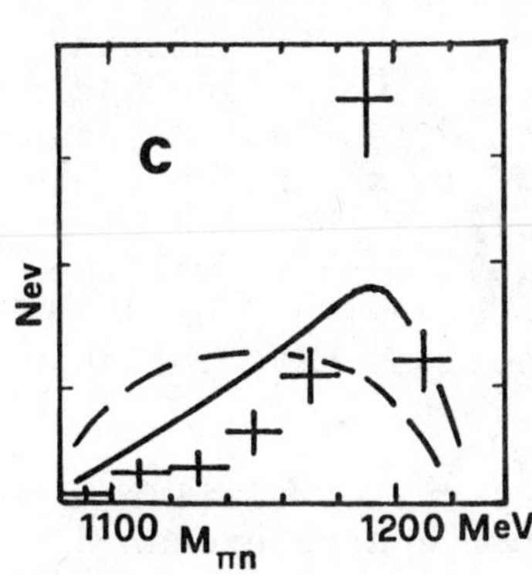

REFERENCES

1. L. Busso et al., Nucl. Instr. and Meth. 102, 1 (1972).
2. F. Binon et al., Nucl. Phys. A298, 499 (1978).
3. E.C. Fowler et al., Phys. Rev. 91, 135 (1953).
4. M.S. Kozodaev et al., Sov. Phys. JETP 11, 300 (1960).
5. D.V. Bugg et al., Nucl. Phys. B26, 588 (1971).
6. F. James, report CERN 68-15.

Measurement of Double Charge Exchange Cross-Section $^{16}O\ (\pi^+,\pi^-)$ at 240 MeV.

J. Bolger, E.T. Boschitz, C.H.Q. Ingram, G. Pröbstle

Kernforschungszentrum und Universität Karlsruhe,
Institut für Experimentelle Kernphysik, Postfach 3640
D-7500 Karlsruhe 1

R. Mischke, P.A.M. Gram

on leave from Los Alamos Scientific Laboratory,
Los Alamos, New Mexico, USA

J. Jansen, J. Zichy

SIN, Schweizerisches Institut für Nuklearforschung,
CH-5234 Villigen

The double charge exchange cross-section $\frac{d^2\sigma}{d\Omega\pi dE\pi}$ $[^{16}O\ (\pi^+,\pi^-)]$ has been measured for 240 MeV incident pions. at 50°, 85° and 130°. The data were taken using the SIN pion spectrometer, with an effective acceptance of 15 msr and 30 % in momentum (π^- in the spectrometer were identified by trajectory analysis and time of flight. Backgrounds were negligible). Results will be presented, covering outgoing pion energies from 40 MeV upto $\sim$ 200 MeV, where the cross-sections have fallen to $<$ 100 nb $(sr.\ MeV)^{-1}$.

The cross-sections at all 3 angles are of similar magnitude, and consistent with a total charge exchange cross-section of a few mb. Comparison will be made with the $^{16}O\ (\pi^+,\pi^{+\prime})$ data also presented to this conference.

ISSN:0094-243X/79/540268-01$1.50

STUDY OF THE INCLUSIVE PION DOUBLE CHARGE EXCHANGE REACTIONS IN ^{12}C AND ^{40,44,48}Ca

J. Davis, J. Källne, J.S. McCarthy, R.C. Minehart
University of Virginia
Charlottesville, Va., 22901

C.L. Morris and H.A. Thiessen
Los Alamos Scientific Laboratory
Los Alamos, New Mexico 87545

G. Blanpied and G.R. Burleson
New Mexico State University
Las Cruces, New Mexico 88001

K. Boyer, W. Cottingame and C.F. Moore
University of Texas
Austin, Texas 78712

C.A. Goulding
Florida A & M University
Tallahassee, Florida 32306

The differential cross section $\frac{d^2\sigma}{d\Omega dT}$ for the double charge exchange (DCE) reactions (π^+,π^-) and (π^-,π^+) have been measured using the EPICS pion spectrometer system of LAMPF. The exit pions were detected over an energy range of T_π = 175-255 MeV at $\theta = 60°$ for an incident energy of 290 MeV. The cross section shows a significant energy dependence in this region and increases with increasing energy loss. Within the (sometimes poor) statistics, the energy dependence is the same for the four nuclei studied and no difference could be distinguished between (π^+,π^-) and (π^-,π^+).

We have examined the cross section magnitudes and find that DCE is one to two orders of magnitude smaller than (π,π) scattering in the same kinematics region which is that of quasi-free scattering. This is not a surprising difference since DCE is a second order process in terms of πN scattering, i.e., two πN single charge exchange scatterings are required in DCE. The systematics of the DCE are not easily reconscilable with a description based on multiple SCE scatterings from individual nucleons. Instead we find, for instance, a correlation between the DCE cross section and the number of nucleon pairs in the target nucleus that form core-excited configurations. It is possible that DCE is sensitive to such configurations and could explain the puzzling observation that $d\sigma(\pi^-,\pi^+) > d\sigma(\pi^+,\pi^-)$ in the self conjugated nuclei ^{12}C and ^{40}Ca. The results obtained for the calcium isotopes points to the possibility that the cross section ratio for (π^+,π^-) and (π^-,π^+) might be sensitive

ISSN:0094-243X/79/540269-02$1.50

to differences between neutron and proton distributions at the nuclear surface. The (π^-,π^+) and (π^+,π^-) reactions procede through the transitions $2p \rightarrow 2n$ and $2n \rightarrow 2p$. They are thus uniquely selective with respect to p and n. The attenuation of the reactions is symmetric with respect to p and n interactions so the attenuation would follow the nuclear matter distribution.

PROPERTIES OF PION SINGLE CHARGE EXCHANGE REACTIONS IN NUCLEI

T. Bowles, D. F. Geesaman, R. J. Holt, H. E. Jackson,
R. M. Laszewski, J. R. Specht, and E. J. Stephenson
Argonne National Laboratory, Argonne, Illinois 60439*

R. E. Segel
Northwestern University, Evanston, Illinois 60201

R. P. Redwine and M. A. Yates-Williams
Los Alamos Scientific Laboratory, Los Alamos, New Mexico 87545

J. Julien
C.E.N. - Saclay

ABSTRACT

Pion single charge exchange reactions initiated by π^+ and π^- beams have been surveyed on targets ranging from Be to Pb with the use of a back angle γ-ray spectrometer. The beam energy varied from 50 to 200 MeV, and the emerging π° spectra were recorded at 6 angles from 25° to 150°. This experiment was an extension of earlier measurements[1] at T_π = 100 MeV and θ = 40° and 120°. The 2γ decay of the π° was detected colinearly with the π° momentum. The substantial Doppler shifts of the two γ-rays allowed the low-energy photon to be detected with good resolution in a large volume NaI(Tl) spectrometer while the presence of the high-energy photon was observed in a Pb-glass Cerenkov counter. The π° spectrum was obtained by transforming the NaI γ-ray spectrum. The efficiency of the detector system was calibrated using the $\pi^- + p \rightarrow \pi^\circ + n$ reaction observed in a CH_2 - C difference spectrum. The effects of the NaI response function were removed from the π° spectra.

The more complete angular distributions determine the total charge exchange cross section, whose dependence on pion charge and target mass can be investigated. As seen in Fig. 1 for ^{16}O, the resemblance of the T_π = 100 MeV measurements to the shape of the free nucleon cross section is confirmed. Compared to the shape of the free nucleon cross section, the observed angular distributions show forward angle enhancement at 50 MeV and lower forward angle values near the 3,3 resonance, indicating the importance of other reaction mechanisms.

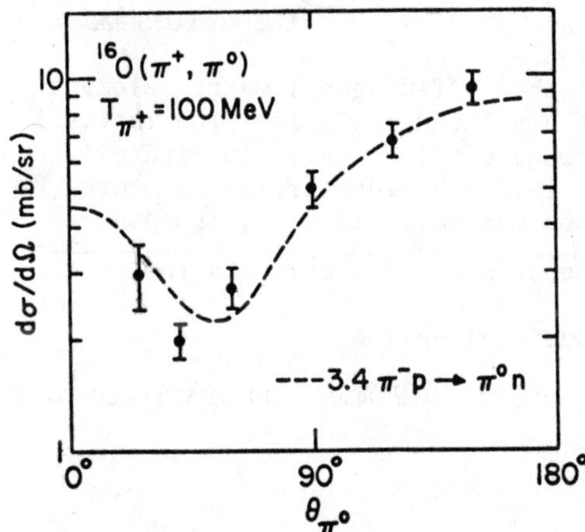

Figure 1. (See text.)

*Work supported by the U. S. Department of Energy.

[1]T. Bowles, et al. Phys. Rev. Lett. 40, 97 (1978).

ISSN:0094-243X/79/540271-01$1.50

MEASURING ISOSPIN MIXING IN NUCLEI USING $\pi^{\pm}$ INELASTIC SCATTERING*

W. B. Cottingame† and W. J. Braithwaite
University of Texas at Austin, Austin, Texas 78712

and

C. L. Morris
Los Alamos Scientific Laboratory, Los Alamos, N. M. 87545

ABSTRACT

A new strongly isospin-mixed doublet has been found in ^{12}C near 19.5 MeV, in a comparison of π^- and π^+ inelastic scattering at 180 MeV. The present techniques may be universally employable, at least in self-conjugate nuclei, in extracting isospin-mixing matrix elements.

Historically, only two examples of strong isospin mixing between two distinct nuclear states have been observed: (1) nearly maximally mixed 2^+ states in ^8Be near 16.5 MeV in excitation[1] and (2) partially mixed 1^+ states in ^{12}C at 12.7 and 15.1 MeV in excitation, respectively.[2]

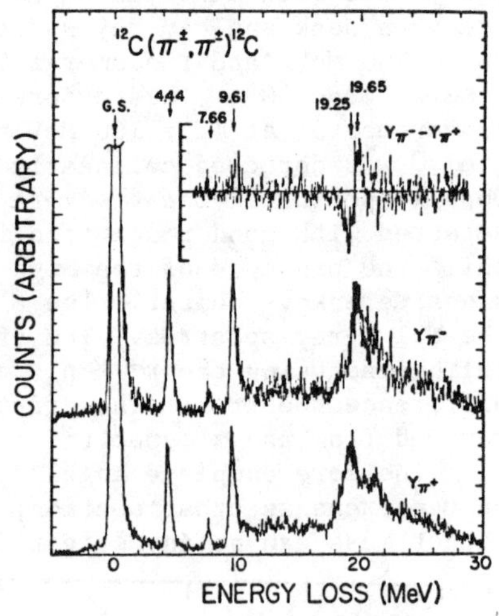

One of the few ways to look for possible charge dependence in the short-range interaction within nuclei is to measure the size of the off-diagonal, charge-dependent matrix elements between known nuclear states, for comparison with calculations based on mixing due to residual Coulomb interactions alone. More strongly-mixed states must be found if short-range nuclear charge dependence is to be established in this fashion. Both the sign and the size of the mixing matrix element may be important, as short-range nuclear charge dependence may result in a change of sign in this off-diagonal matrix element.

Figure 1 exhibits spectra of $\sigma(\pi^-)$ and $\sigma(\pi^+)$ for pion inelastic scattering. The difference spectrum, $\sigma(\pi^-) - \sigma(\pi^+)$, in Figure 1, shows a strong bipolar shape near 19.5 MeV, identifying one more pair of states as strongly, perhaps maximally mixed in isospin.

---

*Supported in part by the U. S. Department of Energy and the Robert A. Welch Foundation.
†Robert A. Welch Foundation Pre-doctoral Fellow.

ISSN:0094-243X/79/540272-02$1.50

When two states are sufficiently close in excitation energy to permit use of the two-state theorem to describe isospin mixing, one state has an increased neutron particle-hole strength and a decreased proton particle-hole strength, with this situation reversed for the other state. Negative-pion inelastic scattering favors excitation of the state with increased neutron particle-hole strength, with positive-pion inelastic scattering favoring the other state due to its increased proton particle-hole strength. Thus, measuring a "dip" in the difference spectrum, $\sigma(\pi^-) - \sigma(\pi^+)$, indicates a larger proton particle-hole strength for the state, whereas a "peak" indicates a larger neutron particle-hole strength for the state.

Peaks and dips in the difference spectra may be used to extract the sign of the isospin-mixing matrix element. However, if there is no isospin mixing, i.e., for state of pure isospin, there will be no favoring of π^- or π^+ in the difference spectrum, $\sigma(\pi^-) - \sigma(\pi^+)$. Thus, all contributions from states of pure isospin will subtract to zero, within statistical errors, reducing the amount of structure in the difference spectrum. Figure 1 shows some examples of this, even in the region of the two states near 19.5 MeV.

The dip in the difference spectrum in Figure 1 occurs for the lower of the isospin-mixed states, whereas the peak is associated with the upper member of these states, thus the lower of these states has an enhanced proton particle-hole strength and the upper of these states has an enhanced neutron particle-hole strength, establishing the sign of the charge-dependent mixing matrix element. This sign is consistent with previous results[1,2] in ^{8}Be and ^{12}C, where the upper member of the doublet, in each case, is associated with enhanced neutron particle-hole strength.

The present work suggests a new method for measuring isospin mixing, and perhaps short-range nuclear charge dependence, using a comparison of negative versus positive pion inelastic scattering to states in self-conjugate nuclei. This method is sensitive to both strength and phases of the off-diagonal mixing matrix elements. For isospin mixing amongst several states, comparisons between inelastic scattering of the two charged pion species to each state should be measureable, and the contribution of isospin mixing should be observable as a distinct peak or a distinct dip in the inelastic scattering $\sigma(\pi^-) - \sigma(\pi^+)$ difference spectrum, associated with each isospin-mixed nuclear state.

REFERENCES

1. J. B. Marion, et al., Nucl. Phys. 77, 129 (1966); J. B. Marion, et al., Phys. Rev. 157, 847 (1967); M. A. Oouthoudt and G. T. Garvey, Nucl. Phys. A 284, 41 (1977).

2. W. J. Braithwaite, et al., Phys. Rev. Letters 29, 376 (1972); E. G. Adelberger, et al., Phys. Rev. C 15, 484 (1977); J. M. Lind, et al., Nucl. Phys. A 276, 25 (1977).

ISOSPIN EFFECTS IN THE REACTIONS $^{12}C(\pi^{\pm},\pi^{\pm}+p)^{11}B$ AND $^{12}C(\pi^{\pm},\pi^{\pm}+D)^{10}B$

R. J. Ellis, H. J. Ziock, K.O.H. Ziock, Univ. of Virginia
J. Bolger, E. Boschitz, Univ. of Karlsruhe
J. Piffaretti, R. Corfu, Univ. de Neuchatel
J. Arvieux, Univ. de Grenoble, France

We have put the magnetic spectrometer, SUSI[5], in the pion channel[6] at SIN, in coincidence with a Si, Ge spectrometer and made measurements of the above reactions in the $(^3/_2,^3/_2)$ resonance region.

There is a well known anomaly in the ratio:

$$R_1 = \frac{\sigma(^{12}C(\pi^-;\pi^-n)^{11}C)}{\sigma(^{12}C(\pi^+;\pi^+n,\pi^0p)^{11}C} = 1.55 \pm 0.10^1;\ 1.57 \pm 0.05^2 \qquad (1)$$

which differs significantly from the value 3 one would expect from isospin conservation in the impulse approximation. Explanations have been proposed[3] which assume that the nucleon leaving the nucleus may charge exchange. Comiso et al.[4] used this to calculate that the nucleon charge exchanges 21% of the time if $R_1 = 1.55$. From this they calculated the ratio:

$$R_2 = \frac{\sigma(^{12}C(\pi^+,\pi^+p)^{11}B}{\sigma(^{12}C(\pi^-,\pi^-p)^{11}B)} = 2.7 \pm 0.68 \qquad (2)$$

instead of the value 9 expected from isospin conservation in the impulse approximation. We have measured this ratio at an incident pion energy of 180 MeV and a proton angle of 30°. SUSI accepted pions in the momentum range 140 to 200 MeV/c. Pions in the beam were selected by time of flight. Protons were selected in the Si, Ge detector by E,ΔE measurements. Our results are:

TABLE I

RUN	Θ_π	$\pi^{\pm}$	N_H/N_L	R_2^A	R_2^L	R_2^H
1	100°	+	1.17 ± 0.79	5.92 ± 0.79	14.6 ± 4.5	3.93 ± 0.59
2	100°	-	4.36 ± 1.5			
3	110°	-	4.87 ± 0.9	6.58 ± 0.52	18.9 ± 3.3	4.04 ± 0.38
4	110°	+	1.04 ± 0.09			
average:				6.25 ± 0.47	16.7 ± 2.8	3.99 ± 0.35
std dvn from 9				5.9	2.7	14.3
std dvn from 2.7 ± 0.68				4.3	4.9	1.7

The superscripts A, L and H refer to all the events, events where the ^{11}B excitation energy is less than 10 MeV, and to events where it is above 10 MeV, respectively. We make these distinctions because R_1 was determined by activation experiments, which are insensitive to particle unstable states. It is R_2^L which should be compared with the above prediction. The above results have been normalized against the number of incident pions. The ratio N_H/N_L is independent of this normalization and it is a measure of the shape of the Boron excitation energy spectrum.

We have also analyzed the deuteron knock-out events for these runs. We define the ratio:

ISSN:0094-243X/79/540274-02$1.50 Copyright 1979 American Institute of Physics

$$R_3 = \frac{\sigma(^{12}C(\pi^+,\pi^+ + D)^{10}B)}{\sigma(^{12}C\ \pi^-,\pi^- + D)^{10}B)} \quad (3)$$

TABLE II

RUN	Θ_π	$\pi^\pm$	N_H/N_L	R_3^A	R_3^L	R_3^H
1	100°	+	0.96 ± 0.2	1.77 ± 0.44	1.64 ± 0.55	1.92 ± 0.70
2	100°	-	0.82 ± 0.37			
3	110°	-	1.04 ± 0.19	1.14 ± 0.19	0.98 ± 0.23	1.34 ± 0.34
4	110°	+	0.76 ± 0.21			
average:				1.46 ± 0.24	1.31 ± 0.30	1.63 ± 0.39
std dvn from 1:				1.9	1.0	1.6

The fact that N_H/N_L and R_3^A, R_3^L and R_3^H are all essentially unity for all these runs enables us to "normalize" the proton data against the deuteron data. We define the ratio:

$$R_4 = \frac{\sigma(^{12}C(\pi^\pm,\pi^\pm + p)^{11}B)}{\sigma(^{12}C(\pi^\pm,\pi^\pm + D)^{10}B)} \quad (4)$$

TABLE III

RUN	R_4^A	R_4^L	R_4^H	R_{2D}^A	R_{2D}^L	R_{2D}^H
1	12.3 ±1.4	10.1 ±1.6	15.1 ±2.6	3.75±1.1	11.9±5.2	1.57±0.79
2	3.28±0.88	0.85±0.35	9.6 ±4.5			
3	2.06±0.28	0.76±0.16	3.19±0.48	5.82±1.1	15.2±4.6	3.93±1.1
4	12.0 ±1.9	11.5 ±2.5	12.6 ±2.8			
average:				4.79±0.79	13.5±3.5	2.75±0.66
std dvn from 9:				5.3	1.3	9.5
std dvn from 2.7 ± 0.68:				2.0	3.0	0.05

The subscript D means that the ratio has been normalized against the deuteron events.

We conclude:

1. The deuteron data is consistent with isospin predictions.
2. The deuteron ratios are independent of ^{10}B excitation energy. Therefore they confirm that the effects seen in the proton data are real physical effects and are not due to systematic effects.
3. The proton data (R_2^L and R_{2D}^L) are completely inconsistent with the predictions of the coherent charge exchange model.
4. There is a significant dependence of the proton ratio (R_2 or R_{2D}) on the ^{11}B excitation energy.
5. We conclude from the ratios N_H/N_L in Table I and R_4 in Table III that the $(\pi^-,\pi^- p)$ interaction-rate is four times greater in the high ^{11}B excitation energy range than the low, and that this causes the decrease by a factor of four in the proton ratios (R_2 and R_{2D}) between the low and high ^{11}B excitation energy ranges.

1. B. J. Dropesky et al., Phys. Rev. Lett. 34, 821 (1975).
2. L. H. Batist et al., Nuc. Phys. A254, 480 (1975).
3. P. W. Hewson, Nuc. Phys. A133, 659 (1969).
4. J. Comiso et al., Z. Physik A285, 101 (1978).
5. J. P. Albanese, Nuc. Instr. and Meth. 158, 363 (1979).
6. R. Balsiger, Nuc. Instr. and Meth. 157, 247 (1978).

THE ^{11}C PRODUCTION YIELD RATIOS FROM ^{12}C VS ^{13}C AND ^{12}C VS ^{16}O FOR INCIDENT $\pi^{\pm}$ FROM 100 TO 250 MeV

K. Aniol, D. Chiang, I. Halpern

University of Washington, Seattle, Wa. 98195

D. Drake, G. Haouat
University of California
Los Alamos Scientific Laboratory, Los Alamos, NM. 87545

ABSTRACT

From an activation measurement we have determined the ratios of ^{11}C production from ^{12}C vs ^{13}C and ^{12}C vs ^{16}O using the $\pi^{\pm}$ beams of LEP at LAMPF. The ratios were determined to within ≤2% by a direct comparison technique for $\pi^{\pm}$ energies between 100 and 250 MeV. In our measurement samples were simultaneously exposed to the same beam and subsequently counted in the same geometry. The ^{12}C vs ^{13}C ratio was measured using interspersed disks of graphite made from ^{12}C and ^{13}C. These disks were transferred to a separate holder for counting. For the ^{12}C vs ^{16}O measurement interspersed chambers of water and octane were bombarded and then the liquid was transferred to separate chambers for counting. For the π^{+} reactions the ratios are nearly constant. The ^{12}C vs ^{16}O ratio remains at about 2.2. The ^{12}C vs ^{13}C ratio increases from 3.5 to 3.8. However, for the π^{-} reaction, the ^{12}C vs ^{13}C ratio decreases monotonically from 8.25 at 100 MeV to 5.45 at 250 MeV, and the ^{12}C vs ^{16}O ratio peaks at 4.6 at 200 MeV. An interpretation of these ratios is currently in progress.

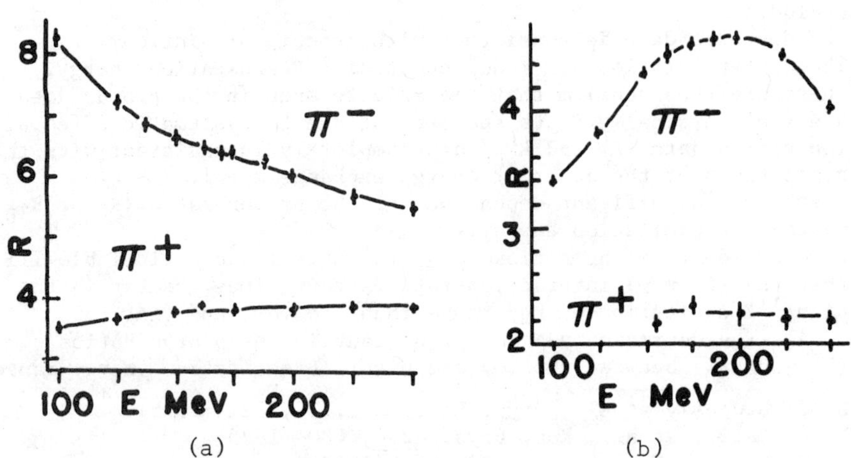

Fig. 1. The ratio R for ^{11}C production from (a) ^{12}C vs ^{13}C and (b) ^{12}C vs ^{16}O for $\pi^{\pm}$ beams.

ISSN:0094-243X/79/540276-01$1.50

THREE NUCLEON EXCHANGE MECHANISM IN π^+ INDUCED KNOCK OUT REACTIONS ON ^{4}He

F. Balestra, M.P. Bussa, L. Busso, R. Garfagnini and G. Piragino
Istituto di Fisica dell'Universitá di Torino, Torino, Italy
I.N.F.N., Sezione di Torino, Torino, Italy

C. Guaraldo, A. Maggiora and R. Scrimaglio
L.N.F. dell'I.N.F.N., Frascati, Italy

R. Mach and M.G. Sapozhnikov
J.I.N.R., Dubna, U.S.S.R.

The total cross sections of the reactions $\pi^+ {}^4He \to \pi^+ p\, {}^3H$ and $\pi^+ {}^4He \to \pi^+ n\, {}^3He$, the total and differential cross sections of the charge exchange reaction $\pi^+ {}^4He \to \pi^o p\, {}^3He$, have been measured with the pion beam of the LNF[1] at the energies $E_\pi = (120 \pm 15), (145 \pm 10), (165 \pm 10)$ MeV. The total cross sections for the three inelastic processes are given in table 1. In the same table we also give the experimental values of the R ratio: $R = \sigma(\pi^+ p\, {}^3H)/[\sigma(\pi^+ n\, {}^3He) + \sigma(\pi^o p\, {}^3He)]$. Our values are compared with those obtained at 273 MeV by Kozodaev et al[2].

Table 1

Experimental total inelastic cross sections and R values

E_π (MeV)	($\pi^+ p\, {}^3H$) (mb)	($\pi^+ n\, {}^3He$) (mb)	($\pi^o p\, {}^3He$) (mb)	R
120 ± 15	74 ± 10	31 ± 5	23 ± 7	1,37 ± 0.36
145 ± 10	73 ± 8	46 ± 5	28 ± 6	0.99 ± 0.26
165 ± 10	107 ± 12	75 ± 8	47 ± 10	0.88 ± 0.20
273 ± 7	50 ± 7	14 ± 4	34 ± 6	1.04 ± 0.35

As can be seen from table 1, the simple plane wave impulse approximation (PWIA) fails to reproduce the experimental values of the ratio R (in fact a value of about 3 is expected). This apparent con-

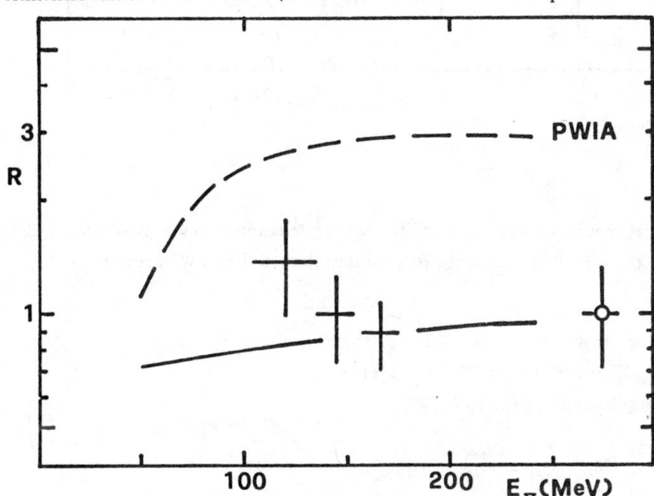

Fig. 1. Energy dependence of the ratio R. Crosses: present work;open circle:ref.2. Full line:three nucleon exchange mechanism; dashed line: PWIA results.

ISSN: 0094-243X/79//540277-02$1.50 Copyright 1979 American Institute of Physics

flict indicates the considerable role of the initial and/or final state interaction in pion induced knock out reactions on ^{4}He. By the aid of a simple theoretical model[3], which incorporates the three nucleon exchange mechanism in the PWIA, in addition to the one nucleon mechanism, it is possible to explain, qualitatively, the main features of the considered knock out reactions. In fig. 1 the experimental ratios are compared with the PWIA calculation and with the model which takes into account also the three nucleon exchange mechanism. The calculated values with the exchange mechanism lie rather close to the experimental ones, much closer than the PWIA values. In fig. 2 is reported the angular distribution of the π^0 in the charge exchange reaction $\pi^+ {}^4$He $\rightarrow \pi^0 p\, {}^3$He at E_π=165 MeV, compared with the results of the three nucleon exchange calculation. The present model describes the shape of the differential cross section much better than the PWIA does (dashed line in the figure). The model predicts correctly the position of the first maximum as well the build up in the backward angular region.

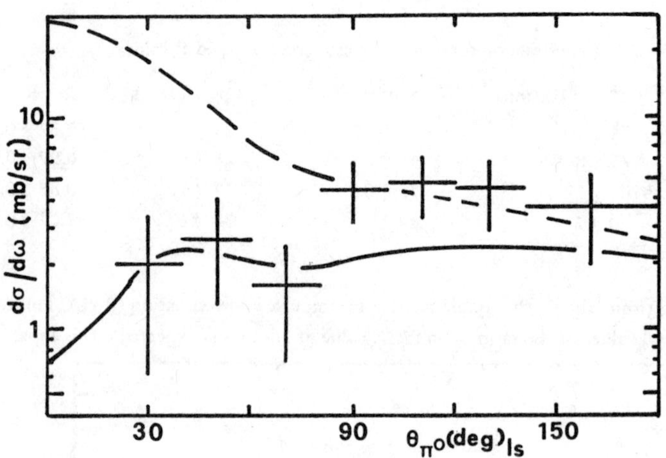

Fig. 2. Differential cross section for the $\pi^+ {}^4$He $\rightarrow \pi^0 p\, {}^3$He reaction for E_π=165 MeV. Crosses: present work; full line: three nucleon exchange mechanism; dashed line: PWIA result.

REFERENCES

1. L. Busso et al., Nucl. Instr. and Meth. 102, 1 (1972).

2. M.S. Kozodaev et al., Sov. Phys. JETP 11, 300 (1960).

3. R. Mach et al., Nuovo Cim. A45, 325 (1978).

MULTIPLE SCATTERING PROCESSES IN PION ^{4}He INELASTIC INTERACTION

F. Balestra, M.P. Bussa, L. Busso, R. Garfagnini and G. Piragino
Istituto di Fisica dell'Universitá di Torino, Torino, Italy
I.N.F.N., Sezione di Torino, Torino, Italy

C. Guaraldo, A. Maggiora and R. Scrimaglio
L.N.F. dell'I.N.F.N., Frascati, Italy

I.V. Falomkin, G.B. Pontecorvo and Yu. A. Shcherbakov
J.I.N.R., Dubna, U.S.S.R.

The total cross section of the reaction $\pi^+ {}^4He \to \pi^+ 2p2n$ has been measured with the pion beam of the Laboratory Nazionali di Frascati[1] at the energies E_π= (120±15), (145±10) and (165±10) MeV.The mechanism of this reaction can be interpreted in two ways. Firstly, the reaction may be the result of simultaneous interaction between the pion and complexes of nucleons. This assumption can be supported by the fact that the main mechanism in non radiative capture of π^+ on ^{4}He does result in quasideuteron absorption, i.e. in processes in which at least two nucleons partecipate. Secondly, this case may be the result of successive collisions of the meson with the nucleons, justifying the terminology adopted tentatively by Kozodaev et al of "multiple scattering processes"[2]. In this last case, since no correlation is effective, the angular distribution of the particles in the final state is expected to be isotropic. Actually, it is probable that both processes are in operation. The comparison among the values for the ^{4}He (π^+,π^+2p2n) cross section and the previous ones obtained in ref.2 and 3, is shown in fig. 1. Fig. 2 shows the distribution of the events for E_π=145 MeV as a function of the opening angle between the two protons. The behaviour is nearly isotropic indicating that in the reaction the cascade process does take place and seems to be the predominant one. This conclusion is also stressed in fig. 3, in which the differential cross section for E_π=145 MeV is reported versus the angle of the scattered pion. In fact this angular distribution is in agreement, as shown in the figure, with the result obtained using the FOWL program[4] to calculate the phase space distribution, with the hypothesis of isotropically scattered particles and without interaction in the final state.

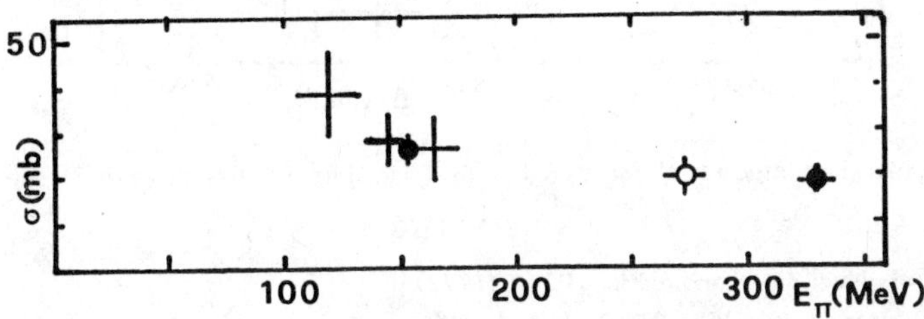

Fig. 1. ^{4}He (π^+,π^+2p2n) reaction cross section. Crosses: present data; open circle: ref. 2 - ^{4}He(π^-, π^-2p2n) reaction cross section. Full points: ref. 2 and 3.

On the other way, reactions as ^{4}He (π^+,π^+p^{3}H) and ^{4}He (π^+,π^+n^{3}He) can be considered as the results of a quasi free scattering by individual bound nucleons, considering that the residual nuclei do not partecipate directly in the interaction with the incoming pions. The relative probability ε of multiple scattering in inelastic scattering on ^{4}He can be determined directly from the number of cases that correspond to the above reported reactions: $\varepsilon=\sigma(\pi^+2p2n)/\left[\sigma(\pi^+2p2n)+\sigma(\pi^+p^3H)+\sigma(\pi^+n^3He)\right]$. The mean value of ε in the energy region considered in our experiment resulted ε= 0.20 ±

ISSN: 0094-243X/79/540279-02$1.50

±0.05 to be compared with: $\varepsilon \simeq 0.24 \pm 0.06$ obtained in ref.2 with π^+ of 273 MeV; $\varepsilon \simeq 0.29 \pm 0.06$ obtained in ref. 3 with π^- of 153 MeV and $\varepsilon \simeq 0.29 \pm 0.05$ obtained in ref.2 with π^- of 330 MeV.

From this, it is possible to conclude remarking the relevant role of multiple scattering processes in pion ^{4}He inelastic interaction, together with their relatively independence on energy and pion charge.

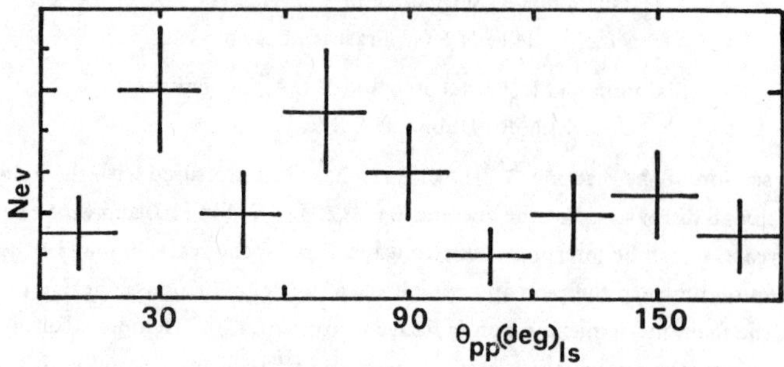

Fig. 2. Distribution of ^{4}He (π^+,π^+2p2n) events versus the opening angle between the two protons. E_π=145 MeV.

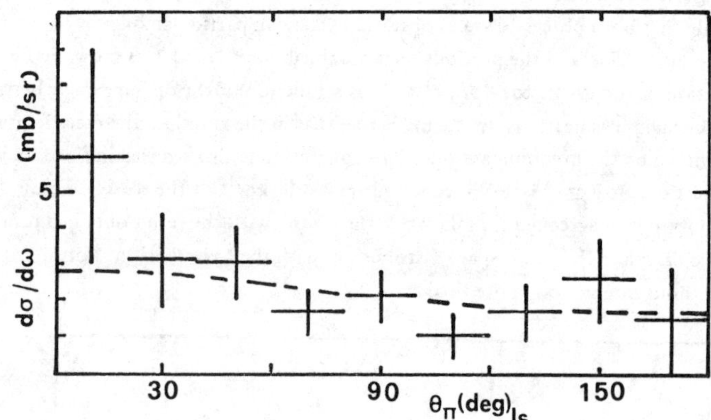

Fig. 3. Angular distribution of the scattered pions. Dashed line: FOWL program[4] result.

REFERENCES

1. L. Busso et al., Nucl. Instr. and Meth., 102 , 1 (1972).
2. M.S. Kozodaev et al., Sov. Phys. JETP 11, 300 (1960).
3. Yu. A. Budagov et al., Sov. Phys. JETP 15, 824 (1962).
4. F. James, report CERN 68-15.

QUASIELASTIC SCATTERING OF PIONS FROM ^{12}C AND $^{40,44,48}Ca$*

G. S. Blanpied and G. R. Burleson
New Mexico State University, Las Cruces, NM 88003

J. Davis, J. S. McCarthy, and R. C. Minehart
University of Virginia, Charlottesville, Va 22901

C. Goulding
Florida A&M University, Tallahassee, Fl 32306

C. L. Morris and H. A. Thiessen
Los Alamos Scientific Laboratory, Los Alamos, NM 87545

W. B. Cottingame, S. Greene, and C. F. Moore
University of Texas, Austin, TX 78712

ABSTRACT

Measurements have been made of the differential cross section $d^2\sigma/d\Omega dE$ for $\pi^{\pm}$ scattered from ^{12}C and $^{40,44,48}Ca$ in the energy region corresponding to quasielastic scattering, using the EPICS facility at LAMPF. Data were taken at 180 Mev, 60°, and 290 Mev, 60° and 120°. Corrections were made for muon contamination, using information found on muons seen in π-p scattering measurements. A sample of the data is shown in Fig. 1. The rest of the data for each of the two angles are very similar in character to this. Within the errors, no simple systematic effects can be seen either in the Ca isotope cross sections or in the π^-/π^+ ratios. These results are being studied using calculations based on pion scattering from a Fermi gas of nucleons.

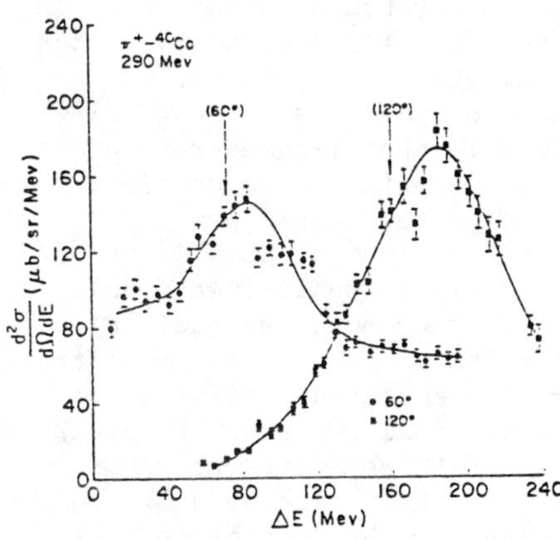

Fig. 1. Experimental values of $d^2\sigma/d\Omega dE$ for π^+ scattered from ^{40}Ca at 290 Mev, 60° and 120°, as a function of pion energy loss ΔE. The arrows indicate the values of ΔE expected for free π-p scattering. The curves shown are to guide the eye.

*Work supported in part by the U.S. Department of Energy
ISSN:0094-243X/79/540281-01$1.50

QUASI-FREE SINGLE NUCLEON REMOVAL IN PI-NUCLEAR SCATTERING

O.H. Otteson, R.E. McAdams, V.G. Lind, W.F. Denig; Utah State University, Logan, Utah 84322: C.A. Goulding, M. Greenfield; Florida A&M, Tallahassee, Florida 32307: H.S. Plendl; Florida State University, Tallahassee, Florida 32306: B.J. Lieb; George Mason University, Fairfax, Virginia 22030: C.E. Stronach; Virginia State College, Petersburg, Virginia 23803: P.A.M. Gram; Los Alamos Scientific Laboratory, Los Alamos, New Mexico 87544: T. Sharma; Oregon State University, Corvallis, Oregon 97330.

ABSTRACT

Proton and pion spectra obtained in coincidence with the prompt gamma rays following the bombardment of ^{27}Al by 190-MeV pions provides evidence for quasi-free single nucleon removal. The single nucleon removal ratios are not in agreement with a single quasi-free scattering model. However, the proton distributions in coincidence with the respective gamma rays are in good agreement. The angular distributions are also in substantial agreement with a quasi-free scattering mechanism.

DISCUSSION

An experiment has been conducted on the P^3 channel at LAMPF in which 190-MeV pions were incident on a 2 gm/cm^3 ^{27}Al target. Gamma-ray spectra and the accompanying particle spectra were obtained simultaneously in coincidence[1]. Figure 1 shows the gamma-ray spectra in coincidence with protons from π^+ incident on the target. Note the strong ^{26}Mg I→o line at 180 kev. In this reaction the proton was detected but the scattered pion was not. The ^{26}Al line at 417 kev represents the removal of a neutron by charge exchange of the π^+. Similar spectra exist for π^- incident on the target except that the ^{26}Al 417 line is of course not seen. Single nucleon removal lines are also observed in the pi-gamma-ray coincidence spectra. Table 1 lists the differential cross sections for single particle removal as obtained in this experiment.

As in single γ-ray experiments, the ratios in table 1 are not in agreement with the expectations of a simple quasi-free (QFS) model (i.e. neglecting charge exchange of the outgoing nucleon.

The data in this experiment was taken in such a way that it could be also sorted to give the particle energy distribution in coincidence with a gamma ray from the residual nucleus. Figure 2 shows the proton energy distributions for three different scattering processes. The dashed line is the prediction of a QFS model. There is good agreement between the data and the prediction. We also sorted our data to give the gamma-ray intensities as a function of particle energy. Again the agreement with QFS is good. A non-quasi-free calculation is much broader and predicts considerably larger intensities at higher proton energies.

ISSN:0094-243X/79/540282-02$1.50

Assuming QFS one can solve the kinematic equation in both the lab and the center of mass frames. From such a calculation it is found that a pion at 35° is equivalent to a proton at 67°. If one also assumes symmetry about the beam axis then there are four points (only two independent) for an angular distribution curve. Figure 3 shows the angular distributor in the center of mass system for the same three scattering reaction. The comparison with the $1 + 3\ \mathrm{Cos}^2$ prediction for free scattering is only fair for the protons from the π^- beam. However, the agreement is excellent for proton removal and good for neutron removal (Via the charge exchange) with the π^+ beam. Taking all evidence into consideration, there is good agreement with a QFS model for single nucleon removal in Pi-nuclear scattering.

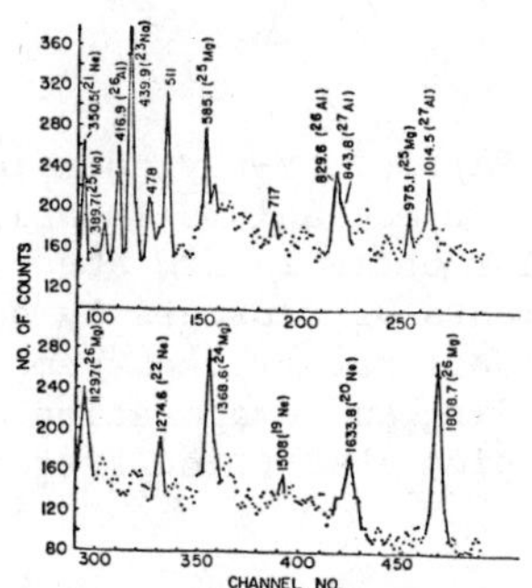

Fig. 1. The gamma-ray spectra in coincidence with the total proton spectra from a π^+ beam.

Table 1. The differential cross-section for single particle removal by 190-MeV pions incident on a 2 gm/cm^2 ^{27}Al target.

Residual Nucleus and Transition	E_γ(keV)	Coincident Particle	$\frac{d\sigma}{dr}(35°\pi^+)$ μb/str	$\frac{d\sigma}{dr}(35°\pi^-)$ μb/str	Ratio π^-toπ^+	QFS Prediction
^{26}Mg I→0	1089.7	P	5790±330	1150±240	0.20±0.04	1/9
^{26}Mg I→0	1809.7	π	2200±200	2380±270[a]	1.08±0.16	1/9
^{26}Al II→0	416.8	π	130±30	430±45[a]	3.23±0.82	9/1

[a]Values reported in reference 1.

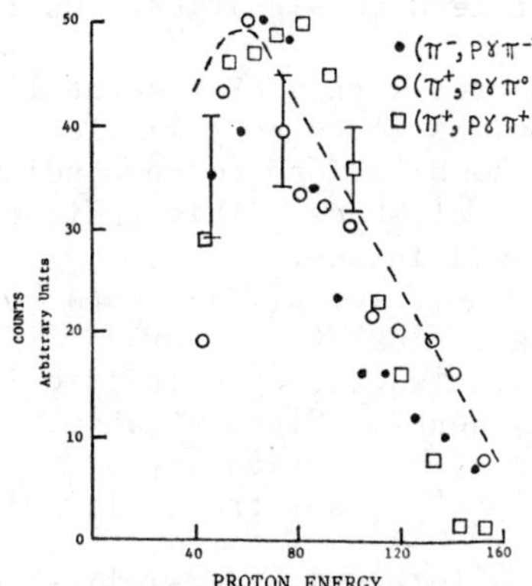

Fig. 2. The proton energy distribution for protons in coincidence with gamma rays from the residual nucleus.

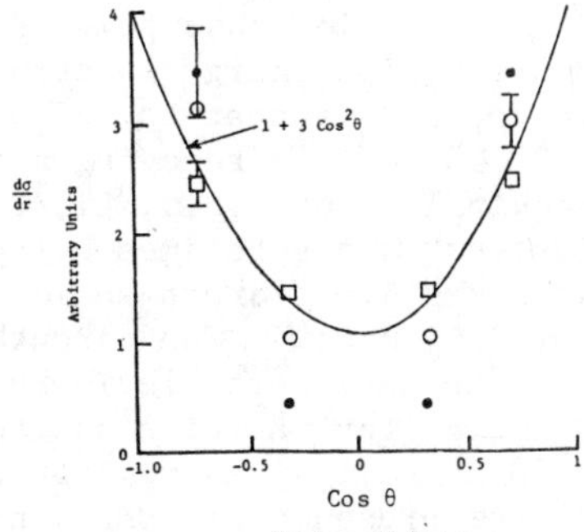

Fig. 3. The angular distribution in the center of mass for ● single Proton Removal π^- beam, □π^+ beam, and O Single Neutron Removal π^+ beam.

REFERENCES

1. V.G. Lind, et al., Phys. Rev. Lett., 41, 1023, (Oct. 9, 1978).

REACTION MECHANISMS IN PI-NUCLEAR MULTI-NUCLEON REMOVAL PROCESSES

V.G. Lind, R.E. McAdams and O.H. Otteson
Utah State University 84322

ABSTRACT

Particle γ-ray coincidence studies illuminate the pi-nuclear reaction mechanisms in which multi-nucleons are removed. In particular there is new strong evidence of the pion interacting with fragments or clusters by sometimes scattering and removing them as a unit rather than removing the nucleons exclusively through cascading and evaporation mechanisms. There is also new evidence that pion absorption frequently occurs on multi-nucleon systems.

INTRODUCTION

Particle γ-ray coincidence studies following pi-nuclear interactions reveal many aspects of the reaction not formerly understood. The present evidence from 190 MeV π^+ and π^- on ^{27}Al[1] points to the following interpretations.

(1) Reactions in which the pion is absorbed constitute $\leq$50% of the reaction cross section even for multi-nucleon removal. This is evident by comparing π-γ coincidence cross sections with γ- singles.

(2) Evaporation energy neutrons are detected by the presence of the 478-keV line in ^{7}Li which results from (n,α) reactions on the boronated paraffin surrounding the γ-detector. The absence and presence of the line in our π-γ data and p-γ data respectively, indicates π-scattering reactions do not lead to significant nucleon evaporation but pion absorption does.

(3) The pion energy spectrum measured with a ΔE-E scintillator telescope shows that the pion sometimes loses very little energy (20-40 MeV) in removing nucleon combinations corresponding to the equivalent of d,t,α,^{6}Li,^{7}Li, etc. clusters. This indicates the "clusters" are sometimes being removed intact.

(4) ^{25}Mg lines are seen in π^--γ, π^+-γ, and p(π^+)-γ data but not in p(π^-)-γ data. (The parenthetically inserted π^+ or π^- signifies the beam particle.) This indicates the np pair is being removed intact for the pion scattering events. Since π^- absorption on an np pair gives no protons but π^+ absorption gives 2 protons, the presence of ^{25}Mg lines in p(π^+)-γ spectra is largely due to π^+ absorption on an np pair.

(5) The ratio of the cross sections $\sigma(p(\pi^-)-\gamma)/\sigma(p(\pi^+)-\gamma)$ for ^{24}Mg is .26±.08. This is accounted for if 33% of the protons contributing to this line are due to break-up of the equivalent "t" cluster and 67% are due to absorption. However, absorption events account for <50% of the total reaction cross section leading to ^{24}Mg[1], therefore intact t removal must be on the order of

ISSN:0094-243X/79/540284-02$1.50

25% or more.

(6) The π^- to π^+ cross section ratios for ^{25}Mg, ^{24}Mg, ^{23}Na and ^{21}Ne are 1.16±.36, 1.56±.31, 1.28±0.21, and 1.30±0.29. These are consistent with the 1, 7/5, 1, 1 ratios expected if the pions are interacting with d,t,α and ^{6}Li sub units respectively.

(7) High energy protons (>165 MeV) arise only from π absorption reactions. Absent from a γ-ray spectrum in coincidence with these protons are the ^{26}Mg lines as expected. Somewhat surprisingly, however, is the near absence of ^{25}Mg as well, whereas ^{24}Mg, ^{23}Na, ^{21}Ne and ^{20}Ne are enhanced (See Fig. 1). These data are understandable if one recognizes that the proton comes off with greater energy when recoiling against the remainder of a cluster fragment rather than against a single nucleon followed by nucleon evaporation. Thus it appears that absorption is occuring on multi-nucleon combinations as well as nucleon pairs. The nucleon pair absorption process generally gives protons with energies less than 175 MeV and in that data ^{25}Mg lines are important.

(8) The ratio of cross sections for the 440-keV line of ^{23}Na for protons with energies 165 MeV, i.e. those known to arise from absorption is 3.47±1.45 for π^+ to π^-. This is in good agreement with what one would expect if the absorption took place on an α cluster, namely a ratio of 3.

FIGURE

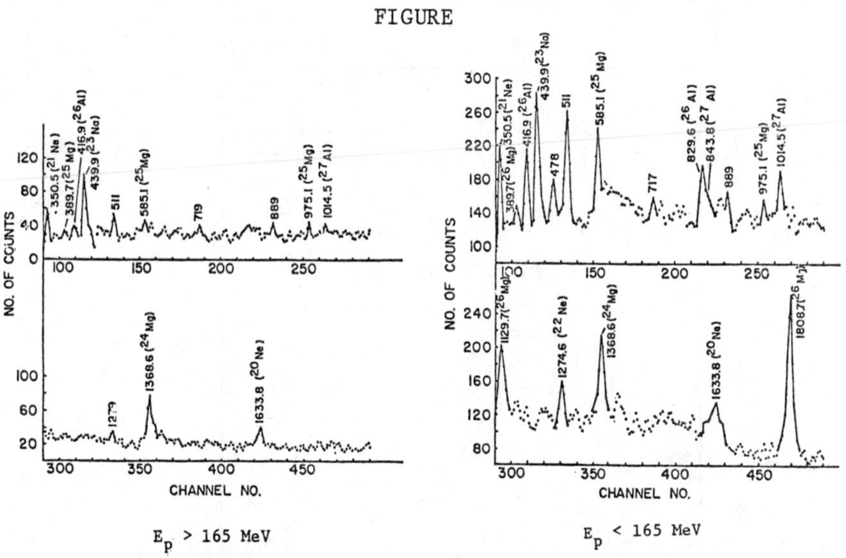

Figure 1. The gamma ray spectra taken in coincidence with protons with energies greater than 165 MeV (left) and with energies less than 165 MeV (right).

REFERENCES

1. V.G. Lind, et al. Phys. Rev. Lett. 41, 1023, (1978).

THE ENERGY DEPENDENCE OF THE (π,N) REACTION IN ^{3,4}HE

Jan Källne and J.S. McCarthy
University of Virginia, Charlottesville, VA 22901

H.A. Thiessen and C.L. Morris
Los Alamos Scientific Laboratory, Los Alamos, N.M. 87544

S.L. Verbeck, G.R. Burleson, and M.J. Devereaux
New Mexico State University, Las Cruces, N.M. 88001

J.E. Bolger and C.F. Moore
University of Texas, Austin, TX 78712

C.A. Goulding
Florida A & M University, Tallahassee, FL 32307

A comprehensive study of the (π^-,n) in ^{3,4}He have been made using the EPICS pion channel of LAMPF. The cross section was measured for θ ≈ 20-150° over the energy range T_π = 50-290 MeV. The characteristic features of the angular distributions have been discussed recently[1] and here we focus our attention on the information contained in the energy dependence.

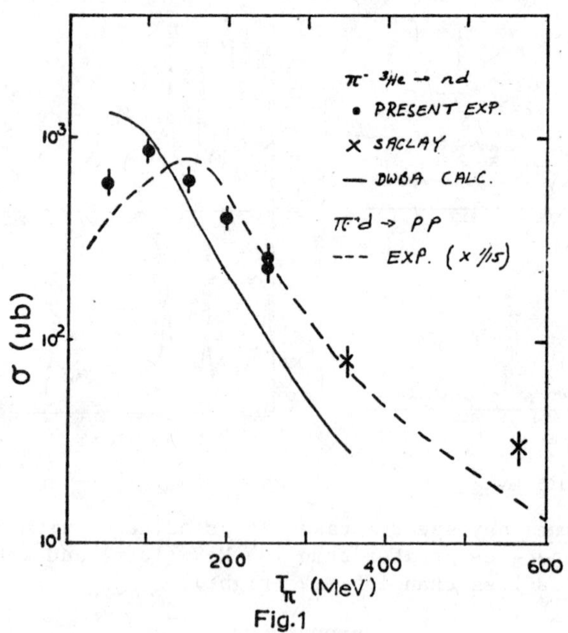

Fig.1

ISSN:0094-243X/79/540286-02$1.50

The (π^-,n) cross section is strongly energy dependent except for the backward hemisphere. This is an intrinsic energy dependence superimposed on that coming from the momentum transfer (and hence nuclear structure dependence) which also varies with T_π. An example is shown in Fig. 1 in form of the angular integrated cross section for ^3He. The two points at higher energies are taken from $^2H(p,\pi^\circ)^3He$ measurements[2] using detailed balance. The cross section maximum just above 100 MeV is associated with the 3,3 resonance which reflect the importance of πN scattering in (π,N). The $\pi^+d \rightarrow pp$ reaction is known to be dominated by the pion exchange (PE)

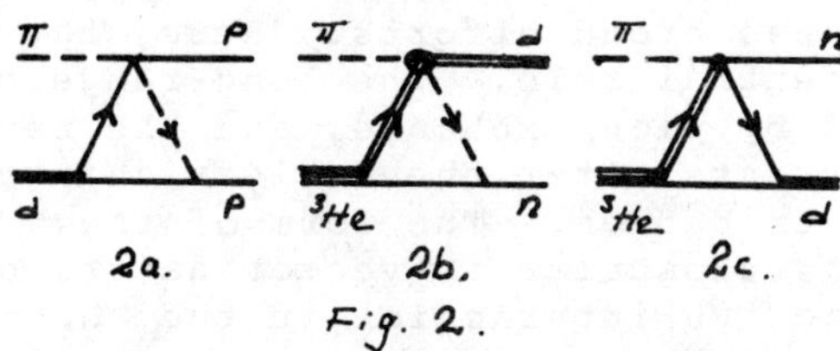

Fig. 2.

diagram (2a) which contains the πN scattering explicitly so nucleon form factors can dominate. Besides πN scattering, the (π,N) reaction for $A > 2$ must involve nuclear form factors as in the pion rescattering diagram (2b) or the triangle diagram using the $\pi d \rightarrow pp$ sub-process (2c). Fig. 1 shows a comparison between the $\pi d \rightarrow pp$ and $\pi^3He \rightarrow nd$ cross sections, and the predicted [3] difference between the two in the form of diagram (2b).

---

1. J. Källne et al., Phys. Rev. Lett. 40, 378 (1978).
2. J. Banaigs et al., Phys. Lett. 45B, 398 (1973).
3. H.W. Fearing, Phys. Rev. C 16, 313 (1977).

PIONS AND ISOBARS IN NUCLEI - AN OVERVIEW*

E.J. Moniz
Center for Theoretical Physics
Laboratory for Nuclear Science and Department of Physics
Massachusetts Institute of Technology
Cambridge, Massachusetts 02139

A basic challenge in nuclear physics has been the investigation both of complicated many-body phenomena (encompassing nuclear structure and reactions) and simultaneously of the underlying two-body force. A special challenge of intermediate energy nuclear physics is the unification of these broad efforts. Here, the pion obviously plays a central role. The long-range nuclear force is mediated by pion exchange, and the resulting charged meson currents alter the nuclear charge and magnetization distribution. The form of the πNN coupling, dictated by the pseudoscalar isovector nature of the pion, leads to an attractive interaction in the πN J=3/2, T=3/2 channel and consequently to a resonance, the Δ(1232) isobar. The Δ in turn plays a crucial role in generating the medium range attraction in the NN force, in contributing to nuclear magnetism, and in serving as the "doorway" for medium energy pion-nucleus absorption reactions. Clearly, we hope to achieve a quantitative understanding of the coupled pion and Δ (more precisely, Δ-nucleon hole) propagators in the nuclear medium. Further, we would like to understand these over a very wide frequency range, from the static limit ($\omega_\pi = 0$) appropriate to studying the structure of low-lying nuclear states up to the much larger energies ($\omega_\pi \sim$ 2 to 3 M_π) appropriate to pion-nucleus reactions in the resonance region. In this brief overview, I will try to assess our present understanding of the pion and Δ-hole propagators in nuclei. I shall start in the resonance region, where I feel we understand the most, and then "work down" to the question of Δ's in nuclear ground state wavefunctions. Much of the discussion will consist of raising questions which hopefully will be answered at least partially in other contributions to this conference.

PION NUCLEUS SCATTERING IN THE RESONANCE REGION ($\omega_\pi \approx 2M_\pi$)

We shall discuss pion-nucleus scattering in the resonance region in the framework of the isobar-hole doorway

* This work is supported in part through funds provided by the U.S. Department of Energy (DOE) under contract EY-76-C-02-3069.

ISSN: 0094-243X/79/540288-18$1.50

formalism[1-5]. Assuming that πN scattering is dominated at medium energies by intermediate Δ-excitation (Fig. 1a), pion-nucleus elastic scattering proceeds through intermediate excitation of Δ-hole states (Fig. 1b).

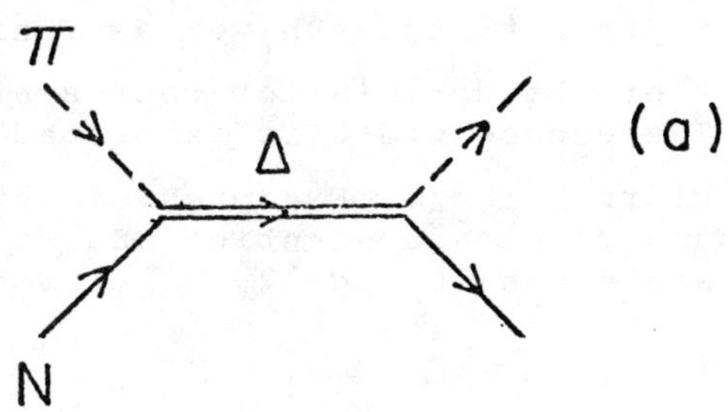

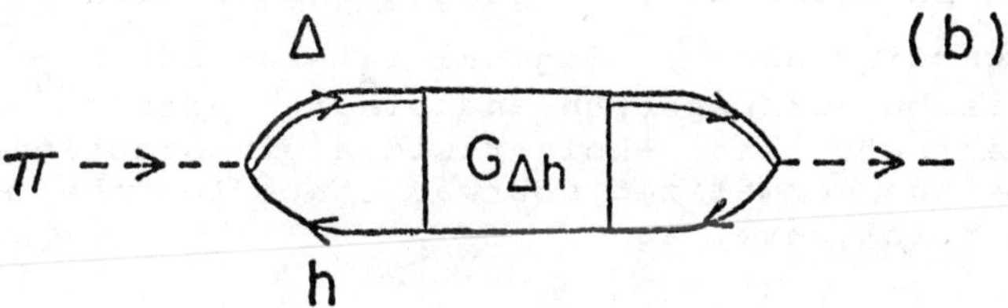

Figure 1. (a) πN scattering via intermedtiate Δ excitation.
(b) π-nucleus elastic scattering via excitation and propagation of Δ-hole states.

The scattering amplitude is given then by the appropriate matrix element of the Δ-hole propagator $G_{\Delta h}$. This can be made explicit in the projection operator formalism. Let the P-space contain all states consisting of the pion plus ground state nucleus; the D-space, all Δ-hole states; and the Q-space, all more complicated channels. With the doorway condition $H_{PQ} = 0$,[1] the pion-nucleus elastic transition matrix is given by

$$T_{PP} = H_{PD}\, G_{\Delta h}\, H_{DP}$$

$$= H_{PD}\, [E-E_R+\frac{i}{2}\Gamma(E)-\mathcal{H}_{\Delta h}]^{-1} H_{DP} \qquad (1)$$

$$\mathcal{H}_{\Delta h} = H_{DD} + H^{\uparrow}_{DD} + H^{\downarrow}_{DD} \qquad (2)$$

$$H^{\uparrow}_{DD} = H_{DP}(E^{+} - H_{PP})^{-1}H_{PD} \tag{3}$$

$$H^{\downarrow}_{DD} = H_{DQ}(E^{+} - H_{QQ})^{-1}H_{QD} \tag{4}$$

Here, H_{DP} represents the $\pi\Delta h$ vertex function and the Breit-Wigner part of the denominator represents the free Δ propagator. The escape width is generated by $H^{\uparrow}_{DD}$, and the spreading width by $H^{\downarrow}_{DD}$. Examples of spreading interactions, corresponding to pion absorption and to pion multiple reflection, are given in Fig. 2. The isobar-hole Hamilton-

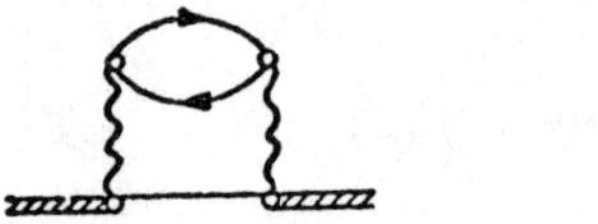

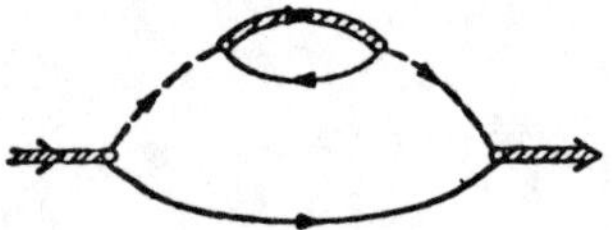

Figure 2. Examples of Δ spreading interactions.

ian $\mathcal{H}_{\Delta h}$ contains also a diagonal interaction H_{DD} which includes isobar propagation and binding effects, Pauli blocking effects, and Δ-hole residual interactions. We can define the normalized doorway state for the L-th pion-nucleus partial wave as

$$|D_{o}^{L}\rangle = N_{L}\ H_{DP}\ |(\vec{q})_{L};\ 0\rangle \tag{5}$$

where $|(\vec{q})_{L};\ 0\rangle$ represents the partial wave projection of the pion plane wave (wavenumber $\vec{q}$) and the nuclear ground state, and where N_L is a normalization factor. The pion-nucleus elastic transition matrix is then the expectation value of $G_{\Delta h}$ in the doorway state

$$T_{L} = N_{L}^{-2}\ \langle D_{o}^{L}|G_{\Delta h}|D_{o}^{L}\rangle \tag{6}$$

In microscopic calculations, this matrix element is evaluated by diagonalizing $G_{\Delta h}$ in a shell model basis[2-5] or in a basis constructed by repeated application of $\mathcal{H}_{\Delta h}$ to the doorway state.

A simple approximation, which we shall call the one-doorway approximation, is obtained by replacing $\mathcal{H}_{\Delta h}$ in Eq.(1) by its doorway expectation value:

$$T_{L}^{\text{Doorway}} = N_{L}^{-2}\ \frac{1}{E-E_{R}+i\Gamma/2\ -\ \langle D_{o}^{L}|\mathcal{H}_{\Delta h}|D_{o}^{L}\rangle} \tag{7}$$

The doorway approximation is useful for discussing quantitatively the relative importance of different aspects of the dynamics; the imaginary part of the expectation value for each term in $\mathcal{H}_{\Delta h}$ corresponds to a partial width for the reaction. In Fig. 3, we show as a function of pion

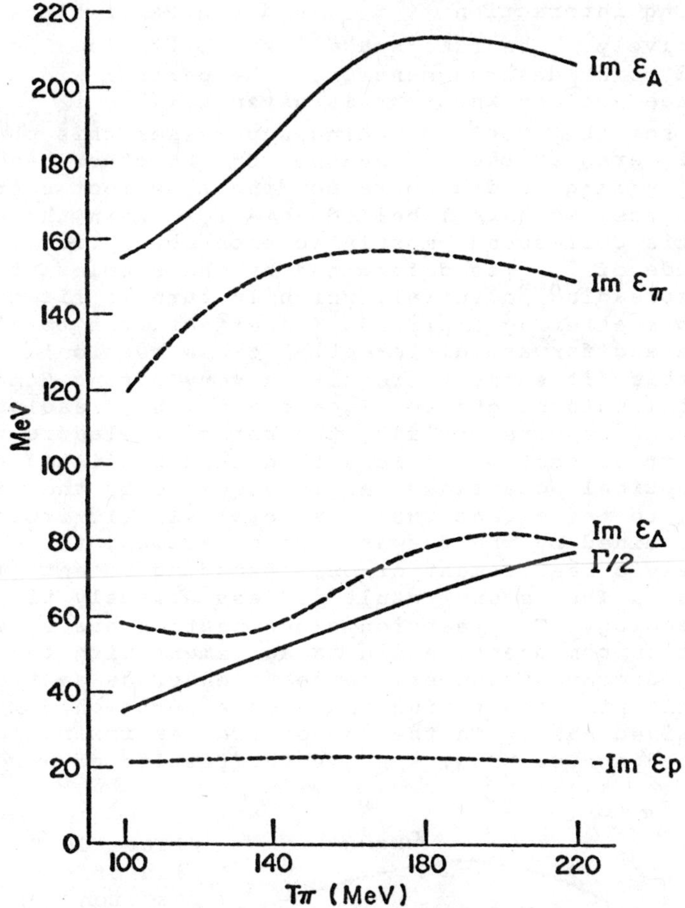

Figure 3. Decomposition of the imaginary part of the doorway expectation of the Δ-hole Hamiltonian $\text{Im}\ \varepsilon_A = -\text{Im}\ \langle D_o^L | \mathcal{H}_{\Delta h} | D_o^L \rangle$ for the $L^\pi = 0^-$ partial wave. The contribution $\text{Im}\ \varepsilon_\pi = -\text{Im}\langle H_{DD}^{\uparrow}\rangle$ and $\text{Im}\ \varepsilon_\Delta = -\text{Im}\ \langle H_{DD}^{\downarrow}\rangle$ are shown separately. The supression of the Δ free space width Γ due to Pauli blocking of the Δ decay is indicated by $\text{Im}\ \varepsilon_P$.

kinetic energy the imaginary part of the doorway expectation value[3] $\mathrm{Im}\ \varepsilon_A = -\mathrm{Im} \langle \mathcal{H}_{\Delta h} \rangle$ for the π- ^{16}O partial wave $L^{\pi} = 0^-$. Also shown are the separate contributions $\Gamma_{Elas}/2$, $\Gamma_{Abs}/2$, and $\Gamma_{Pauli}/2$, corresponding to the imaginary parts of the expectation values of the escape interaction $H^{\uparrow}_{DD}$, the spreading interaction $H^{\downarrow}_{DD}$, and the Pauli blocking term, respectively. The free-space Δ width $\Gamma/2$ is also plotted. Several points deserve mention. The partial width for quasifree nucleon knockout is given by $(\Gamma - \Gamma_{Pauli})$, and we can see that Pauli blocking suppresses this channel strongly even at the resonance. On the other hand, there is a very large width corresponding to direct coupling to the Q-space; we have labelled this $\Gamma_{Abs}/2$ in the expectation that this corresponds mostly to pion absorption. The magnitude of Γ_{Abs} is determined by the complex strength of the Δ spreading potential, which in turn is fit to the forward scattering amplitude (in effect, the total cross section and forward differential cross section). Therefore, this fit seems to require a very strong damping of the Δ-hole states due to Q-space coupling, leading to two nontrivial results: First, the ratio of elastic to reaction cross section is less than that predicted by first order optical potentials, as is required by the data. Of course, to the extent that the total elastic cross section is determined by the forward cross section, this result is merely a restatement of our spreading potential phenomenology. The second result is less directly tied to the phenomenology: The reaction cross section has a very large absorption component, again in agreement with the data. A rough decomposition of the reaction cross section into inelastic pion scattering and pion absorbtion components is obtained easily in the one-doorway approximation, as indicated above. Fig. 4 shows the results of a more

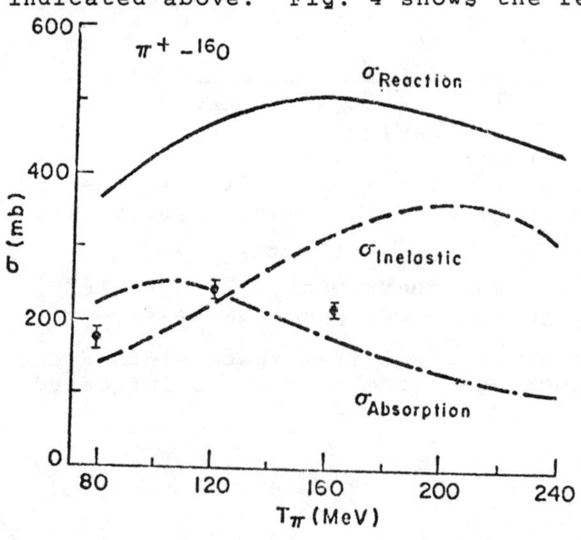

Figure 4. Decomposition of the $\pi^{\pm 16}$O reaction cross section into total absorption and inelastic scattering cross section[3]. The absorption cross section data are interpolations of the results of Navon et al[6].

detailed calculation[3] for π-^{16}O scattering. The absorption cross section data are interpolations of the results of Navon et al[6]. The agreement with the elastic, inelastic and absorption inclusive cross sections indicates that the Δ-hole dynamics shown in Fig. 3 is quite reasonable.

Other tests are provided by the elastic scattering angular distributions. Fig. 5 shows the results for $\pi^{\pm}$-^{16}O

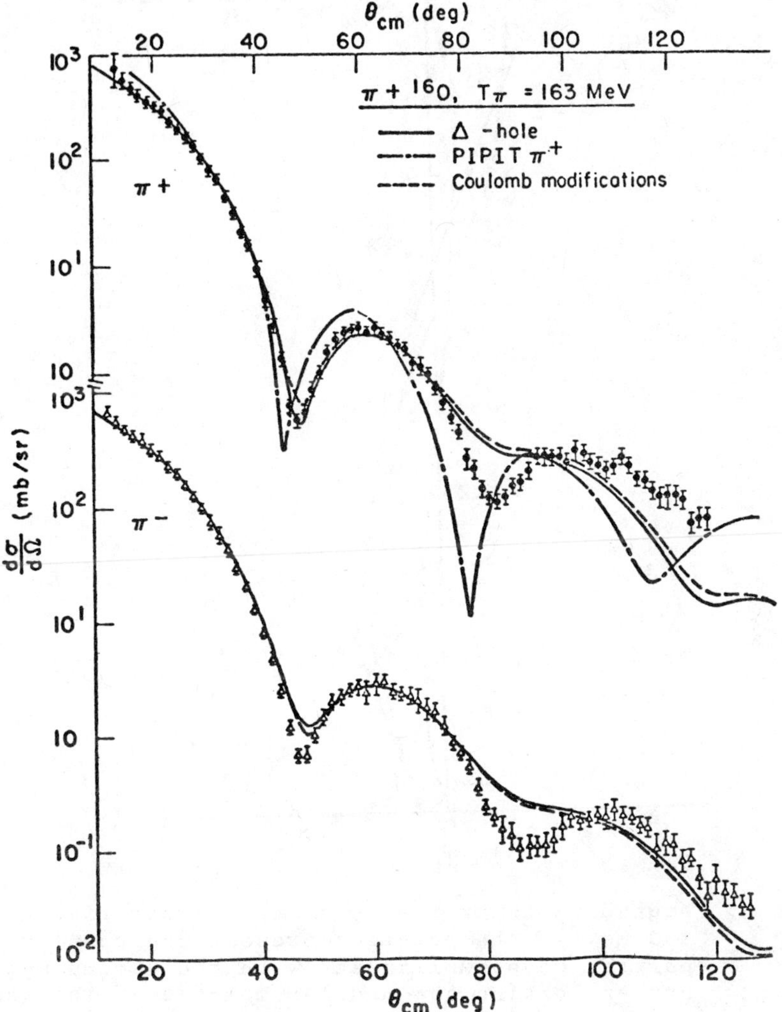

Figure 5. Elastic differential cross section for $\pi^{\pm}$-^{16}O scattering at 163 MeV. Data from Reference 7; Δ-hole calculations from Reference 3; PIPIT calculations according to Reference 8.

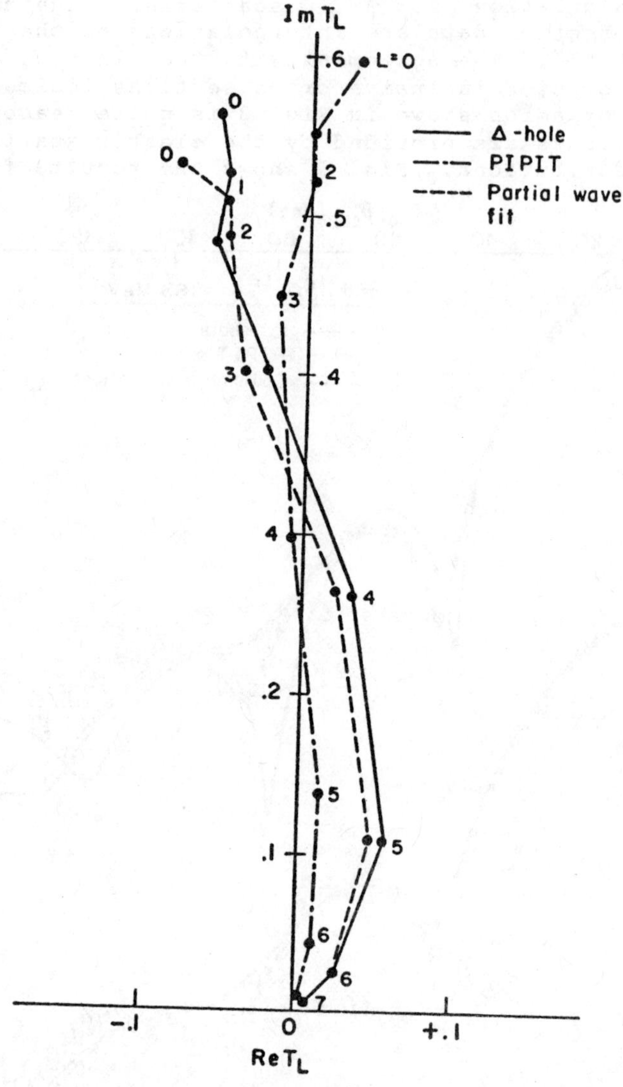

Figure 6. Argand plot for π^+-^{16}O elastic scattering at 163 MeV[3]. The points represent the complex partial wave amplitudes, with the accompanying numbers giving the angular momentum. The lines serve only to guide the eye. Calculations same as those in Fig. 5; dashed line connects partial wave amplitude fit to the data.

scattering at 163 MeV; the data are from S.I.N.[7] and the Δ-hole calculations are those of Hirata et al[3]. The π^+ results calculated with the first order optical potential code PIPIT are also shown. It is interesting that the two calculations, which use exactly the same nuclear density (fit to electron scattering), give substantially different results even close to resonance and in the angular range out to the second maximum ($\theta_{c.m.} < 60^o$). This might lead one to be suspicious of present attempts to extract from pion-nucleus scattering information about neutron radii with sufficient accuracy to make any impact on theories of nuclear structure[9]. Further insight into the difference between the two calculations is obtained upon examination of the Argand diagram shown in Fig. 6. The partial wave fit[3] to the data is not unique, but the uncertainties lie almost entirely with the central partial waves (L<3). The important point is that the peripheral partial waves are reproduced much more closely in the Δ-hole calculation than in the PIPIT calculation. This has comparatively little to do with the spreading interaction. Binding, recoil and Pauli blocking effects play a crucial role in these partial waves, and these effects must be treated carefully in either Δ-hole or optical potential calculations. Without such carefuly treatment, we cannot hope to extract quantitative information about the pion or Δ-hole propagators "inside" the nucleus.

Inelastic pion scattering and pion production/absorption reactions are sensitive to the half-off-shell pion nucleus transition matrix and so may test further theories of pion-nucleus interactions. A particularly "clean" probe of the pion wavefunction would appear to be coherent π^o photoproduction. The only data available in the resonance region are rather old and of dubious quality[10], but we shall use this process to illustrate potential difficulties with DWIA analyses of inelastic reactions. In the isobar-hole formalism, the photoproduction amplitude is[11]

$$T_{\gamma\pi^o} = H_{PD}\, G_{\Delta h}\, H_{D\gamma} \qquad (8)$$

The only difference to the pion elastic amplitude T_{PP} is that the Δ-hole doorway to the reaction is generated by the $\gamma N\Delta$ vertex operator $H_{D\gamma}$. Calculations for $^{16}O(\gamma,\pi^o)^{16}O_{g.s.}$ yield results, shown in Fig. 7, rather close to those obtained previously[12] in standard DWIA calculations with phenomenological first order pion optical potentials. The interesting result emerges from re-writing Eq. (8) in a "DWIA form":

$$T_{\gamma\pi^o} = \{T_{PP} G_P + 1\} \left[H_{PD} \frac{1}{E-E_R + \frac{i}{2}\Gamma \qquad - H^{\downarrow}_{DD}} H_{D\gamma} \right] \qquad (9)$$

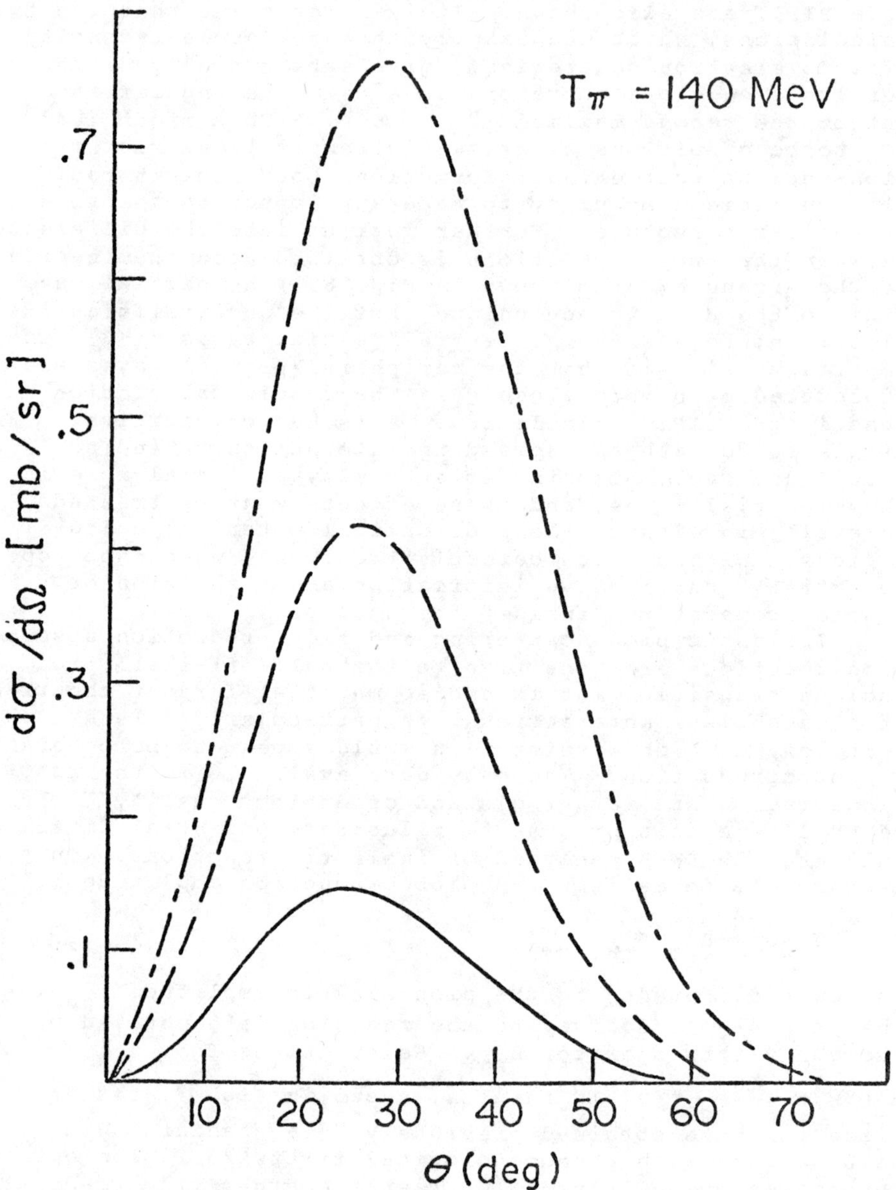

Figure 7. Differential cross section for coherent π^0 photoproduction $^{16}O(\gamma,\pi^0)^{16}O_{g.s.}$ Dot-dash line corresponds to impulse approximation, and solid line, to full Δ-hole results. The dashed line corresponds to a DWIA calculation using the Δ-hole distorted wave (see Eq.(9) and related discussion). All calculations from Reference 11.

The quantity in curly brackets generates the pion distorted wave from the Δ-hole pion nucleus elastic transition matrix. However, the production operator in square brackets, containing the spreading interaction, does not correspond to impulse approximation; in the usual reaction theory language, the modified production operator sums a variety of multistep contributions. We show in[11] in Fig. 7 the results of a DWIA calculation in which the impulse approximation is used for the production operator (i.e, $(H_{DD}+H^{\dagger}_{DD})$ are replaced by only the recoil energy) while the full Δ-hole results are retained for the distortion. The result is an increase in the cross section by more than a factor of two. Clearly, it is very important to treat consistently the dynamical input to the distorted wave and production operator in order to calculate adequately the normalization (note that the angular shape is essentially unchanged). The large effect seen in Fig. 7 points to the importance of a microscopic understanding of the optical potential used in DWIA calculations. Of course, this problem is less severe for inelastic reactions confined to the nuclear surface because of the nuclear transition density involved. Concomitantly, these reactions have less power for teaching us about the pion wavefunction inside the nucleus.

In closing this section, we should remark that incoherent reactions can also be important in learning about Δ-nucleus interactions. The advantage here lies in having considerably more freedom in fixing the kinematics of the outgoing particles so as to emphasize specific reaction mechanisms. As an example, the Saclay group has extracted information about the Δ-N S-wave interaction by looking at $\gamma d \to pp\pi^-$ in a region of phase space suitable to zero relative motion between the isobar and nucleon[13]. Another example is inclusive Δ-excitation in deep inelastic electron-nucleus scattering; available results for the cross section on very heavy nuclei show a surprisingly narrow peak.[14] The latter experiment should be repeated. More generally, many more inclusive and coincidence measurements will be needed for studying the Δ-propagator in nuclei and for understanding better pion-nucleus reaction mechanisms[15].

PIONIC ATOMS ($\omega_{\pi} \approx M_{\pi}$)

The width of pion atomic levels provides a fairly direct measure of the strength of the imaginary part of the pion optical potential at threshold. The only inelastic channel available for strong interaction processes is pion absorption. Consequently, if we assume that p-wave absorption goes through the Δ-doorway, the level widths provide a constraint on the Δ spreading potential discussed in our consideration of the resonance region. The central strength of the Δ absorptive potential was found to be about 50 MeV throughout the resonance region.[3] We shall show that this strength, extrapolated to threshold, is consistent with the observed pionic atom level widths.

The phenomenological p-wave pionic atom optical potential can be written as

$$V = 4\pi\vec{\nabla}\cdot[c_o\rho(r) + C_o\rho^2(r)]\vec{\nabla} \equiv 4\pi\ \vec{\nabla}\cdot\alpha(r)\vec{\nabla} \tag{10}$$

with c_o the πN p-wave scattering volume ($c_o = 0.21\ M_\pi^{-3}$) and C_o a complex constant which includes the effects of absorption. Fits to the pionic atom data give[16] $\delta \equiv (\mathrm{Im}\ C_o\ \rho(o)/c_o) \simeq 0.10$ to 0.17. We want now to calculate δ with a crude model in which the Δ propagator is modified by an absorption width. We write for the first order term in the optical potential $\alpha(r) = \rho(r)\ [c_{nr} + C_\Delta]$ where C_Δ is the contribution to the scattering volume coming from the "Δ-pole". We adopt the value[17] $C_\Delta = 0.55c_o$ (it should be clear that we do not pretend to supply here more than an order-of-magnitude estimate). We now modify the resonant term

$$\alpha(r) = \rho(r)\left[C_{NR} + C_\Delta\frac{E-E_R}{E-E_R+\frac{i}{2}\Gamma_{Abs}\ \tilde{\rho}(r)}\right]$$
$$= \rho(r)\ c_o\left[1 - \frac{C_\Delta}{c_o}\frac{i(\Gamma_{Abs}/2)\ \tilde{\rho}(r)}{E-E_R+\frac{i}{2}\Gamma_{Abs}\ \tilde{\rho}(r)}\right] \tag{11}$$

with $\tilde{\rho}(r) \equiv \rho(r)/\rho(o)$. Taking $\Gamma_{Abs}/2 = 50$ Mev from the central isobar spreading potential, we have $\delta = 0.14$, which obviously is consistent with the phenomenological value. We repeat that this exercise is intended only to demonstrate that the Δ spreading potential strength is in the right ballpark.

Recent experimental developments indicate that the phenomenological optical potential discussed above may be inadequate for describing "hidden" levels. Because of the competition between absorption and radiative decay of the pion atomic levels, only levels with a sufficiently high angular momentum could be observed for a given nucleus. For example, p-levels have been observed directly only for $Z \lesssim 32$. By definition, the pion wavefunction for the observed levels had small overlap with the nucleus, so that the form of the optical potential was not tested severely. One way to get information on "hidden" levels is through nuclear polarization. In the event of a near degeneracy between atomic and nuclear transitions, a "hidden" level may be admixed substantially with an observable level. A number of experiments were performed[18] and seemed to point to major discrepancies with optical potential calculations. In particular, the width of the 3p level in pionic ^{110}Pd seemed to be overestimated by at least a fact or two[18]. A more careful theoretical treatment[19], including the effects of strong coupling between the pion and nuclear vibrational excitations and, in some cases, the effects of mixing more than two levels, led to a different conclusion while still using standard optical potentials. All cases except ^{110}Pd were explained satisfactorily, and a possible explanation for the ^{110}Pd result in terms of still unknown spectroscopic information was given[19]. However, very recent experimental results on "hidden" d-levels in very heavy nuclei[20] have re-opened the issue of possibly serious shortcomings in the optical potential. Konijn et al[20] find shifts and widths a factor of 2 to 3 less than those predicted, reminiscent of the ^{110}Pd nuclear polarization result. In fact, because the nuclear polarization experiments cannot extract the width and shift separately, it is possible that a substantially modified optical potential might not destroy agreement with many of these results. Crude estimates[21] indicate that this may actually be the case. In any event, pionic atom experiments such as these, which probe deeply bound levels, are very important for our understanding near threshold pion propagation in nuclei. With a more refined theory spanning the energy region from threshold to resonance, p-wave absorption will constrain strongly the model for isobar spreading interactions.

MESON EXCHANGE CURRENTS AND Δ's IN THE NUCLEAR GROUND STATE($\omega_\pi \approx 0$)

In discussing meson exchange currents in electromagnetic processess, it is useful to start by classifying the currents as "potential" or "nonpotential". This distinction is drawn easily by resorting to diagrams. The "potential" terms are those for which the corresponding diagram represents a term in the NN force when the photon is removed. For example, the currents related to the static one pion exchange potential (OPEP) are indicated in Fig. 8(a). For consistency, these exchange currents must be treated in a unified way with the nuclear force and wavefunction calculations. For the isovector M1 operator generated by static OPEP, there is essentially no ambiguity in the long range part of the current. The classic tests of this operator are provided by radiative capture of thermal neutrons[22] $n+p\rightarrow d+\gamma$ and by near threshold electrodisintegration of the deuteron[23]. In both cases the OPEP exchange currents have been used to resolve significant discrepancies between theories and experiment, and the same technology has now been applied in numerous calculations.

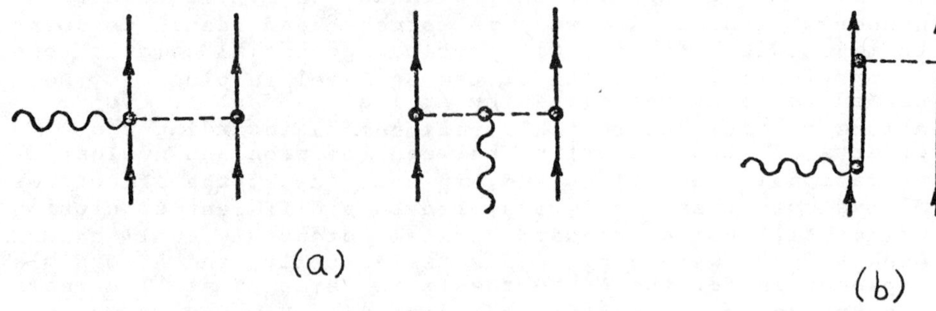

Figure 8. Electromagnetic exchange currents arising (a) from single pion exchange and (b) from intermediate Δ excitation.

In the "nonpotential" diagrams, the photon cannot be removed sensibly. An example of particular relevance for our consideration is given in Fig. 8(b): a two-body current is generated by intermediate excitation of the Δ. This term is by itself gauge invariant and is not tied directly to the theory of the NN force. The Δ exchange current, which is of pionic range, has been incorporated into calculations of electromagnetic processes, such as those mentioned above. However, there is no very convincing result for the magnitude of this contribution. Uncertainties arise in specification of the hadronic form factors used and of the Δ propagator in the nucleus. Uncertainties in the nuclear wavefunction, in the OPEP exchange currents (again, hadronic form factors must be specified), and in the importance of shorter range exchange currents preclude experimental determination of its magnitude.

Further progress on understanding the role of exchange currents in nuclear charge and current densities is dependent on further development of the meson and/or quark theory of nuclear forces. Recent progress on derivation of the intermediate range force due to two pion exchange (with and without Δ excitation) has not yet been reflected in theories of exchange currents. Theoretical work in this direction is certainly very desirable and, at the least, should pin down the Δ exchange current contribution. The short range interactions present a considerably more difficult problem, since even the appropriateness of a meson (versus quark) description of the NN interaction may be questioned.

We turn finally to the question of Δ's in nuclear ground state wavefunctions. We refer here to nonperturbative calculations of nuclear wavefunction components containing at least one Δ or N*[24]. Typically, a quoted result of such calculations is a Δ-probability in the ground state, and several experimental searches based upon the spectator model have been carried out (with little success). There are formidable barriers to realistic theoretical calculations, without which the experiments cannot be interpreted. Relativity provides a central complication. For example, the deuteron ΔΔ wavefunction component should be useful for describing both the ΔΔ contribution to the static magnetic moment and the role of the ΔΔ admixture in intermediate energy breakup experiments with a Δ spectator "left behind"; because of the large ΔN mass difference, the latter necessarily involve large energy transfer. Presumably, such a two baryon relativistic "wavefunction" must retain the relative time variable in a nontrivial way. Thus, simplified relativistic equations, such as the Blankenbecler-Sugar equation, are not adequate[24].

Another difficulty, stressed by Amado[25], lies with the ambiguity in defining a Δ probability. The only "model-independent" way of defining the Δ wavefunction component entails going to the complex energy resonance pole (where the Δ is "on-shell"), but the resulting amplitude is not constrained by normalization[25]. Of course, the probability could still be defined within the context of any specific dynamical scheme. However, theories giving equivalent descriptions of the relevant πN phase shifts would in general give different operational prescriptions for the probability.

A third difficulty, also discussed by Amado[26], lies in disentangling Δ wavefunction effects from final state interactions in breakup experiments. The characteristic falloff distance for a two-body wavefunction is $\alpha=(2\mu B)^{-1}$ where μ is the reduced mass and B is the energy of the state below breakup threshold. For a ΔN "bound state", $\alpha \approx 0.35$ fm, which is short compared to the range of the NN→ΔN "transition potential". Consequently, there is no asymptotic region in which to sample the Δ wavefunction, so that final state interactions are always a serious matter[26]. Again, this does not preclude using Δ wavefunction components as a means for understanding medium energy reactions, but it does imply that information about these components can be extracted from experiments only by comparison with very involved dynamical calculations.

CONCLUDING REMARKS

We have seen that substantial progress has been made in understanding quantitatively the pion and Δ-hole propagators in nuclei. Further theoretical and experimental advances can be expected in the near future. In the resonance region, both elastic pion scattering and various inclusive cross sections are described well with comparatively simple isobar-hole dynamics (analogous statements could be made in the optical potential language). Theoretical tasks which remain are quantitative microscopic evaluation of the Δ-nucleus spreading interaction and investigation of the πΔh vertex corrections in the medium. Further experimental work in measuring inclusive cross sections (e.g., (π,π'),(π$^{\pm}$,π$^{\mp}$) and (π,N) and in determining coincidence cross sections (e.g., (π,πN)) will be very important. We can expect also substantially more data on inelastic pion scattering and on pion absorption/production reactions, but uncertainties about the reaction mechanism will make it difficult to learn quantitatively about the pion wavefunction in the nuclear interior.

Pionic atom level widths appear to be consistent with the Δ spreading interaction discussed in the resonance region. However, recent experimental information on strongly bound levels indicates serious problems for the standard optical potential. There is considerable room for theoretical improvement on the form of the optical potential; for example, nonlocality and proper isospin dependence should be included. The most important task, however, is the experimental one of learning more about the strong interaction shifts and widths of deeply bound pion atomic states.

The role of pion exchange currents in small energy transfer nuclear electromagnetic processes has been clarified greatly in recent years. Improved understanding of the medium range nuclear force should now be reflected in further development of the exchange current technology. In particular, this should establish the importance of the Δ exchange current. However, neither experimental data nor theoretical calculations have, to date, provided any convincing evidence about the (non-perturbative) role of Δ's in the nuclear ground state. Extraction of a Δ "probability" cannot proceed without reference to a detailed dynamical scheme.

As a final note, we observe that progress has been much more rapid, not surprisingly, in the accumulation of high quality data than in the development of comparably sophisticated theories of pion-nucleus dynamics. All signs point to a continuation of this trend.

REFERENCES AND FOOTNOTES

1. L. Kisslinger and W. Wang, Ann. Phys. 99(1976)374.
2. M. Hirata, F. Lenz and K. Yazaki, Ann. Phys. 108(1977) 116.
3. M. Hirata, J.H. Koch, F. Lenz and E.J. Moniz, Phys. Lett. 70B(1977)281; M.I.T. preprint CTP-748, to be published in Ann. Phys.
4. M. Dillig and M.G. Huber, Phys. Lett. 48B(1974)417; K. Klingenbeck, M. Dillig and M.G. Huber, Phys. Rev. Lett. 41(1978)387; invited paper of M. Dillig presented at this conference.

5. E. Oset and W. Weise, Phys. Lett. 77B(1978)159; invited paper of W. Weise presented at this conference.
6. I Navon et al, Tel-Aviv preprint TAUP 701-78 and paper contributed to this conference.
7. J.P. Albanese et al, Phys. Lett. 73B(1978)119; S.I.N. preprint.
8. R.A. Eisenstein and F. Tabakin, Comp. Phys. Comm. 12(1976)237; R. Landau and F. Tabakin, Phys. Rev. D5(1972)2746.
9. It has been argued that isotope differences will be far less sensitive to the higher order dynamics responsible for the differences evident in Fig. 5. However, rather high accuracy in pinning down parameters of neutron distributions has been achieved already from high energy proton scattering and from magnetic electron scattering; further, the excellent agreement found for the proton distributions extracted from Hartree-Fock calculations and from electron scattering data implicity confirms that the Hartree-Fock neutron distribution is quite sensible (see the review of J.W. Negele in the Proceedings of the Conference on Modern Trends in Elastic Electron Scattering, I.K.O., Amsterdam, 1978). Consequently, pion scattering can contribute further only by extracting neutron densities to very high accuracy; for example, the neutron root-mean-square radius should be pinned down to about 0.05 fm. To do this, the isospin dependence of higher order terms in the optical potential must be known very well (note that the difference in position of the minimum in the PIPIT and Δ-hole calculations in Fig. 5 translates naively into a RMS difference of about 0.25 fm). It would seem more profitable at this stage to take the best available Hartree-Fock neutron distributions and to use isotope difference measurements to study the isospin dependence of the optical potential. It is worth noting that, for pionic atoms, the isospin dependence found for the level shifts and widths seems substantially different from that expected from the conventional semi-phenomenological pion optical potential (G.Beer, private communication).
10. G. Davidson, Ph.D. thesis, Massachusetts Institute of Technology (1959).
11. J.H. Koch and E.J. Moniz, M.I.T. preprint CTP-765.
12. L.M. Saunders, Nucl. Phys. B7(1968)293.
13. For a review and further references see the paper of J.M. Laget, in Proceedings of the International Symposium on Photopion Nuclear Physics, P. Stoler, editor (Plenum Press, N.Y., 1979) pg. 255.

14. For a review, see the paper of E.J. Moniz, in Proceedings of the International Symposium on Photopion Nuclear Physics, P. Stoler, editor (Plenum Press, N.Y., 1979) pg. 335.
15. Recent (π,π') measurements are reported in the invited paper of Q. Ingram presented at this conference.
16. J. Hufner, Phys. Reports 21(1975)1.
17. S. Barshay, G.E. Brown and M. Rho, Phys. Rev. Lett. 32(1974)787.
18. For a summary, see the paper of M. Leon, in Proceedings of the LAMPF Summer School on Nuclear Structure with Pions and Protons (June, 1977), R.L. Burman and B.F. Gibson, editors.
19. J.F. Dubach, E.J. Moniz and G.D. Nixon, M.I.T. preprint CTP-749.
20. J.Konijn et al, paper contributed to this conference; J.H. Koch, private communication.
21. J.H. Koch, E.J. Moniz and G.D. Nixon, unpublished.
22. D.O. Riska and G.E. Brown, Phys. Lett. 38B(1972)193.
23. J. Hockert, D.O. Riska, M. Gari and A. Huffman, Nucl. Phys. A217(1973)14.
24. For a review and further reference, see the invited paper of L.S. Kisslinger at this conference.
25. R.D. Amado, Phys. Rev. C19(1979)1095.
26. R.D. Amado, to be published.

Question: A. Rinat, Weizmann Institute

I whole heartedly agree with your remark that the notion of "Δ probability" is model-dependent. However, one should stress that within any given consistent unitary model those probabilities are defined.

Response: Moniz

No response.

Question: W. Weise, University of Regensburg

Do you have any argument for your phenomenological spreading potential being local in r-space? I'm thinking especially about the real part of it, which you have not shown explicitly in your talk, but which seems to be hard to interpret on physical grounds.

Response: Moniz

The Δ spreading potential should, of course, be nonlocal. However the expectation that the precise form would not matter very much was confirmed by repeating some of the calculations after "smearing" the spreading potential via a one Fermi nonlocal range, while maintaining the total cross section constant. The description of elastic scattering was essentially unchanged (see Reference 3). We would certainly agree that a microscopic calculation of the real part of the spreading potential is very difficult.

Question: E. Oset, Universidad de Barcelona

In order to stress the difference between the Δ-h approach to π-A interactions and the standard optical potential approach, it is interesting to separate the effects of the different pieces contributing to the Δ-h interaction with the medium, i.e. Pauli effects, absorption and reflection pieces etc. Sometimes the effects individually are large but there are cancellations and this makes the optical potential approach more successful, but by pure accident. I would like to know if this separation has been done for the photonuclear processes?

Response: Moniz

The point made by Figure 7 is precisely that. In optical potential language, the pion-nucleus (or Δ-hole) dynamics must be handled consistently in calculating the distorted waves and production operator. Eliminating the spreading interaction in the production operator while retaining it in the distortion operator led to significant error.

Comment: L. Kisslinger, Carnegie-Mellon University

It is true that the asymtotic normalization of the N^*-N (virtual N^*) states cannot be identified as the residue of a pole as with the N-N S and D states. However, even the D state of the deuteron might well be an entirely model dependent quality. Yet it is important in nuclear physics. The N^* components are important for high momentum component properties. We must develop reliable models to study them as well as the N-N D state.

Study of Giant Isobaric Resonances in the Isobar Doorway Model

M. Dillig, H.M. Hofmann and K. Klingenbeck
Institute for Theoretical Physics
University of Erlangen-Nürnberg, Erlangen/W.-Germany

I. Introduction

It is well known that the excitation of isobaric resonances - which clearly show up in the elemantary πN amplitude - also dominate most of the medium energy processes on nuclei. In particular the $\Delta(33)$ isobar seems to play a crucial role for a detailed understanding of the dynamics in medium energy physics. As a consequence, a proper microscopic treatment of such processes requires, to extend the conventional optical model or multiple scattering approach and to treat the Δ-isobar propagation explicity as part of the nuclear many body problem on an equivalent footing as the nucleons. The isobar doorway or giant resonance model provides the proper framework for such an approach[1].

II. Basic formalism

The basic ideas and approximations behind the giant resonance model are most clearly revealed in the projection operator formalism[2]. Concentrating for a simpler notation for the moment on elastic π-nucleus scattering it is technically convenient to divide the whole Hilbert space into the elastic channel (P space), the doorway channel with one nucleon excited into an isobar (D space) and all the inelastic channels (R space). Decomposing the π-nucleus scattering amplitude at the scattering energy ω in a nonresonant, $T^{NR}(\omega)$, and a resonant part, $T^{R}(\omega)$,

$$T_{\pi A}(\omega) = T^{NR}(\omega) + T^{R}(\omega) \qquad (1)$$

an appropriate approximation scheme for $T^{R}(\omega)$ is easily worked out from two assumptions:

- the isobar doorway hypothesis, stating that there is no direct coupling between the P and the R spaces; and

ISSN:0094-243X/79/540306-12$1.50

- the giant resonance assumption that the dynamics of the system is dominated by the coupling of the doorway state to the elastic channel.

As a consequence, in the resonant amplitude[3,4)]

$$T^R(\omega) = \langle f| H_{PD}(\omega - H_{DD} - W_{DD} - \widetilde{W}_{DD})^{-1} H_{DP} |i\rangle \qquad (2)$$

(the indices symbolize the projection on the model spaces) the coupling between the doorway state and the elastic continuum due to W_{DD} has to be treated in detail, whereas in a first approximation the Hartree-type self energy correction H_{DD} of the isobar and the coupling of the doorway state to the inelastic channels via $\widetilde{W}_{DD}$ may be

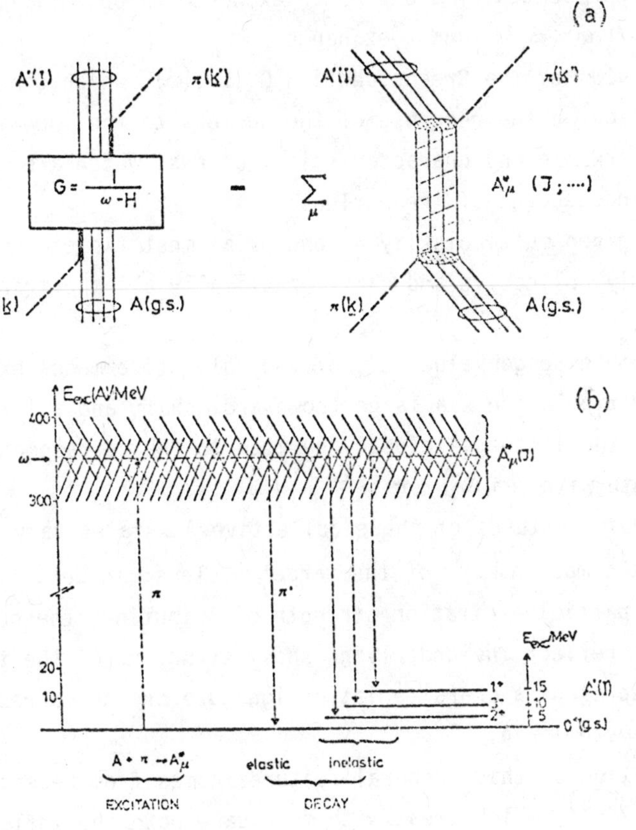

Fig. 1: Schematical representation of the (a) π-nucleus amplitude and (b) of elastic and inelastic π-nucleus scattering in the A* model.

parametrized in a simple form. The resulting truncated A-body system is then solved in the TD approximation yielding for the resonant scattering amplitude

$$T^R(\omega) = \langle f | H_{PD}\, G_{\Delta,A-1}(\omega) H_{DP} | i \rangle \qquad (3)$$

with (Fig. 1a).

$$G_{\Delta,A-1}(\omega) = \sum_{J^\pi} \left\{ \sum_{\mu} \frac{|(A^*)^{J^\pi}_\mu \rangle\langle (A^*)^{J^\pi}_\mu |}{\omega - \varepsilon^{J^\pi}_\mu} \right\} \qquad (4)$$

where for a given multipolarity J^π the eigenstates A^*_μ represent the eigenmodes of the whole nuclear system. For practical calculations W_{DD} is represented by a one boson exchange ΔN potential, with contributions from π, σ, ρ and ω^0 exchange.

For $\omega \lesssim 400$ MeV the Greensfunction $G_{\Delta,A-1}(\omega)$, which contains all the information on the response of the nucleus to the incoming pion, shows some remarkable and characteristic features, which are found to be rather independent of the nucleus:

- for a given multipolarity J^π one or at most two eigenmodes A^*_μ are strongly collective and carry practically all the transition strength S^{J^π} ;
- the complex eigenvalues $\varepsilon^{J^\pi}_\mu$ of the collective modes exhibit for very low multipolariteis a large (downward) shift and a large elastic width; the latter is closely related to the corresponding transition strength with $Im\, \varepsilon^{J^\pi}_\mu \propto |S^{J^\pi}_\mu|^2$;
- the gross features of these collective A^* states vary only smoothly with the mass number of the target nucleus; in detail, however, their partial excitation strength distribution, the positions and widths reflect the underlying shell structure of the target nucleus and represent a characteristic signature of the corresponding many-body system (Fig. 2)[3].

Confirmation of these general features comes from recent model calculations[4,5,6], which treat with more care both the influence of short ranged meson exchange contributions, in particular the effects of two-boson exchange corrections on the isobar self energy (in the calculation by Hofmann the imaginary parts of diagrams listed in Fig. 3a were investigated[5]), while in the work of Oset and Weise also the corres-

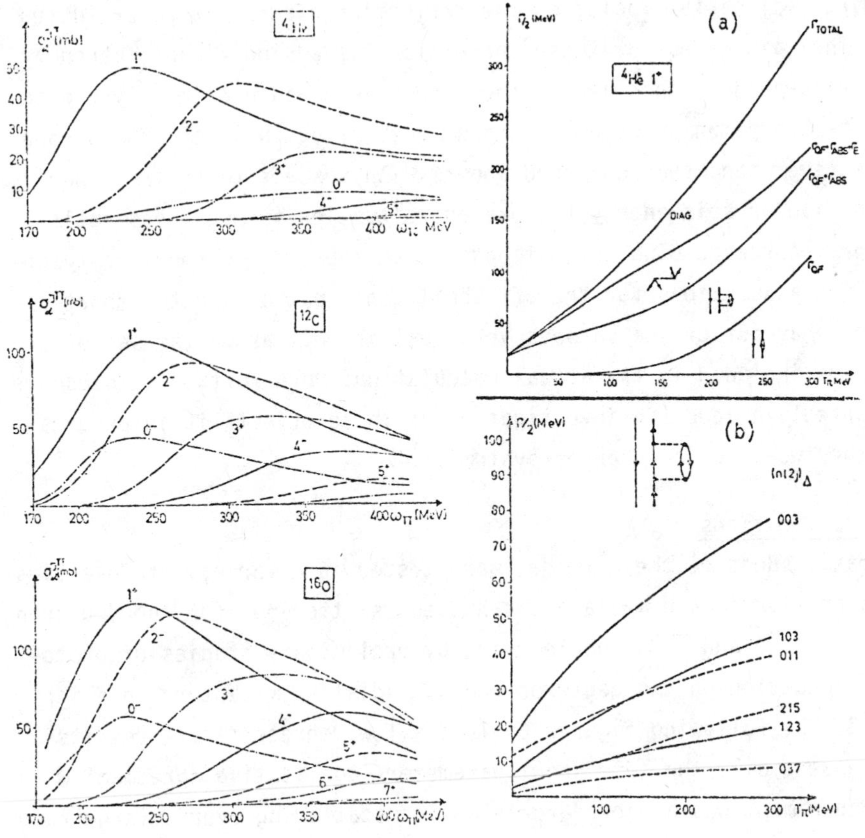

Fig. 2: Contributions of A* resonances of different multipolarity to the integrated elastic π-nucleus cross section on ^{4}He, ^{12}C and ^{16}O (as a function of the pion energy ω).

Fig. 3: a) Representative example for the contribution of the quasi-free decay (QF), true π-absorption (ABS) and the coupling to the elastic continuum (EL) to the total width of the diagonal matrix element of the ^{4}He, 1^{+} state in the isobar propagator (as a function of ω).

b) Contribution of true π-absorption to the width of different isobar configurations (as a function of the pion energy).

ponding real parts, including the reflection diagram, were evaluated [6]). The main effect of these corrections is a significant, strongly channel dependent increase of the effective Δ width primarily due to true π-absorption (Fig. 3b), without destroying, however, the strong collective phenomena mentioned above. With respect to the real part of the isobar self energy the present situation is much less satisfactory: corresponding estimations how an appreciable sensitivity on specific assumptions for the off shell continuation and the short range behaviour of the various vertices, as well as on the set of diagrams included in the actual calculation. Here further studies are undoubtedly needed (an investigation of these effects is in progress for the two- and four-baryon system).

III. Applications

The basic ideas of the A^* model were tested in a variety of investigations on elastic and inelastic π-nucleus scattering, for the deuteron, ^{4}He, ^{12}C, ^{16}O and ^{40}Ca supplemented by preliminary studies of photopion production on the deuteron and of radiative π-capture in flight on ^{12}C; the following figures collect a few representative results. With essentially the same input parameters qualitative agreement with the experiment was achieved for elastic π scattering over a large range of scattering energies: even in its crudest version the model is capable of reproducing the characteristic experimental trends (Fig. 4) [3,7,8]. The inclusion of absorptive corrections on a microscopic level signals an improvement of the theory (Fig. 4b) and stresses the need for their proper treatment in more refined calculations[4,5,6]. Conceptually the transition to inelastic processes is fascinatingly simple: it involves just the appropriate projection of the A^* resonances onto the corresponding entrance and exit channels (Fig. 1b). Again, first results for inelastic π scattering (Fig. 5)[3] and the photoproduction of pions (Fig. 6) show encouraging qualitative agreement with the experiment *).

---

*) We point out that extensions to incorporate in more details the final state interaction of inelastically scattered pions as well as the nonresonant background contributions in the (π,γ) process, are under investigation.

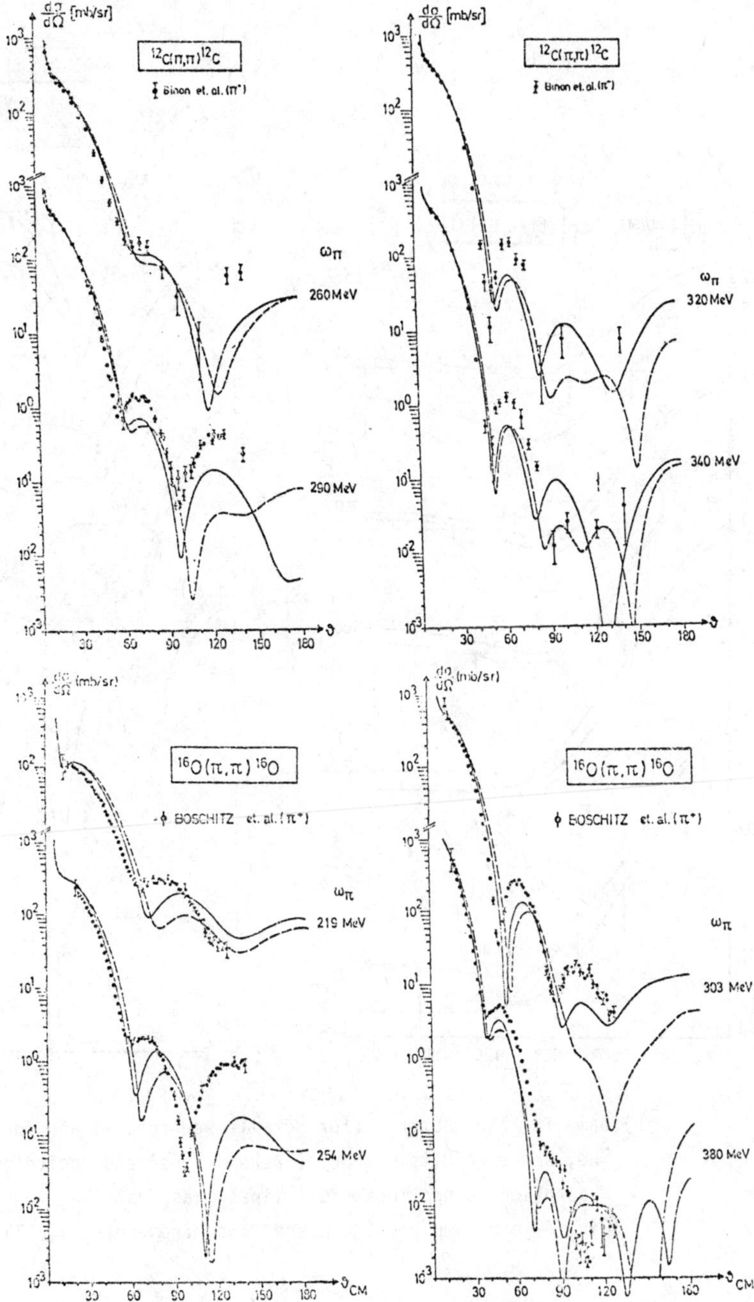

Fig. 4: a) Differential cross sections for elastic π scattering on ^{12}C and ^{16}O; solid (dashed) curves are with (without) nonstatic corrections (data are taken from refs. 10, 11).

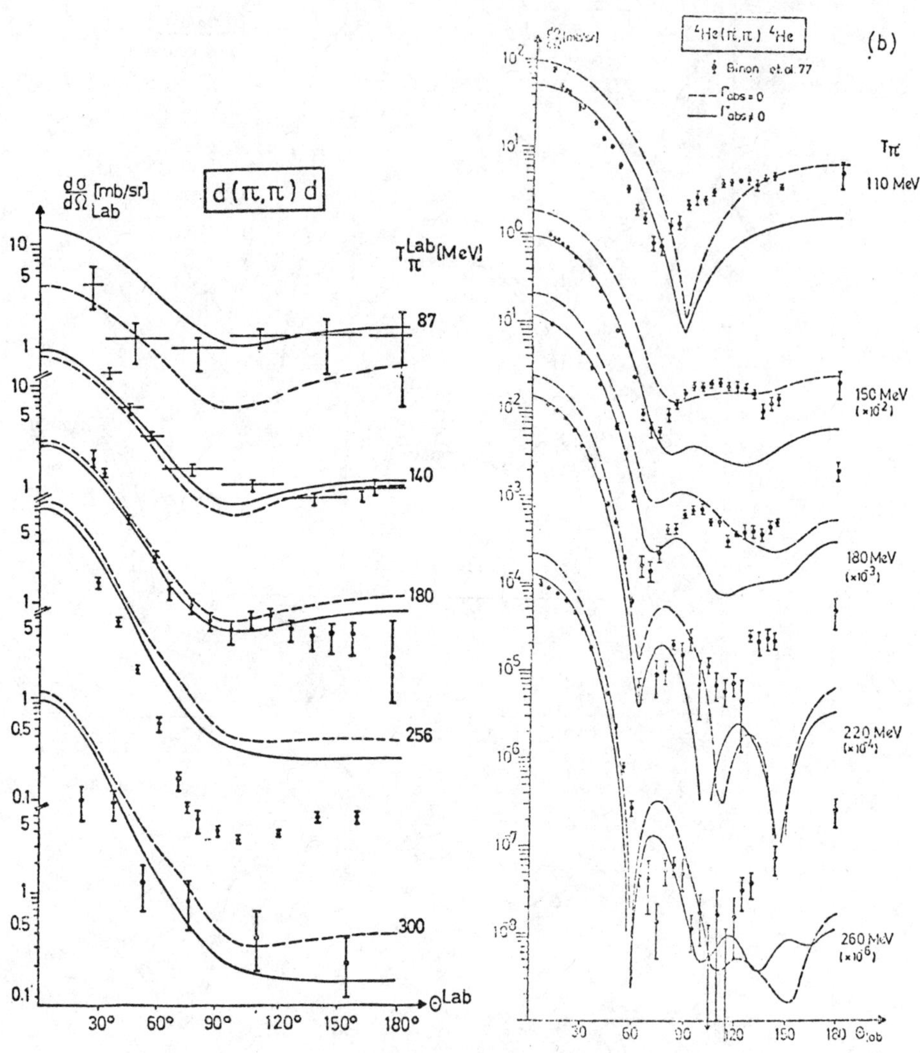

b) Same for the deuteron (for details see ref. 7) and for ^{4}He, where results without (broken lines) and including absorption connections (full lines), as described in Fig. 3, are compared (data are taken from refs. 9, 12).

IV. Conclusions and Outlook

Inspite of its conceptual novelty the isobar doorway approach has already proven as an adequate starting point for a microscopic understanding of the dynamics in isobar dominated medium energy processes on nuclei. In our opinion its two outstanding features are:

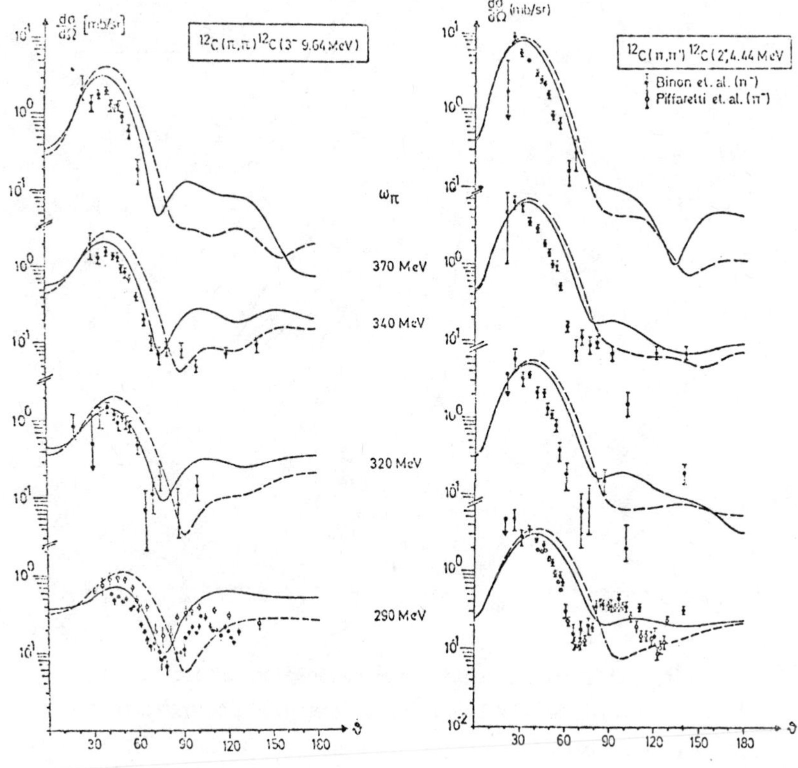

Fig. 5: Differential cross section for inelastic π - ^{12}C scattering into two different final states. The predictions with (full lines) and without nonstatic corrections (dashed lines) are compared with data from ref. 10.

the model accounts for the prominent role of the isobar in the dynamics of medium energy processes on nuclei by generalizing the conventional many-body problem, to account for the internal excitations of the nucleons in a conceptually transparent way which is accessible to specific calculations;
the model provides comprehensive information on the nuclear dynamics in form of a few collective A^* resonances as a characteristic signature of the response of a given nucleus on the interaction with an external probe. If established in detail these giant isobaric resonances should allow a unified description of a wide class of medium energy processes in terms of a few "fundamental" excitation modes.

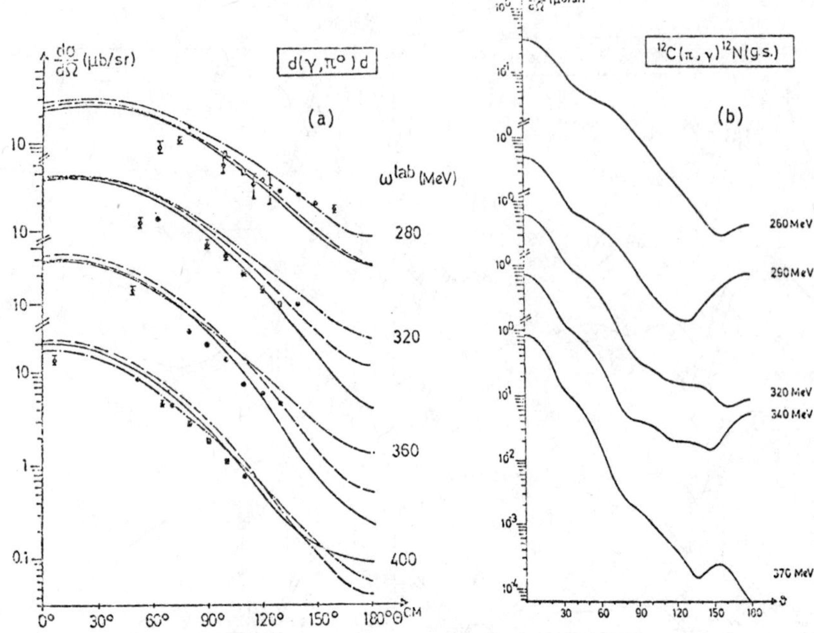

Fig. 6: a) Differential cross section for photo-pion production on the deuteron at different photon energies (compared with data from ref. 14).

b) Same for the inverse reaction (π,γ) on ^{12}C.

For the details, however, we have to keep in mind, that we are just at the beginning of a quantitative understanding of medium energy processes on nuclei. Quite some work, such as the investigation of different pieces in the isobar self energy or the extension of the formalism on other reactions like (π,π N), (p,π) or (γ,p), remains to be done. We feel confident, however, that such a theoretical program, supported by a stream of good and systematic data, will boost us to a far deeper understanding of nuclear dynamics in the near future.

References:

1. L.S. Kisslinger an W.L. Wang, Phys. Rev. Lett. 30 (1973) 1071
M. Dillig and M.G. Huber, Phys. Lett. 48B (1974) 417
F. Lenz, Ann. of Phys. 95 (1975) 348
G.E. Brown and W. Weise, Phys. Rep. 22C (1975) 909

2. H. Feshbach, Ann. Phys. (N.Y.) 5 (1958) 357, 19 (1962) 287

3. M.G. Huber and K. Klingenbeck, Nucleonica (1979, in press)

4. M. Hirata, F. Lenz and K. Yazaki, Ann. Phys. 108 (1977) 116,
M. Hirata, J.H. Koch, F. Lenz and E.J. Moniz, preprint (1978, sub. to Ann. Phys.)

5. H.M. Hofmann, Z. Physik A 289 (1979) 273

6. E. Oset and W. Weise, Phys. Lett. 77B (1978) 159 and preprint (1978, Univ. Regensburg)

7. R. Händel, M. Dillig and M.G. Huber, Phys. Lett. 73B (1978) 4

8. K. Klingenbeck, M. Dillig and M.G. Huber, Phys. Rev. Lett. 41 (1978) 387

9. E.G. Pewett et al., Phys. Rev. 131 (1963) 826
K.G. Gabathuler et al., Nucl. Phys. B55 (1973) 397

10. J. Pifaretti et al., Phys. Lett. 67B (1977) 239, 71B (1977) 324
F. Binon et al., Nucl. Phys. B17 (1970) 168

11. J. Jansen et al., Phys. Lett. 77B (1978) 359
J.P. Albanese et al., Phys. Lett. 73B (1978) 119

12. F. Binon et al., Nucl. Phys. A 298 (1978) 499

13. R. Händel, M. Dillig and M.G. Huber, Proceed. of the Symposium on Photopion-Reactions, Troy (1978) (Plenum press, in press)

14. B. Bouquet et al., Phys. Lett. 41B (1972) 536, Nucl. Phys. B79 (1974) 45
E. Hilger et al., Nucl. Phys. B93 (1975) 7

Question: M. Furic, Rice University

In the 1960's the so called shell model states embedded in the continium were treated in a similar way. Of course your explicit interaction is very specific. Are your "technical steps" identical to previous ones?

Response: Dillig

Yes! Though the force and some technical details are quite different, our model is in principle an extension of the conventional model as introduced around 1962 by Brown and Bolsterli.

Question: A. Rinat, Weismann Institute

There are standare arguments which stress absorption effects on elastic distributions for low incident energies. Your results seem to imply that large effects appear for all angles through the resonance energy. Can you give criteria (for energy and angles) for those corrections to be important?

Response: Dillig

Up to now no, but we are just extending the calculations to cover different nuclei at quite different kinematics.

Question: J. Alster, TRIUMF

You mentioned that you want to use the formulism for other pion reactions, such as single charge exchange. Infact Hirata did such a calculation for $^{13}C(\pi^+,\pi^0)^{13}N$(IAS) and his result is just as bad as the DWIA calculations.

Response: Dillig

I think it fair to argue that the approach is simply too new and needs additional substantial refinements before a reliable statement about its quality can be made.

Question: A. Saharia, TRIUMF

1) In our phenomenological parameterization of the isobar-doorway model [Saharia and Kisslinger, to be published] we find that nonlocality associated with Δ-propogation decreases with energy which is consistent with your observation that the number of doorway states increases as energy. 2) As far as the isobar doorway propogator is concerned, we find that the channel dependent and channel independent parameterizations give very similar results as long as the total width in the peripheral partial wave is the same.

Response: Dillig

It is not clear to me the merit of a simple parameterization (i.e. to use a channel independent width) because if you apply the model to processes involving high momentum transfer on nuclei it is clear that peripheral waves are quite differently damped compared to the central waves.

Question: F. Tabakin, University of Pittsburgh

Would you please indicate if, when, and how your model breaks down as you go either up or down in energy, away from the region of the dominance of the Δ and the Δ-h collective states you discuss? I would just like to know the limits of your formulation.

Response: Dillig

The limits of the model, concerning its validity at very large or very low energies, have not really been exploted so far. However for low energies I am pretty convinced that the model makes sense down to very low energies, as there are indications that even for π-absorption in mesic atoms the isobar excitation plays a significant role (if recoil is properly taken into account). At energies beyond the 33 resonance of course the influence of particle hole resonances principly of the d wave resonance around 1800 MeV have to be studied.

MESONIC EXCHANGE CURRENT EFFECTS IN ELASTIC ELECTRON-DEUTERON SCATTERING*

Earle L. Lomon
Center for Theoretical Physics
Laboratory for Nuclear Science and Department of Physics
Massachusetts Institute of Technology
Cambridge, Massachusetts 02139

ABSTRACT

The features of elastic electron-deuteron scattering for $0 \leq q^2 < 200\ fm^{-2}$ are reviewed. The complete set of three observables, electric and magnetic scattering and tensor polarization is examined. The contribution of a variety of relativistic and meson exchange current corrections (including isobar terms) is considered in conjunction with the impulse approximation from several nucleon force models. A phenomenological way of handling unknown, short range meson exchange current effects is described. It is shown that with the meager data presently available there is some information about which nucleon forces may be consistent over a range of q^2 and what the meson exchange current corrections are that go with such models. Experiments which are most likely to give information on models and meson exchange currents are discussed for several ranges of q^2.

INTRODUCTION

Elastic electron-deuteron scattering is sensitive to:

1. The deuteron wave function (i.e. the nucleon-nucleon force.
2. Electromagnetic form factors (emff) of nucleons
3. Relativistic corrections
4. Meson exchange current (MEC) corrections, including isobar terms.

Given the several undetermined elements which contribute significantly, can one learn anything about any of them, especially the MEC and the NN forces? In fact very little has been learned to date. Can future data improve the situation?

Observables

Because the deuteron and electron have spins of 1 and 1/2 respectively there are three independent amplitudes $G_i(q^2)$, i=0,1,2 which determine the observables - the

*This work is supported in part through funds provided by the U.S. DEPARTMENT OF ENERGY (DOE) under contract EY-76-C-02-3069.

ISSN:0094-243X/79/540318-14$1.50

electric and magnetic scattering and the tensor polarization:

$$A(q^2) = G_o^2 + \frac{8}{9}\eta^2 G_2^2 + \frac{2}{3}\eta\ G_1^{\ 2}$$

$$B(q^2) = \frac{4}{3}\eta\ (1 + \eta)\ G_1^{\ 2}$$

$$P(q^2) = \frac{4\sqrt{2}}{3}\eta\ (G_0 G_2 + \eta\ \frac{G_2^2}{3})/(G_0^2 + \frac{8}{9}\eta^2 G_2^2)$$

The G_i are normalized by

$$G_0(0) = 1 \qquad G_1(0) = \frac{M_d}{M_p}\mu_d \qquad G_2(0) = M_d^2\ Q$$

To determine the G_i all three observables must be measured at a given value of q^2. However P has never been measured except at q^2=0.

EXPERIMENTAL SITUATION

Before 1973

μ_d was known to much higher accuracy than needed for our purposes. The error in Q was estimated from the convergence rate of the binding energy of the molecular electronic wave function as about ± 1%.

Measurements of $A(q^2)$ were available for $q \lesssim 26\ fm^{-2}$ and had an accuracy of ± 5-15%. $B(q^2)$ was only measured for $q^2 < 15\ fm^{-2}$ to an accuracy of ± 15%. There were no measurements of $P(q^2)$, $q^2 > 0$.

Since 1973

A very accurate (±0.4%) measurement of $A(q^2)$ for $q^2 < 0.5\ fm^{-2}$ was made by R. W. Berard et al, Phys. Lett. 47B, 355 (1973). The range of measurements of $A(q^2)$ was very much extended by R.G. Arnold et al., Phys. Rev. Lett. 35, 776 (1975). They obtained measurements of ±13% for $q^2 < 80\ fm^{-2}$ and made measurements up to $q^2 = 150\ fm^{-2}$ where they were able to obtain a useful upper limit. These measurements established that $A(q^2)$ was of the order of the impulse approximation (IA) even at their highest values of q^2. On the other hand previously published values of MEC contributions were two orders of magnitude larger. The surprising results of Arnold et al forced one to consider the importance of hadronic form factors in decreasing MEC effects. Simultaneously those results demonstrated that the IA was of importance at high q^2, and that one may be able to learn something about the short range nucleon force.

The same experimental group, F. Martin et al, Phys. Rev. Lett. 38, 1320 (1977) tried to measure $B(q^2)$ at

$q^2 = 25\ fm^{-2}$. They only succeeded in getting a useful upper limit.

Up to this day there are still no measurements of $P(q^2)$!

THEORY

The S-matrix, or Feynman diagram, approach to MEC contributions is valid if the N-N interaction is described on the same footing. Dispersion theory and field theory NN potentials are successful down to heavy meson ranges, which is good enough for calculating the larger MEC effects. The MEC effects of diagrams A (pair terms, retardation and $\rho\pi\gamma$) and B (iterated isobar intermediate states) in Fig. 1 have been calculated, at least for the Reid Soft Core (RSC) potential. Those of diagrams C (uniterated isobar diagrams) have not been calculated and neither have those of D except at q=0 where they are small as expected. However because of their long range (two-pion exchange) one may expect them to be as large as pion pair terms and $\rho\pi\gamma$ at $q^2 \approx 30\ fm^{-2}$.

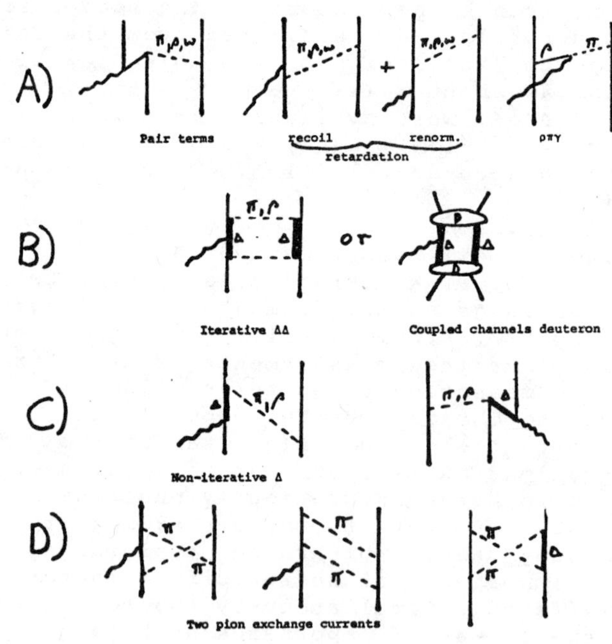

Figure 1

Calculations at $q^2=0$ (μ and Q) have recently been made by E. Hadjimichael, Nuc. Phys. A312, 341 (1978). Included are relativistic corrections, L·S and Q_{12} operator terms, pair (π, ρ and ω exchanges) retardation, $\rho\pi\gamma$ and iterative $\Delta\Delta$ terms of the coupled channel type, all done for RSC, three Feshbach-Lomon (FL), Paris, Stony Brook, Yale, Hamada-Johnston (HJ) and Sprung-deTourreil potentials.

For $q^2>0$ pair and $\rho\pi\gamma$ terms have been computed by M. Gari and H. Hyuga, Nuc. Phys. A264,409 (1976) and the same authors have computed the retardation terms, A278, 372 (1977) with hadronic form factors. Because of the strong coupling to isobars and the ambiguity of short range interactions between baryons, photon insertions on NN and ΔΔ coupled channel wave functions are preferred to a relativistic treatment of the Δ diagrams. This however can only take care of Δ's as iterated by the Schrödinger equation. The contribution of the iterated terms was calculated by H. Arenhövel and H. G. Miller, Z. Phys. 266, 13 (1974), by W. Fabian, H. Arenhövel and H. G. Miller, Z. Phys. 271, 93 (1974) and by M. Gari, H. Hyuga and B. Sommer, Phys. Rev. C14, 2196 (1976). The $q^2>0$ calculations were done over a large q^2 range only for the Reid Soft Core potential, and used either dipole fit or Iachello, Jackson, Lande emff, neither of which is wholly adequate to the elastic and quasi-elastic eN data.

RESULTS

$q^2=0$

As shown in Fig. 2, the "experimental" value of Q prefers a small D state deuteron, such as FL1 (4.57% D state) while μ_d prefers the large D state Hamada-Johnston and Yale potentials. The latter interactions have a smaller μ_D because of their strong long range quadratic spin orbit force. All other models give too large a μ_D even if the ΔΔ contribution is neglected. Therefore no single model with its consistent relativistic and MEC corrections agrees with both μ_d and Q!

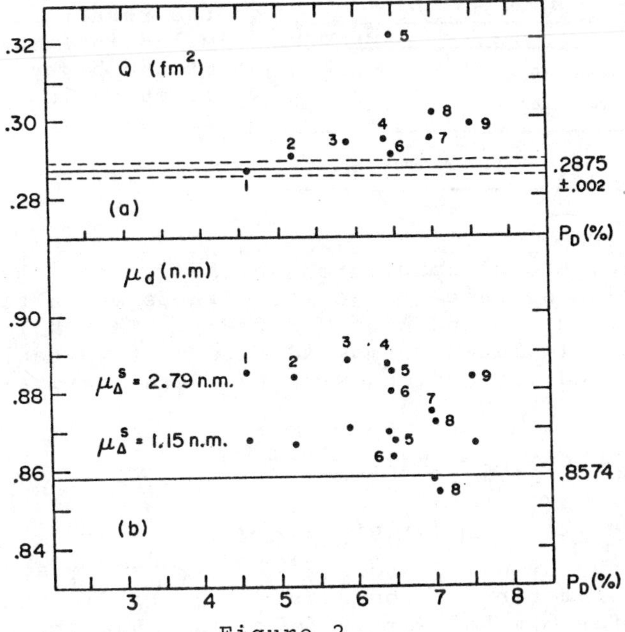

Figure 2

In resolving the above consideration μ_d^{MEC} is suspect because of (a) the ambiguous couplings and magnetic moment, μ_Δ^s, of the isobar term, and (b) the uncalculated non-iterative Δ contributions which may be of the same order as the iterative ones.

The value of Q^{MEC} is less vulnerable, being dominated by the π-pair term. However "Q^{expt}" is obtained by dividing the hyperfine structure by the second derivative at the deuteron of the HD molecule electronic wave function. It is the supposed accuracy of this latter quantity which determines the stated "experimental accuracy". As the variational calculation optimizes the electronic wave function to the energy, which gets negligible contribution from the point position of the deuteron, I feel that our actual knowledge of the value of Q may be somewhat less than stated.

$A(q^2<0\ 5\ fm^{-2}$

The analysis of this high precision data by Berard et al included relativistic corrections and showed significicant correlation with the large P_D wave function of FL15 (7.55% D state). Even with the new value of $G_E^{n'}(0)$ by Koester, Nistler and Waschkowski, Phys. Rev.Lett. 36, 1021 (1976), and the addition of MEC and G_2 contributions the result remains in favor of the higher P_D as shown in Table I.

Table I $G_E^{n'}(o)$ values analyzed from the data of Ref. 44, modified by inclusion of leading MEC term. The model independent relativistic and magnetic scattering correction are kept.

	$G_E^{n'}(o)$ in fm^2	
Curvature C	Model FL 1 or 5	Model FL 15
-.0036	.0257 ± .0023	.0223 ± .0023
0	.0244 ± .0022	.0215 ± .0022
Experimental value[18] of $G_E^{n'}(o)$ = .0199 ± .0003 fm^2		

Phenomenological MEC

It is possible, though considered unlikely, that the uncalculated two-pion or short range MEC effects may alter the comparisons with μ_d, Q and $A(q^2<0.5\ fm^{-2})$. Threshold dependence and the calculated terms indicate that we can treat the uncalculated corrections with the parameterization

$$\Delta G_i^{Phen} = \beta_i\ q^{n(i)}\ e^{-\alpha_i q^2}$$

$n(i) = 2(i=0),\ 0(i=1,2)$ for $q^2 \lesssim 30\ fm^{-2}$

The β_i are determined by μ_d, Q and $A(q^2<.5\ fm^{-2})$. There is then only one parameter per observable, α_i, to fit the three observables for $0.5\ fm^{-2}<q^2<30\ fm^{-2}$, in conjunction with each nucleon-nucleon force model. In this way both MEC and IA can be consistently phenomenologized at short range.

Using emff which fit the elastic and quasi-elastic eN data, I have computed the IA for FL1, FL5, FL15, RSC and HJ for $q^2 < 20\ fm^{-2}$. Predictions are then made in conjunction with the MEC corrections calculated with the RSC potential, or the phenomenological corrections determined by fitting the β_i as above, α_0 and α_1 from A and B at $q^2 = 5\ fm^{-2}$ and assuming $\alpha_2^{Phen} = (15m_\pi^2)^{-1}$.

Some of the results are summarized here.

$q^2 = 5\ fm^{-2}$

Examination of Tables II and III shows that MEC corrections to G_1 and G_2 are important. The iterative ΔΔ contributions make the total MEC effects in G_0 and G_1 highly ambiguous. The central value of β_2 and the assumed value α_2 (Table IV) determine ΔG_2^{Phen}. Table III displays the resulting values of the observables.

For A(5 fm^{-2}) it is seen that the IA is not adequate, that IA + MEC will fit the data if the ΔΔ contribution is reduced, and that using ΔG_2^{Phen} improves the fit. The experimental value of B(5 fm^{-2}) needs the use of the experimentally good emff (Case I which utilizes a dipole form G_m^n) and also requires very small (or negative) isobar contributions. The predicted values of P(5 fm^{-2}) show that it has 10% variations with model and 25% sensitivity to the corrections to IA.

$12\ fm^{-2} < q^2 < 30\ fm^{-2}$

G_0 vanishes for the first time (see Table V) at $q^2 \simeq 21\ fm^{-2}$ in the IA, and MEC corrections indicate that this may occur closer to 16 fm^{-2}. At q_0, $P(q^2) = \frac{1}{\sqrt{2}}$. Because of the rapid variation of $P(q^2)$ in the vicinity of q_0 (see Fig. 3) measurements of $P(q^2)$ can accurately determine q_0. As the deuteron wave function model dependence of $G_0^{IA}(16\ fm^{-2})$ is small we will then have a good determination of $G_0^{MEC}(q_0^2) = -\,G_0^{IA}(q_0^2)$.

At $q=q_0$ we have $B = \frac{4}{3}\,\eta(1+\eta)G_1^2$ and $A = G_2^2 + \frac{2}{3}\eta G_1^2$ providing a relatively "clean" separation of G_1 and G_2 which may then be compared with those calculations that are consistent with $G_0=0$.

At $q^2 = 15\ fm^{-2}$, as shown in Table VI, A(IA + MEC) fits the data better for a small P_D model such as FL1. On the other hand both A(IA + ΔG_i^{Phen}) and B(IA + ΔG_i^{Phen}) fit the data best for large P_D models such as HJ and FL15.

Table II Theoretical contributions to the $G_i(q^2=5\ fm^{-2})$ for various nucleon force models.

Amplitude	Model FL 1	FL 5	FL 15	RSC	HJ
G_o^{IA}	.1331	.1326	.1344	.1336	.1319
G_o^{MEC} (Refs4 and 22)				-.0067	
$G_o^{\Delta\Delta}$ (Ref 36b)[(a)]		.0051		.0057	.0064
$G_o^{Rel}=2\eta G_o^{MEC}$	-.0033	-.0033	-.0034	-.0033	-.0033
G_1^{IA}	.2397	.2393	.2405	.2381	.2370
G_1^{MEC} (Ref4)				.02	
$G_1^{\Delta\Delta}$ (Ref 36b)[(a)]		.0377		.0371	.0403
G_2^{IA}	3.383	3.514	3.804	3.709	3.753
G_2^{MEC} (Refs4 and 22)				.298	
$G_2^{\Delta\Delta}$ (Ref 36b)[(a)]		-.032		-.031	-.032
ΔG_2^{Phen} (see Table IV)	.739	.510	-.125	.343	.116

(a) $G_2^{\Delta\Delta}$ has been scaled down from Ref. 36(b) to correspond to 0.7% ΔΔ component.

Table III Predicted values of A, B and P at $q^2 = 5\ fm^{-2}$ for various corrections and models.

Terms used	Model FL 1	FL 5	FL 15	RSC	HJ
(a) $A(5\ fm^{-2})$	Experimental value (Ref. 48) .0180 ± .0011				
A(IA) [(a)]	.0202(3)	.0202(3)	.0211(1)	.0207(8)	.0203(4)
A(IA,MEC)	.0190	.0190	.0199	.0196	.0193
A(IA,MEC, ΔΔ)	.0204	.0204	.0213	.0209	.0205
$A(IA,MEC_{0,1},R_o,\Delta G_2^{Phen})$	.0186	.0183	.0183	.0186	.0180
(b) $B(5\ fm^{-2})$	Experimental value (Ref. 48) .0011 ± .00015				
B(IA) ff model I	1.07×10^{-3}	1.07×10^{-3}	1.08×10^{-3}	1.06×10^{-3}	1.05×10^{-3}
B(IA) ff model II	1.29×10^{-3}	1.28×10^{-3}	1.27×10^{-3}	1.26×10^{-3}	1.24×10^{-3}
B(IA,MEC,ΔΔ) ffmodel I	1.65×10^{-3}	1.65×10^{-3}	1.66×10^{-3}	1.63×10^{-3}	1.65×10^{-3}
(c) $p(5\ fm^{-2})$					
P(IA)	.667	.693	.734	.722	.738
P(IA,MEC)	.753	.779	.818	.807	.823
P(IA,MEC,ΔΔ)	.722	.747	.787	.775	.791
$P(IA,MEC,R_o)$	.756	.782	.821	.810	.826
$P(IA,MEC,R_o,\Delta G_2^{Phen})$	.847	.833	.764	.832	.810
$P(IA,\Delta G_o^{Phen},\Delta G_2^{Phen})$	.859	.839	.769	.846	.809

(a) Digits in parentheses correspond to replacing ff model I with ff model II.

Table IV. Coefficients of phenomenological correction forms $B_i\ e^{-\alpha_i q^2}$

Coefficients	Model				
	FL 1	FL 5	FL 1 5	RSC	HJ
β_o	$-.0024\pm.0025$	$-.0023\pm.0025$	$-.0058\pm.0025$	$-.0105\pm.0025$	$-.0032\pm.0025$
α_o	$-.005\ {}^{+.22}_{-\infty}$	$-.009\ {}^{+.24}_{-\infty}$	$-.19\ {}^{+.16}_{-.17}$	$.29\ {}^{+.12}_{-.11}$	$.10\ {}^{+.22}_{-.37}$
β_1	.0072	.0144	.0411	.0288	.0345
α_1	$.18\ {}^{+\infty}_{-.3}$	$.13\ {}^{+\infty}_{-.4}$	$.60\ {}^{+\infty}_{-.5}$	$.16\ {}^{+\infty}_{-.06}$	$.16\ {}^{+\infty}_{-.06}$
β_2	$1.44\pm.23$	$.99\pm.23$	$-.24\pm.23$	$.67\pm.23$	$.23\pm.23$
α_2 (a)	.133	.133	.133	.133	.133

(a) Assumed value

Table V Zeros of the Nucleon and Deuteron Form Factors, $0<q^2<200\text{fm}^{-2}$ (Entries are q^2 in units of fm^{-2})

Model / deut-f-f	FL 1	FL 5	FL 15	RSC	HJ
C_E	21,160	20.5, 158	21,160	19.5, 148	20,160
C_Q	86	103	125	118	125
C_L	45,177	57	88	?	88
C_S	30.5, 180	32,180	39.5, 170	?	39,193

Model / Nucleon FF	I	II
G_E^S	193	193
G_M^S	–	71

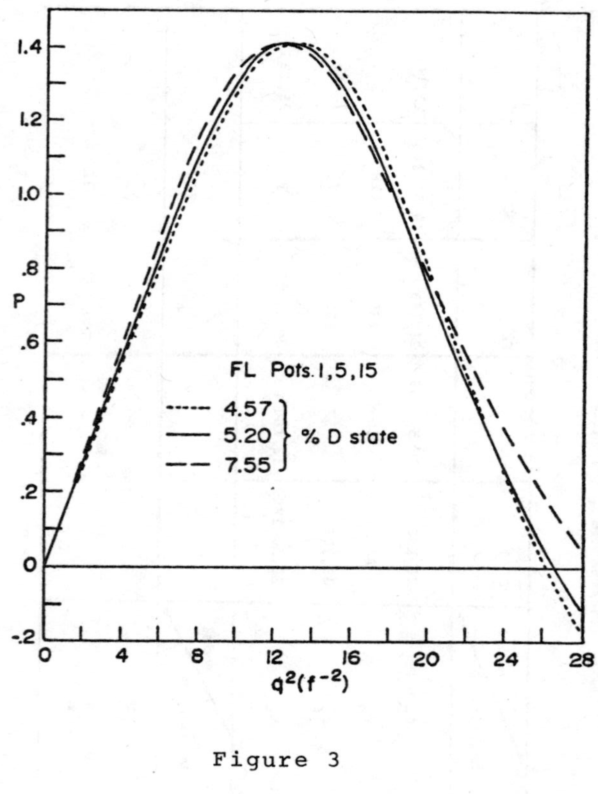

Figure 3

Table VI Amplitudes and Observables for q^2=15 fm^{-2}

Terms	Model				
	FL1	FL5	FL15	RSC	HJ
(a) A(15 fm^{-2})	Experimental value (Ref9) $(44.2\pm4)\times10^{-5}$				
G_o^{IA}	.0084	.0081	.0086	.0072	.0078
G_o^{MEC}(Refs4&36a)				-.0046	
ΔG_o^{Phen}	-.036	-.030	-.005	-.002	-.011
A(IA)	38.5×10^{-5}	42.7×10^{-5}	55.3×10^{-5}	49.5×10^{-5}	51.5×10^{-5}
A(IA,MEC)	42.9×10^{-5}	48.0×10^{-5}	61.6×10^{-5}	56.6×10^{-5}	58.2×10^{-5}
A(IA,ΔG_i^{Phen})	139.5×10^{-5}	106.6×10^{-5}	43.8×10^{-5}	63.3×10^{-5}	52.2×10^{-5}
(b) B(15 fm^{-2})	Experimental value (Ref16) $(80\pm25)\times10^{-6}$				
G_1^{IA}	.0289	.0299	.0341	.0286	.0323
G_1^{MEC}(Refs4&36c)[(a)]				.008	
ΔG_1^{Phen}	.0005	.0021	.0000	.0026	.0032
B(IA)	48.1×10^{-6}	51.5×10^{-6}	67.0×10^{-6}	47.1×10^{-6}	60.1×10^{-6}
B(IA,MEC)	78.4×10^{-6}	82.8×10^{-6}	102.1×10^{-6}	77.2×10^{-6}	93.6×10^{-6}
B(IA,ΔG_1^{Phen})	49.8×10^{-6}	59.0×10^{-6}	67.0×10^{-6}	56.1×10^{-6}	72.6×10^{-6}
(c) P(15 fm^{-2})					
G_2^{IA}	.436	.469	.540	.524	.527
G_2^{MEC}(4,22&35b)				.060	
ΔG_2^{Phen}	.195	.135	-.033	.091	.031
P(IA)	1.36	1.33	1.30	1.25	1.28
P(IA,MEC)	1.06	1.02	1.02	0.92	0.97
P(IA,$\Delta G_{o,2}^{Phen}$)	-0.68	-0.62	1.04	1.09	0.41

(a) To compensate for large ΔΔcomponent in Ref 36c, we use $\frac{1}{2}$ x Ref 36

Table VII Amplitudes and Observables for $q^2 = 26\ fm^{-2}$

Terms	Model				
	FL1	FL5	FL15	RSC	HJ
(a) $A(26\ fm^{-2})$	Experimental value (Ref38a) $(9.5\pm1.3)x10^{-5}$				
G_o^{IA}	$-2.05x10^{-3}$	$-2.18x10^{-3}$	$-2.07x10^{-3}$	$-2.50x10^{-3}$	$-2.14x10^{-3}$
G_o^{MEC}(Refs4&36a)				$-2.4x10^{-3}$	
ΔG_o^{Phen}	$-62.4x10^{-3}$	$-47.4x10^{-3}$	$-1.07x10^{-3}$	$-.145x10^{-3}$	$-6.51x10^{-3}$
A(IA)	$4.06x10^{-5}$	$5.12x10^{-5}$	$7.68x10^{-5}$	$6.92x10^{-5}$	$7.05x10^{-5}$
A(IA,MEC)	$6.96x10^{-5}$	$8.27x10^{-5}$	$11.19x10^{-5}$	$10.45x10^{-5}$	$10.48x10^{-5}$
$A(IA,\Delta G_i^{Phen})$	$423.6x10^{-5}$	$254.0x10^{-5}$	$7.41x10^{-5}$	$9.75x10^{-5}$	$11.84x10^{-5}$
(b)	Experimental value (Ref38b) $(5.9\pm12.0)x10^{-6}$				
G_1^{IA}	$3.04x10^{-3}$	$3.87x10^{-3}$	$7.00x10^{-3}$	$4.88x10^{-3}$	$6.26x10^{-3}$
G_1^{MEC}(Refs4&36c)(a)				$4.3x10^{-3}$	
ΔG_1^{Phen}	$.063x10^{-3}$	$.516x10^{-3}$	$.000x10^{-3}$	$.438x10^{-3}$	$.57x10^{-3}$
B(IA)	$0.952x10^{-6}$	$1.54x10^{-6}$	$5.04x10^{-6}$	$2.48x10^{-6}$	$4.03x10^{-6}$
B(IA,MEC)	$5.54x10^{-6}$	$6.86x10^{-6}$	$13.12x10^{-6}$	$8.66x10^{-6}$	$11.46x10^{-6}$
$B(IA,\Delta G_1^{Phen})$	$0.990x10^{-6}$	$1.98x10^{-6}$	$5.04x10^{-6}$	$2.91x10^{-6}$	$4.79x10^{-6}$
(c) $P(26\ fm^{-2})$					
G_2^{IA}	.0884	.0998	.1236	.116	.118
G_2^{MEC}(Refs4&36a)				.013	
ΔG_2^{Phen}	.0450	.0310	-.0076	.0209	.0071
P(IA)	.021	.057	.201	.065	.161
P(IA,MEC)	-.414	-.361	-.209	-.315	-.250
$P(IA,\Delta G_{o,2}^{Phen})$	-.0180	-.0177	-.079	.094	-.653

(a) To compensate for large ΔΔ component in Ref 36(c), we use $\frac{1}{2}$ x Ref 36.

At $q^2=26\ fm^{-2}$, as shown in Table VII, A(IA + MEC) now fits better with the models of larger P_D, while $A(IA + \Delta G_i^{Phen})$ still fits best for the models with large P_D.

In summary with the presently calculated MEC the predicted $A(q^2)$ is better for large P_D models when G_0 dominates A, and it is better for small P_D models when G_2 dominates A. This is the same source of inconsistency as in Q compared to $A(q^2<0.5\ fm^{-2})$. It is for these reasons that the ΔG_i^{Phen}, with which the larger P_D models are consistent for $q^2\leq30\ fm^{-2}$, do not agree with MEC.

To put the phenomenological approach on a sounder footing calculations of the two-pion exchange MEC and of the non-iterative Δ terms should be done together with other MEC and relativistic terms for each model. With the ΔG_i^{Phen} being taken as only the corrections to MEC and relativistic terms, then the phenomenological forms will be more trustworthy and the magnitudes of the ΔG_i^{Phen} should be much smaller.

In the context of the above information if in the future A is measured to ∿ ±3%, B to ∿ ±5-10% and P to ±.03 then we will probably be able to:

(1) Eliminate many nucleon force models as inconsistent for $q^2<30\ fm^{-2}$.

(2) Determine ΔG_i^{Phen} to compare with the expected magnitudes of uncomputed MEC terms.

Higher q^2, $30\ fm^{-2} < q^2 < 200\ fm^{-2}$

For $q^2>30\ fm^{-2}$ there are several complications that are not so important for smaller q^2.

- Relativistic effects are not expandable in q^2/M^2 and are unknown.
- Hadronic form factors are not well enough understood.
- Threshold form parameterization of unknown MEC effects is invalid.
- Quark aspects of nucleons may be important.

One may still use the following properties to separate MEC and model effects.

- Relativistic and MEC effects are expected to be smooth over scales $q^2 \tilde{\ } M^2$. Sharp structures indicate the IA behavior.
- Hadronic form factors and/or cancellations between MEC terms may keep the MEC contribution from overwhelming the IA. The IA effects, especially where zeros predict substantial dips or shoulders may therefore be measurable up to very high q^2. The interference with the MEC background will be a measure of MEC. In fact the IA form factors have several zeros as shown in Table V.

As shown by Fig. 4 the magnitude of $B(q^2)$ is very sensitive to (a) the choice of emff (much larger for Case I than for IJL) and (b) to the deuteron wave function (larger for small P_D). Because of (a) and maybe with a further boost in IA from (b), dips or shoulders may appear in the vicinity of 40 fm^{-2}. The magnitude of the maximum of $B(q^2)$ after the dip will be a measure of the IA which together with the position of the dip will indicate MEC contributions.

The large value of G_1 also contributes importantly to the size of $A(q^2 \tilde{\ } 80\ fm^{-2})$ as shown in Fig. 5. Then at $q^2 \tilde{\ } 180\ fm^{-2}$ A^{IA} has a very sharp structure because G_0 and G_2 vanish near each other (Table V). The width of the structure is controlled by the space between the zeros of G_0 and G_2 and therefore is a signature of the deuteron model. If the IA is still 20% or more of the cross section in this q^2 region, then this structure may be visible with 10% measurements of $A(q^2)$. The depth of the dip will indicate the size of MEC contributions.

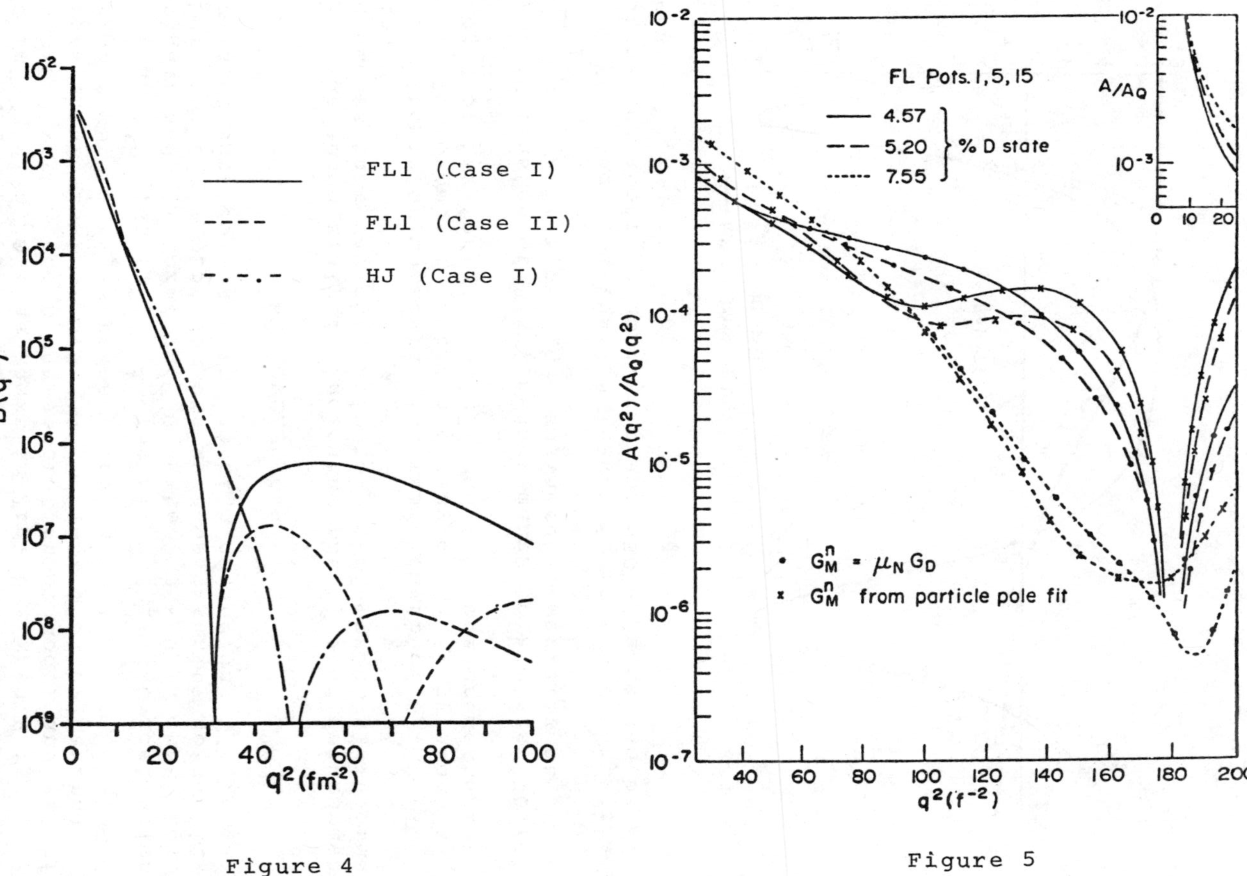

Figure 4

Figure 5

$P(q^2)$ is very sensitive to both model and MEC for $70\ fm^{-2}<q^2<125\ fm^{-2}$ as shown in Fig. 6. When P is combined with A and B the separation of model and MEC effects should be possible. At $q^2 \cong 140\ fm^{-2}$ the maximum value of $P^{IA} = \sqrt{2}$ is reached and G_0 vanishes at $q^2=160\ fm^{-2}$ so that $P^{IA}\ (160\ fm^{-2}) \cong \frac{1}{\sqrt{2}}$ for all models. Cancellation of pair and $\rho\pi\gamma$ terms in G_2 make $P^{MEC} \cong 0$ in that region. Therefore $P(q^2)$ provides one more opportunity to see if MEC or IA effects dominate in the high q^2 region.

CONCLUSIONS

The above analysis indicates the type of future measurements that may most help to provide critical information.

- Atomic or molecular ion measurement of Q.
- Improvements in the accuracy of $A(q^2)$ by about a factor of two for $q^2<100\ fm^{-2}$ and ±10% measurements for $q^2>100\ fm^{-2}$.
- Extension of $B(q^2)$ measurements to at least $50\ fm^{-2}$.
- Measurement of $P(q^2)$ anywhere, especially at $q^2 \cong 15\ fm^{-2}$.

With such observations new information about nucleon forces and about relativistic and MEC effects can be obtained.

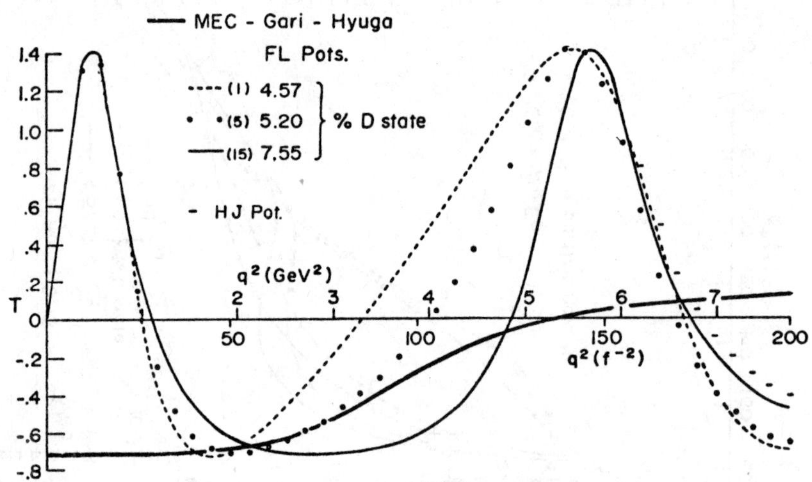

Figure 6

Question: H. Jackson, Argonne National Laboratory

In your view, is P_d, the D-wave probability of the deuteron wave function, a quantity which should be energy independent?

Response: Lomon

The D wave probability is defined as a unique number $\int w^2 d^3r/[\int u^2 d^3w + \int w^2 d^3r]$. The effect of the D state does vary with energy, depending on its detailed range dependence. As the potential is theoretically understood for $r > \frac{1}{2}\mu^{-1}$, P_D (which is determined by the short range tensor force) describes the effect of the D state for $q \lesssim 2\mu/c$. For high momentum transfer the effect of the D state will become sensitive to the detailed form of the short range tensor force.

Question: G. Igo, UCLA

When will the Bates energy doubler be ready to make the measurement of the tensor scattering in e-d scattering?

Response: Lomon

In about two years from now.

ISOBARS AND NUCLEI

Leonard S. Kisslinger
Carnegie-Mellon University, Pittsburgh, Pennsylvania 15213

ABSTRACT

The progress in our understanding of N* components of nuclei during the past two years or so is reviewed. The emphasis is on the treatment of the basic quantities which enter into theoretical calculations and their effect on comparison with experiments.

INTRODUCTION

Baryon Resonances dominate the medium-energy meson-nuclear interaction. From meson-nucleon scattering and reactions a great deal is known about the on-mass-shell formation and decay of isobars, and this has led to at least a semiquantitative understanding of the dynamics underlying "real" isobars. These modes also play an important part in the nucleon-nucleon interaction. At low energies one deals with these same isobars as "virtual" constituents of the two-baryon and many-baryon systems. About ten years ago the first suggestion that N* components of nuclei must be explicitly considered for processes at high momentum transfer was made, and since that time there has been a large body of work. Two years ago three rather complete reviews of the status of this field were made [1, 2, 3]. I shall therefore restrict myself to work appearing since that time except a brief review of older work when necessary for the logic. The major emphasis is on the basic theoretical ingredients, for this is the most important topic for the present stage of this field.

II. THE BARYON-BARYON INTERACTION

In this section we review the effects of N*'s in the two-baryon system, emphasizing the work which has been done during the past two years. The Basic Interaction to be considered is that of Fig. 1a, which when used in a dynamic theory has as its basic ingredients 1) Meson-nucleon vertex functions, 2) meson propagators, and 3) two-baryon propagators. We review these and recent developments in the Baryon-Baryon system.

A. The Meson-Baryon Vertex

The incomplete knowledge of meson-baryon vertex functions, baryon matrix elements of meson currents $<B'|J_m|B>$, is a major problem for accurate calculations of N* effects in nuclei. The form of the meson current for static (no recoil) coupling is [1]

$$V = g_{B'Bm} \; \psi_B^+ \; [\sum \nabla^{\ell}]^{s} \cdot T \cdot \phi_m \psi_B \; , \qquad (1)$$

ISSN: 0094-243X/79/540332-15$1.50

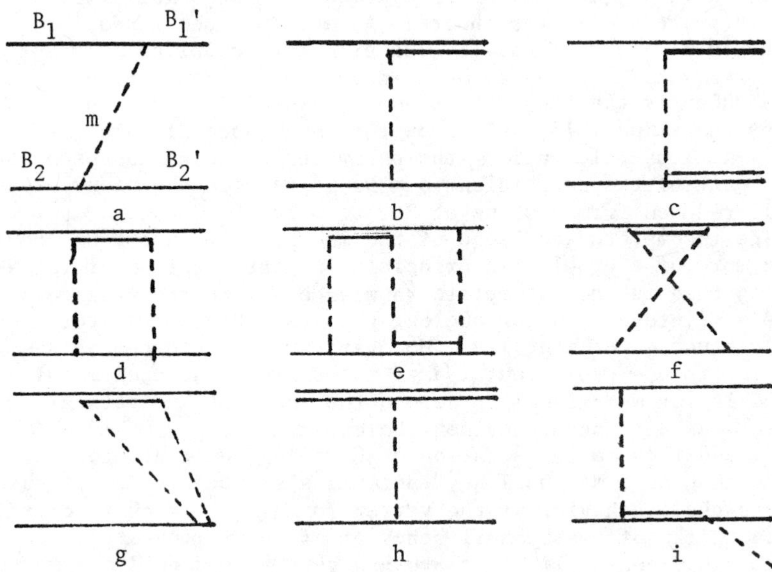

Fig. 1

where $g_{B'Bm}$ is a function of the external [off-shell] masses and the momentum transfer. The on-shell value of $g_{B'Bm}$ is referred to as the coupling constant. The π-N-N* coupling constants are well known, since the pionic widths of N*s are well-known. The relationship is $\Gamma_{N*N\pi} \propto |\langle B'|J_m|B\rangle|^2$, with, e.g., $J_m = g\bar{\psi}_N \gamma_5 \tau \cdot \phi \psi_N$ for the pion-nucleon vertex with p.s. coupling. Thus the longest-range part of the one-boson exchange interaction depicted in Fig. 1 is known not only for the N-N interaction, but also for the NN→NN*, NN→N*N*, etc. processes. However, for the m-N-N* vertex functions for the other mesons the vertices are in most cases completely unknown, and one must rely mainly on theoretical arguments for the coupling constants as well as the off-shell dependence.

The greatest effort has been spent on calculating the effects of the Δ (1232) in nuclear systems. For a quantitative treatment it has been shown [4] that ρ-meson exchange can be very important in determining the NN→NΔ transition potential. Therefore, it is essential to know the $g_{\rho N\Delta}$ vertex function. In the last few years a great deal of theoretical work has been done including ρ-meson effects based on a vector dominance model for the ρNΔ vertex. This involves an extrapolation for which the vector dominance model has not been justified. Recently, a Regge Pole analysis of peripheral reactions has been used [5] to extract the gρNΔ effective coupling constant. This method should be reliable since the ρ-Regge trajectory has been very successfully used for pionic and kaonic reactions, and the extrapolation to the ρ-mass is a standard procedure in the theory. As a result one finds that for the static form of Eq. (1) the effective coupling constant is $g^2_{\rho N\Delta}/4\pi = 3.7$, which can be compared to the range $5.8 \leq g^2_{\rho N\Delta}/4\pi \leq 14.5$ which is the range which one obtains from vector dominance and the quark model [6]. To show the importance of this result in pion-nucleus calculations, the parameter η^2 which has been used [7] as part of the coefficient of the ρ^2 term in the pion potential is reduced from a value of 2.1 to 0.14.

The incomplete knowledge of the mNN* vertex functions should not be more of a problem in principle for the question of the N-N* coupling than is the incomplete knowledge of the short-range part of the N-N interaction for nuclear physics. Some parameters must be determined experimentally. However, a major problem arises in trying to use experimental data to complete the phenomenological models for N-N* coupling due to the inherent theoretical problems, as will become evident below.

In addition to the question of coupling constants for the question of N→N* coupling, there is also the problem of the off-mass-shell behavior of the vertex functions, which is closely related to the off-mass-shell behavior of scattering amplitudes. For the π-nucleon amplitude there has been a good deal of progress recently with a detailed dynamic treatment of the low-energy behavior [8] and studies of the Chew-Low Model [9]. There is also a recent study for the treatment of pp↔πd [38].

The knowledge of the diagonal mN*N* vertices is in a much poorer state. Recently, there have been a number of studies related to the Δ-π-Δ vertex. In a recent analysis [10] of $\pi^- p \to \pi^+\pi^- n$ at resonance energies, an isobar model interpretation corresponding to the process of Fig. 2 was used to extract the Δ-π-Δ coupling constant. It was found that $g^2_{\Delta\pi\Delta}/\psi\pi = 40\pm20$, which can be compared to the SU (6) value of ≈100. It should be noted, however, that when the same diagram is used for internal bremstrahlung, with the wiggly line being a photon, the same interpretation of the $\pi^+ p \to \pi^+ p\gamma$ experiment gives a Δ^{++} magnetic moment in strong disagreement with the quark model. [See, e.g., Ref. [1] for a discussion.] The relevance of this to Δ-interactions is discussed below.

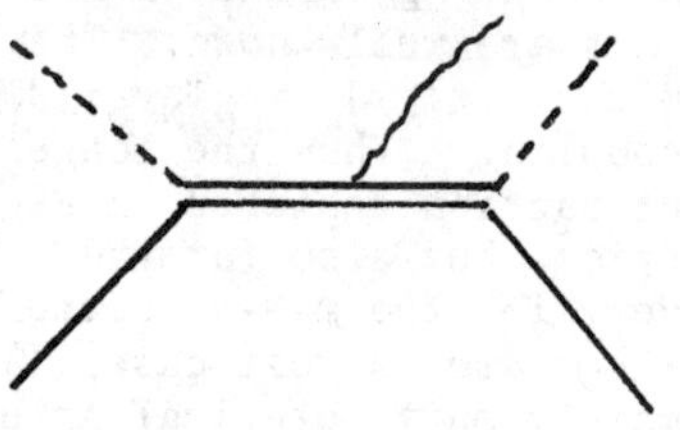

Fig. 2

B. The Coupled-Channel Model

The coupled-channel model for the two-baryon system makes use of the NN→NN* and NN→N*N* coupling potentials as well as the nuclear potential in the N-N subspace. The form of the equations is

$$G_0^{-1}(N_1N_2)\psi_{N_1N_2} = -\sum_{ij}\langle N_1N_2|V|N_iN_j\rangle\psi_{N_iN_j} , \qquad (2)$$

where $\psi_{N_iN_j}$ is the wave function for the two-baryon (N_i,N_j) channel, and $G_0(N_1N_2)$ is the two-baryon propagator in that channel. In the nonrelativistic treatments $G_0(N_1N_2) = \langle N,N_2|(H_0-E)^{-1}|N,N_2\rangle$, where the Hamiltonian H_0 includes the mass operator. In the perturbation theory version only the NN sector is used on the right-hand-side of Eq. (2). The coupling potentials are obtained from one-meson processes depicted in Fig. 1b,c. Applications to two-baryon problems are now reviewed.

1) <u>N-N Scattering</u>

Recently, there has been a good deal of discussion of the form of Green's functions and potentials of Eq. (2) and the appropriateness of the coupled-channel.approach for the effective N-N interaction [1, 2, 3]. I shall concentrate on these aspects. Most of the coupled-channel calculations have been restricted to nucleons and Δ (1232)'s, and I will restrict myself to these baryons in the present section.

The first, iteration of the coupling potential (Fig. 1b) yields the box diagram with an intermediate Δ (Fig. 1d), which was shown to simulate the effects of scalar (σ) mesons for the N-N potential. The early coupled-channel calculation of the N-N phase shifts [11] showed that about half of the effect of the 6 mesons, which were artificially introduced in to N-N one boson exchange potentials in order to fit the experimental phase shifts, could be accounted for. This suggests that at least part of the effect of the correlated two pion exchange processes must be removed from the N-N sector in the coupled channel calculations.

In recent calculations, a good deal of attention has been paid to the problem of altering the N-N interaction to remove "Δ-effects" so that there will hopefully be no double-counting in the coupled-channel calculations, and to the use of nonstatic interactions and propagators. The most elaborate calculations have used the second order form of the coupling potential to get an effective interaction similar to the box diagram [12]. The authors have attempted fits to the N-N phase shifts by adjusting the σ parameters, and have tested sensitivity to the form of the propagator and vertex function. They indeed find that there is great sensitivity. It should be noted, however, that sensitivity is not very useful without a theory which is sufficiently complete to be accurate.

With regard to the use of effective potentials rather than the coupled-channel method, it has been pointed out [11, 13] that there are contributions included in the coupled-channel approach which cannot be obtained with an effective N-N interaction. Of greatest importance is the neglect of "Δ-exchange" processes like those of Fig. 1e. Especially at energies above pion production threshold one can expect these processes to be important. In a recent somewhat schematic calculation [13] the Δ-exchange process was iterated to all orders and shown in that model to give important corrections to the N-N effective interaction. This has also been suggested in the earliest coupled-channel calculation of the D-wave N-N phase shifts [11]. As shown in Fig. 3, the coupled channel model with NN→NΔ coupling (curve cc) is very different from the OBEP model of curve 1, which is the OBEP model without 6's. The calculation leading to curve cc differs from that giving curve 1 only in the inclusion of Δ's. The third calculation, shown by curve 2, is the OBEP calculation with the 6's replaced. Thus it is indicated that for this state the coupled channel result differs from that of the effective interaction. It should be noted that the coupled channel and OBEP results are more nearly consistent for the othe phase shifts. This is consistent with the N-Δ channel being particularly important in the $^1D_2$ state, as is expected.

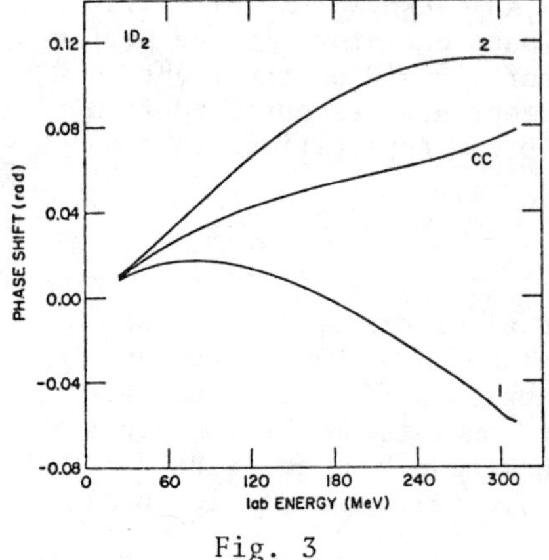

Fig. 3

It should also be noted that the OBEP coupled channel calculations do not include all Δ effects. E.g., the crossed pion process illustrated in Fig. 1f is not included, and must be added as a distinct irreducible diagram. However, a valid theory above the pion threshold must include processes like that of Fig. 1i, so that reliable coupled-channel calculations must be carried out.

Green and his coworkers [14] have recently calculated the imaginary part of the N-N phase shifts by a model which includes the Δ-exchange mechanism and other sources of inelasticity associated with the unstable Δ. Although these calculations are exploratory and do not result in good fits to the data, these are the types of effort which will be needed in the future.

The two most important questions for this area of physics are 1) the development of a truely relativistic theory which can accurately handle this field theoric problem, and 2) the incorporation of all of the physical processes necessary for the treatment of the problem. Most of the recent effort has gone toward the former, and there has been significant progress in testing sensitivity to the forms of the various ingredients. However, all realistic calculations have been done using a variety of three-dimensional forms, which simulate relativity but which are not unique, and which suppress variables important for the problem. One can anticipate progress in the future here. It is only after achieving such progress that one can attempt the second question: the development of a theory sufficiently complete to represent the physics.

2) The Deuteron

Naturally, the deuteron has received most attention in theoretical studies of N* components. This is the simplest relevant system and one should try out one's model here. The older calculations have been thoroughly reviewed [1, 2, 3]. They differ in the various detailed aspects, some of which have been discussed above: perturbation theory vs. coupled-channel treatment, static vs. relativistic forms for the propagators, recoil terms in the operators, the effect of ρ as well as π coupling potentials, the type of N*s included, and the inclusion of the diagonal N*N* interaction (Fig. 1h). The main difference between the results of the various calculations is the "probability" of the Δ-Δ component of the deuteron; i.e., $\int|\Psi_{\Delta\Delta}|^2 d^3x = P_{\Delta\Delta}$. The published values for that quantity have ranged from 0.015% to >2.0%. Recently, there has been an extensive calculation which has incorporated many of the ideas which have been proposed in the past few years [15]. The result is that $P_{\Delta\Delta} \simeq 0.1\%$, value which is lower than most recent calculations, and an order of magnitude smaller than previous results of Green and his coworkers. Both this result and the shape of the wave functions are similar to the earliest calculations of Jena and Kisslinger [11].

There has also been considerable work on the static properties of the deuteron, most of which occurred early in the N* game and has been reviewed. A recent calculation [16] on the effect of Δ's on the moments of the deuteron has included the effects of the Δ along with corrections from meson degrees of freedom and some relativistic effects. All of this is very complicated, but necessary if one is to use the moments as constraints on the N-N interaction.

There have also been predictions for other N* components of the deuteron and other nucleon systems, and the theoretical situation is similar. I shall refer to these below, since the different properties of the various N*s lead to different experimental consequences. In particular, since the deuteron is an isospin zero object, the I = 1/2 N*s are especially important for certain processes even though they are more massive than the $\Delta(1232)$.

My own conclusions for the status of the N-N system and the basic formulation of the theory of N* components of nuclei is that although there has been important progress, we are some distance from being able to calculate a priori and to make predictions of observable effects. But at this time new accurate data for N-N scattering above the pion threshold and related processes is becoming available. What is needed is the interplay between experiment and theory which characterizes nuclear physics. For this, however, one must unravel the reaction mechanism and learn how to make reliable comparison with experiment. I shall spend much of my remaining time on this aspect.

3) Δ-N and Δ-Δ Interactions

The most direct way to explore the interaction of N*s with nucleons is to create them with pions in nuclei. There have been a number of recent investigations related to various aspects of the Δ-N and Δ-Δ interactions, which I now review.

a) The Isobar-Doorway Model

In the Isobar-Doorway Model [17] one introduces directly Δ-components of nuclei as doorway states for pionic scattering and reactions. In the last couple of years there has been good progress both in the phenomenological applications [17, 18] and in the microscopic approaches [19]. This topic is the subject of Dr. Dillig's talk, so I shall only mention the results most directly applicable to the question of N* components of nuclei. In the region of the Δ the closure approximation can be used to find the average properties of the Δ in the nucleus. In this approximation the pion optical potential can be written

$$V^{opt} = V^{N.R.} + \frac{E-M_\Delta+i\Gamma_\Delta/2}{E-M_\Delta-\Delta E+i\beta\Gamma_\Delta/2} \; V^{(1)} , \qquad (3)$$

where $V^{(1)}$ is the first order optical potential derived from pion-nucleus scattering near the Δ in the impulse approximation, β gives the modification of the width of the Δ in nuclear matter, and ΔE is the difference of binding energy of the Δ and the nucleon averaged over states. The present results [18] indicate that $\beta \simeq 1.1$ (a net 10% broadening of the width) and that $\Delta E \simeq 10$ MeV. I.e., the Δ interacts less strongly than a nucleon with nuclear matter.

Therefore, the results of pion elastic scattering do not seem to be consistent with a strong interaction of the type of Fig. 1h. This in turn suggests that the $\Delta\pi\Delta$ vertex is much weaker than the prediction of the quark model [17]. Let us now consider recent work on this problem.

b) The Δ-N and Δ-Δ Systems. The Dibaryon

Since the $\pi N\Delta$ vertex is well known, the one pion Δ-N exchange reaction (Fig. 4) well known. This interaction has been used in the microscopic treatments of Δ-hole states to explore the isobar-doorway model [19]. In a recent study of Δ-N scattering [20] various approximations for the Bethe-Salpeter equation with the process of Fig. 4 as the driving term have been used. The results depend strongly on the assumptions. Once more one finds sensitivity, but has not been able to carry out an accurate calculation. By

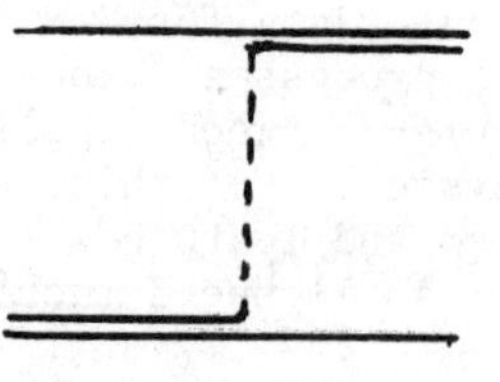

Fig. 4

means of detailed comparison with pion-deuteron scattering at the Δ-resonance, in the future it should be possible to improve the theoretical treatment.

Of great current interest is the possible existence of a 6-quark object - the "dibaryon". In the SU(6) or quark model, the Δ-Δ interaction is very strong. Using the SU(6) coupling constants with pion exchange, Kamae and Fujito [21] find that the Δ-Δ system is bound by some 100 MeV. They offer this dibaryon state as the explanation of the sharp peak seen in $\gamma D \to pn$ polarization [22], shown in Fig. 5. However, the existence of this very strong Δ interaction is not consistent with the results of the phenomenological I D model for pion scattering discussed above. The explanation of this strong polarization peak is now one of the most interesting open problems in the field.

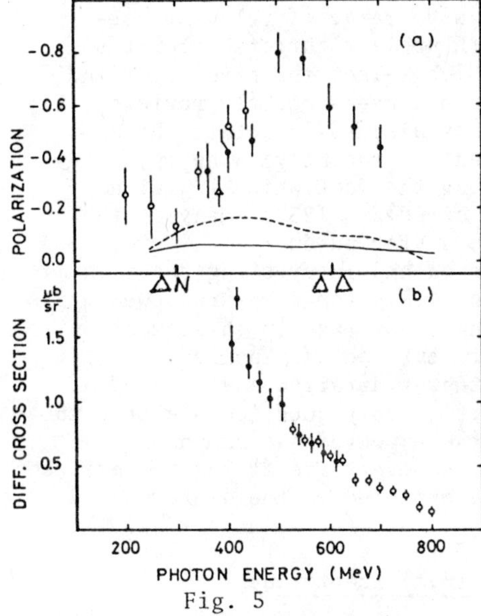

Fig. 5

III. ELECTROMAGNETIC EFFECTS

There is a large body of work on electromagnetic effects of meson exchange currents, and the effect of N*s must be considered to be part of this area of study. The reviews of two years ago [1, 2, 3] have covered this topic quite completely, and not very much has changed with regard to isobar contributions. Among the many significant theoretical problems is the question of adding together contributions from explicit meson degrees and explicit N* degrees. There might be a serious problem of double counting.

One comment I would like to make, is that it is the Δ which has mainly been considered. Since the γNΔ coupling is magnetic, it is unlikely that the Δ will contribute much to the charge form factor, as has been shown. However, the N(1688) has a sizable γN E2 coupling and could play an important role. This has been suggested in a calculation for the ^{3}He charge form factor [23]. It is interesting to note that there seems to be a discrepancy between experiment and theory in the high momentum transfer charge form factor of nuclei [24], and the possible effects of the N(1688) are being explored by Dan Riska and myself.

IV. N* TRANSFER REACTIONS

Transfer reactions provide the most specific tests of particle properties of nuclei. Since one can control the quantum numbers of the transferred system via exclusive reactions, and since nuclear reactions provide a wide range of relevant possibilities, transfer reactions with nuclear targets offer the best possibility of experiments which select the effects of individual N*s. This general question has been recently reviewed [1], so in the present review I will only discuss in detail three reactions which have received attention recently: The (p,d), (p,π^-), and $pp\leftrightarrow\pi d$ reactions. Among the work which I will not discuss is the recent calculation of $pd\to dN^*$ [25]. Many of the theoretical questions are the same as for $pd\to dp$.

One of the most widely discussed and interesting reactions is the spectator isobar production. This topic was reviewed quite beautifully by Weber [26], so I shall not take up this question at the impact of the current experiments on our knowledge of Δ-components. It should be noted, that a detailed analysis by A. S. Goldhaber [26] has shown that it is very questionable that the present experiments could detect the expected Δ-Δ components of the deuteron. Recent experiments, however, are at least consistent with Δ components of the magnitude expected in the deuteron and ^{3}He [27].

A) The (p,d) Reaction. Backward (p,d) Scattering

In the Born approximation the cross section for a (p,d) reaction removing a neutron which had been in the single-particle nucleon configuration α is

$$\left(\frac{d\sigma}{d\Omega}\right)^{PWBA} = H(E)(\Delta^2 + \kappa^2)|\psi_D(\underset{\sim}{\Delta})|^2|\psi_\alpha|(\underset{\sim}{K})|^2, \quad (4)$$

where $\underset{\sim}{\Delta} = \underset{\sim}{p} - \underset{\sim}{d}/2$ and $K = \underset{\sim}{d} - \underset{\sim}{p}\,(A-1)/A$, with $\underset{\sim}{p}$ and $\underset{\sim}{d}$ being the center of mass proton and deuteron momenta, and H(E) is a function of the energy. In fact over a wide range of energies the first three factors are constant and one has the form

$$\left(\frac{d\sigma}{d\Omega}\right)^{PWBA} \cong D|\psi_\alpha(K)|^2 . \quad (5)$$

This form suggests the "K plots" [1]. To the extent that it is valid, (p,d) cross section for different energies and angles for a particular nuclear transfer should lie on a single curve when plotted as a function of $(\underset{\sim}{K})$. At medium energies the distortions are not very energy or angle dependent, and should not change this qualitative observation. It is this approximate relation which shows most clearly how the (p,d) reaction can determine the Fourier components of nuclear single-particle wave functions. A few years ago attempts were made to use the Saturn (p,d) experiments [28] to extract effects of the I = 1/2 N*'s for complex nuclei but the theoretical calculations [29] indicated that there is such a small probability of N*'s in the discrete nucleon state of complex nuclei that they do not show up at energies L|GeV. This low N* probability vs. the much larger N* probability in light nuclei is still a puzzle to me.

For the backward $p + d \to d + p$ scattering the cross section is approximately by Eq. 5 as

$$\left(\frac{d\sigma}{d\Omega}\right)^{PWBA} \cong D|\psi_D(\underset{\sim}{\Delta})|^2 \qquad (6)$$

This suggests that $\underset{\sim}{\Delta} = |\underset{\sim}{p} - \underset{\sim}{d}'/2|$ should be a "scaling variable". I.e., if the I=1/2 baryon transfer reactions (N+N*'s) dominate, a plot of backward differential cross sections for all energies as a function of $(\underset{\sim}{\Delta})$ should lie on a universal curve. As shown in Fig. 6, this seems to be true except for the region of perhaps 4-800 MeV, where the s-channel, crossed pion, Δ-process is very important and adds a bump to the Δ-plot (31), which clearly shows up in 180° experiments (32) and in Fig. 6.

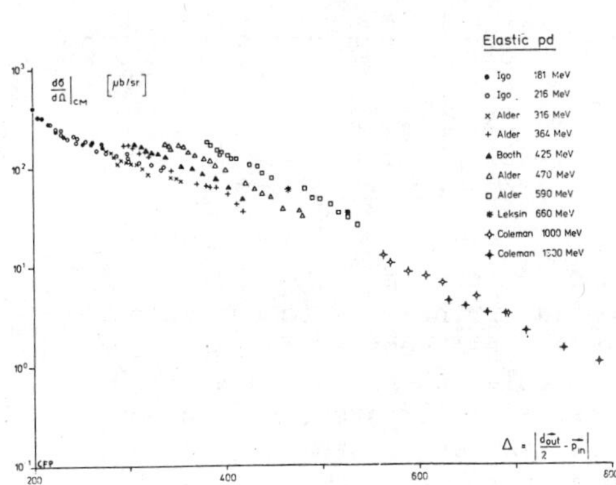

Fig. 6

The most recent theoretical results [30] show that the 1-2.5 GeV backward p-d data is consistent with a 1% $N_{5/2}(1688)$ and 1/2% $N_{9/2}(2220)$ probability. These are the first and second Regge recurrents, and the picture is consistent with that originally proposed [33].

The most series objection to this interpretation is the question of the use of Eq. (4), which includes relativistic kinematics but uses a nonrelativistic wave function eva uated with a nonrelativistic variable, Δ. Noble and coworkers [34] suggest that one should use the variable $\underset{\sim}{q} = \underset{\sim}{p} - \underset{\sim}{d}$, and point out that $|\underset{\sim}{q}|$ in the deuteron rest system is smaller than $(\underset{\sim}{\Delta})$ in the center of mass system, and thus the Kerman-Kisslinger [33] argument that the ordinary N-N wave function does not have enough high momentum components to fit the backward p-d data is not valid. From a practical point of view, this argument cannot be completely correct. First, for Δ(or q) very large ($\gtrsim 5F^{-1}$) the N-N components can decrease with decreasing Δ, and N* components can be just as large vs. N-N components. Thus for $E_{Lab} \lesssim 2$ GeV, the Kerman-Kisslinger argument is not changed dramatically and indeed detailed calculations using the $\underset{\sim}{q}$ variable cannot fit the data [34] without N* effects. Secondly, the $\underset{\sim}{q}$ variable is the three-momentum component of the q = p-d four-vector, and plots with $u = q^2$, the Mandelstam variable. The experimental p-d data simply does not scale in this variable [1].

I believe that the correct variable is the 4-vector $\Delta = p - d'/2 = (\Delta_o, \underset{\sim}{\Delta})$. This is the variable of a relativistic composite model of the deuteron such as the Bethe-Salpeter equation. It is interesting to look at the kinematics given in table 1. The variable $\bar{\Delta}=\bar{q}$

Table I

	T_L = 0.6 GeV			T_L = 1.0 GeV			T_L = 1.5 GeV		
θ	180°	150°	120°	180°	150°	120°	180°	150°	120°
Δ	0.36	.45	.62	.46	.59	.81	.57	.71	.99
$\bar{\Delta}$	.32	.42	.62	.38	.54	.82	.44	.64	1.02
Δ_o	-.13	-.18	-.18	-.28	-.28	-.28	-.38	-.38	-.38
$\bar{\Delta}_o$	-.05	-.09	-.19	-.08	-.14	-.31	-.10	-.20	-.45
$\bar{q}_o$	.86	.79	.63	.86	.79	.63	.84	.73	.49

$= \underset{\sim}{p} - \underset{\sim}{d}'/2$ in the rest system of the deuteron d'. Note that it is smaller than $\underset{\sim}{\Delta}$. However, the essential problem is to evaluate the Bethe-Salpeter amplitude $\psi(\Delta_o, \underset{\sim}{\Delta})$ in the p-d center of mass system. Although Δ_o is a constant over the angular distribution at each energy, the value of Δ_o in the deuteron rest system, $\bar{\Delta}_o$, is not at all constant. The value of q_o in the rest system, $\bar{q}_o$, is very large and not at all constant. Thus, the problem now is to use truely relativistic wave functions, so that one can handle the transformation to the dueteron rest system, which is where the bound-state calculation is done. Brad Keister and I are studying this now, and hope for a "final" solution to this problem. See also the discussion of Smorodinskaya [35].

B) <u>Pion Production and Absorption</u>

1) The pp↔πd Reaction

The reaction π + d↔p + p is the clearest example of virtual N* effects in nuclear physics and has received a great deal of attention lately as a basic process for pion absorption in nuclei. It has long been studied from the point of view of isobar intermediate states. Since the lowest order nucleon exchange process involves high momentum components of the deuteron, the second order process with pion scattering before absorption is favored. Since the kinematics favor backward pi-nucleon backward scattering as the intermediate process, the resonance isobar scattering dominates at suitable kinematic regions for intermediate isobar formation.

The recent work in this area involving explicit isobar components includes a study of effects of intermediate ρ-meson propagation [36], and coupled N-Δ channel calculation of the angular distribution and polarization of the proton in πd↔pp [37].

Recently, a relativistic formulation of this process has incorporated an off-mass-shell treatment [38] of the pion-nucleon amplitude. The results are shown in Fig. 8, where the off-mass-shell effects for the forward angle are given by the difference between the long-dashed and solid curve. This process can be considered to be a laboratory for exploring N* dynamics, and will prove most valuable for this area of physics.

A number of theorists have attempted to use the $\pi d \leftrightarrow pp$ process directly in a DWIA calculation of pion absorption. However, pion absorption and production obviously involves more dynamics than just the πd absorption. The possibility to obtain fits to data probably originates from the fact that the detailed structure of the two-nucleon clusters which mainly account for pion absorption is not important due to the rescattering process which allows low momentum transfer, s-state processes to dominate.

2) The (π^-,p) Reaction

The (π^-,p) process has been suggested as a list of Δ-components of nuclei [39]. Present experiment indicate a very small probability for Δ components in complex nuclei [40], but experimental work in more favorable kinematic region, as well as more detailed theoretical work is necessary before one can draw strong conclusions from this reaction.

V. OTHER PROCESSES AND CONCLUSIONS

In addition to the studies in light nuclei to which I have mainly restricted myself here there also have been calculations related to N*s in light nuclei and nuclear matter [1, 2, 3]. Particularly, in dense nuclear matter the isobars should be important as the Fermi energy approaches the pion mass. Recent calculations of nuclear matter [41] find a Δ probability of the order of 1%, with the effects being much more important for high Fermi momentum, as expected. One of the most important aspects of this is the three-body force which arises from N*s. A realistic coupled-channel calculation involves a self-consistent treatment which removes Δ's from the two-body force and allows the three-body clusters with intermediate N*s. This has not yet been done.

Some of the most interesting aspects of N* constituents of nuclei are effects on weak interactions. Much of this has been previously reviewed [1].

In conclusion, the past two years have seen good progress in developing the relativistic field theoretic methods necessary to treat the question of N* components of nuclei. The relevant experimental information is being provided at even a faster pace. I would expect considerable progress in "hard" results on the effects of N* components of nuclei in the next few years.

* Supported in part by NSF PHY 78-19757.

1. L. S. Kisslinger, "Mesons in Nuclei", M. Rho and D. Wilkinson, Editors (1979).
2. A. M. Green, Rep. Prog. Phys. 39, 1109 (1976).
3. H. J. Weber and H. Arenhövel, Phys. Reports 36C, 277 (1978).
4. P. Haapak sk , Phys. Lett. 48B, 307 (1974).
5. L. S. Kisslinger, Carnegie-Mellon preprint (1979).
6. M. Brack, D. O. Riska, and W. Weise, Nucl. Phys. A287, 425 (1977).
7. M. Thies, Phys. Lett. 63B, 39,43 (1976).
8. M. K. Banerjee and J. B. Cammarata, Phys. Rev. C17, 1125 (1978).
9. G. A. Miller, Phys. Rev. C14, 2230 (1976).
10. D. A. Dicus et al, Univ. Texas preprint (1978).
11. S. Jena and E. S. Kisslinger, Ann. Phys. (N.Y.) 85, 251 (1974).
12. K. Holinde and R. Machleidt, Nucl. Phys. A280, 429 (1977). K. Holinde et al, Phys. Rev. C18, 870 (1978).
13. H. J. Weber, J. M. Eisenberg, and M. D. Shuster, Nucl. Phys. A278, 491 (1977).
14. A. M. Green, J. A. Niskanen, and M. E. Sainio, J. Phys. G. Nucl. Phys. 4, 1055 (1978); A. M. Green and M. E. Sainio, Univ. Helsinki preprint (1978).
15. S. Häkkinen, Univ. Helsinki preprint (1978).
16. E. Hadjimichael, Nucl. Phys. A312, 341 (1978).
17. L. S. Kisslinger and W. Wang, Phys. Rev. Lett. 30, 1071 (1973); Ann. Phys. (N.Y.) 99, 374 (1976).
18. L. S. Kisslinger and A. Saharia, contribution to this conference.
19. M. Hirata, J. H. Koch, F. Lenz, and E. J. Moniz, Phys. Lett. 70 B, 281 (1977); MIT preprint (1978); K. Klingenbeck, M. Dillig, and M. G. Huber, Phys. Rev. Lett. 41, 387 (1978); Evlangen preprint (1978); E. Oset and W. Weise, Regensburg preprint (1978).
20. E. Kapon, E. Piasetzky, and J. M. Eisenberg, Tel Aviv preprint (1978).
21. T. Kamae and T. Fujita, Phys. Rev. Lett. 38, 471 (1977).
22. T. Kamae et al, Phys. Rev. Lett. 38, 468 (1977).
23. Q. Ho-Kim and L. S. Kisslinger, unpublished. Results given in Ref. 1.
24. J. W. Negele and D. O. Riska, Phys. Rev. Lett. 40, 1005 (1978).
25. J. S. Sharma and A. M. Mitra, Nucl. Phys. A292, 437 (1977).
26. H. J. Weber, "Meson-Nuclear Physics - 1976", AIP Conference Proceedings No. 33, P. D. Barnes, R. A. Eisenstein, and L. S. Kisslinger, eds; A. S. Goldhaber, Nucl. Phys. A294, 293 (1978).
27. R. Beurty et al, Phys. Lett. 61B, 409 (1976); B. Tatischeff et al, Phys. Lett. 77 B, 254 (1978).
28. S. D. Baker et al, Phys. Rev. Lett. 32, 839 (1974), Phys. Lett. B52, 57 (1974), See G. J. Igo, Revs. Modern Phys. 50, 523 (1978) for a comprehensive discussion.
29. R. Schaeffer, L. S. Kisslinger; and E. Rost, unpublished.
30. S. Jena and L. S. Kisslinger, to be published.

31. N. S. Craigie and C. Wilkin, Nucl. Phys. B14, 477 (1969).
32. B. E. Bonner et al, P. R. C17, 671 (1978), LASL Report (1978).
33. A. K. Kerman and L. S. Kisslinger, Phys. Rev. 180, 1483 (1969).
34. J. V. Noble and H. J. Weber, Phys. Lett. 50B, 233 (1974); M. Levitas and J. V. Noble, Nucl. Phys. A251, 384 (1975).
35. N. Ya. Smorodinskaya, "Seminar on the Nuclear Theory", ITEP (1978); L. A. Kondratydik and F. M. Lev, Ibid. (1976).
36. M. Brack, D. O. Riska, and W. Weise, Nucl. Phys. A287, 425 (1977).
37. J. S. Niskanen, Nucl. Phys. A298, 417 (1978).
38. B. D. Keister and L. S. Kisslinger, Carnegie-Mellon preprint (1978).
39. L. S. Kisslinger and G. A. Miller, Nucl. Phys. A254, 493 (1975).
40. P. Couvert et al, Phys. Rev. Lett. 41, 530 (1978), private communication.
41. B. D. Day and F. Coester, Phys. Rev. C13, 1720 (1978); W. Manzke and M. Gari Nucl. Phys. A312, 457 (1978); M. R. Anastasio et al, Jülich and Bonn preprint (1978).

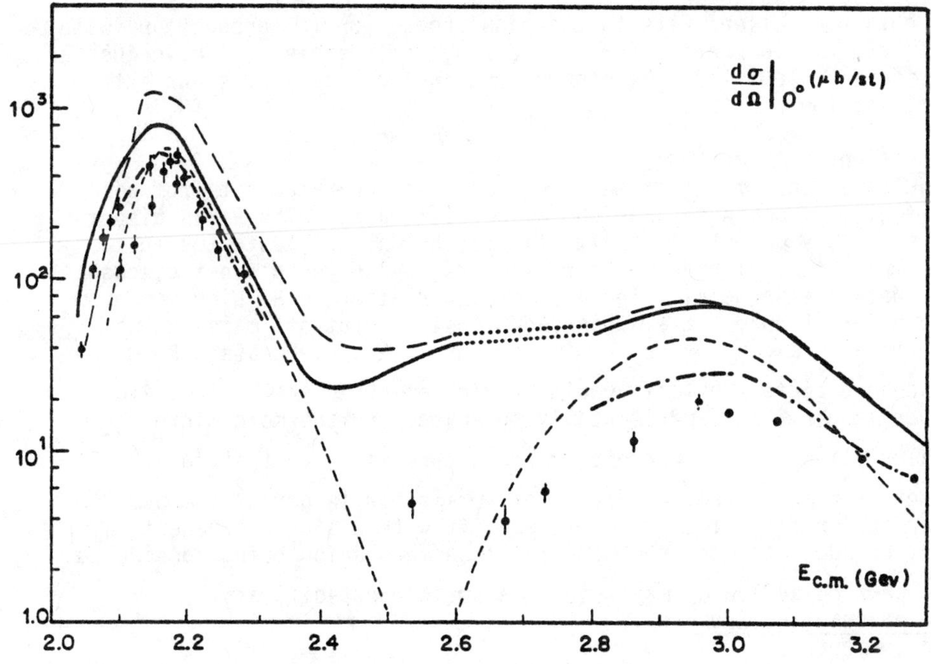

Fig. 8

Question: G. Stephenson, Jr, LASL
I have a comment on your remarks about the effect of Δ's on double beta decay if lepton number is conserved. While it is true that $\Delta^- \to p\beta^-\beta^-$ could go, the operator must involve a spin vector, hence the matrix element must vanish for any $0^+ \to 0^+$ transition, which, unfortunately includes all of the cases of interest.

Response: Kisslinger
I thought that in the geochemical studies there were cosmic ray processes which allowed the process to take place.

Question: M. Banerjee, University of Maryland
I was particularly intrigued by your determination of ρNΔ coupling constant through Regge analysis. This requires a model for example the Zacharaisen-Gell-Mann model, which in turn requires a scale parameter. What did you use for the scale parameter?

Response: Kisslinger
I used 1 GeV.

Question: Banerjee, continued
In 1967 Levinson and I determined these scale parameters from elastic scattering data and found it to be 200 MeV.

Response: Kisslinger
I obtain excellent fits to the pion production (Δ production) with π's and K's on protons for a wide range of energies. I have just repeated calculations by others done in the late 1960's and extrapolated the $g_{\alpha\rho n}$.

Question: E. Lomon, MIT
1) The effect of Δ intermediate states is separated out from core effects by looking at peripheral partial waves. The Paris dispersion theory potential, in its fit to the data, illustrates this.
2) The effect of hadronic form factors, which you did not discuss in detail is crucial. The elastic e,d scattering at high momentum transfer (Arnold, *et al*) has shown that the cut off mass in the form factors can not be much more than ~1 Gev/c, because the Meson exchange current effects would predict electric scattering, $A(q^2)$, much larger than experimentally observed. Furthermore microscopic calculations of the hadronic form factors indicate that for $q^2 < 25f^{-2}$ the σ-vertex correction dominates, while for larger q^2 the ρ-vertex correction dominates. Therefore a fit with a single exponent may not be adequate over the momentum transfer region being considered. In particular low q^2 may require a smaller cut-off mass.

Response: Kisslinger
No comment.

THE PROBLEM OF EXCHANGE CURRENTS IN THE TRANSITION : $^{16}O(0^+) \longleftrightarrow {}^{16}N(0^-)$

P. A. M. Guichon et M. Giffon
Institut de Physique Nucléaire, Université Claude Bernard Lyon-I
43, Bd du 11 Novembre 1918, 69621 Villeurbanne, France

C. Samour
DPh. N/HE, CEN Saclay, B. P. 2, 91190 Gif sur Yvette, France

In a previous work[1] we have proposed that the one pion exchange (O. P. E.) correction could solve the longstanding problem of the two inverse processes :

$$^{16}O(0^+) + \mu^- \longrightarrow {}^{16}N(0^-) + \nu_\mu \qquad (\text{rate}^2\ \Lambda_\mu = 1560 \pm 108\ s^{-1}),$$

$$^{16}N(0^-) \longrightarrow {}^{16}O(0^+) + \bar{\nu}_e + e^- \qquad (\text{rate}^3\ \Lambda_\beta = 0.43 \pm 0.10\ s^{-1}).$$

Indeed we were able to reproduce the experimental ratio $\rho = \Lambda_\mu/\Lambda_\beta$ after correcting the impulse approximation (I. A.) with the O. P. E. current as given in ref. 4. We recall that the large O. P. E effect for this transition is considered as reliable because it concerns the time-part of the axial current and thus can be deduced directly from current algebra and P. C. A. C. hypothesis. Since then we have shown[5] that core deformation in ^{16}O could have an effect similar to the O. P. E. current in what concerns ρ. Consequently the O. P. E. correction may be smaller than the one calculated in ref. 4. This point we investigate here.

We do not want to commit ourselves in incorporating short range effects (form factors, short range correlations) which are well known to reduce appreciably the size of the O. P. E. current. The reason is that, at present time, it is not possible to give a consistent treatment of these high pion momentum effects. We incorporate them in a phenomenological coefficient φ that multiplies the O. P. E. current of ref. 4, and we allow φ to run from 0 to 1. On the other side, we treat long range effects consistently, namely we calculate I. A. and O. P. E. matrix elements with wave functions that describe core deformation. The 0^+ state is taken as a combination of 0p - 0h and 2p - 2h states. We adopt the admixture coefficients of BGN, BGK (see ref. 6) and Walker[7]. We describe the 0^- state with p-h excitation in the (2s - 1d) shell. The effect of more complex excitations (3p - 3h for instance) has been found to be negligible. In order to simplify the presentation

ISSN: 0094-243X/79/540347-02$1.50

of the results we give them for the pure configuration (1p 1/2) (2s 1/2). It is well known that ρ is practically independent of the admixture of (1p 3/2) (1d 3/2) states. To be complete we have included some refinements (as compared with ref. 1 and ref. 5) in the calculation, particularly in the evaluation of Coulomb corrections.

Fig. 1

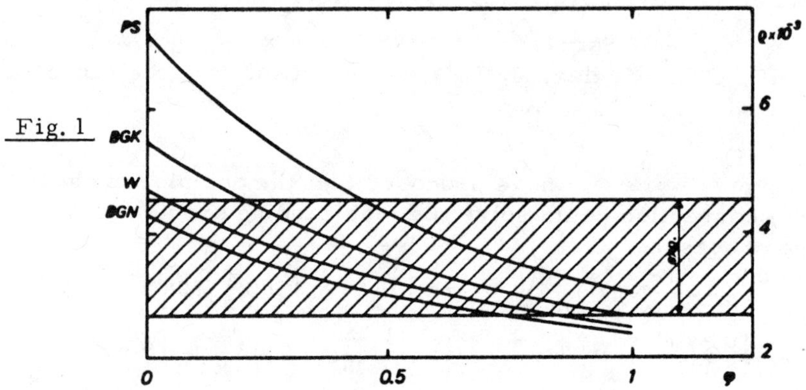

In Fig. 1 we give the curve $\rho = f(\varphi)$ for BGN, Walker (W) and BGK. For purpose of comparison, we also plot the curve (PS) corresponding to no core deformation. It shows that the latter is much more effective at $\varphi = 0$ (I. A.) than at $\varphi = 1$ (I. A. + O. P. E.). More important is the evident ambiguity arising from experimental uncertainties : indeed, there is overlap between theory and experiment both at $\varphi = 0$ and $\varphi = 1$. Even if we take BGN and BGK as limiting cases, we get : $0.2 < \varphi < 0.7$. It is a satisfying result for supporters of I. A. as well as for supporters of a strong O. P. E. effect. The only reasonable conclusion is that without a new (i. e. different experimental conditions) and more accurate (at least 10% precision instead of 23 %) measurement of Λ_β, the problem of exchange currents in the $(0^+ \leftrightarrow 0^-)$ transition cannot be solved.

REFERENCES

1. P. A. M. Guichon et al., Phys. Lett. 74B, 15 (1978).
2. P. A. M. Guichon et al., Phys. Rev. (on press).
3. L. Palffy et al., Phys. Rev. Lett. 34, 212 (1975).
4. P. A. M. Guichon et al., Z. Physik A 285, 183 (1978).
5. P. A. M. Guichon and C. Samour, Phys. Lett. B (on press).
6. A. M. Green et al., Nucl. Phys. A 130, 112 (1969).
7. G. E. Walker, Phys. Rev. 174, 1290 (1968).

CONTRIBUTION OF MESONIC EXCHANGE EFFECTS TO THE PION ^{3}He SCATTERING

I.V. Falomkin, V.I. Lyashenko, R. Mach, V.A. Panyushkin, G.B. Pontecorvo, M.G. Sapozhnikov, and Yu. A. Shcherbakov
J.I.N.R., Dubna, U.S.S.R.

F. Balestra, L. Busso, R. Garfagnini and G. Piragino
Istituto di Fisica dell'Universitá di Torino, Torino, Italy
I.N.F.N., Sezione di Torino, Torino, Italy

T. Angelescu, A. Mihul, and F. Nichitiu
Institute of Nuclear Physics, Bucharest, Romania

The differential cross section of (π , ^{3}He) elastic scattering at 145, 180 and 195 MeV has been measured at the Laboratory of Nuclear Probles of the JINR. The experiment has been carried out by means of a high pressure self shunted streamer chamber, filled with helium and triggered by an hodoscope of scintillation counters[1]. The experimental results have been compared with optical model calculations. In constructing the optical potential the strong interaction nuclear form factors have been derived from the electrical and magnetic form factors, taking into account the mesonic exchange effects in the electron nucleus scattering. The contribution due to the mesonic exchange effects has been substracted from the electric and magnetic nuclear form factors obtained in electron scattering experiments[2]. The mesonic exchange can contribute to the magnetic form factors of ^{3}He and ^{3}H up to $10 \div 15^o/_o$. It is natural to assume that the contribution of mesonic exchange effects are most effective, when the virtual mesonic exchange takes place between two nucleons located in close space points. So we supposed that the mesonic exchange corrections are proportional, in the coordinate space, to the nucleon density squared. The admixture of S' and D states to the ground state wave function of ^{3}He has also been taken into account.

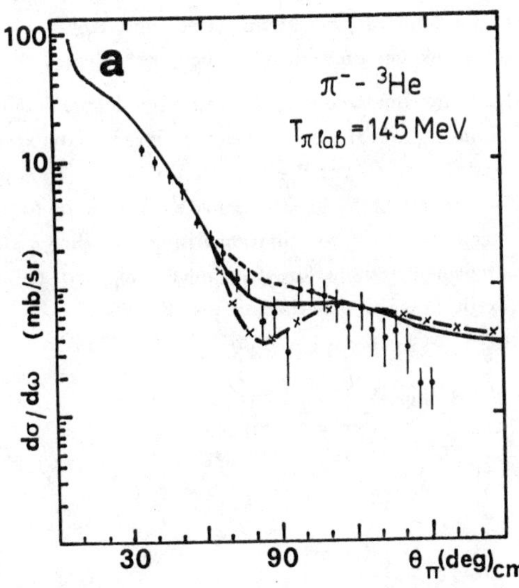

ISSN:0094-243X/79/540349-02$1.50

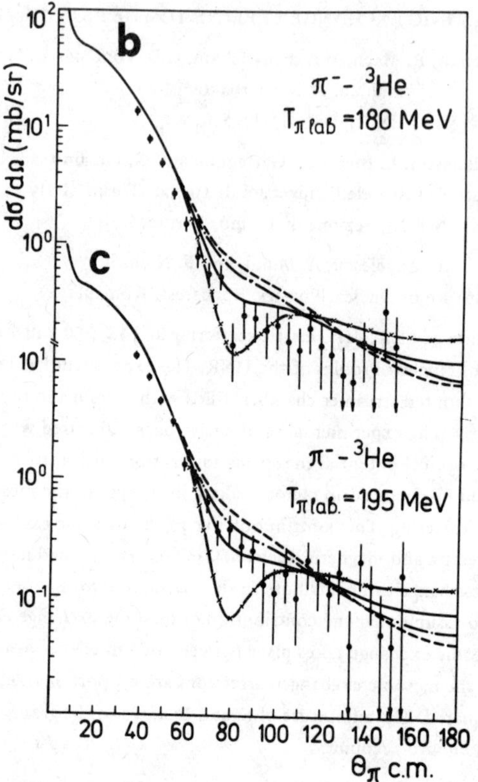

Fig. 1a, b, c. Differential cross section of (π , ^{3}He) elastic scattering at 145, 180, 195 MeV. Cross dashed line: R_m=1.73 fm; full line: R_m=1.95 fm; dashed line: R_m= 2.17 fm. Dot dashed lines show the result for R_m= 1.95 fm,when meson exchange effects are neglected.

In fig. 1 the experimental data are compared with the optical model predictions, for different values of the nuclear magnetic radius R_m of ^{3}He. As one can see, there is better agreement for R_m values in the (1.73 ÷1.95)fm interval, than for the R_m values in the (1.95 ÷ 2.17)fm. On the other hand, it must be noted that the mesonic exchange effects give a contribution to the differential cross section of the same order of magnitude as the contribution arising from the variation of the R_m within 10%. So it seems that a reliable determination of the nuclear magnetic radius from pion nucleus scattering data is rather a theoretical than an experimental problem.

REFERENCES

1. I.V. Falomkin et al., Lett. Nuovo Cim. 5, 757 (1972).
2. J.S. McCarthy et al., Phys. Rev. Lett. 25, 884 (1970) and H. Collard et al., Phys. Rev. 138, B57 (1965).

EFFECTS OF THE MESON EXCHANGE CURRENTS ON THE PION ABSORPTION

Il-T. Cheon

Yonsei University, Seoul 120-00, Korea

ABSTRACT

Effects of meson exchange currents have been investigated in the pion absorption process $\pi^+ + d \rightarrow p + p$.

INTRODUCTION

Since it has already been established that meson exchange currents were very important for explanation of the nuclear electromagnetic structure, one may expect that these currents might contribute to the processes in which the pion is involved. So far, only the effects of rescattering and $\Delta(1236)$-resonance were taken into account in the pion absorption processes. In the present paper, we investigate effects of the meson exchange current $\rho\pi\pi$, which is dominant among the currents, $\pi\omega\gamma$, $\pi\rho\gamma$ and $\rho\pi\pi$, on the pion absorption by the deuteron.

THEORY

The cross section for the process considered here is given by calculating the Feynman diagramshown in Fig. 1. With the conventional expressions for each vertexes, $f_{\rho\pi\pi}(p+q)_\mu$, $g_\pi \gamma_5$ and $\frac{i}{2} f^V_{NN\rho}\gamma^\mu + \frac{1}{2M} f^T_{NN\rho}\, l_\nu \sigma^{\mu\nu}$, we can easily obtain the S-matrix in the following form

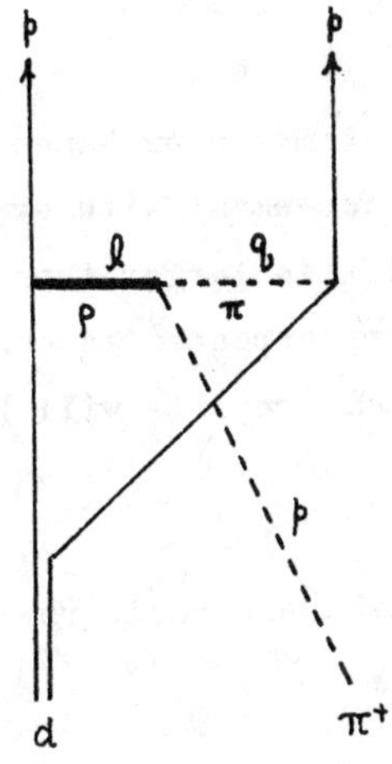

Fig. 1. Feyman graph.

$$
\begin{aligned}
S_{\pi\rho} = \int d^4 y_p\, d^4 y_n\, \chi^+_f (y_p\, y_n) \\
&* \left(\frac{i}{2} f^V_{NN\rho}\gamma^\mu + \frac{1}{2M} f^T_{NN\rho}\right)\left[\frac{d^4 l}{(2\pi)^4}\frac{e^{-il(y_n - x)}}{l^2 - m_\rho^2 + i\varepsilon}\right] \\
&* \left[f_{\rho\pi\pi}(p+q)_\mu\right]\left[\frac{d^4 q}{(2\pi)^4}\,\frac{e^{-iq(y_p - x)}}{q^2 - m_\pi^2 + i\varepsilon}\right] \\
&* g_\pi \gamma_5\, \chi_i (y_p\, y_n)\, \phi_\pi (px)\, (2\pi)^4\, \delta(q + l - p) + \text{Exchange} \qquad (1)
\end{aligned}
$$

ISSN:0094-243X/79/540351-02$1.50

where $\chi(y_p, y_n)$ is the wave function of the two nucleon system. The coupling constants are $f^2_{\rho\pi\pi}/4\pi = 2.1$, $g^2_\pi/4\pi = 14.64$, $(f^V_{NN\rho})^2/4\pi = 3.87$ and $(f^T_{NN\rho})^2/4\pi = 13.07$. Considering only the dominant term, we can rewrite eq. (1) as follows,

$$S_{\pi\rho} = \frac{3\pi E_\pi}{2M} f_{\rho\pi\pi}\, g_\pi\, f^V_{NN\rho} \left[\int e^{ipr}\phi_\pi(pr)\, d^4r\right] L(p,k) \, , \qquad (2)$$

where

$$L(p,k) = \int d^3y\, d^3\tau\, \varphi_f^+(\vec{k}\vec{y}) \frac{\langle X_{1m} | \vec{S}\cdot(\frac{\vec{p}}{2} - \vec{\tau}) | X_{1m'} \rangle\, e^{-i\vec{\tau}\cdot\vec{y}}}{[(\frac{\vec{p}}{2}+\vec{\tau})^2 + m_\rho^2 - \frac{E_\pi^2}{4} - i\varepsilon][(\frac{\vec{p}}{2}-\vec{\tau})^2 + m_\pi^2 - \frac{E_\pi^2}{4} - i\varepsilon]} \varphi_i(\vec{y}) \, . \qquad (3)$$

Here X_{1m} and $\varphi(\vec{y})$ are spin state and the spatialwave function of the two nucleon system. $\vec{S}$ denotes the two nucleon spin operator.

After carrying out some algebra, we obtain the cross section for the process $\pi^+ + d \rightarrow p + p$ in the form

$$\sigma = \frac{E_\pi}{|\vec{p}|} \frac{9\pi^2\sqrt{E_\pi^2 - 4M_p^2c^4}}{8(\hbar c)^4} \left(\frac{f^2_{\rho\pi\pi}}{4\pi}\right)\left(\frac{g^2_\pi}{4\pi}\right)\left(\frac{f^{V2}_{NN\rho}}{4\pi}\right)\left(\frac{\hbar}{M_p c}\right)^2 E_\pi \frac{M_p c^2}{2}\, \xi \, , \qquad (4)$$

$$\xi = \left[\int e^{ipr}\phi_\pi(pr)\, d^4r\right] \frac{1}{3}\sum_{AV} \int |L(p,k)|^2 d\Omega_k \, , \qquad (5)$$

where $d\Omega_k$ denotes the solid angle of the relative momentum k between the outgoing protons.

CONCLUSION

Without meson exchange currents, the cross section is generally in a good agreement with experimental results[1)] except the energy region a little larger than the (33)-resonance. Meson exchange currents are expected to make some contributions to this region. The numerical results will be presented at the Conference.

REFERENCES

1) J.A. Niskanen, Nucl. Phys. A298, 414 (1978).

AXIAL MESON EXCHANGE CURRENTS IN ANTINEUTRINO DISINTEGRATION OF THE DEUTERON

J. Hošek and E. Truhlík
Institute of Nuclear Physics, 250 68 Řež near Prague, Czechoslovakia

ABSTRACT

It is shown that the accurate measurement of the cross section for the disintegration of the deuteron by reactor antineutrinos may provide an essential information on the existence of axial meson exchange currents (MEC) in nuclear physics.

The reaction

$$\bar{\nu}_e + d \rightarrow n + p + \bar{\nu}_e \quad (1)$$

with the reactor antineutrinos is in the Weinberg-Salam theory (see, e.g., ref. [1]) determined by the axial vector isovector coupling constant. This constant being known [1], we calculate the axial MEC effects for reaction (1) in the framework of the model [2] based on the hard pion method. In reaction of muon capture in deuterium this model predicts a MEC effect of 8.5% [3] for the doublet transition rate. Using the soft pion theorems, the MEC corrections have recently been calculated [4] for β-decay and μ-capture in light nuclei (^{12}C, ^{12}B etc.). However, the calculations of nuclear matrix elements, in comparison with reaction (1) are much more uncertain. We used the antineutrino spectrum from ref. [5], but our $G = 2.874 \times 10^{-22}$ cm/MeV. Employing eq.(3) of ref. [5] we have $\bar{\sigma} = 8.00 \times 10^{-45}$ cm^2 in zero range approximation. In table I, we present the results of our calculations with the Yamaguchi potential ($a_{np} = -23.7$ f, $r_{np} = 2.8$ f) and Úlehla [6]

Table I Cross section (in cm^2) for reaction (1)

n-p interaction	Y	U
$\bar{\sigma} \times 10^{45}$	7.43	7.22
$\bar{\sigma}_M \times 10^{45}$	1.05	1.04

ISSN:0094-243X/79/540353-02$1.50

realistic interaction, which fits n-p and p-p data in the $^1S_0$ channel separately. We employed the RSC deuteron [7]; $\bar{\sigma}$ is the cross section for the single particle current and $\bar{\sigma}_M$ represents the MEC effects ($\sim$ 15%).

CONCLUSION

A carefully measured cross section for reaction (1) with low-energy antineutrinos may provide an essential information about the existence of axial MEC effects in nuclear physics.

REFERENCES

1. J. J. Sakurai, preprint UCLA/78/TEP/18, 1978.
2. E. Ivanov and E. Truhlík, to be published in Nucl. Phys.
3. E. Ivanov and E. Truhlík, to be published in Nucl. Phys.
4. P. A. M. Guichon, M. Giffon and C. Samour, Phys. Letters 74B, 15 (1978); K. Kubodera, J. Delorme and M. Rho, Phys. Rev. Letters 40, 775 (1978).
5. F. T. Avignone III and Z. D. Greenwood, Phys. Rev. D17, 154 (1978).
6. I. Úlehla, in The Nuclear Many-Body Problem (ed. by F. Calogero and Cioffi degli Atti, Editrice Compositori, Bologna, 1973), v.1, p. 145.

MESON EXCHANGE CURRENTS AND THE EXTRACTION OF a_{nn} AND g_P FROM THE REACTION $\mu^- d \rightarrow nn\nu_\mu$

G. E. Dogotar
Institute of Applied Physics, Kishinev, USSR

R. A. Eramzhyan
Laboratory of Theoretical Physics, JINR, Dubna, USSR

E. Truhlík
Institute of Nuclear Physics, Řež, Czechoslovakia

ABSTRACT

The influence of the meson exchange currents (MEC) on the neutron spectra and the ratio of the doublet Λ_d and quadruplet Λ_q transition rates is investigated.

EXTRACTION OF THE NEUTRON-NEUTRON SCATTERING LENGTH a_{nn}

It was shown in refs.[1, 2] that the normalized differential neutron spectra for reaction

$$\mu^- + d \rightarrow 2n + \nu_\mu \tag{1}$$

are suitable for the extraction of the a_{nn}. Here we report on the study of the influence of the MEC on those spectra, using a consistent approach[3] to the axial MEC problem in nuclear physics based on the hard pion method. The axial MEC operator is reasonably given in the tree approximation and it automatically satisfies the PCAC constraint. In this approach, the contribution of the MEC effect to the Λ_d is ~8.5%[4]. Here we have found that in the range $0^\circ < \theta_{12} < 15^\circ$ and for the values of $p_1 < 70$ MeV/c ($t < 190$ nsec) the normalized spectra are changed not more than by 0.01%. It results in an uncertainty of 0.05 f for a_{nn}.

EXTRACTION OF g_P

It was noted in ref.[5] that the ratio Λ_q/Λ_d is sensitive to the value of g_P. In order to calculate this ratio with the MEC effects taken into account, we evaluated them also for the Λ_q in the

ISSN:0094-243X/79/540355-02$1.50

same model [3] as used before for the Λ_d. The calculations show [6] that the MEC effects are negligible in this case. The reason is that now $j_2(\nu r)$ (ν is the neutrino momentum) enters the nuclear matrix element instead of $j_o(\nu r)$ as is the case for the Λ_d.

In table I we present the value of this ratio $R(R_o)$ with (without) MEC effect. We see that the measurement of R within 10% would considerably reduce the uncertainty in our knowledge of g_P.

Table I Values of Λ_q/Λ_d for various g_P

g_P/g_A	5	6	7	8	9	10	11	12
$100 \times R_o$	1.47	1.86	2.33	2.88	3.51	4.23	5.06	6.00
$100 \times R$	1.36	1.72	2.14	2.64	3.22	3.88	4.63	5.48

REFERENCES

1. G. E. Dogotar and R. A. Eramzhyan, Journal of Nucl. Phys. 26, 483 (1977).
2. B. F. Gibson and G. J. Stephenson, Jr., preprint LA-5424-MS, 1973.
3. E. Ivanov and E. Truhlík, to be published in Nucl. Phys.
4. E. Ivanov and E. Truhlík, to be published in Nucl. Phys.
5. G. E. Dogotar and R. A. Eramzhyan, Journal of Nucl. Phys. 25, 1042 (1977).
6. G. E. Dogotar, R. A. Eramzhyan and E. Truhlík, to be published.

THE ROLE OF THE Δ(1232) ISOBAR IN THE $^{16}O(\gamma,p)$ PROCESS

M. J. Leitch, J. L. Matthews, D. Ingham, W. W. Sapp, and C. P. Sargent
Massachusetts Institute of Technology, Cambridge, MA 02139*

D.J.S. Findlay and R. O. Owens
Glasgow University, Glasgow, Scotland+

B. L. Roberts
Boston University, Boston, MA 02115++

ABSTRACT

A recent measurement of $^{16}O(\gamma,p_o)$ cross sections for $E_\gamma \geq$ 100MeV is compared with two theoretical calculations which involve the excitation of the Δ(1232) isobar in an intermediate state.

INTRODUCTION

The cross section for the $^{16}O(\gamma,p_o)^{15}N$ reaction for E_γ=40-100MeV is observed to decrease approximately exponentially with increasing photon energy.[1] This behavior can be understood in terms of a simple one-step direct knockout mechanism, for which the differential cross section can be shown to be proportional to the square of the initial proton wave function in momentum space.[2] As the initial proton momentum increases with E_γ, this direct amplitude will decrease, and for higher E_γ one might expect two-step mechanisms such as an intermediate (virtual) excitation of the Δ(1232) isobar to play a role.

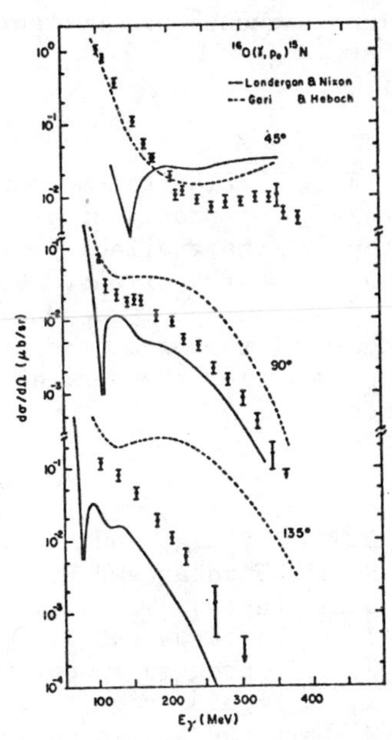

Fig. 1 Differential cross section for $^{16}O(\gamma,p_o)^{15}N$ vs. photon energy for proton angles 45°, 90° and 135°. Curves are described in the text.

EXPERIMENT

The $^{16}O(\gamma,p)$ process in the energy range E_γ=100-380MeV has been investigated at the M.I.T. Bates Linear Accelerator, for proton emission angles 45°, 90° and 135°.

* Supported in part by DOE Contract #76-C-3069
+ Supported in part by the Science Research Council
++ Supported in part by the National Science Foundation
ISSN:0094-243X/79/540357-02$1.50

The relatively high excitation energy (≈5.3MeV) of the first excited state of the residual nucleus ^{15}N allows the cross section for the (γ,p_0) reaction (leaving ^{15}N in its ground state) to be obtained directly from the highest 5MeV of a bremsstrahlung-produced proton spectrum. Some preliminary results have been published.[3] The most recent data are shown in Fig. 1.

THEORY

The solid lines in Fig. 1 represent the calculation of Londergan and Nixon.[4] The Hartree-Fock wave function given by Negele[5] is used for the initial bound state, and the distortion of the outgoing protons is included by an approximate method. The isobar width is taken as the free value (Γ=115MeV), and center-of-mass corrections are made.

The dashed lines represent the calculation of Gari and Hebach,[6] in which the isobar is treated as a giant M1 resonance state with width 100 MeV at energy 300 MeV, and initial- and final-state interactions and exchange currents are included in a phenomenological manner.[7]

CONCLUSION

Fig. 1 illustrates that whereas the $\Delta(1232)$ excitation mechanism is capable of providing a large amplitude in the E_γ>100MeV region, neither of the available calculations agrees quantitatively with the measured cross sections. Both theories roughly reproduce the observed photon energy dependence, but do not predict the angular dependence of the data, and are seen to fail in very different ways. A more detailed theoretical treatment of this problem is clearly indicated.

REFERENCES

1. D.J.S. Findlay and R. O. Owens, Nucl. Phys. A279, 385(1977).
2. D.J.S. Findlay and R. O. Owens, Phys. Rev. Lett. 37, 674(1976), Nucl. Phys. A292, 53(1977).
3. J. L. Matthews *et al*., Phys. Rev. Lett. 38, 8(1977).
4. J. T. Londergan and G. D. Nixon, Phys. Rev. C (to be published).
5. J. W. Negele, Phys. Rev. C1, 1260(1970).
6. M. Gari and H. Hebach, preprint (1978).
7. H. Hebach, A. Wortberg, and M. Gari, Nucl. Phys. A267 425(1976).

INVESTIGATION OF POSSIBLE DIBARYON RESONANCES IN THE ΔΔ SYSTEM

M. Dillig
University of Erlangen-Nürnberg, Erlangen, W-Germany

Recent experimental investigations of various scattering processes involving two baryons, such as polarized pp scattering[1] or the photo-induced deuteron break-up[2], have raised some speculation about a possible formation of two-baryon resonances at CM-energies between 2.1 and 2.4 GeV with a typical decay width of 100 MeV. Theoretically a formation of such resonances is in general studied either in the quark model[3] or on the basis of one-boson-exchange potential models (OBEP)[2,4,5]. In the present investigation the latter approach is pursued: as suggested from the position of candidates for di-baryon resonances the interaction of the ΔN and the ΔΔ system is studied in the potential model. As a major extension of the

- OBE ansatz (including π, η, σ, ρ and ω exchange, supplemented by vertex corrections and short range correlations)
- TBE diagrams (induced by $\pi\pi$ and $\pi\rho$ exchange with intermediate NN or NΔ excitation)

are included in the model. It is found that the TBE pieces are of crucial importance for the investigation of di-baryon resonances far above the NN-threshold (due to the coupling to other baryonic and mesonic decay channels); they strongly modify the real part of the interaction potential (dispersive corrections) and introduce in particular strong imaginary components into the two-particle potential.

The concept sketched above is applied for the study of the ΔN and the ΔΔ interaction. Quoting one result we

ISSN:0094-243X/79/540359-02$1.50

find that the inclusion of TBE contributions strongly suppresses the formation of a T = L = 0, J = 3 resonance, as suggested from an unexpected behaviour of the proton polarization in $\gamma d \rightarrow np$ at $E_\gamma \cong 600$ MeV. Unfortunately, irreducible corrections of even higher order seem to be important; their investigation as well as the extension of the approach to other two-baryon systems is under way.

1. I. P. Auer et al., Phys. Rev. Lett. 41, 354 (1978)
2. T. Kamae et al., Phys. Lett. 38, 471 (1977)
3. P. J. G. Mulders, A. Th. Aerts and J. J. de Swart, Phys. Rev. Lett. 40, 1543 (1978)
4. H. Arenhövel, Nucl. Phys. A 247, 473 (1975)
5. R. Händel, M. Dillig and M. G. Huber, Phys. Lett. 73 B, 4 (1978)

PRECRITICAL PHENOMENA CLOSE TO PION CONDENSATION IN FINITE NUCLEI

H. Toki and W. Weise
Institute of Theoretical Physics, University of Regensburg
D-8400 Regensburg, W. Germany

Although pion condensation is unlikely to appear around or below normal nuclear matter density in symmetric nuclear matter, the relative proximity of the pion condensation threshold may show up in precritical behaviour in ordinary nuclei. For this purpose, we have studied the nature of low frequency pion-like excitations in finite nuclei [1], as produced, for example, via inelastic proton scattering.

As shown in Fig. 1, this process is accompanied by various intermediate particle-hole (including Δ isobar-hole) excitations due to many body polarisation effects caused by the interaction between nucleons in pion-channel;

$$V(\vec{q}) = \frac{f^2(q^2)}{m_\pi^2}\left[\vec{\sigma}_1\cdot\vec{q}\, D_o(q)\, \vec{\sigma}_2\cdot\vec{q} + g'\vec{\sigma}_1\cdot\vec{\sigma}_2\right]\vec{\tau}_1\cdot\vec{\tau}_2 \tag{1}$$

The first term denotes the one pion exchange interaction and the second the short range correlations in terms of the Landau-Migdal interaction. The model for the interaction of eq. (1) is the same as the one used in considerations of pion condensation in infinite nuclear matter.

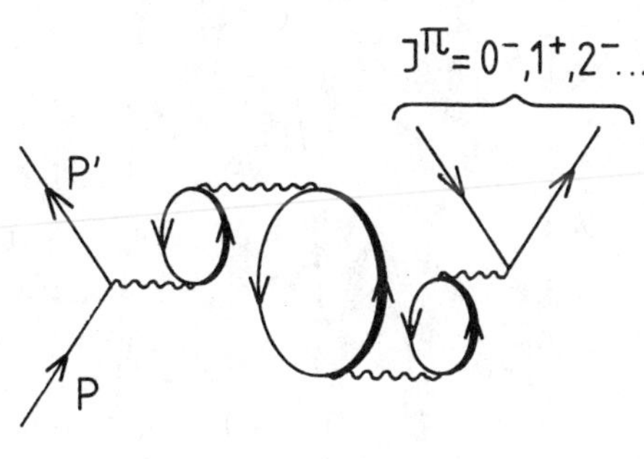

Fig.1: Proton inelastic excitation of unnatural parity states. The thick line denotes both nucleon and Δ isobar states, whereas the thin line nucleon hole states.

In finite nuclei, momentum conservation is lost in the summation over intermediate particle-hole excitations (see Fig. 1) in contrast to the infinite matter case, and then one is led to the integral equation for the response function R_J;

$$R_J(q,q') = \Pi_J(q,q') + \int_0^\infty \frac{q''^2 dq''}{(2\pi)^3}\, \Pi_J(q,q'') D(q'') R_J(q'',q') \tag{2}$$

where Π_J denotes the pion self-energy with partial wave J. Furthermore, the propagator D contains the short range correlations (g') in addition to the free pion propagator, $D_o(q)$. The response function is then used as the interaction for the inelastic proton excitation of unnatural parity states.

The pion self-energy $\Pi_J(q,q')$ is shown in Fig. 2. One observes that while convergence is obtained already at the principal quantum number N=10 for $q=m_\pi$, 30 shells are required to collect the full

ISSN:0094-243X/79/540361-02$1.50

strength for q=2~3 m_π.

As an particular case, we show the inelastic proton scattering into $J^\pi = 1^+$ states in ^{208}Pb. The '0$\hbar\omega$' excitations are taken into account explicitly, whereas all others are treated in terms of the pion self-energy. Thus most of the polarization strength is contained in the effective interaction. The result for $d\sigma/d\Omega$ in the plain wave Born approximation for the lowest 1^+ state is presented in fig. 3 as a function of momentum transfer q. The various curves show the g' dependence, which is strong at q=2-3 m_π, while only small changes appear for q < 1.5 m_π. We note that the precritical behaviour in this 1^+ excitation becomes critical for g'=0.40 which is somewhat smaller than the critical g'(=0.45) in nuclear matter.

1) H. Toki and W. Weise; preprint Regensburg (1978)

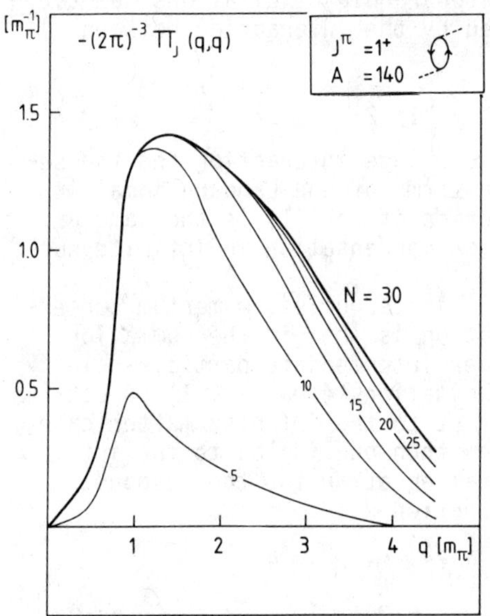

Fig. 2: Diagonal part of partial-wave pion self energy $\Pi_J(q,q')$ for $J^\pi = 1^+$ with A = 140. Particle states are only nucleons. Numbers denote the maximum principal quantum number which truncates the particle space

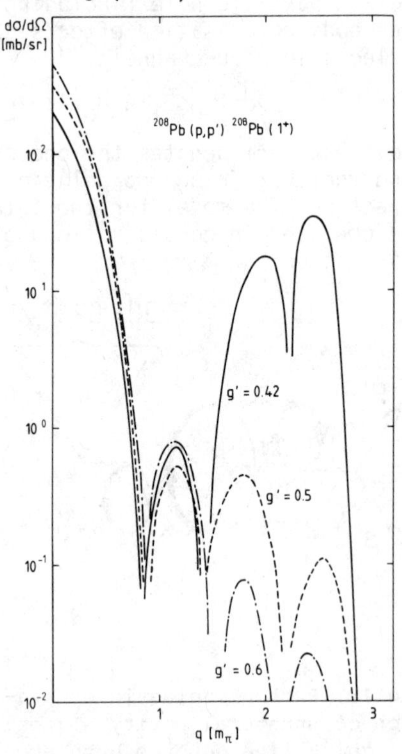

Fig. 3: Proton inelastic scattering cross section to the lowest 1^+state in ^{208}Pb. The various curves show the dependence on the Landau parameter g'.

PRODUCTION OF ISOBARS VIA RELATIVISTIC HEAVY-ION COLLISIONS

P. A. Deutchman
The University of Idaho, Moscow, Idaho 83843

ABSTRACT

A theoretical estimate of the total cross-section for the production of isobars in an ^{16}O projectile and the concomitant excitation of a ^{12}C target to a giant resonance via peripheral relativistic collisions at 2.1 GeV/nucleon is presented.

TEXT

Recently[1], it has been pointed out that the one-pion-exchange interaction is ideal for producing an isobar in a projectile ion, which subsequently decays emitting a pion concomitantly exciting a target to a giant resonance state. Following an approach similar to Feshbach and Zabek[2], the frequency spectrum for the production of the $\Delta(3,3)$ isobar is given by time-dependent perturbation theory as

$$F(\omega) = \frac{1}{2\pi} \int dt \, \langle \Phi_{\beta f} \Psi_{\beta} \chi_f | V u^2 | \Phi_{\alpha i} \Psi_{\alpha} \chi_i \rangle, \qquad (1)$$

where the motion of the impact parameter $\underset{\sim}{b}_{cm}$ is given by the wave functions $\Phi_{\alpha i}(\underset{\sim}{b}_{cm})$ and $\Phi_{\beta f}(\underset{\sim}{b}_{cm})$. The target states are described by space-spin states $\chi_i(\underset{\sim}{\xi}_t, \underset{\sim}{\sigma}_t)$ and $\chi_f(\underset{\sim}{\xi}_t, \underset{\sim}{\sigma}_t)$ and the space-spin projectile states from nucleon to isobar is given by $\Psi_\alpha(\underset{\sim}{\xi}_p, \underset{\sim}{\sigma}_p)$ and $\Psi_\beta(\underset{\sim}{\xi}_p, \underset{\sim}{S}_p)$ with the $p^{\underline{th}}$ nucleon having transition spin $\underset{\sim}{S}_p$[3]. The straight line assumption $u^2 = \delta(z-vt)$ is used. Only the triplet ispin and spin-spin part of the one-pion-exchange potential is considered between projectile-target nucleons and is given by

$$V = \sum_{pt} \underset{\sim}{S}_p \cdot \underset{\sim}{\sigma}_t v_{pt}(\zeta, b), \qquad (2)$$

where longitudinal and transverse separation distances are ζ and b and v_{pt} is chosen to be separable gaussian forms. The following assumptions are also made:

The interactions between target-nucleons and projectile-nucleons among themselves are neglected; the assumption of peripherality involves interactions only between p-state nucleons; the independent-particle model is assumed for target and projectile with no $\underset{\sim}{\ell} \cdot \underset{\sim}{s}$ interaction; the projectile and target internal wave functions are taken to be unsymmeterized products of space and spin functions; Fermi motion is ignored; and lower energy n-p triplet scattering parameters are used. The estimate for the total cross-section is about 2 millibn.

REFERENCES

1. G.E. Brown and P.A. Deutchman, Workshop on "High Res. Heavy Ion Phys," Saclay, France, May 31-June 2, (1978).
2. H. Feshbach and M. Zabek, Ann. Phys. **107**, 110 (1977).
3. H. Sugawara and F.V. Hippel, Phys. Rev. **172**, 1764 (1968).

ISSN:0094-243X/79/540363-01$1.50

PION NUCLEUS SINGLE CHARGE EXCHANGE IN THE ISOBAR-HOLE MODEL

E. Oset
Department of Physics, University of Barcelona, Spain

ABSTRACT
A microscopic calculation on the pion-nucleus single charge exchange reaction is performed in the framework of the isobar-hole model. Results are given for $^{15}N(\pi^+,\pi^0)^{15}O$, where a flat cross section of about $\sigma = 0.4$ mb is obtained in the resonance region once all the pieces contributing to the Δ selfenergy in the medium are included.

A new approach to pion scattering in the resonance region, based on the isobar-hole excitation and posterior propagation in the medium has become available [1] , which incorporates Pauli and binding effects, true pion absorption, quasielastic and reflection contributions, and which gives a good description of the π elastic scattering in the resonance region and of the absorptive part of the low energy pion nucleus optical potential [2].

The important medium corrections appear through selfenergy contributions to the Δ states as shwon in fig.1.

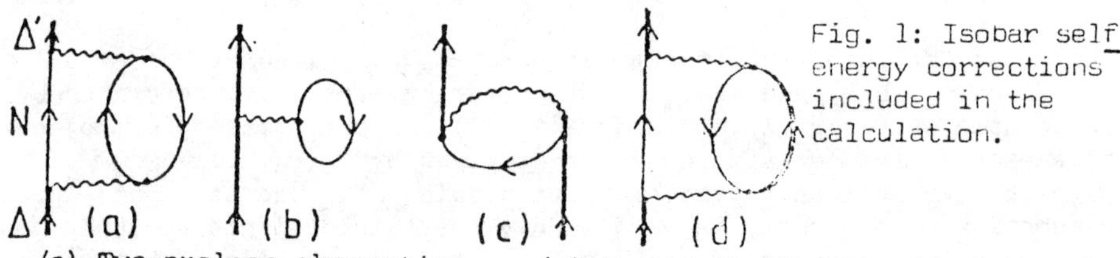

Fig. 1: Isobar selfenergy corrections included in the calculation.

(a) Two-nucleon absorption (b) Hartree binding) term (c) Fock term (Pauli effect) (d) Reflection correction.

We have used the same model to study the pion SCE reaction for closed shell nuclei plus and minus one nucleon. Diagramatically the process is shown in fig. 2, where we consider direct and exchange terms and the multiple scattering of the pion before and after the change exchange takes place. The multiple scattering is treated as in [1], with the Δ selfenergies of fig. 1 and π and ρ meson exchange plus additional short range correlations for the Δ-hole interaction.

The results can be seen in fig. 3 for the $^{15}N(\pi^+,\pi^0)^{15}O$ reaction in three steps. Free Δ means no Δ selfenergies included (it would be the closest result to the standard optical potential calculations). The next step includes Pauli and binding effects plus the exchange term of fig. 2. The cross section gets tremendously enhanced. The third step includes the remainder of the Δ selfenergies, absorption and reflection terms, and they produce a drastic change in the cross section to the point of being practically flat in the resonance region around the value of $\sigma = 0.4$ mb (results below 100 MeV should not be taken too strongly since the background potential, not included in this calculation, becomes important there).

ISSN:0094-243X/79/540364-02$1.50

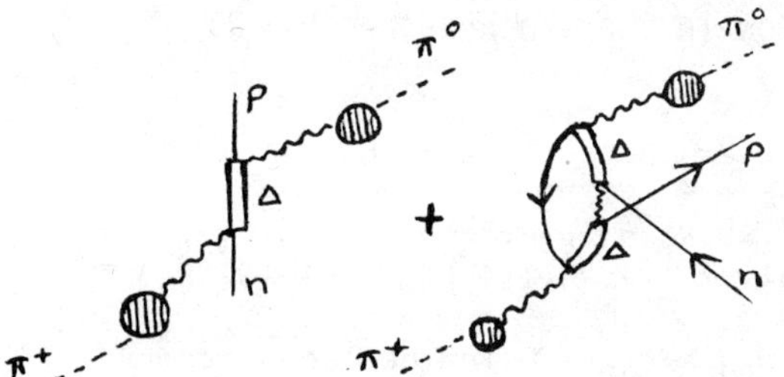

Fig.2: Contribution to SCE process. The stripped circle stands for multiple scattering and the wavy line for the different ingredients of the Δ-h interaction.

The SCE reaction depends on a critical balance of the Born pieces, which have a resonant structure, and the absorption of the pion wave due to multiple scattering, which drastically reduces the cross section in the resonance region. The true pion absorption and reflection pieces of the Δ selfenergy produce essentially a negative shift of the resonance position and an enchancement of the Δ decay width and thus they alter substantially these two pieces and their balance. We thus think that the experimental results of this reaction, to be measured at Los Alamos, will be of great interest to clarify the reaction mechanisms in π nucleus scattering.

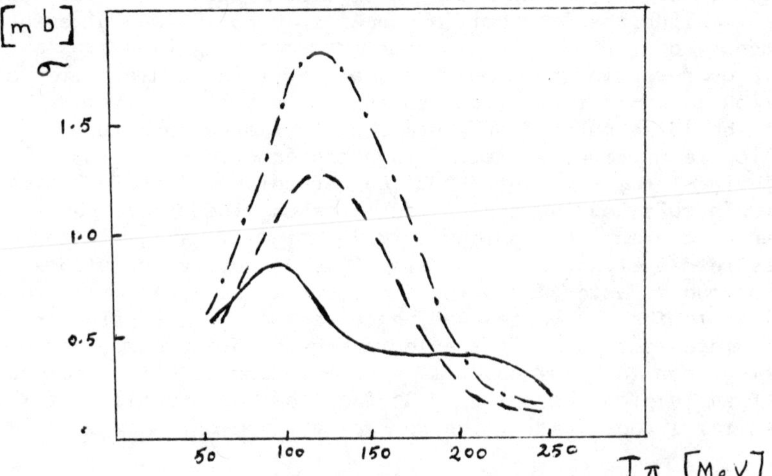

Fig. 3: SCE integrated cross section for $^{15}N(\pi^+,\pi^0)^{15}O$. — — — free delta. — . — Pauli plus binding. Solid line total selfenergy. A closed shell with $(1p_{\frac{1}{2}})^{-1}$ nucleon hole has been taken for the nuclear wave function.

1 E. Oset and W. Weise, Phys. Lett. 77 B(1978) 159 and Nucl. Phys. A (1979), in print.

2 E. Oset, W. Weise and R. Brockmann, submitted to Phys. Lett.

PION TRUE-ABSORPTION IN THE ISOBAR-DOORWAY-MODEL

A.N. Saharia
TRIUMF, Vancouver, B.C., Canada V6T 1W5

L.S. Kisslinger
Carnegie-Mellon University, Pittsburgh, PA 15213

The doorway assumption $H_{pq} = H_{qp} = 0$ in the isobar-doorway theory enables us to identify the non-resonance part of the pion optical potential in isobar-doorway model as the conventional optical potential with the interaction in π-N 3-3 channel switched off.[1] The non-vanishing of the coupling H_{pq} leads to terms in optical potential which correspond, in the lowest order, to pion true-absorption with intermediate states being 2 nucleon-2 hole states. [These absorption terms are to be distinguished from the ones included in the resonance part of the optical potential which would be generated by coupling <u>both the incident and final channels</u> to the eigenstates of $\mathcal{H}_{dd}$.[1]] To estimate these terms we take H_{pq} as the πNN vertex. Writing the nuclear ground state in terms of independent pair approximation, we find that the matrix elements $\langle Q|H_{qp}|P\rangle$ are given by nucleon-nucleon residual interaction (remember H_{pq} is a single particle operator for nucleons). As a result the contribution of the absorption terms to the optical potential is small.

It should be pointed out here that the large magnitude of the absorption terms found in Ref. 2 is a consequence of including rescattering terms in the two-nucleon mechanism. As Ref. 2 uses the t-matrix formulation the absorption terms obtained there should not be taken as corrections to the optical potential as the latter includes rescattering to all orders. Also in our calculations we find that the "s-wave absorption" term arises as a result of recoil correction to the πNN vertex and hence is small. The large s-wave repulsion needed to explain the low-energy pion-nuclear elastic scattering data thus may have its source in some other phenomena rather than in pion absorption. On the other hand small contribution of these terms does justify the doorway assumption $H_{pq} = 0$ of Ref. 1.

REFERENCES

1. L.S. Kisslinger and W.L. Wang, Ann. Phy. 99, 374 (1976).
2. G.F. Bertsh and D.O. Riska, Phys. Rev. C18, 317 (1978).

ISSN:0094-243X/79/540366-01$1.50

PHOTO- AND ELECTRO-PRODUCTION OF CHARGED PIONS FROM NUCLEI AT THRESHOLD

T. W. Donnelly
Institute of Theoretical Physics, Department of Physics, Stanford University, Stanford, CA 94305*

ABSTRACT

A brief introduction to the subject of electromagnetic interactions in meson-nuclear physics is presented followed by a more detailed discussion of threshold photo- and electro-production of charged pions from nuclei.

INTRODUCTION

This section involves a very brief summary of the general field of electromagnetic interactions in meson-nuclear physics. The subject is very broad and diverse and here I hope only to place in context the rest of the present talk on photo- and electro-production of $\pi^{\pm}$ near threshold with the other talks in this session and in the rest of the conference.

One subject which falls in the category under discussion is the study of mesic atoms (π,K,Σ) where the electromagnetic interaction is involved in considering atomic cascades and in discussions of radiative capture of the mesons by the nucleus (see following papers). One may for instance consider processes such as (π^-,γ) (see ref. 1) where the nuclear cross section comes mainly from the radiative capture of a pion on individual nucleons. All processes in this category involve a meson in a bound (nearly zero energy) atomic state. Or, more exotic processes may be studied such as (π,2γ) where the pion may annihilate on another (virtual) pion already present in the nucleus and so probe the mesonic content of the nucleus (see following paper).

Alternatively one may study the subject of electromagnetic interactions in meson-nuclear physics by using photons (real or virtual) to produce mesons in the final state. For example the reactions (γ,π) and (e,e'π) are of this sort.[2] Here, in addition to the bound state of the meson-nuclear system as above, one may study the electromagnetic production of unbound final states of the meson-nuclear system as a function of energy. The nature of the physics

*Research supported by the National Science Foundation, Grant NSF PHY 77-16188.

ISSN:0094-243X/79/540367-14$1.50

involved may vary as the energy is varied. For example, for photo- and electro-production of charged pions near threshold (say within a few MeV of threshold) the reasonably well-known s-wave production mode is dominant and all other partial waves are suppressed by the angular momentum barriers. On the other hand, at higher energies away from threshold the resonant p-wave production through the N^* may dominate and lead to a different set of problems to be addressed: What is the production mechanism? How does the N^* propagate through the nucleus? What is the nature of the final state (N^* particle-N hole, etc.)? Some of these topics have already been discussed by previous speakers.

Of course in all cases one is confronted with some of the same questions which must be addressed when studying other reactions involving mesons and nuclei, such as scattering and hadronic production: How do the mesons interact with the nucleus? How are they absorbed? Can we employ an optical model discription of the meson-nucleus interaction? What aspects of nuclear structure enter and how may the knowledge of specific nuclear transitions be used to disentangle the various contributions to the meson-nucleus problem? What can one learn about nuclear physics <u>per se</u> using mesons as probes? Again these topics have also been discussed in other sessions in this conference.

I wish to turn now from the general to the very specific and discuss in a little more detail the topic of threshold photo- and electro-production of charged pions from nuclei. This is a continuing program of study at Stanford beginning with work on the $(\gamma,\pi^{\pm})$ reactions [3,4] and presently involving the $(e,e'\pi^{\pm})$ coincidence* reactions[5]. In addition to these exclusive processes we have been interested in the inclusive electron scattering cross section above pion threshold where both real pion production and virtual pion meson-exchange-current contributions play a role.[6,7]

THE REACTIONS $(\gamma,\pi^{\pm})$ AND $(e,e'\pi^{\pm})$ ON NUCLEI NEAR THRESHOLD

We begin the discussion by considering first electromagnetic interactions with nuclei[8,9] where pions are not produced in the final state. These are represented by the diagrams in Fig. 1.

*By $(e,e'\pi)$ we mean the coincidence reaction; some authors use this notation to refer to the process (e,π), i.e., not a coincidence involving detection of the scattered electron. The information obtained in the case of the (e,π) reaction is more akin to what is learned from the (γ,π) reaction as in the former the virtual photon spectrum is integrated over. In contrast the coincidence reaction $(e,e'\pi)$ yields more information as it involves measurements point by point across the virtual photon spectrum.

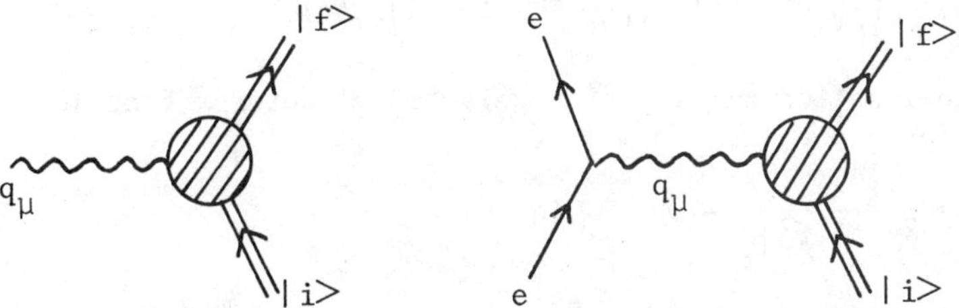

Fig. 1. Electromagnetic processes not involving produced pions.

The left-hand figure corresponds to a variety of real-photon processes such as γ-decay and photoexcitation where the photon momentum is on-shell, $q_\mu^2 = q^2 - \omega^2 = 0$. Here $q = |\vec{q}|$ is the three-momentum and ω is the energy. The right-hand figure corresponds to electron scattering where an electron initially of energy ε is scattered through an angle θ to final energy ε' and exchanges a virtual photon with the nuclear system. Here the photon momentum may be off-shell, $q_\mu^2 = q^2 - \omega^2 \geqslant 0$. Thus it is possible to fix the energy transfer $\omega = E_f - E_i$ to look at a given nuclear excitation from state $|i\rangle$ to state $|f\rangle$ and yet to vary the momentum transfer q over all values such that $q \geqslant \omega$.

Let us consider electron scattering in more detail. The interaction of the photon with the nucleus goes like $\hat{J}_\mu A_\mu$, where $\hat{J}_\mu$ is the nuclear current density operator (the caret signifies a second-quantized operator) and where A_μ is the vector potential of the electromagnetic field. All of the nuclear physics is buried in the matrix elements $\langle f|\hat{J}_\mu|i\rangle$ of the current operator. For reference later we note that this is a vector current, that it is conserved ($\partial_\mu \hat{J}_\mu = 0$) and that it contains isoscalar (S) and isovector third component (V_3) isospin contributions. It contains a three-vector piece $\hat{\vec{J}}$ (convection current and magnetization current operators) and a time component $\hat{\rho}$, the charge density operator. We may make multipole projections of $\hat{J}_\mu$ to define operators with good angular momentum quantum numbers[8,9] $\hat{M}_{JM}$ (projections of ρ), $\hat{L}_{JM}$ (longitudinal projections of $\vec{J}$) and $\hat{T}^{el}_{JM}$ and $\hat{T}^{mag}_{JM}$ (transverse projections of $\vec{J}$). In fact, using current conservation we may eliminate L and write for the electron scattering cross section[8,9]

$$\frac{d\sigma}{d\Omega} = 4\pi\sigma_M F^2(q,\theta), \qquad \omega = E_f - E_i, \tag{1}$$

where σ_M is the Mott cross section[8,9] and $F^2(q,\theta)$ is the form factor in which is buried the nuclear physics in the problem. A Rosenbluth separation of F^2 can be made to yield

$$F^2(q,\theta) = \left(\frac{q_\mu^2}{q^2}\right)^2 F_L^2(q) + \left(\frac{1}{2}\frac{q_\mu^2}{q^2} + \tan^2\frac{\theta}{2}\right) F_T^2(q), \tag{2}$$

defining the longitudinal and transverse form factors (functions only of q),

$$F_L^2(q) = \frac{1}{2J_i+1} \sum_{\mathcal{J}\geq 0} |\langle f||\hat{M}_{\mathcal{J}}(q)||i\rangle|^2 \tag{3a}$$

$$F_T^2(q) = \frac{1}{2J_i+1} \sum_{\mathcal{J}\geq 1} \left\{ |\langle f||\hat{T}_{\mathcal{J}}^{e\ell}(q)||i\rangle|^2 + |\langle f||T_{\mathcal{J}}^{mag}(q)||i\rangle|^2 \right\} \tag{3b}$$

where the matrix elements are reduced in angular momentum. By mapping out the q-dependence of these form factors one is in effect mapping out the Fourier transforms of the charge and current density distributions for the nuclear transition being considered. A similar analysis may be repeated[8] for real photon processes (the left-hand part of Fig. 1) to yield: for γ-decay the rate

$$\omega_\gamma = 8\pi\alpha\omega\, F_T^2\,(q=\omega) \tag{4a}$$

and for photoexcitation, integrating in energy over the spectrum for the transition $i \to f$,

$$\int_{\text{peak}} \sigma(\omega')d\omega' = (2\pi)^3 \frac{\alpha}{\omega} F_T^2\,(q=\omega). \tag{4b}$$

With this summary of the usual electromagnetic interactions with nuclei in mind let us turn to electromagnetic processes where pions are present, as in Fig. 2.

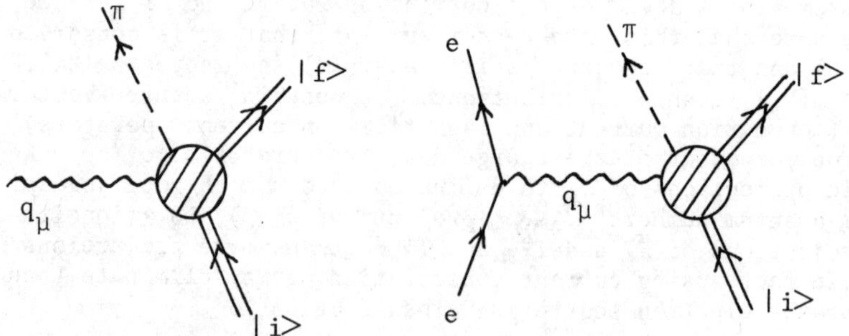

Fig. 2. Electromagnetic process involving produced pions.

The labelling is as in Fig. 1 except now one has a pion in the final state. The photon carries energy ω where

$$\omega = m_\pi + \omega_{fi} + E_\pi^{CM} + E_{recoil}\,(q), \tag{5}$$

where m_π is the pion rest mass, ω_{fi} is the difference in energy of the states $|i\rangle$ and $|f\rangle$, E_π^{CM} is the π-nucleus energy in the CM system and E_{recoil} (q) is the recoil energy of the entire final state. As before we have $q = \omega$ for real photons (left-hand diagram) and $q \geqslant \omega$ for virtual photons (right-hand diagram). While we have written the real photon diagram to indicate pion photoproduction, time-reversing everything will yield the diagram for radiative pion capture; that is, it is not really a distinct process, it differs only in that the pion is captured from a bound state of the pion-nucleus system.

To describe the processes in Fig. 2 we must consider the photon-pion-nuclear vertex. At this stage we consider only charged pion production and so the states $|i\rangle$ and $|f\rangle$ differ in charge by one unit (see Fig. 4 below). Two steps are performed (see ref. 5 for the details): (1) Using PCAC the pion field may be replaced by the divergence of the axial-vector current $\partial_\mu \hat{J}_\mu{}^5$ (we use the notation from past work, see refs. 3,4,5,9,10); (2) Using current algebra we combine this axial-vector current (A) with the vector electromagnetic current (V) to obtain the axial-vector current again. These steps are indicated schematically in Fig. 3.

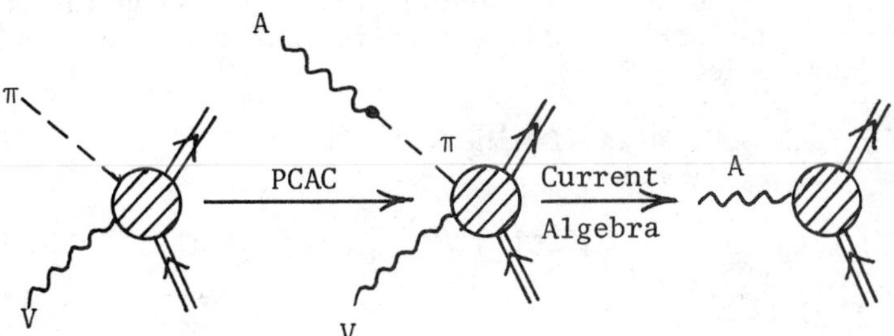

Fig. 3. Reduction of the current using PCAC and current algebra.

We then must deal with the axial-vector current $\hat{J}_\mu^5$ which is not conserved and has isospin structure $V_\pm$ (charge-changing). As above we may make multipole projections of the time-piece, the longitudinal piece and the two transverse pieces to define $\hat{M}^{5\prime}_{\mathcal{JM}}$, $\hat{L}^{5\prime}_{\mathcal{JM}}$, $\hat{T}^{e\ell 5\prime}_{\mathcal{JM}}$ and $\hat{T}^{mag 5\prime}_{\mathcal{JM}}$ respectively (here as above the "5" indicates the extra γ_5 which enters in the elementary axial-vector currents). The prime is used to indicate that the actual spatial form of the pion wave function must be included in the current operators. We return to this point below. In the case of s-wave pion production, as we will restrict ourselves to at threshold, this means multiplying the nuclear densities by the s-wave radial wave function of the pion.

We may proceed to express the $(e,e'\pi^{\pm})$ cross section in terms of matrix elements of these multipole operators[5]. For a given nuclear transition $|i\rangle \to |f\rangle$ and for a fixed energy E_π^{CM} (the π-nucleus energy in the CM system) and hence a fixed pion wave function we have

$$\frac{d^3\sigma}{d\Omega dE' d\Omega_\pi^{CM}} = \frac{4\pi\sigma_M}{M_T} G^2(q,E_\pi^{CM},\theta), \tag{6}$$

where the electron variables are in the lab system and the pion variables refer to the final-state CM system. The energy transfer ω is given by Eq. (5) and varies with momentum transfer q in such a way as to keep E_π^{CM} fixed. In Eq. (6) M_T is the target mass which has been included so that G^2 is dimensionless. We again have a Rosenbluth decomposition

$$G^2(q,E_\pi^{CM},\theta) = K_L\left(\frac{q_\mu^2}{q^2}\right)^2 G_L^2(q,E_\pi^{CM}) + K_T\left(\frac{1}{2}\frac{q_\mu^2}{q^2} + \tan^2\frac{\theta}{2}\right) G_T^2(q,E_\pi^{CM}), \tag{7}$$

where K_L and K_T are composed of kinematic factors and coupling constants used in our definitions of the longitudinal and transverse form factors (independent of θ):

$$G_L^2(q,E_\pi^{CM}) = \frac{1}{2J_i+1} \sum_{J\geq 0} |\langle f||\hat{M}_J^{5\prime}(q)||i\rangle|^2 \tag{8a}$$

$$G_T^2(q,E_\pi^{CM}) = \frac{1}{2J_i+1} \sum_{J\geq 1} \left\{|\langle f||\hat{T}_J^{el_5\prime}(q)||i\rangle|^2 + |\langle f||T^{mag_5\prime}(q)||i\rangle|^2\right\} \tag{8b}$$

As noted above the prime in the operators indicates that the pion wave function is to be included with the nuclear distributions. Thus measurements of the q-dependence of these form factors yield information on the current distributions which are combinations of the nuclear distributions and the pion wave function at CM energy E_π^{CM}. If we were to have already determined the former, the nuclear (axial-vector) distributions, we would then be sensitive directly to the spatial distribution of the pion wave function. We return to discuss the implications of this statement at the end.

We may obtain expressions for the real photon process (the left-hand diagram in Fig. 2) just as we obtain Eqs. (4) in concert with Eqs. (1-3). Thus the photoproduction cross section is proportional to $G_T^2(q=\omega, E_\pi^{CM})$. Note that, if one wishes to keep the energy E_π^{CM} fixed and hence keep the pion wave function fixed, there is only one value of $q=\omega$ possible for a given E_π^{CM} from Eq. (5). That is, it requires a virtual photon as in electroproduction to enable one to map out the pion wave function. If $q=\omega$ were to vary in photoproduction then so must E_π^{CM} vary and, for instance, it would not be possible to restrict one's attention to s-wave pions, but would soon require s- and p-waves over a range of energies.

The question remains: Can we determine the nuclear distributions in the absence of pions? For this we are led to consider the weak interactions, since the axial-vector current required above is the same as the axial-vector part of the weak current aside from the inclusion of the pion wave function. We have been engaged in a continuing study of semi-leptonic weak interactions with nuclei[9,10,11] including β-decay, μ-capture, ν-reactions and ν-scattering. These interactions and the interactions considered above mediate nuclear transitions as shown in Fig. 4, where we take the A=12 system as an example.

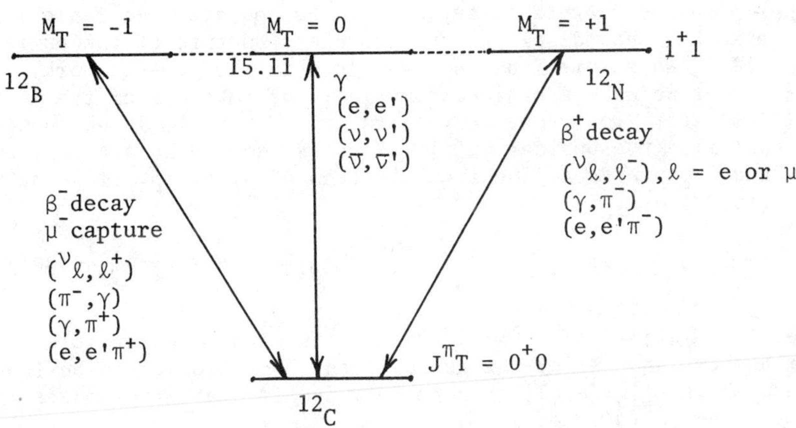

Fig. 4. Transitions considered in the A=12 system.

In studying the weak interaction the procedure is similar to the one outlined above: The current-current form of the weak interaction is employed with vector and axial-vector currents; the leptonic matrix elements are computed and, after making multipole projections of the nuclear current operators, we may write the cross sections in terms of reduced nuclear matrix elements of these multipole operators (see refs. 10,11 for details). Table 1 contains a summary of all of the multipole operators involved.

Table 1: Multipole Operators

	Electromagnetic	Weak		Pion Production[a]
$\mathcal{J}\geqslant 0$	$\hat{M}$	$\hat{M}$	$\hat{M}^5$	$\hat{M}^{5'}$
$\mathcal{J}\geqslant 0$	$(\hat{L})^b$	$(\hat{L})^b$	$\hat{L}^5$	$(\hat{L}^{5'})^b$
$\mathcal{J}\geqslant 1$	$\hat{T}^{e\ell}$	$\hat{T}^{e\ell}$	$\hat{T}^{e\ell 5}$	$\hat{T}^{e\ell 5'}$

(cont'd next pg)

$\mathcal{J} \geq 1$	$\hat{T}^{mag}_V$	$\hat{T}^{mag}_V$	$\hat{T}^{mag5}_A$	$\hat{T}^{mag5'}_A$

a) The prime indicates that the pion wave function is to be included with the nuclear distributions.
b) Current conservation may be used to eliminate these multipoles.

In fact we usually invoke the CVC hypothesis to relate the vector part of the weak interaction current to the electromagnetic current.

To obtain cross sections for all of the processes under discussion we require nuclear matrix elements of multipole operators $\hat{T}_{\mathcal{J}\mathcal{M}_{\mathcal{J}};\mathcal{T}\mathcal{M}_{\mathcal{T}}}(q)$, where $\hat{T}$ is any one of the operators in Table 1 and where we have now displayed the isospin structure of the current as well. (we assume isospin invariance in the present work) At this point we make a basic assumption: we take all of the multipole operators to be one-body operators ($\hat{T} \sim a^+a$), in which case we can write the nuclear matrix elements reduced in angular momentum and isospin as a linear combination of single-particle matrix elements[9,11]

$$\langle J_f;T_f \,\|\, \hat{T}_{\mathcal{J};\mathcal{T}}(q) \,\|\, J_i;T_i\rangle = \sum_{a\,a'} \langle a;\tfrac{1}{2} \,\|\, T_{\mathcal{J};\mathcal{T}}(q) \,\|\, a';\tfrac{1}{2}\rangle \psi^{(fi)}_{\mathcal{J};\mathcal{T}}(aa'), \quad (9)$$

where $a = \{n\ell j\}$, $a' = \{n'\ell'j'\}$ are sets of single-particle quantum numbers and all of the nuclear many-body problem is buried in the numbers $\psi^{(fi)}_{\mathcal{J};\mathcal{T}}(aa')$, the one-body density matrix elements. The procedure followed is:

(1) Take electron scattering data to determine the many-body matrix elements of the electromagnetic multipole operators in Table 1 (that is, determine the left-hand side of Eq. (9)).
(2) Expand the right-hand side of Eq. (9) using the (reasonably) well-known single-particle matrix elements of these multipole operators, including their q-dependence. Hence, determine the set of density matrix elements. Of course the sums over a and a' will have to be truncated to some model space and this constitutes a second assumption.
(3) Using these determined ψ's, turn the relationship around and for given single-particle matrix elements for another process (perhaps the weak interaction matrix elements to test the procedure, perhaps the single-nucleon pion production matrix elements to make predictions for that process) predict the many-body matrix elements and hence the cross sections.

By using experimental data as input hopefully the predictions will be relatively model independent, certainly much less subject to model assumptions then shell model calculations for instance. In other words our procedure is to "calibrate" a given nuclear transition (and its isospin analogs) using one probe (electron scattering) and then to use this "calibrated" system to predict the rates for other less well-known processes. Some examples may help to tie together all of the discussion presented thus far.

First we consider the A=3 system, the $J^{\pi}T = \frac{1}{2}^{+}\frac{1}{2}$ ground states of ^3He and ^3H. Let us truncate the sums over single-particle quantum numbers in Eq. (9) to only the $1s_{1/2}$ shell. Then there are only four density matrix elements required to determine all of the processes under discussion, $\psi_{0;0} = 3$, $\psi_{1;0} = \psi_{0;1} = 1$ and $\psi_{1;1} = -1$ (see refs. 11,12). The electron scattering and weak interaction matrix elements all involve only one single-particle radial matrix element, $\langle 1s_{1/2}| j_0(qr)| 1s_{1/2}\rangle$, and this may be taken directly from electron scattering by fitting the elastic charge form factor of ^3He (see refs. 12,13). As a first test of the procedure we may predict the elastic magnetic form factor of ^3He and the elastic charge and elastic magnetic form factors of ^3H; all are quite satisfactorily reproduced. As a second test we may predict the $^3H \rightarrow {}^3He$ β-decay rate and the $^3He \rightarrow {}^3H$ μ-capture rate; again both are in excellent agreement with experiment. Now to predict the $^3He(\gamma,\pi^+)^3H$ and $^3He(e,e'\pi^+)^3H$ cross sections near threshold we require the pion wave function at energy E_{π}^{CM} as now the radial matrix elements involve $\langle 1s_{1/2}| j_0(qr)\phi_{\pi}^{s\text{-wave}}(r)| 1s_{1/2}\rangle$. We use optical model wave functions which employ the Krell-Ericson potential to estimate the cross sections. In Fig. 5 we show results for $^3He(e,e'\pi^+)^3H$ with $E_{\pi}^{CM} = 3.2$ MeV as a function of momentum transfer q. The point marked "γ" corresponds to the real-photon point for the reaction $^3He(\gamma,\pi^+)^3H$ and agrees well with the most recent experimental value[14]. A series of measurements at moderate to high values of q where one is sensitive to fine spatial variations in the pion wave function (in contrast to the photon point where only a relatively low moment of the spatial distribution is required) would help decide which optical potential is correct.

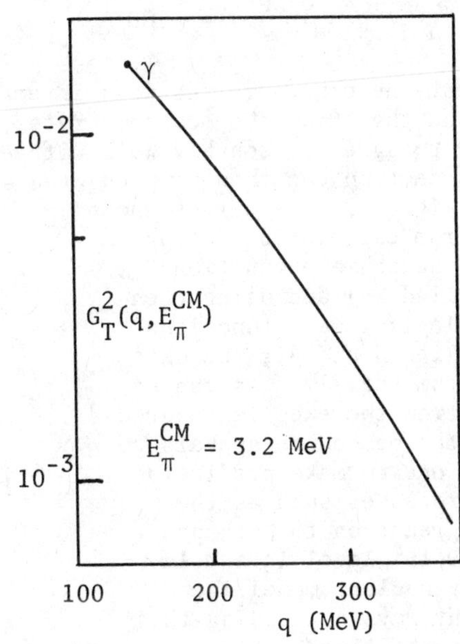

Fig. 5. Transverse form factor for the reaction $^3He(e,e'\pi^+)^3H$.

As a second example we consider the A=6 system consisting of the ground state of $^6Li(1^+0)$, the second excited state of $^6Li(0^+1, 3.562$ MeV) and its isospin analog the ground state of $^6He(0^+1)$. An analysis of the electromagnetic and weak interactions in this system was performed some time ago[15] and used in predicting the $^6Li(\gamma,\pi^+)^6He$ rate[3,4,16]. Here the model space was taken to be a closed $1s_{1/2}$ shell plus two nucleons in the 1p-shell. The wave functions can then be written

$$\begin{aligned} |1^+0\rangle &= A|1p^2_{3/2},1^+0\rangle + B|1p_{3/2}1p_{1/2},1^+0\rangle + C|1p^2_{1/2},1^+0\rangle \\ |0^+1\rangle &= D|1p^2_{3/2},0^+1\rangle + E|1p^2_{1/2},0^+1\rangle, \end{aligned} \tag{10}$$

where we use some basis of single-particle wave functions such as harmonic oscillators[15], or Woods-Saxon wave functions[4] or another phenomenological set[15]. There are two normalization conditions, $A^2+B^2+C^2=1$ and $D^2+E^2=1$ and by fitting elastic and inelastic electron scattering data (including the magnetic dipole and electric quadrupole moments of the ground state of ^6Li) all of the parameters can be determined. From this one may immediately obtain the one-body density matrix elements for the $1^+0 \rightarrow 0^+1$ transition:

$$\begin{aligned} \psi_{1;1}(1p_{3/2}1p_{3/2}) &= -\frac{1}{\sqrt{2}}AD \\ \psi_{1;1}(1p_{3/2}1p_{1/2}) &= -\frac{1}{2}BD \\ \psi_{1;1}(1p_{1/2}1p_{3/2}) &= \frac{1}{\sqrt{2}}BE \\ \psi_{1;1}(1p_{1/2}1p_{1/2}) &= -CE \end{aligned} \tag{11}$$

To test the procedure the weak interaction rates, the β-decay of ^6He and the $^6Li \rightarrow {}^6He$ μ capture rates, may be predicted and are found to agree reasonably well with experiment. The $^6Li(\gamma,\pi^+)^6He$ cross section can then be predicted and compared with experiment (see Fig. 6). The curves shown are from calculations using Woods-Saxon wave functions[4] (labelled WS) and using phenomonological wave functions[16] (labelled A and B); the data are from ref. 17. As can be seen from the excellent agreement the procedure appears to allow one to make predictions for processes such as the (γ,π) reaction to perhaps the 10-15% level in a relatively nuclear model independent way. Extending this to the reaction $^6Li(e,e'\pi^+)^6He$ we obtain results as shown in Fig. 7 where again for definitness Krell-

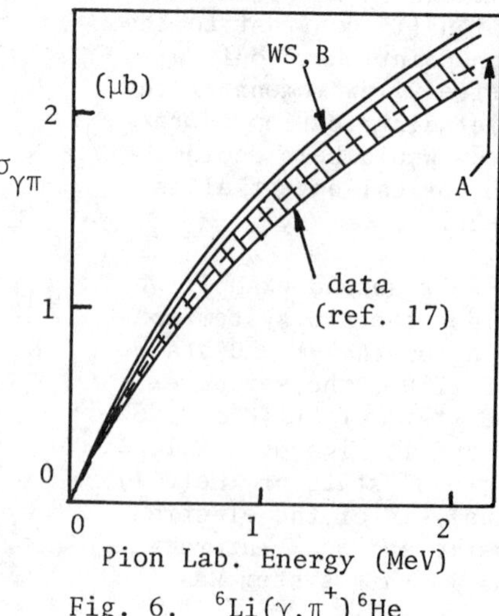

Fig. 6. $^6Li(\gamma,\pi^+)^6He$

Ericson potential pion wave functions were employed at energy E_{π}^{CM} = 3.2 MeV. At intermediate values of q these predictions should be more sensitive to changes in the pion optical potential assumed than the predictions given in Fig. 6 (which were rather insensitive to the choice of optical potential employed).

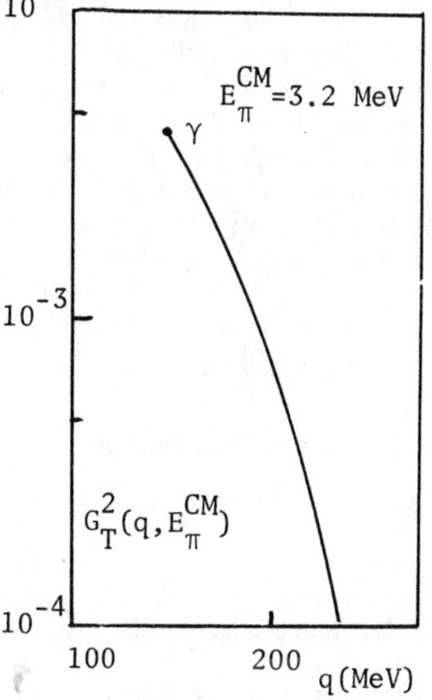

Fig. 7. $^6Li(e,e'\pi^+)^6He$

The above two simple examples serve to illustrate the procedures used. Of course other systems are of interest and some have already been explored in photoproduction such as the A=12 system where considerable theoretical work has also been undertaken[3,11,18]. Rather than summarize these other photoproduction examples I wish to conclude this talk by collecting together some of our observations thus far on the as-yet-to-be-measured electroproduction process:

(1) The electroproduction form factors contain information on the spatial distribution of the pion wave functions. We can determine the axial-vector nuclear distributions by studying electromagnetic and weak interactions with the system as functions of momentum transfer. Our analyses proceed through the approach using one-body density matrices to represent the nuclear many-body physics in the problem. The q-dependence of the electroproduction form factors then yields the Fourier transforms of the combined pion-plus-nuclear distributions from which we can extract information on the pion wave function. The hope is that this new information may help to define which is the correct pion optical potential from among the choices that are consistent with pion scattering data. We note that such a measurement involving the spatial distribution of a given pion partial wave at a given energy is not possible with real photons (photoproduction) where $q=\omega$.

(2) In a coincidence experiment such as $(e,e'\pi)$ the center-of-mass angular distribution may be measured. One may then obtain a separation of s-wave (isotropic) from p-wave production even at energies away from threshold where the two partial waves compete.

(3) In photoproduction we have $q=\omega$ and so the momentum in the final state is relatively low. However in electroproduction $q \geqslant \omega$ and so q may be large enough that $E_{recoil}(q)$ (see Eq. (5)) may be significant. For sufficiently light nuclear systems this opens the possibility of detecting not the pion but the recoiling

daughter nucleus and so going quite close to threshold. An example of such a reaction is provided by $^3He(e,e'{}^3H)\pi^+$ which involves a coincidence between the scattered electron and the recoiling triton. The recoiling daughter has an angular distribution which is highly peaked along the direction of momentum transfer $\vec{q}$ and so for practical sized detector systems one has an effective 4π geometry. Finally, to give a feeling for the typical values of the cross sections we are dealing with in studying the reaction $^3He(e,e'{}^3H)\pi^+$, we take ε = 265 MeV, $\theta = 45^o$ and ω = 149 MeV corresponding to q = 200 MeV and obtain

$$\frac{d^3\sigma}{d\Omega d\varepsilon' dE_t^L} = 1.1x10^{-35}\ cm^2\ sr^{-1}\ MeV^{-2},$$

where E_t^L is the triton energy in the lab. system.

REFERENCES

1. H. W. Baer, K. M. Crowe and P. Truöl, Adv. in Nucl. Phys. 9, 177 (1977).
2. International Symposium on Photopion Nuclear Physics, Rensselaer Polytechnic Institute, August 10-12, 1978.
3. J. H. Koch and T. W. Donnelly, Nucl. Phys. B64, 478 (1973); Phys. Rev. C10, 2618 (1974).
4. J. B. Cammarata and T. W. Donnelly, Nucl. Phys. A267, 365 (1976)
5. T. W. Donnelly and J. B. Cammarata (to be published).
6. T. W. Donnelly, J. W. van Orden, T. deForest, Jr. and W. C. Hermans, Phys. Lett. 76B, 393 (1978).
7. J. W. van Orden (Ph.D. Thesis, Stanford, 1978, unpublished).
8. T. deForest, Jr. and J. D. Walecka, Adv. in Phys. 15, 1 (1966).
9. T. W. Donnelly and J. D. Walecka, Ann. Rev. Nucl. Sci. 25, 329 (1975).
10. J. D. Walecka, in Muon Physics, vol. 2, ed. V. W. Hughes and C. S. Wu (Academic Press, N.Y. 1975) p. 113.
11. T. W. Donnelly and R. D. Peccei, Phys. Rep. 50, 1 (1979).
12. T. W. Donnelly and J. D. Walecka, Nucl. Phys. A274, 368 (1976).
13. T. W. Donnelly and J. D. Walecka, Nucl. Phys. A201, 81 (1973).
14. P. Argan et al. (preprint).
15. T. W. Donnelly and J. D. Walecka, Phys. Lett. 44B, 330 (1973).
16. J. C. Bergstrom, I. P. Auer and R. S. Hicks, Nucl. Phys. A251, 401 (1975).
17. G. Audit et al., Phys. Rev. C15, 1415 (1977).
18. W. C. Haxton, Phys. Lett. 76B, 165 (1978).

Question: K. Amos, University of Melbourne

I have a comment and a question. I agree that high spin states of natural parity, such as the 6^- states in ^{28}Si, are of simple nature, but in a transition you must know well the ground state occupancy of particles whose evaluation leads to the particle-hole prescription for the high spin unnatural parity states. This critically effects cross section magnitudes. The question I have concerns the use of transition data (such as (e,e')) to determine, in general, nuclear transition density matrixes. Given that one must truncate a single particle basis space, how can you be sure that the extracted density matrix elements coincide with normalized wave functions of the nuclear states? Furthermore, how unique are the set of numbers that result?

Response: Donnelly

The results are testable by increasing the size of the model space. In several cases we have reasonable confidence in our one-body density matrix analysis of the high spin states. As to uniqueness we attack this problem by using different basis sets of single-particle wave functions (harmonic oscillators, Woods-Saxon wave functions, Nileson wave functions, etc.). Any complete set will do, and within the context of one-body operators the answers must be equivalent. The approximation tested is again the truncation scheme. That is one asks, are the results more convergent with one basis set or another? We are studying this at present.

Question: K. Seth, Northwestern University

The outgoing pion energies you talked about were 1-10 MeV. It would appear that in this case all that you will be sensitive to is Couloumb distortions. If so, your main objective, finding the pion wave function or the strong interaction distortions to it, will not be realized.

Response: Donnelly

The Couloumb distortion is smooth, whereas the strong interaction distortions may be more rapid in coordinate space and hence lead to significant effects at large q where such Fourier components are probed selectively. Again, this is one of the prime reasons for using virtual photons as in electroproduction.

Question: L. Kisslinger, Carnegie-Mellon University

We are fairly content using the concept of a pion wave function for elastic and inelastic pion scattering, which involves mainly the nuclear surface. However, in electroproduction you will be studying the pionic field deep inside the nucleus. The pion interacts largely by being destroyed and created. I believe the healthiest attitude one may profess is that you will be testing whether the concept of a pion wave function is a good one. Perhaps many-body, nonlocal forms must be used.

Response: Donnelly

I agree. I feel that this is one of the merits of electroproduction where probing the spatial distributions, including the effects of the pion, is an attainable goal.

Comment: G. Walker, Indiana University

The phase relations between such matrix elements as $\langle j_p|o|j_h\rangle$ and

$<j_h|o|j_p>$ which enter in evaluating particle-hole creation or destruction diagrams, is the same for the comparable electron, pion, and kaon matrix elements. It is different for distorted protons due to the presence of the antisymmetrization induced exchange term. This is a useful difference in trying to deduce what corrections must be made in either nuclear structure or reaction mechanisms when theory and experiment do not agree. Note that in some cases the exchange term is very important and in other cases unimportant in (p,p') and (p,n) reactions.

DOUBLY RADIATIVE PION-CAPTURE : A STATUS-REPORT

J. Deutsch
Institut de Physique Corpusculaire
Université Catholique de Louvain
Chemin du Cyclotron, 2
B - 1348 LOUVAIN-LA-NEUVE, Belgium

The field of doubly radiative pion-capture attracted much attention these last years because of its possible connection to eventual pion-condensates in nuclei, idea first introduced by Migdal [1].

The situation is far from being settled and so we hope that this "mid-stream" status-report will be out-dated quite soon...

1) The aim of doubly-radiative pion-capture experiments was originally to probe the pion-field in nuclei [2]. More specifically, it was considered that the "annihilation-graph" represented in Fig. 1.a, dominating the process in the Coulomb-gauge [2], shall provide a more selective probe of the pion-field, then e.g. the well-known "induced pseudoscalar" graph (Fig. 1.b), which contributes only %-effects to muon-capture. The process could have mapped out, moreover, the q^2-dependence of the pion-nucleon coupling constant $g_{\pi NN}$ [2].

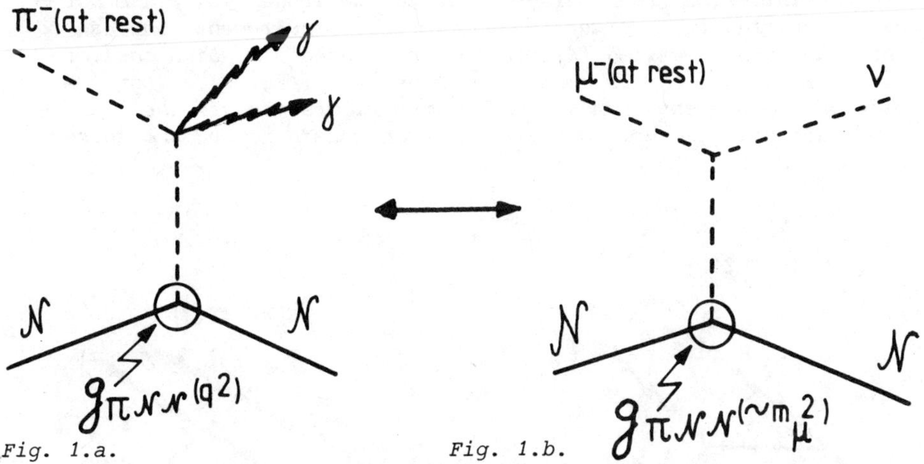

Fig. 1.a. *Fig. 1.b.*

Even if it was not expected that the pion-condensation would manifest itself at normal nuclear matter-densities, it was hoped that "pre-critical" enhanced fluctuations of the pion-field may show up in the process under investigation [3]. (For another probe of such a "pre-critical" situation, cfr. also the contribution of Toki and Weise to this conference).

ISSN: 0094-243X/79/540381-09$1.50 Copyright 1979 American Institute of Physics

2) These aims in mind, several _experimental investigations_ *of the process were initiated. A Virginia Polytechnic Institute/Indiana University Group working at SREL obtained an upper limit to the branching ratio* [4]*. A group from the Louvain University, working at the CERN SC reported recently the first observation of the process* [5] *at a rate confirmed by the preliminary result contributed to this conference by a University of British Columbia/Université de Montréal collaboration working at TRIUMF* [6]*. Finally, a Argonne-M.I.T. Northwestern University-University of Illinois collaboration intents to take up the investigation of the process in hydrogen and also in heavier nuclei.*

Before entering into the details of these experiments and the discussion of the results, let us comment on the theoretical developments which followed the first theoretical papers in the field.

3) The first _numerical treatment_ *of the doubly radiative process* [2] *presented, basically, the following* _two shortcomings_ *: the branching ratio was estimated considering only s-wave capture on free nucleons and the "annihilation-graph" alone. The effect of the Fermi-motion of the nucleons and the Pauli-blocking was hoped to account for phenomenologically using the experimental nuclear branching-ratio for singly radiative capture and using the free-nucleon estimate for the double-to-single radiative capture ratio only :* $[N\gamma\gamma/N(\pi\text{-stop})] = [N\gamma\gamma/N\gamma]_{free}\,[N\gamma/N(\pi\text{-stop})]_{exp}$, *but the overwhelmingly p-wave capture in moderately heavy nuclei such as* ^{9}Be *and* ^{12}C *was not accounted for. Moreover, the consideration of the "annihilation graph" alone leads to amplitudes which are not gauge-invariant. In order to achieve this requirement, one has to introduce the so-called "Bremmstrahlung-graphs" in which one (or both) of the photons is emitted by a nucleon (Fig. 2). (For the ease of the presentation we omit on the figure the off-shell pion-lines and photon-coupling to the neutron considered by the authors we shall quote).*

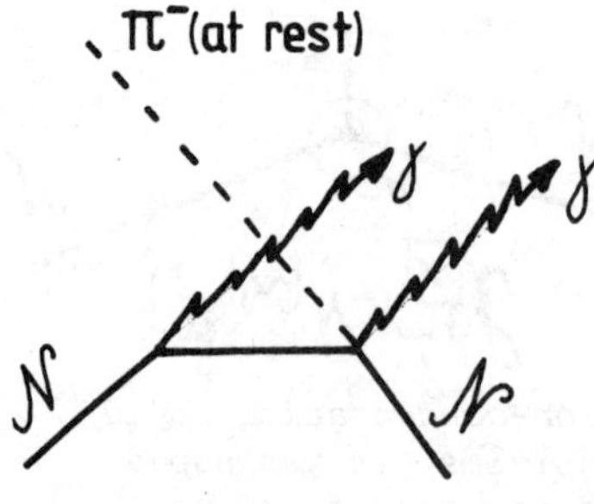

Fig. 2.a.

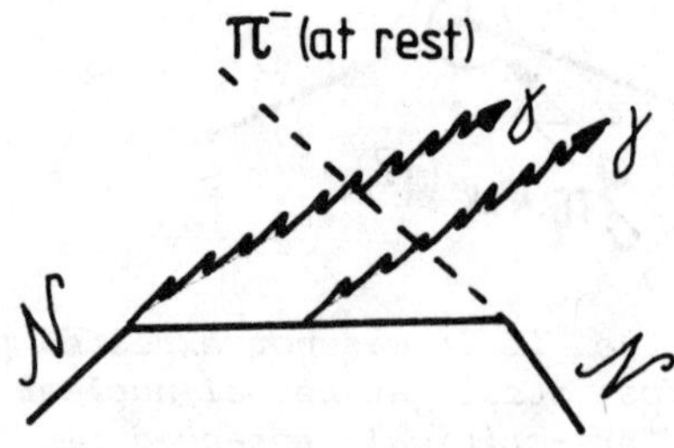

Fig. 2.b.

To improve upon some of these shortcomings S. Barshay introduced [7] both s-wave and p-wave pions in the evaluation of the doubly-radiative capture amplitude and D. Beder [8] considered the effect of the Bremmstrahlung graphs. Finally, P. Christillin and T. Ericson in a contribution to this conference, evaluate the doubly radiative branching ratio considering simultaneously the annihilation and Bremsstrahlung-graphs, p-wave pions and the excitation spectrum of the final states in ^{12}B using energy-weighted sum-rules.

We shall compare the numerical predictions later ; let us note already, however, that contrarily to the early believe : 1) in principle the "annihilation-graph" cannot be singled out because its amplitude depends on the particular gauge used and 2) a comparaison with experiment requires the evaluation of the Bremsstrahlung-graphs (enhanced - even in the Coulomb-gauge - for p-wave capture). As all the graphs imply one pion-nucleon vertex, the total amplitude is still proportional to the pion-nucleon coupling $g_{\pi NN}$ in the nucleus, but because of the contribution of different off-shell pion - or nucleon-lines - its momentum-dependence cannot be infered from the distribution of the doubly radiative photons.

4) The experiments will not be discussed here in detail ; for these details the reader may consult refs. 4 and 5 as well as the contribution of Bassalleck et al. to this conference. We shall, however, describe briefly how the basic challenge of the experiment was met by the two successful experiments performed up to now.

The basic challenge of the experiment is the observation of the decay-channel π^-(stopped) + (Nucleus) $\rightarrow$ $(\text{Nucleus})^\star$ + γ + γ of expectedly very low branching-ratio (∿ 3 10^{-6} per stopped pion) in the presence of a 10^4 - 10^6 times more abundant background from $(\pi^-,2n)$-, $(\pi^-,n\gamma)$-, (π^-,γ)-reactions, the last (single radiative) decay appearing as a $(\pi^-,\gamma\gamma)$-reaction through accidental coincidences or inter-detector cross-talk. (π^-, π^o) charge-exchange reactions from pions in flight (or stopped in materials where this reaction is energetically not forbidden), followed by the π^o-decay, may also lead to a strong two-photon background, especially for inter-photon angles around 180°.

The photon detector of the Louvain-group consisted in a hodoscope of 20 lead-glass truncated prisms surrounding the target in a plan perpendicular to the pion-beam axis. The angular acceptance of a pair of detectors is 13° FWHM, inter-detector cross-talk is negligible for inter-photon angles $\theta_{12} \geq 50°$. Around 100 MeV photon-energy the energy-resolution is 45 % FWHM and the total acceptance is about 20 %. The detector is heavily shielded against photons originating from the degrader and the beam-telescope. The beam-telescope allows to reject pairs of particles arriving within the time resolution of the photon detectors and to select the pions which enter the target with an energy lower than the charge-exchange threshold (∿ 9 MeV in ^9Be and ^{12}C). An event is retained if two and only two lead-glass cells sensed a signal in time-coincidence with one from the telescope selected according the above-mentioned criteria. From the photon energy- and angular-

distribution the background is substracted. Measurements were performed both with ^{9}Be and ^{12}C targets.

The set-up of the U.B.C.-Montréal group consisted basically in two large NaI crystals of good energy-resolution placed at inter-photon angle $\theta \approx 120°$. The neutron-rejection is provided by the time-of-flight information which is used also. Two supplementary lead-glass detectors are used to explore other inter-photon angles. The energy of the incoming beam is low enough (20 MeV) to minimize the background from (π^-,π^o)-reactions. The experiment was performed on ^{12}C alone.

5) The <u>*results*</u> *are essentially three-parametric distributions $N(\theta_{12}, E_{\gamma 1}, E_{\gamma 2})$, functions of the inter-photon angle and both photon-energies. For easier discussion we shall consider only some secondary distributions one can deduce from these results.*

The distribution $N(E_{\gamma 1} + E_{\gamma 2})$ integrated over all observed interphoton angles gives the excitation-spectrum of the residual nucleus. In both experiments ($50° \leq \theta_{12} \leq 180°$ for Louvain and $\theta_{12} \approx 120°$ for U.B.C.-Montréal) this distribution is very similar to the one observed in single radiative pion capture. This is in good agreement with the expectations of P. Christillin and T. Ericson in a paper submitted to this conference.

The angle-integrated bi-parametric distribution $N(E_{\gamma 1}, E_{\gamma 2})$ is also instructif because it allows to visualize the energy-sharing distribution between the two photons. Though no theoretical computations are presented as yet in this form, one may expect intuitively that the Bremsstrahlung-graphs (single-radiative capture + radiation of a nucleon) favour unequal energy-partition between the two photons. The results do not bear out this expectation and favour a more homogenous filling of the $N(E_{\gamma 1}, E_{\gamma 2})$-distribution.

The energy-integrated inter-photon angular distributions obtained by the two experiments in ^{12}C are compared in Fig. 3. (The distribution observed in ^{9}Be [5] is quite similar). They are in satisfactory agreement especially if we note the data-points of ref. 5 for $\theta_{12} > 140°$ may contain a contribution from contaminating gases of low charge-exchange threshold absorbed in the graphite target. The angular distributions are in agreement with the ones predicted for free nucleons [7,8], but seem to disagree with the more realistic but as yet preliminary results of P. Christillin and T. Ericson contributed to this conference. These authors predict a distribution which decreases with increasing θ_{12}.

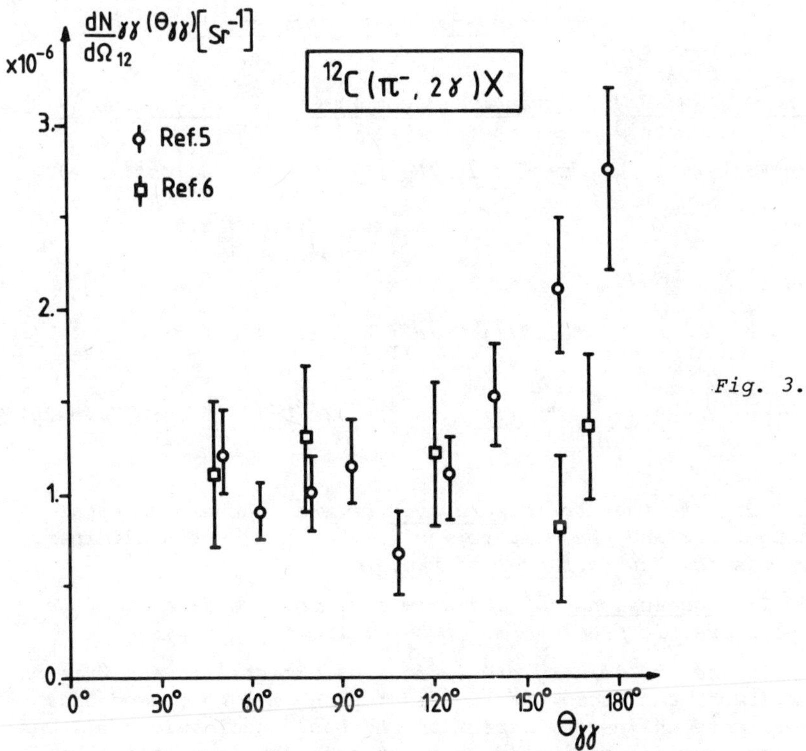

Fig. 3.

Finally, let us quote the total branching ratios obtained for doubly radiative capture. The Louvain-group obtains [5], *with an energy-cut of* $E_\gamma \geq 25$ *MeV and* $50° \leq \theta \leq 180°$, *a branching ratio of* $(10 \pm 1)10^{-6}$ *for* ^9Be *and* $(14 \pm 2)10^{-6}$ *for* ^{12}C. *This value changes only slightly if the suspect data-points above* $\theta_{12} = 140°$ *are neglected :* $(12 \pm 2)10^{-6}$. *The Louvain-result is in excellent agreement with the preliminary result* $(14 \pm 4)10^{-6}$ *of the U.B.C.-Montréal collaboration contributed to this conference. This result was obtained with an energy-threshold of* $E_\gamma \geq 10$ *MeV, but an inspection of the* $N(E_{\gamma 1}, E_{\gamma 2})$*-distribution shows that the bulk of the events is situated above the* $E_\gamma \geq 25$ *MeV cut ; so the two results are directly comparable.*

The different <u>*theoretical predictions*</u> *of the branching-ratio are resumed in the table. We indicate also the major ingredients of these predictions as discussed already in § 3.*

Table

Branching-ratios $N_{\gamma\gamma}/N(\pi\text{-stop})$

Author	*$\ell(\pi)$*	*graphs*	*computed*	*$N_{\gamma\gamma}/N_{\pi}(10^{-6})$*
Ericson/Wilkin	0	"π-π" only	$(N_{\gamma\gamma}/N_{\gamma})$(free)	3 (estimate)
Barshay	0	"	$(N_{\gamma\gamma}/N_{\pi})$(free + opt^{al} pot.)	3.5
Barshay	1(!)	"	"	5
Beder	0	"all"(!)	$(N_{\gamma\gamma}/N_{\pi})$(free)	2.5
Christillin/ Ericson	1(!)	" (!)	$(N_{\gamma\gamma}/N_{\pi})$(^{12}C:!)	6 (preliminary)

Let us note that the discrepancy between the experimental branching-ratio and the most reliable (though as yet preliminary) predictions is about a factor of two.

6) The conclusions one can draw from this "mid-stream" status-report are, it seems to me, two-fold.

A) If the preliminary predictions of Christillin and Ericson are confirmed, the factor-of-two discrepancy may be traced back either to some unexpected aspect of the basic nucleonic amplitude or to some specific collective nuclear contributions not taken into account by the nucleon-graphs (Fig. 2) used by Christillin and Ericson. The first explanation will be tested by the planed $\pi^- + p \to n + \gamma + \gamma$ experiment of the Argonne/MIT/Northwestern/ Illinois collaboration. As for the second one, let us note that e.g. graphs implying the Δ-resonance contribute only ∿ 7 % to the amplitude in the case of the nucleon [8]. They may be enhanced in nuclei, where, moreover, genuine two-nucleon graphs may also contribute to the doubly radiative capture. A numerical estimate of this possibility should be certainly worked out. On the experimental side, one may trace down also the A-dependence of these collective effects, measured as yet only for A = 9 and 12.

B) Because of the small mean nuclear density and the relativily small momentum-transfer of the reaction which was investigated, the identification of the pion-condensates may be performed with a reaction which can map out the momentum-transfer region $q^2 = (2.5\, m_{\pi})^2$ [9]. Of course, in-flight doubly radiative capture could be considered for this purpose (cfr. also the contribution of B. Cammarata to this conference), but the strong photon background from charge-exchange reactions would impose stringent requirements on the precision with which the photon's momentum should measured. Other reactions may turn out to be preferable in this respect.

Discussions with D. Beder, P. Christillin, J. Pestieau and A. Weise clarified my understanding of the theoretical aspects of the problem ; I am indebted also to M. Hasinoff and M. Salomon for informations on the TRIUMF-experiment.

REFERENCES

1. A.B. Migdal, Rev. Mod. Phys. 50, 107 (1978) and refs. cited therein
2. T. Ericson and C. Wilkin, Phys. Letters 57B, 345 (1975) ; the search for pion-capture analogous to induced pseudo-scalar muon-capture was first suggested by L. Grenacs
3. J. Delorme and M. Ericson, Phys. Letters 76B, 182 (1978)
4. P.L. Roberson et al., Phys. Letters 70B, 935 (1977)
5. J. Deutsch et al., Phys. Letters 80B, 347 (1979)
6. B. Bassaleck et al., contribution to this conference
7. S. Barshay, Phys. Letters 78B, 384 (1978) and 79B, 514 (1978)
8. D. Beder, Univ. of British Columbia preprint, November 1978 and private communication
9. A. Weise, private communication and Toki and Weise, contribution to this conference

Question: M. Furic, Rice University

Instead of concentrating on an absolute value discrepancy between the theory and the experiment, you may compare the trends in some specific variables. This may discriminate between various calculations especially those which use different pionic fields.

Response: Deutsch

I completely agree. I mentioned already the angular distribution of the two photons as one of these specific variables, but there are many others; e.g. the inter-photon energy-sharing. The predictions however, are not yet available.

Question: H. Jackson, Argonne National Lab.

The statement which Christillin and Ericsen make in their contribution to this conference is very strong. The Bremsstrahlung process should overwhelm the pion annihilation process. It would appear that your experimental results contradicts their conclusion. What is your feeling about their calculation?

Response: Deutsch

The main contradiction between our result and their preliminary branching ratio using all graphs is a factor of 2. There is another, indirect, contradiction between their (and other) evaluations of the annihilation graph which alone should lead to a branching ratio much smaller than $6x10^{-6}$ according to their contribution. One should keep in mind, however, that; (1) they are the first to compute in a finite nucleus; and (2) the relative contribution of the different graphs is not guage invariant.

Question: H. Jackson, Argonne National Lab.

It seems there is another background reaction that must be considered. If you look at the one photon pion capture for carbon there is an appreciable probability of emitting a capture photon, leaving the residual nucleus in a highly excited state, perhaps $E_{ex} \approx 40-50$ MeV. Such a state could decay by emitting a second decay photon. What is your reasoning for neglecting this two photon branch of the reaction?

Response: Deutsch

As radiative capture decreases with increasing excitation energy of the residual nucleus, and the gamma-width of the highly excited levels produced is typically 10^{-3} or less of the corresponding particle width, I would not expect this background to exceed a branching ratio of 10^{-6}. Effectively, we found no indication for a trend toward unequal energy sharing behavior of the two photons, which would be the signature of such contributions.

Question: K. Seth, Northwestern University

The main physics interest which motivated all these very difficult experiments, was annihilation γ's and not the Bremmstrahlung γ's. If the two can not be meaningfully separated, either from the theoretical or the experimental point of view, what is the interest left? What does, even a very successful experiment tell?

Response: Deutsch

The interest of the experiment was to test the pion content in nuclei, or if you prefer, to measure the pion nuclear coupling in a nuclear medium. As all the graphs mentioned introduced this coupling linearly in the amplitude, this interest still prevails. The theoretical treatment of the problem becomes, however, more complicated. One should note also that this coupling can be approached by easier probes, such as single-radiative pion capture. So if the discrepancy between the results and the theory remains, we may have come upon an unexpected effect of collective nature.

Question: C. Wilkin, University College London

Do the present theoretical calculations include your experimental cuts?

Response: Deutsch

As far as I know for Christillin and Ericsen the answer is yes.

STRONG INTERACTIONS IN STRANGE HADRONIC ATOMS

S.Wycech
Institute for Nuclear Research, Warsaw, Poland

ABSTRACT

Strong shifts and widths of atomic levels in the K^- and Σ^- atoms are discussed together with their decay modes. An extraction of the elementary subthreshold ΣN scattering amplitudes from nuclear effects is attempted. Introduction to the subject is made in the cases of hadronic hydrogen and deuterium.

INTRODUCTION

Experimental studies of strange hadrons (K,Σ) scattering by nuclei are difficult at low energies due to their short lifetimes. Similar information may be obtained, in part, by studies of K^-, Σ^- stopped in matter and bound into atomic orbits. Details of the formation and electromagnetic cascade processes in atoms may be found in reviews[1]. Here, we recall that the cascade transitions push the orbital hadron to more and more circular orbits. In these states hadron is likely to reach low lying levels where nuclear interactions emerge and may be observed as shifts and widths of the electromagnetic levels. If electromagnetic and strong decay widths are comparable for a given level, the X-ray experiments may detect the strong width (called Γupper) by the intensity loss and the shift and width of the next lower level Γ_L, ΔE_L by the line shape measurements. Then cascade terminates due to the nuclear absorption. These favourable conditions are met only in a few regions of the periodic table. As the nuclear interactions of K and are very strong, these states have high angular momenta.L. Since the hadron wave functions behave as r^L at the origin the maximum overlap of nuclear and atomic densities occurs at nuclear surfaces. This raised high hopes for studies of the nuclear surface structures.

Absolute intensities of the kaonic X rays were reported by Wiegand[2] and showed that about half of mesons reach the bottom of cascade. The relative yields (Γ_U) and the lower shifts and widths were measured by several Laboratories[3]. Usually a simple optical potential $-2\pi/m_{HN}\, a\, \rho$ is used and the parameter a is extracted from the best fit. It is about .4+i.8 in contrast to the average scattering length -.15+i.68[4]. The difference in the sign of the

ISSN:0094-243X/79/540390-14$1.50

real parts is attributed to the $Y_0(1405)$ I=0 quasibound KN state decaying into $\Sigma\pi$ channels.

The first observations of hyperonic X-rays by Backenstoss [5] established constraints for Im a, but complete determination was made only recently by Batty [6]. Surprisingly, the interaction parameter and especially the absorptive part turns out very small a = .35(4) + i.18(3) fm in view of the expected $\Sigma^- p$ quasibound state the scattering length average 0.05(20) + i.49(20) and a_p = .43 + i.68 deduced from the final state interactions of hyperons produced in K^- capture [8, 12]. We argue in the following that the atomic parameters a are related to the hadron-nucleon (HN) scattering amplitudes in the subthreshold energy region and may be a test for those. There is a number of limitations, due to the nuclear physics involved, and we attempt to discuss some of them.

Emulsion techniques provide additional data on the decay modes of hadronic atoms. These are initiated by inelastic elementary reactions.

$$K^- p \rightarrow \begin{matrix} \Sigma^- \pi^+ \\ \Sigma^+ \pi^- \end{matrix} \qquad KNN \rightarrow N\Sigma(\Lambda) \qquad \Sigma^- p \rightarrow \Lambda n$$

$$K^- n \rightarrow \begin{matrix} \Sigma^0 \pi^- \\ \Sigma^- \pi^0 \end{matrix} \tag{1}$$

and momentum distributions of fast emitted particles are measured together with the branching ratios. Two long standing problems are: the indication of a thick neutron skin inferred from the ration Kn/Kp [9] and the hypothesis of strong correlation at the nuclear surface from the KN/KNN ratio [10]. Another useful ratio is $\Sigma^-\pi^+/\Sigma^+\pi^- = \gamma$ which allows to test the Yo behaviour in nuclear matter [11]. These questions cannot be convincingly answered before the understanding of the capture mechanism is reached.

To describe both the elastic and inelastic data, it is profitable to use the multichannel scattering or reaction matrix adapted to the nuclear medium conditions. We discuss this in the following, comparing our own calculations with related approaches [12-18]. The kaonic and hyperonic atoms are related by the strangeness. Full understanding of the K atom decays requires description of the hyperon final state interactions which are tested in the hyperonic atoms.

There exist other descriptions of the $\bar{K}$ nucleus scattering based on: folding of density and KN potential [19], black sphere model [20], and so far not applied to atomic problem dispersion approach [21] and isobar-doorway theory [22]. We touch some of these methods very briefly.

In the next section we explain some problems in the 2 and 3 body case and go over to construction of the nuclear optical model.

HADRONIC HYDROGEN AND DEUTERIUM

We describe briefly the two cases of simple nuclear systems. The aim is to find a relation of the atomic level shifts and widths to the hadron nucleon scattering matrix and expose some features characteristic of the "scattering at negative energies." For this purpose we develop a multiple scattering expansion for the reverse level shift. Imagine hadronic hydrogen first. It is described by 3 length parameters: Bohr radius B, radius of the HN potential r_o and HN scattering length A. The ratio $r_o/B \approx 10^{-3} - 10^{-2}$ is equivalent to small overlaps of the atomic and nuclear densities. We assume A/B to be comparably small. This means (see late) small shift/level spacings. Let A be real (no decay channels) for a moment. Equation for the n-th bound state is

$$|\Psi_n\rangle = G(E)\, V|\Psi_n\rangle \qquad (2)$$

where V is the strong potential and Coulomb interactions are included into the Green's function $G = (E - H_o - V_c)^{-1}$. Let us split it into three parts $G = G_o + G_n + g$ where G_o is the free propagator, $G_n = (E - E_n)^{-1}|n\rangle\langle n|$ projects on the given unperturbed atomic state n with the main q. number n and energy E_n and g is a small correction. Instead of eq 2 we formulate an equation for $|R_n\rangle = V|\Psi_n\rangle$. Recalling that the strong scattering matrix is $T = (1 - VG_o)^{-1}\, V$ we have

$$|R_n\rangle = \frac{T(E)}{E-E_n}|n\rangle\langle n|R_n\rangle + T\, g|R_n\rangle \qquad (3)$$

Perturbative expansion in g produces eq. for the level shift

$$\frac{\langle n|T(E)|n\rangle}{E - E_n} = 1 - \frac{\langle n|TgT|n\rangle}{\langle n|T|n\rangle} + \cdots \qquad (4)$$

with strong cancellations in the higher order terms. To the first order in A/B, we obtain old result [23] that level shifts are equivalent to scattering lengths. For S states

$$E - E_n \approx 4\bar{E}_n\, n^{-1}\, A/B\ . \qquad (5)$$

If strong potential is attractive enough to produce a bound state ($A < 0$) eq 5 yields repulsive shifts. This apparent paradox is due to the node of the wave function induced by the strongly boundstate. The kinetic energy gained in this way and an increase of the atomic radius off set the attraction. The case of $|A| > r_0$ typical for kaons is drawn in fig 1 for 1 S state. The strong accumulation of mesons around the centre is the main difficulty of the scattering on many body systems. Let there be another nucleon at a distance d. Three cases arise: 1° $|A| \ll d$ typical for extreme nuclear surfaces where quasi-free scattering applies 2° $|A| > d$ typical for nuclear centre where we expect average field approximations to work, 3° $|A| \approx d$ the most difficult situation of a partly dissolved HN bound state. The case of ΣN interaction differs slightly since $|A| \approx r_0$ and off shell extensions in momentum variables are more important.

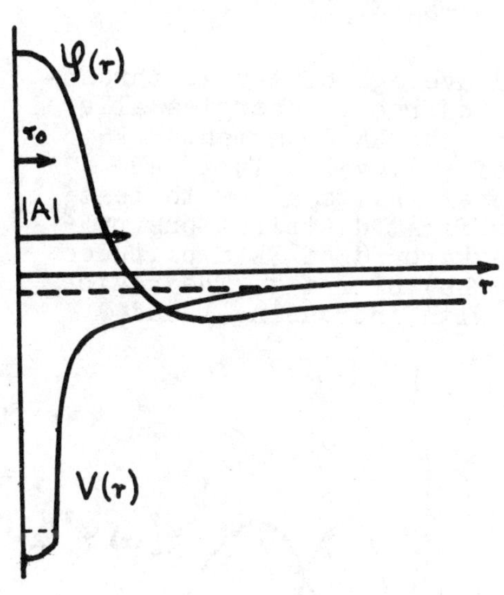

Fig 1. Wave function of an idealized atom.

When decay channels are switched on, the decay probability is determined via the unitarity by im A. Atomic wave functions become complex, but the effect discussed is similar. Eqs 4 and 5 determine also widths of the atomic levels. However, as shown by Ericson and Hambro, the line shapes for transition to the S states asymptotically differ from Lorentzian [24].

Some useful results and intuitions may be obtained in the deuterium case which we discuss in the limit of zero range S wave HN interactions. Eq (3) is now a 2 x 2 matrix one. The unperturbed state consists of deuteron and atomic state $|n\rangle = |\varphi_n \varphi_D\rangle$. The Green function g contains now free hadron propagator and projection on the deuteron $g_D = |\varphi_D\rangle\langle\varphi_D| (E-E \;\; -H^0_H)^{-1}$ and residual corrections disregarded for simplicity [25]. As usual in the 3 body problem, the energy argument of the scattering matrices involved is shifted from the energy shell $E_n + E$ by the HN centre of mass energy E . The average of the elastic one is

$$\langle n|T|n\rangle = \Theta \cdot \bar{T} \qquad \text{where} \qquad (6)$$

$$\Theta = \int \varphi_D^2(r)\varphi_n^2(r/2)d\vec{r} \;;\; \bar{T} = \int T(E - E_n - E_{CM}(p))F^2(p)d\vec{p}$$

are the overlap integral and average of the subthreshold amplitude over the range of recoil energies allowed by normalized distribution of the HN CM momentum F^2. In the first approximation of (4), valid for high atomic angular momenta L, level shifts allow to test subthreshold amplitudes. Few first distributions p^2F^2 are drawn in Fig 2. on the background of $\bar{K}N$ amplitudes $-m_{KN}T/2\pi$ (normalized to lengths) of ref [4]. The region of allowed energies narrows with increasing L and $\bar{T} \to T(E)$.

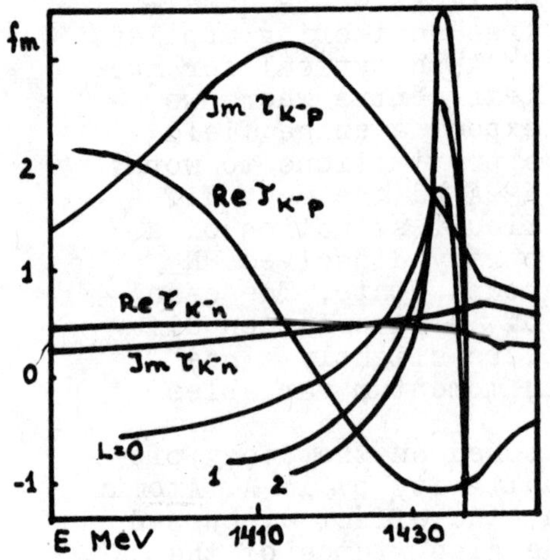

Fig 2. $\bar{K}N$ scattering matrices and averaging functions.

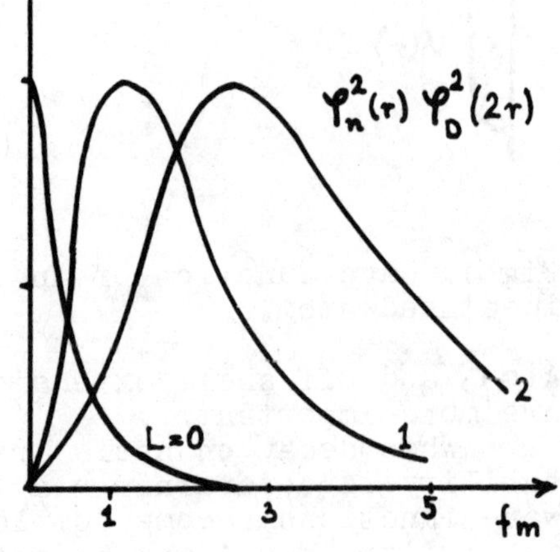

Fig 3. Overlap densities.

We also draw the overlap densities in fig 3. The L = 0 case is close to the deuteron density in CM system, L = 1 is typical for lower levels in heavier atoms, L = 2 is typical for the upper levels of kaons or lower levels of hyperons.

The next term of expansion (4) sums a whole class of multiple scattering processes so that the subsequent contributions are small. It contains scattering on nucleons frozen in the deuteron state, due to g , and scattering with free nucleons in the intermediate states. The level shift is

$$\frac{\Theta}{E-E_n} = \frac{1}{\bar{\bar{T}} + T_n} - \frac{1}{\Theta} \iint \bar{\varphi}_n(x)\, \varphi_D^2(x) g_A(x-y)\, \varphi_D^2(y) \varphi_n(y) - \tag{7}$$

$$- \frac{2\bar{T}_p\bar{T}_n}{(T_p + T_n)^2}\, \frac{1}{\Theta} \int \bar{\varphi}_n(x/2) g_D(x)\, \varphi_n(-x/2) \varphi_D^2(x)$$

where $g(r) = -m_{HD}\exp(-kr)/2\pi$ is the hadronic propagator at negative energy. In the first "frozen" part (we call it S) the damping factor k is related to the atomic binding energy. In the second, free scattering part it is given by the nuclear binding and two recoil energies $k^2 = 2m_{HD}|E_D - 2\bar{E}_{CM}|$. The damping is quite strong in this case.

In the particular case of $T_n = 0$, i.e., scattering on a bound proton the first contribution S may be interpreted as a dynamical shift of the resonance. If $T_p(E) = r^2(E - E_R + i\Gamma/2)^{-1}$ than from eqs 5 and 7 the effective scattering matrix on the bound proton $T_{BP}(E) = r^2(E - E_R + E_D - \bar{E}_{CM} - Sr^2 + i\Gamma/2)^{-1}$. The kinematical correction due to binding and recoil energies, $E_D - \bar{E}_{CM}$ is repulsive and pushes the position of the resonance closer to the threshold. For negative energies of hadron and in some region above threshold we see from (7) that S is negative and the dynamical shift is always attractive. This effect is due to the energy gap between the on shell conditions and the free spectrum in the intermediate states of HN intercations. It may damp completely the resonance behaviour of the scattering length $T_{BP}(o)$ considered as a function of binding energy E. For 1S states of kaons this was found by Revai [26] in a numerical way. However, for higher L states the resonance shift S falls down by order of magnitude for each L and resonance behaviour is recovered [25].

We are going to extend these intuitions to cases of heavier atoms.

FIRST ORDER OPTICAL POTENTIAL

We expect the first order optical potential to be approximately correct in the region of extreme nuclear surface. Provided there are no multinucleon correlations there the justification stems from the 3 body model of scattering by a bound nucleon. It is equivalent to the two first terms in eq 7. The next term describing virtual excitations of nucleon pairs is reduced by the exponential damping of the hadron propagator. This happens in the region of small average nuclear potentials where the damping distance k^{-1} is given by negative single particle and recoil energies. For nuclear densities of interest in the kaonic case (10% of the central one) the k is about 1 fm^{-1} and the pair correction due to the Pauli principle must be added to that.

In general, the H-N scattering is described by a matrix in channel indices of a four point functions. The elastic part produces optical potential related to eq (6)

$$V(\vec{x}\vec{x}'E) = \sum_n \int d\vec{r}d\vec{r}\ \tilde{T}_n(\vec{r}\vec{r}',\vec{R}-\vec{R}',E)\bar{\varphi}_n(\vec{x}-\vec{r})\varphi_n(\vec{x}'-\vec{r}') \qquad (8)$$

with

$$\tilde{T}_n(\vec{r}\vec{r}',\vec{R}-\vec{R}',E) = \int \frac{d\vec{p}}{(2\pi)^3} e^{i\vec{p}(\vec{R}-\vec{R}')} T(\vec{r},\vec{r}',E-E_n-E_{CM}(p))$$

where x denotes the hadron coordinates, r is the H-N relative, $R = x - rm_N/m_N + m_H$ is the HN centre of mass coordinate and φ_n, E_n are the single particle nucleon wave functions and energies. The r,r′ dependence is due to the range of the HN forces while R-R′ dependence describes a propagation of the HN pair relative to the centre of mass of the whole atom and is given by the Fourier transform over the HN momentum. If a resonance arises, a complex pole appears in T giving oscillatory behaviour and Yukawa-like damping in the space coordinates. The range of R-R′ allowed is about 1fm in the Y_0 case and is comparable to the thickness of the nuclear surface.

Level shifts and widths are obtained by solving the Schrodinger or Klein-Gordon equations with the electromagnetic and strong potentials. Technical problems arise as the potential is nonlocal due to the force range and the resonance propagation. A number of local equivalents may be defined. It was shown by Alberg [13] that first order of the standard one

$$V_L(\vec{x},\vec{x}') = \delta(\vec{x}-\vec{x}')\int d\vec{x}V(\vec{x},\vec{x}') + \vec{\partial}\delta(x-x')\int d\vec{x}V(\vec{x}\vec{x}')\vec{x}\ldots \qquad (10)$$

gives satisfactory results for short range S wave KN interactions. Another simple way is to compare Born terms of the nonlocal and local equivalent potentials as done by Bardeen and Torigoe [14]. This offers the best average over the resonance momentum but runs into difficulties when the optical potential changes sign. Although differences of various approximations are at 5% level we discuss it since we need an extension to longer range forces. A factorization of nonlocality and nonlinear effects may be obtained with the help of separable scattering matrix with Gauss farmfactors $T = v(m,r)T(E)v(m,r')$; $v = \exp(-m^2r^2)(m/\sqrt{\pi})^3$. The single nucleon mixed density appears in V of eq (8) if an average E_n is introduced. It may be factorized by $\rho(r,r') = \rho(R)J_R(r-r')$ where ρ is the density at $R = (r + r')/2$ and J is a function given by Campi [27]. The Fourier transform of J characterizes local distribution of nucleon momenta. In the surface region this differs strongly from the local Fermi gas counterpart. Application of both the factorization and the approximation (10) produces local potential

$$V(\vec{x}E) = -\frac{2\pi}{m_{HN}} a_x(E) f_x \rho_F(\vec{x}) \qquad \text{where}$$

$$\begin{aligned} \rho_F &= \int d\vec{y}\,\rho(\vec{x}-\vec{y})v(y,m\sqrt{2}) \\ f_x &= \int d\vec{y} J_x(y)v(y,m/\sqrt{2}) \\ a_x &= -\frac{m_{HN}}{2\pi}\int d\vec{p}\,T(E-E_x-E(p))W_x(p) \end{aligned} \tag{11}$$

and the momentum distribution w is determined by normalized product of the Fourier transforms of J and v. The average of single nucleon energies E_n weighted by the contribution of levels to the total nuclear density is denoted by E_x. This point is of some importance since in the kaonic case ReT crosses zero in the subthreshold region and a may change sign at the nuclear surface [12,13].

Optical potential is strongly nonlinear in density at the nuclear surface due to folding over the potential formfactor. This effect is very strong in case of hyperonic atoms which involve very low density region ($<$5%) and relatively long forces. The table produces the overlap integrals $\Theta' = 2\pi\Theta/m_{HN}$ for the ^{32}S, 4f states. The three values of range parameter m correspond to zero range, two pions and one pion exchange. Substantial differences of overlaps are fortunately balanced by the factor f of eq (11). The last column shows the a fitted to the experimental widths and shifts. It is quite stable

$m fm^{-1}$	Θ' KeV/fm	, f	Θ'f	a , fm
	1.21	1	1.21	.28 + i.15
.65	1.71	.87	1.49	.27 + i.18
.32	3.90	.59	2.29	.16 + i.16

but from the upper 5g level width we [31] obtain different values Im a = .14 , .12 or .08 fm. The upper levels are practically given by Born term and the dependence on force introduces considerable uncertainty. On the other hand, with the long range forces the differences due to various nuclear density shapes are considerably reduced. It is expected that KN potential is dominated by exchanges of heavy vector mesons and thus very short ranged. Similar estimation follows from the S wave dominance of the $\bar{K}$N scattering [4]. The overlap question is less severe but a few per cent uncertainty arises.

A choice must be made what the object of our study is: nuclear distributions or interaction parameters. In both the K and sigma cases interactions with protons are dominant. The second choice is favoured. In hyperon case we look for qualitative effect which we try to find later. In kaonic case one hopes for a more quantitative result. The calculations of Bardeen [14] show 10-20% differences of complex level shifts caused by various KN reaction matrix parameters. With increased experimental accuracy (10-20% errors are typical nowadays) one would hope for a strong test. This unfortunately is hampered by second order corrections.

We analyse the hyperon case in some detail since the topic is relatively new. Experiments of Batty [6] produced line shapes in five cases of light nuclei indicating very small absorption and an attractive optical potential. The attraction has been expected in many models but the absorption due to $\Sigma \rightarrow \Lambda$ conversion on protons was expected to be strongly enhanced by the $\Sigma^- p$, I = $^1/_2$, J = 1^+ quasibound state [28]. We analyse the data in terms of the potential (11) and deduce phenomenological values of a, for the likely two pion force range. In fig 4 Im a for 4f and 5g levels in S and Si are indicated at the average subthreshold energies calculated as in ref [14]. The subthreshold extrapolation of several I = $^1/_2$ absorptive amplitudes is given. The A, B, C are based on Alexander [7] scattering length for 3 channel $\Sigma^- p$, $\Sigma^0 n$, Λn problem and averaged over spin states. The extrapolation so far below the

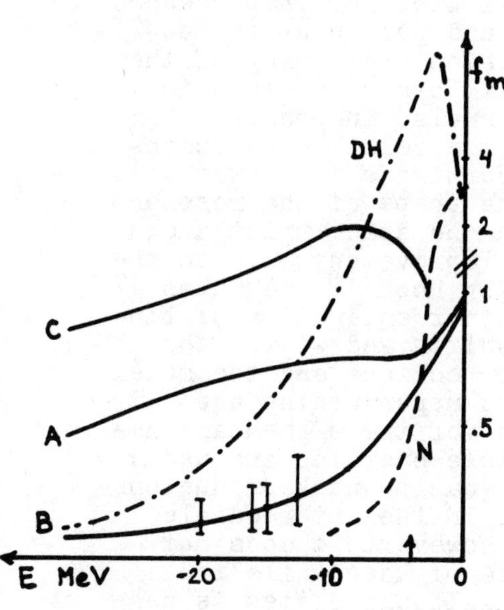

Fig 4. Absorptive ΣN amplitudes and atomic results.

threshold is quite uncertain, however. The DH is given by Dosch and Hepp [29] in J = 1 case and N is the dominant J = 1 amplitude of the sophisticated multichannel calculation by Nagels [30]. The last two resonant amplitudes fall by order of magnitude and as well as the nonresonant solution B are consistent with the atomic data. We do not feel able to make a definite decision in favour of one or the other description. A special difficulty concerns strong cancellation between the scattering amplitudes when calculating the value of Re a. To the available accuracy the results are consistent with the work of Nagels or Dosch. Direct extrapolation of B solution seems to be excluded as it gives repulsive potential. Stability of these results against higher order corrections is discussed in the next section.

HIGHER ORDER CORRECTIONS

Our aim is to find the hadron nucleon scattering matrix in the nuclear medium T. On the basis of previous discussion we reduce nuclear effects to the Pauli principle and HNN excitations. Within this the equation for T is

$$T = V_{HN} + V G_{HN} D T \qquad (12)$$

where G is the nucleon and D the hadron propagator in the nuclear medium. Optical potential is given by eq. 8 and in the next step it produces the hadron propagator D and the set of equations closes. To solve eq. 12 for T we need a multichannel potential. The phenomenological reaction matrix parameters of Dalitz and Tuan [32,4] fix the potential parameters pretty well and the off shell corrections are small. The program outlined was performed in refs 12, 13, 16, 17, 31. Two physical effects lie behind it: 1° the Pauli blocking that reduces the strength of KN interactions and pushes up the position of Y_0 resonance 2° collision broadening of the resonance due to interactions with other nucleons in the elastic and the decay $\Sigma\pi$ channels. In practice this problem reduces to finding the relative HN momentum in the nuclear medium. The Fermi gas model was used in early calculations [12, 13]. The shape of the resonance curve of fig 1 depends now on the space point and the total momentum of the pair. Improvements due to the shell structure were introduced by Rook [16] and Thies [17]. Changes are mainly due to the better estimation of binding effects. For lower levels shifts and widths the medium effects ammount to 25% corrections and the model dependence is at a 10% level. The upper widths are fairly insensitive and may be a test for the elementary amplitudes if correct single particle energies are used. This scheme may be inaccurate if quasibound kaon nucleus states arise [33] and would appear in the intermediate states of the interaction (12). However, it does not happen for higher angular momenta of kaon while the coupling to a lower L nuclear quasibound states is negligible.

The resonance Y_0 is treated as an elementary object in papers by Eisenberg [18] and Brockman [15]. However, it is allowed to dissolve into $\Sigma\pi$ [18] and KN [15] states. Nuclear corrections, expressed as resonance shifts are calculated in this way and turn out bigger than in the potential approach (20-30 MeV shift [16]). Although the result may be correct these models are not consistent with the Dalitz Tuan description as they treat the interactions in the intermediate state in a different way.

Important corrections due to KNN$\rightarrow\Sigma$N virtual absorptions were treated systematically by Rook [17]. He shows that due to propagation of resonance the correction is linear rather than quadratic in densities. The widths are enhanced by 30% (lower) and 10% (upper).

The antisymmetrization of nucleon states in the course of HNN interactions should be introduced [12]. The physical effect is that around an interacting proton it is less likely to find another one and the collision broadening is reduced. It is observed in the decay pion hyperon channels by Vander Velde - Wilquet [8, 11]. The absorption of amitted Σ^- in the parent ^{12}C nucleus is 38% while that of Σ^+ is 62% and the difference is mostly attributed to correlation effects [12]. Another emulsion experiment [11] determines the branching ratio γ as discussed in the introduction. The results in heavy emulsion nuclei are drawn in fig 5. Corresponding curves are due to the free ratio [4] and to those with the

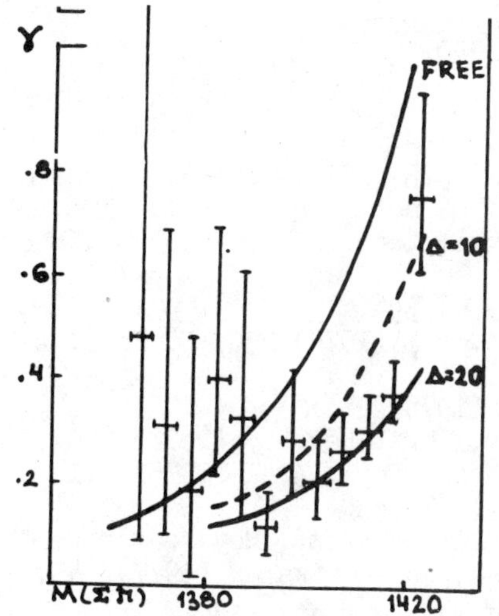

Fig 5. The γ branching ratio

resonance shifted by 10 and 20 MeV, all corrected for the final state interactions. It is likely that the resonance is shifted in nuclei but the conclusion is weak due to the errors and uncertain final state absorptions.

The medium effects on the hyperon nucleon resonant amplitude are quite strong [31], but atomic experiments involve energies below the narrow resonance and corrections are small in this region.

The conclusions of the previous section are thus maintained.

CONCLUSIONS

Kaons and hyperons provide means to study the structure of nuclear surface. However, hopes have not been fully accomplished yet. Instead, the atomic physics of kaons is rather concerned with understanding the interaction mechanisms. It seems also that the three experimental atomic numbers are not restrictive enough.

More precise and perhaps additional information on level shifts is needed.One of the methods proposed to obtain it is to use the E 2 mixing technique [34]. The studies of the decay products are not fully convincing yet as the interactions of the outgoing hyperons is not well controlled and the state of capture not well known. The measurement of X rays in coincidence with the decay products would be very useful here. Testing the KN elastic amplitudes seems uncertain on the theory side at a 20% level. The experiment in deuterium would be helpful for theorists. The measurement of of the KP scattering length in atomic or standard scattering experiments would fix some of the reaction matrix parameters.

The hyperonic atoms brought data on the ΣN interactions not accessible by other techniques. The attractive optical potential and the observed reduction of the sigma lambda coupling in the subthreshold region suggest existence of relatively long-lived sigma hypernuclei

REFERENCES

1. R. Seki, C. E. Wiegand, Ann. Rev. Nucl. Sci. 25 (1975) 241.
2. C. E. Wiegand, G. L. Godfrey, Phys. Rev. A9 (1974) 2282.
3. G. Backenstoss et all, Nucl. Phys. B73 (1973) 189; R. Kunselman, Phys. Lett. 34B (1971) 485; P. D. Barnes et all, Nucl. Phys. A231 477(1974); C. J. Batty et all, Phys. Lett. 60B (1976) 355. Nucl. Phys. A282 (1977) 487.
4. A. D. Martin, Phys. Lett. B65 (1976) 346.
5. G. Backenstoss et all, Phys. Lett. 33B (1970) 230, Z. Phys. A273 (1975) 137.
6. C. Batty et all, Phys. Lett. 74B (1978) 27.
7. G. Alexander, Y. Gell, I. Sturmer, Phys. Rev. D6 (1972) 2405.
8. C. Vander Velde-Wilquet et all, Nucl. Phys. A241 (1975) 511.
9. E. H. S. Burhop, Nucl. Phys. B1 (1967) 438.
10. D. H. Wilkinson in Proc. Rutherford Jub. Conf. Heywood co London 1961
11. D. H. Davis, D. N. Tovee, R. Nowak, Nukleonika 22 (1977) 845.
12. S. Wycech, Nucl. Phys. B28 (1971) 541.
13. M. Alberg, E. Henley, L. Wilets, Phys. Rev. Lett. 30 (73) 245, Ann. Phys. 96 (76) 43.
14. W. A. Bardeen, E. W. Torigoe, Phys. Rev. C3 (71) 1785, Phys. Lett. 38B (72) 135.
15. R. Brodkman, W. Weise, L. Tauscher, Nucl. Phys. A308 (1978) 365.
16. J. R. Rook, Nucl. Phys. A 249 (1975) 466.
17. M. Thies, Nucl. Phys. A 298 (1978) 344.
18. J. M. Eisenberg, Phys. Rev. C14 (1976) 2343.
19. A. Deloff, J. Law, Phys. Rev. C10 (1974) 1668; M. Atawashi et all, Progr. Th. Phys. 60 (1978) 209; R. Seki, Phys. Rev. C5 (1972) 1196.
20. H. Pilkuhn, Z. Phys. 1977.
21. O. Dumbrajs, reported to this conference.
22. M. Huber, Nukleonika 1/2 (1979); L. Kislinger, W. L. Wang, Ann. Phys. 99 (1976) 374.
23. T. L. Treuman, Nucl. Phys. 20 (1961) 57.
24. T. E. D. Ericson, Hambro L. Cern Th. Preprint 1977.
25. S. Wycech, Karlsruhe Univ. preprint 1974.
26. J. Revai, Phys. Lett. 33B (1970) 587.
27. X. Campi, A. Bouyssy, Phys. Lett. 73B (1978) 263.
28. Tai Ho Tan, Phys. Rev. D7 (1973) 600; P. Cline, R. Lauman, J. Mapp, Proc. Hyp. Phys. Argonne 1969, p. 62.
29. H. G. Dosch, V. Hepp, CERN Report TM 2310 (1977).
30. M. M. Nagels, T. A. Rijken, J. J. de Swart, Phys. Rev. D15 (1977) 2547.
31. S. Wycech, W. Stepien-Rudzka, J. R. Rook, to be published.
32. R. H. Dalitz, S. F. Tuan, Ann, Phys. 10 (1960) 307.
33. T. E. O. Ericson, F. Scheck, Nucl. Phys. B19 (1970) 450.
34. M. Leon, Phys. Lett. 53B (1974) 141.

Question: S. Hoath, Rutherford Laboratory

Comment: Data on K^- atoms measured at the Rutherford Laboratory should achieve approximately 10% uncertainties on basic scattering lengths. This analysis is being written up now.

Response: Wycech

This is to be very helpful, especially in cases of like nuclei.

PIONIC ATOMS AND STRONGLY BOUND STATES

E. Friedman
The Racah Institute of Physics
The Hebrew University of Jerusalem, Jerusalem, Israel

ABSTRACT

Recent precision measurements of strong interaction level shifts and widths in pionic atoms serve as the basis for new sets of parameters for the zero energy pion-nucleus potential. Different versions of the non-local component are tried with limiting forms of the Lorentz-Lorenz effect and of the dispersive term. Very good fits to the data are obtained throughout the periodic table. Many nuclei are found to meet the criticality condition for the existence of strongly bound pionic states. Some properties of these states are discussed with emphasis on the sensitivity to the range of the p-wave interaction.

INTRODUCTION

In this talk, I will present results of a recent analysis of strong interaction level shifts and widths for pionic atoms. The experimental basis of this analysis is provided by the precision measurements by Batty et al[1, 2]. The acuracy of these results is generally an order of magnitude better than the accuracies of previous measurements and it will be shown that the improved accuracy reveals several effects which were not observed before. The strong interaction level shift and width are described in terms of an effective zero range non-local pion-nucleus potential and parameters of that potential have been obtained from fits to the experimental data. The potentials so obtained are non-unique and several of them pose the unusual characteristic of having a "binding" kinetic energy in their non-local component, thus raising the question of the possible existence of strongly bound pionic states. Some properties will be discussed with special attention paid to the role played by the finite range of the non-local interaction.

EXPERIMENTAL

The experiments were performed using the stopping meson beam from the 7 GeV proton Synchrotron Nimrod. Fig. 1 shows the beam line which was originally designed for kaons[3], but was tuned for 200 MeV/c pions in the present experiments. Stopping pions were identified, as usual[4], with a counter telescope but an unusual feature of the beam line was a beam separator, based on crossed electrostatic and magnetic fields, which greatly reduced the electron and muon contamination of the beam, thus contributing significantly to the quality of the present data in terms of peak to background ratios. The detectors used were a planar 5 cm^3 Ge(Li) and a co-

ISSN:0094-243X/79/540404-14$1.50

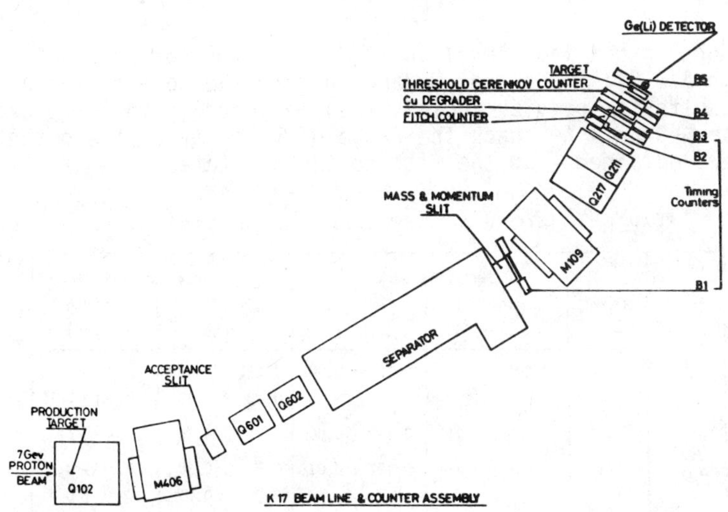

axial 70 cm^3 Ge(Li) detectors with 600 eV resolution at 122 keV and 1.68 keV resolution at 1.33 MeV, respectively. The experimentally determined response function of the detector was folded[5] with a Lorentzian and fitted to the strong interaction broadened peaks to obtain the energies and widths of the X-ray transitions. As an example, fig. 2 shows a nuclear gamma ray peak and a broadened pionic transition peak together with a fit to the latter. Small peaks due to gamma rays and muonic transitions are also shown in the vicinity of the pionic peak.

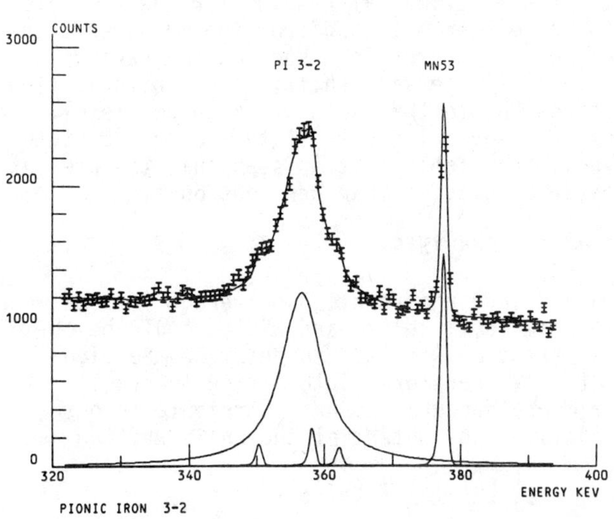

Fig. 2. An example of an X-ray spectrum and a fit to a broadened line.

RESULTS

Strong interaction level shifts and widths for 2p pionic states in Aℓ, Si, S, Ca, Fe, Cu and Zn form the core of the present work; results for heavier nuclei, also measured in the same experiments, served only to check the range of validity of the potential parameters obtained from the fits to the 2p states.

Table I Strong interaction shifts and widths

Element	Level	present work shift (keV)	present work width (keV)	previous work shift (keV)	previous work width (keV)
Aℓ	2p	0.201 ± 0.009	0.120 ± 0.007	0.212 ± 0.023	
				0.12 ± 0.10	0.36 ± 0.15
Si	2p	0.308 ± 0.010	0.192 ± 0.009	0.29 ± 0.15	
S	2p	0.635 ± 0.016	0.422 ± 0.018	0.54 ± 0.10	0.79 ± 0.15
				0.7 ± 0.3	0.8 ± 0.4
				0.502 ± 0.035	0.50 ± 0.06
Ca	2p	1.929 ± 0.019	1.590 ± 0.023	1.97 ± 0.18	2.00 ± 0.25
				1.6 ± 0.3	2.1 ± 0.6
Fe	2p	4.368 ± 0.113	6.87 ± 0.21	4.0 ± 0.3	8.7 ± 0.6
				4.4 ± 1.0	6.0 ± 2.5
Cu	2p	6.67 ± 0.28	11.4 ± 0.8	7.0 ± 2.0	15.9 ± 4.0
Zn	2p	6.44 ± 0.38	12.4 ± 1.4	8.0 ± 3.0	16.8 ± 6.0

References to previous work can be found in Ref. 2.

Table I summarizes the experimental results for the 2p states. These shifts and widths have been obtained from the measured transitions[2] by applying a host of corrections such as the various electromagnetic terms (e.g. finite size shift, vacuum polarization) and also small corrections due to the width of the upper level. In some cases transitions from several higher states to the 2p state have also been observed. From Table I it is seen that the present results have much greater accuracies than previous ones.

ANALYSIS

It has been shown a decade ago[6] that good average fits between calculated and experimental level shifts and widths could be obtained if an effective potential was used to describe the pion-nucleus interaction. It is, therefore, interesting to see if the greatly improved experimental accuracies can contribute to our knowledge of that potential. The potential chosen is written as

$$V(r) = \frac{1}{2\mu}\left[q(r) + \vec{\nabla}\alpha(r)\cdot\vec{\nabla}\right] \qquad (1)$$

where μ is the reduced mass, q is the local component and α the non-local component of the potential. The latter results from the predominant p-wave term $\vec{k}\cdot\vec{k}'$ in the pion-nucleon interaction. This potential is inserted linearly into the Klein-Gordon equation which is solved numerically including, of course, the potential due to the realistic charge distribution and also vacuum polarization. Calculated level shifts and widths are obtained by comparing binding energies calculated with and without the strong interaction potential (1). The local potential q(r) is parameterised in terms of the neutron (ρ_n) and proton (ρ_p) density distributions as follows

$$q = -4\pi(1+\frac{\mu}{m})\ [b_o(\rho_n+\rho_p) + b_1(\rho_n-\rho_p)]\ -4\pi(1+\frac{\mu}{2m})\ 4\ B_o\rho_n\rho_p \quad (2)$$

where m is the nucleon mass and b_o, b_1 and B_o are parameters determined from fits to the data. The parameter B_o is a complex number whose imaginary part represents the absorption of pions on pairs of nucleons.

The non-local part of the potential is usually written as

$$\alpha = \alpha_o/(1 + \frac{1}{3}\ \xi\ \alpha_o) \quad (3)$$

where ξ represents the Lorentz-Lorenz (LL) effect[6]. The term α_o is parameterised as follows

$$\alpha_o = 4\pi(1+\frac{\mu}{m})^{-1}\ [c_o(\rho_n+\rho_p) + c_1(\rho_n-\rho_p)] + 4\pi(1+\frac{\mu}{2m})^{-1}\ 4C_o\rho_n\rho_p \quad (4)$$

where again c_o, c_1 and C_o are parameters determined from fits to the data. The parameter C_o is a complex number in analogy with B_o. The data was also analyzed with a different form of the non-local component of the potential where the LL effect is included only in the term linear in the nuclear density. In this case α is written as

$$\alpha = \alpha_1/(1 + \frac{1}{3}\ \xi\ \alpha_1) + \alpha_2 \quad (5)$$

where

$$\alpha_1 = 4\pi(1 + \frac{\mu}{m})^{-1}\ [c_o(\rho_n + \rho_p) + c_1(\rho_n - \rho_p)] \quad (6)$$

$$\alpha_2 = 4\pi(1 + \frac{\mu}{2m})^{-1}\ 4\ C_o\ \rho_n\rho_p. \quad (7)$$

The 2p states cover the part of the periodic table which is most suitable for investigating the pion-nucleus interaction. This is due to:

(a) For 2p states the sensitivity of the shifts and widths to the strong interaction is greater than for any of the states with larger quantum numbers.

(b) The 1s states in lighter nuclei have very small sensitivity to the non-local component of the interaction. Also, the whole concept of an effective potential might be inappropriate for very light nuclei.

(c) The neutron density distribution which forms an essential ingredient of the potential is generally not very well known for heavier nuclei. It can be assumed to be equal to the corresponding proton distributions only up to about the Ca region.

It was considered preferable to start with nuclei where one may safely assume that the neutrons and protons have the same density distributions. Consequently, fits to the 2p data were made in two steps. First the parameters were adjusted to fit the Aℓ, Si, S and Ca data and then fits were made to the Fe, Cu and Zn data. Experience showed that any set of parameters which fitted the Ca data always gave very good agreement with the experimental results for Aℓ, Si and S. Therefore, the first step was essentially a fit to the Ca data. As the data consists of just two numbers (level shift and width) we could only adjust two parameters in the first step of the fitting procedure. The choice was to vary two out of the three parameters, c_o, ReC_o and ImC_o of the non-local potential, as these are the more relevant ones for the 2p state. Note that b_1 and c_1 are not effective for nuclei with equal numbers of neutrons and protons and have negligible effects in the case of Aℓ. The parameters b_o and B_o of the local potential were taken from a fit[7] to 1s states.

Four different sets of parameters were used in the analysis, which represent limiting forms of the non-local component of the potential, as described by eq. (3) and (5). (a)c_o and ImC_o were adjusted, keeping ReC_o=0. (b)c_o was held constant at its theoretical value[8] of $0.21m_\pi^{-3}$ and the complex number C_o was adjusted. In both cases the conventional form of the LL effect was chosen, (Eq. (3)) with $\xi = 1$. (c)For the third parameter set the LL effect (with $\xi = 1$) was included only in the term which is linear in the density (Eq. (5)) and again C_o was adjusted whilst keeping c_o constant at $0.21m_\pi^{-3}$. Finally, parameter set (d) was obtained with $\xi = 0$ (i.e. no LL effect) and again adjusting C_o only. For ρ_n and ρ_p the charge density distributions were used without unfolding the finite size of the proton. Repeating the analysis with density distributions where the finite size of the proton was unfolded leads to equally good fits to the data, although with slightly different potential parameters. The present approximation , in particular, should not affect our ability to compare neighboring elements or different isotopes of the same element. Figure 3 shows that a precise fit to the data for Aℓ, Si, S and Ca has been obtained.

In the second stage of the fitting procedure the data for Fe, Cu and Zn were used and only the coefficients of $(\rho_n-\rho_p)$(i.e. b_1 and c_1) were adjusted. For these nuclei, which have an excess of neutrons, the radial parameters of the neutron density distribution could no longer be considered as identical to those for the protons and the radius parameter of the neutron distributions was

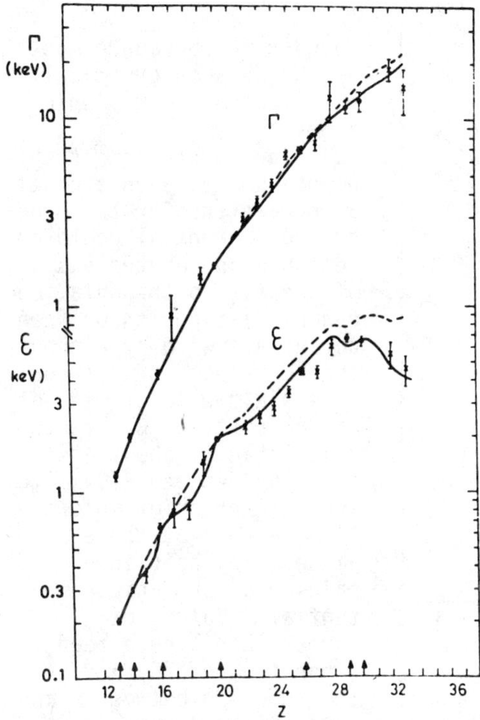

Fig. 3. Fits to the 2p data (continuous curve). The dashed curve is for $b_1 = -0.087m_\pi^{-1}$.

allowed to vary. Figure 4 shows an example for Fe where the correlation between r_n-r_p, the difference between neutron and proton rms radii, and the parameter b_1 is evident. An interesting conclusion is that $|b_1|$ had to be increased by 40% compared to its previous value in order to fit the data with reasonable values of r_n-r_p. The dashed curves are for a different version of the potential where $\rho_n\rho_p$ in the imaginary terms are replaced by $\frac{1}{2}\rho_p\rho_m$. It is seen that the $\rho_n\rho_p$ version is preferred; a conclusion which is obtained also for other nuclei. Results for all the 2p states are shown in Fig. 3 and the parameters obtained from the above fits are summarized in Table II. The extra structure displayed by the smooth curve in Fig. 3 as compared to the dashed curve is due to the increase in $|b_1|$. The improved agreement with experiment is evident.

Figure 5 shows experimental[10-14] and calculated shifts and widths for 1s states, using, as an example, parameter set (b) for the calculations, without any further adjustments. An excellent agreement is obtained, with the exception of ^{4}He and ^{7}Li. (Note that the shift for ^{3}He is negative).

A further check on the range of validity of the present parameters is provided by comparing predictions with the experimental results[2] for Ag, Cd, Pb, Bi and U, and good agreement is observed. However, unlike the lighter nuclei, it was found that whilst the predictions for parameter sets (a), (b) and (c) were very similar to each other, the predictions of parameter set (d) differ by several standard deviations. Consequently, whereas agreement with experiment was achieved for sets (a), (b) and (c) using values r_n-r_p in the range of 0 to 0.1fm, it was necessary to increase r_n by a further 0.15 to 0.2 fm in order to get agreement for parameter set (d). The present potential parameters were also

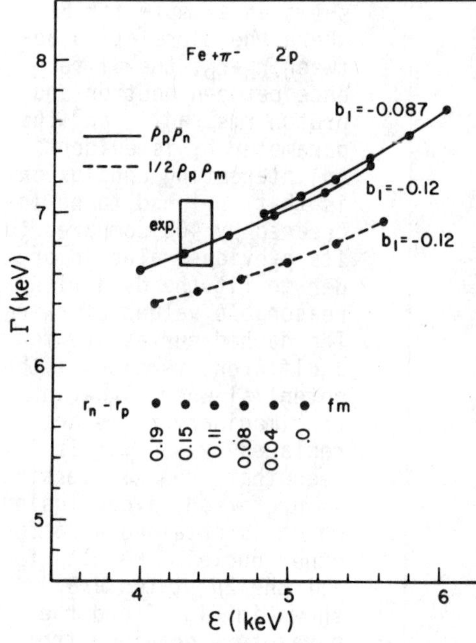

Fig. 4. Fits to the Fe data. The dashed curves are for $\frac{1}{2}\rho_p\rho_m$ replacing $\rho_p\rho_n$ in the potential.

found[15] to reproduce most of the data on dynamical polarization in pionic atoms.

The conclusion at this point is that a unique set of parameters for the pion-nucleus potential could not be found and on the basis of the fit to the data, one cannot distinguish between potentials with or without the LL effect and between potentials with or without a real part for C_o. Note that Table II shows a coupling between ReC_o, ImC_o and ξ. Any prior information on some of these parameters may help in obtaining a more unique potential. Values of ξ larger than 1 have been suggested before[17,18] and the coupling between ξ and ReC_o observed here is in good agreement with that observed in a recent microscopic calculation[19]

Table II Parameters for effective pion-nucleus potential

The conventional units of m_π^{-1} = 1.4138fm are used

	(a)	(b)	(c)	(d)
b_o (m_π^{-1})	-0.017	-0.017	-0.017	-0.017
b_1 (m_π^{-1})	-0.12 ± 0.02	-0.12 ± 0.02	-0.12 ± 0.02	-0.13 ± 0.02
ImB_o (m_π^{-4})	0.0475	0.0475	0.0475	0.0475
ReB_o/ImB_o	-1.0	-1.0	-1.0	-1.0
c_o (m_π^{-3})	0.250 ± 0.003	0.21	0.21	0.21
c_1 (m_π^{-3})	0.17	0.17	0.17	0.17
ImC_o (m_π^{-6})	0.090 ± 0.005	0.080 ± 0.005	0.0425 ± 0.005	0.0425 ± 0.005
ReC_o/ImC_o	0.0	1.7 ± 0.1	1.95 ± 0.05	-0.80 ± 0.05

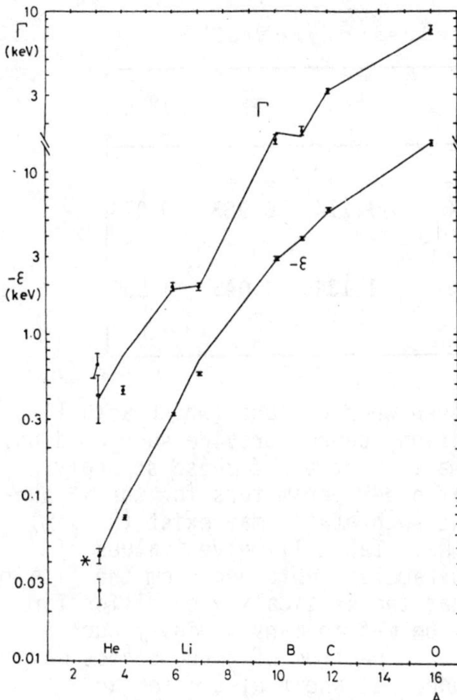

where for a large value of ξ a $ReC_0 > 0$ is obtained. The present type of potential (eq. (1)-(7)) is capable of successfully reproducing the experimental results for pionic atoms throughout the periodic table and therefore it may have some predictive power.

Fig. 5. Comparisons between predictions and available data for 1s states.

STRONGLY BOUND PIONS

An interesting phenomenon in connection with the zero-energy pion-nucleus interaction has been pointed out very recently by Ericson and Myhrer[20]. The Schrödinger equation with the potential (1)

$$(\nabla^2 - \kappa^2)\psi = [q + \vec{\nabla}\boldsymbol{\alpha}\cdot\vec{\nabla}]\psi \tag{8}$$

with $\kappa^2 = 2\mu B$, B being the binding energy, can be rewritten as follows

$$\vec{\nabla}\cdot(1-\alpha(r))\vec{\nabla}\psi - \kappa^2\psi = q(r)\psi. \tag{9}$$

Multiplying on the left by ψ^* and integrating to infinity, assuming ψ represents a bound state, we find

$$2\mu B \int|\psi|^2 d\vec{r} = -\int q(r)|\psi|^2 d\vec{r} - \int(1-\alpha(r))|\vec{\nabla}\psi|^2 d\vec{r}. \tag{10}$$

The second term on the right shows that if over some part of the nucleus 1-Re $\alpha<0$, then the $|\vec{\nabla}\psi|^2$ term can contribute towards

Table III Values of Reα(0)

Nucleus	^{27}Al	^{40}Ca	^{44}Ca	Pb	^{209}Bi
set (c)	1.068	1.028	1.114	0.983	0.974
set (d)	1.063	1.009	1.134	1.045	1.036

the strong binding of the pion even when q alone (which actually contains also the Coulomb attraction) cannot provide such binding. Ericson and Myhrer[20] discuss some properties of these strongly bound states, and on the basis of older parameters for the π^- nucleus potential they conclude that such states may exist (Re $\alpha>1$) only in exotic nuclei such as ^{32}Na. Table III gives values of Re $\alpha(0)$ for some nuclei, using parameters obtained from the fits to pionic atom data. It is seen that the criticality condition for the existence of such states may be met in many ordinary nuclei.

Some general properties of the spectrum of bound states may be obtained from a simplified model, as shown by Ericson and Myhrer[20]. In this model α = const. for $r\leq R$ and $\psi=0$ for $r\geq R$. Eq. (10) can then be written as

$$(\alpha-1)K^2 - q - \kappa^2 = 0 \tag{11}$$

where K is an "inner" wave number such that $\nabla^2\psi = -K^2\psi$. With the above boundary conditions K is determined from $j_\ell(K_{\ell n}R) = 0$ and then $K_{\ell n} = \pi\beta_{\ell n}/R$ where $\pi\beta_{\ell n}$ is the n-th zero of the Bessel function j_ℓ. For large n, $\beta_{\ell n} \to (n + \ell/2)$. The binding energy is given, within this model, by

$$B = -\frac{q}{2\mu} + (\alpha - 1)\frac{\pi^2\beta^2_{\ell n}}{2\mu R^2} \tag{12}$$

and it is clear that for Reα > 1 there is an infinite number of bound states in a "reversed order", i.e. B increases with increasing n and ℓ. These results, and many others, are derived in a more general way[21] where it is also shown that the WKB scheme can provide meaningful approximations.

Calculations of strongly bound pionic states have been made[21,22] using parameter sets (c) and (d) and realistic nucleon distributions. Table IV shows some examples of numerical results, where n_i is the number of nodes of the wave function in the internal region, i.e. where Reα > 1. The general picture obtained from

Table IV Strongly bound pionic states in ^{209}Bi

n_i	ℓ	B(MeV)	Γ(MeV)
0*	0	1.73	25.3
	1	1.78	34.4
	2	2.04	45.6
1	0	1.01	50.7
	1	2.10	67.1
	2	3.22	85.4
2	0	4.41	89.8
	1	6.19	113
	2	7.87	138

*for n_i=0 the value of b_1 was $-0.11m_\pi^{-1}$

the above model is indeed observed. It appears that heavy nuclei form the best candidates for such states to occur although the very large widths will make an experimental identification difficult.

The infinite number of bound states is a consequence of the zero range of the non-local interaction. The effects of a finite range can be easily studied[21] within the above simple model. For simplicity we drop now the local potential q. In momentum space, the $\vec{k}\cdot\vec{k}'$ term is replaced by

$$\vec{k}\,\frac{\Lambda^2}{\Lambda^2 + k^2}\cdot\frac{\Lambda^2}{\Lambda^2 + k'^2}\,\vec{k}' \tag{13}$$

where the form factor $\Lambda^2/(\Lambda^2 + k^2)$ suppresses high-momentum terms and ensures that for $\Lambda \to \infty$ the zero-range limit is obtained. With the Fourier transform of this form factor,

$$g(r) = \frac{1}{(2\pi)^3}\int \frac{\Lambda^2}{\Lambda^2 + k^2}\, e^{i\vec{k}\cdot\vec{r}}\, d\vec{k} = \Lambda^2\,\frac{e^{-\Lambda r}}{4\pi r}\,, \tag{14}$$

the $\vec{\nabla}\alpha\vec{\nabla}\psi$ of the zero range version is replaced by

$$\vec{\nabla}\int g(\vec{r} - \vec{r}')\,\alpha(r')g(\vec{r}' - \vec{r}'')\vec{\nabla}''\psi(\vec{r}'')d\vec{r}'\,d\vec{r}'' \quad . \tag{15}$$

We now operate twice with $(\nabla^2 - \Lambda^2)$ on the Schrödinger equation which contains (15). Noting that g(r) (eq. (14)) satisfies the equation

$$\nabla^2 g - \Lambda^2 g = -\Lambda^2\delta(\vec{r}) \tag{16}$$

one arrives at the equation

$$(\nabla^2 - \Lambda^2)^2\,\nabla^2\psi - \Lambda^4\vec{\nabla}\alpha\vec{\nabla}\psi - (\nabla^2 - \Lambda^2)^2\kappa^2\psi = 0 \tag{17}$$

which can be rewritten as

$$\frac{1}{\Lambda^4}[\nabla^6 - \kappa^2\nabla^4]\psi + \frac{1}{\Lambda^2}[-2\nabla^2 + 2\kappa^2\nabla^2]\psi + [\nabla^2 - \vec{\nabla}\alpha\vec{\nabla} - \kappa^2]\psi = 0. \quad (18)$$

When $\Lambda \to \infty$ the zero range is obtained and the last term is indeed eq. (9) (where q has been dropped). The first finite range correction is obtained by retaining only the term Λ^{-2}. The wave equation is then

$$\frac{2}{\Lambda^2}(-\nabla^2 + \kappa^2)\nabla^2\psi + \vec{\nabla}(1 - \alpha)\vec{\nabla}\psi - \kappa^2\psi = 0 \; . \quad (19)$$

Within the simplified model which leads to eq.(11) for $\Lambda \to \infty$ we write $\nabla^2\psi = -K^2\psi$ for $r \leqslant R$ and $\psi = 0$ for $r \geqslant R$. Eq. (19) then becomes

$$(\frac{2K^2}{\Lambda^2} + 1)\;(\kappa^2 + K^2) - \alpha K^2 = 0 \quad (20)$$

Rewriting eq. (20) as

$$\kappa^2 = -K^2 + K^2\,\frac{\alpha\,\Lambda^2}{2K^2 + \Lambda^2} \quad (21)$$

the condition for a positive binding energy is

$$(\alpha - 1)\Lambda^2 > 2K^2 \; . \quad (22)$$

With

$$K_{\ell n} = \frac{\pi\beta_{\ell n}}{R} \simeq \frac{\pi}{R}\,(n + \frac{\ell}{2}) \quad (23)$$

there results an upper limit on the radial quantum number n for a bound state. For example, for $\ell = 0$ one gets

$$n < \frac{R\Lambda}{\pi}\,(\frac{\alpha - 1}{2})^{1/2} \; . \quad (24)$$

From eq. (21) it is also found that the maximum binding energy is obtained for $K^2/\Lambda^2 = (\sqrt{\alpha} - 1)/2$ and that maximum is given by $(\sqrt{\alpha} - 1)^2\Lambda^2/4\mu$. Table V shows some examples for $\alpha - 1 = 0.1$ and $R = 5$ fm, which is typical of heavy nuclei. It is seen that only 1-2 states survive. Recent numerical results by de Takascy[23], who used a finite range potential, support this conclusion. However, he did not check whether his potential reproduced the results for pionic atoms.

The obvious next step in the problem of the possible existence of strongly bound pionic states is to analyze the pionic atom data with a finite range potential or with an alternative form of the

Table V Results for finite range uniform model
(α-1=0.1,R=5fm)

Λ (Mev/c)	500	800	1000	1200	1500
n_{max}	0.90	1.44	1.80	2.16	2.71
B_{max} (Mev)	1.07	2.73	4.27	6.14	9.60

potential, such as the Laplacian form. A good fit to the recent precision results for pionic atoms is a prerequisite for any potential aimed at describing the zero-energy pion nucleus interaction.

I wish to thank the many colleagues who contributed so much to the work described here and in particular to the team at the Rutherford Laboratory headed by C.J. Batty and to A. Gal and V.B. Mandelzweig at the Racah Institute of Physics.

REFERENCES

1. C.J. Batty, S.F. Biagi, E. Friedman, S.D. Hoath, J.D. Davies, G.J. Pyle and G.T.A. Squier, Phys. Rev. Lett. 40, 931 (1978).
2. C.J. Batty, S.F. Biagi, E. Friedman, S.D. Hoath, J.D. Davies, G.J. Pyle, G.T.A. Squier, D.M. Asbury and A. Guberman, Nucl. Phys. A, to be published.
3. M.K. Craddock, C.J. Reason and R.A.J. Riddle, Rutherford Laboratory report RL-76-091 (unpublished).
4. C.J. Batty, S.F. Biagi, M. Blecher, R.A.J. Riddle, B.L. Roberts, J.D. Davies, G.J. Pyle, G.T.A. Squier and D.M. Asbury, Nucl. Phys. A282, 487 (1977).
5. C.J. Batty, S.D. Hoath and B.L. Roberts, Nucl. Inst. Methods 137, 179 (1976).
6. M. Ericson and T.E.O. Ericson, Annals of Physics 36, 323 (1966); M. Krell and T.E.O. Ericson, Nucl. Phys. B11, 521 (1969).
7. M. Tauscher and W. Schneider, Z. Phys. 271, 409 (1974).
8. J. Hüfner, Phys. Rep. 21C, 1 (1975).
9. C.W. deJager, H. deVries and C. deVries, Atomic Data and Nucl. Data Tables 14, 479 (1974).
10. R. Abela, G. Backenst oss, A. Brandao d'Oliveria, M. Izycki, H.O. Meyer, I. Schwanner, L. Tauscher, P. Blüm, W. Fetcher, D. Gotta, H. Koch, H. Poth and L.M. Simons, Phys. Lett 68B, 429 (1977).
11. G.R. Mason, G.A. Beer, D.A. Bryman, M.S. Dixit, S.K. Kim, J.A.

MacDonald, A. Olin, R.M. Pearce, M. Krell, J.S. Vincent, Phys. Lett. 74B, 179 (1978).

12. G. Backenstoss, J. Egger, T. von Egidy, K. Hagelberg, C.J. Herrlander, H. Koch, H.P. Povel, A. Schwitter and L. Tauscher, Nucl. Phys. A232, 519 (1974).
13. G. Backenstoss, I. Bergström, J. Egger, R. Hagelberg, C.J. Herrlander, H. Koch, H.P. Povel, R.H. Price, A. Schwitter and L. Tauscher, Nucl. Phys. B66, 125 (1973).
14. G. Backenstoss, S. Charalambus, H. Daniel, H. Koch, G. Poelz, H. Schmitt and L. Tauscher, Phys. Lett. 25B, 365 (1967).
15. J.F. Dubach, E.J. Moniz and G.D. Nixon, submitted to Phys. Rev. (1978).
16. M. Leon, Nucl. Phys. A260, 461 (1976).
17. G. Baym and G.E. Brown, Nucl. Phys. A247, 395 (1975).
18. J. Delorme and M. Ericson, Phys. Lett. 60B, 451 (1976).
19. E. Oset, W. Weise and R. Brockman, preprint.
20. T.E.O. Ericson and F. Myhrer, Phys. Lett. 74B, 163 (1978).
21. V.B. Mandelzweig, A. Gal and E. Friedman, to be published.
22. E. Friedman, A. Gal and V.B. Mandelzweig, Phys. Rev. Lett. 41, 794 (1978).
23. N. deTakacsy, preprint 1978.

<u>Question: E. Oset, University of Barcelona</u>

Is there any reason to use a term $\rho_p\rho_n$ instead of the standard ρ^2 term to account for pion absorption in pionic atoms?

<u>Response: Friedman</u>

As negative pions are absorbed predominantly by proton-neutron pairs the $\rho_p\rho_n$ parameterization is more appropriate than the ρ_m^2 one. Indeed we found that with the former we could not fit simultaneously self-conjugate nuclei and nuclei with an excess of neutrons whereas we succeeded when using the latter. The factor of 4 was included to enable direct comparisons between our C_0 and the old one.

<u>Question: R. Seki, California State University, Northridge</u>

I think that an order-of-magnitude measurement in π^--atoms are revolutionary. What has made youre measurements so accurate compared to the previous measurements?

<u>Response: Friedman</u>

The order-of-magnitude improvement in the accuracy of the experimental results is due to the combined effects of; (1) the resolution of the 70 cc GeLi detector (1.68 keV at 1.33 MeV); (2) the use of an electrostatic beam separator which effectively removed particles that may contribute to the background; and, (3) a more sophisticated curve fitting program which folded a Lorentzian line shape into an experimentally determined response function while handling simultanelusly several background peaks. One should also mention the stability of ADC's that are now available.

OBSERVATION OF 2γ AND 2e EMISSION IN NUCLEAR PION CAPTURE*

E. Mazzucato, B. Bassalleck, M.D. Hasinoff, T. Marks and M. Salomon
University of British Columbia, Vancouver, B.C., Canada, V6T 1W5

J.M. Poutissou
TRIUMF and Université de Montréal, Montreal, P.Q., Canada, H3C 3J7

INTRODUCTION

The possibility of the existence of rare electromagnetic processes in pionic atoms was suggested by Ericson and Wilkin[1] in 1975. Their estimate of the branching ratio (B.R.) for the $(\pi^-,2\gamma)$ reaction in a nucleus was $\approx 10^{-5} - 10^{-6}$ assuming that the reaction was due mainly to the annihilation of the π^- with a "virtual" π^+ inside the nucleus. A branching ratio $\approx 10^{-4}$ for electron pair emission at small opening angles (due to the internal conversion of radiative capture photons) and $\approx 10^{-5}$ at larger opening angles was also given.

The $(\pi^-,2\gamma)$ process has been suggested by other authors[2] as a method of testing for a pion condensate in nuclei.

An experimental upper limit for this process has been published by a SREL group[3] and the first observation of the γγ pairs was reported recently by Deutsch et al.[4] In this last paper the angular distribution of the γγ pairs emitted after π^- capture in C and Be was observed with a total B.R. = $(1.0 \pm 0.1) \times 10^{-5}$ and $(1.4 \pm 0.2) \times 10^{-5}$, respectively.

EXPERIMENTAL METHOD

We have looked for γγ and ee pairs at TRIUMF using 2 large NaI crystals and 2 lead glass Cerenkov counters as detectors. They were placed at +55°, +105°, -55°, and -135° with respect to the incident pion beam. The pion beam from the stopped π/μ channel at TRIUMF (M9) was tuned for 20 MeV pions; the average rate was $8 \times 10^5 \pi^-/\text{sec}$. A 3 counter telescope, defined the incoming beam. Counter 3 defined the time of the incoming pions and the energy deposited in it was recorded for each event.

A 1 cm thick carbon target was used to reduce the production of neutral pions since the charge exchange reaction in ^{12}C has a Q value of -9.3 MeV.

The γ counters had charge identification counters placed in front of them to recognize e^+e^- pairs. A veto counter was used to measure the actual π stop rate but it was removed during the data collecting part of the run, since the hydrogen in it was found to produce unwanted π^0 background. Time of flight was used to separate γ's and n's in the NaI crystals. The timing resolution between the beam counters and the NaI crystals was 2 ns FWHM. Large plastic counters covered the NaI crystals to reduce the cosmic ray background. Steel shielding between the large crystals was used to reduce cross-talk and other souces of background.

A valid event was defined as $(S1 \cdot S2 \cdot S3) \cdot (N_i \cdot N_j)$ where N_i are any of the γ ray counters. For each event, we recorded on magnetic tape

ISSN:0094-243X/79/540418-02$1.50

the energy and time of all N_i counters, the energy and time of S3, the signals in the various charge identification counters, S5 to S10, the signal in the cosmic ray counters, and the time of the proton beam pulse. This last signal was used to separate the pions (45%) from the electrons (50%) and muons (5%) in the beam.

The data were recorded over a 2 week running period with a total number of π stops = 2.6×10^{11}.

RESULTS

The analysis of the data is still in progress but some preliminary results are available.

Events were classified in charged-charged and neutral-neutral groups and N_iN_j pairs for different counters. The main sources of background (after all the soft-ware cuts have been applied) are random coincidences of several types. In all cases they amount to less than 30% of the real events. In particular the e^+e^- pairs have a very low background of less than 5% of the real events.

Our preliminary results for $\gamma\gamma$ pairs indicate branching ratios in reasonable agreement with Deutsch's results for a C target. The summed energy distributions for both $\gamma\gamma$ and ee pairs peak at approximately 115 MeV indicating that the residual nucleus is not highly excited. The widths of the distributions are reasonably narrow ($\Gamma \approx 35$ MeV). The branching ratio for ee pairs is of the same order of magnitude as that for the $\gamma\gamma$ pairs. Detailed results for the energy and angular distributions will be presented at the conference.

CONCLUSIONS

Our results, together with those of Deutsch et al[4)] indicate that the $(\pi^-,2\gamma)$ process certainly exists at approximately the level predicted by Ericson and Wilkin.[1)] It is not so clear, however, whether or not the only contribution to the process is due to $\pi^+\pi^-$ annihilation or virtual π° decay. Recent calculations for this process in hydrogen by D. Beder[5)] indicate that only $\sim$ 70% of the rate is due to annihilation and the remainder is due to single radiative capture plus bremsstrahlung diagrams. The importance of the condensate mechanism proposed by Barshay[2)] is not clear at the present time.

REFERENCES

1. T.E.O. Ericson and C. Wilkin, Phys. Lett. 57B (1975) 345.
2. S. Barshay, Phys. Lett. 78B (1978) 384 and references therein.
3. P. L. Roberson et al, Phys. Lett. 70B (1977) 35.
4. J. Deutsch et al., submitted to Phys. Lett.
5. D. Beder, submitted to Nucl. Phys. B.

*Work supported in part by the Natural Sciences and Engineering Research Council of Canada.

TWO PHOTON EMISSION IN PION CAPTURE ON ^{12}C

P. Christillin and T.E.O. Ericson
CERN - Geneva

ABSTRACT

The 2γ branching ratio, energy distribution and angular correlation are derived from a gauge invariant effective Hamiltonian using energy weighted sum rules.

Some time ago it was suggested[1] that doubly radiative capture of pions in nuclei could be a sensitive probe of the nuclear pionic degrees of freedom. In fact, in the long wave length limit the dominant mechanism is the annihilation of a π^- from an atomic 1s orbit with a virtual π^+ in the nucleus, so that the measurement of the detailed energy and angular distribution of the photons gives information on the virtual π^+.

The first experimental observation[2] of the process has recently been repeated in ^{9}Be and ^{12}C. This has prompted us to a detailed theoretical investigation of the process assuming a conventional nuclear picture, in the hope that any major discrepancy would signal anomalies in the nuclear meson field. In the nuclei in question, total 2p absorption is known to be important. Therefore, the ratio of emitted γ's per stopped pion is (for details and notation see Ref. 3):

$$R_{2\gamma} = \frac{\lambda_{2\gamma}(1s)}{\lambda_a(1s)} \omega_s + \frac{\lambda_{2\gamma}(2p)}{\lambda_a(2p)} \omega_p \quad (1)$$

We now construct an effective Hamiltonian for the process by a non-relativistic reduction of gauge invariant Feynman diagrams with 2γ emission. The rate, the 2γ angular correlations and the γ energy distribution is obtained using energy weighted sum rules over the nuclear excitation spectrum.

A verification of the procedure is the excellent agreement with experiment for the corresponding $R_{1\gamma}$.

The results may be summarized as follows: the doubly radiative capture in the 2p state is dominated by bremsstrahlung diagrams of single radiative capture which overwhelm the $\pi^-\pi^+$ annihilation graphs by close to a magnitude. Unless the annihilation process is not substantially enhanced little information on pionic structure can be directly extracted.

ISSN:0094-243X/79/540420-02$1.50

The present situation concerning pionic degrees of freedon is therefore similar to that of exchange effects in intermediate energy Compton scattering.[4] The total 2γ branching ratio is $R^{2\gamma} \simeq 6 \times 10^{-6}$ with a decreasing angular correlation in the backward direction.

Relevant experiments should be performed in lighter elements or for very low energy pions in flight.

REFERENCES

1. T. Ericson and C. Wilkin, Phys. Lett. 57B, 345 (1975).
2. J. Deutsch et al., Université de Louvain, preprint (1978).
3. H.W. Baer et al., Adv. in Nucl. Phys. 9, 177 (1977).
4. P. Christillin and M. Rosa-Clot, Nuovo Cimento 43A, 172 (1978).

TWO-PHOTON EMISSION IN NEGATIVE PION CAPTURE BY PROTONS

J. B. Cammarata*
Department of Physics
Virginia Polytechnic Institute
Blacksburg, Virginia 24061

ABSTRACT

Using a chiral-invariant Lagrangian we have evaluated the tree diagrams for $\pi^- p \to \gamma\gamma n$. With this model we have studied the feasibility of using an experiment which detects only the final state γ's to resolve two matters of theoretical interest: the off-mass-shell behavior of the πN amplitude and the sign of the $\pi^0 \to \gamma\gamma$ amplitude.

DISCUSSION

Recent experiments [1] have detected the simultaneous emission of two photons in the capture of negative pions by nuclei. The nuclear $(\pi^-, 2\gamma)$ process is interesting for several reasons. As discussed by Ericson and Wilkin[2], this reaction can serve as a means to probe the pion content of the nucleus, and it can provide a means to study π-nucleon interactions and pion propagation in the nuclear medium. Barshay[3] has shown further that, with appropriate kinematics, this π-capture process can provide a definite signal for the presence of a pion condensate in a nucleus.

In order to use effectively the $(\pi^-, 2\gamma)$ reaction to study these various nuclear effects it is essential that one have an accurate model of the single-nucleon reaction $\pi^- p \to \gamma\gamma n$. Given the success of a chiral-invariant Lagrangian model for pion photoproduction[4], it is natural to generalize this approach to the 2γ process. We have done this and have obtained an expression for the two-photon amplitude which is similar to the one derived by Lapidus and Musakhanov[5], who followed the method of Adler and Dothan[6].

The main diagram of interest in our study (which included all 10 diagrams of order α) describes the charge exchange process $\pi^- \to$ virtual $\pi^0 \to \gamma\gamma$. We have evaluated this diagram using a scattering length approximation and various dynamical models of the off-mass-shell πN amplitudes[7]. Our numerical work has been based on an assumed experiment in which only the final-state gammas are detected, and we have explored the feasibility of using such an experiment to resolve two matters of theoretical interest: (1) whether such an experiment can be used to determine

*Work supported in part by the National Science Foundation under Grant PHY 77-04408.

ISSN:0094-243X/79/540422-02$1.50

the sign of the $\pi^0 \to \gamma\gamma$ amplitude, since knowledge of the sign (and magnitude) can be used to distinguish between different quark models of hadrons[8], and (2) the sensitivity of the reaction at different energies to the off-mass-shell variation of the πN amplitude, to determine whether this process can serve as a test to differentiate between various models of this amplitude.

Our principal conclusions are that: (1) only in an experiment in which the polarizations and energies of the gammas are measured can the contribution of the virtual charge-exchange process be large relative to other processes when $q^2_{\pi^0} < 0.9m^2_\pi$, (2) this reaction is not a practicable way to distinguish between different models of the off-mass-shell amplitude since cross sections are ~1 picobarn/MeV for kinematic configurations in which the charge exchange process with virtual π^0's plays an appreciable role, and (3) a reasonably accurate experiment could determine the sign of the π^0 decay amplitude.

REFERENCES

[1]J. Deutsch et al., Proc. 7th Intern. Conf. on High-Energy Physics and Nuclear Structure, Zurich (Birkhauser, Basel, 1977); P. L. Roberson et al., Phys. Lett. 70B, 35 (1977).
[2]T. E. O. Ericson and C. Wilkin, Phys. Lett. 57B, 345 (1975).
[3]S. Barshay, Phys. Lett. 78B, 384 (1978).
[4]R. D. Peccei, Phys. Rev. 181, 1902 (1969).
[5]L. I. Lapidus and M. M. Musakhanov, Sov. J. Nucl. Phys. 15, 558 (1972).
[6]S. L. Adler and Y. Dothan, Phys. Rev. 151, 1267 (1966).
[7]M. K. Banerjee and J. B. Cammarata, Phys. Rev. C17, 1125 (1978).
[8]F. J. Gilman, Phys. Rev. 184, 1964 (1969); S. L. Adler, Phys. Rev. 177, 2426 (1969); S. Okubo, Phys. Rev. 179, 1629 (1969).

COLLECTIVE RESONANCES WITH ISOSPIN T=2,1 AND 0 IN PHOTOREACTIONS, ELECTRON SCATTERING AND RADIATIVE PION CAPTURE ON A=14 NUCLEI

H. R. Kissener
Zentralinstitut für Kernforschung, Rossendorf, DDR

R. A. Eramzhyan
Joint Institute for Nuclear Research, Dubna, USSR

ABSTRACT

Shell model predictions are given for photo- and low-q 180° electroexcitation and (π^-,γ) on A=14 nuclei. Strong T=2 resonances $(2^-,1^-)$ are expected at $E_x(^{14}C) \approx$ 24, 26 and 30 MeV and for the analog $^{14}C(\pi^-,\gamma)$ transitions. The (π^-,γ) g.s. branch and the analog M2 electroexcitation in ^{14}C should be weak.

INTRODUCTION

The A=14 isobars are well suited for search for collective states with isospin $T=T_{min}$, $T_{min}+1$ and $T_{min}+2$. Shell model calculations predicted strong T=2 resonances at rather low energy [1,2]. But neither the $^{13}C(p,\gamma_1)^{14}N(0^+,T=1)$ data above $E_x(^{14}N) \approx 25$ MeV [3] nor the ^{14}C electroexcitation function around the lowest T=2 state [4] shows indications of narrow resonances.

PHOTO EXCITATION

The ^{14}N g.s. photoresonance $(0^-1^-2^-,T=1)$ shows only little structure, peaking at $E_x \approx 23$ MeV (2^-). The T=0 and T=2 branches of the dipole resonance above the first 0^+,T=1 state are predicted to be separated in energy $(E_x(T=0) \lesssim 25$ MeV). The main T=2 dipole state is expected at $E_x \approx 28$ MeV, being shifted up by more than the energy of the initial level, and to contain a larger fraction of the corresponding dipole sum, as compared with the g.s. resonance peak. The T=2 resonance should not be very narrow (thresholds for isospin-allowed nucleon decay are about 23 MeV); excitation of this state by the (p,γ) reaction on ^{13}C (see ref. [3]) is isospin-forbidden and suppressed by the structure of the resonance state. Hence, T=2 resonances should rather be looked for in reactions on ^{14}C (see ref. [1] for predictions of partial $^{14}C(\gamma,n)$ cross sections).

ISSN:0094-243X/79/504424-02$1.50

ELECTROEXCITATION OF ^{14}C

The calculated ^{14}C electroexcitation function at $\theta = 180^{o}$ and $q \approx m_{\pi_+}$ agrees well with recent data [5] up to $E_x \approx 15$ MeV; the 1^+ assignment to the resonance at 11.3 MeV is consistent with the predicted M1 strength in this region. Strong T=2 resonances are expected around 24 MeV (M2) and 30 MeV (transversal E1). The lowest T=2 state (2^-, $E_x \approx 22.5$ MeV; not seen in ref. [4]) and the photo dipole state (1^-,2; 26 MeV) should be weakly excited at these values of q and θ.

THE $^{14}C(\pi^-,\gamma)$ REACTION

The calculated excitation function shows close analogy to the $T_>$ branch in the backward electroexcitation of ^{14}C at $q \approx m_\pi$. The main part of the $^{14}C(\pi^-,\gamma)$ yield is due to spin-dipole resonances 2^- and 1^-, with dominant branches to two states (2^-, 1.8 MeV and 1^-, $E_x \gtrsim$ 5 MeV) in ^{14}B which should show up in the neutron decay of the resonance. The g.s. branch is predicted to be weak. Since M1, $\Delta T=1$ excitations and 1s hole states in ^{14}C are suppressed, the (π^-,γ) strength distribution should be rather narrow, except for some background from the spin-quadrupole mode as in the ^{16}O case [6].

REFERENCES

1. H. R. Kissener, R. A. Eramzhyan and H. U. Jäger, Nucl. Phys. A207, 78 (1973)
2. J. D. Vergados, Nucl. Phys. A239, 271 (1975)
3. P. Paul, H. M. Kuan and E. K. Warburton, Nucl. Phys. A254, 1 (1975)
4. F. J. Kline, H. Crannell, J. M. Finn, P. L. Hallowell, J. T. O'Brien, C. Werntz, S. P. Fivozinsky, J. W. Lightbody Jr. and S. Penner, Il Nuovo Cim. 23A, 137 (1974)
5. Hall Crannell, J. M. Finn, P. L. Hallowell, J. T. O'Brien, N. Ensslin, L. W. Fagg, E. C. Jones Jr. and W. L. Bendel, Nucl. Phys. A278, 253 (1977)
6. R. A. Eramzhyan, M. Gmitro, R. A. Sakaev and L. A. Tosunjan, Nucl. Phys. A290, 294 (1977)

PHOTOPRODUCTION MEASUREMENTS OF π^+ NEAR THRESHOLD ON ^{3}He AND ^{2}H

P. Argan, G. Audit, A. Bloch, N. de Botton, C. Schuhl,
G. Tamas, C. Tzara, E. Vincent
DPh-N, CEN Saclay, BP 2, 91190 Gif-sur-Yvette, France

J. Deutsch, D. Favart, R. Prieels, B. Van Oystaeyen
Institut de Physique Corpusculaire, Université Catholique de Louvain
B-1348 Louvain-la-Neuve, Belgium

The two-and three-nucleon systems offer the ground for a significant study of nuclear structures generated by the nucleon-nucleon interaction. We present here a contribution to the knowledge of these systems, based on the fact that charged pion photoproduction at threshold is a weak probe measuring the spin-flip form factor of the nuclear target. We have thus measured[1,2] the cross-sections for reactions :

$$\gamma + {}^3\text{He} \to {}^3\text{H} + \pi^+, \qquad (1)$$

$$\gamma + {}^2\text{H} \to n + n + \pi^+, \qquad (2)$$

relatively to the elementary one :

$$\gamma + p \to n + \pi^+, \qquad (3)$$

at energies ranging from 1 to 5 MeV above threshold.

From these data, we extract the ratio of the slopes of the cross sections at threshold for reactions (1) and (3) :

$$\frac{a({}^3\text{He})}{a(q)} = \lim(q = 0)\ \frac{\sigma({}^3\text{He})\ k/qs}{\sigma(p)\ k/q} = 0.62 \pm 0.02.$$

As for reaction (2), because of the three-body final state, the cross section has a more complicated shape, but it can nevertheless be characterized by one number playing a role equivalent to the slope a(A), and which was measured with a ± 2.5 % accuracy.

In order to extract from the data a form factor, it was necessary to estimate correctly the modification to a plane wave calculation due to the pion rescattering. Using a multiple scattering expansion[3], we found a small effect (-1 %) for reaction (1) and a larger one (+ 10 %) for reaction (2). By applying the correction to the data on reaction (1), we obtain

$$|F(Q^2)|^2 = 0.52 \pm 0.02. \qquad (4)$$

This is 20 % lower than the form factor squared extracted from the magnetic electron scattering on ^{3}He and ^{3}H. Calculations based on realistic wave functions[4] yield form factors in general agreement

ISSN:0094-243X/79/540426-02$1.50

with the experimental one (4). These calculations corrected by the effect of the nucleon fermi-motion should result in a slightly lower threshold cross section improving the overall agreement with the experiment.

The cross section for reaction (2) has been computed[5] including the effect of the fermi motion of the nucleon, but with less scruting concerning the 2N wave functions. Once corrected for the pion multiple scattering, they fall 10 % above the measured cross section. The cross section for the threshold electrodisintegration :

$$e + d \rightarrow e' + n + p$$

(which measures the spin flip transition form factor). When computed in impulse approximation is about 20 % lower than the measured one.

Thus we observe in the case of the two- and three-nucleon targets that their spin-flip form factor as measured by threshold photoproduction is 10-15 % lower than the one measured by magnetic electron scattering, and agrees better with impulse approximation calculations. It indicates the following trend :

- pion photoproduction measures essentially the one-body spin flip form factor ;
- magnetic electron scattering is more sensitive to many-body corrections.

REFERENCES

1. P. Argan et al., to be published.
2. E.C. Booth et al., Phys. Lett. 66B, 236 (1977).
   G. Audit et al., Phys. Rev. C16, 1517 (1977).
3. N. de Botton and C. Tzara, Rapport interne DPh-N/HE/78-06.
4. A.C. Phillips and F. Roig, Nucl. Phys. A234, 378 (1974)
   B. Goulard, A. Laverne and J.D. Vergados, Phys. Rev. C18, 944 (1978).
5. J.V. Noble, Phys. Lett. 67B, 39 (1977).
   J.M. Laget, Nucl. Phys. A296, 388 (1977).

MEASUREMENT OF π° PHOTOPRODUCTION NEAR THRESHOLD ON LIGHT NUCLEI

P. Argan, G. Audit, N. de Botton, J.L. Faure, C. Schuhl, G. Tamas, C. Tzara, E. Vincent
DPh-N, CEN Saclay, BP 2, 91190 Gif-sur-Yvette, France

J. Deutsch, D. Favart, R. Prieels, B. Van Oystaeyen
Institut de Physique Corpusculaire, Université Catholique de Louvain B-1348 Louvain-la-Neuve, Belgium

ABSTRACT

π° photoproduction on ^{1}H, ^{2}H, ^{3}He and ^{4}He has been measured in the region of 1 to 10 MeV above threshold. Our measurement confirms the importance of pion rescattering for the reactions on ^{2}H and ^{3}He, and yields information on the threshold amplitudes on the nucleons.

The basic motivations of π° photoproduction measurements near threshold on light nuclei are twofold : i) obtain information on the so far poorly determined π° production amplitudes on the nucleons ; ii) learn about photoproduction mechanism in nuclei, especially in the case of deuterium[1] and helium-3 where large contributions of pion rescattering terms dominate the one-body amplitude, whereas they affect weakly the reaction on ^{4}He.

The experiment consists in the comparison of the π° photoproduction yields on ^{1}H, ^{2}H, ^{3}He and ^{4}He. Measurements are made for several end-point energies E_e of the bremsstrahlung spectrum ranging up to approximately 10 MeV above threshold. The two gammas from the π° decay (emitted almost back to back) are converted in a lead foil and subsequently detected in two Čerenkov telescopes placed symmetrically at 90° about the photon beam direction. The measured yields are related to the π° photoproduction cross section $d\sigma/d\Omega$ by the relation

$$Y(E) = \int_{E_o}^{E_e} \int_{\Omega_d} B(E,E_e)\, C\varepsilon(E,\theta)\, \frac{d\sigma}{d\Omega}\, dE d\Omega,$$

where $B(E,E_e)$ is the photon spectrum, $\varepsilon(E,\theta)$ the geometrical detection efficiency for photon energy E and π° emission angle θ, E_0 the threshold energy and Ω_d the solid angle of the detection system ; the constant factor C accounts for the efficiency of the Cerenkov detectors.

In a preliminary analysis we have compared our data to the predictions of a simple PWIA theoretical model allowing, in addition, for pion rescattering effects in the s-wave.

We have restricted the elementary nucleonic amplitudes to the dominant E_{0+} and M_{1+} multipoles ; the rescattering effects were estimated in a very simple model[2] involving only π production amplitu-

ISSN:0094-243X/79/540428-02$1.50

des and scattering lengths on the nucleons, and the average value of the inverse nucleon separation in ^2H and ^3He. The resulting cross-sections were folded with the Jabbur and Pratt bremsstrahlung shape and the Monte Carlo simulated detection efficiency ε. For each reaction the ratio of s wave (including rescattering) to p wave production amplitude was left as a free parameter to be adjusted on the data. Assuming the values in units $m_\pi = \hbar = c = 1$,

M_{1+} ($\pi^\circ p$) = 0.0085 qk and
M_{1+} ($\pi^\circ n$) = 0.0077 qk deduced of the multipolar analysis[3] of photoproduction at higher energies, we obtained from the general fit (see fig.) of all data points up to 6 MeV above threshold the following s-wave amplitudes :

$E_{0}+$ (π° p) = - 0.0020 ± 0.0002

$E_{0}+$ (π° d) = - 0.0056 ± 0.0003

$E_{0}+$ (π° ^3He) = - 0.0037 ± 0.0003

The $E_0+(\pi^\circ p)$ value is in agreement with existing data[4]. As for the $E_0+(\pi^\circ n)$ determination, it depends critically on the calculated values of the rescattering amplitudes for ^2H and ^3He which are presently affected by large theoretical uncertainties. Our preliminary analysis illustrates that using a very simple model we have an overall understanding of the four studied reactions. A more thourough theoretical treatment is now needed to extract from our data a reliable $E_0+(\pi^\circ n)$ value.

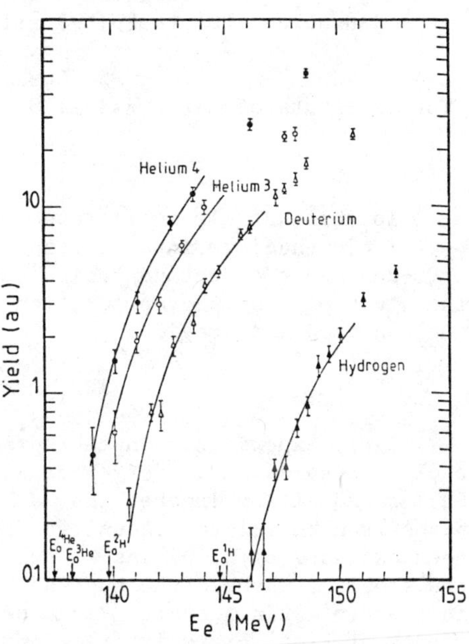

Fig. - The measured photoproduction yields as a function of the bremsstrahlung end-point energy E_e. Curves are theoretical yields adjusted as described in the text.

REFERENCES

1. P. Argan et al., Phys. Rev. Lett. 41, 629 (1978).
2. N. de Botton and C. Tzara, Rapport interne DPh-N/HE/78-06.
3. P. Noelle, W. Pfeil and D. Schwela, Nucl. Phys. B31, 1 (1971).
4. B.B. Govorkov et al., Proceedings (Trudy) of the P.N. Lebedev Physics Institute 54, 1 (1974).

AN IMPROVED SURFACE PRODUCTION MODEL FOR THE STUDY OF PION PHOTOPRODUCTION FROM NUCLEI

K. Srinivasa Rao
Department of Physics
Rensselaer Polytechnic Institute, Troy, N.Y. 12181

S. Susila
The Institute of Mathematical Sciences, Madras-600020, India

ABSTRACT

In this article an improved surface production model, where the p-wave pion-nucleus interaction is taken into account through an energy-dependent effective radius cut-off parameter, is suggested for the study of pion photoproduction from nuclei in the first pion-nucleon resonance region.

THE MODEL

Early experiments on photoproduction of charged pions from nuclei revealed an almost exact $A^{2/3}$ dependence for the cross section. Butler and Wilson invoked the mechanism of surface production[1] of pions from nuclei to account for this gross feature. This idea has been extended by Laing and Moorehouse[2] to a shell model treatment of the problem, wherein the nucleus makes a specific transition from an initial to a final state. In a series of papers[3], we simulated the effect of Final State Interaction (FSI) of the outgoing pion with the residual nucleus, by the use of the surface production model and studied charged pion photoproduction from closed, as well as non-closed shell nuclei and found that in the absence of detailed information about the initial (ground) and final (low-lying excited) states, the phenomenological surface production model is reasonably successful in accounting for the experimental data in the first pion-nucleon resonance region.

Recently, Freed et al.[4] have shown that the results for charged pion photoproduction obtained with the phenomenological surface production model resemble those obtained in the DWIA with pions distorted by the s-wave component of the optical potential alone. However, the use of the complete optical potential to study the effect of FSI, exhibits a very strong energy dependent p-wave absorption which drastically reduces the cross section to values far below the experimental data in the resonance region.

In their studies on pion-nucleus scattering at medium energies, Silbar and Sternheim[5] find that a simple optical model, in which the p-wave pion-nucleon interaction is taken into account through an energy dependent effective radius, fits the pion-nucleus scattering data. The effective nuclear radius R^* is defined by: $R^* = R^2 + 3/2\cdot\ell(\ell+1)k_{c.m.}^{-2}$, where R is the r.m.s. nuclear radius, ℓ is the ℓth partial wave which dominates the pion-nucleon amplitude and $k_{c.m.}$ is the c.m. pion momentum.

ISSN:0094-243X/79/540430-02$1.50

In Table 1 below, we give the total cross section for $^{16}O(\gamma,\pi^+)^{16}N$ - treating the ^{16}O ground state to be spherical and the low-lying states of $^{16}N(2^-,0^-,3^-,1^-)$ to be particle-hole states[6] whose mixing coefficients are given by Elliot-Flowers (EF) and Gillet-Vinh Mau (GV), with GV in the Tamm-Dancoff (TDA) and Random Phase (RPA) approximations - for an incident photon energy of 260 MeV, as a function of the surface production cut-off parameter R.

Table 1. Cross sections in μb for $^{16}O(\gamma,\pi^+)^{16}N$

R	IPM	EF	GV(TDA)	GV(RPA)
0.0	32.33	26.39	26.68	25.77
1.6	19.43	15.04	15.33	14.84
1.9	11.39	8.79	8.97	8.72
2.2	4.9	3.77	3.83	3.74

Note: R is in pion Compton wave lengths and IPM stands for the Independent Particle Model for ^{16}N states.

The cross section is extremely sensitive to the cut-off parameter, as seen from Table 1. In view of the fact that $R^*/R \sim 2$ at 170 MeV and it decreases to $\sim$ 1.1 at 420 MeV incident photon energy for ^{16}O and since the (γ,π) reaction is sensitive to the exact nature of the pion wave function inside the nucleus (than elastic pion-nucleus scattering) an improved surface production model for pion photoproduction is one in which the cut-off parameter is energy dependent (R^*) instead of the energy independent R. Results obtained using this model for specific (γ,π) reactions will be reported.

REFERENCES

1. R. R. Wilson, Phys. Rev. 86, 125 (1952); S. T. Butler, Phys. Rev. 87, 1117 (1952).
2. E. W. Laing and R. G. Moorehouse, Proc. Phys. Soc. 70, 629 (1957).
3. For a list of references see K. Srinivasa Rao, Proc. Tamil Nadu Acad. Sci. (India) 1, 127 (1978).
4. N. Freed and P. Ostrander, Phys. Rev. C11, 805 (1977); Phys. Letts. 61B, 449 (1976); I. Blomqvist et al., Phys. Rev. C15, 988 (1977).
5. R. R. Silbar and M. M. Sternheim, Phys. Rev. Letts. 31, 941 (1973); Ann. Rev. Nucl. Sci. 23, 249 (1974).
6. J. P. Elliot and B. H. Flowers, Proc. Roy. Soc. A242, 57 (1957); V. Gillet and N. Vinh Mau, Nucl. Phys. 54, 321 (1964).

$^{12}C(\gamma,\pi^-)^{12}N_{g.s.}$ FROM THRESHOLD TO RESONANCE

G.N. Epstein*
Center for Theoretical Physics, Laboratory for Nuclear Science, MIT, Cambridge, MA 02139

M.K. Singham and F. Tabakin
Department of Physics and Astronomy, University of Pittsburgh, Pittsburgh, PA 15260

ABSTRACT

We calculate both total and differential cross-sections for the reaction $^{12}C(\gamma,\pi^-)^{12}N_{g.s.}$ for pion laboratory kinetic energies ranging from 0 - 180 MeV. We pay particular attention to the role of non-static terms in the photoproduction operator.

INTRODUCTION

In previous work[1,2] we have calculated $^{12}C(\gamma,\pi^-)$ $^{12}N_{g.s.}$ for pion kinetic energies from 0 - 50 MeV. We have now extended the analysis to include the full range of pion kinetic energy from threshold to the resonance region. Here we draw attention to the main features of our approach. Complete results for both total and differential cross-sections will be presented in the appropriate conference session. There are essentially three elements in the analysis: the photoproduction operator ($H_{\gamma\pi}$); nuclear structure; pion distortions.

First of all we have designed our calculation to handle the complete photoproduction operator $H_{\gamma\pi}$. In particular we include the non-static pieces, which play an increasingly important role as the pion energy approaches resonance. We use a simplified version of the Blomqvist-Laget model[3] for the $\gamma N \to \pi N$ amplitude (N=nucleon) and convert this into a coordinate space operator for use in the nucleus. In this way we are able both to maintain reasonable agreement with the $\gamma N \to \pi N$ data up to the resonance region and to conveniently handle the

*This work is supported in part through funds provided by the US DEPARTMENT OF ENERGY (DOE) under contract EY-76-C-02-3069.

ISSN:0094-243X/79/540432-02$1.50

transformation from πN centre-of-mass system to π-nucleus centre-of-mass system.

For a reliable analysis it is vital to use a good quality nuclear structure description. We use Haxton's shell model wave functions[4] which are fitted to M1 form factors obtained from recent electron scattering data from $^{12}C_{g.s.}$ to $^{12}C(1^+, 15.11\ MeV)$, where the latter state is taken to form an isotriplet with $^{12}N_{g.s.}$ and $^{12}B_{g.s.}$. To estimate the sensitivity to the nuclear structure description we have also used the configurations of O'Connell, Donnelly and Walecka.[5]

Finally, to handle the final state pion distortions we have used pion optical potential parameters consistent with the available pion elastic scattering and pionic atom data. Sensitivity to the distorted waves was investigated.

OUTLOOK

We view our calculation of $^{12}C(\gamma,\pi^-)^{12}N_{g.s.}$ as a "case-study" in which we make a careful analysis of a photopion reaction which we should be able to understand. We examine in detail the sensitivities to all of the ingredients. It is our hope that once we have this case-study reaction "under control", then we can proceed to systematically study other photopion reactions to build a complete picture of (γ,π) processes both near threshold and in the important isobar region. We have developed all of our analysis of $^{12}C(\gamma,\pi^-)^{12}N_{g.s.}$ in a general framework so that the extension to handle other photopion reactions is simple.

REFERENCES

1. G.N. Epstein, M.K. Singham and F. Tabakin, Phys. Rev. C<u>17</u>, 702 (1978).
2. G.N. Epstein, M.K. Singham and F. Tabakin, Proc. Int. Symposium on Photopion Reactions R.P.I. 1978.
3. I. Blomqvist and J.M. Laget, Nucl. Phys. A<u>280</u>, 405 (1977).
4. W.C. Haxton, Phys. Lett. <u>76B</u>, 165 (1978).
5. J.S. O'Connell, T.W. Donnelly and J.D. Walecka, Phys. Rev. C<u>6</u>, 719 (1972).

SHIFT AND WIDTH OF THE PIONIC *3d* STATE IN Ta, Re AND Bi

J.Konijn,J.K.Panman,J.H.Koch,W.van Doesburg,G.T.Ewan,
T.Johansson,G.Tibell,K.Fransson and L.Tauscher
(IKO,Amsterdam-CERN-Kingston-Stockholm Collaboration)

ABSTRACT

Strong interaction monopole and quadrupole shifts and widths of the *4f* and *3d* levels have been measured in Ta, Re and Bi, by observing the pionic *4f*→*3d* X-ray transition. Thus in addition to the strongly shifted and broadened *5g*→*4f* transitions a second strongly affected line is available for these elements. In contrast to the pionic *4f* level data, the strong interaction shifts and widths of the deeper lying *3d* level differ typically by a factor two or more from the standard optical potential predictions.

The experiment was performed at the muon channel of the CERN Synchro-Cyclotron (SC) stopping pions in Ta, Re and Bi targets. The experimental set-up was especially designed to detect weak transitions by sacrificing some of the resolution in energy to obtain a higher photo peak efficiency, achieved by replacing the usual LEPS small volume detectors by a highly efficient large volume Ge(Li) diode with a high peak to Compton ratio. A further reduction of the background was accomplished by incorporating this detector into a Compton suppression system [1]). We were therefore able to separate from the background the broad line corresponding to the pionic *4f*→*3d* transition.

In Table 1 the observed shifts and widths due to the strong interaction are presented, obtained by subtracting from the experimental values the calculated electromagnetic values, which include corrections for finite size,

ISSN:0094-243X/79/540434-02$1.50

vacuum polarization to higher orders, Lamb shift and electron screening. The experimental data are also corrected for the shift and width of the upper level and for the radiative widths of both upper and lower levels.

The theoretical strong interaction level shifts and widths are calculated with the optical potential of Tauscher [2]). The calculations for the pionic *4f* level agree reasonably well with experiment. However, ε_0, ε_2 and Γ_0, calculated for the pionic *3d* level are larger than the experimental value by a factor of two or more. Different optical potential parameters give similar results. Other refinements such as a different neutron distribution are not sufficient to explain the discrepancies found for the *3d* orbit which has a large nuclear overlap.

REFERENCES

1. R. Beetz et al., Nucl.Instr.Meth. 145(1977)353
2. L. Tauscher, Proc.Int.Sem. on π-meson-nucleus interactions, Strasbourg,1971(CNRS,Strasbourg,1971) p.45

Table I Strong interaction monopole and quadrupole, ε_0 and ε_2, shifts and widths, Γ_0, in keV for the pionic *4f* and *3d* levels.

	ε_0		ε_2		Γ_0	
Nucleus	Exp.	Calc.	Exp.	Calc.	Exp.	Calc.
4f level						
^{181}Ta	-0.540 ± 0.100	-.533	0.128 ± .031	0.034	0.225 ± .057	0.322
Re	-0.673 ± 0.050	-.690	0.064 ± .037	0.028	0.547 ± .050	0.430
^{209}Bi	-1.79 ± 0.15	-1.83			1.157 ± .070	1.216
3d level						
^{181}Ta	-13.5 ± 1.0	-24.5	-0.94 ± .60	-3.81	17.7 ± 3.3	41.9
Re	-15.2 ± 1.0	-28.4	-1.56 ± .73	-3.06	16.8 ± 2.0	50.2
^{209}Bi	-20.6 ± 1.3	-44.5			25.4 ± 5.3	90.5

COHERENT NUCLEUS APPROXIMATION IN KAONIC ATOMS THEORY

A. Deloff*
University of Guelph, Guelph, Ont., Canada N1G 2W1

ABSTRACT

We show that the phenomenological kaon-nucleus optical potential is derivable from microscopic theory.

INTRODUCTION

Kaonic atoms level shifts and widths can be quite satisfactorily explained by employing a phenomenological optical potential of the form

$$2\mu V(r) = -4\pi A(1 + m_K/m_N)\bar{a}\,\rho(r) \quad ; \tag{1}$$

(m_K, m_N are kaon and nucleon masses, μ is K-nucleus reduced mass, $\rho(r)$ is nuclear density normalised to one). When the complex depth parameter $\bar{a}$ is naively identified with the $\bar{K}$-N scattering length, the value $\bar{a} = (-0.15 + i\,0.68)$fm is obtained (for $2Z = 2N = A$) and the theory disagrees with experiment. If instead $\bar{a}$ is adjusted to the data, a two parameter fit yields[1] the value $\bar{a} = (0.44 + i\,0.82)$fm and the agreement is restored. Our goal is to answer two vital questions: (i) how (1) can be derived from theory, and (ii) why $\bar{a}$ is not just the scattering length.

THEORY

In Watson formulation the exact expression for the optical potential is

$$V = \langle 0|\sum_{\alpha=1}^{N}\tau_\alpha + \sum_{\alpha\neq\beta}\tau_\alpha G_o Q \tau_\beta + \ldots|0\rangle \tag{2}$$

where Q is excited states projector, G_o is the full many-body Green's function and the scattering operator τ_α in medium is related to the free scattering operator t_α, by the equation

$$\tau_\alpha = t_\alpha[1 + (G_o Q - g_o)\tau_\alpha] \quad , \tag{3}$$

where g_o is a two-body Green's function. The usual procedure is to retain only the first term in (2) setting $\tau_\alpha \simeq t_\alpha$ (impulse approximation - IA). Since the kaon energy is by an order of magnitude smaller than the nucleon binding energy, IA is unlikely to hold. To go beyond IA, we suggest a coherent approximation (CA) which consists in neglecting excitations ($Q\to 0$). Assuming CA, eq. (3) becomes $\tau_\alpha = t_\alpha(1 - g_o\,\tau_\alpha)$ with the solution $\tau_\alpha = (1 + t_\alpha g_o)^{-1} t_\alpha = v_\alpha$ where v_α is the $\bar{K}N$ interaction. Representing v_α by a local potential, the folding model[2] follows. With a range small in comparison with the spread of $\rho(r)$, CA model yields

*From: Institute for Nuclear Research, Warsaw, Poland

ISSN:0094-243X/79/540436-02$1.50

$$V(r) \simeq N \int \rho(\vec{r} - \vec{x})\, v(x) d^3x \simeq N \int v(x) d^3x\, \rho(r) \qquad (4)$$

and eq. (4) answers our first question. To answer the second question, the volume integral in (4) is not just proportional to the scattering length a, unless Born approximation were to hold. Actually, to a good approximation[3] $2\mu_{KN} \int v(x) d^3x = -4\pi a(1 + 1/bq)^{-1}$, where b is the intrinsic range and q is a parameter (related to the shape) of the order of one. Since $|a| \simeq 1$ fm and all known mechanisms indicate that $b \lesssim 1$ fm, the denominator $(1 + a/bq)^{-1}$ gives rise to an important renormalisation of the IA value of $\bar{a}$. Taking the same shapes for $\rho(r)$ and $v(x)$ for both neutrons and protons, (4) goes into (1) with $\bar{a}$ given as

$$\bar{a} = (N/A)a_n(1 + a_n/bq)^{-1} + (Z/A)a_p(1 + a_p/bq)^{-1} \;. \qquad (5)$$

In (5) the only unknown quantity is bq (rather than two Re $\bar{a}$ and Im $\bar{a}$) and by adjusting it to the data, one obtains $bq \simeq 1.05$ fm. Setting $2Z = 2N = A$, from (5) one has $\bar{a} = (0.42 + i\,0.74)$fm, not far from the estimated value.[1] Qualitatively, the Q dependent terms hitherto neglected, generate corrections of presumably shorter range and might be relatively unimportant owing to large ℓ values.

REFERENCES

1. J. H. Koch and M. M. Sternheim, Phys. Rev. Lett. 28, 1061 (1972)
2. A. Deloff and J. Law, Phys. Rev. C10, 1688 (1974)
3. A. Deloff, Nucl. Phys. A303, 412 (1978)

STUDY OF A CLOSED SHELL PLUS ONE PARTICLE USING POLARIZED μ-ATOMS+

P.Quarati , A.Pompei
Istituto di Fisica dell'Università, Cagliari (Italy)

M.Guidetti
Istituto di Fisica Sperimentale del Politecnico, Torino(Italy)

ABSTRACT

We propose to analyze X-rays emitted from μ-atoms with nucleus polarized in the same direction of the spin of the μ-beam to study the structure of a nucleus consisting f a closed shell plus one valence nucleon.

In recent measurements of cross sections for magnetic scattering of electrons from a nucleus consisting of a closed shell plus one valence nucleon (VN) a problem has arisen [1]: what is the relative importance of core-polarization and exchange-currents in nuclei as e. g. ^{17}O and ^{41}Sc. However, cross sections containing both the effects (see ref.2), although representing an improvement respect to the HOSM, don't fit the experimental resuts. On the other hand a good fit is obtained using a smaller HO parameter for the VN (b_v). Recently Zamick [3] has shown physical reasons for lowering b_v respect to b of the core (b_c), but the exact amount of this lowering still remains a problem. A few properties of the nuclei deduced from e-scattering cross sections can be deduced as well by analyzing the X-rays emitted from μ-atoms. For the solution of the above mentioned problem we also propose to analyze the X-rays emitted from the corresponding polarized μ-atoms. In the evaluation of the energy difference between two μ-levels one has to take care of the shift due to nuclear polarizability (NP) which has to be calculated using some nuclear model. NP shifts for this particular problem are calculated below.

As we have already shown [4], in a decoupled scheme, the μ-energy-level-shifts (ELS) depend not only on the double-energy-weighted-sum rule (spherical term), but also on direct and exchange expectation values of one- and two-body operators, calculated between core-nucleon and VN states : these non-spherical (ns) terms are directly related to the VN orbit. The described physical scheme can be performed by using nuclei polarized in the same direction of the incident μ-beam. All angular momenta involved are, in this case, parallel to the μ-spin (z-axis) and the total spin has its maximum projection. In practice, the expressions valid for the decoupled scheme can be used here. As a consequence the ns terms enter into the expression of the ELS and all the calculations can be explicitely and very easily developed. The expression for the shift of the μ-level n,l,m, $j=l+1/2$, $m_j=l+1/2$ due to E1 NP for the ^{41}Sc (the VN is the 0f7/2 proton) is :

+ Supported in part by I.N.F.N.

ISSN:0094-243X/79/540438-02$1.50

$$\Delta E_{nlm}(E1) = -\frac{e^4}{3\hat{E}} \langle r_\mu^{-4} \rangle_{nl} \left\{ A_s + A_{ns} - C_s - C_s(0f7/2) - C_{ns} \right\}$$

A_s is the one-body s term, A_{ns} is the one-body ns term related to the VN, C_s is the two-body exchange s term, $C_s(0f7/2)$ is the exchange s term calculated between the 0f7/2 and the 0d states, C_{ns} is the two-body ns exchange term. The quantities A_{ns}, $C_s(0f7/2)$, C_{ns} and the contribution of the orbit 0f7/2 to A_s are directly connected to the VN. Our calculations, using an HOSM with $b_v \neq b_c$, show that, in a nucleus of 41 nucleons, the 15% of the NP effect is due to the VN. This percentage is 50% larger than that for non polarized nuclei. b_v can be determined with the same precision as the transition energy between two μEL. In the following table we write, for two μ-levels, the s and ns contributions to E1 NP ($b_v \neq b_c$) :

Table: Values of s and ns contributions to E1 NP

	A_s	A_{ns}	$-C_s$	$-C_s(0f7/2)$	$-C_{ns}$
n=4,l=3,m=3	$60b_c + 9/2b_v$	$+5/3\ b_v$	$-15b_c$	$+48/\sqrt{7}\ R$	$+16/\sqrt{7}\ R$
n=5,l=4,m=4	$60b_c + 9/2b_v$	$+20/11b_v$	$-15b_c$	$+48/\sqrt{7}\ R$	$+(198/\sqrt{7}\cdot 11)\ R$

$$R = b_c^{-7/4}\, b_v^{-9/4} \,/\, (b_c^{-1} + b_v^{-1})^{9/2}$$

Other calculations of NP, for different μEL have been performed, using the Zamick deformed w.f. with different $b_{xv}, b_{xc}, b_{zv}, b_{zc}$ and zero range interaction between core and VN [5].

As it appears from the table, the ns terms are very important in the evaluation of the energy difference between two μEL, thus giving a very precise determination of the parameters b . This will be possible when the needed experimental data will be available.

REFERENCES

1. I.Sick et al.:Phys.Rev.Lett.38,1259(1977);P.K.A.deWitt Huberts et al.:Phys.Lett.71B,317(1977); M.V.Hynes et al.:preprint.

2. A.Arima et al.:Phys.Rev.Lett.40,1001(1978); J.Dubach,W.C.Haxton: Phys.Rev.Lett.41,1453(1978).

3. L.Zamick: preprint.(One of us (P.Q.) thanks L.Zamick for private communication and discussions)

4. M.Guidetti,G.Puddu,P.Quarati: Lecture Notes in Physics, Springer Verlag, Heidelberg, incourse of publication.

5. L.Zamick: Nucl.Phys.A260,241(1976).

AN OVERVIEW OF π-NUCLEUS SCATTERING

R. A. Eisenstein[+]
Carnegie-Mellon University, Pittsburgh, PA 15213

ABSTRACT

A concise examination of the π-nucleus elastic and inelastic scattering problem is made. A brief discussion of some of the ambiguities in current interpretations of the data is presented. The influence of such ambiguities on nuclear matter density determinations and other topics of nuclear structure interest is examined. Such uncertainties may be more harmful to elastic than inelastic scattering analysis.

INTRODUCTION

One of the principal aims of π-nucleus physics is to understand π-nucleus (πA) scattering in terms of the elementary π-nucleon (πN) amplitude. Should this goal be realized, it would be a major step forward in our understanding of many-body processes, and would perforce involve a detailed knowledge of both the interaction mechanisms and the nuclear structure involved in the process. The fact that the basic πN interaction is fairly well understood (and relatively simple) on shell below T_π = 300 MeV makes the prospects of success more tantalizing.

Since our goal is such an ambitious one, it is well to begin our studies with the simplest of all πA interactions, elastic and inelastic scattering. These processes are interesting not only in their own right, but also because they provide phenomenological information necessary for better understanding of other processes. It may be true that we do not need a detailed dynamical understanding of the process in order to make useful contributions to such reaction studies as (p,π) or (K,π), or to extractions of neutron-proton matter radius differences. Of course, since the more exotic reactions such as (K,π) are quite sensitive to parts of the πA interaction not well sampled by scattering experiments, information of that kind will be an important input to our overall understanding.

In what follows, I would like to indicate briefly, in the nature of an overview, the salient features of πA scattering processes. Because many of the observed phenomena are to be discussed by Drs. Redwine and Ingram, I will mention only illustrative examples here. As the discussion progresses, I would like to focus on one problem: the role of ambiguities in the theory and how they hinder our attempts to learn detailed nuclear structure information from πA scattering. This is important to do because πA physics is entering its quantitative phase.

ISSN:0094-243X/79/540440-15$1.50

OUTLINE OF πA SCATTERING

We are all aware that the basic feature of the elementary amplitude below 300 MeV is the Δ_{33} resonance in the πN p-wave at $T_\pi \sim 180$ MeV. The resonance is so strong that the π^+p total cross section equals the unitary limit at this energy.[1] The s-wave is non-resonant and is weak at low energies, in accordance with current algebra predictions.[1] Other partial waves are unimportant[2] for our purposes. The off-shell behavior of the πN amplitude is not determined experimentally and must be surmised from other physical considerations.[3] Our knowledge of the on-shell problem has been summarized in an energy-dependent phase shift analysis by Rowe, *et al.*[4]

The traditional approach in making a theory of πA scattering has been to imbed the elementary πN t-matrix into conventional multiple scattering theory (MST), the goal being the replacement of a many body problem by a two-body one.[5] From the πA T-matrix, an optical potential representation of the π scattering from the nucleus can be obtained. The nucleus is now an elementary object, since all possible virtual excitations and internal degrees of freedom are accounted for in the optical potential. Because of the simple nature of the πN interaction, the πA scattering problem has long been expected to be one of the best MST "laboratories" we are likely to find.

However, the usual treatments of the multiple scattering series, such as Watson theory or KMT theory[6] do not include pion production or absorption, and thus neglect mesonic degrees of freedom (such as ρ meson exchange[7]) in nuclei as well as true meson absorption.[8] Such phenomena require the use of π-nucleus field theories for consistent treatment. (See G. Miller, this Conference.)

The latter effect, in which the incoming π is annihilated (principally on two nucleons) is known to be important, especially at low π energies.[8] In addition to this omission, there are other effects, which although in principle are included in MST, are difficult to calculate in practice. These effects include such first order phenomena as proper averaging over nucleon Fermi motion,[9,10] and high-order effects such as nucleon binding, and nucleon correlations (both Pauli and LLEE types).[11] Even such an apparently straightforward problem as the proper kinematic transformation of the πN t-matrix (from the πN to the πA center of mass frame) is ambiguous when treated non-covariantly.[12] All of these effects can be quite sizeable.[9,10,13,14]

Assuming for the moment that a suitable optical potential can be constructed, it may then be inserted into a wave equation and π wavefunctions obtained. Current lore has it that a reasonable wavefunction to use is a truncated (omission of the term $V^2_{OPTICAL}$) Klein-Gordon equation or perhaps the relativistic Schrodinger equation. This topic is currently under discussion in the literature.[15]

Let us turn to a simple form for the optical potential as a guide to the phenomenology. In the limit of low nuclear density, free but static nucleons, and a sharply peaked πN amplitude, the πA optical potential in momentum space would be:

$$2E_\pi V_{\pi A} \sim Ak^2\ t_{\pi N}\ \rho_A(q)$$

with (1)

$$t_{\pi N} \sim [b_0 + b_1 \cos\theta]$$

The Kisslinger variation[2] of this is, in co-ordinate space:

$$2E_\pi V_{\pi A} = -Ak^2\ b_0\ \rho_A(r) + Ab_1\ \vec{\nabla}\cdot\rho_A(r)\vec{\nabla} \quad (2)$$

Here $\rho_A(q)$ is the Fourier transform of the nuclear matter density normalized to unit r-space volume, while b_0 and b_1 are complex numbers computed from πN phase shifts. In Born approximation, neglecting the Coulomb interaction for the moment, the πA cross section is proportional to $|V_{\pi A}|^2$. Thus, we see that minima in the angular distribution will arise from two sources:[16] the zero in $t_{\pi N}$ (an "elementary particle" minimum) and the zeros due to the change in sign of $\rho_A(q)$. The process therefore will manifest both scattering from the elementary scattering centers as well as from the distribution as a whole. These minima appear also in detailed calculations (see fig. 1). When the Coulomb interaction is included an additional dip appears at forward angles due to Coulomb-nuclear interference. Note that the t-matrix minimum stays relatively fixed in position as the energy increases, while the nuclear size minima move into smaller angles and deepen as the energy approaches the resonance value.

On resonance, the nucleus is very absorptive and the π mean free path is $\sim$0.5 fm. Here, the angular distribution resembles the diffractive pattern observed for scattering from a strongly absorbing black disk. The parameter b_0 remains small, while b_1 becomes purely imaginary. The real part of b_1 is passing through zero with negative slope as we pass upward in energy across the resonance; the interaction is then quite surface-peaked. This can be seen from eq. (2).

In contrast, at low energies the π can penetrate much more deeply into the nuclear interior because the πN resonance is less dominant; here the mean free path is $\sim$5 fm. In this case, we expect the effects of nuclear structure to manifest themselves via both first and second order corrections to the optical model.

In figure 2, the Kisslinger potential with free πN parameters is compared to π^+ elastic scattering data from ^{12}C at 50 and 150 MeV. The calculation is seen to reproduce the resonance region qualitatively; the minima are nearly in the right place and the cross section agrees pretty well in the regions near maxima. Presumably this is because the calculation is mostly sensitive to the strength of the absorption and the nuclear size. At the lower energy, the Kisslinger result is considerably worse, failing not only to predict the location of the minimum but the size of the cross section as well. However, in both energy regions there is room for improvement over this simple model. An examination of the data from both of these regions should help us decide in what energy region the multiple

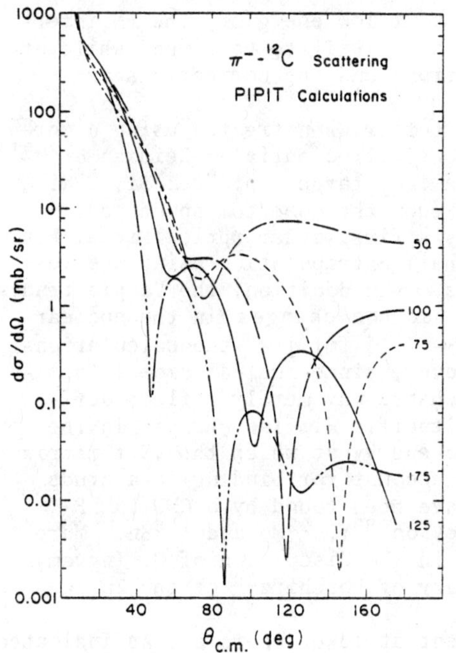

Fig. 1

PIPIT calculations of $\pi^- - ^{12}C$ scattering at various energies. The Coulomb-nuclear interference minimum is at ∿10°; the "elementary particle" minimum is at ∿65°. Other minima are due to zeros of ρ(q). The elementary and ρ(q) minima become diffractive as the πN resonance is approached.

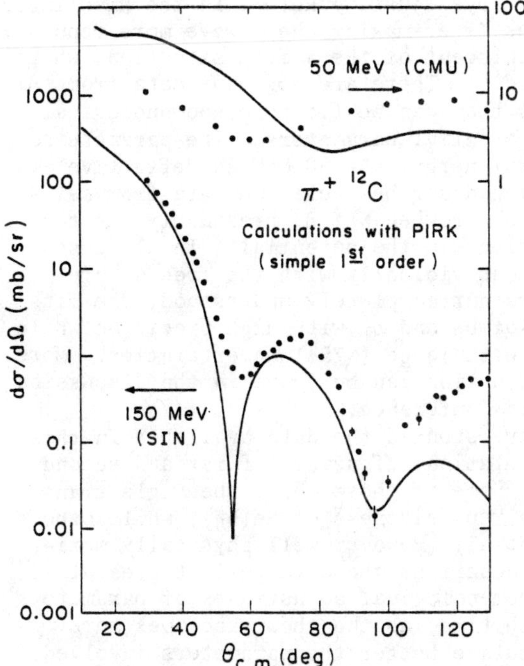

Fig. 2

PIRK calculations using free πN information compared to data from refs. 22 and 39.

scattering theory is most reliable. At low energies, the πN interaction is weak but the corrections are likely to be large, while at high energies the interaction is strong and the corrections are expected to be smaller.

The situation on resonance is better when treated using a more realistic, but still first-order, optical potential. Zeidman *et al*[17] have compared their $\pi^{\pm}$ 162 MeV data from targets of ^{9}Be, Si, ^{58}Ni and ^{208}Pb to calculations obtained from the momentum space code PIPIT.[18] This program incorporates a first-order optical model but with a realistic (separable) off-shell extrapolation using the results of Londergan, McVoy and Moniz.[3] In addition, the "angle transformation" is included. If we make ad hoc changes in the nuclear size parameter, $\langle r\rangle^{1/2}_{mod} = [\langle r^2\rangle_{chge} - (1.3\ \mathrm{fm})^2]^{1/2}$, the calculations reproduce the data very well. However, since nuclear size information is reasonably reliable, this change may merely reflect deficiencies in the optical potential itself. Similar changes in the calculations can be effected if the energy at which the πN t-matrix is evaluated is allowed to vary to account for binding in a crude sense. Similar results (fig. 3) have been found by a CMU-LAMPF-NMSU experiment[19] with $\pi^{\pm}$ at 180 MeV on ^{89}Y, ^{92}Mo and ^{118}Sn. More details on this topic can be found in the discussion of Q. Ingram at this Converence and in the summary of Boschitz[20] at the Zurich Conference.

The situation is quite different at lower energies, as indicated in fig. 2 and discussed briefly above. The minimum, which is due to the πN t-matrix, can be moved[16] by adjusting Re(b_0) in the Kisslinger model to a large negative value (i.e. making the s-wave more repulsive). This situation is reminiscent of the π atom situation, where a similar effect has been noted.[21] There are now many data from this energy region[22-25] and most of them can be fit by phenomenological adjustment of the Kisslinger potential parameters. The parameters are found to vary smoothly with energy. At 50 MeV an "effective interaction" has emerged[22] which describes well most data from self-conjugate nuclei. Deviations occur when $N > Z$, presumably due to the entrance of an isovector piece in the potential. In all cases, it is Re(b_0) which disagrees most violently with the free πN predictions. For reasons that are not completely understood, the fits determine the real parameters of b_0 and b_1 with high precision while the imaginary parts are found with large ($\sim$75%) uncertainties. More details of the experimental situation can be found in the discussion presented by R. Redwine at this Conference.

Many workers[9,10,13,14] have studied the data available in this energy region and careful calculations of several first and second order effects have been made. Some of these (e.g. the angle transformation)[12] will increase the repulsiveness of Re(b_0), while others involve new terms in the potential. However well physically motivated, none of these theories describes the data well at present without appealing to some phenomenological adjustments of parameters. This situation may improve with time, as the theory becomes more sophisticated and able to calculate better the parameters involved. The work in this area has shown one thing very clearly: all of these effects are large (see, e.g. ref. 13), both in first and second order

Fig. 3

Data from ref. 19 compared to PIPIT calculations. The matter radii have been reduced as outlined in the te t and ref. 17.

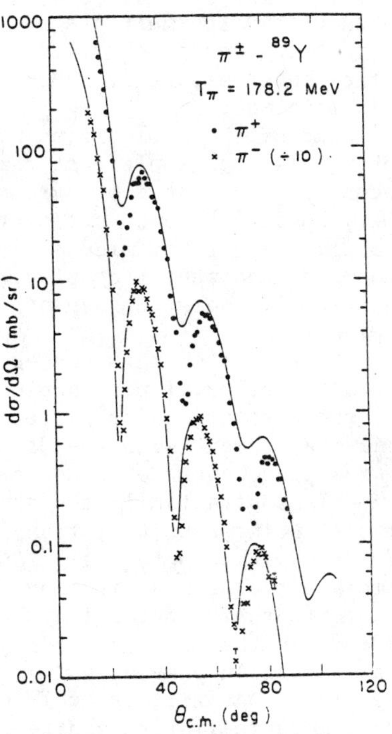

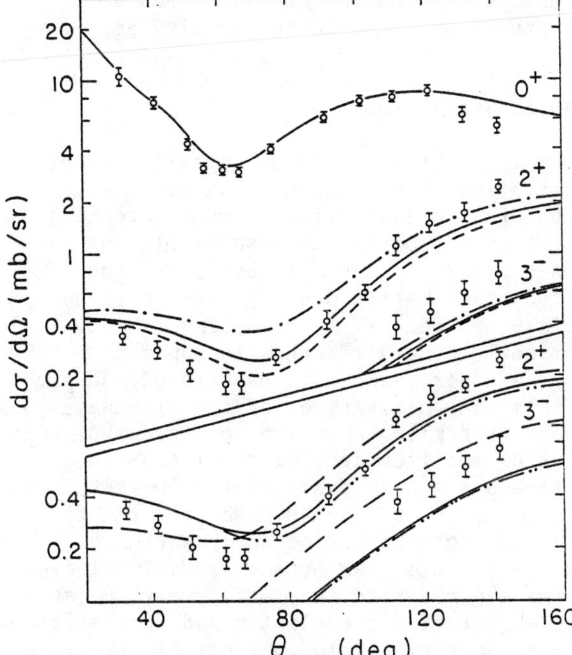

Fig. 4

The π^{+}-^{12}C data of ref. 22 compared to the calculations of Keister (ref. 26). The inelastic calculations are done using various phase-shift equivalent potentials, all of which reproduce exactly the elastic data.

as well. From this state of affairs, one might pessimistically conclude that the multiple scattering series is *not* converging, at least at low energies. This situation is examined more critically in the contributions of McManus, Moniz, Dillig and Kisslinger to this Conference.

What of the ambiguities mentioned earlier? One source of such problems is the simple fact that effects due to one type of physical process can be masked by parameter adjustments in terms representing other phenomena. Such correlations are to some extent statistical in origin, and to my mind are somewhat less important than those due to a more fundamental problem, namely the fact that the elastic data alone specify only the *asymptotic behavior* of the outgoing pion wavefunction.

The result of this can be seen in the fact that there exist several theories of pion-nucleus interactions whose underlying physical contents are quite different, yet which achieve qualitatively similar agreement with the data. This is true for data both at low and resonance energies. Recently, Keister[26] has shown it possible to modify substantially the short range behavior of π-nucleus interactions, using a unitary transformation, while producing a *phase-shift equivalent* potential. Such a unitary transformation can generate additional contributions to the potential which are identical in form to terms that arise when higher order corrections are included. In the case of the Kisslinger model, such terms include a local Laplacian p-wave, $\rho^2(r)$ contributions to both the s- and p-wave pieces, as well as modifications to the imaginary part of b_0. When a more realistic, finite range, non-local potential is considered, a more general transformation can be applied; the result is an ambiguity in *non-locality* in addition to the density ambiguity mentioned above.

INELASTIC SCATTERING

Such ambiguities can hopefully be resolved by appealing to processes such as inelastic scattering which sample the nuclear interior more effectively. If the elastic scattering optical potential is given crudely by eq. (1), then the potential responsible for inelastic scattering will be $A\ t_{\pi N}\ \delta\rho_A$. Here $\delta\rho_A$ represents the difference between the excited state matter density and the ground state; this function is surface peaked. If the Kisslinger model (eq. 2) is taken as the form for the t-matrix then the optical potential will contain derivatives of $\delta\rho(r)$, and will be strongly surface dependent, in addition to containing derivatives of the pion wavefunction. We therefore expect that mild variations of the wavefunction can produce substantial modifications of the inelastic cross section. This is confirmed by an analysis of the inelastic scattering results of Dytman, *et al*[22] at T_π = 48.5 MeV for the first 2^+ and 3^- states in ^{12}C. Indeed, both the shape and position of the minimum in the inelastic calculation can be considerably altered, especially when the transformation modifies the surface region (see fig. 4). However, in this energy region the elastic and inelastic data, even when taken together, do very little to constrain the wave function in the central region. One must then rely on more stringent tests (via reactions such as (p,π) or (γ,π)) to limit wave function variations in the nuclear interior. Similar fundings have been discussed by Yoo and Sternheim.[27] However, the ambiguity is not sufficient to cloud the fact that the DWIA with no adjustable parameters does well in predicting the inelastic scattering for ^{12}C once the elastic channel is properly described [22,28] (see fig. 5).

Discrepancies have been observed in places where the DWIA works well for other probes. Examples are the $\pi^{\pm}$ - ^{24}Mg results obtained[29] at SIN, and recent studies[19] with π^+ on the lowest 2^+ states in ^{92}Mo and ^{118}Sn. In these cases, the collective deformation parameter β_L required to fit the data is significantly larger than obtained with other probes. This may reflect the usual sort of discrepancy associated with this parameter, or else indicate a need for more careful (microscopic?) treatment for some of these excitations. In both cases, the angular distribution is well reproduced (see fig. 5).

In many cases of interest however, the DWIA will fail, as in the case of multistep excitations. Examples of this are seen in 50 MeV π^+ scattering[22] to the 0^+ (7.6 MeV) state in ^{12}C and in the transition to the 4^+ state (2.25 MeV) in ^{118}Sn observed with $\pi^{\pm}$ at 182 MeV[19] (see fig. 6). In the former case, a coupled channels analysis[30] has been shown to account for the data well.

Other effects due to nuclear structure can be expected to show up in π scattering studies. The usefulness of such work for distinguishing between T=0 and T=1 excitations has been emphasized by Walker,[31] who discusses this topic in his contribution to this Conference.

MATTER DENSITY STUDIES WITH PIONS

As a last topic for discussion, I would like to consider briefly the perennial question of neutron radius determinations using π-scattering. Studies of this type seek to exploit the differences between π^+ and π^- scattering on resonance, where $\sigma_{\pi N}(T=3/2)/\sigma_{\pi N}(T=1/2)$ is of order 9. A major hope of pion physics has been that such large isospin selectivity would provide unambiguous information about the neutron distribution in nuclei for both ground and excited states.

For the case of elastic scattering, recent measurements have been done on the oxygen isotopes $^{16,18}O$ (refs. 32, 33) and on the calcium isotopes $^{40,48}Ca$ (ref. 34). On one issue there is no argument: the results are very sensitive to the *number* of neutrons in the particular isotope under study. For an isotope with $N \neq Z$, π^+ and π^- elastic scattering show differences which should be well described in terms of an isovector part of the πA interaction, a major part of which can be accounted for once neutron number is taken into account.

Difficulties arise, however, when one seeks to interpret such observations in terms of the *matter distributions* involved. One problem is that predicted[35] distributions for neutrons and protons are not very different; in addition, we suffer from the fact that we do not understand the connection between the nuclear densities

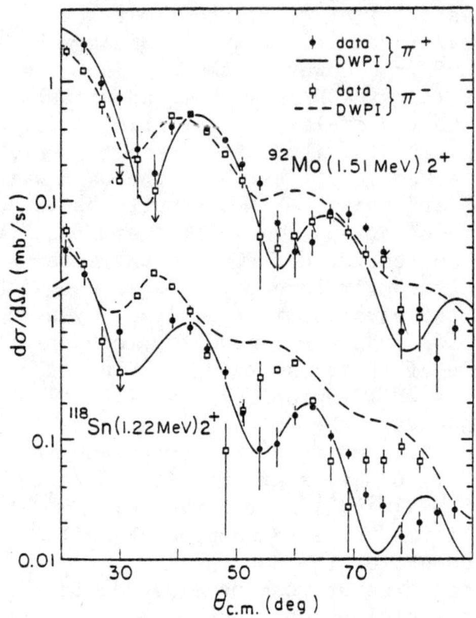

Fig. 5

Data from ref. 19 at 180 MeV. Basic agreement with the DWBA is seen. The β_L values agree fairly well with those obtained with other probes.

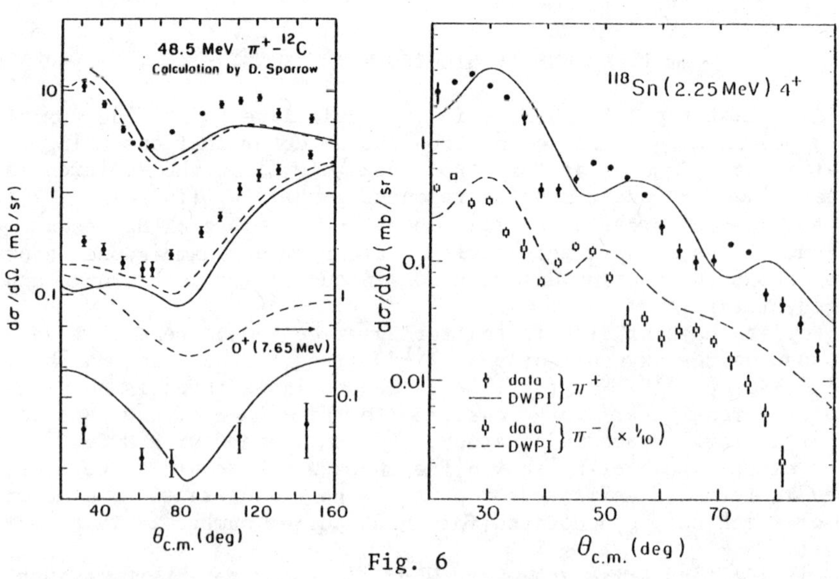

Fig. 6

Data from refs. 19 and 22 compared to DWBA and coupled channels calculations. DWBA fails due to the multistep nature of the excitations. The CC calculation of Sparrow and Gerace (ref. 30) obtains the correct magnitude for the 0^+ state in ^{12}C.

and the optical potential well enough. Unfortunately, this situation is not like electron scattering where these matters are much clearer. That ambiguities in the optical potential can mask differences due to distribution size parameters has been shown by Sternheim and Yoo,[36] who conclude that with the present state of the art, one cannot make definitive statements based on elastic scattering alone. A possible remedy is to treat both elastic and total cross section data via a common potential description.[36]

An interesting addendum to this discussion is to consider neutron distribution determinations at low energies ($T_\pi \sim 50$ MeV).[37,38] One would try to exploit the large s-wave isovector part of the πN interaction in an energy region where the isoscalar part is small. The method would work best in a study using π^- scattering on neighboring isotopes (e.g. $^{12,13}C$). In a well chosen case, the ratio of cross sections for the two isotopes should involve a cancellation of effects due to scattering from the core, leaving only the extra neutron to account for the observed ratio. In a study at 50 MeV with π^+, such a description was found to be qualitatively correct. However, the measurement was not accurate enough to determine quantitative details of the neutron distribution. The situation should be much clearer with π^- projectiles.[38]

In the case of inelastic scattering, the situation may not be as discouraging. This is because one is often dealing with excitations which primarily involve either neutrons or protons, and so the excitation probabilities should differ markedly when one uses π^- or π^+ as the probe. There is some evidence for the general effect; the most thoroughly investigated[32,33] examples are the 2^+ state (1.98 MeV) and 3^- state (5.10 MeV) in ^{18}O. Other cases have shown up in our recent $\pi^\pm$ resonance measurements[19] on ^{89}Y, ^{92}Mo and ^{118}Sn. In particular, the first excited state of ^{89}Y ($^9/_2$, 0.91 MeV) is excited about three times more strongly with π^+ than π^-; qualitatively this is to be expected because the transition is a pure single particle promotion of a proton from the $2p_{1/2}$ orbit to the $1g_{9/2}$. The strength for π^+ is in agreement with the single particle prediction.

In the case of the ^{18}O data, we again expect differences between π^+ and π^- scattering, for both elastic and inelastic processes, simply because N/Z = 10/8. Such differences are clearly observed in the LAMPF 164 MeV data[33] of Iverson. For the elastic scattering, N≠Z accounts for most of the observed difference, leaving only $0.3 \pm 0.3°$ of minimum shift to be accounted for by a difference in rms radius for neutrons and protons. Thus, the Δ(rms) obtained is consistent with zero.

For the inelastic scattering to the 2^+ state, we are dealing with states thought to consist mainly of neutron $2s_{1/2}$ and $1d_{5/2}$ orbitals. The transition will thus involve mostly the valence neutrons with little core polarization. Thus, we expect π^- inelastic scattering to be enhanced over π^+ by the ratio N/Z = 1.25. In fact, a much larger ratio (1.86 ± 0.16) is found,[33] not only at 164 MeV but at 230 MeV as well. This ratio is interpreted as evidence for an enhanced ratio of $(\beta r)_n/(\beta r)_p = 1.68 \pm 0.13$. If this analysis withstands further scrutiny, it would be the first clear separation of neutron and proton deformation lengths as far as I am aware. In closing, it should be noted that the SIN results[32] do not support this finding, indicating instead equal deformation lengths.

CONCLUSION

As described above, experimental findings are in qualitative agreement with simple theories of πA interactions which are based on multiple scattering theories and the known properties of the πN amplitude. However, a much better understanding of both the amplitude and the various terms of the MST must be reached before agreement with the data can be obtained. This is true both on resonance and at low energies. While some progress has been made in this direction, a fuller understanding will bring not only a better knowledge of the nuclear structure input, but also the ability to measure elusive quantities such as neutron matter distributions and deformation parameters.

ACKNOWLEDGEMENT

I would like to record once again my thanks to my colleagues S. Dytman, B. Keister, F. Tabakin and W. Wharton for many helpful discussions of the above material.

REFERENCES

+ Work supported by the USDOE.
1. Robert J. Cence, Pion-Nucleon Scattering, Princeton University Press, Princeton NJ (1969). B. H. Bransden and Moorhouse, R. G., The Pion-Nucleon System, Princeton University Press, Princeton NJ (1973).
2. D. S. Koltun, in Advances in Nuclear Physics, (vol. 3), edited by M. Baranger and E. Vogt, Plenum Press, New York (1969).
3. J. Londergan, K. McVoy and E. Moniz, Ann. Phys. 86, 147 (1974).
4. G. Rowe, M. Salomon and R. Landau, Phys. Rev. C18, 584 (1978).
5. J. Hüfner, Physics Rpts. 21, 1 (1975).
6. A pedagogic discussion of these theories is given in N. Austern, F. Tabakin and M. Silver, Am. J. Phys. 45, 361 (1977). See also L. Foldy, Phys. Rev. 67, 107 (1945) and A. Kerman, H. McManus, and R. Thaler, Ann. Phys., New York 8, 551 (1959).
7. G. Brown, B. Jennings, and V. Rostokin, "The Pion-Nucleus Many Body Problem", in Lectures from the LAMPF Summer School on Nuclear Structure with Pions and Protons, R. Burman and B. Gibson, editors, Los Alamos Report LA-6926-C (1977).
8. F. Hachenberg and H. J. Pirner, Ann. Phys. 112, 401 (1978) and R. Rockmore, E. Kanter and P. Goode, Phys. Lett. 77B, 149 (1978).
9. L. Liu and C. Shakin, Phys. Rev. C16, 1963 (1977) and references therein.
10. R. Landau and A. Thomas, Nuc. Phys. A302, 461 (1978).
11. M. Ericson, T. Ericson, Ann. Phys. 36, 323 (1966).
12. D. Ernst and G. Miller, Preprint, (1979).

13. N. DiGiacomo, A. Rosenthal, E. Rost and D. Sparrow, Phys. Lett. 66B, 421 (1977).
14. K. Stricker, H. McManus, J. Carr, Preprint, (1978).
15. E. Siciliano and R. Thaler, Phys. Rev. Lett. 41, 927 (1978).
16. M. Cooper and R. Eisenstein, Phys. Rev. C13, 1334 (1976).
17. B. Zeidman, C. Olmer, D. Geesaman, R. Boudrie, R. Siemssen, J. Amann, C. Morris, H. Thiessen, G. Burleson, M. Devereux, R. Segel, L. Swenson, Phys. Rev. Lett. 40, 1539 (1978). See also Phys. Rev. C18, 2223 (1978).
18. R. Eisenstein and F. Tabakin, Computer Phys. Comm. 12, 237 (1976).
19. M. Doss, R. Eisenstein, W. Wharton, J. Amann, G. Burleson, G. Blanpied, Exp. 23 LAMPF. Analysis in Progress.
20. E. Boschitz, in Seventh International Conference on High Energy Physics and Nuclear Structure, edited by M. P. Locher, Birkhauser Verlag, Basel (1977).
21. M. Krell and T. Ericson, Nuc. Phys. B11, 521 (1969).
22. S. Dytman, J. Amann, P. Barnes, J. Craig, M. Doss, R. Eisenstein, J. Sherman, W. Wharton, G. Burleson, S. Verbeck, R. Peterson, H. Thiessen, Phys. Rev. C (in Press).
23. M. Moinester, R. Burman, R. Redwine, M. Yates-Williams, D. Malbrough, C. Darden, R. Edge, T. Marks, S. Dam, B. Preedom, F. Bertrand, T. Cleary, E. Gross, C. Ludemann, M. Blecher, K. Gotow, D. Jenkins and F. Milder, Phys. Rev. C18, 2678 (1978) and D. Malbrough *et al*, Phys. Rev. C17, 1395 (1978).
24. R. Johnson, T. Masterson, K. Erdman, A. Thomas, and R. Landau, Nuc. Phys. A296, 444 (1978), and R. Johnson, B. Bassalleck, K. Erdman, B. Gyles, T. Marks, T. Masterson, D. Gill and C. Sabev, Phys. Lett. 78B, 560 (1978).
25. C. Guaraldo, A. Maggiora, R. Scrimaglio, F. Balestra, L. Busso, R. Garfagnini, G. Piragino, F. Cannata, Phys. Lett. 80B, 203 (1979).
26. B. Keister, Phys. Rev. C18, 1934 (1978).
27. K.-B. Yoo and M. Sternheim, Phys. Rev. Lett. 40, 498 (1978).
28. E. Rost, in Lectures from the LAMPF Summer School on the Theory of Pion-Nucleus Scattering, W. Gibbs and B. Gibson, editors, Los Alamos Reprot LA-5443-C (1973).
29. C. Wiedner, J. A. Nolen, W. Saathoff, R. Tribble, J. Bolger, J. Zichy, K. Stricker, H. McManus, and J. Carr, Phys. Lett. 78B, 26 (1978).
30. D. Sparrow and W. Gerace, Phys. Rev. Lett. 41, 1101 (1978).
31. G. Walker, in ref. 7.
32. J. Jansen, J. Zichy, J. P. Albanese, J. Arvieux, J. Bolger, E. Boschitz, C. Ingram, and L. Pflug, Phys. Lett. 77B, 359 (1978).
33. S. Iversen, H. Nann, A. Obst, K. Seth, N. Tanaka, C. Morris, H. Thiessen, K. Boyer, W. Cottingame, C. Moore, R. Boudrie, and D. Dehnhard, to be published in Phys. Lett. See also Iversen, *et al*, Phys. Rev. Lett. 40, 17 (1978).
34. J.-P. Egger, R. Corfu, P. Gretillat, C. Lunke, J. Piffaretti, E. Schwarz, C. Perrin, J. Jansen, and B. Preedom, Phys. Rev. Lett. 39, 1609 (1977).

35. J. W. Negele, Phys. Rev. C1, 1260 (1969).
36. M. Sternheim and K.-B. Yoo, Phys. Rev. Lett. 41, 1781 (1978).
37. S. Dytman, J. Amann, P. Barnes, J. Craig, M. Doss, R. Eisenstein, J. Sherman, W. Wharton, G. Burleson, S. Verbeck, R. Peterson, H. Thiessen, Phys. Rev. C18, 2316 (1978).
38. E. Rost, R. Johnson, and T. Marks, Univ. Colorado Progress Report (1978).
39. J. Piffaretti, R. Corfu, J.-P. Egger, P. Gretillat, C. Lunke, E. Schwarz, and C. Perrin, Phys. Lett. 67B, 289 (1977).

Question: M. Banerjee, University of Maryland

Early in your talk you referred to the question of whether to use the so called Klein Gordon wave function or the Schrodinger wave function. This question was anticipated and studied by Cammarata and myself in 1973. We pointed out in a reaction theory calculation one should use the Schrodinger wave function. But this is the correct Schrodinger wave function and not the incorrect one discussed yesterday.

Response: Eisenstein

No Comment.

Question: Q. Ingram, Karlsruhe

I think it appropriate to make a comment on the ^{18}O inelastic scattering data of Jansen, et al. Since the discrepancies with the data of Iverson, et al were pointed out, we have looked at our data analysis once more. It does appear that the method by which we estimated the background from the tail of the elastic peak was unsatisfactory. We are currently reanalyzing the data, and in the meantime I do not think that you should put too much emphasis on the discrepancies with Iverson, et al.

Response: Eisenstein

No comment.

Question: F. Tabakin, University of Pittsburgh

Concerning the phase shift equivalent of Keister, I'd like to mention that these changes in the pion distorted waves have been tested (for up to 50 MeV pions) in the $\gamma\pi^-$ reaction on ^{12}C by Singham, Epstein and myself. It turns out that the surface nature of the pion production in $\gamma\pi$, makes that reaction seem to be quite insensitive to volume changes in the distorted waves. We're looking at this question some more.

Response: Eisenstein

No Comment.

Question: M. Cooper, LAMPF

Why is the ratio for $\beta_{\pi-}/\beta_{\pi+}$ excitation of the 2^+ state in ^{18}O expected to be 1.25 in view of the fact that not all the nucleons in ^{18}O are simply coupled in first order to this state?

Response: Eisenstein

The ratio of 10/8 for inelastic scattering to the inelastic state would be correct if the neutrons and protons participate equally. Since the state in question is reasonably collective (B(E2)≈7spu) such a ratio may not be too bad an approximation.

Question: C. Shakin, Brooklyn College

I understand the motivation in attempting to fit the elastic data with a first-order potential, however, from a fundamental point of view such fits are inappropriate. We know that if one uses a first-order potential the implication is that the inelastic flux is going into the quasi-elastic knockout channels. There is no description of the absorption process which is known to generate a very large part of the reaction cross section. This is one of the reasons that various unphysical modifications of the first-order theory are necessary. Because of the very large effects due to the second-order potential (which has a different geometry than the first-order potential) it is unlikely that we will learn anything concerning the matter distribution.

Response: Eisenstein

Your comment linking first-order optical potentials and the depletion of flux into quasi-elastic channels is well taken. However, when the parameters are allowed to vary freely, the effect is to mock up phenomenologically the second-order effects that have been omitted. Therefore, I would think it a little pessimistic to conclude that the reaction cross sections will be wrong or that we will be unable to extract neutron radii. I do agree that our understanding of the phenomenology is not all that it could be.

Question: J. Alster, TRIUMF

Two comments on the question of βR determinations. 1) The βR description has meaning only for strongly absorbed particles, thus one should not attempt to use it for 50 MeV pion scattering. 2) The first excited state of ^{18}O is somewhat collective and it would be good to analyze the elastic and inelastic scattering in a coupled channel calculation. The values of R and βR could change and consequently modify the difference between π^+ and π^- results.

Response: Eisenstein

Both of these points are well taken.

Question: J.-P. Egger, University of Neuchatel

At SIN, the Neuchatel-Grenoble-SIN collaboration measured the π^-/π^+ cross section ratio for the first excited 2^+ state in ^{18}O (neutron excitation) and ^{52}CR (proton excitation at 180 MeV. $\frac{d\sigma}{d\Omega}(\pi^-)/\frac{d\sigma}{d\Omega}(\pi^+)$ is 1.58±0.15 for ^{18}O and smaller than 1 for ^{52}Cr.

Response: Eisenstein

No comment.

PION-NUCLEUS SCATTERING AROUND THE 3-3 Resonance

C.H.Q. Ingram

Institut für Kernphysik des Kernforschungszentrums und I.E.K.P. der Universität, 7500 Karlsruhe, Fed. Rep. Germany

The experimental knowledge of pion-nucleus scattering from 80 to 300 MeV is reviewed. The accuracy and the extent to which we may learn about nuclear structure from it are discussed. The recent improvement in our quantitative knowledge of the systematics of pion-nucleus reactions is indicated and the implications for an improved understanding of the pion- and Δ-nucleus interaction explored.

INTRODUCTION

In this talk I shall attempt to review the progress of our knowledge of pion-nucleus scattering in the past couple of years, as seen by an experimentalist, rather than to give an exhaustive review of the data. This will, I hope, help to give coherence to my talk, particularly since what data have appeared in print or pre-print form which were not presented at the Zurich conference in 1977 have little natural coherence in their approach to problems in pion-nucleus physics.

So I shall attempt to emphasize the new types of data, and especially those which are sufficiently systematic that we are able either to learn something from them immediately about the pion-nucleus interaction or to hope that comparison with theoretical calculations will be meaningful. On the other hand, I will use the existence of other sessions and workshops as an excuse to say relatively little about nuclear structure studies using pions.

My talk is divided into about five sections. First, I shall make some introductory remarks whose purpose is to establish the point of view from which I shall discuss the data. Then I shall make a few technical comments before starting on elastic scattering. This will be followed by inelastic scattering to discrete nuclear states, while the last part of my talk will be concerned with deep inelastic and other reactions.

Despite all the pre-, over- and other-views given at this conference already, I would like to briefly outline my own historical perspective as a background for reviewing recent progress. Before the meson factories

ISSN:0094-243X/79/540455-46$1.50

began producing large quantities of data, say some three or four years ago, our knowledge of the pion-nucleus interaction could perhaps be rather briefly summarized as follows:

1) Total cross-sections were known, over a respectable range of pion energies and nuclei, and were seen to be essentially geometrical, with an $A^{2/3}$ dependence[1-4].

2) The elastic differential cross-sections on a few nuclei were known (for example refs.[5,6,7]) and shown to be diffractive in nature, i.e. also substantially geometric (except for very light nuclei). Indeed there was initially some puzzlement about how calculations based on Glauber theory, for example, reproduced the data so well[8]. However, it was soon noted that any model which takes proper account of the strength of the pion-nucleon interaction must reproduce the optical blackness of the nucleus for pions. Thus the diffractive nature of the elastic scattering can be reproduced fairly well with little sensitivity to the microscopic processes or modes of inelasticity contained in a given model.

3) True pion absorption, on two nucleons predominantly, was known to be an important reaction mechanism.

In addition, there was an accumulation of data on specific reactions or nuclei, whose net effect was almost more puzzling than helpful, partly because of the unreliability of the data themselves, and partly because isolated data on reactions forming only a small part of the π-nucleus reaction cross-sections can be misleading when the overall systematics are unknown. For less cryptic summaries of the state of our knowledge at this period, review articles exist[9,10].

A couple of years later, this knowledge had been confirmed in greater detail and accuracy, by the new generation of experiments at the meson factories (refs. 11-16). In addition, qualitatively new data became available in the detailed comparison between the scattering of π^+ and π^-, both elastically and inelastically. On N=Z nuclei, differences between π^+ and π^- cross-sections must be attributed to Coulomb effects. This is discussed in some detail in Wilkin's talk at this conference, and I will merely comment that since the Coulomb interaction can be relatively well described, π^+/π^- differences can be used to gain information on the strong amplitudes of the pion-nucleus interaction. I will make more detailed comments on the interactions with N≠Z nuclei later in my talk.

Also confirmed by this new data was the difficulty of learning about the pion-nucleus interaction from this type of data: in elastic scattering the reaction mechanisms are hidden in the small deviations from pure diffraction scattering (or in the backward angle region where the cross-sections are small and contain too extreme a sensitivity to many details of the pion-nucleus interaction for easy interpretation).

There was, however, sufficient and precise enough data to make clear that deficiencies and problems existed in theoretical models when detailed and careful comparisons are made, in particular when all the available data, including the total cross-sections, are compared to the theoretical predictions[17,18].

I will not attempt to summarise the progress of the large amount of theoretical work done up to this time, notably that based on microscopic optical models, (see, e.g. refs. 9 and 10), but will merely point out that no quantitatively reliable description of the π-nucleus interaction existed (nor does now). This was undoubtedly partly due to the ambitious nature of many attempts to build microscopic potentials, while completely new techniques are needed to include the interesting and important phenomena of Δ-formation and pion absorption. However, the lack of data to provide insight into the reaction mechanisms must have also played an important role.

As recently as 1976, our knowledge of the decomposition of the pion-nucleus total cross-section into partial cross-sections relied on a substantial amount of intuition in extracting the information from the available data[9]. Our knowledge of the differential cross-sections in the major reaction channels was virtually non-existent. An important part of this talk will be to show the progress being made in the establishment of the gross systematics of the pion-nucleus interaction.

The experiments of Navon et al.[19] (to which I shall return later) have provided us with partial cross-sections on a range of nuclei from ^{7}Li to ^{209}Bi. Their preliminary results at 165 MeV are shown in fig. 1. The charge exchange cross-sections are estimates based on the inelastic $(\pi^+,\pi^{+\prime})$ cross-sections. Let me note the important increase in our quantitative knowledge this represents, and extract from it the information summarised in Table I, where the partial cross-sections are described as fractions of σ_{TOT} as A increases from 7 to 209.

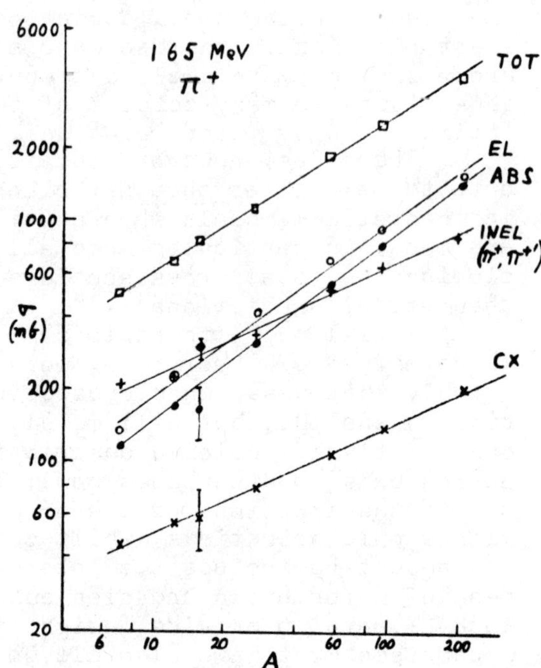

Fig. 1. Decomposition of the π^+-nucleus total cross-section from Li to Bi, at 165 MeV (from ref. 19, preliminary results). The data for ^{16}O are from ref. 52.

Table I Partial π-nucleus cross-sections as % of σ_{TOT} at 165 MeV

	Li → Bi
$\sigma_{elastic}$	∿ 35 - 40 % (∿ const.)
$\sigma_{(\pi,\pi')}$ without charge exchange	35% → 20 %
$\sigma_{absorption}$	20% → 35 %
$\sigma_{charge\ exchange}$ (estimated)	7% → 4 %

Of the two major reaction channels, the absorption cross-section is harder to decompose into differential cross-sections than the pion inelastic cross-section, $\sigma(\pi,\pi')$. This is substantially due to the fact that there is so much energy deposited in the nucleus that there are a large number of channels available. Measurements of the spectra of heavy particles emitted in π-nucleus collisions either do not clearly separate processes with pion absorption from those where a pion remains in the final state, or are obliged, to apply cuts to the data either in the kinematic regions examined, or on the number or type of particles emitted, or usually, both. Theoretical studies have similar problems in that even if a π-2N process is assumed to be the elementary absorption mechanism, the possibilities of what may be called initial and final state interactions are very large and diverse. Statistical cascade descriptions have been used to try to reproduce the data[20,21].

The other major reaction channel, inelastic pion scattering, which proceeds mainly with the knock out of one or more nucleons, is more amenable to systematic investigation. The spectrum of scattered pions must reflect the elementary pion-nucleon collisions inside the nucleus, and also be strongly influenced by the equally important and competing pion absorption process. There is now progress in examining this reaction channel, and I shall discuss this and what we may learn from it in detail later.

The single charge exchange reaction is presumably also dominated by quasi-free processes involving nucleon knockout.

Inelastic scattering to discrete (bound) nuclear states forms a small part of the π-nucleus reaction cross-section, and the body of data available is too small for us to learn about the reaction mechanisms from systematic examination.

To conclude these introductory remarks, I will remind you that the phenomena of Δ-production and propagation and true pion absorption, are not only very interesting aspects of the pion-nucleus system, but also dominating ones. Thus in studying the pion-nucleus interaction in this energy range we may hope to find that the nuclear medium affects the behaviour of a strong resonance (the Δ) in a measurable way. A complete understanding of the role of true absorption at the microscopic level implies an understanding of pion exchange in the nuclear field, since the processes are frequently indistinguishable. Although understanding the pion-nucleus system is not an easy task, it does, however, provide us with these unique opportunities in nuclear physics.

TECHNICAL COMMENT

I will now make a few technical comments on pion-nucleus scattering experiments. Fig. 2 shows a spectrum of pions scattered by ^{12}C taken at Los Alamos (EPICS), and a similar spectrum from π-^{13}C scattering taken by the

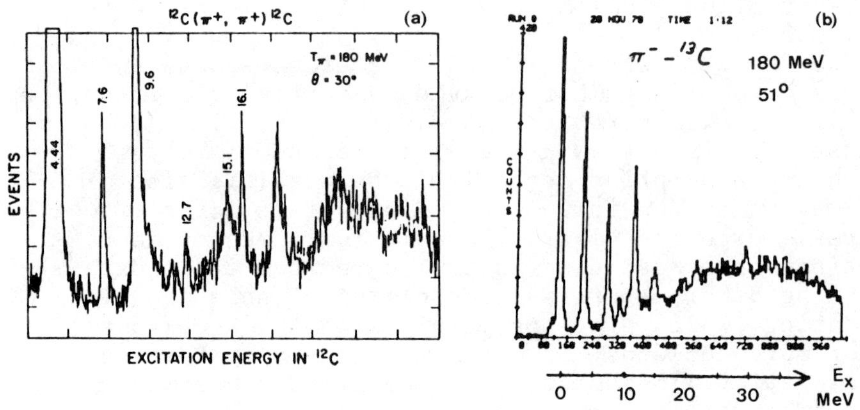

Fig.2 Spectra of pions scattered from ^{12}C, taken at Los Alamos (EPICS) (a); and from ^{13}C taken at SIN (b).

Neuchâtel group with the SIN pion spectrometer. These two spectrometers remain the principal sources of data in this field. I think that the EPICS spectrum is impressive not only in the ability it shows to separate inelastic peaks, but also in the peak-to-background ratio it manifests. The elastic peak, which is off-scale but close, would rise to a height of about 2000 on the scale indicated in the figure. Although the background level at SIN is comparable, the working resolution is in the range 400-450 keV FWHM compared to 250-270 keV at EPICS. The states in the spectrum from ^{13}C have not yet been identified unambiguously.

Considering the complexity of both of these spectrometers, it is perhaps unfortunate, if understandable, that there has so far been very little data available to make direct comparisons possible. In Fig. 3 are shown the differential cross-sections for π^+-^{12}C elastic scattering at 163 MeV, as measured at the two facilities. Although casual inspection indicates fairly good agreement, there are significant differences: First, the depth of the first minimum of the EPICS data is deeper, indicating better angular resolution or a smaller band in the incident momentum. More important is a difference in the position of the minimum, that of the EPICS data being at an angle at least 1/2° smaller than that of the SIN data. Finally, at larger angles, there is significant difference in normalisation of the data, with that from EPICS systematically lower than the SIN data.

Fig. 3. Comparison of π^+-^{12}C elastic scattering cross-sections at 163 MeV, as measured at Los Alamos and at SIN.

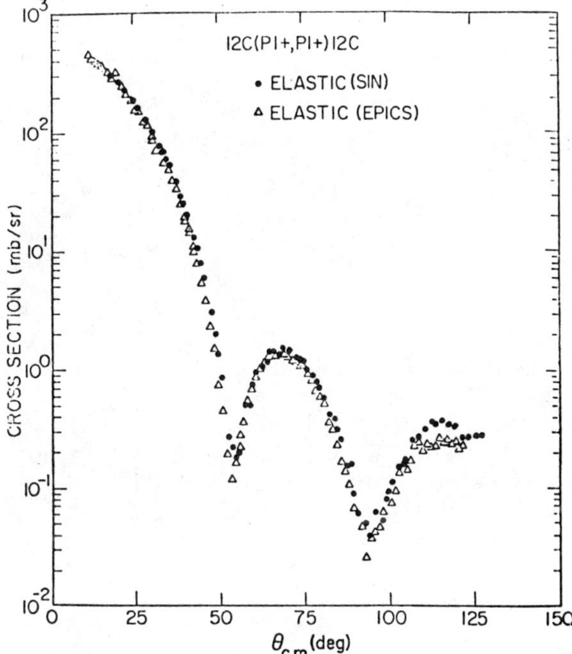

In view of the less-than-exhaustive discussion of possible systematic errors in the published presentation of recent data, this comparison (together with that I shall show later for π-^{18}O scattering) may be an indication of the accuracy of the data which is achieved. How serious such discrepancies are is not readily seen (at least, so far as the backward angle normalisations are concerned), as there is a parallel lack of discussion of sensitivity to errors in theoretical papers. I can only point out that in view of the substantially diffractive nature of elastic scattering, close comparison with very accurate data may prove to be necessary to extract useful information.

ELASTIC SCATTERING

I will begin my discussion of elastic scattering with the special case of deuterium, where a diffractive description is clearly inappropriate. As the simplest case, π-d scattering may be a useful place to gain insight into microscopic models of the π-nucleus interaction, despite the exceptionally low density, of the

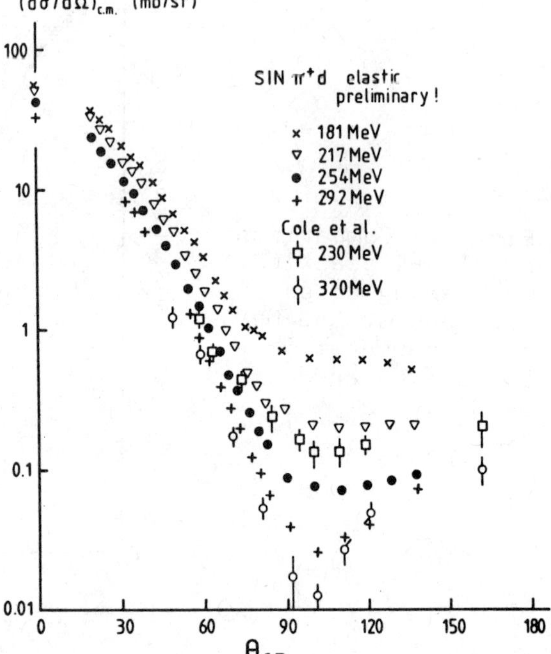

Fig. 4. Preliminary results for π^+-d elastic scattering from ref. 22, with data from ref. 23 for comparison.

deuteron. Fig. 4 shows preliminary results for π^+-d elastic scattering, from Gabathuler et al.[22], at four incident pion energies at and above resonance. These data have been taken with the SIN spectrometer, carefully calibrated by measuring π^+-p scattering under identical spectrometer conditions at each angle. Thus the greatest systematic uncertainty may prove to be in the subtraction of background from the $\pi d \rightarrow \pi p n$ breakup.

Also shown in fig. 4 are the recent data of Cole et al.[23], where the energies are comparable. The agreement is mainly good, with the quite sudden development of the minimum at 100° above 250 MeV being apparent in both sets of data. Cole et al. detected the recoil deuteron to identify elastic scattering, which is a technique particularly suited to backward (pion) angles. This may be connected to their forward point at 320 MeV being lower than the new SIN data, and which is also difficult to reproduce theoretically[24].

Agreement with the bubble-chamber data of Norem et al.[25] at 180 MeV (not shown) is good. The previous experiment of Gabathuler et al. at 256 MeV[26], however, appears to have given too low cross-sections at the backward angles, the new data being up to 60% higher. Recent 3-body calculations of Rinat and Thomas[27] still do not agree with the new data at 254 MeV, although the higher backward cross-sections reduce the discrepancies.

Turning to more "normal" nuclei, the data of Zeidman et al.[28] shown in fig. 5 demonstrate the essentially

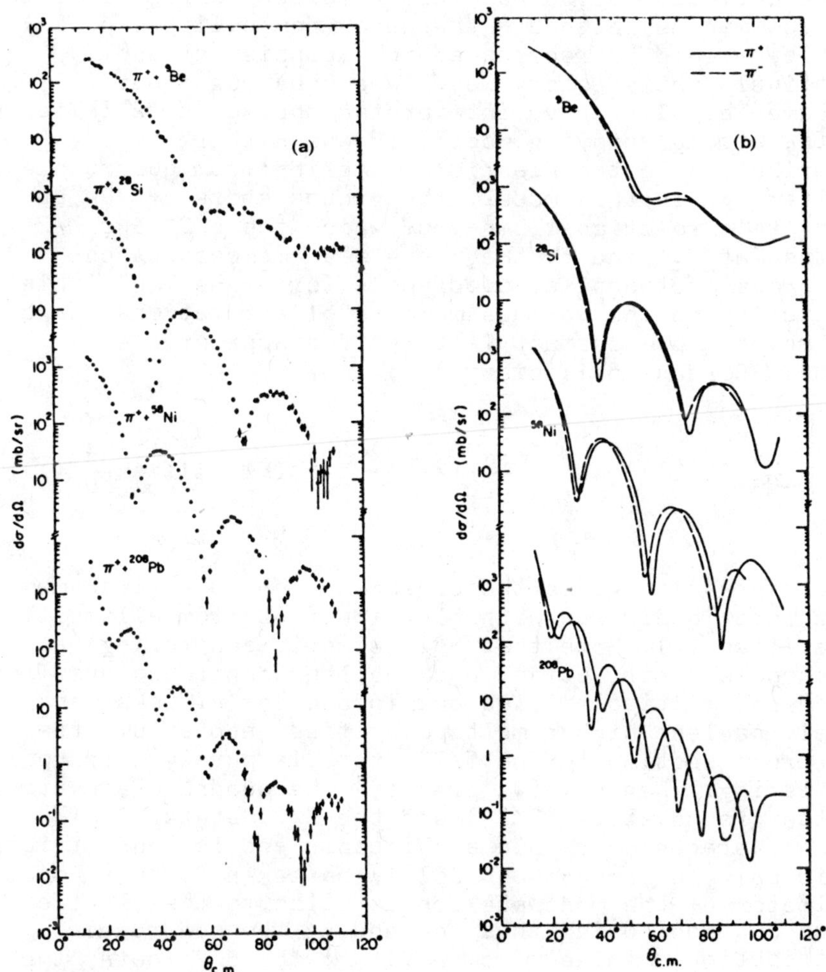

Fig. 5. Elastic pion scattering from ^9Be, ^{28}Si, Ni and ^{208}Pb at 163 MeV, from ref. 28. (a) π^+ data (b) fits to the π^+ and π^- data for comparison.

diffractive nature of elastic pion scattering near resonance (163 MeV), at least for the three heavier nuclei measured. Before discussing the ^{9}Be data, I will just comment that optical model codes show that most of the shift in the positions of the minima between π^+ and π^- on ^{208}Pb is a result of Coulomb effects, with only about 25% of the shift being due to the excess of 44 neutrons.

Zeidman et al. noted that the minima in the elastic distributions from ^{9}Be were much shallower than in the cases of the heavier nuclei (and also than from ^{12}C [12]). In addition, calculations[28] using the first order momentum space optical model code PIPIT[29] failed to reproduce the shallow minima, although the data from silicon, nickel and lead could be reproduced in a consistent and relatively satisfactory way. In a subsequent paper, Geesaman et al.[30] have interpreted the ^{9}Be data in terms of the strong coupling model, in which there is a contribution to the elastic scattering from quadrupole scattering. In this model, the ground state of ^{9}Be belongs to a rotational band, of which the $5/2^-$ and $7/2^-$ states (at 2.43and 6.76MeV) are also members. Then the cross-sections for quadrupole (or higher multipole) scattering to the various members of a rotational band are related by constant factors (the appropriate Clebsch-Gordon coefficients)[31]:

$$\frac{d\sigma}{d\Omega}(I\rightarrow I') = \delta_{I,I'}\frac{d\sigma}{d\Omega}(\lambda=0) + \sum_{\substack{\lambda\neq 0\\ \text{even}}} \langle IK\lambda 0|I'K\rangle^2 \frac{d\sigma}{d\Omega}(\lambda) \qquad (1)$$

This expression gives the cross-section to a given member of a rotational band with contributions from all multipolarities (λ). Note that only elastic scattering has a monopole contribution, and that the contributions from different multipolarities add incoherently. Geesaman et al. neglect higher multipolarities and assume that the cross-section to the $5/2^-$ state is purely quadrupole scattering. Then eq. (1) predicts the quadrupole scattering to the ground state ($3/2^-$) and the $7/2^-$ state. PIPIT, which averages over spin and isospin and is thus strictly valid only for even-even nuclei, is taken to be a good estimator of the monopole contribution to the elastic cross-section; using this and adding the quadrupole contribution obtained from a fit to the $5/2^-$ data, the dashed curves of fig. 6 are obtained (the full lines are the PIPIT calculations alone). The agreement is impressive, although, as the authors point out, possibly fortuitous.

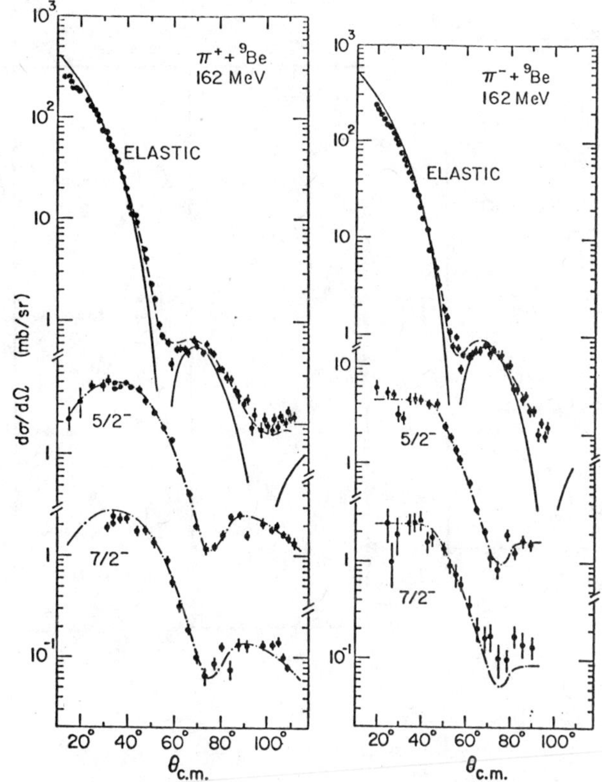

Fig. 6. Elastic and inelastic pion scattering from ^{9}Be at 163 MeV, with calculated curves as described in the text (from ref. 30).

In electron scattering, the ^{9}Be form-factor's minimum is similarly washed out by quadrupole scattering[32] so that the interpretation in terms of nuclear deformation might seem reasonable. However, with pions, effects due to spin-flip and the behaviour of the pion-nucleon amplitude should be considered. Geesaman et al. estimate spin-flip effects to be negligible, but do not appear to have addressed the question of whether PIPIT satisfactorily describes scattering from a light nucleus, which may not be optically "black" for pions. It is well known that in π^--^{4}He elastic scattering[33], there is a minimum near 80° which does not move very much with pion energy, and appears to be a manifestation of the minimum in the π-N cross-section (a minimum at larger angles has a momentum-dependent position more appropriate for diffraction). Furthermore, the minimum at 80° is not very deep for ^{4}He near 163 MeV. Fig. 7 shows the π^--^{4}He data, with PIPIT calculations (dashed lines) using the r.m.s. charge radius of ^{4}He as the matter radius (without correction for the proton form-factor). At least at energies below

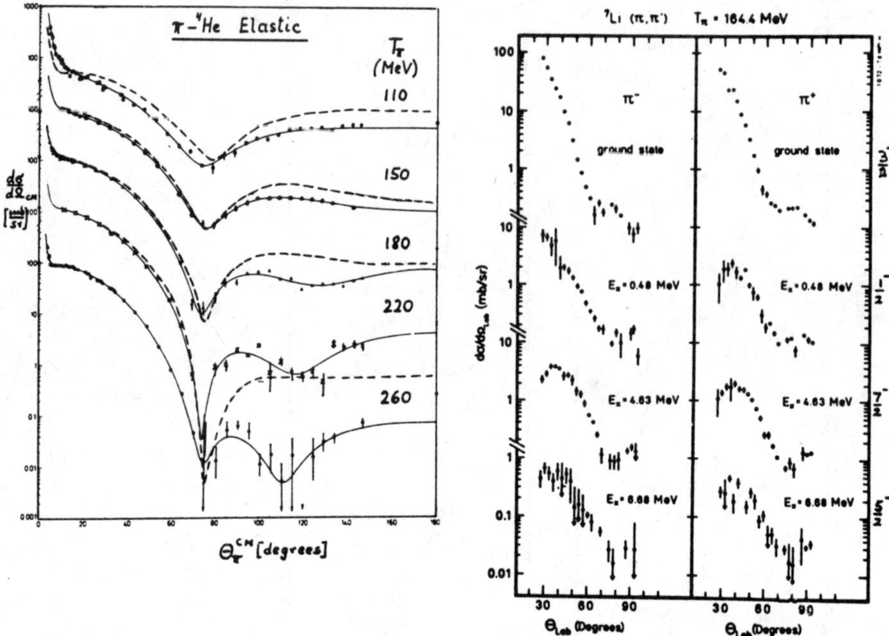

Fig. 7. Elastic $\pi^- - {}^4He$ data[33]. The full lines are fits to the data, dashed lines PIPIT calculations using $\langle r^2\rangle^{1/2} = 1.71$ fm.

Fig 8. Elastic and inelastic $\pi - {}^7Li$ cross-sections at 163 MeV[34].

200 MeV, the agreement is impressively good, and implies that PIPIT ought to satisfactorily reflect any increased transparency of ^9Be for pions (compared to ^{12}C).

At 163 MeV there is also preliminary data on two other light, deformed nuclei, ^7Li[34] (shown in fig. 8) and ^6Li[35]. The authors of these works have not attempted to analyse the depths of the minima in these elastic distributions, and in the contribution to this conference on ^7Li the comparison of the inelastic states with nuclear models does not indicate any simple interpretation. However, at this one energy, the available pion scattering data now forms a significant body of information on these light nuclei.

ELASTIC SCATTERING FROM ISOTOPES

I shall now turn my attention to the comparison of π^+ and π^- elastic scattering from pairs or groups of

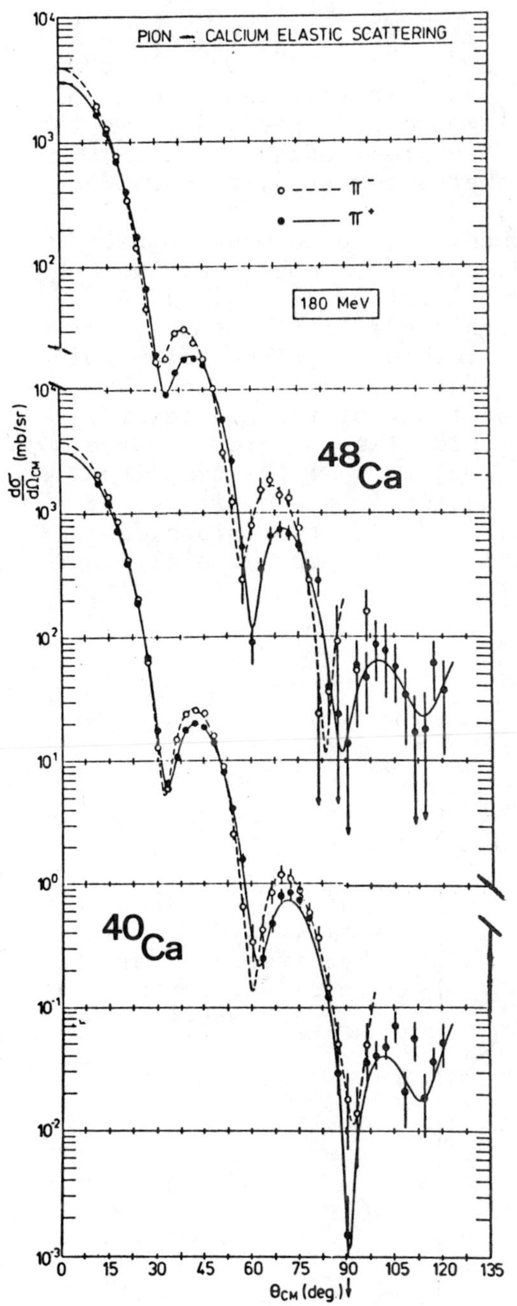

Fig. 9. Elastic scattering of π^+ and π^- from ^{40}Ca and ^{48}Ca at 180 MeV[36].

isotopes. The purpose of these experiments is to investigate the neutron and proton wave functions in the nuclear ground states, the π^- and π^+ being the respective quasi-independent neutron and proton probes (with relative sensitivities of 9 to 1).

At present, the experimental and theoretical uncertainties appear to be sufficiently large as to make the extraction of useful information possible only in careful comparative studies of isotopes or isotones. Typically an N=Z nucleus is chosen as the reference since the π^+/π^--nuclear symmetry is then brokeh by only the Coulomb interaction; even Coulomb effects in the pion-nucleus interaction are sufficiently large, however, to require very careful calculation.

Fig. 9 shows preliminary results for π^+ and π^- elastic scattering on ^{40}Ca and ^{48}Ca at 180 MeV from Egger et al.[36]. These data illustrate the features common to this type of data: notably, the shift in the positions of the diffraction minima between the π^+ and the π^- data for ^{48}Ca is greater than that for ^{40}Ca. There are now data on these two isotopes at anumber of pion energies[37,38], while the other "favoured" isotopes, ^{16}O and ^{18}O, have been studied together at only one energy, 163 MeV[39].

Preliminary data also exist, in the resonance region, on ^{12}C and ^{13}C[36], and on the isotones ^{89}Y and ^{92}Mo[40]. In order to illustrate how these data may be interpreted, I will discuss the ^{18}O and ^{16}O data in some detail. Fig. 10 shows the elastic cross-sections at 163 MeV from Jansen et al. [39] (with the data for each pion charge compared, rather than for each isotope, as in fig. 9).

As the ^{18}O data is the second instance where direct comparison of LAMPF[41] and SIN results is possible, the published cross-sections are shown together in fig. 11. As with the ^{12}C data of fig. 3, there are significant systematic differences, with the minima in the LAMPF data again coming at smaller angles ($\sim 1/2^o$); but in this case the LAMPF data give higher cross-sections at the subsidiary maximum than the SIN data. I would like to underline again, the experimental uncertainity implied by this comparison. On the other hand, the discrepancies in the positions of the minima are eliminated in π^+ and π^- comparisions, as indicated in Table II ($\Delta q = \cdot 001\ fm^{-1}$ is approximately equivalent to $\Delta\Theta = \cdot 045^o$).

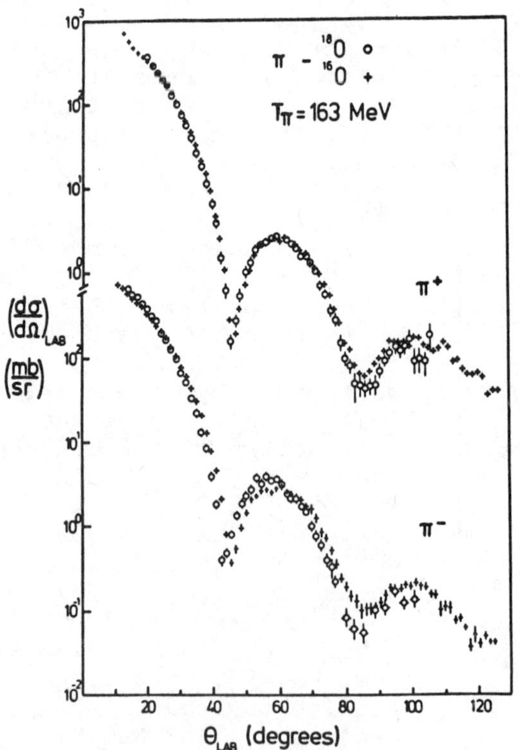

Fig. 10. Elastic scattering from ^{16}O and ^{18}O of π^+ and π^- at 163 MeV[39].

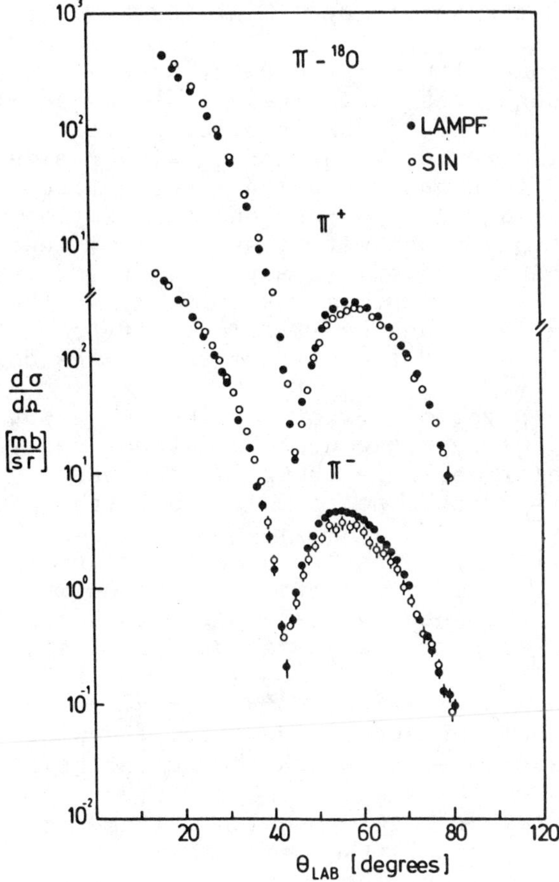

Fig. 11. π^+ and π^- cross-sections at 163 MeV as measured at Los Alamos[41] and at S.I.N.[39].

Table II Shifts in position of first diffraction minima between π^+ and π^- elastic scattering. $\Delta q = q_{\pi^+} - q_{\pi^-}$, where q_π ($\simeq 1.05\ fm^{-1}$) is the momentum transfer at the minimum

	Nucleus	T_π(MeV)	Δq (fm^{-1})
Jansen et al.[39]	^{18}O	163	0.050 ± .006
Iverson et al.[41]	^{18}O	164	0.053 ± .007
Lunke et al.[42]	^{18}O	180	0.049 (no quoted error)
Jansen et al.[39]	^{16}O	163	0.019 ± .006

Included in the table is the shift for ^{18}O at 180 MeV from Lunke et al.[42] and the shift for ^{16}O at 163 MeV from ref.[39].

The crudest interpretation of the data, the black disc diffraction model, concentrates exclusively on these shifts: for true diffraction, the product (qR) = 3.83, where R is the equivalent black disc radius. Approaches using realistic nuclear matter distributions and eikonal[43,39] or optical model[39, 41] descriptions of the pion-nucleus interaction have also been used in comparison with these shifts. There is, of course, significantly more information in the data than just the position of one minimum but this may be the most important characteristic and so I will use it to illustrate how we can interpret the differences in the data from the isotopes.

Before going into details of what conclusions about the neutron matter of ^{18}O may be deduced from these data, I will make some general remarks. Although it has been stated before, there seems to be no harm in repeating, as my first point, that

I. Since the nucleus is optically black to pions, elastic scattering is only sensitive to the nuclear surface. Hence low moments of the density such as the r.m.s. radius cannot be well measured by elastic pion scattering.

The data can always be related to the r.m.s. radius, however, through models of the nuclear density. The wiser course at present appears to be to check the compatibility of the data with models of the nuclear shape derived by independent means such as Hartree-Fock calculations, so that uncertainties are limited to the description of the pion-nucleus interaction. There appears to be a dispute in the literature, whether the pion effectively interacts with the nuclear matter near the half-density radius[44] or about half a fermi further out[43].

II. Coulomb effects in the pion-nucleus interaction must be carfeully handled.

Coulomb effects in ^{18}O will not be completely identical to those in ^{16}O, and since they are large, this should not be overlooked. Nor is there an entirely satisfactory way of checking calculations, since the data on ^{16}O do not distinguish between Coulomb effects in the pion interaction, and those intrinsic to ^{16}O (i.e. differences between neutron and proton matter distributions in ^{16}O). Careful calculation should reduce errors from this source to small if not negligible uncertainties.

III. Last, but not least, is the asymmetry of charge exchange and true pion absorption in $N \neq Z$ nuclei. In a neutron-rich nucleus these routes out of the elastic channel are more available to π^+ than to π^-.

A model of the interaction which uses the free π-nucleon amplitudes as input ought to take care of the charge exchange, but there are no good ways to handle absorption. The effect of this asymmetry will undoubtedlv be reduced by the preference of pions to absorb on an n-p pair rather than an n-n or p-p pair and by the fact that absorption is probably a relatively high density (central nucleus) phenomenon. Since absorption provides nearly half the reaction cross-section, however, to ignore it totally and then draw detailed conclusions about matter distributions (as is done, for instance, in ref.[41] is, at best, sloppy.

Returning to the data, I must continue by pointing out that the larger part of the shifts of the minima shown in figs. 9 and 10 and table II have as their origin two essentially uninteresting effects. These are the Coulomb interaction and the increase in absolute neutron density caused by the 25% excess in the number of neutrons (in ^{18}O). The shift of 0.19 fm^{-1} between π^+ and π^- in ^{16}O implies that 40% of the shift in ^{18}O is of Coulomb origin (with the caveats mentioned after point II above).

We may consider the position of a diffraction minimum to be related to the impact parameter (radius) of a pion's trajectory where there is some probability (~50%) of the pion being scattered out of the elastic channel. An increase of the absolute neutron matter density, within an identical shape, causes this probability to be reached at a greater radius. We can estimate the magnitude of this second uninteresting contribution to the shifts by doing optical model calculations with $\rho_n(^{18}O)$ = 1.25 x $\rho_n(^{16}O)$. With such densities, PIPIT (and ref.[44]) gives a shift between π^+ and π^-, about .02 fm^{-1} greater for ^{18}O than for ^{16}O, which indicates that roughly a further 40% of the shift for ^{18}O shown in Table II derives from this second effect. That around 80% of the observed shift between π^+ and π^- on ^{18}O is accounted for so trivially should be borne in mind when considering comparison between calculations and the data.

The important question to answer is whether the apparent sensitivity of the pion to the nuclear matter distribution is sufficient to overcome the uncertainties of interpretation. Maillet et al.[44] have considered the ^{16}O and ^{18}O data of ref.[39]in considerable detail, using a first-order momentum space optical model code and various published models of the ^{16}O and ^{18}O matter distributions. They find their calculations to be sensitivite to the nuclear form-factor $F^2(q)$ only in the region $q < 2\ fm^{-1}$. Surprisingly, they also find that they are sensitive to the radius parameter, but not to the diffuseness parameter, α, in distributions such as[45]:

$$\rho(r) = \rho_0\ [1+\alpha\ (\frac{r}{a})^2]e^{-(\frac{r}{a})^2}$$

They find that the ^{18}O data alone is completely unable to distinguish between the four matter distributions they used. Indeed not only is the comparison with ^{16}O essential, but they also find that inclusion of the total cross-section[46] and Coulomb-nuclear interference data provide important additional constraints. They have investigated some of the theoretical uncertainties, for example the choice of range for the π-N interaction, and conclude that they are unable to distinguish between models with differing radii, such that they find:

$$r_p^{18} - r_p^{16} \leq 0.05\ fm$$

$$r_p^{16} - r_n^{16} \leq 0.06\ fm$$

$$r_n^{18} - r_n^{16} \leq 0.15\ fm$$

(Notation: r_p^{18}=proton r.m.s.radius in ^{18}O)

Now, while these conclusions are drawn after a refreshingly thorough discussion of the difficulties of interpretation, that due to pion absorption does not appear to have been quantitatively included. The possibility of experimental error which has not cancelled in the ^{16}O comparison has also been ignored, but this seems to be a relatively small worry, at the moment.

In an analysis of the ^{40}Ca-^{48}Ca case, Sternheim and Yoo[47] also found sufficient model dependence that they were essentially unable to draw any reliable conclusion, although they state that one of their analyses gives $r_n^{48} - r_n^{40} \simeq 0.2$ fm

I think it is fair to conclude from this discussion, that the accuracy with which information on nuclear matter distributions may be deduced from these data is not yet sufficient to allow us to claim that pions are significantly advancing our knowledge. Theoreticians need a lot more (and accurate) data to reduce the ambiguities: elastic, including the interference region, and total cross-sections, over a significant range in energies to vary the pion-nucleus interaction, appear to be necessary and always for pairs or groups of isotopes or isotones. The more general problem of how to include pion absorption awaits solution, but it would be pleasant to see attempts at quantitative estimates of its effects on these specific cross-sections.

INELASTIC SCATTERING TO DISCRETE STATES

I will cover this topic only very briefly since it will be discussed by a number of other speakers at this conference. There is also a lot less published than promised, or surreptitiously circulating, data, and so I shall confine myself to a couple of examples.

It is still too early to say whether pions have an unusual facility for exciting certain types of states, where they would prove useful as a spectroscopic tool, although there is still speculation concerning the fairly high excitation region (see G. Walker's and F. Bertrand's talks). And if pions are going to be interesting in the excitation of well-known states, then they must, of course, be able to tell us something new about those states; or else the inelastic scattering must teach us about the pion-nucleus interaction.

A popular goal of inelastic scattering studies is to measure the extent to which a particular state is a neutron (or proton) excitation, rather than a collective state, by using, again, the π^- and π^+ as neutron- and proton-sensitive probes. Excitation of the 2^+ (1.98 MeV) state of ^{18}O, which has been measured together with the elastic data, provides a convenient example. Again the problem of interpretation arises, with neither a sound theory nor a well-developed phenomenology available. Comparative studies can assist us, however, and in the data of Iverson et al.[41] of fig. 12, for example, the 2^+ state in ^{18}O can be readily seen to be more excited by the π^- than by the π^+, while the state or states around 5 MeV are almost equally excited by the π^- and the π^+. They claim that this shows that the 2^+ state contains significantly more contribution from excitation of the valence neutrons than the state(s) around 5 MeV. On the other hand, it is suggested[39, 41] that excitations of these 5 MeV state(s) is reduced, for π^-, by the blocking of excitations of core neutrons by the valence neutrons. Presumably both the effects play a rôle, but we really need a much wider phenomenology of inelastic pion scattering before we can deduce details with confidence.

Cross-sections for π^+ and π^- scattering by ^{28}Si have been measured at 130, 180 and 226 MeV, and this allows a phenomenological separation of structure and reaction mechanism effects to be attempted; Preedom et al.[48] have analysed their data at the three energies together, using the DWIA code DWPI[49] to calculate inelastic cross-sections. The optical potential used was of the

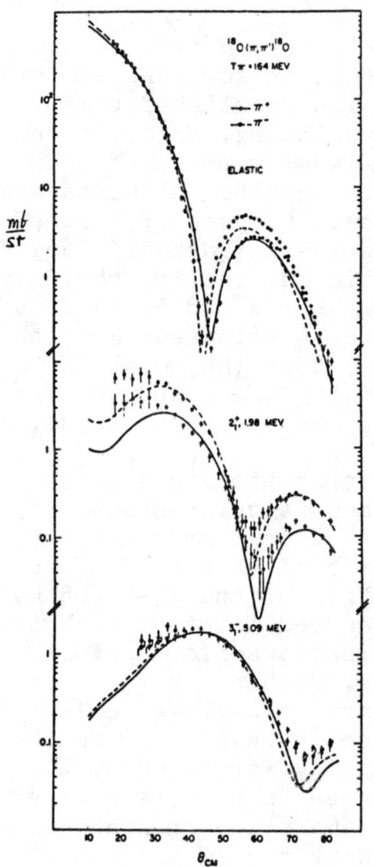

Fig. 12. Elastic and inelastic scattering of π^+ and π^- from ^{18}O at 163 MeV. The curves are from Kisslinger optical potential and DWIA analysis. From ref. 41.

Kisslinger type, $U(\underline{k}, \underline{k}') \sim [b_0 + b_1 \underline{k}\cdot\underline{k}']$, and the nuclear density of the Woods-Saxon form: $\rho(r) = \rho_0 (1 + \exp[(r-c)/a])^{-1}$. Taking the potential parameters b_0 and b_1 from the free π-nucleon cross-sections, and the radius and shape parameters from the electron scattering results (c = 3.1 fm, a = 0.54 fm), cross-sections for inelastic scattering to collective states were calculated. Deformation lengths, $\beta_L c$, were then deduced by normalising the curves to the 2^+ (1.78 MeV), the 4^+ (4.62 MeV) and the 3^- (6.88 MeV) states'measured cross-sections. Only the 2^+ state was unambiguously resolved experimentally.

By fitting to the ^{28}Si elastic data, phenomenological values of b_0 and b_1 were determined, independently for π^+ and π^-, at each energy; in this case, the radius parameter, c, was adjusted to 2.5 fm, to improve the quality of the fits. The deformation lengths were then

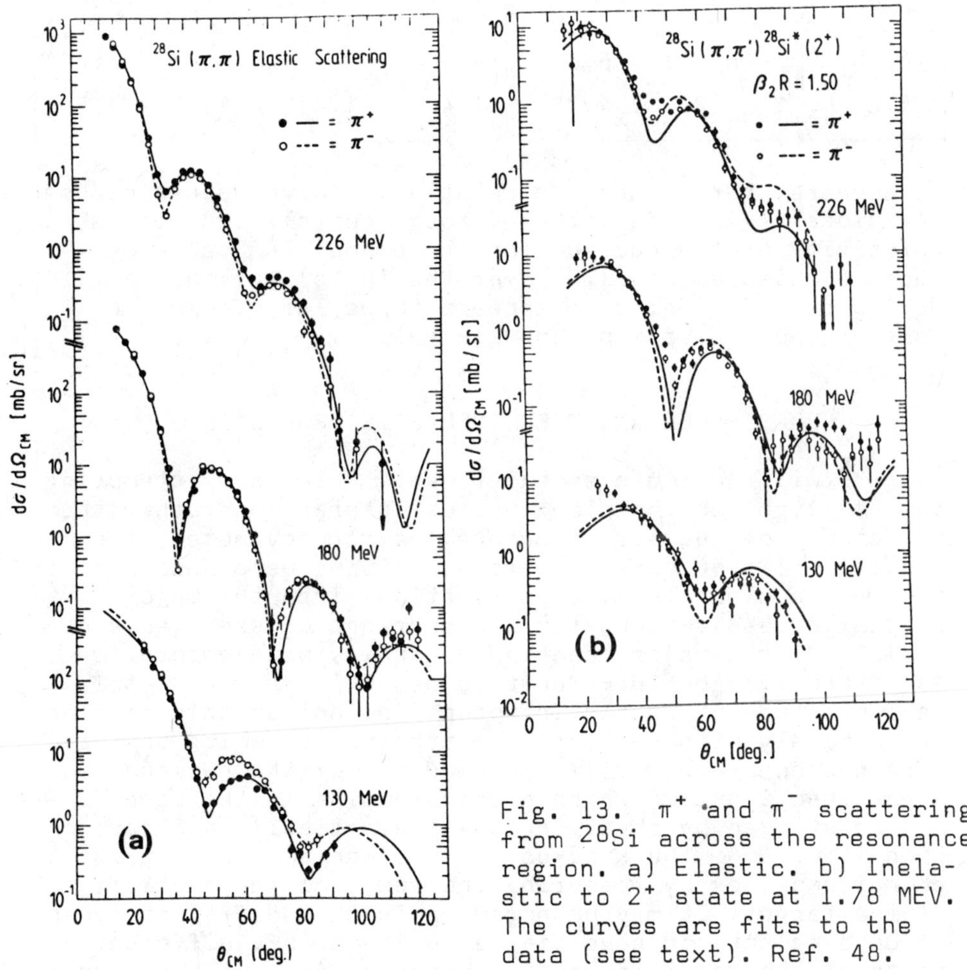

Fig. 13. π^+ and π^- scattering from ^{28}Si across the resonance region. a) Elastic. b) Inelastic to 2^+ state at 1.78 MEV. The curves are fits to the data (see text). Ref. 48.

determined once more using the phenomenological potential; the fits to the elastic scattering and the 2^+ state cross-sections are shown in fig. 13. The authors note that the deformation lengths determined were independent of the incident pion energy (over a range where the π-nucleon cross-sections change by a factor of 2), for both potentials used, and also equal for both pion charges. This is an encouragingly successful test of DWIA analysis in pion scattering, although the deformations deduced appear to be rather large. In Table III, the values of (β_2 c) deduced by various probes are given (from ref.[48]):

Table III (β_2 c) for ^{28}Si (1.78 MeV)

Method	(e,e')	(p,p')	(n,n')	(d,d')	(α,α')	(π,π')
$\beta_2 c$	1.22	1.28	1.47	1.41	1.50	1.50

It appears, here, that the relatively large values deduced for pions is in line with the rough correlation with absorptivity of the nucleus for the probe. The value deduced by Wiedner et al.[50], for the (β_2 c) for the 2^+ (1.4 MeV) state of ^{24}Mg, also appears to be large compared to that deduced from proton scattering.

QUASI-FREE AND DEEP INELASTIC PROCESSES

I will now turn to experiments which are attempting to shed light on the pion-nucleus interaction rather than on details of nuclear structure. As already noted, the results of Navon et al.[19] are now giving us a decomposition of the π-nucleus total cross-sections into the major partial cross-sections for a wide range of energies and nuclei. Of the major reaction channels, pion absorption is interesting but difficult to study in greater detail, as I pointed out in the introduction, and so this section deals mainly with inelastic scattering, in which one or more nucleons are usually knocked out of the nucleus. These reactions, in which a pion remains in the final state, may be studied by observing the inelastically scattered pion itself, or the knocked out nucleon(s), or both in coincidence, or by measuring the residual radioactivity of the target. It may be noted again that different types of data select, or have biases in favour of, different parts of the total inelastic cross-section, with varying amounts of background from pion absorption.

In "one-arm" experiments, (π, π'), in which only the pion is observed in the final state, practically the whole differential cross-section can be measured, without bias and the only background, in this case from pion production, is not a serious difficulty at these energies. Consequently, the data can give a full picture (integrated over all final nuclear and nucleon states), where none of the features are manifestations of natural biases of the experiment. Such unbiased and complete pictures should be useful as guides to reaction mechanisms, and also as providers of phenomenological frameworks within which restricted but more detailed data (such as coincidence experiments) must be viewed. Until recently, however, the experimental data available on (π, π') cross-sections have been limited and of poor quality (for a review, see ref.9).

Since people may not be very familiar with this type of data, before showing the new results I will discuss briefly what one might expect to see in a (π, π') experiment in which the full energy spectrum of the scattered pions is measured.

The pion-nucleon interaction near resonance is so strong that the mean free path between pion collisions inside the nucleus (defined as $\lambda = 1/\sigma x \rho$, where ρ is the nuclear density),is $\leq$ 1 fm. This implies that the spectrum of scattered pions might indicate a dominance of multiple collisions within the nucleus; such a spectrum would lack any notable structure in angle or energy. It also implies that single, quasi-free (π-nucleon) scattering would be visible only from peripheral interactions, in which the pion is scattered <u>out</u> of the nucleus. (In this type of experiment, what happens to the struck nucleon, in this case probably passing through the nucleus, is of no consequence as only the pion is observed).

However, although incident pions near resonance have kinetic energies much larger than nucleon separation energies, their momenta are comparable to the nuclear Fermi momentum. After knocking out a nucleon, a pion's momentum may be substantially reduced (typically by $\sim$ 100 MeV/c) such that in a second collision it might have difficulty in ejecting the struck nucleon out of the Fermi sea. This "Pauli blocking" may be expected to give some reduction to multiple collisions, and also to inhibit single-nucleon knock out at small momentum transfers (i.e. at small scattered pion angle).

Since true pion absorption is a very important process, it must compete with the (π, π') reaction to a considerable extent. If, as a result of a collision within the nucleus, a pion is absorbed, it is lost more or less irrevocably from the (π, π') channel (unless the probability of subsequent pion production is large, which seems unlikely).Thus pion absorption must take strength from the (π, π') reaction, particularly from the multiple collision contribution since the probability of absorption must increase with the number of pion interactions. Furthermore, if the elementary absorption process is a π-2 nucleon mechanism, then the

(π, π') spectrum from single π-nucleon collisions will be affected little by this competing channel. Note that pion absorption is almost immune from "Pauli blocking" because of the availability of the energy from the pion's mass.

Thus we have two mechanisms which might preferentially weaken the multiple scattering contributions to the (π, π') spectra, making them less important compared to that from single quasi-free π-nucleon scattering than is implied by the strength of the π-nucleon interaction.

As I shall compare the nuclear (π, π') cross-sections with the free π-nucleon one, I shall define, here, the quantity N_{eff}, the effective number of nucleons participating in the reaction. This is defined simply as the ratio of the nuclear cross-section to the average π-nucleon elastic cross-section at the same incident energy, where the average is taken over N neutrons and Z protons:

$N_{eff} = \sigma_A\ (\pi,\ \pi') \times A/[Z*\sigma_{\pi p}(el) + N*\sigma_{\pi n}(el)]$ where $A = N + Z$. There are some obvious reasons why we would expect this N_{eff} to be less than A. These are shadowing of the nucleons in the nucleus by each other, true pion absorption, and Pauli blocking. In addition, the π-nucleon average cross-section should be "Fermi-averaged", and finally, there are kinematical and technical reasons (the nuclear "response function" integrated over the experimental variables does not equal A[51]) why N_{eff} would differ from A. Thus, although I shall use this quantity later, we should remember that it is a very crude parametrisation of the data.

Measured doubly differential cross-sections for the reaction ^{16}O (π^+, $\pi^{+\prime}$) for 115, 163 and 240 MeV incident pions, for angles between 30° and 134°, are shown in figs. 14a-c[52]. These are preliminary results, but the systematic errors including normalisation are thought to be about ± 10%. The data were taken with a water target, and, except for the data for 30° or 35°, the π^+– p elastic peak has been omitted from the data shown in the figures, and the gap filled by dashed lines to guide the eye.

At all three incident energies these data show a large, broad peak, occurring at the kinematically appropriate position for quasi-free π-nucleon scattering: the peak centroid is close to the peak from free π^+-p scattering (whose position may be deduced from the gaps in the data), and moves with it to larger energy-loss (smaller scattered pion energy) with increasing scattering angle. For quasi-free scattering from a stationary nucleon, one would expect the scattered pion's energy to be shifted

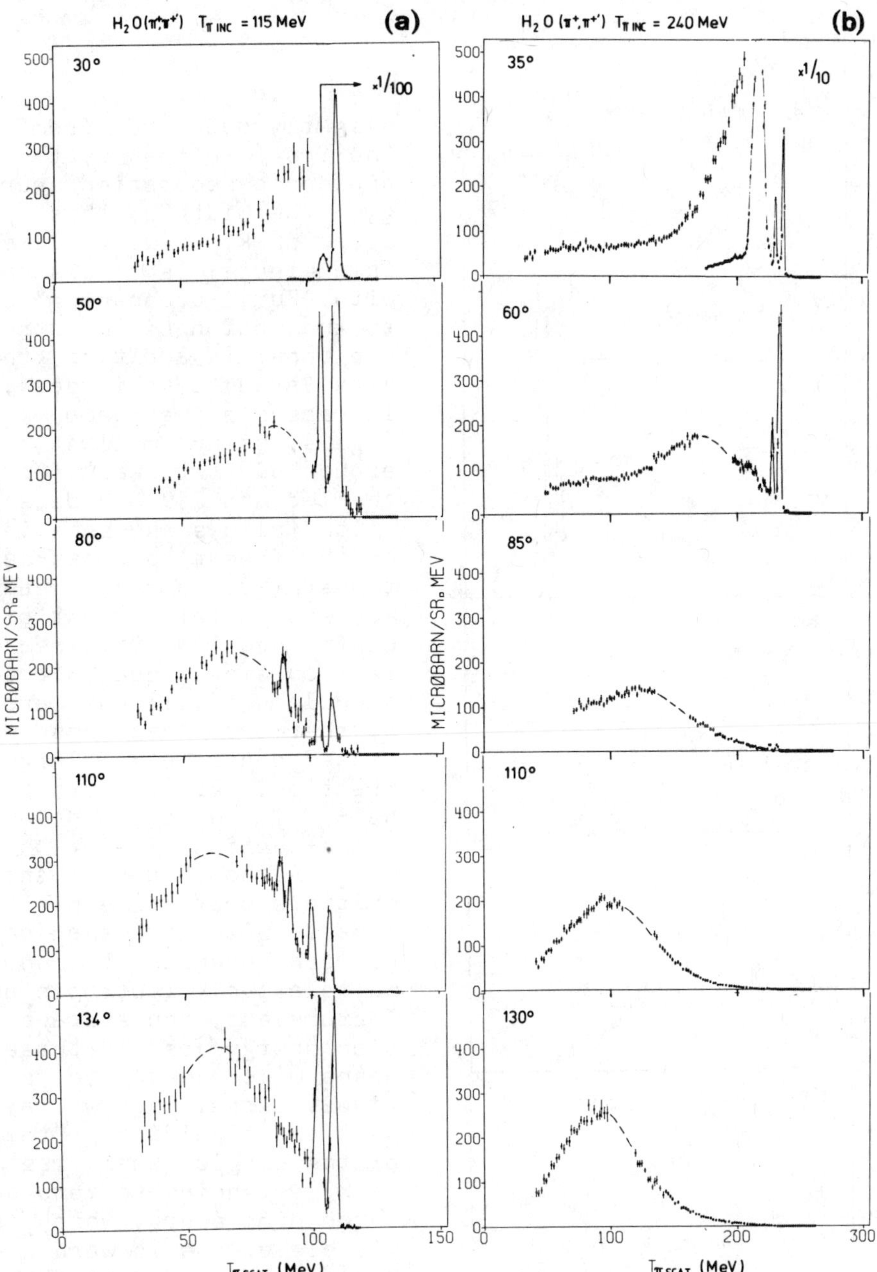

Fig. 14. Doubly differential cross-sections for the reaction $^{16}O(\pi^+,\pi^{+'})$ for incident pion energies below and above resonance: (a) 115 MeV, (b) 240 MeV, and at resonance: (c) 163 MeV on next page, from ref 52. The scales for the kinetic energy of the scattered pions are different for (a), (b),(c).

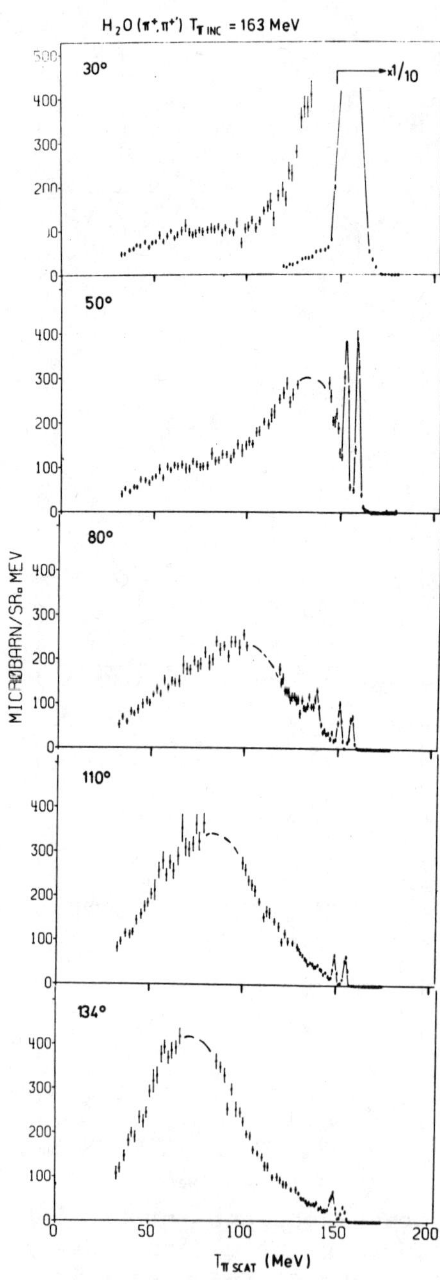

Fig. 14 (c). See caption under Figs 14 (a) and (b). The narrow peaks are from scattering to discrete states.

slightly downwards from the free π-p peak by the nucleon's separation energy. The width of this broad peak, as we shall see in a minute, is consistent with "Fermi-broadening" due to the motion of the struck nucleons. In addition, the size (area) of this peak, increases at backward angles, being smallest around 80° (at least for the 163 and 240 MeV data), which follows the behaviour of the free π^+-p cross-sections. Thus we can already say with a fair amount of confidence that this peak is from <u>single</u> quasi-free π-nucleon collisions, and furthermore, that these processes contribute a large fraction, perhaps more than half, of the total inelastic cross-section of ^{16}O.

Although double and multiple scattering processes (with more than one nucleon knocked out) should be relatively isotropic and featureless, the average pion energy loss in these cases will clearly be relatively large. Thus we may expect the significant "tail" of the single quasi-free peak, extending to zero scattered pion energy which is visible at the forward angles, to be due to these processes. At backward angles where the kinematical recoil shifts the single-scattering peak to the large energy loss end of the spectra, this kinematic separation of the single- and multiple scattering regions is lost, and we must assume that the single-scattering peak contains some multiple-scattering background.

The data show another feature which both confirms the single-scattering nature of the large peaks and illustrates how the data are sensitive to the nature of the π-nucleon interaction within the nucleus. Fig. 15 shows the most backward angle spectrum at each energy, with the positions of the free π-p peaks (H) and the ^{16}O (π, π') peak centroids (located by eye) explicitly marked. It can be seen that the ^{16}O (π, π') centroid moves relative to the free π-p peak, from smaller to larger energy-loss, as the incident pion energy increases - or, more pertinantly, as it passes across the resonance.

To see a possible explantion for this, consider first the 240 MeV data of fig. 15, in the region where

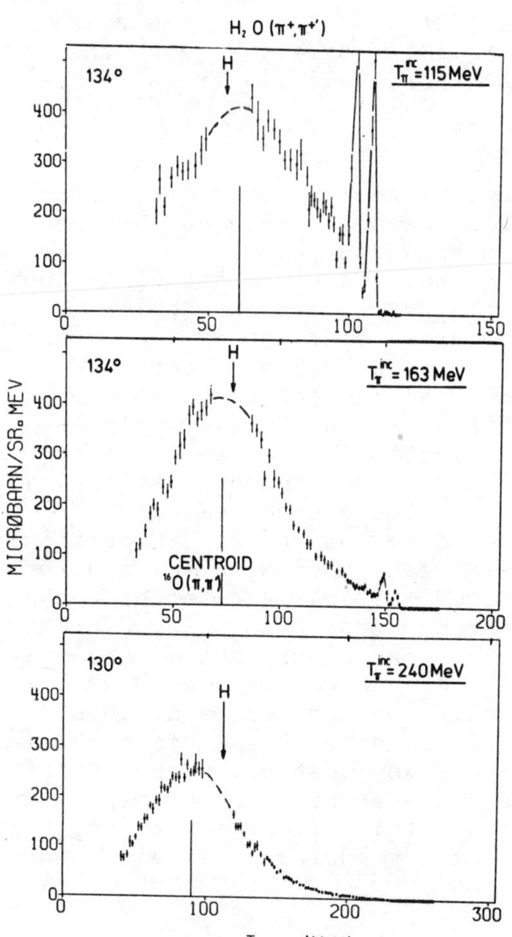

Fig. 15.
^{16}O $(\pi^+,\pi^{+\prime})$ at backward angles across the resonance [52]. Note the different scales for scattered pion energy. See text.

the scattered pion has lost less energy than it does in a collision with a stationary nucleon (indicated by the position of the free π-p peak) i.e. for $T_{\pi'} > 120$ MeV. Pions from single quasi-free collisions will fall into this kinematical region if the momentum vector of the struck nucleon is towards the incoming pion. In this case, the total centre of mass energy of the pion-nucleon system is greater than that in a free π-nucleon collision. For 240 MeV pions, the free π-N cross-section is already well down from the resonance peak (at 180 MeV), and increasing the total energy of the π-nucleon collision reduces it further. Thus, the quasi-free cross-section in this part of the spectrum should be depressed by the kinematics and energy-dependence of the π-nucleon interaction. For scattered pions on the other side of the free π-p peak ($T_{\pi'} < 120$ MeV) the kinematical situation is reversed, the π-nucleon collision is closer to resonance, and the quasi-free cross-section is enhanced. The relevant change in the π-nucleon total energy, W, over the peak of the 240 MeV spectrum in fig. 15, is given approximately by

$$\Delta W = 2\, P_{\pi}\, P_F\, /W \quad \simeq 120 \text{ MeV}$$

(where P_{π}, P_F are the momenta of the incident pion and the struck nucleon). The free π-N cross section changes by a factor of 5 over this energy change.

For 115 MeV incident pions, the kinematic situation is, of course, the same, but since we are here below resonance, the effect on the cross-sections is the reverse: those at high energy-loss ($T_{\pi'} < 60$ MeV) are depressed, those at low energy-loss ($T_{\pi'} > 60$ MeV) are enhanced.

The shift in the position of the $^{16}O\ (\pi, \pi')$ peak relative to that of free π-p scattering, as the incident pion energy crosses the resonance, as indicated in fig. 15, may be seen as a consequence of these enhancements and depressions of the cross-sections. The shapes, in particular the curvatures, of the quasi-free peaks may also be noted. If one considers the effects of multiple scattering backgrounds under the peaks (probably greatest in the 240 MeV data) these effects may become less clear, but if one takes the level of the tail at 35°, or the double charge exchange data I will show shortly, as indicators of this background, the qualitative features of these effects remain. Calculations which employ the fixed scattering approximation, with or without the usual Fermi averaging procedures, should not be able to reproduce this effect since they hold the centre-of-mass energy of the π-nucleon collision constant.

Fig. 16 which shows calculations of Thies, Horikawa and Lenz[51], gives these interpretations some more solid support. These calculations use the plane-wave impulse approximation for the π-nucleon collision within the nucleus, consider only the P_{33} partial wave, and represent single π-nucleon collisions only. A single particle model of the nucleus was used, with final state interactions of the outgoing nucleon (only) included; the prominant peak at about 85 MeV scattered pion energy in the 115 MeV calculations result from the single-particle resonances (analogues of the giant resonances of real nuclei) of the model; this overestimate of this nuclear structure

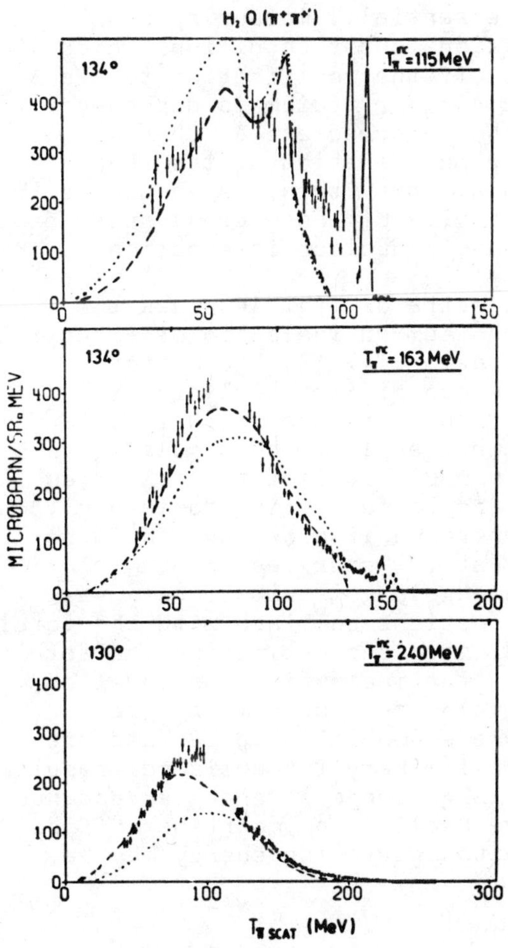

Fig. 16. The data of fig. 15 compared with calculations of ref. 51. Each of the calculated curves is arbitrarily normalised. See text.

effect is not important for this discussion.

The two curves in fig. 16, each of which is arbitrarily normalised, represent different ways of treating the microscopic π-nucleon dynamics. The dotted line uses the fixed-scatterer approximation, while the dashed line includes the variation of the Δ kinetic energy over the energy-loss spectrum - that is, the energy variation of the resonant πN amplitude is included. First, note that these curves confirm that the widths of the peaks in the data are appropriate for Fermi-broadening of quasi-free π-nucleon scattering. Second, the qualitative effect of the shift of the peak in the data relative to the free π-p peak is reproduced in the dashed line, where the π-nucleon amplitude's energy variation is included.

Now this is more than just a pleasing effect. It shows that these data are essentially measuring single quasi-free π-nucleon collisions under conditions which vary with pion energy-loss. If the modification to the π-nucleon amplitude by the nuclear medium is dependent on the kinetic energy of the intermediate Δ, then we must expect a corresponding modification of the shape of these energy-loss spectra. This, then, is an explicit demonstration of the sensitivity of these cross-sections to the dynamics of the π- and Δ-nucleon interaction within the nucleus.

Integration of the spectra of fig. 14 gives the angular distribution for the summed inelastic differential cross-sections on ^{16}O, shown in fig. 17. The dotted lines are the average π-nucleon differential cross-sections, multiplied by an arbitrary factor (N_{eff}), to give agreement at the backward angles. The angular distributions are seen to broadly reflect the π-N cross-sections (consistent with single-scattering dominance), except at forward angles where Pauli blocking should become important, especially at low energies.

Integration of these angular distributions gives the total inelastic cross-sections: combined with the total cross-section, the total elastic and estimating the total charge-exchange as 1/5th of the inelastic, the total absorption cross-section may also be deduced. The preliminary results of this are shown in Table IV, and are broadly in line with the preliminary (unpublished) results of ref. 19, which cover a large range of energies and nuclei. In particular, the relative decline in importance of the absorption cross-section with increasing energy may be noted.

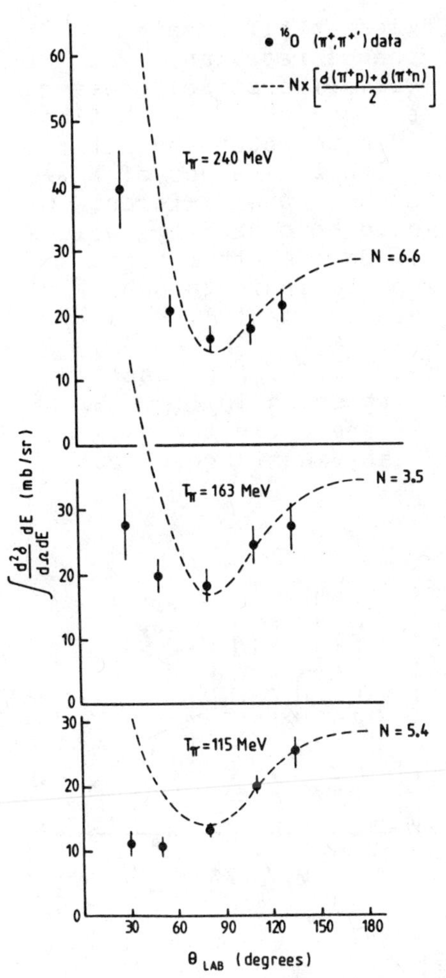

← Fig. 17. Summed inelastic cross-sections [52] for the reaction ^{16}O $(\pi^+,\pi^{+'})$, compared to the average π-nucleon elastic cross section.

Fig. 18. Summed charge exchange cross-section [53] at 100 MeV for the reaction ^{16}O (π^+,π^0).

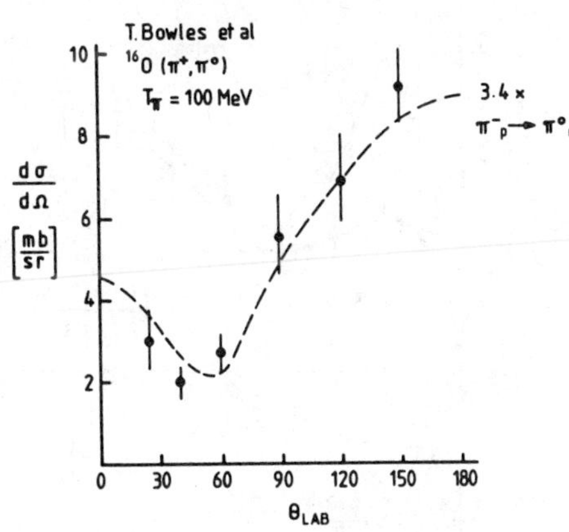

Table IV Preliminary partial cross-sections for π^+ - ^{16}O in mb

	115 MeV	163 MeV	240 MeV
σ_{TOT}[2,3]	735 ± 15	807 ± 15	676 ± 15
σ_{el}[13,18]	280 ± 20	297 ± 20	270 ± 20
$\sigma_{(\pi^+,\pi')}$[52]	224 ± 30	291 ± 40	284 ± 40
σ_{cx}(est)	45 ± 15	60 ± 20	57 ± 20
σ_{abs}	186 ± 40	159 ± 50	65 ± 50

In a similar experiment, Bowles et al.[53] have measured the quasi-free charge-exchange reaction, ^{16}O (π^+, π^0), at 100 MeV, and the summed differential cross-sections are shown in fig. 18. The $\pi^- p \to \pi^0 n$ distribution has again been arbitrarily normalised to match the data; the required factor of 3.4 agrees with the values of N_{eff} shown in fig. 17, if account is taken of the different definition (the N_{eff} of fig. 17 would have half the values shown, in a comparable definition). On the other hand, the charge exchange data do not seem to indicate any strong Pauli blocking effect as seen in the $(\pi^+, \pi^{+'})$ data.

The data of Gismatullin et al.[54], for the reaction $\pi^+ + {}^{12}C \to \pi^+ + p + {}^{11}B(*)$, which are shown in fig. 19, may shed light on this apparent difference in the oxygen data[51]. At 112 MeV Pauli blocking is evident again, but

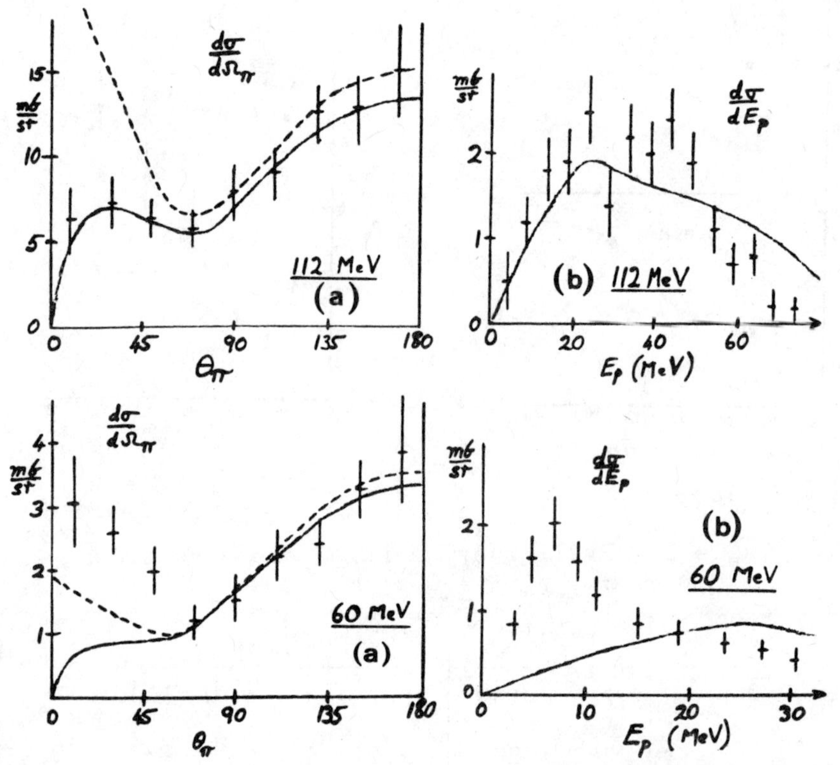

Fig. 19. Data for the reaction $\pi^+ + {}^{12}C \to \pi^+ + p + {}^{11}B(*)$ from ref. 54. (a) integrated over scattered pion energy. (b) integrated over knocked out proton angle. The full curves are from a PWIA Fermi gas model calculation of Lenz[9], and the dashed curves the π^+-p laboratory cross-section, arbitrarily normalised.

at 60 MeV there is actually a relative enhancement of the π - ^{12}C cross-section at the forward pion angles. The spectrum of knocked out protons (which is integrated over all angles) at 60 MeV contains a pronounced peak near 8 MeV, which is absent however, in the 112 MeV data. Linking these two qualitative differences in the 60 and 112 MeV data, we can speculate that the enhancement of the forward pion cross-section at 60 MeV may be due to strong excitation of particle unstable states in this region. In other words, at small momentum transfers enhancements of the cross-section by strong nuclear resonances may become greater than the reductions from Pauli blocking.

It seems appropriate to consider, here, one of the puzzles in this field. Fig. 20 shows the data of Binon et al.[4] for the summed inelastic cross-sections on ^{12}C at 0°, together with points deduced by extrapolating the data of fig. 17 and of fig. 18 (multiplied by the appropriate isospin factor of 5). The puzzle in the ^{12}C data is the sudden large rise in cross-section as the incident energy falls below 150 MeV. This seems to imply either a rapid increase in multiple scattering processes, or of some non-locality of the interaction, overcoming the Pauli blocking. The new oxygen results at 115 MeV and 100 MeV disagree with the ^{12}C data and indicate a steadily falling cross-section at lower energies, consistent with what one might expect as the π-nucleon cross-section falls, and Pauli blocking effects increase.

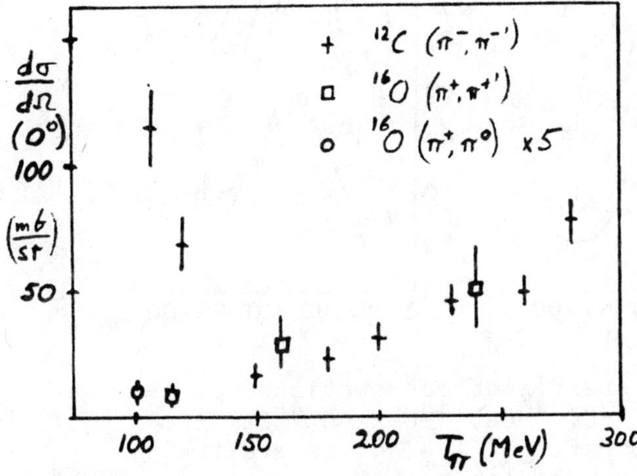

Fig. 20. Summed inelastic cross-sections at 0° on ^{12}C, and on ^{16}O. The ^{16}O points are extrapolations to 0°, the point at 100 MeV is for the charge exchange reaction, multiplied by 5. Data from refs. 4, 52 and 53.

In a contributed paper to this conference, Blanpied et al.[55] show (π, π') spectra from ^{12}C, ^{40}Ca, ^{44}Ca and ^{48}Ca, at 180 and 290 MeV, for both π^+ and π^-. Fig. 21 shows their results at 290 MeV and 60°. Isotopic comparisons in this type of experiment tell us about the dynamics of the interaction, since the addition of neutrons alters the balance between the reaction channels differently for π^+ and for π^-. For the π^-, the neutrons are the principal source of quasi-free elastic collisions, and so we might expect a corresponding enhancement of the (π^-, $\pi^{-'}$) cross-section in the neutron-rich isotopes. For the π^+, however, additional neutrons may primarily enhance the probability of leaving the (π^+, $\pi^{+'}$) channel through charge exchange or pion absorption:it is possible to envisage a reduction of the (π^+, $\pi^{+'}$) cross-section due to increased competition. The consequences of nuclear

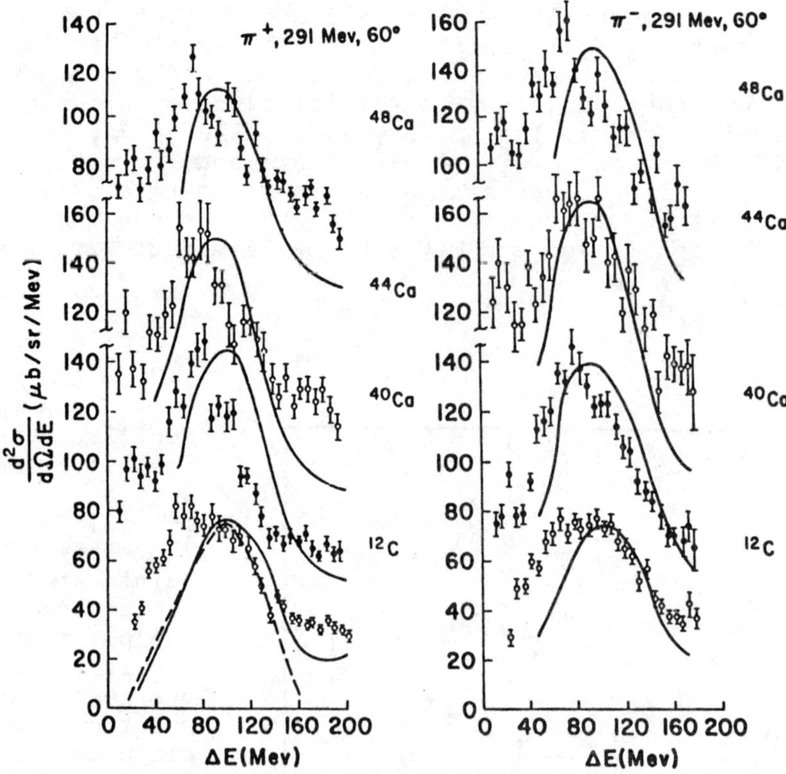

Fig. 21. Doubly differential cross-sections for the ($\pi^{\pm}$,$\pi^{\pm'}$) reaction on ^{12}C, ^{40}Ca, ^{44}Ca and ^{48}Ca at 291 MeV and 60° from ref. 55. The curves are the results of Monte Carlo calculations.

structure should, of course, be considered - for example a change in binding energy might shift the spectrum a few MeV.

I shall return now to the relative importance of single and multiple scattering processes. I have indicated the kinematic separation of the two processes in the forward angle data, and will now examine this in more detail. Fig. 22a shows the 240 MeV ^{16}O (π. π') data at 60° in comparison to the 360 MeV ^{12}C (e, e') data of Mougey et al.[56] at 60° with an arbitrary normalisation. The momentum transfer for these two spectra are equal within 5% throughout, but it should be noted that the electrons have an incident energy 120 MeV higher. The rise in the (e, e') cross-section at energy losses greater than 250 MeV is from pion production. If we subtract the (e, e') cross-section (as normalised in fig. 22a from the (π, π') data, we have a rough estimate of the multiple scattering contribution to the ^{16}O (π, π') cross-section.

In fig. 22b the same (π, π') data is compared to the calculation of Thies et al.[51], once more arbitrarily normalised. We may once more assign at least most of difference between the calculation and the data to multiple scattering processes.

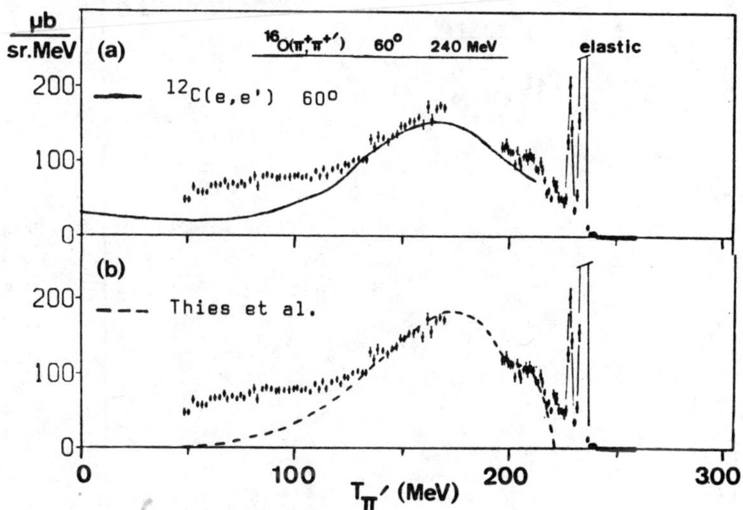

Fig. 22. The ^{16}O ($\pi^+,\pi^{+\prime}$) data [52] at 240 MeV (355 MeV/c) and 60° compared to a) the ^{12}C (e,e') data [56] at 360 MeV and 60°; b) calculations of single pion quasi-free scattering from Thies et al. [51] with arbitrary normalisation.

Pion double charge exchange can also tell us something about the importance of multiple scattering processes, since it cannot proceed via a single π-nucleon interaction. The free π-nucleon cross-section, at resonance, goes in the ratio 9 : 2 : 1 for π^+-p elastic: π^+-n→π^0-p: π^+-n elastic interactions. If one takes equal numbers of neutrons and protons in the nucleus, we find that multiplying the isospin factors gives us relative probabilities of 26 : 9 : 1 for finding a $\pi^+ : \pi^0 : \pi^-$ in the final state, for a π^+ undergoing <u>two</u> quasi-free interactions. Thus if the (π, π') reaction proceeds via largely unperturbed π-nucleon collisions at resonance we would expect the quasi-free double charge exchange cross-section to be just 1/26 of the double scattering contribution to ^{16}O (π^+, π^+).

Fig. 23a shows the measured double charge exchange cross-sections on ^{16}O at 240 MeV, at 50°, 85° and 130°[57]. Note that these are quite isotropic, with a much smaller shift to greater energy loss with increasing angle than for the single-scattering dominated (π^+, $\pi^{+\prime}$) data. Using this isotropy, and extrapolating to 0 MeV T_{π}', a total double charge exchange cross-section of 4.6 ± 0.7 mb is obtained.

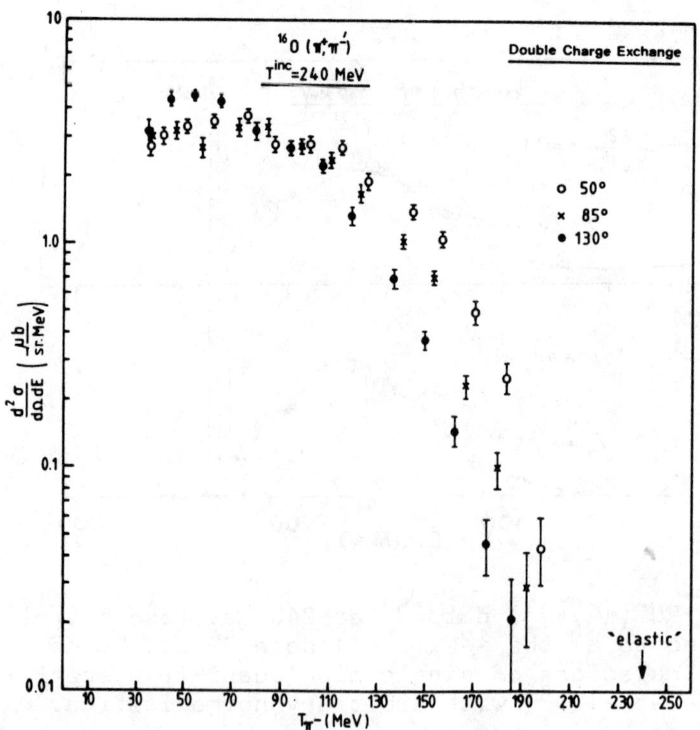

Fig. 23a. Pion Double Charge Exchange on ^{16}O at 240 MeV

Pion production could contribute significantly to these (π^+, π^-) spectra as an unwanted background. The total cross-section on emulsion nuclei has been reported[59] as about 150 μb at 240 MeV, for the reaction $\pi^- + A \rightarrow \pi^+ + \pi^- + A'$, one tenth of the corresponding double charge-exchange cross-section. If we assume isotropy, with the pion spectrum peaking at 35 MeV $T_{\pi'}$, then we may get an estimate of the peak cross-section of 35 $\mu b.sr^{-1}.MeV^{-1}$. This is a 10% background. On the other hand, the total cross-section for the elementary cross-section $\pi^- + p \rightarrow \pi^+ + \pi^- + n$ has recently been reported[60] to be 150 μb also, equal to that of emulsion nuclei, implying that this background could be higher. However, bearing this in mind, and that at the detailed level Coulomb barrier and nuclear separation effects may perturb the π^+/π^- symmetry, we may now compare the double charge exchange with the (π^+,$\pi^{+'}$) results.

In fig. 23b, the estimates for multiple scattering at 60° obtained from the comparisons of figs. 22a and 22b, are compared to the double charge exchange data at 50°. The estimates have been divided by 17. The agreement is very satisfactory considering the approximations used to obtain the estimates, and implies that the double charge exchange spectra may be used to indicate the level of multiple scattering in the (π^+, $\pi^{+'}$) spectra at larger angles.

Concluding this discussion we may say with some confidence that at 240 MeV, multiple scattering processes with more than one nucleon knocked out of the nucleus, appear to make up less than half the ^{16}O (π^+, $\pi^{+'}$) cross-

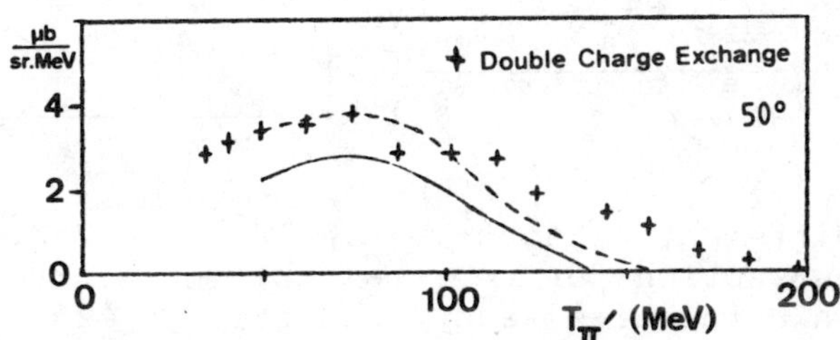

Fig. 23b. The ^{16}O (π^+,π^-) data at 50° compared to estimates of multiple scattering in the ^{16}O (π^+,$\pi^{+'}$) reaction at 60°, divided by 17. These estimates are obtained by using the ^{12}C (e,e') data (full line), and the calculation of ref. 51 (dashed line) as indicators of the single scattering (see fig. 23a).

section. At lower energies we may expect multiple scattering to be even less important. It is clear from the size of the pion absorption cross-section that this must be an important competing process throughout the resonance region, and that this may be expected to reduce the multiple-scattering contributions in particular. Thus we may hope that those data reflect not only the pion-nucleon interaction inside the nucleus, but also the mechanisms of pion absorption; that is, the shape of the spectra should reflect the whole pion-nucleus dynamics.

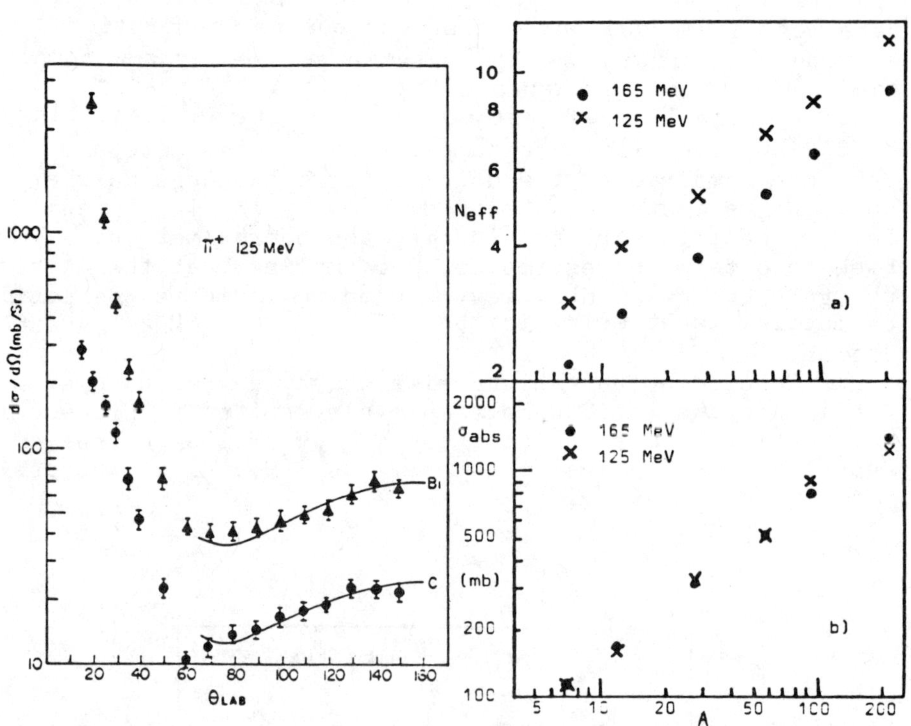

Fig. 24. Differential scattering cross-section for 125 MeV π^+ on C (circles) and Bi (triangles). The solid line is the free πN cross-section normalized to the data.

Fig. 25a and 25b.
a) Neff obtained for π^+ energies of 125 MeV and 165 MEV vs. atomic mass no. of target A.
b) Cross-section for absorption of 125 MeV and 165 MeV pions, vs. atomic mass no. of target A.

To end my talk I shall return to the experiments of Navon et al.[19], which were composed of two types. In one, scattered pions were identified as such but neither their energy nor charge (π^+ or π^-) determined. Thus the data produced, such as that in fig. 24, are angular distributions of the summed elastic, inelastic and double charge exchange cross-sections. In the forward region these are essentially elastic cross-sections, in the backward region, (π,π') cross-sections.

The second type of experiment was a standard transmission experiment giving a total cross-section, from which the results of the first experiment are subtracted, to give the summed absorption and single charge exchange cross-section. By doing the subtraction in the raw data, the difficulties of extrapolating to zero solid angle, and of the Coulomb effects are removed. To obtain the total absorption cross-section, the single charge-exchange cross-section must, however, be estimated at present.

Only the 165 MeV results of fig. 1, and 125 MeV results are presently available, and these are summarised in figs. 25a and b. The N_{eff} of fig. 25a is obtained by fitting the average π-nucleon elastic cross-section to the backward part of the data of the measured [elastic+(π,π')] cross-sections.

CONCLUSION

To conclude I will point to the steady flow of new pion-nucleus data, only part of which I have been able to show here, and to the increasing diversity of the types of data. We now have significant amounts of data not only on elastic and total cross-sections, but also on inelastic reactions, both to bound and unbound final states and on the partial reaction cross-sections,while there is a visible growth of interest in the single and double charge exchange reactions. There are still plenty of obvious gaps, partly because these experiments are really not very easy to do well. There is still a need for more systematic studies. An isolated angular distribution of the excitation of a state for example, is unlikely to yield useful information unambiguously, and a collection of such distributions on a variety of nuclei at a variety of energies is not a great deal better unless the choices are the product of some unifying motivation. The ablity to vary the pion interaction substantially by varying the pion energy across the resonance is a very powerful diagnostic tool which is discarded in a measurement at a single energy.

However, the increasing variety of data is not only increasing the constraints of theoretical interpretation, but in its growing emphasis on defining the systematics of the reaction mechanisms should help to point out how theoretical models of the pion-nucleus interaction may be improved. This is encouraging both for those whose primary interest lies in the pion-nucleus system itself, with its own unique degrees of freedom, and for those who wish to use the pion's well known special properties as a probe for nuclear structure studies.

It is a pleasure to thank members of and visitors to the S.I.N. theory group for many helpful discussions, and also many colleagues who have assisted me in preparing this talk.

REFERENCES

1 C. Wilkin, C.R. Cox, J.J. Domingo, K. Gabathuler, E. Pedroni, J. Rohlin, P. Schwaller and N.W. Tanner, Nucl. Phys. B62 (1973) 61

2 N.D. Gabitzsch, G.S. Mutchler, C.R. Fletcher, E.V. Hungerford, L. Coulson, D. Mann, T. Witten, M. Furić, G.C. Phillips, B. Mayes, L.Y. Lee, J. Hudomalj, J.C. Allred and C. Goodman, Phys. Lett. 47B (1973) 234

3 A.S. Clough, G.K. Turner, B.W. Allardyce, C.J. Batty, D.J. Baugh, W.J. McDonald, R.A.J. Riddle, L.H. Watson, M.E. Cage, G.J. Pyle and G.T.A. Squier, Nucl. Phys. B76 (1974) 15

4 A.S. Carrol, I.H. Chiang, C.B. Dover, T.F. Kycia, K.K.Li, P.O. Mazur, D.N. Michael, P.M. Mockett, D.C. Rahm and R. Rubinstein, Phys. Rev. C14 (1976) 635

5 F. Binon, P. Duteil, J.P. Garon, J. Görres, L. Hugon, J.P. Peigneux, C. Schmit, M. Spighel and J.P. Stroot, Nucl. Phys. B17 (1970) 168

6 J. Rohlin, S. Rohlin, B.W. Allardyce, J.J. Domingo, C.H.Q. Ingram, N.W. Tanner, E.M. Rimmer and J.P. Girardeau-Montaut, Nucl. Phys. B37 (1972) 461

7 M. Blecher, K. Gotow, D.K. Anderson, R. Kerns, R. Minehart, K. Ziock, R.W. Bercaw, J.S. Vincent and R. Johnson Phys. Rev. C10 (1974) 2247

8 C. Schmit, Nuovo Cimento Letters 4 (1970) 454; C. Wilkin, ibid, p. 491

9 F. Lenz, Proceedings of the Topical Meeting on Intermediate Energy Physics at Zuoz (1976), published by SIN

10 J. Hüfner, Phys. Reports 21C (1975) 1

11 J. Piffaretti, R. Corfu, J.P. Egger, P. Gretillat, C. Lunke, E. Schwarz and C. Perrin, Phys. Lett. 67B (1977) 289

12 J. Piffaretti, R. Corfu, J.P. Egger, P. Gretillat, C. Lunke, E. Schwarz, C. Perrin and B.M. Preedom, Phys. Lett. 71B (1977) 324

13 J.P. Albanèse, J. Arvieux, E. Boschitz, C.H.Q. Ingram, L. Pflug, C. Wiedner and J. Zichy, Phys. Lett. 73B (1978) 119

14 C.H.Q. Ingram, E. Boschitz, L. Pflug, J. Zichy, J.P. Albanèse and J. Arvieux, Phys. Lett. 76B (1978) 173

15 E. Boschitz in Proceedings of the Seventh International Conference on High Energy Physics and Nuclear Structure, Zurich ed. M.P. Locher, Birkhäuser, 1977

16 J. Arvieux, J.P. Albanèse, J. Bolger, E. Boschitz, C.H.Q. Ingram, L. Pflug, J. Jansen, J. Zichy, E. Rost and A.S. Rosenthal, to be published in Nucl. Phys.

17 J.P. Maillet, J.P. Dedonder and C. Schmit, Nucl. Phys. A316 (1979) 267

18 M. Hirata, J.H. Koch, F. Lenz and E.J. Moniz, SIN preprint 78-011, to be published in Annals of Physics

19 I. Navon, D. Ashery, G. Azuelos, H.J. Pfeiffer, H.K. Walter and F.W. Schleputz, to be published, contribution to this conference, and private communication

20 J.N. Ginocchio, Phys. Rev. C17 (1978) 195

21 J. Hüfner and M. Thies, submitted to Phys. Rev.

22 K. Gabathuler, J.J. Domingo, J. Arvieux, J. Bolger, P.A.M. Gram, W. Hirt, C.H.Q. Ingram, R.E. Mischke, P. Schwaller and J. Zichy, private communication

23 R.H. Cole, J.S. McCarthy, R.C. Minehart and E.A. Wadlinger, Phys.Rev. C17 (1978) 681

24 A.S. Rinat and A.W. Thomas, Nucl. Phys. A282 (1977) 365

25 J.H. Norem, Nucl. Phys. B33 (1971) 512

26 K. Gabathuler, C.R. Cox, J.J. Domingo, J. Rohlin, N.W. Tanner and C. Wilkin, Nucl. Phys. B55 (1973) 397

27 A.S. Rinat, private communication

28 B. Zeidman, C. Olmer, D.F. Geesaman, R.L. Boudrie, R.H. Siemssen, J.F. Amann, C.L. Morris, H.A. Thiessen, G.R. Burleson, M.J. Devereux, R.E. Segel and L.W. Swenson, Phys. Rev. Lett. 40 (1978) 1316

29 R.A. Eisenstein and F. Tabakin, Comp. Phys. Comm. 12 (1976) 237

30 D.F. Geesaman, C. Olmer, B. Zeidman, R.L. Boudrie, R.H. Siemssen, J.F. Amann, C.L. Morris, H.A. Thiessen, G.R. Burleson, M.J. Devereux, R.E. Segel and L.W. Swenson, Phys. Rev. C18 (1978) 2223

31 J.S. Blair and I.M. Naquib, Phys. Rev. C1 (1970) 569

32 M. Bernheim, T. Stovall and D. Vincinguerra, Nucl. Phys. A97 (1967) 488

33 F. Binon, P. Duteil, M. Gouanére, L. Hugon, J. Jansen, J.P. Lagnaux, H. Palevsky, J.P. Peigneux, M. Spighel and J.P. Stroot, Nucl. Phys. A298 (1978) 499

34 J. Bolger, E. Boschitz, R. Mischke, A. Nagel, W. Saathoff, C. Wiedner and J. Zichy, contributed paper at this conference

35 J. Zichy, J.P. Albanèse, J. Arvieux, J. Bolger, E.Boschitz, C.H.Q. Ingram and E. Rost, contributed paper at this conference

36 R. Corfu, J.P. Egger, F. Goetz, P. Gretillat, C. Lunke, J. Piffaretti, E. Schwarz, C. Perrin and R. Mischke, contributed paper at this conference and private communication

37 J.P. Egger, R. Corfu, P. Gretillat, C. Lunke, J. Piffaretti, E. Schwarz, C. Perrin, J. Jansen and B.M. Preedom, Phys. Rev. Lett. 39 (1977) 1608

38 C.L. Morris, private communication

39 J. Jansen, J. Zichy, J.P. Albanèse, J. Arvieux, J. Bolger, E. Boschitz, C.H.Q. Ingram and L. Pflug, Phys. Lett. 77B (1978) 359

40 W.R. Wharton, J.F. Amann, G. Blanpied, G.R. Burleson, K.G.R. Doss and R.A. Eisenstein, private communication

41 S. Iversen, H. Nann, A. Obst, K.K. Seth, N. Tanaka, C.L. Morris, H.A. Thiessen, K. Boyer, W. Cottingame, C.F. Moore, R.L. Boudrie and D. Dehnhard, Phys. Lett. 82B (1979) 51

42 C. Lunke, R. Corfu, J.P. Egger, P. Gretillat, J. Piffaretti, E. Schwarz, J. Jansen, C. Perrin and B. Preedom, Phys. Lett. 78B (1978) 201

43 M.B. Johnson and H.A. Bethe, Comments on Nucl. and Particle Phys. 8 (1978) 75

44 J.P. Maillet, J.P. Dedonder and C. Schmit, Orsay preprint 79-8

45 R.P. Singhal, J.R. Moreira and H.S. Caplan, Phys. Rev. Lett. 24 (1970) 73

46 M.D. Cooper in Meson.Nuclear Physics, ed. P.D. Barnes (A.I.P., N.Y. 1976) 237

47 M.M. Sternheim and K.B. Yoo, Phys. Rev. Lett. 41 (1978) 1781

48 B.M. Preedom, J.P. Egger, P. Gretillat, C. Lunke, J. Piffaretti, E. Schwarz, J. Jansen and C. Perrin, to be published

49 R.A. Eisenstein and G.A. Miller, Comp. Phys. Comm. 11 (1976) 95

50 C.A. Wiedner, J.A. Nolan, W. Saathoff, R.E. Tribble, J. Bolger, J. Zichy, K. Stricker, H. McManus and J.A. Carr, Phys. Lett. 78B (1978) 26

51 M. Thies, Y. Horikawa and F. Lenz, private communication

52 C.H.Q. Ingram, J. Bolger, E.T. Boschitz, G. Pröbstle, J. Jansen, J. Zichy, J. Arvieux, P.A.M. Gram and R.E. Mischke to be published

53 T. Bowles, D.F. Geesaman, R.J. Holt, H.E. Jackson, J. Julien, R.M. Laszewski, J.R. Specht, E.J. Stephenson, L.L. Rutledge, R.E. Segel, R.P. Redwine and M.A. Yates-Williams, presented at LAMPF Workshop on Pion Single Charge Exchange, Jan. 1979, and contributed paper at this conference

54 Yu. R. Gismatullin and V.I. Ostroumov, Sov. J. of Nucl. Phys. II (1970) 159;
Yu. R. Gismatullin, I.A. Lautsev, V.I. Ostroumov and A. Ya. Smelyanshii, ibid 19 (1974) 22; Yu. R. Gismatullin, J.A. Lautsev and V.I. Ostroumov, ibid 21 (1975) 488; as quoted in ref. 9

55 G.S. Blanpied, G.R. Burleson, J. Davis, J.S. McCarthy, R.C. Minehart, C. Goulding, C.L. Morris, H.A. Thiessen, W.B. Cottingame, S. Greene and C.F. Morris private communication

56 J. Mougey, M. Bernheim, D. Royer, D. Tarnowski, S. Turck, P.D. Zimmerman, J.M. Finn, S. Frullani, D.B. Isabelle, G.P. Capitani, E. de Sanctis and I. Sick, Phys. Rev. Lett. 41 (1978) 1645

57 J. Bolger, E.T. Boschitz, C.H.Q. Ingram, G. Pröbstle, R. Mischke, P.A.M. Gram, J. Jansen, J. Zichy, contributed paper at this conference

59 Yu. A. Batusov, S.A. Bunyatov, N. Dalkhazhav, G. Ionice, E. Losneanu, V. Mihul, V.M. Sidorov, D. Tuvdendorzh and V.A. Yarba, Sov. J. of Nucl. Phys. 9 (1969) 221

60 P.A.M. Gram, F.T. Shively, C.W. Bjork, T.R. King, A.J. Oyer, G.A. Rebka, J.B. Walter, C.A. Bordner, R. Carawon, E. Lomon, private communication

Question: M. Sternheim, University of Massachusetts

There is a qualitative difference in scattering on calcium isotopes at 130 and 180 MeV in discussing neutron distributions. K. Yoo and I find that for a given optical potential, the χ^2 value depends only on the RMS neutron radius at 130 MeV. At 180 MeV, the shape is also quite important.

Response: Ingram

This does not appear to be the conclusion of Maillet, *et al.* who also investigated ^{40}Ca around the resonance.

Comment: B. Zeidman, Argonne National Laboratory

I have two comments. (1) With DWPI, a space coordinate calculation for inelastic scattering, we find βR in agreement with other hadronic probes. With the momentum space code, however, β is in agreement with electromagnetic values. Since the effective radius of the potentials is similar to the nuclear size. This implies that β(DWPI) is about 1.4 x β(momentum space). The reason for this discrepency is not clear, nor are the fortuitous agreements with alternative ways of describing collective excitations. (2) We also find, for some strange reason, an RMS radius dependence for the oscillating frequency in elastic scattering. I will elaborate privately.

Response: Ingram

No reply.

Question: K. Seth, Northwestern University

A short comment on the absolute value of $\beta_2 R$, especially in relation to the factor of 1.2 between pion scattering and for example (e,e') for ^{28}Si. We have played around with DWIA calculations a bit and find that Sternheim and Yoo's comments about ambiguities in R determination from pion elastic scattering apply equally well to βR determination from inelastic scattering. With all the uncertainty about the amplitudes b_o,b_1 etc., the potential formulation, (Kisslinger, Laplace or LMM) and the second order effects, a 20% or even greater uncertainty in absolute value of βR is very easy to obtain. What does remain relatively stable is ratios of βR for different states of the same L in a given nucleus or for π^+ and π^- in the same state.

Response: Ingram

No reply---See Preedom's comment.

Comment: B. Preedom, University of South Carolina

I certainly agree with Professor Seth that the DWIA collective model analysis has uncertainties on the order of 20% and deformation lengths varying within those uncertainties should be considered to be in agreement. Also, I would like to stress the energy-independence of the deformation lengths that we obtained. We measured both π^+ and π^- inelastic scattering from ^{28}Si at 130, 180, and 226 MeV and found that we could use the same deformation length at all energies and for both charges. We conclude that this fact indicates that the DWIZ is correctly accounting for the reaction dynamics.

Response: Ingram
No reply.

Comment: H. Jackson, Argonne National Laboratory
With regard to your remarks concerning Pauli blocking in our $^{16}O(\pi^+,\pi^\circ)$ data, there is no evidence from our data that Pauli blocking is not operative. If you look at the differential cross section, say at 40°, in the region corresponding to quasi-elastic scattering there does appear to be a blocking effect. Most of the cross section at forward directions appears to come from more complicated processes such as two step scattering.

Response: Ingram
This appears to imply that Pauli blocking is less important for multiple scattering processes than for single quasi-free scattering which is the opposite of what I would expect. One might note that the ratio between charge exchange and elastic scattering for two quasi-free collisions in an N=Z nucleus is nearly twice as large as for a single collision, so that (π^+,π°) spectra should show more multiple scattering than (π^+,π^+) ones.

Comment: A. Thomas, TRIUMF
In reply let a theorist make a request for data of good quality! In particular, if one ever wants to learn about the πN interaction inside the nuclear medium, one must have good momentum resolution on both the outgoing pion and nucleon. Examples of typical pion and nucleon angles are given in tables in a recent preprint by Jackson, Ioannides and myself (TRI-PP-78-29, to be published). In fact, all of the very interesting medium effects which we show there, occur in one or two of the angular bins (±5°) of the recent experiment of Ziock, et al. Finally, let me point out, that although we did find interesting effects from the Pauli blocking of the πN interaction, "classical" off-shell effects are negligible in this reaction.

Response: Ingram
Thank you for your advertisement. I fully endorse your request for good data, in the sense both of accuracy and of providing sufficiently detailed or comprehensive information to be really useful. I don't see that Pauli blocking effects are especially interesting in themselves, except that if they are not observed this would provide evidence of a very non-local pion-nucleon interaction in the nucleus.

LOW ENERGY PION-NUCLEUS SCATTERING

R. P. Redwine
Los Alamos Scientific Laboratory †, Los Alamos, N. M. 87545

and

Department of Physics †, Massachusetts Institute of Technology, Cambridge, MA 02139*

ABSTRACT

Recent experiments on low energy pion-nucleus scattering are reviewed. There has been a significant increase in π^- scattering data at energies below 80 MeV. Analysis of an impressive amount of π^+ scattering data yields interesting systematic variation in optical model parameters. New experimental techniques and prospects for the near future are discussed.

INTRODUCTION

It is my intention in this talk to review recent experimental developments in the field of low energy pion-nucleus scattering. The last such review was given by Barry Preedom at the Zurich Conference.[1] While I will attempt to give a general overview of the field for the non-expert, I will concentrate on developments that have occurred since the time of the Zurich Conference. In particular, while I will show a large fraction of all of the data that have been taken in this field, I am not attempting to show in fact all of it. It is perhaps worth emphasizing that this review will concentrate on experimental developments, certainly more so than did that of Preedom. Theoretical developments will obviously be addressed by a number of other speakers at this conference.

For the purposes of this review "low energy" is defined as <80 MeV. This definition is arbitrary, of course, but seems justified not only from physics considerations, but also from the differences between facilities in use for the study of pion scattering. The high resolution pion spectrometers, such as SUSI at SIN and EPICS at Los Alamos, are not well suited to operate below 80 MeV because of their length and the resulting loss of pions due to decay. The low energy experiments have, therefore, been done by groups working on shorter pion channels and with more modest detection schemes. I will be rather loose with the word "nucleus" as it refers to the scope of the review, including in one case the deuteron as a nucleus. This is partially to insure that these new data are brought to the attention of the conferees and partially to discuss the relevance of the experimental techniques.

† Supported by the United States Department of Energy
*Permanent Address

ISSN:0094-243X/79/540501-14$1.50

Motivations for the study of low energy pion-nucleus scattering are by now familiar. The expectation that data in the low energy region would emphasize different aspects of the pion nucleus interaction than would data in the (3,3) resonance region has indeed been realized. This comes, of course, from the fact that the pion-nucleus interaction is weaker at low energies (i.e., the nucleus looks less "black"). From the point of view of nuclear structure studies this means that scattered low energy pions probe more of the nuclear volume than do pions in the (3,3) resonance energy region. Conversely, from the point of view of pion-nucleus reaction mechanism studies, this means that a richer variety of phenomena may be available for study. The problem of obtaining both reaction mechanism information and nuclear structure information simultaneously (since the result of an experiment is in general a function of both) is one that is of concern in all of pion physics. Clearly we must understand pion-nucleus scattering over a broad energy range if we are to achieve the goal of using it as an effective probe of the nucleus.

The remainder of the talk is organized in the following way. In the next section an overview of the new data is given. This includes final versions of data that may have been quite preliminary at the time of the Zurich Conference, as well as data that have been taken since that time. Inelastic π^+-nucleus scattering data are discussed briefly. The next section examines in some detail the π^--nucleus scattering data that have been obtained recently. Next, phenomenological fits to π^+-nucleus elastic scattering data are discussed, especially in terms of the energy and mass number dependence of the fitted parameters. The concluding section attempts to foresee developments in this field in the near future.

OVERVIEW OF NEW DATA

A significant amount of low energy π^+-nucleus elastic scattering data has been taken in recent years by each of three groups: 1) a collaboration of physicists from the University of British Columbia and TRIUMF, working at TRIUMF, 2) a collaboration of physicists from Carnegie-Mellon University, the University of Colorado, New Mexico State University, and LAMPF, working at LAMPF; and 3) a collaboration of physicsts from the Virginia Polytechnic Institute and State University, LAMPF, Oak Ridge National Laboratory, the University of South Carolina, and Tel Aviv University, working at LAMPF. In all three cases the detection technique consisted of stopping the π^+'s in solid state detectors and depending upon pion dE/dx for identification and the total stopping energy signal for energy resolution. In some circumstances it was also necessary to detect the μ^+ signal from π^+ decay as an aid to pion identification.

The UBC-TRIUMF group has concentrated on taking angular distributions for π^+ - ^{12}C elastic scattering at pion energies between 29 MeV and 50 MeV.[2,3] They have in addition, using a different technique, taken data for π^- - ^{208}Pb elastic scattering at 29 MeV,[4] π^- - ^{12}C and π^- - ^{13}C elastic scattering at 29 MeV and 49 MeV,[5] and π^- - ^{16}O and π^- - ^{18}O elastic scattering at 29 MeV.[5] These latter data will be described in more detail in the next section.

The CMU-CU-NMSU-LAMPF group has completed analysis on data that were available in preliminary form at the time of the Zurich Conference. These data include angular distributions for 50 MeV π^+ - elastic scattering on ^6Li, ^7Li, ^{12}C, ^{13}C, ^{16}O, ^{28}Si and ^{56}Fe.[6,7,8,9]

Fig. 1 shows their measured angular distributions for even-even target nuclei.[9] The minimum which occurs at ~ 65° has been interpreted as being due to interference between the s- and p- waves of the π-nucleon potential. The minimum is most evident for light nuclei. The curves are the result of a Kisslinger potential optical model fit with the s- and p- wave pieces (b_o and b_1) of the π-N t-matrix constrained to be the same for all four nuclei. It is seen that the data are represented quite well by this phenomenological potential.

Data for 50 MeV π^+ elastic scattering on ^{12}C and ^{13}C [8] are shown in Fig. 2. The curve through the ^{12}C data is the result of a best fit Kisslinger potential.[8] The curve through the ^{13}C data is the result of a core-nucleus-plus-valence-nucleon calculation. The core is represented by a Kisslinger potential which fits the ^{12}C data; the valence neutron is described by a free π-N t-matrix operating on a neutron matter density calculated from shell model harmonic oscillator wave functions.[8] This prescription seems to do rather well in describing the ^{13}C data.

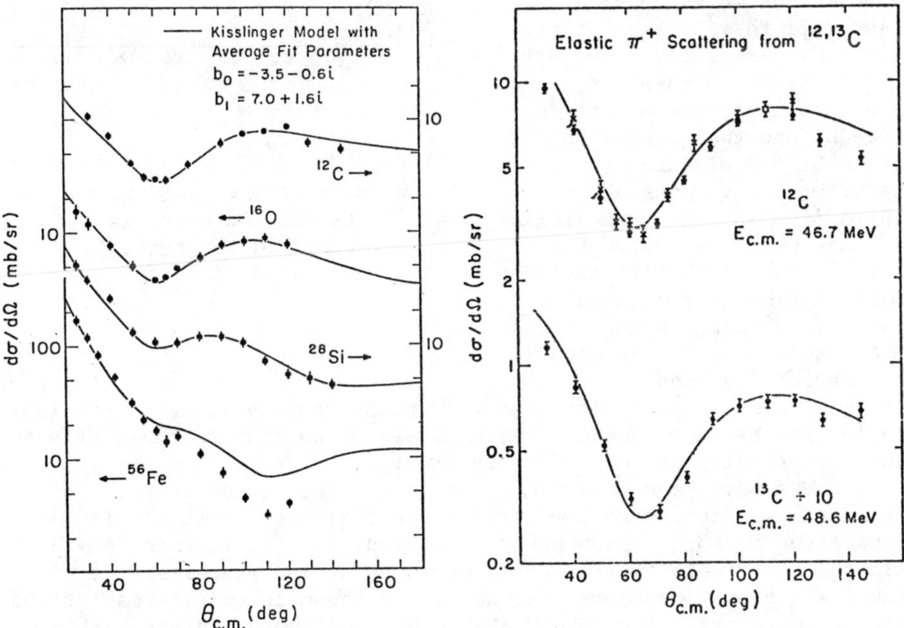

Fig. 1. 50 MeV π^+ nuclear elastic scattering cross sections (Ref. 9).

Fig. 2. π^+-$^{12,13}C$ elastic scattering data (Ref. 8). The calculated curves are described in the text.

The isotopic dependence of the elastic differential cross section for π^+ scattering is of interest in itself. Experimentally, the measurement of the ratio can involve fewer uncertainties than an absolute measurement. Theoretically, there is hope that a calculation of an isotopic ratio or difference of cross sections may be less model dependent than an absolute calculation. This latter point has in fact already been demonstrated for pion total cross sections.[10] Fig. 3 shows data taken by the CMU-CU-NMSU-LAMPF group[8] on the ^{13}C - ^{12}C differential cross section ratio for 50 MeV π^+ scattering. The curve is a core-nucleus-plus-valence-neutron calculation as described above. The calculation can at best be said to describe roughly the trend of the data.

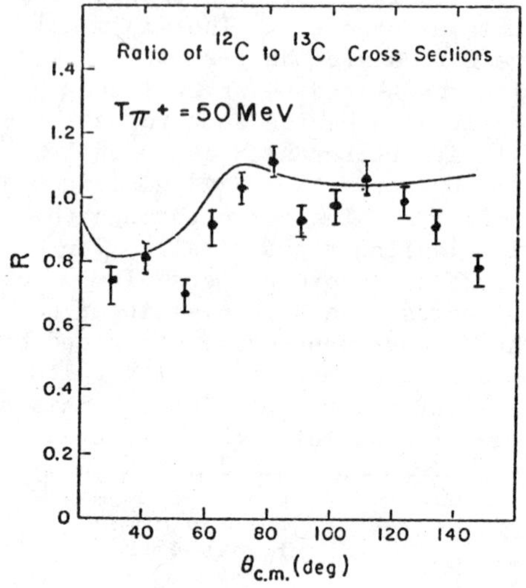

Fig. 3. Ratio of π^+-$^{13}C/^{12}C$ cross sections (Ref. 8). The calculated curve is described in the text.

Similar 50 MeV π^+ data and calculations[8] for the isotopic pair ^6Li, ^7Li are shown in Fig. 4 and Fig. 5. It is clearly seen in Fig. 4 that the core-nucleus-plus-valence-neutron calculation works considerably less well for ^7Li than it did for ^{13}C. The authors suggest that failure to take into account the quadrupole moments of ^6Li and ^7Li and the general failure of a ^6Li + n picture of ^7Li may be primarily responsible. The solid line in Fig. 5 is similar to that described above for Fig. 3, except that it includes only an isovector part calculated from free π-N information.[8] The agreement with the data is not especially good.

It is worth emphasizing at this point the magnitude of the isotopic effects shown above. We shall see below that isotopic effects in π^- scattering are considerably greater.

There have been very few attempts to look at low energy pion inelastic scattering to low-lying nuclear states, certainly in comparison to (3,3) resonance region studies. The reasons for this are the relatively smaller cross sections at low energies, the somewhat lower flux beams available, and the more modest resolutions of the detectors. The CMU-CU-NMSU-LAMPF collaboration has obtained data for a few low-lying states in ^{12}C, ^{28}Si, and ^{56}Fe.[7,9] Fig. 6 shows data for 50 MeV π^+ excitation of the 2^+ (4.44 MeV) and 3^- (9.64 MeV) states of ^{12}C. The elastic angular distribution is also shown. The curves in Fig. 6 are Kisslinger potential optical model calculations. The inelastic calculations are DWBA predictions based on the elastic optical potential. The solid curves are based on free π-N information, while the dashed curves use parameters obtained from a phenomenological fit to the elastic data. As has been found in other studies,[11,12,13] the free π-N parameter calculation does not do very well in describing the data. It has been

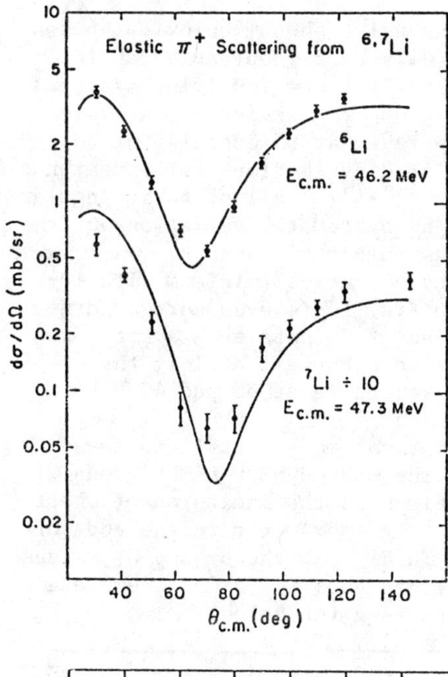

Fig. 4. π^+-^{6,7}Li elastic scattering data (Ref. 8). The calculated curves are described in the text.

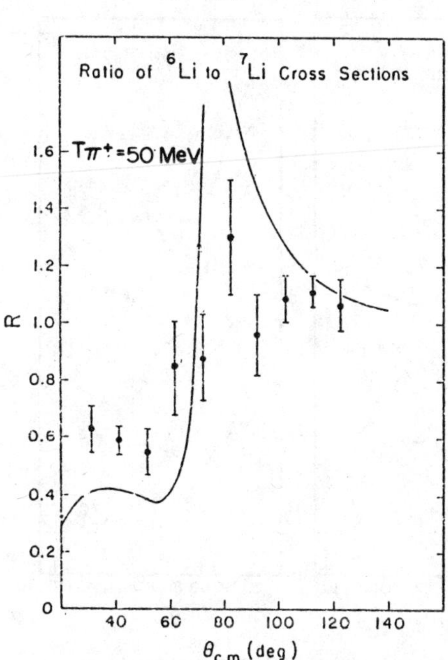

Fig. 5. Ratio of π^+-Li^7/^{6}Li cross sections (Ref. 8). The calculated curve is described in the text.

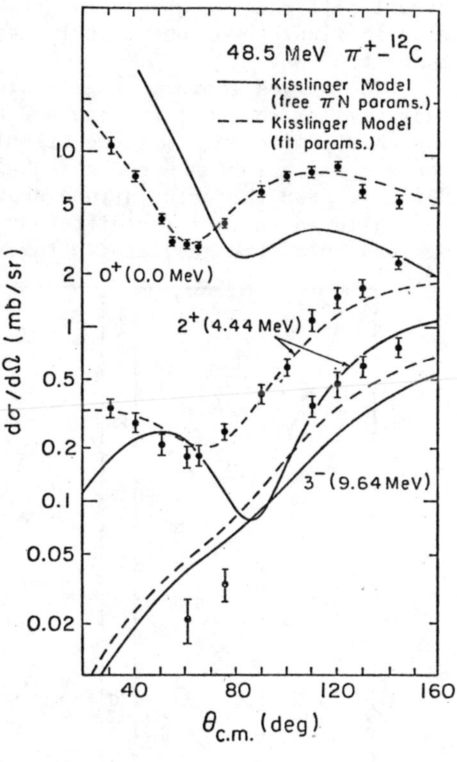

Fig. 6. Elastic and inelastic scattering data for π^+-^{12}C (Ref. 7,9). The curves are described in the text.

found, however, that DWBA calculations using phenomenological potentials that fit the elastic data generally do a creditable job in describing inelastic scattering to the first few low-lying states.[7,9]

The VPI-LAMPF-ORNL-USC-Tel Aviv group is completing analysis of the elastic scattering data they have taken at π^+ energies of 30, 40, and 50 MeV. They have obtained angular distributions for target nuclei of ^{12}C, ^{16}O, ^{40}Ca, ^{90}Zr, and ^{208}Pb.[14,15,16] All of these angular distributions are shown in Fig. 7. The systematic variations of the angular distributions with target mass number and with π^+ energy are interesting. At 50 MeV, the washing out of the 65° minimum with increasing mass number, already seen in Fig. 1, is even more evident here. In addition, the 50 MeV ^{90}Zr and ^{208}Pb data show a more obviously diffractive angular distribution. For all nuclei, the distributions show considerably less structure at 30 and 40 MeV than at 50 MeV.

The curves shown in Fig. 7 are optical model fits[17] to these data similar to the fits obtained by the CMU-CU-NMSU-LAMPF group. The chief difference in the calculations is the inclusion of short range correlations and a π-N off-shell range parameter in the code of Ref. 17. For the calculations shown in Fig. 7, the b_o and b_1 values were allowed to vary for different target nuclei. A more complete discussion of the resulting parameters is given below.

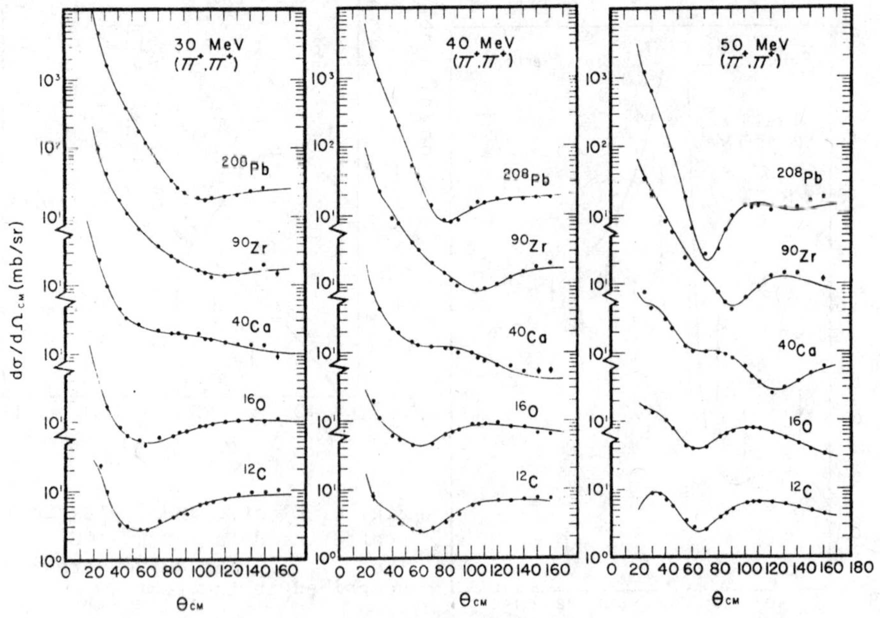

Fig. 7. Elastic scattering angular distribution for various π^+ energies and targets (Ref. 16). The optical model fits use a modified Kisslinger type potential (Ref. 17).

The VPI-LAMPF-ORNL-USC-Tel Aviv group has also begun a program of π^+ and π^- elastic scattering in the region from 50 MeV to

100 MeV.[18] These measurements are done with the so-called BCS magnet on the Low Energy Pion Channel at LAMPF. The BCS system has a resolution of the order of 1 MeV. The first (preliminary) results of this work are shown in Fig. 8. The 65 MeV $\pi^{\pm}$ - ^{40}Ca elastic scattering angular distributions were taken under the same conditions and with a relative monitor that is essentially sensitive only to the pions (π^+ or π^-) in the incident beam.[19] The curves in Fig. 8 are drawn to guide the eye. As expected, differences between π^+ and π^- scattering on ^{40}Ca are considerably greater at 65 MeV than at (3,3) resonance energies.

Enhanced Coulomb effects are in fact one of the exciting reasons for wanting to study low energy π^- (as well as π^+) elastic scattering. The requirement that a given model fit both π^+ and π^- data is a strong constraint on the model. If the data are in a region where π^+, π^- differences are large, then a failure of that type will be more obvious. A clear example of just how different π^+ and π^- angular distributions can be is shown in Fig. 9. The 30 MeV π^+ - ^{208}Pb data are from the VPI-LAMPF-ORNL-USC-Tel Aviv group[16] and the 29 MeV π^- - ^{208}Pb data are from the UBC-TRIUMF group.[4] Once again, the curves are drawn to guide the eye.

As will become more clear in the next section, the use of pion range telescopes is increasingly important in the field of low-energy pion nucleus scattering. Because the stopped π^- very quickly loses a fair fraction of its mass in star fragments, a total stopping energy technique (such as that used for π^+) is not suitable. Besides the magnetic spectrometer discussed above, the technique that has been used is a pion range telescope. In general, the telescope

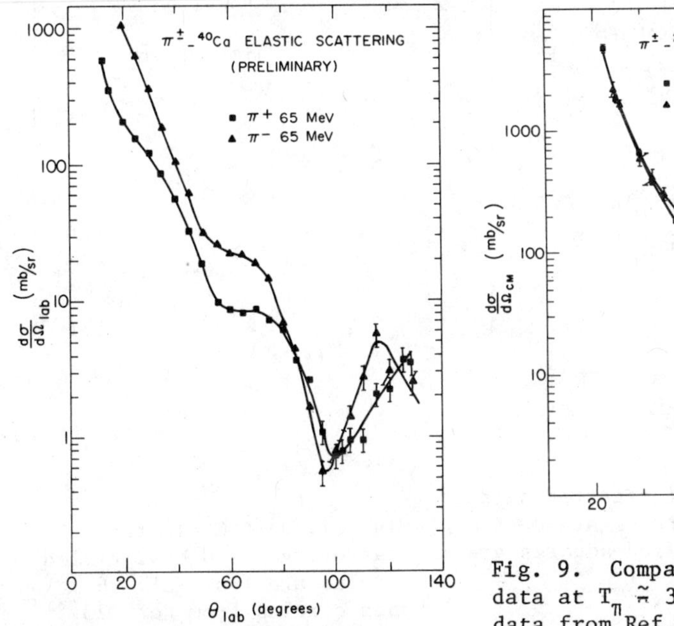

Fig. 8. Preliminary $\pi^{\pm}$-^{40}Ca data taken at LAMPF (Ref. 18). Curves are drawn to guide the eye.

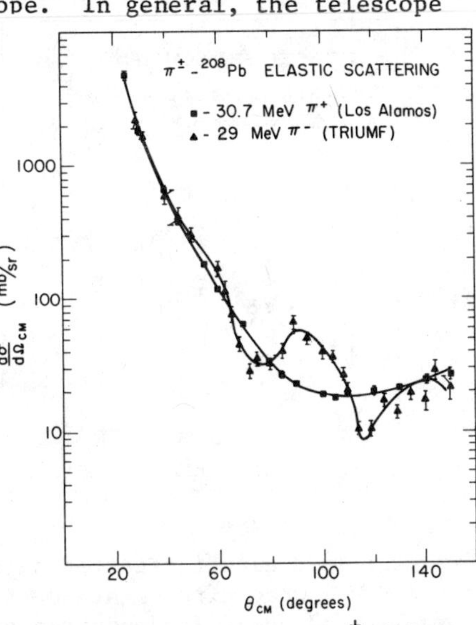

Fig. 9. Comparison of $\pi^{\pm}$-^{208}Pb data at $T_{\pi} \approx$ 30 MeV. Squares are data from Ref. 16 and triangles are data from Ref. 4. Curves are drawn to guide the eye.

consists of several thin counters in which the pions stop, as well as thicker counters for dE/dx identification. The technique is useful for both π^+ and π^-. The efficiency is not seriously degraded by the stopping π^- fragments. Intrinsic resolutions (determined by the range straggling) of 1.3 MeV at 30 MeV can be obtained.

To my knowledge, the first group to make general use of this technique was at Saclay.[20] This group has so far been primarily concerned with π-d and π-p scattering. They have obtained data on $\pi^\pm$-d scattering for a pion energy of 47 MeV and lab angles of 40° to 90°.[21] The π^+-d data, along with older data from Axen et al.,[22] are shown in Fig. 10. There is significant disagreement between the two experiments. More measurements are planned by the Saclay group. The use of pion range telescopes is discussed in more detail below.

It has become traditional in a talk such as this to include a figure with the world's supply of 50 MeV π^+ - ^{12}C elastic scattering data. This shows not only the impressive amount of data, but also the agreement (or lack of it) between different experiments. I have decided to follow the trend of the field and show the same thing for 30 MeV. Fig. 11 contains data from four groups: the open circles are from an earlier Rochester experiment,[23] the closed circles are from the VPI-LAMPF-ORNL-USC-Tel Aviv group,[16] the closed squares are data from the UBC-TRIUMF group,[3] and the open squares are from a recent streamer chamber experiment at Frascati.[24] The π^+ energy is in all cases within 1.5 MeV of 30 MeV. While the amount of data, especially the angular range covered, is impressive, there are some significant disagreements between experiments.

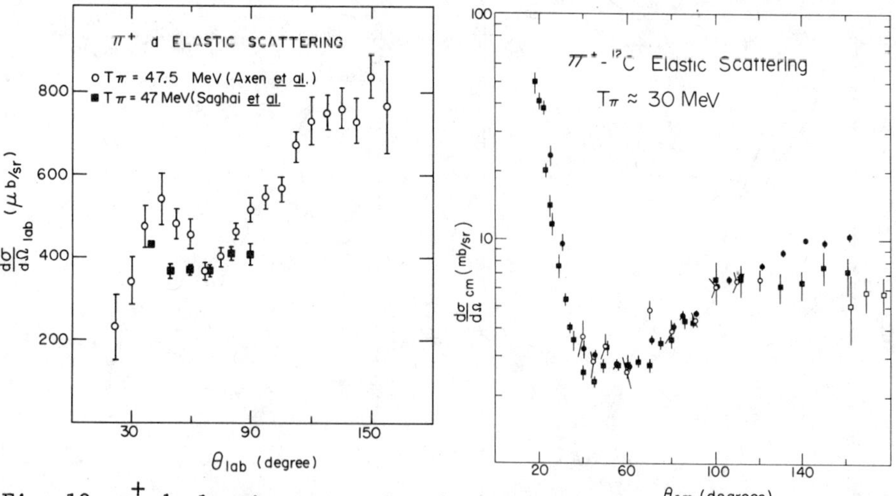

Fig. 10. π^+-d elastic scattering data at 47 MeV. Open circles are data from Ref. 22 and filled squares are data from Ref. 21.

Fig. 11. π^+-^{12}C elastic scattering at 30 MeV. Filled circles are from Ref. 16, open circles from Ref. 23, filled squares from Ref. 3, and open squares from Ref. 24.

ISOTOPIC COMPARISONS OF π^- SCATTERING

The UBC-TRIUMF collaboration has recently completed a series of measurements on 29 and 49 MeV π^- scattering on $^{12,13}C$ and 29 MeV π^- scattering on $^{16,18}O$,[5] as well as the ^{208}Pb data already shown. The measurements were done with the apparatus shown in Fig. 12. Two independent range telescopes (one with 5 counters in the stopping region and one with a single counter in the stopping region) took data simultaneously. Monitoring was done in three ways: 1) an in-beam ion chamber, 2) an in-beam scintillator coincidence (S1.S2), and 3) a thick detector fixed at 115°. It is worth noting that other existing pion range telescopes contain on the order of 6 to 10 counters in the stopping region.[25] The UBC-TRIUMF group feels that they have a sufficient understanding of their efficiency. The 29 MeV $^{12,13}C$ data are shown in Fig. 13.

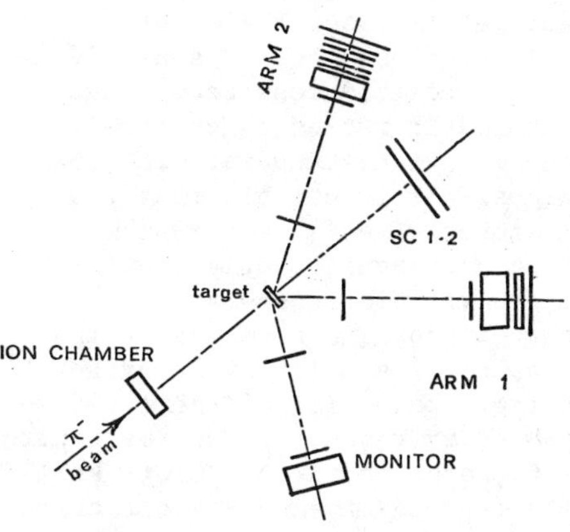

Fig. 12. Pion range telescope setup used by UBC-TRIUMF group (Refs. 4,5) for π^- scattering measurements.

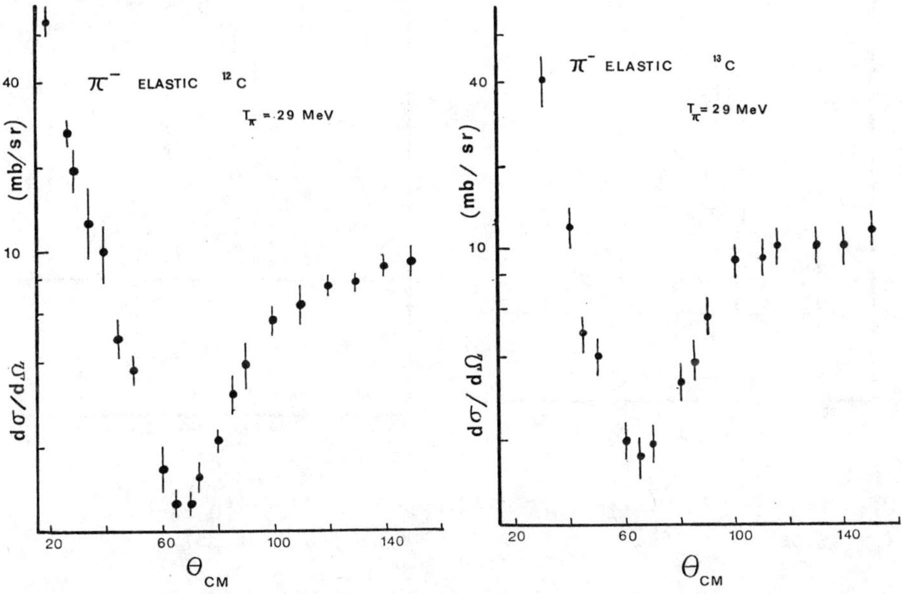

Fig. 13. 29 MeV π^--^{12}C, ^{13}C angular distributions (Ref. 5).

The isotopic cross section ratios for 29 MeV $^{12,13}C$, 49 MeV $^{12,13}C$, and 29 MeV $^{16,18}O$ are plotted in Figs. 14, 15 and 16. As in the earlier π^+ case, the experimental values of the $^{13}C/^{12}C$ and $^{18}O/^{16}O$ ratios contain less uncertainty than would ratios calculated from the individual angular distributions. It is notable that the isotopic effects seen in the π^- data are significantly greater than those visible in the π^+ data. The curves shown in Fig. 14 are the results of a calculation by Rost[26] using a potential of the type employed by Thies.[27] The rms neutron and proton radii of ^{12}C are kept constant and the curves are calculated for three different values of the rms neutron radius of ^{13}C. One should not take the absolute values shown seriously, but rather view the curves as a measure of the sensitivity of the data to the ^{13}C neutron radius. The sensitivity is considerable indeed. Questions

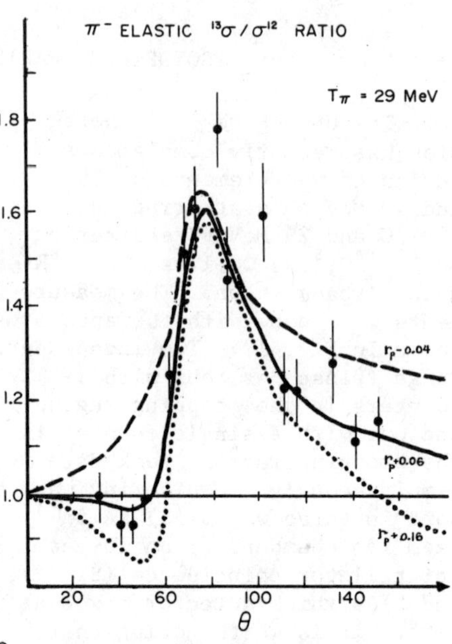

Fig. 14. ^{13}C, ^{12}C ratio for T_π=29 MeV (Ref. 5). Curves are calculations by Rost (Ref. 26) and are labelled by the size of the rms neutron radius r_n used.

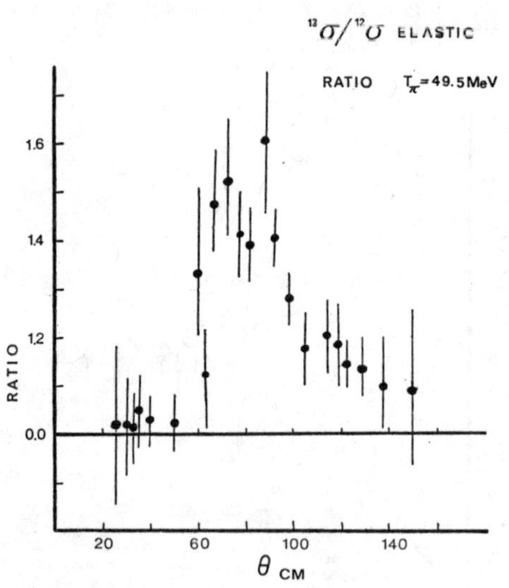

Fig. 15. ^{13}C, ^{12}C ratio for T_π= 49 MeV (Ref. 5).

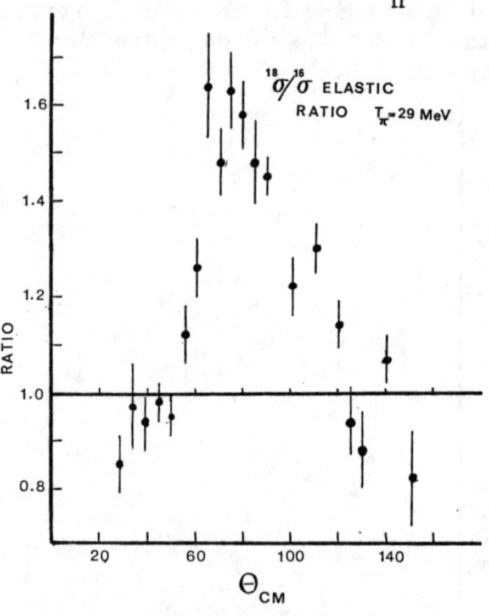

Fig. 16. ^{18}O, ^{16}O ratio for T_π= 29 MeV (Ref. 5).

of model dependence and dependence on the form of the matter distribution are under active investigation by the UBC-TRIUMF group.[28] If these can be resolved, low energy π^- scattering can clearly be a powerful tool in studying neutron distributions.

SYSTEMATIC VARIATIONS IN PHENOMENOLOGICAL PARAMETERS

From the beginning, one of the aims of the field of pion scattering was to obtain optical model parameters from elastic scattering that could then be used to distort incoming and outgoing pion waves in other calculations. While we know now that the simplest optical models are not at all successful at predicting low energy elastic scattering, we have also seen that simple phenomenological potentials are successful at describing inelastic scattering to low-lying states.

Both the VPI-LAMPF-ORNL-USC-Tel Aviv[16] and CMU-CU-NMSU-LAMPF[9] groups have fit their data with such phenomenological potentials to investigate the systematic dependence of the parameters on energy and mass number. The fits of the first group are shown in Fig. 7. The second group has done similar fits (allowing b_o and b_1 to vary for different nuclei) to their data shown in Fig. 1. The resulting parameters are plotted as a function of mass number A in Fig. 17. Uncertainties on the parameters are difficult to assess because of correlations between the parameters themselves and between parameters and other inputs to the models. It is probably reasonable to say that the size of the uncertainties is sufficiently large so that there is no significant difference between the different energy parameters, i.e., no obvious energy dependence.

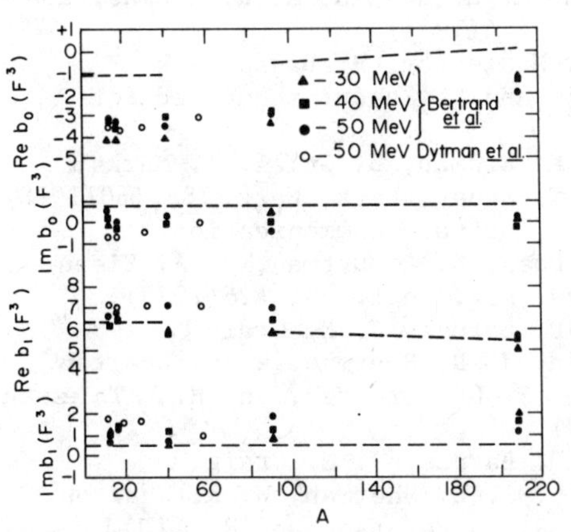

Fig.17. Fitted optical model parameters using phenomenological Kisslinger-like potentials. Filled points are from Ref. 16 and open circles are from Ref. 9. Dashed lines correspond to prediction using free π-N information (Ref. 29).

The dashed lines drawn on Fig. 17 correspond to the b_o and b_1 values from calculations using free π-N data.[29] There is uncertainty in these values as well, coming from the ambiguities in π-N phase shift solutions. The greatest uncertainty is in Re b_1 and is of the order of $\pm$ 0.5 F^3. The rough isospin dependence of the free π-N predictions for A>40 is shown.

One sees immediately that there is good agreement between the parameters as found by the two experiments. Comparing to the free π-N predicted values, only Re b_o is greatly different. The isospin effect on Re b_o and Re b_1 seems to be seen in the fitted parameters. It is reasonable to conclude that these parameters are describing fairly well the general features of low-energy pion nucleus elastic scattering.

CONCLUSIONS

We have seen that there is an increasing body of data on low energy pion nucleus elastic scattering. In particular, good quality π^- data are now appearing. Systematic variations in the data are being probed, as well as aspects of the reaction unique to low energy pions. In the next year or two there should be a large increase in the amount of π^- data available[30] as well as π^+ data at even lower energies.[31]

I wish to thank the theorists and experimentalists who provided me with calculations and data prior to publication. I appreciate useful conversations with J. Alster, M. Blecher, R. Burman, R. Eisenstein, K. Gotow, B. Preedom and B. Saghai. By far the major portion of my contributions to this field was accomplished while I was a member of the staff of the Los Alamos Scientific Laboratory. It is a pleasure to thank Louis Rosen and Darragh Nagle for a most useful and enjoyable association.

REFERENCES

1) B. M. Preedom, in Seventh International Conference on High Energy Physics and Nuclear Structure, ed. by M. P. Locher, Birkhauser Verlag-Basel, 1977, p. 119.

2) H. Dollard, K. L. Erdman, R. R. Johnson, H. R. Johnston, T. Masterson, and P. Walden, Phys. Lett. 63, 416(1976).

3) R. R. Johnson, T. G. Masterson, K. L. Erdman, A. W. Thomas, and R. H. Landau, Nucl. Phys. A296, 444(1978).

4) R. R. Johnson, T. Marks, T. G. Masterson, B. Bassalleck, K. L. Erdman, W. Gyles, D. Gill and C. Sabev, submitted for publication.

5) R. R. Johnson, B. Bassalleck, K. Erdman, B. Gyles, T. Marks, T. Masterson, D. R. Gill, and C. Sabev, Phys. Lett. 78B, 560(1978) and J. Alster and R. R. Johnson, private communication.

6) J. F. Amann, P. D. Barnes, M. Doss, S. A. Dytman, R. A. Eisenstein, and A. C. Thompson, Phys. Rev. Lett. 35, 426(1975).

7) S. A. Dytman, J. F. Amann, P. D. Barnes, J. N. Craig, K. G. R. Doss, R. A. Eisenstein, J. D. Sherman, W. R. Wharton, R. J. Peterson, G. R. Burleson, S. L. Verbergk, and H. A. Thiessen, Phys. Rev. Lett. 38, 1059(1977).

8) S. A. Dytman, J. F. Amann, P. D. Barnes, J. N. Craig, K. G. R. Doss, R. A. Eisenstein, J. D. Sherman, W. R. Wharton, G. R. Burleson, S. L. Verbeck, R. J. Peterson and H. A. Thiessen, Phys. Rev. C18, 2316(1978).

9) S. A. Dytman, J. F. Amann, P. D. Barnes, J. N. Craig, K. G. R. Doss, R. A. Eisenstein, J. D. Sherman, W. R. Wharton, G. R. Burleson, S. L. Verbeck, R. J. Peterson, and H. A. Thiessen, to be published.

10) M. J. Jakobson, G. R. Burleson, J. R. Calarco, M. D. Cooper, D. C. Hagerman, I. Halpern, R. H. Jeppeson, K. F. Johnson, L. D. Knutson, R. E. Marrs, H. O. Meyer, and R. P. Redwine, Phys. Rev. Lett. 38, 1201(1977)

11) R. Landau, S. Phatak, and F. Tabakin, Ann. Phys. 78, 299(1973).
12) G. Faldt, Phys. Rev. C5, 400(1972) and Nucl. Phys. A206, 176(1973).
13) K.-B. Yoo and M. Sternheim, Phys. Rev. Lett. 40, 498(1978).
14) D. J. Malbrough, C. W. Darden, R. D. Edge, T. Marks, B. M. Preedom, R. L. Burman, M. A. Moinester, R. P. Redwine, F. E. Bertrand, T. P. Cleary, E. E. Gross, C. A. Ludemann, and K. Gotow, Phys. Rev. C17, 1395(1978).
15) M. A. Moinester, R. L. Burman, R. P. Redwine, M. A. Yates-Williams, D. J. Malbrough, C. W. Darden, R. D. Edge, T. Marks, S. H. Dam, B. M. Preedom, F. E. Bertrand, T. P. Cleary, E. E. Gross, C. A. Ludemann, M. Blecher, K. Gotow, D. Jenkins, and F. Milder, Phys. Rev. C18, 2678(1978).
16) F. E. Bertrand, T. P. Cleary, E. E. Gross, C. A. Ludemann, M. Blecher, K. Gotow, D. Jenkins, F. Milder, M. A. Moinester, R. L. Burman, R. P. Redwine, M. A. Yates-Williams, D. J. Malbrough, C. W. Darden, R. D. Edge, T. Marks, S. H. Dam, and B. M. Preedom, in preparation.
17) W. R. Gibbs, B. F. Gibson, and G. J. Stephenson, Jr. private communication.
18) S. H. Dam et al., and LAMPF Proposal #333, K. Gotow, Spokesman.
19) E. A. Wadlinger, Nucl. Inst. Meth. 134, 243(1976).
20) B. Saghai, talk given at IKO Summer Study on Physics with Low Energy Beams of Pions and Muons, Amsterdam, 1978, and B. Saghai, private communication.
21) B. Saghai, private communication.
22) D. Axen, G. Duesdieker, L. Felawka, Q. Ingram, R. Johnson, G. Jones, D. Lepatourel, M. Salomon, and W. Westlund, Nucl. Phys. A256, 387(1976).
23) J. F. Marshall, M. E. Nordberg, Jr., and R. L. Burman, Phys. Rev. C1, 1685(1970).
24) C. Guaraldo, A. Maggiora, R. Scrimaglio, F. Balesta, L. Busso, R. Garfagnini, G. Piragino, and F. Cannata, Phys. Lett. 80B, 203(1979).
25) B. Saghai, private communication, and LAMPF Proposal #316, K. K. Seth and G. R. Burleson, Spokesmen.
26) E. Rost, from J. Alster, private communication.
27) M. Thies, Phys. Lett. 63B, 43(1976).
28) J. Alster, private communication.
29) M. Blecher, private communication.
30) LAMPF Proposal #316, K. K. Seth and G. R. Burleson, Spokesmen, and LAMPF Proposal #333, K. Gotow, Spokesman.
31) LAMPF Proposal #29/54, B. M. Preedom, Spokesman.

Question: G. Igo, UCLA

In view of the uncertainty in the interaction, are you being realistic in your statement that low energy pions (~30 MeV) can determine the RMS radius of neutron mass distributions with anything like the sensitivity of intermediate energy protons?

Response: Redwine

To repeat what I did state, the use of low energy pions to determine RMS neutron radii awaits a convincing demonstration of model independence and independence of the form of the neutron distribution. If these requirements can be met, then the sensitivity already shown would make low energy negative pions a quite powerful tool.

Question: C. Guaraldo, INFN Frascati

The Frascati experiment, which is a Torino-Frascati collaboration, has already performed 16 MeV and 20 MeV measurements at 180°. The preliminary results will be presented in the TRIUMF Conference. The reaction is $(\pi^+, {}^{12}C)$.

Question: Q. Ingram, Karlsruhe

I think it should be pointed out that the true pion absorption, which is asymmetric for π^+ and π^- or $N \neq Z$ nuclei, is likely to be an even more serious difficulty at low energies than at resonance, since it is predominantly a central nucleus (high nucleon density) phenomenon.

Response: Redwine

Certainly true pion absorption is more important at low energies. However, my own opinion is that in the next few years any convincing use of pions to determine neutron distributions will depend upon a demonstration of model independence rather than a demonstration of a correct model.

Question: A. Thomas, TRIUMF

I hesitate to inject a note of optimism into the proceedings, but as will be seen in the workshop (when Johnson presents the details of the analysis) a considerable amount of model independence has already been demonstrated. On the question of ambiguities in the inclusion of absorption it is worth stressing that we do have a good understanding of the relative importance of ρ and ρ^2 terms at low energy from considerations of reaction content.

$\pi^{\pm}$-DEUTERON DIFFERENTIAL ELASTIC SCATTERING CROSS-SECTION AT 47 MeV

B. Balestri†, P.Y. Bertin††, B. Coupat††, G. Fournier†, A. Gérard†, L. Guechi†, E.W.A Lingemann†††, J. Miller†, J. Morgenstern†, J. Picard†, B. Saghai†, K.K. Seth††††, C. Tzara† and P. Vernin†

Pion-deuteron interactions, because of their particular features, might be of special interest in understanding of various aspects in pion-nuclei interactions like : pion absorption, off-shell effects, multiple scattering, distortion of strong interaction by Coulomb field, Fermi motion and possibly the admixture of baryonic resonances in the ground state of finite nuclei. On the other hand our comprehension of nucleon-nucleon interactions might also be improved.

Using the facilities of low energy pion channel of Saclay's Linear Accelerator[1], we have measured the differential cross-sections of $\pi^{\pm}$-d elastic scattering and π^{+}-d non radiative capture, $\pi^{+}d \to pp$, at T_{π} = 24 and 47 MeV. The detection system consists essentially of a range spectrometer, designed and constructed at Saclay[2,3]. The reason of this choice was to get identical performances in π^{+} and π^{-}[3,4], in fact the measured resolution is about 2 MeV and the efficiency to pions of $T_{\pi^{\pm}}$ = 20 to 80 MeV varies from 93 % to 75 %.

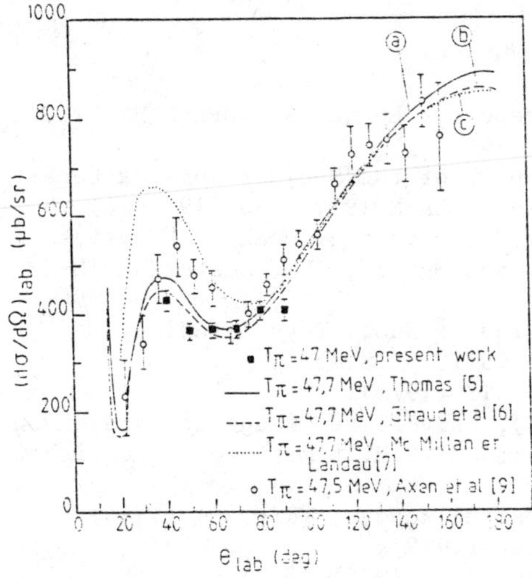

Liquid deuteron targets from 10 to 25 mm thick have been used. A relative monitoring system independent of electron and muon contaminations of the beam, has allowed us to determine the number of incident pions with $\sim$ 2 % accuracy.

Fig. 1 - Differential cross-section of π^{+} elastic scattering : $\pi^{+}d \to \pi^{+}d$. Curves (a) and (b) are 3-body calculations using a Yamaguchi potential and a deuteron wave function with 7 % D-state component, curve (c) is a single scattering calculation.

† DPh-N/HE, CEN Saclay, BP 2, 91190 Gif-sur-Yvette, France.

†† IN2P3, Laboratoire de Physique Corpusculaire, Université de Clermont, BP 45, 63170 Aubière, France.

††† IKO, Oosterringdijk, 18A Posthus 4395, Amsterdam 1006, The Netherlands.

†††† Department of Physics and Astrophysics, Northwestern University, Evanston, Illinois 60201, U.S.A.

On fig.1 and 2 our data at 47 MeV for forward angles in π^+ and π^-, and the results of recent calculations are given[5-8]. The only other enough accurate data at low energy are also reported on fig.1[9]. The theoretical curves issued from the three body formalism are in fairly good agreement with our data.

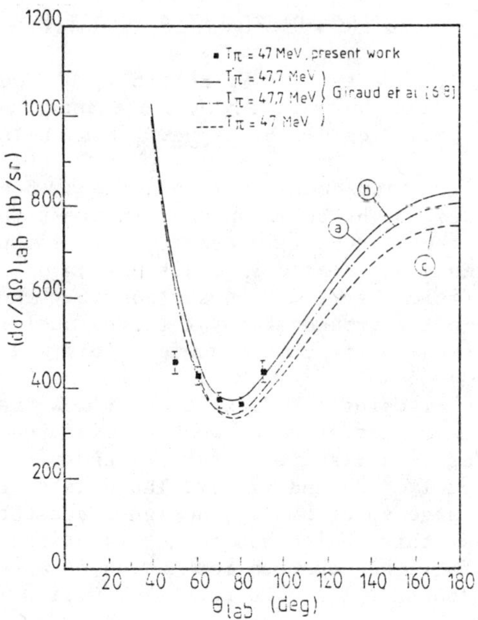

Fig. 2 - Differential cross-section of π^- elastic scattering : $\pi^- d \to \pi^- d$. Curves are 3-body calculations using Yamaguchi (a, b)[6] and Pieper (c)[8] potentials with respectively 4 %, 7 % and 6.7 % for P_D.

References

1. P.Y. Bertin et al., Internal report, CEN Saclay, DPh-N/HE 71/3 (1971).
2. Y. Baillon et al., Progress Report of the Nuclear Physics Department, CEN Saclay, 1975-1976, Note CEA-N-1959, 180 (1975-76).
3. B. Saghai, Talk given at the IKO Summerstudy Week on "Physics with Low Energy Beams of Pions and Muons", IKO, Amsterdam 18-22 September 1978.
4. B. Saghai, Fourth biennal session of Nucl. Phys. studies, La Toussuire, 28 février - 4 mars 1977, p. S.15.1.
5. A.W. Thomas, Nucl. Phys. A258, 417 (1976).
6. N. Giraud, Thesis, n°683 (1978), Université Claude Bernard, Lyon. N. Giraud, Y. Avishai, C. Fayard and G.H. Lamot, to be published in Phys. Rev.
7. M. McMillan and R.H. Landau, TRIUMF Report, TRI-74-1 (1974).
8. N. Giraud, private communication (1979).
9. D. Axen et al., Nucl.Phys. A256, 243 (1976).

MEASUREMENT OF THE BACKWARD ELASTIC SCATTERING OF PIONS ON DEUTERONS IN THE REGION OF THE (3,3) RESONANCE

K.Bos,[ϕ] T. Bressani,[‡] E. Chiavassa,[*] S. Costa,[*] J.D. Davies,[††] G. Dellacasa,[*] W. van Doesburg,[†] M. Gallio,[*] G. Kernel,[∫] E. G. Michaelis,[ϕ] A. Musso,[*] M. Panighini,[*] A. Stanovnik,[∫] N.W. Tanner[ι]

Omicron Collaboration, CERN, Genéva, Switzerland

Preliminary results of the measurement of elastic scattering of pions on deuterons at backward angles between 130° and 175° and pion laboratory energies 141, 177, 260 MeV for π^+ and 151, 185, 189 MeV for π^- are presented and compared to other existing data.

The experiment was performed at the CERN 600 MeV synchrocyclotron. Beams from an external production target were transported to the apparatus by two QQDQQ systems. Pions were identified by means of a telescope consisting of two scintillation counters and two Cherenkov detectors. A gas Cherenkov detector (CO_2 at 10 atm) was used to veto electrons and a liquid freon detector (FC75) served for off-line determination of the muon contamination.

Identification of elastic events rests upon the detection of the recoil particle direction and its pulse heights in a scintillation counter telescope. A heavy water target was located between two pairs of multiwire proportional chambers; the first pair measuring directions of incident and scattered pions, the second pair measuring the direction of recoil. The solid angle of the apparatus was about 1 s.r.

By requiring that pulse heights in the recoil counter telescope are in the range of the recoil deuteron and that the relation between directions of the three tracks is within the πd elastic kinematical relation most of the inelastic background was eliminated. The remaining background under the deuteron peak in the pulse height distribution was then subtracted.

Preliminary results are shown in Fig.1. where they are also compared with existing data[1-4]. The indicated error bars refer to statistical errors which include background subtraction. An estimated systematic error smaller than 15% has been confirmed by

† Free University, Amsterdam, The Netherlands
††Rutherford Laboratory and Birmingham University, United Kingdom
‡ Universita de Cagliari, Cagliari, Italy and INFN, Sez.de Torino.
ϕ CERN, Genéva, Switzerland
∫Jozef Stefan Inst. and Univ. of Ljubljana, Yugoslavia
ιOxford University, United Kingdom
*Universita de Torino, Torino, Italy and INFN, Sez.de Torino.

ISSN:0094-243X/79/540517-02$1.50

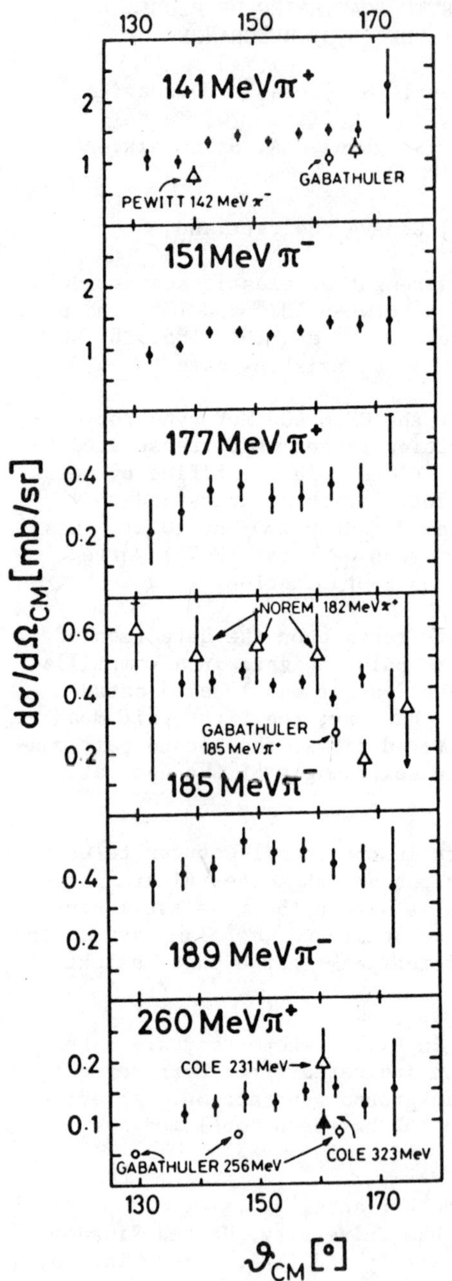

Fig.1. Preliminary cross sections for elastic scattering of pions on deuterons.

comparison of measured 141 MeV π^+p and 185 MeV π^-p differential cross sections with calculations from phase shifts.

All our results are high compared to the data of Gabathuler *et al.*[2] This disagreement is especially pronounced at 260 MeV where our results seem to be in agreement with the new data of Cole *et al.*[1]

Assuming a smooth energy dependence of the backward cross section, our π^-d results appear to be higher than π^+d. Until the analysis is complete and our results are confirmed, we would not like to draw any conclusions about this difference.

REFERENCES

1. R.H. Cole *et al.*, Phys.Rev. *C17*(1978) 681.

2. K.Gabathuler *et al.*, Nucl.Phys.*B55* (1973) 397.

3. J.H. Norem, Nucl.Phys. *B33* (1971) 512.

4. E.G. Pewitt *et al.*, Phys.Rev. *131*(1963) 1826.

ELASTIC AND INELASTIC SCATTERING OF PIONS FROM ^{7}Li AT $T_\pi = 164$ MeV

J. Bolger, E. Boschitz
Institut für Kernphysik des Kernforschungszentrums und
Institut für experimentelle Kernphysik der Universität
Karlsruhe, D-7500, Karlsruhe, Fed. Rep. of Germany

R. Mischke
Clinton P. Anderson Meson Physics Facility, Los Alamos
Scientific Laboratory, Los Alamos, New Mexico, U. S. A.

A. Nagel, W. Saathoff, C. Wiedner
Max-Plank-Institut für Kernphysik, Heidelberg, W. Germany

J. Zichy
Swiss Institute for Nuclear Physics, Villigen, Switzerland

The elastic and inelastic scattering of positive and negative pions from ^{7}Li has been investigated at $T_\pi = 164$ MeV. Angular distributions were measured between 27° and 93° in 3° steps. The ratios of the π^-/π^+ cross sections have been determined for several states.

The data were taken with the SUSI pion spectrometer at SIN and the details of the experimental apparatus are described elsewhere.[1] The target consisted of natural lithium of thickness 0.159 gm/cm^2 with which we obtained an energy resolution of 600 keV FWHM. This energy resolution was not sufficient to fully resolve the elastic scattering from the first excited state at 0.48 MeV; however, we were able to decompose the observed peak into its two constituents by a chi-squared fitting procedure[2] that accurately describes the observed straggling distribution. Contamination due to ^{6}Li and other, heavier elements were eliminated by a background analysis. For the absolute normalization we measured the elastic scattering

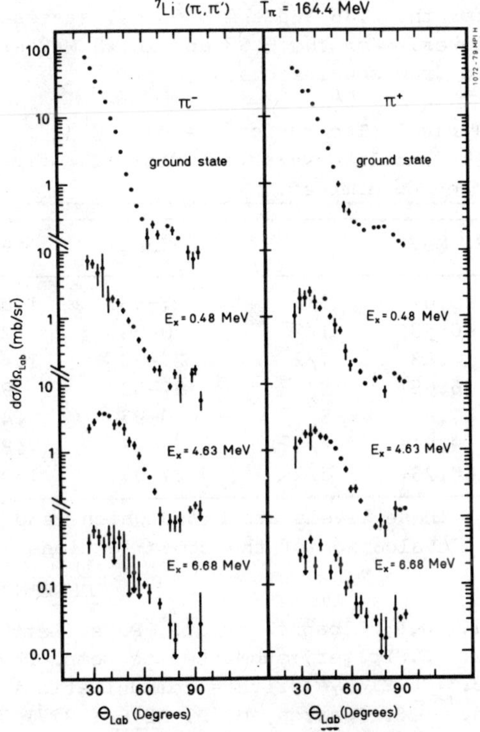

Fig. 1 Angular distributions

ISSN:0094-243X/79/540519-02$1.50

from a pure ^{6}Li target under identical experimental conditions as were used for measurements of scattering from the natural lithium target; these results were compared to a previous measurement.[3] We estimate an absolute error of ± 20%.

Angular distributions for the four lowest excited states of ^{7}Li are shown in Fig. 1. Predictions of D. S. Sparrow[4] may be directly compared to our results. The predicted slope at forward angles for the elastic angular distribution is reproduced by the data, however the data exhibit a much shallower first minimum and a flatter second maximum.

In Table 1 the ratios, $R = \sigma(\pi^-)/\sigma(\pi^+)$, are contained together with the angular interval where the averaging has been done. The relative strengths of isoscaler and isovector excitation have been investigated by writing the ratio of the cross sections, $\sigma(\pi^-)/\sigma(\pi^+)$, as[5]

$$R = \frac{\sigma(\pi^-)}{\sigma(\pi^+)} = \left| \frac{A_0 + \frac{1}{\sqrt{6}}A_1}{A_0 - \frac{1}{\sqrt{6}}A_1} \right|^2 \tag{1}$$

where A_0 and A_1 denote the isoscalar and isovector amplitudes. For the 4.63 and 7.47 MeV states the ratio agrees with the value of 0.25 ±0.05 which the ^{18}O pion data demand for the 3^- state;[6] however, for the 0.48 and the 9.61 MeV states the ratio is significantly lower. For the 6.68 and 10.25 MeV states the errors are too large to draw conclusions.

Table 1 The ratios $R = \sigma(\pi^-)/\sigma(\pi^+)$ for states in ^{7}Li. $\langle\Theta\rangle$ denotes the angular averaging interval. From eq. (1) the ratios A_1/A_0 have been determined.

E_x (MeV)	J^P	$\langle\Theta\rangle$ (deg)	R	ΔR	A_1/A_0	$\Delta(A_1/A_0)$
0.0	$3/2^-$	27-93	1.041	±.056	.024	±.033
0.48	$1/2^-$	30-90	1.206	±.072	.115	±.037
4.63	$7/2^-$	27-93	1.475	±.074	.238	±.031
6.68+	$5/2^-$	27-93	.988	±.096	.007	±.060
7.47	$5/2^-$	30-93	.458	±.097	.473	±.125
9.61+	$7/2^-$	27-93	.805	±.088	.132	±.162
10.25+	$3/2^-$	27-93	1.010	±.060	.006	±.037

\+ These levels are broad which lead to uncertainties in the evaluation of the cross sections

REFERENCES

1. J.P. Albanese et al., Phys. Lett. 73B (1978) 119.
2. J. Bolger, submitted to Comp. Phys. Comm.
3. J. Zichy (private communication).
4. D.S. Sparrow, N. P. A276 (1977) 365.
5. D.S. Koltun, Advances in Nuclear Physics, Vol. 3(1969) 71.
6. S. Iversen et al., Phys. Rev. Lett. 40 (1978) 17.

FORWARD AMPLITUDES FOR $\pi^{\pm}$ ON Al, Ca, Cu, Sn, Ho, and Pb IN THE ENERGY RANGE 65-215 MEV†

R. H. Jeppesen, M. J. Jakobson
University of Montana, Missoula, MT 59812

M. D. Cooper, D. C. Hagerman, M. B. Johnson, R. P. Redwine
Los Alamos Scientific Laboratory, Los Alamos, NM 87545

G. R. Burleson, K. F. Johnson*
New Mexico State University, Las Cruces, NM 88003

I. Halpern, L. D. Knutson**
University of Washington, Seattle, WA 98195

R. E. Marrs, H. O. Meyer
Indiana University, Bloomington, IN 47401

ABSTRACT

Measurements have been made of the attenuation cross section, $\sigma(\Omega)$, for both π^+ and π^- projectiles on Al, ^{40}Ca, Cu, Sn, Ho, and Pb nuclei. The measurements were made over a range of energies including 63, 115, 140, 165, and 215 MeV. A new method of analyzing transmission experiment data, proposed by Cooper and Johnson[1], has been used to extract both the real and the imaginary parts of a forward scattering amplitude $f_N(0)$, where $f_N(0)$ is defined to be the limit as $\Omega \to 0$ of $F(\Omega)-f_c(\Omega)$. $F(\Omega)$ is the full elastic scattering amplitude and $f_c(\Omega)$ is the relativistic point Coulomb scattering amplitude[2]. The values of $f_N(0)$ may be compared with forward amplitudes from theoretical models.

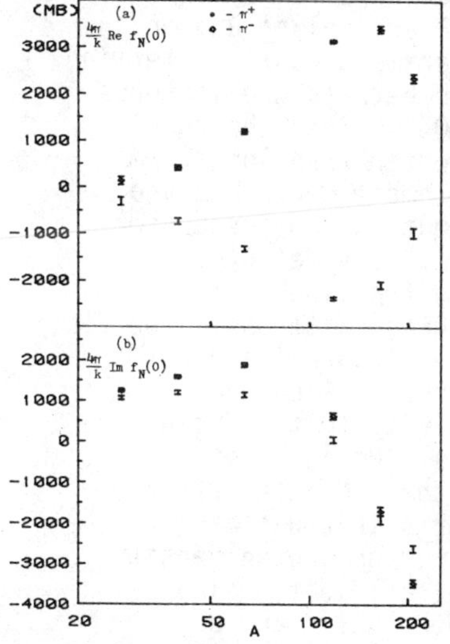

Fig. 1. Forward π-nucleus amplitudes at T_π = 165 MeV

A sample of the data is shown in Fig. 1 for T_π=165 MeV. Figures 1a and 1b are the A dependence of the real and imaginary parts of $f_N(0)$, respectively. Due to isospin invariance, the following relations might be expected to hold: Re $f_N^{\pi^+}(0)$ = Re $f_N^{\pi^-}(0)$ and Im $f_N^{\pi^+}(0)$ = Im $f_N^{\pi^-}(0)$. These symmetries are broken by the unequal number of protons and neutrons in the targets as well as the Coulomb interaction.

† Work supported in part by the U.S. Department of Energy

1. M. D. Cooper and M. B. Johnson, Nucl. Phys., A260, 352 (1976)
2. M. D. Cooper, et al., to be published

*Present address: Swiss Institute of Nuclear Research, Switzerland
**Present address: University of Wisconsin, Madison, WI 53706

ISSN:0094-243X/79/540521-01$1.50

π^- ELASTIC SCATTERING ON $^{12,13}C$ AND $^{16,18}O$ AT LOW ENERGIES

W. Gyles, R.R. Johnson, T. Masterson, B. Bassalleck, T. Marks
K.L. Erdman: Physics Department, University of British Columbia;
A.W. Thomas, D.R. Gill: TRIUMF; C. Sabev**: CERN; J. Arvieux**:
Univ. Grenoble; E. Rost, J.J. Kraushaar**: Univ. Colorado;
J. Alster**: Tel Aviv Univ; M. Krell**: Univ. Sherbrooke

ABSTRACT

π^- elastic scattering experiments on $^{12,13}C$ at 29. and 49.5 MeV and on $^{16,18}O$ at 29. MeV are presented. The ratio of ^{13}C differential cross section to ^{12}C differential cross section as well as the ratio of ^{18}O differential cross section to ^{16}O differential cross section have been measured. Two optical potentials have been used to examine the sensitivity of the ratios to neutron-proton radius differences, model dependence and energy dependence.

We have completed a series of π^- elastic scattering experiments on $^{12,13}C$ at 29.0 and 49.5 MeV and $^{16,18}O$ at 29.0 MeV. The experiment used apparatus that has been reported elsewhere[1] and was performed on the TRIUMF Biomedical channel*** and on the TRIUMF M9 channel. The object of the experiment was to determine the elastic scattering cross section ratio between the isotopes over scattering angles between 30° and 150°. Since the short term stability of the apparatus must be better than long term stability, the isotopes of a particular element were observed at a single angle together with a target out run in succession before a new scattering angle was selected. The detector efficiency and solid angle as well as pion fraction in the incident flux cancel in the ratio of the cross sections. The results of the experiment are presented in figure 1 and 2.

The ratio of the π^- differential cross sections are certainly large and may indicate a sensitivity to the nuclear structure differences between the isotopes. Notice that a large s-wave iso-vector contribution to the π-N interaction at low energies gives large values for terms in π-nuclear potentials that are proportional to the neutron-proton density difference[2]. We have begun to study the ratio data using two optical potentials, the "Colorado" potential[3] and a potential based on a Stricker, McManus, and Carr parameter set[4]. Figure 2 shows a set of curves based on the Colorado potential and demonstrates the sensitivity of the calculations to the neutron-proton radius difference. Likewise, the energy

*Research supported in part by NSERC
**TRIUMF Visitors
***We are indebted to the BATHO Memorial Biomedical Facility for the use of their pion channel.

ISSN:0094-243X/79/540522-02

dependent effects of the calculations seem to be small since the ratio calculations at both 29 and 49.5 MeV for carbon using a neutron-proton radius difference of 0.06 fm at both energies gives the respective curves presented in figure 1.

These results indicate a large sensitivity to nuclear structure details in low energy π^- elastic scattering though more model dependence studies must be done before definitive conclusions can be made.

REFERENCES

1. R.R. Johnson et al. to be published in Canadian Journal of Physics (1979).
2. M. Ericson and T.E.O. Ericson, Ann. Phys. (N.Y.) 36, 323 (1966)
3. A.S. Rosenthal, Ph.K. dissertation, Univ. Colorado (1978) unpublished.
4. K. Stricker, H. McManus, and J. Carr, to be published in Phys. Rev. (1979).

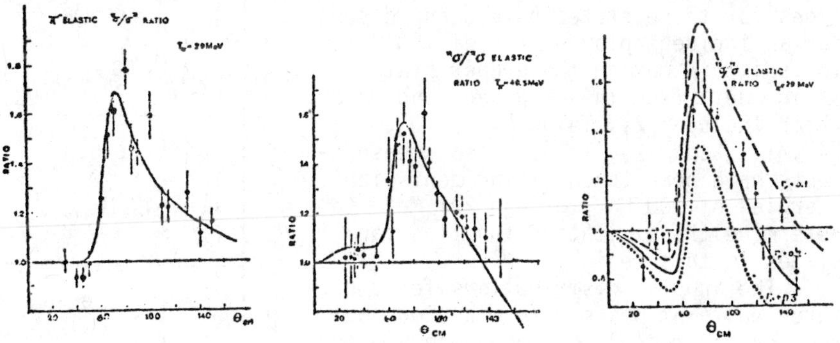

Figure 1. π^- Elastic Scattering differential cross section ratios of ^{12}C and ^{13}C at 29 and 49.5 MeV. The ratios are compared to optical potential calculations with the neutron-proton radius difference of 0.06 f.

Figure 2. π^- Elastic Scattering differential cross section ratios for ^{16}O and ^{18}O. The ratio is compared with calculations where the neutron-proton radius is varied.

ENHANCEMENT OF ^{12}C INELASTIC SCATTERING NEAR THE (3,3) RESONANCE*

W. B. Cottingame, I. B. Moore, W. J. Braithwaite, S. J. Greene, C. J. Harvey, D. B. Holtkamp, R. J. Joseph, C. Fred Moore and H. J. Ward, Univ. of Texas; N. S. King, C. L. Morris and J. Piffaretti, LASL; R. L. Boudrie, J. J. Kraushaar, R. J. Ristinen and G. R. Smith, Univ. of Colorado.

Pion inelastic scattering on ^{12}C exhibits strong enhancement for excitation energies in the region of 19.3 to 21.6 MeV, for incident pion energies near the pion-nucleon (3,3) resonance.

Although both total elastic nuclear cross sections and differential cross sections for inelastic scattering to the 2_1^+ collective state of ^{12}C show only a slight increase near the T=3/2, J=3/2 pion-nucleon resonance, the strong set of states near 20 MeV are observed to have a yield more strongly dependent upon the incident pion energy. The (3,3) resonance may, therefore, provide a mechanism for selective enhancement of pion inelastic scattering to these states. Angular distributions for these states have been measured for an incident pion energy of 180 MeV, to assist in making the proper state assignments from the proposed candidates, which include [2^-, T=1], [4^-, T=0], [4^-, T=1] and [2^+, T=0]. Also, measurements have been taken at incident pion energies of 100 MeV through 291 MeV, to determine the excitation function for states in this region.

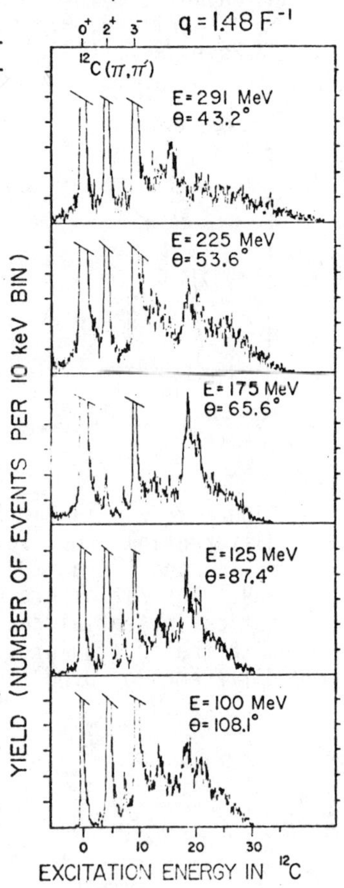

The angular distributions for the three dominant peaks were measured, but the shapes are unlike those observed for the 2_1^+ and 3_1^- states. The 19.3, 20.6 and 21.5 MeV states are more backward peaked than either the 2_1^+ or the 3_1^- states. The angular distributions for the ground, 2_1^+ and 3_1^- states of ^{12}C agree with those of Piffaretti, et al. [Phys. Lett. 67B, 289 (1977)], which are adequately fit by optical model and DWBA calculations. The three states near 20 MeV have angular distributions consistent with large angular-momentum transfer (calculated using a collective form factor), where none of these three states are fit using angular-momentum transfer less than 4, nor by any reasonable sum of two.

*Supported in part by the U. S. Department of Energy and the Robert A. Welch Foundation.

ISSN;0094-243X/79/540524-01$1.50

π^+ AND π^- ELASTIC SCATTERING FROM ^{40}Ca AT 65 MeV

S. H. Dam, R. D. Edge, and B. M. Preedom
University of South Carolina, Columbia, SC 29208*

F. E. Bertrand and E. E. Gross
Oak Ridge National Laboratory, Oak Ridge, Tennessee 37830†

M. Blecher and K. Gotow
Virginia Polytechnic Inst. & State Univ., Blacksburg, VA 24061*

R. L. Burman, R. Carlini, M. E. Hamm, R. P. Redwine, and M. A. Yates-Williams
Los Alamos Scientific Laboratory, Los Alamos, New Mexico 87545†

M. Moinester
Tel Aviv University, Ramat Aviv, Israel

Angular distributions for the elastic scattering of 65 MeV π^+ and π^- beams from ^{40}Ca have been measured utilizing the Bicentennial Spectrometer (BCS) at the Low Energy Pion Channel of the Clinton P. Anderson Los Alamos Meson Physics Facility (LAMPF). The scattered pions were detected by two helical wire chambers near the focal plane of the BCS. The relative beam intensity for each angle was measured by detecting decay muons from the beam in two scintillator counters placed at an angle less than the Jacobian peak angle.

The data, which at the time of this presentation should still be regarded as preliminary, along with some optical model fits to be described below are presented in Figure 1. The absolute normalization was obtained from the π^+p scattering data of Bertin et al.[1] Only relative errors are indicated and the forward angle points have been corrected for the finite angle acceptance of the spectrometer.

The angular distributions show a considerable amount of structure which indicates that the pion-nucleus interaction is going through a transition in this energy region. Both the Coulomb-nuclear and s-p wave interferences are clearly visible, as in lower energy elastic scattering. At the same time the deep diffractive minima characteristic of higher energy data are present, the minimum being at $\sim 100^o$ for π^+ and $\sim 95^o$ for π^-. It is thus expected that the present data will provide a rather stringent test of current pion-nucleus optical models.

Preliminary optical model fits, shown in Figure 1, were generated using a first-order optical model code[2] that utilizes a non-local, off-shell Kisslinger potential for the elementary π-nucleon interaction. Fitting is accomplished by an internal χ^2 search on the complex s and p wave strength parameters (b_0 and b_1) of the potential. Two sets of fits are shown which correspond to using different two-parameter Woods-Saxon distributions for the nuclear matter density. For Set 1, the values for the half-radius (c) and diffuseness (a) were taken from electron scattering data. The curves labeled Set 2 were generated using those values of c and a which gave the best overall χ^2 fit to the π^+ data. The values for the

ISSN:0094-243X/79/540525-02$1.50

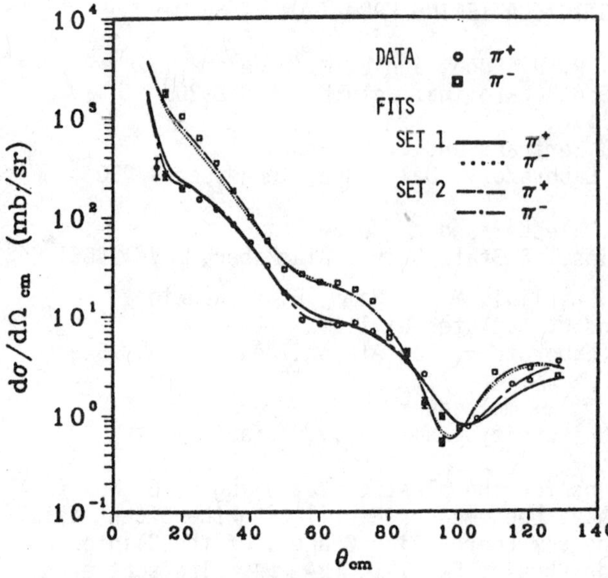

FIGURE 1, $\pi^{\pm}$ - ^{40}Ca elastic scattering at 65 MeV.

the geometry parameters as well as the resulting values of b_0 and b_1 obtained for the various fits are given in Table I. In all cases the value of the off-shell parameter α was taken to be 500 MeV/c. External χ^2 searches showed little sensitivity to changes in this parameter.

As is evident, a better overall fit to the data could be obtained if the shape of the nuclear matter density was allowed to change from that obtained from electron scattering. It should be noted, however, that the values of c and a which gave the best fit to the data correspond to the same rms radius as the electron scattering values. Hence, both densities tend to give similar fits in the diffraction minimum region. This would seem to indicate that some effect not included in the first-order treatment is being mocked up by an effective change in the nuclear geometry. This is not simply a Lorentz-Lorenz effect, since both the half-radius and diffuseness had to be changed to get the best fit to the data. Further optical model studies will be made when the data are finalized.

Table I. OPTICAL MODEL PARAMETERS

Set	Re b_0 (fm^3)	Im b_0 (fm^3)	Re b_1 (fm^3)	Im b_1 (fm^3)	c (fm)	a (fm)
1, π^+	-3.54	-.991	7.42	3.01	3.68	.58
1, π^-	-3.79	-.843	7.64	2.56	3.68	.58
2, π^+	-3.19	-1.16	7.34	2.52	3.95	.48
2, π^-	-2.75	-1.11	7.05	2.67	3.95	.48

† Work supported by D.O.E.
* Work supported by N.S.F.

1. P. Y. Bertin et al., Nucl. Phys. B106, 341 (1976).
2. W. R. Gibbs, B. F. Gibson, and G. Stephenson, private communication.

PION INELASTIC SCATTERING OFF ^{12}C, ^{18}O, ^{28}Si AND ^{52}Cr AROUND THE $\pi N(3,3)$ RESONANCE

C. Lunke, R. Corfu, J.-P. Egger, P. Gretillat, J. Piffaretti, E. Schwarz
Institute of Physics, University, CH - 2000 Neuchâtel (Switzerland)

J. Jansen
S.I.N., CH - 5234 Villigen (Switzerland)

C. Perrin
I.S.N., University of Grenoble, F - 38044 Grenoble-Cedex (France)

B.M. Preedom
University of South Carolina, Columbia SC 29208 (U.S.A.)

The excitation of nuclear states by inelastic scattering has yielded much information about reaction mechanisms and nuclear structure. Since pion beams are now available with better resolution and higher intensities than in the past, it is possible to test mechanisms for the excitation of nuclear states by pion inelastic scattering. However it is necessary to choose targets whose ground state and excited state wave functions are well known. Given this condition, it may be possible to isolate effects resulting from the reaction mechanism. Two types of states that have been studied in sufficient detail both experimentally and theoretically are (a) states whose wave functions contain a minimum number of shell model configurations and (b) collective states. The lowest 2^+ states in ^{18}O (involving neutrons) and ^{52}Cr (involving protons) were chosen in the first group and the lowest 2^+ levels in ^{12}C and ^{28}Si were selected as examples of collective states.

Angular distributions for inelastic scattering to the first excited 2^+ state with π^+ and π^- beams were measured on ^{12}C at 148, 162, 180 and 226 MeV, on ^{28}Si at 130, 180 and 226 MeV and on ^{18}O and ^{52}Cr at 180 MeV. The ^{12}C and ^{28}Si data show small $\pi^+ - \pi^-$ differences which can be explained if Coulomb effects are treated in detail [1]. The ^{28}Si data were analysed using a collective model and the DWIA with a phenomenological potential whose parameters were determined by fitting the elastic scattering angular distribution. This gave deformations independent of pion bombarding energy and in agreement with deformations obtained using other probes. On the other hand, the ^{18}O and ^{52}Cr cross section values for the first excited 2^+ level are significantly different for π^+ compared to π^- [2]. This difference is due to the fact that π^+ or π^- are differently sensitive to protons or neutrons. Our experimental results on ^{18}O and ^{52}Cr imply a neutron deformation length $(\beta R)_n$ different from $(\beta R)_p$.

1. J.-F. Germond and C. Wilkin, to be published.
2. C. Lunke et al., Phys. Lett. 78B, 201 (1978);
B.M. Preedom et al., to be published.

ISSN:0094-243X/79/540527-01$1.50

ELASTIC SCATTERING OF 180 MeV π^+ AND π^- ON NUCLEI OVER A WIDE MASS RANGE

R. Corfu, J.-P. Egger, F. Goetz, P. Gretillat, C. Lunke, J. Piffaretti, E. Schwarz
Institute of Physics, University, CH - 2000 Neuchâtel (Switzerland)

C. Perrin
I.S.N., University of Grenoble, F - 38044 Grenoble-Cedex (France)

R.E. Mischke
L.A.S.L., Los Alamos, NM 87545 (U.S.A.)

A comparison of π^+ and π^- scattering in the region of the N(3,3) resonance is an ideal tool for investigating neutron radii of nuclei since π^+ are believed to interact preferentially with the protons and π^- with the neutrons of the nucleus. However, in order to extract r.m.s. radii, one must first disentangle the effects of matter distributions and of π-nucleus interaction. To clarify the situation, we have measured elastic scattering distributions of $\pi^\pm$ over a wide mass range : $^{12,13}C$, ^{18}O, ^{28}Si, $^{40,48}Ca$ and ^{52}Cr at 180 MeV, where Coulomb effects are expected to be minimum. As a preliminary analysis, a simple black disk calculation was performed, connecting the position of the first minimum of the angular distribution with the black disk radius R. For the N = Z nuclei ^{12}C, ^{28}Si and ^{40}Ca, the differences $\Delta R = R(\pi^-) - R(\pi^+)$ are small and explained with Coulomb effects [1]. However the other nuclei show a significant difference between $R(\pi^-)$ and $R(\pi^+)$, as can be seen on fig. 1, where $\bar{R} = \frac{1}{2}(R(\pi^+) + R(\pi^-))$ has also been plotted in function of $A^{1/3}$.

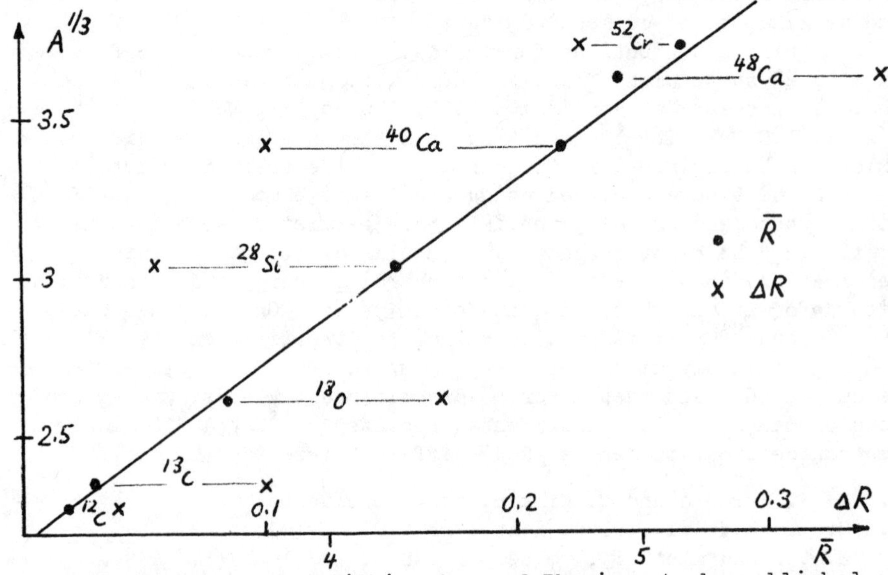

1. J.-F. Germond and C. Wilkin, Ann. of Physics, to be published.

ISSN:0094-243X/79/540528-01$1.50

PAULI BLOCKING IN THE CALCIUM ISOTOPES USING $(\pi^{\pm},\pi^{\pm})$ SCATTERING*

K. G. Boyer, W. J. Braithwaite, W. Cottingame, S. J. Greene, L. E. Smith, and C. Fred Moore, Univ. of Texas; C. L. Morris and H. A. Thiessen, LASL; G. S. Blanpied and G. R. Burleson, New Mexico State Univ.; J. Davis, J. S. McCarthy and R. C. Minehart, Univ. of Virginia; and C. A. Goulding, Florida A&M University.

Differential cross-sections for π^+ and π^- inelastic scattering at 180 MeV from the lowest-lying collective 3^- states in ^{40,42,44,48}Ca are shown in the Figure on the right.

In the simplest shell-model picture, ground state ^{40}Ca fills the s,d shell (pos. parity) in both neutrons and protons, and, as neutrons are added to Calcium, they enter the f-7/2 subshell (neg. parity), filling it in ^{48}Ca. As the next higher orbitals are in the remaining f,p shell (neg. parity), valence neutron transitions to negative parity states (e.g., 3^- states) are inhibited increasingly as one moves to increasingly neutron-rich Calcium targets, due to Pauli blocking. No such inhibition is present for valence proton transitions, thus, since π^+ are known to interact more strongly with protons and π^- are known to interact more strongly with neutrons, one would expect an enhancement of π^+ over π^- in inelastic scattering to negative parity states, as one changes targets from ^{42}Ca to ^{48}Ca. This qualitative effect is seen in the data on the right.

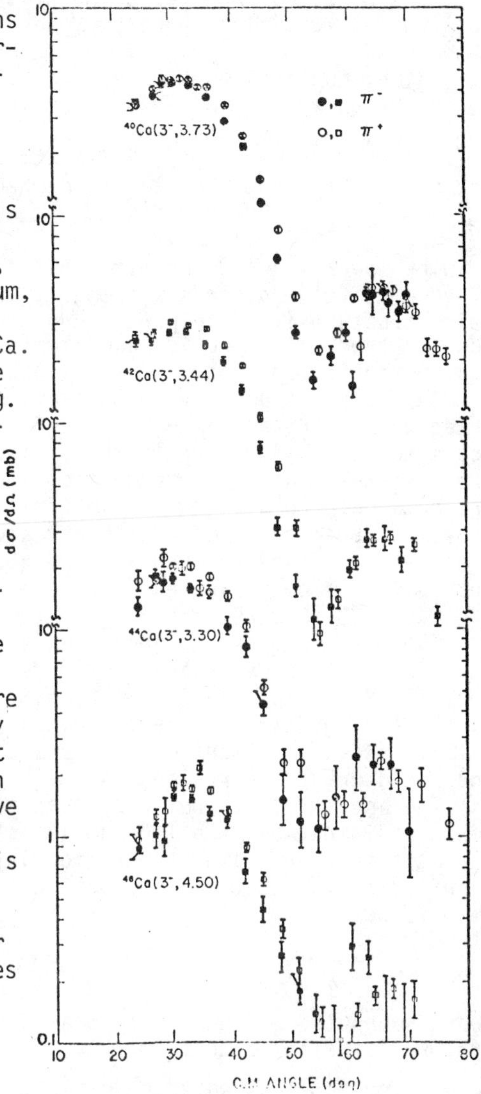

Inelastic $(\pi^{\pm},\pi^{\pm})$ data for lower-lying collective 2^+ states and for higher-lying 3^- and 4^+ states will also be presented for targets of Calcium and for a target of ^{54}Fe.

*Supported in part by the U. S. Department of Energy, the U. S. National Science Foundation and the Robert A. Welch Foundation.

ISSN:0094-243X/79/540529-01$1.50

SPIN-FLIP AND ISOSPIN-FLIP TRANSITIONS IN ^{12}C BY POSITIVE-PION SCATTERING*

R. L. Boudrie, N. J. DiGiacomo, J. J. Kraushaar, R. J. Peterson, R. A. Ristinen, J. R. Shepard, G. R. Smith
University of Colorado, Boulder, Colorado 80309

N. S. P. King, C. L. Morris, H. A. Thiessen
Los Alamos Scientific Laboratory, Los Alamos, New Mexico 87545

W. B. Cottingame, C. Fred Moore, W. J. Braithwaite
University of Texas, Austin, Texas 78712

J. Piffaretti
Institut de Physique, Universite de Neuchâtel, Rue A.-L Breguet 1, CH-2000 Neuchâtel, Switzerland

The lower 1^+, T=0, and T=1 states in ^{12}C at 12.71 MeV and 15.11 MeV respectively, possess a symmetry and simplicity of great appeal. Their structure is quite purely due to the $(p_{3/2}^{-1}, p_{1/2})$ configuration and the states provide good tests of M1 transitions induced by electromagnetic, proton, and alpha particle interactions.

An inelastic positive-pion scattering experiment recently completed with EPICS at LAMPF has provided well-resolved, good statistics spectra showing clearly the excitation of the 1^+ (as well as other) states. At a beam energy of 180 MeV, angular distributions were obtained from 15° to 60° in 5° steps and from 60° to 110° in 10° steps. At beam energies of 140 MeV and 260 MeV, angular distributions were obtained between 20° and 90° in steps of 10°. In addition, a few selected angles were taken at beam energies of 100, 116, 160, 200, 230, and 291 MeV.

The most interesting angular distributions are shown in Figure 1 for incident energies of 116 MeV and 180 MeV. The ratio of the 12.71 MeV (T=0) state to the 15.11 MeV (T=1) state is 1.4 ± 0.5 for the 180 MeV data, while the ratio is ≈3 for the 116 MeV data. A similar wave function for each state and a simple isospin dependence would predict a ratio of 4.

The angular distributions at an incident energy of 140 MeV show a ratio between the T=0 and T=1 states to be around 1.4. Data at the other incident energies is less complete, but there is an indication that for incident energies above 180 MeV the ratio is near 1.

* Work supported by the United States Department of Energy and the R. A. Welch Foundation

ISSN:0094-243X/79/540530-02$1.50

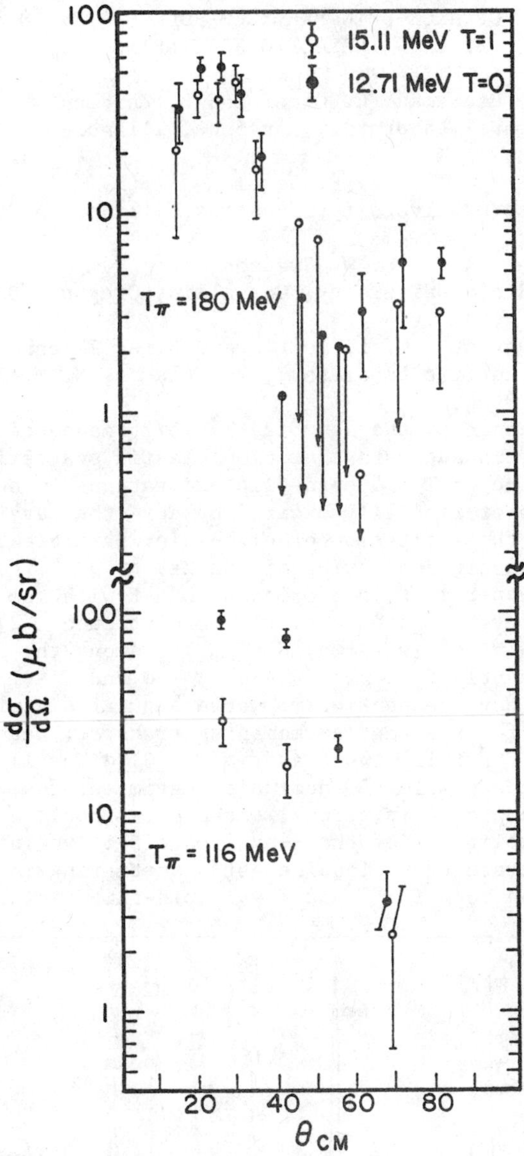

Figure 1. Angular distributions for the lower 1^+, $T=0_1T=1$ states in ^{12}C at incident pion energies of 116 MeV and 180 MeV.

EXCITATION OF HIGH-SPIN PARTICLE-HOLE STATES IN ^{28}Si BY PION INELASTIC SCATTERING

D. F. Geesaman, C. Olmer and B. Zeidman
Argonne National Laboratory, Argonne, Illinois 60439*

R. E. Segel
Northwestern University, Evanston, Illinois 60201

L. W. Swenson
Oregon State University, Corvallis, Oregon 97331

R. L. Boudrie,** C. L. Morris and H. A. Thiessen
Los Alamos Scientific Laboratory, Los Alamos, N. M. 88001

The dominance of the J = 3/2, T = 3/2 resonance in the pion-nucleon interaction suggests that pion elastic scattering should be an excellent probe of T = 0 spin-flip excitations in nuclei. Such states are not preferentially excited by any other available probe. We have studied the excitation of high-spin, particle-hole states in ^{28}Si by the inelastic scattering of 162-MeV π^+ at large momentum transfer. The experiment, performed at LAMPF with the use of the EPICS spectrometer, achieved a system resolution of ~270 keV. A spectrum at θ_{LAB} = 70° is shown in Fig. 1. Among the states observed are the predominantly $f_{7/2} \times d_{5/2}^{-1}$ 6^- T = 0 and T = 1 states at 11.58 and 14.36 MeV, respectively, whose angular distributions are characteristic of large angular momentum transfer, peaking at q ~ 300 MeV/c. For these two states $\sigma(T = 0)/\sigma(T = 1) = 1.7 \pm .4$ as compared to the simple P33 dominance estimate of 4. With detailed microscopic calculations, these data should help define the spin-flip amplitudes of the π-nucleon effective interaction. The data demonstrate that pion inelastic scattering is an excellent tool for studying both T = 0 and T = 1 spin-flip excitations in nuclei.

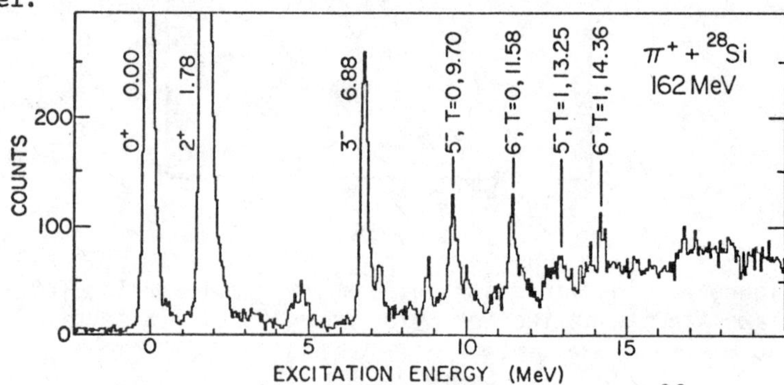

Fig. 1. Spectrum of π^+ inelastic scattering from ^{28}Si at θ_{LAB} = 70°

*Work supported by the U. S. Department of Energy.
**Also at the University of Colorado, Boulder, CO 80302.

ISSN: 0094-243X/79/540532-01$1.50

OCTUPOLE TRANSITIONS IN THE ^{92}Mo($\pi^{\pm}$,$\pi^{\pm}$) REACTIONS

K. G. R. Doss, R. A. Eisenstein and W. R. Wharton
Carnegie-Mellon University, Pittsburgh, PA 15213

G. Blanpied and G. R. Burleson
New Mexico State University, Las Cruces, NM 88003

J. F. Amann
Los Alamos Scientific Lab., Los Alamos, NM 87545

ABSTRACT

The ^{92}Mo($\pi^{\pm}$,$\pi^{\pm}$) reactions at T_{π} = 180 MeV populating octupole states are examined to suggest the (π,π') reaction as an interesting probe of nuclear structure.

TEXT

Angular distributions of states in the ^{92}Mo($\pi^{\pm}$,$\pi^{\pm}$) reactions at T_{π} = 180 MeV have been measured for $15° < \theta \leq 100°$ using the EPICS spectrometer at Los Alamos. A spectrum of π^{+} scattering at 54° is shown in figure 1. The 2.8 MeV and 5.6 + 5.8 MeV doublet octupole states are dominant in the spectrum but there is little or no indication of the low-energy octupole resonance, LEOR, between 5 and 8.5 MeV. This is true for all of our spectra. To be consistent with the (α,α')[1], (e,e')[2], and a recent ^{90}Zr(p,p') measurement[3] about 70% of the events between 5 and 8.5 MeV in figure 1 would have to be attributed to the LEOR. For both (π^{+},π^{+}) and (π^{-},π^{-}) we find that the energy integrated cross sections between 5 and 8.5 MeV and between 8.5 and 12 MeV have nearly the same magnitude and same angular distributions. Normalizing the 2.8 MeV state to have 6% of the L=3 T=0 energy weighted sum rule (EWSR), we place an upper limit of 12% of the EWSR for the LEOR. This is a factor of 2 weaker than measured in (α,α') and (p,p'). In contrast the ratio of the cross section of the 5.7 peak to the 2.8 MeV state is found to be consistent with (α,α').[4]

A feature of π scattering near the 3-3 resonance is that π^{+} scattering is dominated by the proton part of the form factor and π^{-} by the neutron part. RPA calculations[1] attribute the partitioning of the EWSR strength between the LEOR and the 2.8 MeV 3^{-} state to the energy separation of the $(g_{9/2},p_{3/2}^{-1})_{3^-}$ proton particle-hole (p-h) below the other 1ħω p-h. This implies that the 2.8 MeV state is predominately a proton excitation and should have much larger cross section for (π^{+},π^{+}) compared to (π^{-},π^{-}). We find that the π^{+} and π^{-} cross sections are equal to about 5% for both the 2.8 MeV and 5.7 MeV doublet suggesting these states have equal amounts of proton and neutron excitation. This presents a problem because, to our knowledge, there is no other explanation as to why the low lying 3^{-} state with about 20% of the 1ħω EWSR is split apart from the LEOR with the remaining 80% of the strength.

ISSN:0094-243X/79/540533-02$1.50

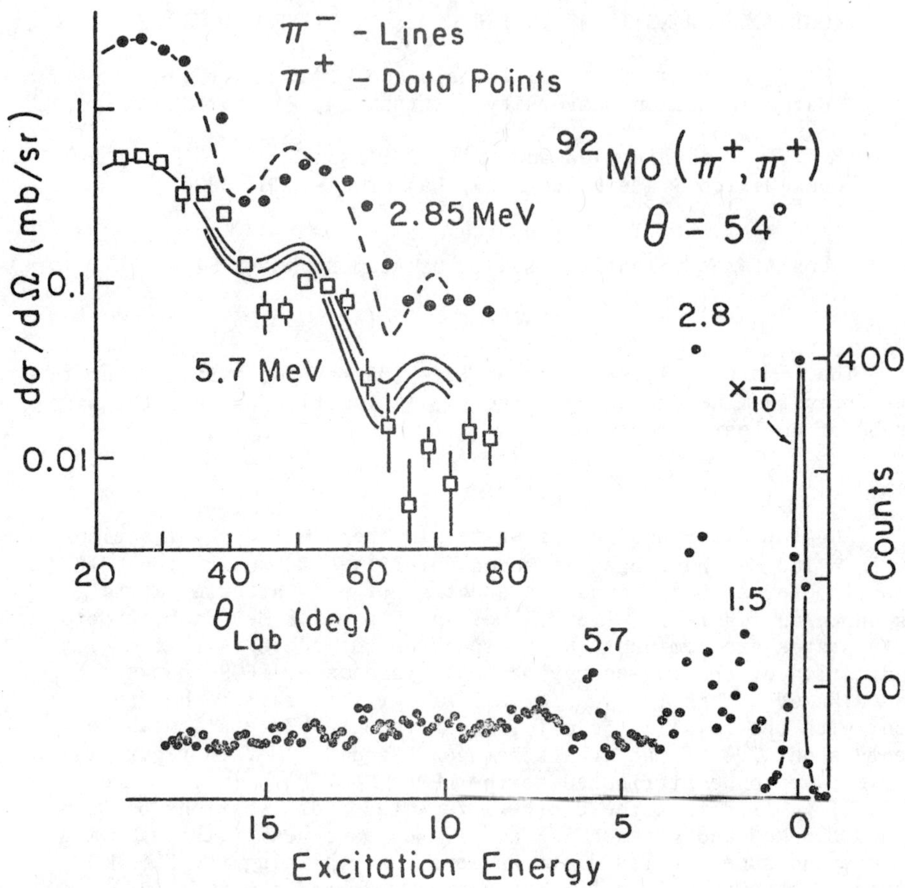

Fig. 1. Excitation spectrum and angular distributions.

We are theorizing that the slopes of the angular distributions shown in the figure are influenced by the radial form factor and that proton and neutron form factors have different radial shapes for the 3^- states. If true, this could reveal a fundamental difference between these octupole states. The ratio of the cross section of the 2.8 MeV to the 5.7 MeV peak at each maximum is 4.4, 4.5, 5.9 (±1.0) for (π^+,π^+) and 5.2, 4.1, 3.5 (±.5) for (π^-,π^-). We are now examining this question.

REFERENCES

1. J. M. Moss et. al., Phys. Rev. C18, 741 (1978).
2. R. Pitthan, H. Hass, D. H Meyer, F. R. Buskirk and J. N. Dyer, to be published.
3. N. J. DiGiacomo, private communication.
4. A. A. Bernstein, Adv. Nucl. Phys. 3, 325 (1969).

DECOMPOSITION OF THE TOTAL PION NUCLEUS CROSS SECTION AT 125 MeV.

I. Navon, D. Ashery, G. Azuelos
Tel-Aviv University, Ramat Aviv, Israel

H. J. Pfeiffer, F. W. Schlepuitz, H. K. Walter
S.I.N, ETH and University of Zurich, Switzerland

The total cross section for pion-nucleus interaction below the threshold for pion production is composed from cross sections for (several channels:) elastic scattering(σ_{el}), inelastic scattering, including to unbound states (σ_{inel}), single and double charge exchange (σ_{scx},σ_{dcx}) and absorption (σ_{abs})

$$\sigma_{tot} = \sigma_{el} + \sigma_{inel} + \sigma_{scx} + \sigma_{dcx} + \sigma_{abs}. \tag{1}$$

While systematic studies exist for the total cross section[1] and elastic scattering[2], little is known about the cross sections for the total inelastic scattering and the absorption processes. We report here results of experiments in which the absorption cross section was measured at 125 MeV as part of a systematic study carried out at bombarding energies in the 3:3 resonance region.

The absorption cross sections were obtained by combining the results of two experiments carried out at the S.I.N. Laboratory. The targets studied were: Li, C, Al, Fe, Nb, and Bi. The first experiment was done using the transmission technique: with several disc-shaped plastic scintillators positioned behind the target, the removal of charged pions from the incident flux was measured. Charged pions are removed through: a) pion absorption process, b) single charge exchange reaction and c) scattering of charged pions (elastic and inelastic) to angles larger than those subtended by the disc counters. (The result of the first experiment is the sum of cross sections for these three processes.) In the second experiment the angular distribution for scattering of charged pions was measured. The scattered pions were detected with plastic scintillator telescopes and were identified and separated from other charged particles through measurement of the energy loss in the detectors. Pions that were stopped in the counter were also identified by the signal generated by the muon decay after the pion came to rest. The measured angular distribution was then integrated over the angular range from 180° to the angle subtended by the disc counters used in the first experiment. After subtracting the results of the second experiment from those of the first, the sum of cross sections for absorption and single charge exchange were obtained.

The charge exchange cross section was calculated using results of measurements of this cross section at 100 MeV on many nuclei[3]

ISSN: 0094-243X/79/540535-02$1.50

together with the measurement for ^{12}C at 130 MeV[4]. The results varied between 25% of the measured sum of cross sections for 12C to 12% for Bi. Therefore, errors in this estimate do not affect much the precision of the absorption cross sections which are obtained by subtracting the single charge exchange contribution from the results of the experiments

By combining the results of the present work with measurements of the total cross section[1] and calculations of elastic scattering, it is possible to obtain the value of the total inelastic cross section (eq. 1) if we neglect the small contribution of the double charge-exchange reaction. Since the elastic cross section is not strongly energy-dependent around 125 MeV, Lenz(5) calculation at 150 MeV was used. Fig 1 shows how the total cross section is decomposed into its major constituents.

REFERENCES

1) A. S. Carroll, et al. Phys Rev C14, 635 (1976)

2) E. Boschitz, Proceedings of the 7th Int. Conf. on High Energy Physics and Nuclear Structure, 1977, Zurich, Switzerland, p. 133 ed. M. P. Locher.

3) T. Bowles, et al. Phys. Rev. Lett 40, 97 (1978)

4) E. Bellotti, D. Cavalli and C. Matteuzzi, Nuovo Cimento 14A, 567 (1973)

5) F. Lenz. Proceedings of the Topical Meeting on Intermediate Energy Physics, 1976, Zuoz, Switzerland. Vol. 2, p. 319; published by SIN, Villigen, Switzerland

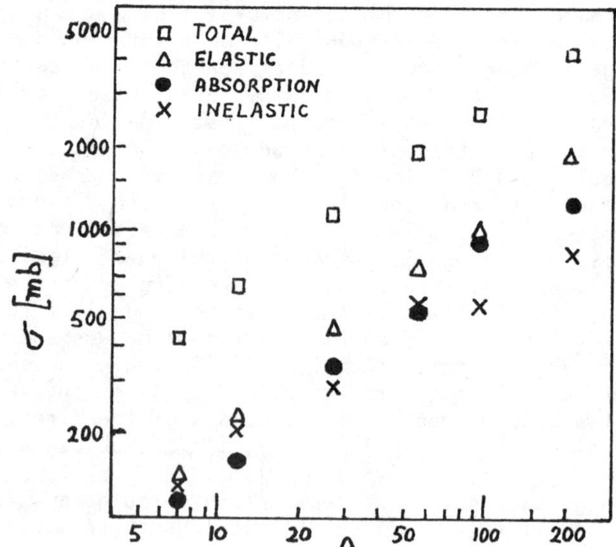

AMPLITUDE ANALYSIS IN PION-NUCLEUS SCATTERING

Colin Wilkin
University College London, WC1E 6BT

ABSTRACT

A simple ansatz for the elastic pion-nucleus scattering amplitude is employed which permits the extraction of an amplitude from the differential and total cross sections. Some of the ambiguities among the parameters are eliminated by looking at the effects of the Coulomb distortion on the combined π^+, π^- data. A generalisation of the prescription for inelastic transitions is proposed though here the influence of the Coulomb field is much reduced. The use of these amplitudes in DWIA calculations is discussed.

Theoreticians would have more insight into the physics if experimentalists could furnish us with amplitudes rather than cross sections. The most ambitious amplitude analysis consists of doing a partial wave decomposition and fitting to data which has the advantage that unitarity is diagonal in ℓ. However, because of the many inelastic channels, unitarity is a very weak constraint for medium energy pion-nucleus scattering and the high number of relevant phase shifts makes this technique too ambiguous to be profitable. We must look elsewhere.

Near the 3-3 resonance the middle of the nucleus is black due to inelastic processes such as nucleon knockout and pion absorption. To a first approximation the scattering from a medium nucleus, such as ^{28}Si shown in Fig. 1, is that of a black disc. Here the elastic scattering is $\{kRJ_1(qR)/q\}^2$ and the excitation of the 2^+ level proportional to $\{J_0(qR)\}^2+3\{J_2(qR)\}^2$ where R is the effective nuclear radius, k the incident momentum and q the momentum transfer. Some allowance for the surface diffuseness has been made by multiplying these forms by an Inopin[2] fuzziness factor $\exp(-dq^2)$.

Although the representation of such elastic and inelastic data in terms of a total of three parameters is impressive this kind of parameterisation is not flexible enough to yield good fits at large angles or lighter nuclei. The crucial features of the Bessel function we want to retain are a few zeros in the physical region modulating a smooth fall-off in q^2. Away from resonance the zeros are are displaced into the complex q^2 plane so that we [3,4] assume that the π-A amplitude can be written in the form

$$F_N(q) = F_N(0) \prod_{i=1}^{M}(1-q^2/q_i^2) \exp(-aq^2/2) \qquad (1)$$

where M is chosen to be the number of observed or inferred minima (or shoulders) in the measured angular distribution. From the optical theorem

$$F_N(0) = ik\sigma_T(1-i\rho)/4\pi \qquad (2)$$

ISSN:0094-243X/79/540537-13$1.50

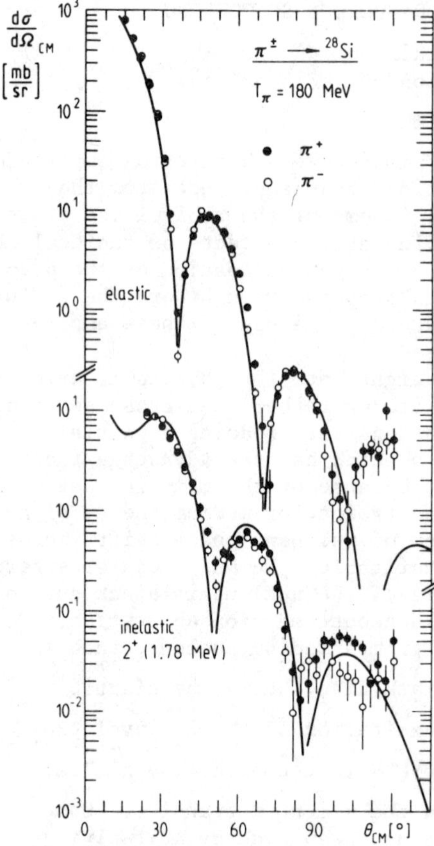

Fig. 1. Elastic $\pi^{28}Si$ scattering and 2^+ excitation at 180 MeV[1] compared with the predictions of the fuzzy black disk model.

where σ_T and ρ are the total cross section and relative real part of the forward amplitude respectively. The Gaussian factor has no theoretical justification and can be replaced by another smooth function if that fits better. The crucial hypothesis is that a is taken to be real. This parameterisation, when applied to data such as that in Fig. 1, yields χ^2 per degree of freedom typically of the order of 1 to 2. In the absence of the Coulomb field the π^+ and π^- amplitudes for the scattering from an isoscalar target should be the same and the signs of ρ and of the imaginary parts of the q_i^2 are not determined by the differential cross section, a 2^{M+1}-fold ambiguity. As is well known ρ may be deduced from Coulomb-nuclear interference (CNI) measurements and we shall see later that the Coulomb field also determines the signs of the $Im(q_i^2)$. What is the behaviour of these in a black disk model where the real part of the amplitude originates from the nuclear surface?

$$F_N(q) \approx ikR^2\left[\frac{J_1(qR)}{qR} - i\rho J_0(qR)/2\right] \tag{3}$$

which for small ρ may be wrttten as

$$F_N(q) \approx ik\bar{R}^2 J_1(q\bar{R})/q\bar{R} \tag{4}$$

with a complex effective radius $\bar{R} = R(1-i\rho/2)$. The imaginary parts of the zeros are then related to the forward real part of the amplitude

$$\nu_i \equiv Im(q_i^2)/Re(q_i^2) \approx \rho \tag{5}$$

and this simple phase rule seems to hold quite well throughout the resonance region as can be seen from Fig. 2. An unexpected result is that all the ρ and ν_1 seem to fall in a narrow band about 0.1 in

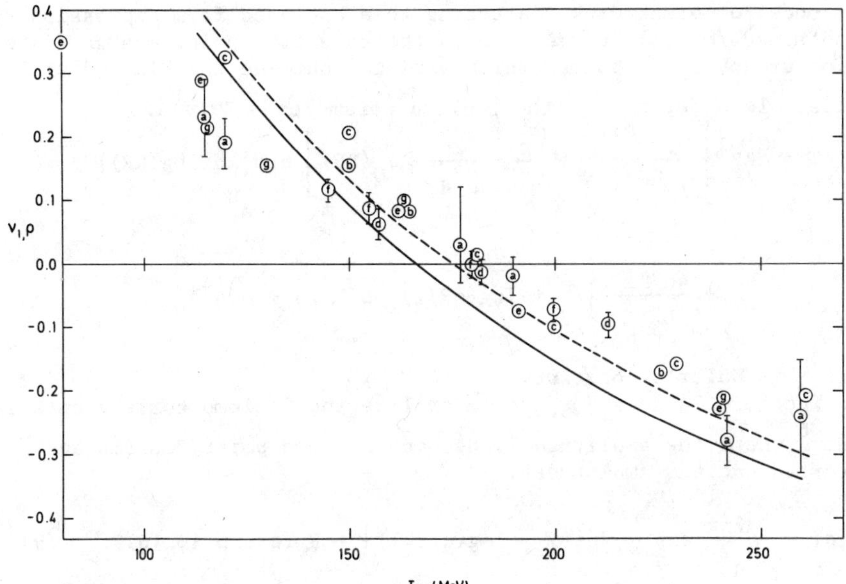

Fig. 2. Plot[4] of ρ for (a) ^{12}C, (d) ^{16}O, (f) ^{40}Ca and ν_1 for (b,c) ^{12}C, (e) ^{16}O and (g) ^{40}Ca against pion kinetic energy. The curves are are dispersion relation calculations of ρ for ^{12}C.

height and this is true for all the isoscalar nuclei thus far analysed. The agreement is not as good for the second zero for the light nuclei but then neither are the data.

Because the parameters of Eq. (1) describe different pieces of the data their errors are largely uncoupled. The fitted parameters therefore vary smoothly with energy and for example the trajectory of q_1^2 in the complex plane can be peacefully continued from the resonance region down towards threshold[3,5].

Coulomb corrections can now be very easily introduced in the eikonal model where use is made of the impact parameter representation

$$F(q) = ik \int_0^\infty J_0(qb)\, bdb \left\{ 1-\exp(i\chi(b)) \right\} \tag{6}$$

Working to first order in the Coulomb parameter $n = Z\alpha/v$ there are three distinct Coulomb modifications to the nuclear phase $\chi_N(b)$.

$$\chi(b) = \chi_N(b) + \chi_C(b) + \chi_{TD}(b) + \chi_{ES}(b) \tag{7}$$

Here $\chi_C(b)$, the phase induced by the charge distribution of the nucleus, is the usual Coulomb correction which accounts for example for most of the Bethe[6] phase. If we go back to our simple illustrative model of black disk scattering then a closed form expression can be obtained for this correction in the case of a point charge distribution <u>or</u> any distribution which vanishes outside the black disk radius. To order n^2 (n the Coulomb parameter = $Z\alpha/v$)

$$F(q) = ikR^2\left[\frac{J_1(qR)}{qR} - i\left\{\frac{\rho}{2} - \frac{2n}{q^2R^2}\right\}J_0(qR)\right]\exp\{2ni\log(kR)\}$$

$$+ \frac{4n^2ki}{q^2}\left[\gamma + \log(qR/2) - q^2R^2/4 + O(q^4)\right] \tag{8}$$

where γ is Euler's constant.

The formula does not look much like the Coulomb correction à la Bethe[6] where the amplitude is decomposed into point Coulomb and distorted nuclear amplitudes.

$$F(q) = -\frac{2nk}{q^2}\exp\left\{-2ni(\gamma + \log(q/2k))\right\} + F_N(q)\exp(i\phi_B(q)) \tag{9}$$

with a Bethe phase given[7] as an integral over the nuclear amplitude.

$$\phi_B(q) = -2n\gamma + n\int_0^{4k^2}\frac{dq'^2}{|q^2-q'^2|}\left[1 - \frac{F_N(q')}{F_N(q)}\right] \tag{10}$$

In fact if the whole nuclear amplitude, the real and imaginary parts of Eq. (3), are substituted into Eq. (10) then for moderate q we find the same answer as that of Eq.(8) apart from corrections which are negligible in our case of large kR. This has been shown for more general nuclear amplitudes by Franco[8]. The inclusion of a form factor in the last term of the integrand of Eq.(10) does not change the result providing, as is a very good approximation in our case, that the charge distribution is restricted to distances less then R. This is easily shown using the properties of the Weber-Schafheitlin discontinuous integrals[9]!

In summary the West-Yennie approach gives essentially the same answer as the simplest eikonal model but in retrospect this is not surprising since they are summing ladder diagrams using techniques which are known to lead to eikonalisation. However it is crucial to feed in all the information you have about F_N to get the imaginary part of the Bethe phase or you will not get the correct π^+, π^- cross section difference which in this case varies like $\{J_0(qR)\}^2/q^2R^2$. It can also be seen from Eq. (8) that in the vicinity of the first minimum the zero is displaced in the complex plane by an amount

$$\delta\nu_1 = 4n/(3.83)^2 \tag{11}$$

so whether the π^+ or π^- minimum is deeper determines the sign of ν_1. For all the isoscalar nuclei examined ν_1 changes sign around the resonance.

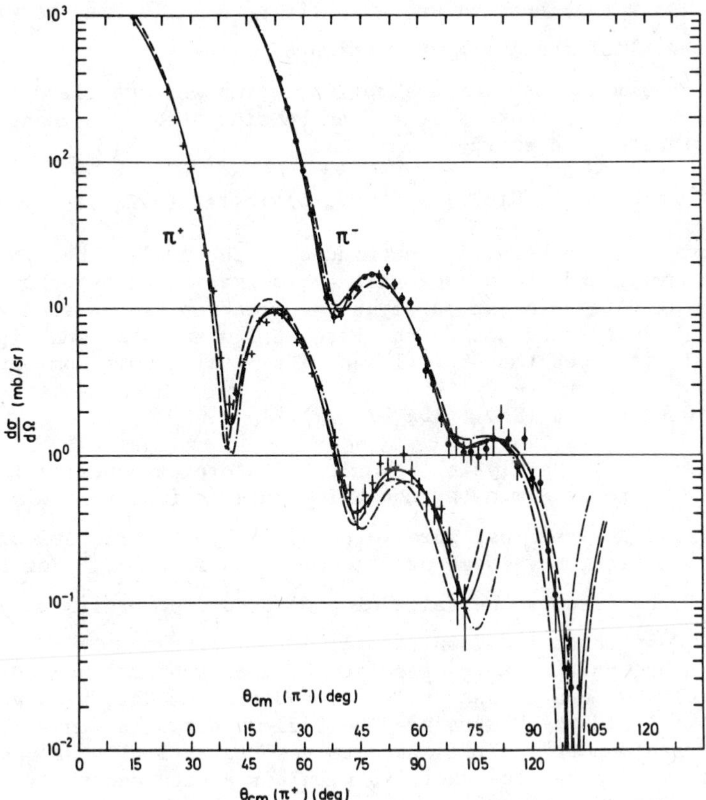

Fig. 3. Elastic π^+ and π^- scattering[10] from ^{40}Ca at 130 MeV. The effect of the Coulomb phase is shown as the dashed curve, with trajectory distortion added as the dot-dashed and with the energy shift also as the solid curve.

Though the black disk shows the correct qualitative features of the data, for accurate comparisons we have only to introduce our parameterisation of F_N, (Eq. (1)), into Eq. (6) and perform the inverse Bessel transform to obtain $\chi_N(b)$. After adding the Coulomb phase and re-transforming, separate predictions for π^+ and π^- can be given and a typical example is shown in Fig. 3. The other corrections drawn are the trajectory distortion (TD) and energy shift (ES) contributions of Eq. (7). Because of the Coulomb field the π^- sees the nucleus through a convex lens, the π^+ through a concave. This focussing, corresponding to a Wallace[11] correction to eikonal propagation, leads to an extra phase $\chi_{TD}(b)$ which can be expressed as an integral over $\chi_N(b)$. Similarly the Coulomb barrier repels (attracts) a $\pi^+(\pi^-)$ and the energy shift term $\chi_{ES}(b)$ may be written as an integral of the energy variation of $\chi_N(b)$

In the limit of strong absorption, which we have close to resonance, almost all of these effects can be simulated by a change of impact parameter and energy.

$$\chi(E{:}b) \approx \chi_C(E{:}b) + \chi_N(\{E-V_C(b)\}{:}b\{1+V_C(b)/kv\}) \tag{12}$$

This form may be doubtful in the middle of the nucleus but that part is black anyway and the use of this approximation rather than the more complex integrals can hardly be detected on graphs such as those of Fig. 3. This would not be the case at low energies and light nuclei. It is hoped that Eq. (12) or its partial wave analogue [5]

$$\delta_N(E{:}\ell) \to \delta_N(E-V_C((\ell + 1/2)/k){:}\ell) \tag{13}$$

will be used in the analysis of Coulomb distortion experiments such as those used to determine ρ from CNI. This is important since a π^+, π^- difference ascribed purely to CNI may, with problems of experimental resolution, have some of its origin in TD or ES. For example at 160 MeV they reduce the value of ρ deduced from preliminary [12] data on ^{12}C by about 15% down to 0.1.

The usual criticism levelled at the approach outlined here is that "surely one can go the whole way and fit the data with an optical potential and then introduce the Coulomb potential in a completely pedestrian manner". It is hard to get statistically as good fits with an optical potential as it is with our ansatz and the resulting parameters are then strongly correlated. One cannot then alter one feature of the data, such as the signs of one of the zeros, at will. Furthermore the Coulomb potential must be introduced consistently into the energy dependence of the potential parameters by $E \to E-V_C(r)$ which is non-trivial for a potential given in momentum space. The optical model is giving you a <u>constrained</u> amplitude analysis and if the constraint is false through, for example, the neglect of true pion absorption or having the wrong off-shell dependence that is unfortunate.

We should now like to extend this study of Coulomb corrections to inelastic transitions. In eikonal DWIA, quantising along the beam direction, the amplitude for exciting a state of angular momentum L and projection M from a spin zero ground state is

$$F_L^M(\vec{q}) = \sqrt{\frac{4\pi}{2L+1}}\; Y_L^M(\hat{q}) \int_0^\infty b\,db\; J_M(qb)\; \Gamma_M(b)\exp\{i\chi(b)\} \tag{14}$$

where for a $\Delta T=0$ transition the transition profile function $\Gamma_M(b)$ is given in terms of the isoscalar pion-nucleon amplitude $f(q)$ and nuclear transition form factor $S_L(q)$ by

$$\Gamma_M(b) = A \int_0^\infty qdq\ J_M(qb)\ f(q)\ S_L(q) \tag{15}$$

Certain simplifications have been made here such as the neglect of spin-flip and excitation energy and differences in the distortion between the initial and final states. In this limit $\chi(b)$ is the eikonal phase for elastic π-A scattering and we have just determined this from the elastic cross section!

The Coulomb potential now enters in two distinct ways, either as a modification of the transition amplitude $\Gamma_M(b)$ or in the distortion phase $\chi(b)$ but its effects are much smaller than for elastic scattering. Close to resonance the transitions are very much peaked near the nuclear periphery so that the main influence of the Coulomb phase $\chi_C(b)$ is as an inconsequential phase factor $\exp\left\{i\chi_C(R)\right\}$. In this region it will also be very hard to detect CNI for inelastic scattering. In the black disk model the distorted Coulomb excitation amplitude is zero unless M=L. This can be proved using our old friends the Weber-Schafheitlin integrals [9] but the result can be seen more clearly if we make a point charge model of an electric multipole (the black disk keeps the probe outside the details of the charge distribution). In Born approximation the only excitations occur for a linear multipole perpendicular to the beam direction and in fact the same result is true for the classical scattering from an electric multipole field. The only nuclear amplitude which does not vanish in the forward direction has M=0 so that there is no CNI here in this simple model unless a significant fraction of the charge lies outside the black disk radius which, I should remind you, is larger than the real nuclear radius. This is also true for monopole excitations where the transition Coulomb potential is essentially zero outside the nuclear region. This is unfortunate since such CNI experiments would have been challenging.

Because of the contributions of the different M values we can't fit an almost model-independent parameterisation to the inelastic scattering as we did for the elastic in order to demonstrate the importance of Coulomb corrections. However rather than presenting fitted DWIA results let me show you another use of our elastic parameterisation. Looking specifically at quadrupole excitations, note from Fig. 1 that the deformed black disk model fits the 2^+ excitation in ^{28}Si very well. More generally [13] in such collective models

$$F_2^0(q) = -\frac{\beta}{\sqrt{4\pi}} \exp(-dq^2/2) \frac{d}{dq^2}\left\{q^2 F_{el}(q)\exp(dq^2/2)\right\}$$

$$F_2^2(q) = -\beta\sqrt{\frac{3}{8\pi}} \exp(-dq^2/2)\, q^2 \frac{d}{dq^2}\left\{F_{el}(q)\exp(dq^2/2)\right\} = F_2^{-2}$$

$$F_2^1(q) = F_2^{-1}(q) = 0 \qquad (16)$$

where β is the deformation parameter and $F_{el}(q)$ the elastic amplitude. To add a little flexibility to the ansatz an Inopin fuzziness factor $\exp(-dq^2/2)$ is removed from the elastic amplitude and replaced in the transition. In the fit to the 162 MeV $\pi\,^{12}C$ excitation in Fig. 4 the

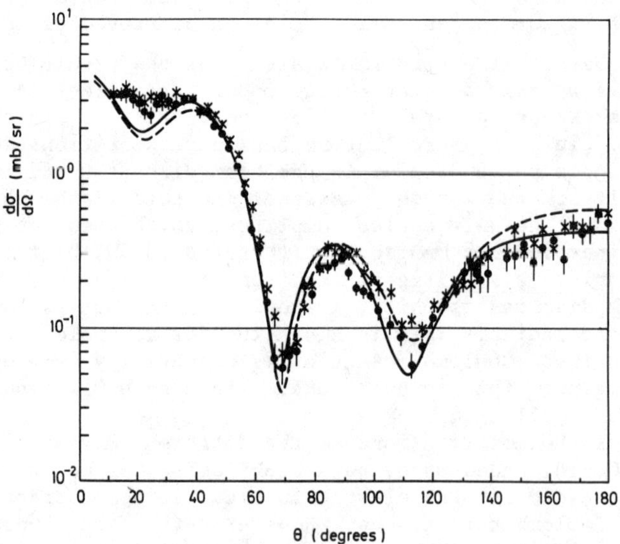

Fig. 4. Excitation of the 2^+(4.4 MeV) level in ^{12}C at 162 MeV [13] with π^+ (crosses) and π^- (circles). The corresponding dashed and solid curves result from introducing Coulomb distortion into the average amplitudes of Eq. (16) with β=0.33 and d=0.01 fm^2.

deduced value of d is reassuringly small. The effects of the Coulomb phase, TD and Coulomb excitation can now all be easily calculated but in order to estimate the ES change we must also fit the same formula of Eq. (16) to data at other energies to find the energy dependence of β and d. As we suspected the most important charge dependence comes now from the trajectory distortion but given the accuracy of the data it is really only significant in the region of the second maximum. The model presented here is not pathological and the conclusions on Coulomb effects are confirmed by DWIA fits where the parameters of the π-nucleon amplitude f(q) of Eq. (15), are allowed to vary.

Some of the ideas presented here are nicely illustrated in a paper [14] on pion scattering from ^{9}Be at 162 MeV. At this energy the diffraction minima should be very pronounced but because of the high target spin (3/2) these are filled in by quadrupole elastic scattering which reorients the nucleus. This adds incoherently and in a strong coupling model its contribution can be related by Clebsch-Gordan coefficients to the cross section for the production of other members of the ground state rotational band. Using measured values for the J=5/2 and 7/2 states this model reproduces the observed shoulders quantitatively [14]. Although there is an extra neutron in this nucleus this interpretation seems to be borne out by the charge dependence. If the minima in elastic scattering were shallow due to large $\mathrm{Im}(q_i^2)$ we would expect large Coulomb correction in that region but if this is due to quadrupole excitation the charge effects, as we see in Fig. 4, should be small as indeed they are.

Now that more precise data are becoming available one can start seriously to question the limits of simple DWIA analyses of inelastic scattering. One interesting step in this direction was carried out by Sparrow and Gerace [15] who considered the excitation of the 0^{+*} (7.6 MeV) level in ^{12}C both as a direct transition and as a two-step process passing via the 2^+ state at 4.4 MeV. Now the 0^* level is hard to describe microscopically but there are extensive measurements of the form factor for the $0 \rightarrow 0^*$ transition as well as that for exciting the 2^+. The B(E2) for the $0^* \rightarrow 2$ transition is known from the de-excitation to the latter state and if we assume that the $0^* \rightarrow 2$ and $0 \rightarrow 2$ form factors are proportional we have enough nuclear information to estimate the relative strengths of the single and two-step mechanisms. There is however an ambiguity of sign between the two contributions and in certain deformed nuclear models this phase information has a

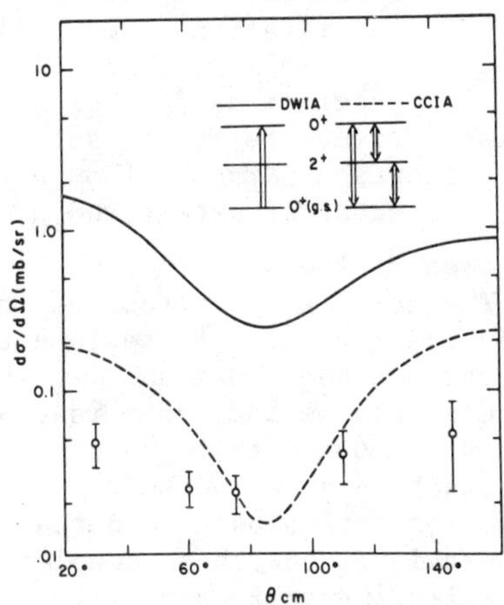

Fig. 5. Effect of two-step processes [15] on the excitation of the 0^+ (7.6 MeV) level in ^{12}C with 50 MeV pions.

direct nuclear structure interpretation [15]. It can be seen from Fig. 5 that at 50 MeV the inclusion of a destructive two-step contribution reduces the predictions by an order of magnitude and gives a respectable agreement with experiment. In their estimates Sparrow and Cerace used a Kisslinger optical potential which gives qualitatively correct results for elastic scattering and 2^+ excitation.

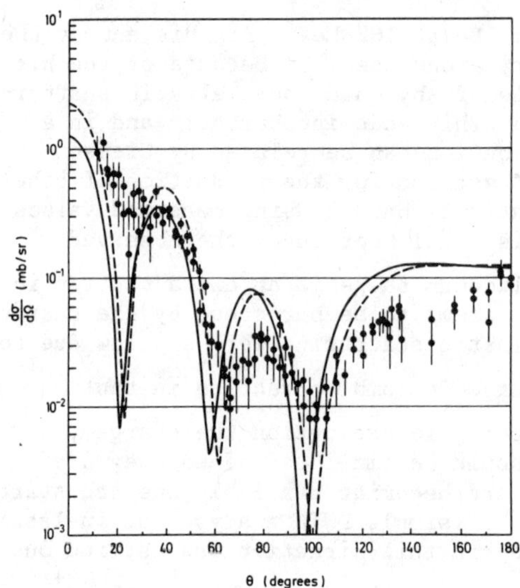

Fig. 6. Eikonal DWIA prediction for exciting the 0^+ (7.6 MeV) level in ^{12}C with 162 MeV pions [16] showing the effect of varying the sign of the two-step amplitude.

At higher energies we investigate the same model in the eikonal approximation where the two-step contribution through a state of angular momentum L to the transition profile function of Eq. (14) is

$$\Gamma^2(b) = \frac{2\pi i}{(2L+1)k} \sum_M |Y_L^M(\pi/2)|^2 \Gamma_{0\to 2}\Gamma_{0*\to 2} \tag{17}$$

The individual profile functions can be evaluated using a parameterisation of the pion-nucleon amplitude f(q) of Eq. (15) which fits the 2^+ transition at that energy and the elastic distortion obtained from our parameterisation of Eq. (1).

In Fig. 6 are plotted predictions for 162 MeV with the two different signs of the interference term, the solid line corresponding to the choice of Sparrow and Gerace. Though the data in the minima may be a bit suspect because of background problems [16] it is clearly out of the question to look for the small difference predicted at the maxima. The bulk of the effect of the two-step processes occurs at relatively small values of the impact parameter and the distortion blots these out near resonance. The result looks like a slight change in disk radius and normalisation. Consequently to search for such effects we must work in a region where the nucleus is much more transparent, e.g. at 50 MeV.

Sparrow and Gerace also find some reflection back into the elastic channel but here I must urge caution since these coupled channel calculations have been done neglecting self-correlations. It can be read in many papers that the imaginary part of the lowest order optical potential is due to nucleon knock-out but in this they also include, via closure, some contribution from the excitation of nuclear levels. The true feed-back really depends upon nuclear correlations which you can't know from just one transition form factor. In the absence of correlations an excited state corresponds to just taking one nucleon out of the core and to end up with elastic scattering the nucleon has to be put back. However in the Glauber limit the pion isn't supposed to hit the nucleon twice so that this contribution should then vanish. The same is true for two-step excitations if it is the same nucleon which is being promoted in the two cases. In practice these states are very collective but in a careful analysis of medium energy proton scattering from ^{12}C in a deformed harmonic oscillator model, Abgrall et al.[17] still find significant modification to these dispersive effects due to the self correlations. It is not clear whether the above dispersive effects or the ones due to pion absorption are the larger, but perhaps the later speakers will tell us.

Talking of Houston, my guide book [18] says it has "more than usually extravagant hopes of future development". Perhaps that is the attitude we should bring to pion-nucleus scattering.

Any unpublished work mentioned here is probably translated from the French of Jean-François Germond.

REFERENCES

1. C.Lunke et al., Phys. Letters 78B, 201 (1978)
2. E.V.Inopin and Yu.A. Berezhnoy, Nucl. Phys. 63, 689 (1965)
3. J.-F. Germond and C. Wilkin, Nucl. Phys. A237, 477 (1975)
4. J.-F. Germond and C. Wilkin, Ann. Phys. (in press).
5. R. Jäckle, H. Pilkuhn and H.G. Schlaile, Phys. Letters 76B, 177 (1978).
6. H.A. Bethe, Ann. Phys. 3, 190 (1958).
7. G.B. West and D.R. Yennie, Phys. Rev. 172, 1413 (1968); M.D. Cooper, M.B. Johnson and G.B. West, Nucl. Phys. A292, 350 (1977)
8. V. Franco, Phys. Rev. D7, 215 (1973)
9. M. Abramowitz and I.A. Stegun, Handbook of Mathematical Functions (N.B.S., Washington, 1964), Chap. 11.
10. J.P. Egger et al., Phys. Rev. Letters 39, 1608 (1977)
11. S.J. Wallace, Phys. Rev. C8, 190 (1973).
12. G.S. Mutchler (private communication).
13. W.H. Bassichis and A. Dar, Ann. Phys. 36, 130 (1960)
14. D.F. Geesaman et al., Phys. Rev. C18, 2223 (1978).
15. D.A. Sparrow and W.J. Gerace, Phys. Rev. Letters 41, 1101 (1978)
16. B. Chabloz et al.,Phys.Letters(in press); private communication.
17. Y. Abgrall, J. Labarsouque and B. Morand, Nucl. Phys. A271, 477 (1976); Y. Abgrall, J. Labarsouque, B. Morand and E. Caurier, Nucl. Phys. A (in press).
18. K. Baedeker, The United States (Baedeker, Leipsic, 1899), p.522.

Question: G. Epstein, MIT
Why can't you tell the relative sign of the one-step and two-step amplitudes?
Response: Wilkin
Sparrow and Gerace use this sign in a collective model to deduce properties of the two 0^+ and 2^+ levels.
Question: K. Amos, University of Melbourne
How seriously do you believe the results of a calculation, such as you described for the $0_1^+ \rightarrow 0_2^+$, that involve the subtraction of two large numbers?
Response: Wilkin
I don't believe that the 50 MeV agreement is very important yet but it is nice that pion scattering does show such a large sensitivity to nuclear structure.
Question: A. Saharia, TRIUMF
Eisenstein told us yesterday that one of the minima in angular distribution for pion elastic scattering occurs because of the p-wave nature of pion-nucleon interaction whereas according to you all the minima occur because of diffractive effects. Can you comment on that?
Response: Wilkin
Bob is an experimentalist so that he tries to be ambitious with the theory and give a microscopic explanation of the phenomena. I am much more modest and am merely trying to determine the pion-nucleus amplitude from experimental data. For this purpose I don't really care if the minima are primarily due to nuclear or nucleon profiles.
Question: A. Gal, Hebrew University
It may be premature to deduce that the $2^{+12}C^*$ dominates the multi-step contributions to the $0^+ \rightarrow 0^{+*}$ pion inelastic excitation before other intermediate states are investigated as to their potential contributions. In the closure approximation, when all such intermediate states are summed on, one obtains a mixed correlation function (between the 0^+ and 0^{+*} states) for the description of two-step contributions. It is in this summation that self-correlations must be treated with the utmost care; however, this is not the case with only one intermediate state, where trivial correction factors like (A-1)/A take care of the worries expressed by you when mentioning the "self-correlations".
Response: Wilkin
I agree entirely with this viewpoint. To calculate the two-step contributions completely we need the two-particle correlation functions and as I stressed we can't deduce these from a couple of transition form factors. It may still however be a reasonable first approximation to keep just the first 2^+ intermediate state because of the strong deformation. There are no clean calculations for the $0^+ \rightarrow 0^+$ transition but for the $0^+ \rightarrow 4^+$ excitation Abgrall *et al.* have shown that the 2^+ is the only significant intermediate state.

Question: K. Seth, Northwestern University

I am quite confused by your statement about self-correlations and double counting. It is true that in the classical Glauber calculation one does not allow scattering on the same nucleon twice. But that is a defficiency of the Glauber model, because it certainly happens in real life. So if calculations are to attempt to describe reality, means should be found to include scattering twice, rather than to throw it out as not appealing or convenient.

Response: Wilkin

As I replied to Gal, two-step processes do exist but their clean evaluation requires knowledge of correlation functions. This statement is true for optical potential approaches as well as in the eikonal model. If we have a bag of free nucleons without any correlations the $0^+ \to 2^+$ and $2^+ \to 0^{+*}$ will both be non-zero. The naive calculation I have outlined will therefore give a non-zero two-step contribution. However, it comes from exciting the same nucleon twice. To a good approximation we shouldn't count this because we have used a π-N amplitude as an input and therefore summed many diagrams. We are however very far from the no-correlation limit because of the strong deformation in ^{12}C. The simple model may therefore give a reasonable answer.

PION-NUCLEUS ELASTIC SCATTERING

C. M. Shakin

Department of Physics and Institute for Nuclear Theory
Brooklyn College of the City University of New York
Brooklyn, New York 11210

ABSTRACT

We describe a covariant theory of pion-nucleus scattering. A parameter-free calculation of the first-order optical potential is performed. A second-order potential is introduced and the parameters of the second-order potential are determined by requiring that the resulting potential provides a fit to the elastic scattering data. The parameters of the second-order potential exhibit a marked resonance behavior.

An approach to the theory of the pion-nucleus interaction based upon unitarity and analyticity is advocated.

In this presentation we will describe the work that has been carried out at Brooklyn College over the past few years. We have developed a covariant theory of the pion-nucleus optical potential.[1] I will describe the essential features of the analysis. We begin with a four-dimensional equation which describes pion-nucleus scattering,

$$M = K + K\mathcal{G}M \; . \tag{1}$$

Here $\mathcal{G}$ is the propagator for the pion and the ground state of the target nucleus and K is an irreducible interaction which does not contain the pion and the target ground state as an intermediate state. This equation may be replaced by a set of two equations, if we use the method Blankenbeckler and Sugar,

$$M = U + UgM \quad , \tag{2}$$

$$U = K + K(G-g)U \quad . \tag{3}$$

Here, g is a Green's function that has the same right-hand cut and discontinuity as G . We may choose g to restrict the heavy nucleus to be on its mass shell. In this fashion Eq. (2) may be reduced to a <u>three</u>-dimensional (covariant) equation,

$$T = V + VG_o T \; . \tag{4}$$

The matrices T and V of Eq. (4) are related to the M and U of Eq. (2) by specific kinematical factors. For example, in the center-of-mass system,

ISSN:0094-243X/79/540550-11$1.50

$$<\vec{k}|T(W)|\vec{k}'>=R^{\frac{1}{2}}(\vec{k})<\vec{k},k^0=k^0(\vec{k})|M(W)|\vec{k}',k'^0=k'^0(\vec{k}')>R^{\frac{1}{2}}(\vec{k}') \quad . \tag{5}$$

In this reduction, the fourth component of the relative four-momentum becomes a specific function of the magnitude of the relative three momentum. For example, if the heavy nucleus of mass M_A is kept its mass shell, $k^0 = W - (\vec{k}^2+M_A{}^2)^{\frac{1}{2}}$. **See Fig. 1.**

The details of the procedure are given in the literature.[1]

As a next **step** we must choose the form of the pion-nucleus optical potential, $<\vec{k}|V(W)|\vec{k}'>$, **which is to be used in** Eq. (4). We choose to determine a first-order optical potential through the evaluation of the diagram shown in Fig. 1. In this figure the **heavy** lines indicate a nucleus of A or (A-1) nucleons. The dashed lines denote

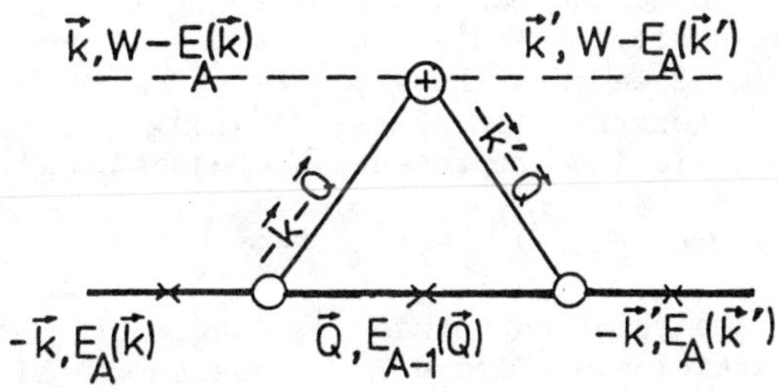

Fig. 1

Diagrammatic representation of the process which determines the first-order optical potential.

pions, the open circles are nuclear vertex functions and the circle containing the plus sign is a pion-nucleon (off-shell) scattering amplitude. The crosses are a reminder that the heavy particles are kept on their mass shells. Note that W is the energy in the center-of-mass of the pion-nucleus system and that $\vec{Q}$ is the momentum of the spectator nucleus of (A-1) nucleons. The fourth component of the momentum vector of the struck nucleon is $E_A(\vec{k}) - E_{A-1}(\vec{Q})$ and therefore the energy in the center-of-mass for the pion-nucleon collision is,

$$\sqrt{s} = \{ [W-E_{A-1}(\vec{Q})]^2 - \vec{Q}^2 \}^{\frac{1}{2}} \quad . \tag{6}$$

Now, in the evaluation of the diagram shown in Fig. 1 we are required to carry out an integral over the values of $\vec{Q}$, the momentum of the spectator nucleus.[2] This must be done for each value of k, k' and $\cos\theta = \vec{k}\cdot\vec{k}'/kk'$, and therefore the calculations involved are rather lengthy. Such integrals are usually avoided by making the fixed-scatterer approximation (FSA), however because of the very rapid variation of the pion-nucleon T matrix with changes in $\sqrt{s}$, the FSA is not a good approximation.

In the evaluation of Fig. 1 we use a separable model for the pion-nucleon T matrix.[3] (The separable potentials models which appear in the literature may be given an interpretation which allows their use in the field theoretic analysis discussed here.[1]) Specification of the other elements of Fig. 1 requires some knowledge of the nuclear density matrix. One may use wave functions that provide good fits to electron scattering data to construct this density matrix.

At this point we may further clarify our observation that the FSA is not useful for this problem. For simplicity, consider the evaluation of the first-order optical potential for forward scattering, $\langle\vec{k}|V^{(1)}(W)|\vec{k}\rangle \equiv \langle k|V^{(1)}(W)|k\rangle$.
The calculation of this quantity involves an integration over various values of the pion-nucleon T matrix, $\langle k_c|t_{\pi N}(\sqrt{s})|k_c\rangle$.
Here k_c is used to denote the magnitude of the relative momentum of the pion and the nucleon. (This quantity also depends on the value of Q) In Fig. 2 we show those values of k_c and $\sqrt{s}$ that are important in the construction of $\langle k|V^{(1)}(W)|k\rangle$.
In this figure the solid line represents the equation

$$\sqrt{s} = (k_c^2+m_\pi^2)^{\frac{1}{2}} + (k_c^2+m_N^2)^{\frac{1}{2}} \tag{7}$$

Equation (7) gives the relation between $\sqrt{s}$ and k_c for scattering of on-mass-shell particles.[4] Figure 2 represents the situation for a pion laboratory momentum of approximately 220 MeV/c. The quantity $\bar{k}_c$ (See Fig. 2) is the relative momentum for a collision in which the nucleon is at rest in the laboratory frame of reference. The shaded area in the figure denotes the kinematic region that is important for the calculation of the first-order optical potential.[4] Note that the values of k_c are centered about $\bar{k}_c$ with a spread of about 100 MeV/c. This spread of values is due to the Fermi motion of the nucleons in the target. If $\vec{p}_N{}^{LAB}$ denotes the nucleon momentum we have, approximately,

$$\vec{k}_c = \frac{m_N\vec{k}^{LAB} - \omega\vec{p}_N{}^{LAB}}{\omega + m_N}, \tag{8}$$

and

$$\bar{k}_c = \frac{m_N k_\pi^{LAB}}{\omega + m_N}. \tag{9}$$

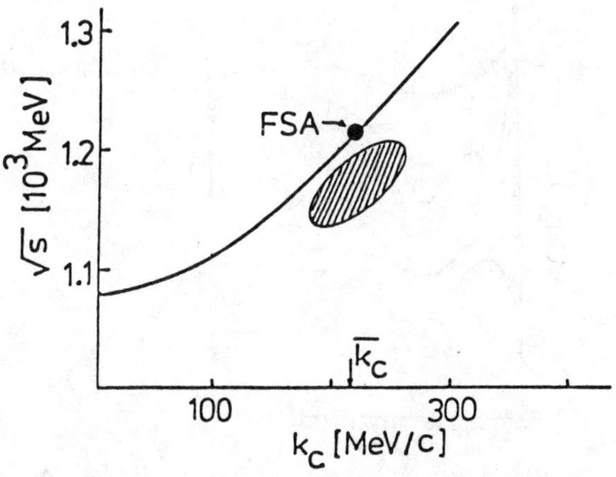

Fig. 2

Kinematical region that is important for the evaluation of the first-order optical potential when $k_\pi^{LAB} \sim$ 220 MeV/c.

In the figure we also show the FSA point at which the π -N T matrix is evaluated if one makes the simplest approximation. One can also consider an empirical shift of the FSA point downward (about 20-30 MeV) into the kinematically significant region, if one insists upon attempting to fit the data without doing the elaborate calculations we are advocating.[5] I believe the limitations of such a procedure are apparent.

In Figs. 3-5 we show the results of calculations of pion-oxygen scattering using the first-order optical potential (dashed lines).[6] In general, the agreement with the data is poor, but these deficiencies are masked to some degree by the customary representation of the data on semi-log graph paper. In order to fit the data it is necessary to consider the addition of a quite large second-order potential. We choose the following phenomenological form[2,7]

$$\langle\vec{k}|V^{(2)}(W)|\vec{k}'\rangle = R^{\frac{1}{2}}(\vec{k})\left[B(W) + \vec{k}\cdot\vec{k}'C(W)\right]R^{\frac{1}{2}}(\vec{k}') \times (2\pi)^3(2M_A)\,A(A-1)\,G(\vec{k}-\vec{k}'), \tag{10}$$

where

$$G(\vec{k}-\vec{k}') = \int e^{i(\vec{k}-\vec{k}')\cdot\vec{r}}\rho^2(\vec{r})d\vec{r} \quad . \tag{11}$$

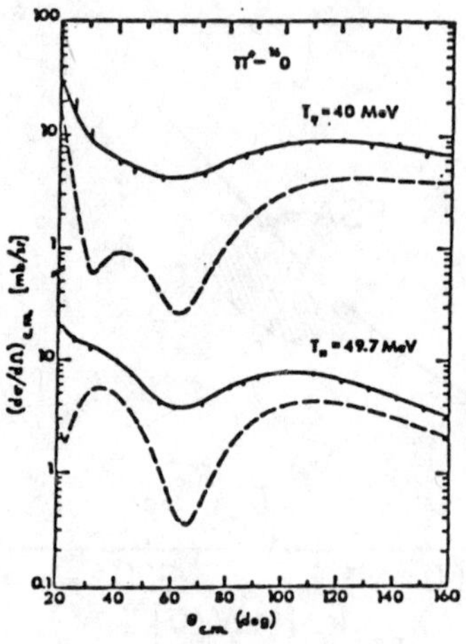

Fig. 3

Comparison of theory and experiment for pion-oxygen scattering at 40 and 49.7 MeV. The dashed curve is the result of the calculation using only the first-order optical potential.

Here B(W) and C(W) are complex phenomenological constants. Thus there are four parameters to determine at each energy. This determination is made in an energy-dependent manner by performing a χ^2 fit to the elastic scattering data.[6,7] The result of this analysis is shown as the solid lines in Figs. 3-5. (Similar results are obtained for π - ^{4}He and π - ^{12}C scattering[7].) It is clear that the inclusion of a large second-order optical potential allows one to make excellent fits to the data at the expense of introducing several new parameters. The parameters determined in this analysis are given in the literature.[6,7] The values of B and C are quite similar for π-^{16}O and π-^{12}C scattering over a large range of energies. The essential features of the energy dependence of Re k^2C and Imk^2C are shown in Fig. 6. This figure represents a somewhat idealized composite of the results obtained for ^{16}O and ^{12}C. The values of B are generally much smaller than k^2C, except at low energies (T_π<80 MeV). The energy dependence of ReB is sketched in Fig. 7. It is possible that the region above 100 MeV in Fig. 7 reflects an "s-wave" parameter, which is induced by the transformation of the fundamental interaction from the pion-nucleon center-of-mass to the pion-nucleus c.m. frame. (We remark that since the quantity $G(\vec{q})$ depends on the angle between $\vec{k}$ and $\vec{k}'$, the parameters B and C appear in all the quantities $< k_c|V_\ell^{(2)}(W)|k_c'>$ which are determined when a partial-wave decomposition of the

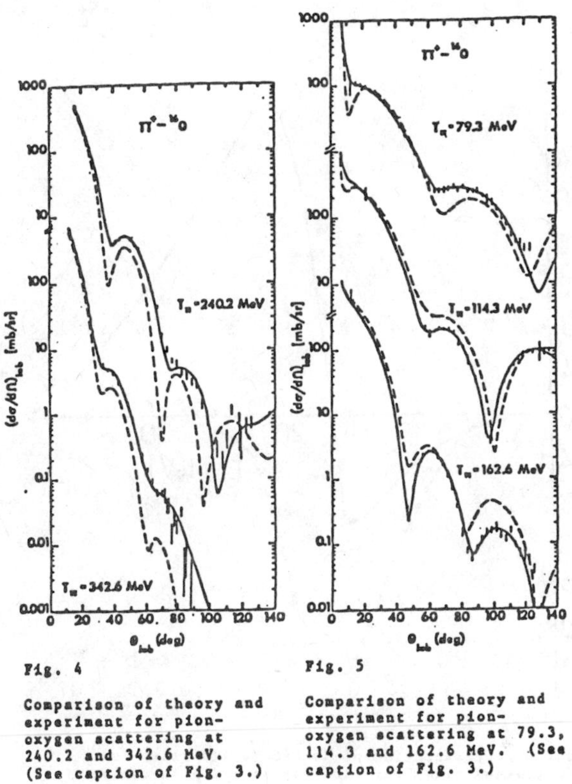

Fig. 4

Comparison of theory and experiment for pion-oxygen scattering at 240.2 and 342.6 MeV. (See caption of Fig. 3.)

Fig. 5

Comparison of theory and experiment for pion-oxygen scattering at 79.3, 114.3 and 162.6 MeV. (See caption of Fig. 3.)

second-order pion-nucleus interaction is carried out. At low energies, i. e. for small $\vec{q}$, B and k^2C contribute mainly to the s and p-wave pion-nucleus optical potentials, respectively.) For low energies ($T_\pi \sim 50$ MeV) ReB provides a large repulsion in the s-wave pion-nucleus interaction. The origin of this repulsion is not fully understood at this time.

It is clear that the parameter $k^2C(W)$ exhibits a remarkable resonance behavior. We conjecture that this is mainly due to a resonance in what has been termed the "true" pion absorption process.[8] Further calculations are needed to support this conjecture. It is well known that the true absorption process is a very large part of the reaction cross section and therefore it is not surprising that large second-order potentials are needed in the calculation of the pion-nucleus interaction. Of course, other processes may give rise to second-order effects and microscopic calculations are needed to elucidate their role in the construction of $V^{(2)}$.

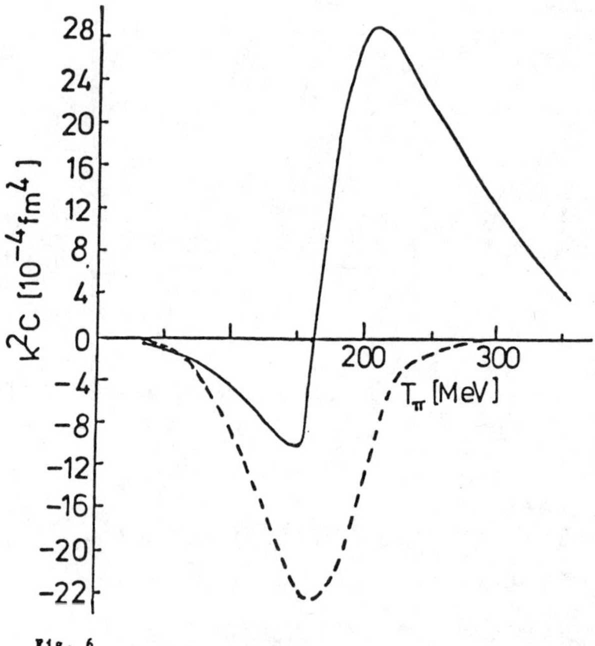

Fig. 6

Schematic representation of Re k^2C (solid line) and Im k^2C (dashed line) for pion-oxygen and pion-carbon scattering.

While most researchers would agree as to the nature of the first-order optical potential, there is no consensus as to the form of the theory beyond the first order. Some workers would propose the use of some variant of the Watson or Kerman, McManus and Thaler (KMT) versions of multiple scattering theory, adopted to describe the pion-nucleus interaction. We do not believe these modified multiple-scattering theories are of value for this problem. Instead, we have recently proposed a new theory of the optical potential for strongly interacting particles which is based upon considerations of unitarity and analyticity.[9] In this short space we are not able to describe this theory in any detail. However, we will remark on the difference in the treatment of the absorption process in a multiple-scattering theory and in our analysis. For example, we indicate, in Fig. 8a), one term of a multiple-scattering series that is related to the absorption process. Again the circles containing the plus signs are π-N T matrices. In general the imaginary part of the potential generated by the process depicted in Fig. 8a) has no definite sign. This is in contrast to the calculation indicated in a schematic fashion in Fig. 8b). This figure is a representation of a calculation of $\langle\vec{k}|V^{(2)}_{abs}(W)|\vec{k}'\rangle$ which follows from the considerations given in ref. 9. For example, we may write

$$\langle\vec{k}|V^{(2)}_{abs}(W)|\vec{k}'\rangle = \sum_n \frac{\int \langle\vec{k},0|T(W'+i\eta)|\vec{p}_1\vec{p}_2,n\rangle d\vec{p}_1 d\vec{p}_2 \langle\vec{p}_1\vec{p}_2,n|\left[T(W'+i\eta)\right]^{\dagger}|k',0\rangle}{W - W' + i\eta} \quad (12)$$

where $T(W + i\eta)$ is the amplitude for pion absorption on a pair of nucleons leading to continuum states of the nucleons labelled by $\vec{p}_1$ and $\vec{p}_2$. Further,

$$W' = (\vec{p}_1^{\,2} + m_N^2)^{\frac{1}{2}} + (\vec{p}_2^{\,2} + m_N^2)^{\frac{1}{2}} + \left[(\vec{p}_1 + \vec{p}_2)^2 + (M^{(n)}_{A-2})^2\right]^{\frac{1}{2}} \quad . \quad (13)$$

In Fig. 8b), the circle containing the minus sign denotes the π-N T matrix, $t_{\pi N}$ $(\sqrt{s} - i\eta)$ in accordance with the structure of Eq. (12). From Eq. (12) we see that $\mathrm{Im}\langle\vec{k}|V^{(2)}(W)|\vec{k}\rangle \leq 0$ so that we restricted our phenomenological parameters such that Im $B(W) \leq 0$ and Im $k^2 C(W) \leq 0$. These considerations have been extended to construct a theory for the calculation of those aspects of the second-order potential that are unrelated to the true absorption process.[9]

REFERENCES

1. L. Celenza, L. C. Liu and C. M. Shakin, Phys. Rev. C11, 1593 (1975). Erratum, Phys. Rev. C12, 712 (1975).
2. L. C. Liu and C. M. Shakin, Phys. Rev. C16, 333 (1977).
3. J. T. Londergan, K. W. McVoy and E. J. Moniz, Ann. Phys.(N.Y.) 86, 147 (1974).
4. L. Celenza and L. C. Liu, Phys. Rev. C12, 1983 (1975).
5. L. C. Liu and C. M. Shakin, Phys. Rev. C16, 1963 (1977).
6. L. C. Liu and C. M. Shakin, "Role of the Second-Order Optical Potential in Pion-Oxygen Scattering", to be published in the Physical Review C.
7. L. C. Liu, Phys. Rev. C17, 1787 (1978).
8. L. C. Liu and C. M. Shakin, Phys. Letters 78B, 389 (1978).
9. L. C. Liu and C. M. Shakin, "Theory of the Optical Potential for Strongly Interacting Particles", submitted to the Physical Review C.

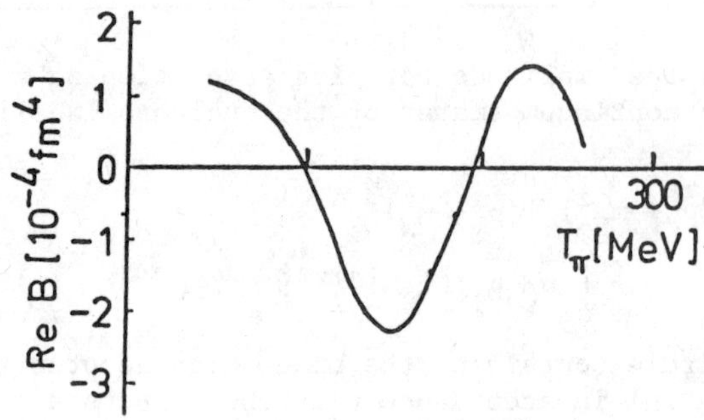

Fig. 7

Schematic representation of Re B determined in the study of pion-oxygen and pion-carbon scattering.

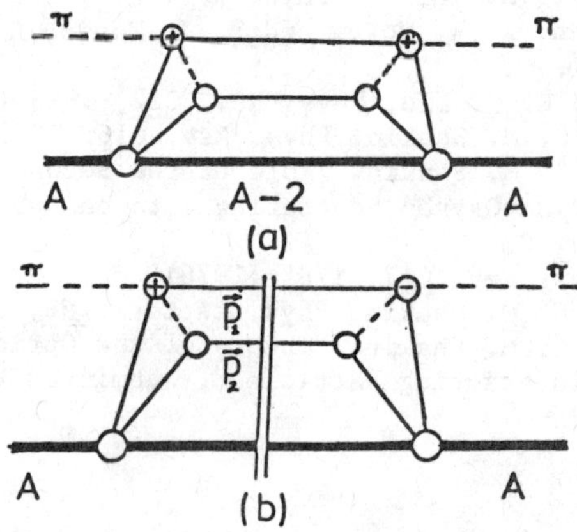

Fig. 8

(a) One term of a "multiple-scattering" series that is related to the true absorption process.
(b) Calculation of the true absorption process in a theory based upon analyticity and unitarity. [See Eq. (12).]

Question: W. Wharton, Carnegie-Mellon University

Have you used your B(w) and C(w) parameters to predict $\sigma_{Absorption}$? (The percentage of $\sigma_{Reaction}$ which is due to absorption, the energy and A dependence of $\sigma_{Absorption}$).

Response: Shakin

Some estimates have been given for $\sigma_{Absorption}$ for pion scattering from ^{12}C and ^{16}O at 30, 40, and 50 MeV in L. Liu and C. Shakin, Phys. Letters 78B,389(1978).

Question: D. Geesaman, Argonne National Laboratory

What you have described is essentially a 4-parameter model. It is not clear to me why this is intrinsically better than fitting a 4-parameter first order potential and then expanding the differences between these parameters and the best calculations of a first order potential in powers of the density. After this proclamation, my second question is: in the resonance region, we know the quasi-free scattering is so strong that the pion never penetrates the nucleus, and ρ^2 terms may not be as important. What I do not understand is, does a simple ρ^2 potential really express all the effects of integrating over the large energy and momentum transfer in the first interaction, as well as other complexities in the complete calculation of true absorption?

Response: Shakin

The procedure you suggest which involves an eight-parameter fit is of no theoretical interest. The goal is to understand how a pion interacts with a nucleus and not to fit the experimental data. We hope to be able to calculate the second-order potential in some approximation in the future. Our calculation of the first-order potential is a first step in our theoretical program. I hope we can go beyond the present semi-phenomenological stage. It is quite possible that a calculation of the second-order potential will show that the simple local form we are using is not realistic. At this point I can only remark that the form we have used is highly suc-successful in providing a universal parameterization for ^{12}C, ^{16}O and ^{40}Ca. Indeed the use of B(w) and C(w) determined for pion ^{12}C scattering, for example, yield very good fits for pion ^{16}O and pion ^{40}Ca scattering. Finally, we note that the pion does penetrate into the nuclear surface region and the ρ^2 term has definite effects on the shape of the differential cross sections. In some cases one may be able to fit the cross sections with a first-order potential but this is often done by making unjustifiable changes in the nuclear radius.

Question: R. Rockmore, Rutgers University

I question the reality of the residue of your dispersion form for the true absorption potential. You will not be able to account for this term as isobar self energy in, say, the isobar model. For reference see the contribution by Brown, Jennings, and Rockmore submitted to this conference.

Response: Shakin

We have not studied the precise relation of our second-order potential to the isobar self-energy. Since our theory is self-contained it is not necessary to determine this relationship although the question is of some interest. As may be seen from inspection of Eq. (12), the absorptive part is indeed real and positive for $\vec{k}=\vec{k}'$.

Question: A. Thomas, TRIUMF

I must object to the somewhat oversimplified picture of the first order optical model which you have presented. In fact, in a complete theory the nucleon which is hit by the pion also interacts (through a binding potential) with the nuclear core. In a very nice piece of analytic work (for the πD system) Fäldt has shown that to a good approximation the three-body energy shift and the so called "TS" correction compensate each other. The compensation is exact in the limit $B \to 0$, although in practice ($B \neq 0$, finite range interactions) one should do a full three body calculation (for πD). In the nucleus case the evaluation of $U^{(1)}$ really requires the solution of a "simple" three body problem. The reason Landau and I introduced the binding energy as a parameter was not that we could not do the fermi average over $D(E,|\vec{Q}|)$(in fact we did this!), but rather our realization that (i) we could not solve the three body problem for $U^{(1)}$ and then calculate π nucleus scattering in a finite amount of time and (ii) Fäldt had shown that "TS" effectively counters the three-body shift at low energy. We certainly do not claim this approximation contains <u>all</u> the physics of the problem, but it makes a good deal of sense (e.g. in terms of reaction content) as a one parameter theory.

Response: Shakin

I agree that there are corrections of the type you suggest, however, I do not believe that they should be represented by an energy shift. Consider the following simple example. Suppose we have a 15 MeV pion incident upon a nucleus which contains a nucleon with a separation energy of 20 MeV. The pion does not have enough energy to knock out the nucleon and therefore the optical potential should be real. In our analysis the t-matrix would be evaluated at negative energies ($\sim$5 MeV and below) yielding a real optical potential. If you introduce a positive energy shift taking you back to the fixed-scatterer value of 15 MeV the t-matrix becomes complex and so does the optical potential. The complex part generated upon shifting the energy is spurious. In our model, however, the analytic structure is given correctly with the thresholds for inelastic scattering at the right energies. More properly the interactions with the spectator particle which you consider should lead to the use of intermediate state distorted waves rather than the plane waves which are presently used in generating the t-matrix via the summation of ladder diagrams.

A FIELD THEORY OF PION-NUCLEUS INTERACTIONS

G. A. Miller*
University of Washington, Seattle, WA 98195

ABSTRACT

Some approximate solutions of the Low equations for pi-nucleon scattering, pi-nuclear scattering and the $pp \rightarrow d\pi^+$ reaction are discussed. Our results differ from those of other treatments by the inclusion of propagator modifications which act as cutoffs for terms involving pions of relatively small momenta. Because of these propagator modifications the use of a pi-nucleon T-matrix of long range is perfectly consistent with the idea that the fundamental pi-nucleon form factor is of very short range. The origin of these propagator modifications is shown to be a proper evaluation of the energy dependence of the off-shell pi-nucleon scattering amplitude. As a consequence of our theory, good agreement with $pp \rightarrow d\pi^+$ data is obtained without the inclusion of large effects from rho-meson exchange terms.

INTRODUCTION

At the present time there is a crisis in the theory of pion-nucleus interactions. There is great disagreement regarding the range of the pion-nucleon interaction. Many groups (see, e.g., Hirata et al[1]) use a pion-nucleon interaction of relatively long range, $\sim$1 fm, but many other groups use an interaction of essentially zero range (see, e.g., McManus[2]). From a simple consideration of the sizes of the pion and proton one would expect that the range of interaction should be fairly long.[3] However, a short range of interaction is to be preferred from field-theoretic arguments.[4] This difference in starting point leads to enormous differences in the tools necessary for computing the cross sections for various reactions. For example, only an essentially zero-range interaction is amenable to a coordinate space treatment. Furthermore one expects to obtain vastly different calculated results using the two different models. However in many cases the two approaches give equally good descriptions of experimental data. For example, a good representation of the $pp \rightarrow d\pi^+$ data can be obtained using a pi-nucleon interaction of long range,[5] or with a short range, if one also includes the effects of rho exchange.[6] Thus I would like to suggest a somewhat unusual approach to the question of determining the range of the pi-nucleon interaction.

The problem of determining the range or, equivalently for separable interactions, off-energy-shell behavior of the pi-nucleon T-matrix, is a difficult one. We cannot consider pi-nucleon data alone, and we must also examine the many-body processes in which the pi-nucleon amplitude is to be used. Furthermore, the dynamics

---

*Supported in part by the U.S. Department of Energy.

ISSN:0094-243X/79/540561-15$1.50

employed in handling pion-nuclear processes should be consistent with those employed in treating pi-nucleon processes. That is, insofar as possible, the same Hamiltonian should be used in handling both problems. In principle, one may use the technique of summing perturbation graphs to deal with both pi-nucleon and pi-nuclear scattering. However the strong interaction is involved, so a non-perturbative technique is necessary.

We adopt the procedure of deriving and attempting to solve the Low equation (non-relativistic dispersion relation) for the various reactions. So far we have studied pi-nucleon scattering,[7,8] pion-nucleus scattering,[9,10] and the $pp \to d\pi^+$ reaction.[11]

Our result is that the use of a pi-nucleon amplitude of long range is totally consistent with having a fundamental pion-nucleon form factor of short range. The use of the T-matrix of long range has its origin in the proper evaluation of the crossed pi-nucleon Born term at off-shell energies.

In this talk we discuss the problems of pi-nucleon scattering, pi-nuclear scattering, and the $pp \to d\pi^+$ reaction in turn.

PI-NUCLEON SCATTERING

Let us consider pi-nucleon scattering within the framework of the Chew-Low[12] model. We work in the static limit, although some of our results are more general.[7] We restrict ourselves to the (3,3) channel only and consider energies in the resonance region or below, so that effects of inelastic channels may be neglected. The pi-nucleon left-hand-cut term is also neglected. It is well known that the attractive nature of the pi-nucleon Born term, $\lambda v/E$,

$$\frac{\lambda}{E}\, v \equiv \qquad (1)$$

combined with the requirement of unitarity, leads to a (3,3) resonance at the correct energy. The parameter λ is proportional to the square of the pi-nucleon coupling constant, f^2, and is negative. The solution of the pi-nucleon Low equation, under the stated assumptions, is

$$t_E(q,k) = D(E)\,\frac{\lambda}{E}\,v(q,k) \quad , \qquad (2)$$

where

$$v(q,k) = \frac{u(q)\,u(k)}{\sqrt{4\omega_q \omega_k}}\,P_{33}(q,k) \qquad (3)$$

and

$$D^{-1}(E) = 1 - \frac{\lambda}{\pi E}\int_0^\infty \frac{p^4 dp}{\omega_p}\,\frac{E^2}{\omega_p^2}\,\frac{u^2(p)}{E^+ - \omega_p} \quad , \qquad (4)$$

in which $t_E(q,k)$ is the off-shell pi-nucleon T-matrix, $P_{33}(q,k)$ is the necessary channel projection operator, $u(q)$ is the fundamental pi-nucleon form factor, and $\omega_p = (p^2 + m_\pi^2)^{\frac{1}{2}}$.

The resonance energy, E_r, is determined from the condition

$$\mathrm{Re}[D^{-1}(E_r)] = 0 \quad , \tag{5}$$

and is a function of $u(q)$. The function $u(q)$ is obtained from the requirement that E_r be the experimental resonance energy. It is typically found that $u(q)$ is essentially constant for small q, but drops rapidly for $q \sim 1$ GeV.[13,8]

Suppose I consider an alternative procedure in which I take the crossed Born term to be the driving term in a Lippmann-Schwinger equation. I would obtain the resulting pi-nucleon T-matrix, t^{LS}, with

$$t_E^{LS}(q,k) = D_{LS}(E)\lambda v/E \quad , \tag{6}$$

and

$$D_{LS}^{-1}(E) = 1 - \frac{\lambda}{\pi E}\int_0^\infty \frac{p^4 dp}{\omega_p}\frac{u^2(p)}{E^+ - \omega_p} \quad . \tag{7}$$

The functions $D^{-1}(E)$ and $D_{LS}^{-1}(E)$ differ only in that the integral over p in the former function contains an additional factor of E^2/ω_p^2. If eq. (7) were used to determine $u(q)$ one would find that $u(q)$ would be very small for $q \gtrsim 350$ MeV (if one obtained the correct resonance energy). The use of the Lippmann-Schwinger equation leads to a cutoff that is important at energies very close to those of interest.

If the Low equation is used one is able to fit the pi-nucleon data with a reasonably high cutoff. The same data is fit for both the Lippmann-Schwinger equation and Low equations because of the presence of the E^2/ω_p^2 factor in eq. (4).

To show that $E^2/\omega_p^2 u^2(p)$ is a reasonable square of a pi-nucleon cutoff function I plot this function, as well as the cutoff function $u_{LS}^2(q)$ used by Moniz,[14] in Fig. 1. There is a striking similarity

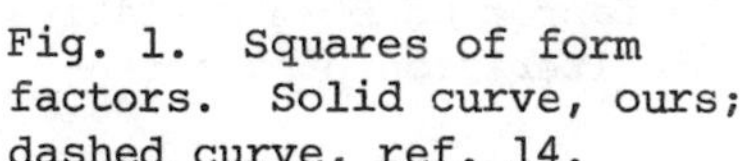
Fig. 1. Squares of form factors. Solid curve, ours; dashed curve, ref. 14.

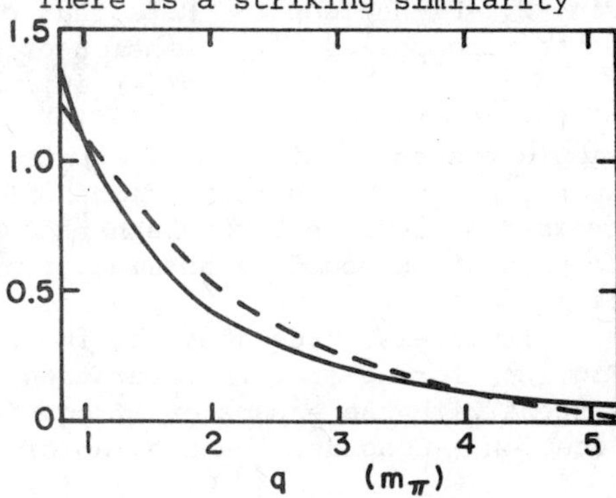

between the two functions. (It is the squares of cutoff functions that enter into calculations.)

For our later purpose in summing the pi-nuclear multiple scattering series, it is desirable to recast the Low equation results of eqs. (2)-(4) in the form of a linear equation. We find that the Chew-Low T-matrix is also the solution of the equation

$$t(E) = \frac{\lambda v}{E}\left(1 + \frac{E^2}{h_0^2}\frac{1}{E^+ - h_0}\,t(E)\right) . \tag{8}$$

This may easily be verified by using (3) in (8) and noting that h_0 is diagonal in momentum space with $(p|h_0|p) = \omega_p$. The eq. (8) is very similar to the Lippmann-Schwinger equation. The only difference is the E^2/h_0^2 factor in the propagator. This factor comes from the E^2/ω_p^2 factor in eq. (4).

The next step is to understand the origin of the E^2/h_0^2 factor. The work that I now discuss has been done in collaboration with E. M. Henley and will appear, in greater detail, elsewhere.[15] Consider the crossed Born graph and the series formed by iterating this graph, Fig. 2. We calculate these graphs using a perturbation theory in which each energy denominator is given by

$$E - \sum_{\text{all pions}} \omega_\pi .$$

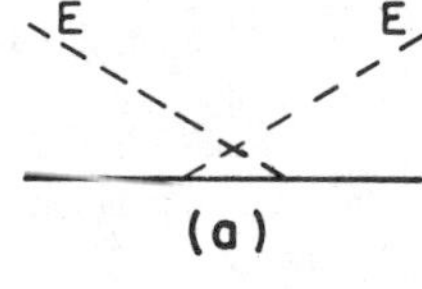

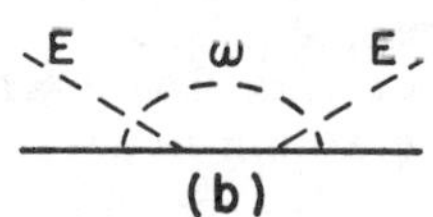

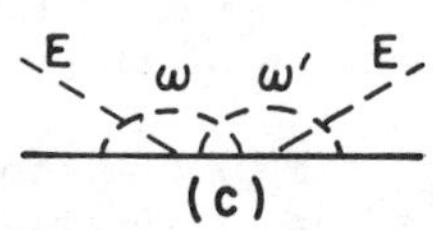

Fig. 2. Pi-nucleon scattering series.

(Note that there are no backward going pions in the static model treatment of pi-nucleon scattering.) We consider only the relevant energy denominators and do not write the spin, isospin and momentum functions. The diagram of Fig. 2a is proportional to $-1/E$. The diagram of Fig. 2b is proportional to

$$\frac{1}{-\omega}\,\frac{1}{E^+ - \omega}\,\frac{1}{-\omega} , \tag{9}$$

in which ω is the energy of the virtual pion, and an integral over ω must be performed. Note that eq. (9) is consistent with the expansion of t, as given by eq. (8), to second order in λ. Note especially the appearance of the two factors of $-1/\omega$ in (9), which replace the expected (from a simple treatment of $\lambda v/E$ as a potential) factors of $-1/E$. Hence it is the proper evaluation of the off-energy shell behavior of the crossed pi-nucleon Born graph that gives the factor of E^2/ω^2 relative to that which would be obtained from a simple iteration of a potential.

It is also true that the infinite sequences of graphs, such as Fig. 2c, formed from the iteration of the crossed Born graph is consistent with the expansion of eq. (8). There is, however, one complication. Consider the diagram of Fig. 2c. The relevant energy denominators are

$$\frac{1}{-\omega}\,\frac{1}{E^+ - \omega}\,\frac{1}{E^+ - \omega - \omega'}\,\frac{1}{E^+ - \omega'}\,\frac{1}{-\omega'}\quad , \tag{10}$$

which is not quite the term of third order in λ of eq. (8). The problem is that the graph of Fig. 2c contains a term which has a two-pion cut. The Chew-Low solution, under the one-meson truncation, does not contain such terms. However, the two- and one-pion cuts may be separated by using the propagator decomposition

$$\frac{1}{E^+ - \omega - \omega'} = \frac{-E}{\omega\omega'} + \frac{(E-\omega)(E-\omega')}{(E^+ - \omega - \omega')\omega\omega'}\quad . \tag{11}$$

The use of the first term on the right-hand side of eq. (11) in eq. (10) gives a result consistent with the third order (in λ) term of eq. (8). The second term on the right-hand side of eq. (11) contains only a two-meson cut. Hence, it should be negligible at the (low) energies of our present interest, and furthermore should not be included in an expansion that reproduces the Chew-Low result. (Sample numerical calculations suggest that the contribution of the second term of (11) is much smaller than the contribution of the first term.)

So far I've told you that $(E^2/\omega_q^2)u^2(q)$ plays the role of the square of a long-range cutoff function in pi-nucleon scattering. The next question to be answered is: what happens when a pion propagates from one nucleon to another? Within the framework of a theory involving potentials, the answer would be clear. We would use the $E^2/\omega_q^2\, u^2(q)$ factor everywhere, including in the propagation of a pion between nucleons. Within our field-theoretic framework, we must study the pi-nucleus Low equation in order to answer this question.

THE PION-NUCLEUS LOW EQUATION

Consider pion-nucleus scattering in which a pion of quantum number k is incident upon a nucleus of quantum number A. Let the final state consist of a pion, k', and a nucleus, A'. We represent the quantum numbers of the corresponding pi-nuclear eigenstates by single lower case letters. The quantum numbers of the initial, final and intermediate pi-nuclear states are designated by the letters i, f and n respectively. With this notation the Low equation for pi-nuclear scattering is given by[8]

$$T_{fi}(E) = \sum_N \frac{\Gamma^*_{Nf}(E_N)\Gamma_{Ni}(E_N)}{E^+ - E_N} + \sum_N \frac{\Gamma_{A',Nk}(E_N)\Gamma^*_{A,Nk'}(E_N)}{-E - E_N}$$

$$+ \sum_N \frac{T^*_{nf}(E_n)T_{ni}(E_n)}{E^+ - E_n}\quad . \tag{12}$$

The matrix element $\Gamma_{Ni}(E_N)$ represents the off-shell probability amplitude for the process $k+A \to N$. If $E_N = E_i$, $\Gamma_{Ni}(E_N)$ is the matrix element for the pion absorption reaction. The first term on the right-hand side of eq. (12) represents the process of pion absorption, leading to a zero-meson nuclear eigenstate, followed by meson emission. The second term on the right-hand side of eq. (12) is the crossed version of the first term. The sum of the first and second terms on the right-hand side of eq. (12) is defined to be the pi-nucleus Born term, $T_{fi}^{(B)}(E)$. The third term on the right-hand side of eq. (12) represents the effects of intermediate pion-nucleus eigenstates. Terms involving the crossed version of the third term of eq. (12) and terms involving intermediate states with more than one asymptotic pion are ignored here.

The amplitudes Γ also satisfy a Low equation, namely[11]

$$\Gamma_{Ni}(E_N) = B_{Ni} + \sum_n \frac{\Gamma^*_{Nn}(E_n)T_{ni}(E_n)}{E^+ - E_n} + \Gamma^L_{Ni} \quad , \tag{13}$$

in which B_{Ni} is the single-nucleon absorption term, Fig. 3. The second term on the right-hand side of eq. (13) represents the effects of pion-nuclear scattering prior to absorption. The term Γ^L_{Ni} has a cut for pions of negative energy.[11]

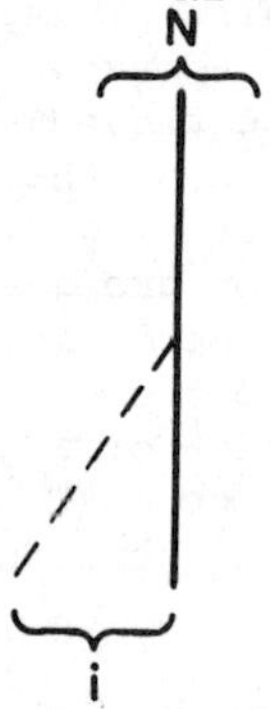

N

i

Fig. 3. Single-nucleon pion-absorption term.

Although we would like to solve the set of coupled equations, eq. (12) and eq. (13), we make a modest first step and keep only the first term on the right-hand side of eq. (13), i.e., we take

$$\Gamma_{Ni}(E) \approx B_{Ni} \quad . \tag{14}$$

As a further approximation we neglect the energy denominators, E_N, that appear in $T_{fi}^{(B)}(E)$, i.e. we take

$$E_N \approx 0 \quad . \tag{15}$$

Nuclear excitation energies are still included in the quantities E_n.

If we make the approximations (14) and (15), the Born term of (12) takes the particularly simple form

$$T_{fi}^{(B)}(E) = \frac{V_{fi}}{E} \quad , \tag{16}$$

in which V_{fi}/E is simply the nuclear matrix element of the sum over nucleons of the pi-nucleon Born terms of Fig. 4a.

Note that, in general, $T_{fi}^{(B)}$ does include terms in which a meson is absorbed on one nucleon and (following some interaction effects in the intermediate states) emitted on another nucleon, Fig. 4b. However, under the approximations of (14) and (15) the sum of the crossed and uncrossed two-nucleon terms of $T^{(B)}$ vanishes. We call

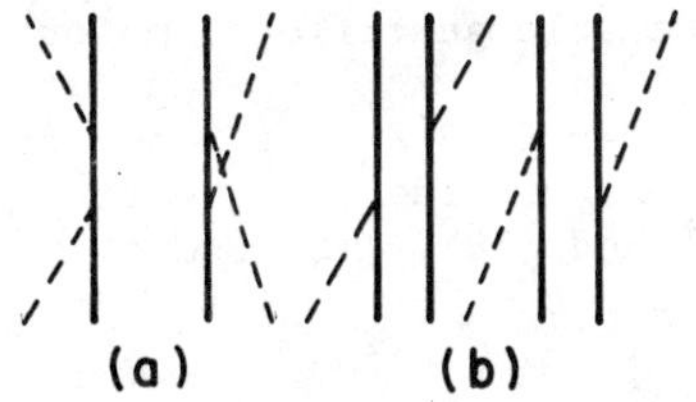

Fig. 4. Some terms appearing in $T^{(B)}$.

the two approximations (14) and (15) the multiple scattering approximation. In using them we ignore one of the major motivations of a field theory, namely the consistent inclusion of true meson absorption. When $E_N = E$, $T_{fi}^{(B)}$ acquires an imaginary part due to physical processes in which the entire energy of the pion is deposited into excitation energy of the nucleus. The approximation of (15) prevents the inclusion of such imaginary terms. Note that because meson absorption processes are dominated by terms involving pion prescattering, the validity of (14) would insure the validity of (15). Hence we group the two approximations together as a single one. However we felt that the use of (14) and (15) does allow us to compute the multiple scattering series. Those terms which are generated by the pi-nucleon Born terms and their iterates are included within the framework of our multiple scattering approximation. Hence we will proceed.

The ultimate inclusion of true meson absorption is one of our main goals, and I will discuss some of our progress towards this goal at the end of the talk.

For now we use eq. (16) in eq. (12) with the result

$$T_{fi}(E) = \frac{V_{fi}}{E} + \sum_n \frac{T^*_{nf}(E_n)T_{ni}(E_n)}{E^+ - E_n} \quad . \tag{17}$$

Eq. (17) is the one we wish to solve. The solution of (17) by iteration or by matrix inversion is impossible because of the presence of the strong interaction and because there are an infinite number of pi-nuclear eigenstates. To handle (17) we use a techniqe of introducing a set of pseudodynamics in which pi-nuclear scattering is treated in a manner very similar to ordinary potential scattering. Consider the Hamiltonian H_0 which has zero- and one-meson eigenstates. The zero-meson eigenstates of H_0 are the ordinary eigenfunctions of nuclear physics. These consist of nucleons which move non-relativistically and which interact via a two-nucleon potential. It is assumed that there is an H_0 that accurately reproduces the spectrum and wave functions of nuclear states. The one-meson eigenstates of H_0 are constructed by creating a single meson on a zero-meson eigenstate. The Hamiltonian H_0 contains no interactions between the pion and the nucleons. Now I claim that the equation

$$T(E) = \frac{V}{E}\left(1 + \frac{E^2}{H_0^2}\frac{1}{E^+ - H_0}T(E)\right) \quad , \tag{18}$$

in which <u>all</u> matrix elements are to be taken between one-meson eigenstates of H_0, is equivalent to eq. (17). That is, the T-matrix elements resulting from eq. (18) obey the relationship of eq. (17).

This equivalence between (18) and (17) is the sole justification for using (18), which is much easier to solve.

The proof of this equivalence is given in ref. 8 and will not be reproduced here. One can make a quick check by considering the terms of second order in V obtained from both (17) and (18). The use of $T_{nf}(E_n) = V_{nf}/E_n$ in the second term of (17) gives

$$T_{fi}^{(2)}(E) = \frac{V_{fi}}{E} + \sum_n V_{fn} \frac{1}{E_n^2} \frac{1}{E^+ - E_n} V_{ni} \quad . \tag{19}$$

The second-order term of (18) is

$$T^{(2)}(E) = \frac{V}{E} + V \frac{1}{H_0^2} \frac{1}{E^+ - H_0} V \quad , \tag{20}$$

and by taking the matrix elements of eq. (20) between one-meson eigenstates of H_0 we have

$$T_{fi}^{(2)}(E) = \frac{V_{fi}}{E} + \sum_n V_{fn} \frac{1}{E_n^2} \frac{1}{E^+ - E_n} V_{ni} \quad , \tag{21}$$

which is the same as eq. (19).

Let's consider the implications of (18) for the question of how a pion propagates from one nucleon to another. As we have stated, the operator V/E is simply the pi-nucleon Born term summed over nucleons and is given in terms of $\lambda v/E$ as

$$\frac{V}{E} = \sum_i \frac{\lambda v_i}{E} \quad , \tag{22}$$

in which the sum over i is over all of the nucleons in the nucleus. The consequence of (22) taken together with (18) is that the factor E^2/H_0^2 appears in the propagation of a pion from one nucleon to another. For example, the use of (22) in (20) leads to the result

$$T^{(2)}(E) = \frac{V}{E} + \sum_{i,j} \frac{\lambda v_i}{E} \frac{E^2}{H_0^2} \frac{1}{E^+ - H_0} \frac{\lambda v_j}{E} \quad , \tag{23}$$

and terms with $i \neq j$ are included as well as terms with $i=j$. The factor E^2/H_0^2 reduces the effects due to intermediate pi-nuclear states of high energy. One then finds that the <u>same</u> cutoff function appears in the propagation between nucleons that appears in the propagation on a single nucleon.

The eq. (18) still needs to be solved, and our technique is the standard one of expressing T(E) as an expansion in terms of pi-nucleon T-matrices and then obtaining an optical potential. The two relevant equations are

$$T(E) = \sum_i \frac{\lambda v_i}{E}\left(1 + \frac{E^2}{H_0^2}\frac{1}{E^+ - H_0}T(E)\right) \tag{24a}$$

and

$$t_i(E) = \frac{\lambda v_i}{E}\left(1 + \frac{E^2}{h_0^2}\frac{1}{E^+ - h_0}t_i(E)\right) \quad . \tag{24b}$$

The use of (24b) in (24a) leads to a multiple scattering series with all of the problems of the conventional one. For example, binding corrections are included via the difference between H_0 and h_0. Note that the Chew-Low theory can be extended so that the effects of nucleon recoil are included in h_0. Such effects are important in pion-nuclear scattering (see, e.g., ref. 1, Shakin and Liu,[16] and Landau and Thomas[17]).

The main difference betwen eq. (24a) and corresponding equations of other theories is the appearance of the E^2/H_0^2 (cutoff) factor in the propagator. As an example of how this cutoff works consider the second-order optical potential (which is also known as the Lorentz-Lorenz effect[18]). Under certain standard manipulations this potential, $U^{(2)}$, may be shown to be given, in our theory, by

$$U^{(2)} = \sum_{i\neq j} t_i(E)\,\frac{E^2}{H_0^2}\,\frac{Q}{E^+ - H_0}\,t_j(E) \quad , \tag{25}$$

in which Q is a projection operator out of the state of one meson plus target ground state. In the usual derivation of the Lorentz-Lorenz effect one assumes that the intermediate pion energies are larger than the nuclear excitation energies so that the approximation, $H_0 \approx h_0$, may be used in (25). The result is

$$U^{(2)} = \sum_{i\neq j} t_i(E)\,\frac{E^2}{h_0^2}\,\frac{Q}{E^+ - h_0}\,t_j(E) \quad . \tag{26}$$

When one takes the matrix element necessary to compute (26) the E^2/h_0^2 factor becomes just E^2/ω_p^2, which acts as a cutoff. It is well known that the appearance of such cutoffs greatly inhibits the Lorentz-Lorenz effect.[19]

The next problem that we face is the one of determining the origin of the E^2/H_0^2 factor in (18). This is more difficult for pi-nuclear scattering than for pi-nucleon scattering because of the appearance (even in the static model) of graphs involving backward going pions.

To enforce some order amongst the various graphs we use and test an approximation scheme due to Chew.[20] According to him, one should apply, after renormalization, ordinary perturbation theory except when intermediate states occur which may have an energy equal to the initial energy. Such terms give rise to poles. For each pole the order of the diagram is to be considered as lowered by one power of f^2. For example, under these criteria all of the graphs of Fig. 2

are of order f^2. This is the prescription that gives the Chew-Low result for pi-nucleon scattering and we will apply it for pi-nuclear interactions.

Let us now apply Chew's idea to pi-nuclear scattering. Consider the time-ordered graphs of Fig. 5. We use the static model in the present exposition. The term of Fig. 5a is of order f^2, but the term of Fig. 5b is of order f^4 (within Chew's criteria) and is ignored to order f^2. The relevant energy denominators of the term of Fig. 5a are

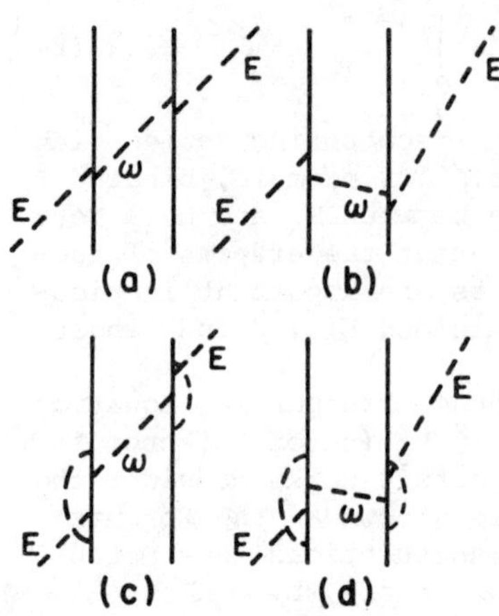

Fig. 5. Some pi-nucleus scattering terms.

$$\frac{E}{-\omega}\frac{1}{E}\frac{1}{E^+ - \omega}\frac{1}{E}\frac{E}{-\omega} \quad , \tag{27}$$

in which two factors of $-1/\omega$ appear. Thus the graph of Fig. 5a is consistent with the corresponding second-order term of eq. (18).

To complete our treatment we must include, for each nucleon, that sequence of graphs that lead to the pion-nucleon (3,3) resonance. However, we use Chew's criteria to eliminate many terms. For example, the graph of Fig. 5c is of order f^2, but the term of Fig. 5d is of order f^6 and is ignored. If we include the sequence of graphs that lead to the (3,3) resonance on one nucleon, allow the nucleon to propagate from one nucleon to another (but apply Chew's criteria), then include the sequence of graphs leading to the pi-nucleon resonance with the second nucleon, we obtain the term in the multiple-scattering series of the form

$$\sum_i t_i^{(E)} \frac{E^2}{h_0^2} \frac{1}{E^+ - h_0} t_j(E) \quad . \tag{28}$$

(Note that we may have to use approximations similar to that of keeping only the first term on the right-hand side of eq. (11).)

Observe that the E^2/h_0^2 factors of (27) and (28) obtain from an evaluation of the pi-nucleon scattering amplitude at the proper off-shell energies.

The approximation scheme of Chew is valid for pi-nucleon scattering, but has never been tested for pi-nuclear scattering. We have made some estimates that suggest that, within the framework of the static model, this approximation scheme is also valid for pi-nuclear scattering. Note however that the use of the static model prevents the inclusion of the effects of true meson absorption.

In order to include effects of true meson absorption we must first learn how to treat pion absorption or production within the Low equation framework. To take the simplest starting point, we study the reaction, $pp \to d\pi^+$.

LOW EQUATION TREATMENT OF THE $pp \to d\pi^+$ REACTION (IN THE Δ-REGION)

This pion production reaction presents an interesting challenge. Since the work of Mandelstam[21] it has been known that reaction proceeds by the exchange of "something" between nucleons, which leads to the production of a Δ on one of them, which is followed by a pion emission from the Δ, e.g., Fig. 6a. Most workers take this graph to be proportional to $e^{i\delta(\varepsilon)}\sin\delta(\varepsilon)$ where $\delta(\varepsilon)$ is the (3,3) phase shift and ε is the sum of the pion and nucleon energies in the center-of-mass frame of the final (interacting) pion and nucleon. There is an interesting question here. The pi-nucleon amplitude appearing in Fig. 6a is an off-shell one, but why do we evaluate it at the final pi-nucleon energy instead of the energy of the initial pi-nucleon state consisting of a pion of high momentum and nucleon of high energy? Within our Low equation framework we find that it is consistent to evaluate δ at the energy ε, but that our treatment of the energy dependence of the off-shell pi-nucleon T-matrix leads to a propagator modified by the factor E/h_0.

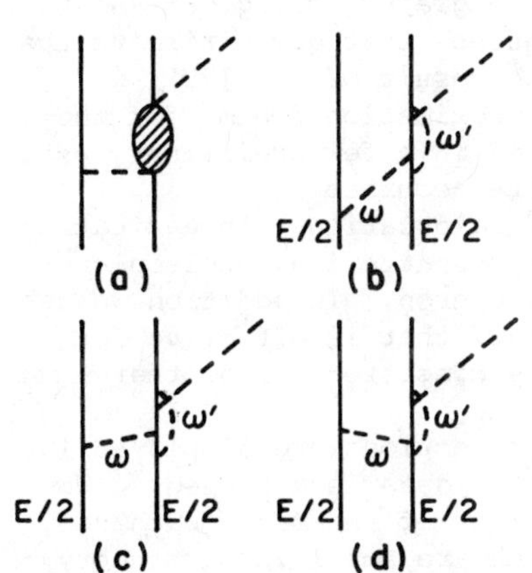

Fig. 6. Mechanisms for pion production.

We must solve the version of the Low equation (13) that is appropriate for the $pp \to d\pi^+$ reaction. The details of the approximation procedure necessary to obtain a solution are complicated, so here we will only state our results.[11] Let $\tilde{\Gamma}(E)$ be the operator for pion production in proton-proton collisions leading to a final pion-deuteron state. We have found that

$$\tilde{\Gamma}(E) = \sum_{i \neq j} t_j(E)\, \frac{E}{h_0}\, \frac{2h_0}{-h_0^2}\, B_i \quad . \qquad (29)$$

(A term, $\sum_i B_i$, has been ignored.) The operator B_i produces a pion from nucleon i which propagates until scattered by the nucleon j. The $-1/h_0^2$ term is the standard Yukawa propagator of the pion. The $2h_0$ factor is absorbed by the $1/\sqrt{2\omega}$ factors appearing in B_i and $t_j(E)$. The major modification of our treatment is the E/h_0 factor which acts as a cutoff of terms involving mesons of high momenta.

As a mathematical indication of the origin of the E/h_0 factor consider the limit of $\tilde{\Gamma}(E)$ as $E \to 0$. The matrix $\tilde{\Gamma}(E)$ has no pole at

E=0, but t(E) does have such a pole. An omission of the E/h_0 factor of (29) leads to a pole in $\tilde{\Gamma}(E)$, a result that is inconsistent with the Low eq. (13). (Note that $\tilde{\Gamma}(E) = \Gamma^{\dagger}(E^{*})$.)

One can also obtain the result (29) from a proper evaluation of the energy dependence of the pi-nucleon scattering amplitude combined with an application of Chew's approximation scheme. Consider, for example, the graphs of Fig. 6b, c and d. For the $pp \rightarrow d\pi^{+}$ reaction each nucleon initially has an energy of E/2. We assume, however, that the intermediate nucleons (on the right-hand lines) have negligible kinetic energy. The graphs of Fig. 6b and 6c are of order f under Chew's criteria but the graph of Fig. 6d is of order f^3 and is ignored to order f. The evaluation of the graphs of Fig. 6b and 6c as well as the inclusion of the full sequence that gives rise to the (3,3) pion-nucleon resonance leads to the result of eq. (29).

The numerical accuracy of Chew's approximation scheme for the meson production reaction has been studied in a few preliminary estimates. So far, this procedure seems quite accurate.

It is now time to discuss an actual calculation. In evaluating (29) we have neglected s-wave pi-nucleon rescattering, nucleon-nucleon scattering and the d-state of the deuteron. In addition we use a fundamental pi-nucleon form factor, u(q), that is effective for momenta at about 1 GeV. This function is consistent with other determinations.[4,13,8]

The results for total $pp \rightarrow d\pi^{+}$ cross sections are displayed in Fig. 7. A quite good representation of the data is achieved. The effects of rho-meson exchange, which are small in our theory,[10] have been ignored. We can easily explain the data with a reasonable fundamental form factor and with the E/h_0 factor. A typical angular distribution is shown in Fig. 8. A qualitative representation of this data is also achieved.

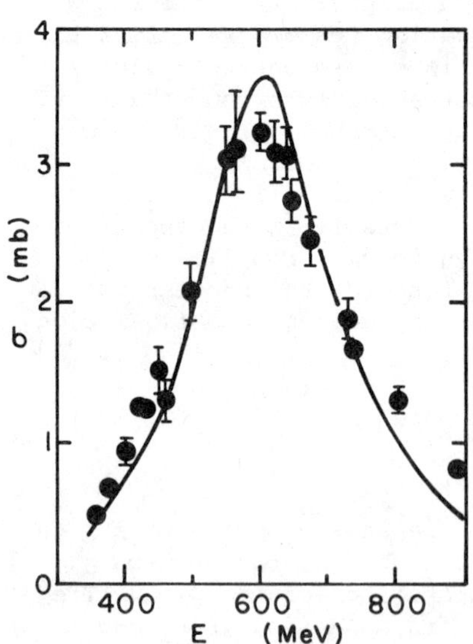

Fig. 7. Total cross section for $pp \rightarrow d\pi^{+}$.

SUMMARY AND CONCLUSIONS

I'd like to briefly review the main points of this talk. For pion-nucleus multiple scattering the factors E^2/H_0^2 or E^2/h_0^2 play the role of a cutoff. Thus the use of a fundamental pi-nucleon form factor of short range is perfectly consistent with the use of a pi-nucleon T-matrix of long range. The origin of these factors is the proper evaluation of the off-energy-shell behavior of the pi-

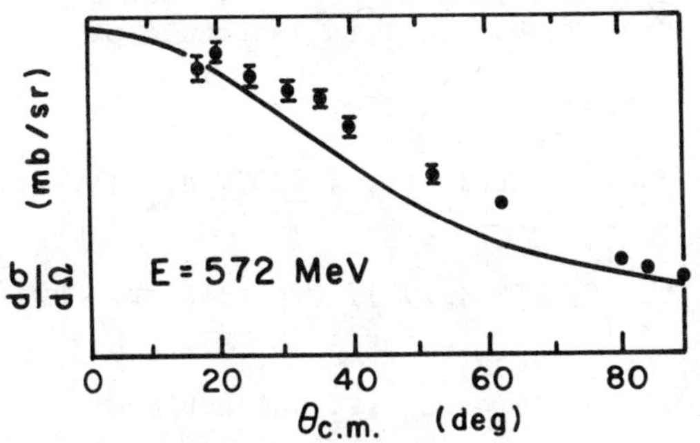

Fig. 8. Angular distribution for pp → $d\pi^+$.

nucleon scattering amplitude. For the pp → $d\pi^+$ reaction a pi-nucleus propagator that is modified by a factor E/h_0 is obtained. This factor also has its origin in a proper evaluation of the energy dependence of the off-shell pi-nucleon T-matrix.

Finally I'd like to mention our plans to include true meson absorption. Our goal has been to solve the full set of coupled Low equations (12) and (13) without making the approximations (14) and (15). If we do not make these approximations the energy dependence of $T^{(B)}$ is quite complicated, and we have been unable to obtain a solution. However an alternate procedure, which makes use of our present approximate solution of the Low equation, is being developed. For multiple scattering terms, the main result of solving the Low equation is that one uses a pi-nucleon T-matrix of long range. Motivated by this, we introduce a new Hamiltonian consisting of an energy independent pi-nucleon potential of long range (effective in the (3,3) channel only), a single nucleon operator which connects the zero and one pion subspaces, and a nucleon-nucleon potential. The set of coupled Low equations for this Hamiltonaian is derived and is very similar to the set obtained without the introduction of the pi-nucleon and nucleon-nucleon potentials. We are presently studying these equations and hope to have some concrete results by the time of the next Meson-Nuclear Physics Conference.

I would like to respond to some criticism raised by Schmit and Maillet.[22] These authors claim that my E^2/h_0^2 effects occur only because of my omission of the uncrossed two-nucleon diagram of Fig. 4b. However, there are no crossed graphs in the theory of ref. 22. In my theory the sum of the two graphs of Fig. 4b vanishes. Furthermore we have seen how E^2/h_0^2 factors arise from the proper evaluation of the energy dependence of the crossed pi-nucleon Born graph. That graph does not even occur in ref. 22. Hence the criticism raised by Schmit and Maillet does not apply to my work.

I thank E. M. Henley for many useful discussions. I thank Chi-Yee Cheung for pointing out an algebraic error in the results of ref. 11.

REFERENCES

1. M. Hirata, J.H. Koch, F. Lenz and E.J. Moniz, 1978 MIT preprint, CTP #748.
2. H. McManus, this conference.
3. R.J. Cence, "Pion-Nucleon Scattering," Princeton University Press, Princeton, New Jersey, 1969.

4. G.E. Brown, in "Nucleon-Nucleon Interactions - 1977," ed. H. Fearing, D. Measday and A. Strathdee, American Institute of Physics, New York, 1978.
5. B. Goplen, W.R. Gibbs and E. Lomon, Phys. Rev. Lett. 32, 1012 (1974).
6. M. Brack, D.O. Riska and W. Weise, Nucl. Phys. A287, 425 (1977).
7. G.A. Miller, Phys. Rev. C14, 2230 (1976).
8. G.A. Miller, Phys. Rev. C18, 914 (1978).
9. G.A. Miller, Phys. Rev. Lett. 38, 753 (1977); G.A. Miller, Phys. Rev. C16, 2325 (1977).
10. G.A. Miller, Z. Physik A287, 387 (1978).
11. M.A. Alberg, E.M. Henley, G.A. Miller and J.F. Walker, Nucl. Phys. A306, 447 (1978).
12. G.F. Chew and F.E. Low, Phys. Rev. 101, 1570 (1956).
13. D.J. Ernst, this conference.
14. E.J. Moniz, in "Meson-Nuclear Physics - 1976," ed. P.D. Barnes, R.A. Eisenstein and L.S. Kisslinger, American Institute of Physics, New York, 1976.
15. E.M. Henley and G.A. Miller, to be published.
16. L.C. Liu and C.M. Shakin, Phys. Rev. C16, 333 (1977).
17. R.H. Landau and A.W. Thomas, Nucl. Phys. A305.
18. M. Ericson and T.E.O. Ericson, Ann. Phys. (N.Y.) 36, 323 (1966).
19. J. Hüfner, J.M. Eisenberg and E.J. Moniz, Phys. Lett. 47B, 381 (1974).
20. G.F. Chew, Phys. Rev. 94, 1755 (1954).
21. S. Mandelstam, Proc. Roy. Soc. A244, 491 (1958).
22. C. Schmit and J.P. Maillet, Nucl. Phys. A312, 236 (1978).

Question: A. Thomas, TRIUMF

While the Chew-Low model is a useful model of the P_{33} resonance, it is very dated. Since then we have discovered SU(3), quarks etc. In that model there is unambiguously an elementary $\Delta \equiv (qqq)$ state. The $\pi N \rightarrow (qqq)_{\Delta}$ vertex function will undoubtedly have a very short range. Is it not possible that the truth about the πN interaction is that the elementary Δ contributes a short-range piece, while the πN rescattering (as many of us have discussed) results in a relatively long range piece of the interaction? On a more philosophical level, why must physics be split into two non-overlapping camps, each refusing to see the merits of the opposition (not to mention the defects of their own approach)?

Response: Miller

The answer to the first question is yes it is possible. However a very large part of the physics of the Δ is its one pion decay. This part I've kept. I've also been thinking about the elementary Δ. I really can't answer the second question, but it does seem to be true that there is a division. The foregoing has been my attempt to connect the two viewpoints.

Question: D. Ernst, Texas A & M

Since the Chew-Low model is only capable of producing the P_{33}, P_{13} and P_{31} two-body data, one needs to extend the model to include other terms to be consistent with the two-body data. Do you have any comments on how you might do this?

Response: Miller

This is a low energy theory, considering energies up to, and slightly greater than the (3,3) resonance. The P_{11} phase shift is small, at such energies, so I'll do what everybody else does.

MANY-BODY APPROACH TO PION-NUCLEUS SCATTERING

W. Weise
Institute of Theoretical Physics, University of Regensburg
D-8400 Regensburg, W. Germany

1. INTRODUCTION

The analysis of presently available π nuclear elastic scattering data at low energy and in the 3.3 resonance region has clearly led to the conclusion that multiple scattering approaches based on first order optical potentials are insufficient to obtain a satisfactory description. Much recent work has therefore been concerned with the treatment of higher order corrections to the pion-nuclear optical potential, in particular with the influence of true pion absorption and of other reaction channels on the elastic channel. Calculations have proceeded along different lines:

(a) In the 3.3 resonance region, the pion optical potential becomes highly non-local because of the propagation of Δ-isobars, once excited by the incoming pion. Methods to handle the pion-nuclear scattering amplitude in the presence of a strongly resonating elementary amplitude have recently become available. We mention here the isobar-hole or isobar-doorway models [1-8] which treat pion scattering as an excitation and propagation of Δ-isobar-hole states in the nuclear environment, thus facing the full complexity of the many-body problem at excitation energies around twice the pion mass.

(b) At low energy ($T_\pi \leq 50$ MeV) phenomenological optical potentials similar to those used in the description of pionic atoms have been applied successfully [9,10]. Terms of second order in density were found to be important. There is general agreement that true absorption channels play an essential role [11,12,13]. The problem is then to understand the sign and size of the parameters associated with such higher order terms on a microscopic level.

(c) In addition, there has been growing interest in the field theoretical foundation of the pion-nuclear scattering problem [14,15,8]. Attempts are being made to establish connections between field theory approaches and the more phenomenologically oriented models mentioned earlier. We refer to the presentation of G.A. Miller for further details and related problems.

Our main emphasis here will be on an approach to pion-nuclear scattering as a many-body response problem, with special emphasis on a microscopic treatment of reaction channels (true absorption, quasi-free scattering) together with binding effects, reflection terms and other many-body degrees of freedom. We shall point out that a large part of the uncertainty occuring in the calculation of such many-body corrections is due to our limited knowledge about the off-shell properties of meson-baryon couplings in general and of the πNΔ and ππNΔ

ISSN:0094-243X/79/540576-14$1.50

(or $\rho N\Delta$) vertex functions in particular. It is for this reason that few-body processes like $\pi D \leftrightarrow NN$, which provide such off-shell information, require a careful discussion.

An important aspect (usually disregarded in theories of pion-nucleus scattering) is the low frequency ($\omega << m_\pi$) limit of the pion-nuclear response problem, as it determines, for example, the renormalization of axial currents in nuclei [16,17]. Various complications associated with strong absorption channels dissappear in this limit. On the other hand, those parts of the relevant ΔN correlations which are dominated by short range dynamics will survive to the extent that they are only smoothly dependent on energy and momentum transfer. For example, the controversal question about the size of the Lorentz-Lorenz correction should also be looked at from this point of view.

2. ISOBARS AND THEIR INTERACTIONS

We specify the basic input in our many-body problem with mesons, nucleons and isobars in some detail since there exist diverging opinions in the literature about the definition of Δ-isobars. We start from a crossing symmetric p-wave π-nucleon K-matrix shown in Fig. 1.

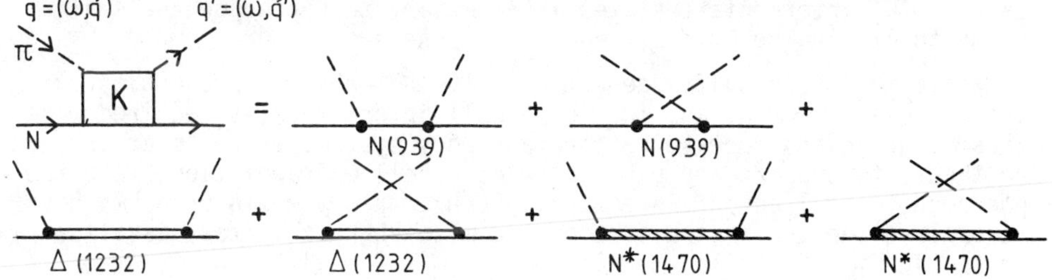

Figure 1: P-wave π-nucleon K-matrix through nucleon, Δ(1232) and N*(1470) intermediate states. The in- and outgoing pion 4-momentum is denoted by $q = (\omega, \vec{q})$ and $q' = (\omega, \vec{q}\,')$.

We then have

$$(q'|K(s)|q) = \sum_x \frac{(qN|\delta H_x|x)\ (x|\delta H_x|q'N)}{M_x - \sqrt{s}} + \text{crossed terms}, \tag{1}$$

where M_x are the masses of the corresponding intermediate states (nucleon, Δ(1232) or N*(1470)) and $\sqrt{s}$ is the total πN c.m. energy. Eq. (1) defines effective meson-baryon Lagrangians δH_x for the transitions $\pi N \to x$ ($x = N,\Delta,N^*$). For example,

$$\delta H_\Delta\ (q) = i\ \frac{f^*(q^2)}{m_\pi}\ \vec{S}\cdot\vec{q}\ T_\lambda\ [\pi_\lambda(q) + \pi^+_\lambda(-q)]\ + \text{h.c.} \tag{2}$$

in the standard notation of ref. [5]. For application in the π-nucleus c.m. system, the πN c.m. momentum $\vec{q}$ has to be transformed into this frame of reference. The πNΔ vertex function $f^*(q^2)$ includes relativistic corrections and a phenomenological monopole form factor:

$$f^*(q^2,s) = f^*\left[\frac{M_\Delta}{\sqrt{s}}\left(\frac{2M_\Delta}{M_\Delta + \sqrt{s}}\right)\right]^{1/2} \frac{\Lambda^2 - m_\pi^2}{\Lambda^2 - q^2} \tag{3}$$

where $f^{*2}/4\pi = 0.32$. The model K-matrix of Fig. 1 then generates p-wave eigenchannel amplitudes which fit to the improved phase shift analysis of ref. [18] within error bars over the entire range of c.m. momenta $|\vec{q}| \leq 3m_\pi$, without the requirement for additional long range cutoffs. Note that the crossed nucleon Born term appears explicitly in the P_{33} channel. We emphasize this point because in models which try to reproduce the 3.3 amplitude by a pure Δ-isobar pole, an additional long range cutoff has to be introduced in order to account for the absence of the crossed nucleon pole. This cutoff has however nothing to do with the range of the πNΔ vertex. Similar problems in connection with πD scattering are pointed out in ref. [19]. The inclusion of N*(1470) intermediate states is necessary to reproduce the small phase shifts in the P_{11}, P_{13} and P_{31} channels.

While the phase shifts determine only the on-shell properties of the effective Lagrangians δH_x, off-shell information is required once these Lagrangians operate in the many-body system. There is at present no theory to support any particular off-shell extrapolation, like the phenomenological cutoff in eq. (3). Such a theory would probably have to explore the quark structure of the meson-baryon vertex, as illustrated in Fig. 2. While the MIT bag model favours rather large baryon sizes (R ≳ 1 fm) [20], recent developments incorporating arguments of chiral symmetry [21] indicate that the bag might be as small as $R \sim M_N^{-1}$. In fact, the whole concept of isobars would be rather meaningless at nuclear densities unless their size (and consequently, the range of the πNΔ vertex) is sufficiently small.*)

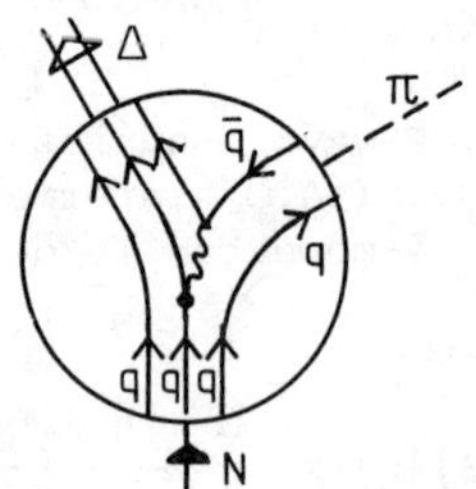

Figure 2:
The quark content of the πNΔ vertex

In the absence of a theory of the πNΔ vertex, one has to rely upon a careful analysis of processes such as $\pi^+ D \leftrightarrow pp$, where there is evidence that the basic πNN two-body absorption in the 3.3 resonance region involves rescattering mechanisms like those shown in Fig. 3, where off-shell momenta $|\vec{q}| \gg q_0$ appear.

---

*) Model calculations [42, 43] find cutoff masses Λ for πNN and πNΔ formfactors of the order of 1 GeV.

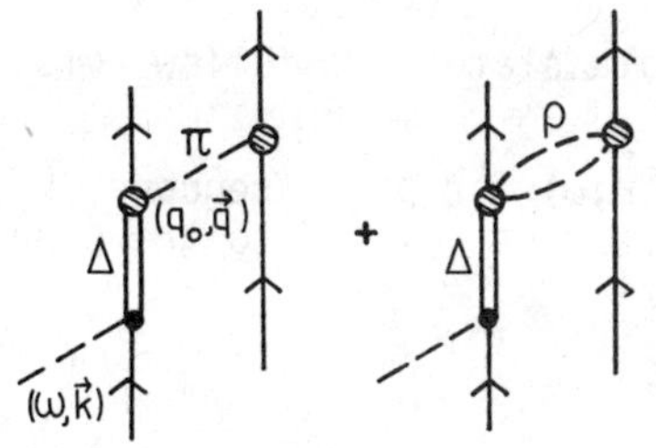

Figure 3:
Rescattering mechanism for p-wave πNN absorption through intermediate Δ-isobar, involving π and ρ exchange.

It was shown in ref. [22] that the total cross section for $\pi^+D \rightarrow pp$ is well reproduced by cutoff masses $\Lambda \gtrsim$ 1 GeV in the πNΔ and πNN vertex together with a rather strong ρ-exchange NΔ → NN interaction. Here ρ exchange is to be understood as correlated 2π exchange with a continuous mass distribution. In the NN interaction, this mass distribution is determined by the f_-^1 helicity amplitude in the $\pi\pi \rightarrow N\bar{N}$ channel [23]; the integral over $|f_-^1|^2$ essentially determines the ρNN coupling constant f_ρ which, according to [23], receives a large value, $f_\rho{}^2/4\pi \simeq 4.5$. We note that the helicity amplitude framework incorporates in principle the finite range structure of the ρ-baryon vertex.

The question is whether there is some simple scaling between f_ρ and the ρNΔ coupling strength f_ρ^*. Within SU(2) x SU(2) quark symmetry one finds $f_\rho^* = \alpha_\rho\ f_\rho$ with $\alpha_\rho^2 =$ 72/25, which has been used in refs. [22,24]. However, it is argued that a Regge model determination of f_ρ^* would give $f_\rho^{*2}/4\pi = 3.7$, thus $\alpha_\rho < 1$ [25]. Furthermore, a field theory treatment of $pp \rightarrow D\pi^+$ based on the Low equation [26] seems to provide long range cutoffs such that a reduction of the one-pion exchange tensor force due to ρ exchange appears unnecessary. On the other hand, such long range cutoffs would have severe consequences in the rather well established boson exchange treatments of the NN interaction [27]. The situation is further complicated by the sensitivity of the $\pi D \leftrightarrow pp$ calculations to the short range part of the deuteron wave function [28, 29]. Reid potential is outdated and is to be replaced by interactions which have a stronger foundation, like the one of Vinh Mau and collaborators [30].

It is hoped that $pp \rightarrow D\pi$ measurements with polarized protons [31,32] will resolve some of these presently unsettled questions.

3. THE PION-NUCLEAR MANY-BODY PROBLEM

We now turn to a description of pion-nucleus scattering in the 3.3 resonance region in terms of the isobar-hole model. Many-body effects appear in this model through self-energy modifications of the Δ-isobar Green's function and through vertex corrections in the isobar-hole interaction, which can be interpreted as a generalized form of the Ericson-Ericson-Lorenz-Lorentz (EELL) correction. The combined effect of all many-body corrections shows up in the energy dependent widths and shifts of the relevant isobar-hole states. We follow refs. [5,6] in the presentation of the theory.

The pion-nuclear optical potential $U(\vec{k}',\vec{k};\omega)$ for pion frequency ω and momenta $\vec{k}$, $\vec{k}'$ in the π-nucleus c.m. system is related to the pion self-energy Π by

$$(\vec{k}'|\Pi(\omega)|\vec{k}) = 2\omega\ U(\vec{k}',\ \vec{k};\omega)\ , \tag{4}$$

such that the T-matrix is

$$2\omega\ (\vec{k}'|T(\omega)|\vec{k}) = (\vec{k}'|\Pi(\omega)|\vec{k}) + 2\omega\int\frac{d^3q}{(2\pi)^3}\ \frac{(\vec{k}'|\Pi(\omega)|\vec{q})(\vec{q}|T(\omega)|\vec{k})}{\omega^2 - \vec{q}^2 - m_\pi^2 + i\delta}\ . \tag{5}$$

It is convenient to split the pion self-energy into a resonant piece Π_Δ involving Δ-isobars and a background contribution which includes the small parts of the p-wave interaction together with s-wave interactions:

$$\Pi = \Pi_\Delta + \Pi_{\text{background}}\ . \tag{6}$$

Let us neglect the background contribution for the moment. In the isobar hole model, the incoming pion excites Δ-hole states, $|\alpha) = |(\Delta h)J^\pi,\ T = 1\rangle$ coupled to angular momentum $J^\pi = 0^-, 1^+, 2^-, \ldots$; a harmonic oscillator basis is usually chosen for these states. The lowest order self-energy $\Pi_\Delta^{(o)}$ is then identified with

$$(\vec{k}'|\Pi_\Delta^{(o)}(\omega)|\vec{k}) = \sum_{\alpha\beta}(\vec{k}'|\delta H_\Delta|\beta)\ [G^{(o)}_{\alpha\beta}(\omega) + G^{(o)}_{\alpha\beta}(-\omega)]\ (\alpha|\delta H_\Delta|\vec{k}), \tag{7}$$

where $G^{(o)}_{\alpha\beta}$ is the free isobar-hole Green's function,

$$G^{(o)}_{\alpha\beta}(\omega) = \frac{\delta_{\alpha\beta}}{\omega - \varepsilon_\alpha}\ ;\quad \varepsilon_\alpha = M_\Delta + \varepsilon_\Delta - \frac{i}{2}\ \Gamma_\Delta(\omega+\varepsilon_h) - M_N - \varepsilon_h \tag{8}$$

where $\Gamma_\Delta(E)$ is the energy dependent free isobar width, ε_h is the single particle energy of the nucleon hole and ε_Δ the isobar kinetic energy. Eq. (7) holds for N = Z spin-zero nuclei; both direct and crossed isobar terms are included. Note also that nucleon binding effects already appear at this level, while the isobar propagates freely.

Iteration of $\Pi_\Delta^{(o)}$ in eq. (5) generates the multiple scattering series in the RPA approximation, where the direct isobar-hole interaction is given by non-static one-pion exchange (OPE). The on-shell propagation of the pion leads to the well known elastic broadening of isobar-hole states. The real part of OPE is attractive and results in a downward shift of Δ-hole states in all partial waves. If small background pieces to $\Pi_\Delta^{(o)}$ are added, then this procedure is equivalent (up to nucleon binding effects) to a standard first order optical potential calculation. The failure of such a calculation (e.g. PIPIT) to reproduce the total cross section, and consequently the requirement to go beyond this first order treatment, becomes obvious from Fig. 7.

In the isobar-hole model, higher order many-body corrections enter at two places:

(a) The non-static OPE isobar-hole interaction in the elastic channel must be replaced by a more realistic interaction, incorporating two-pion exchange degrees of freedom and other short range correlations ("vertex corrections").

(b) The isobar, once excited, experiences interactions which couple it to reaction channels (quasifree, absorption etc.). Such interactions modify the mass and width in the isobar-hole propagator $G^{(o)}$ for each individual partial wave J ("self-energy corrections").

The authors of refs. [1,2] have simulated the full diversity of these many-body corrections phenomenologically by a local complex isobar spreading potential, the parameters of which were adjusted to reproduce the total cross sections. These results have provided clear evidence for the importance of true absorption channels. The position taken in refs. [5,6,7] was instead to calculate these corrections microscopically in order to explore their sensitivity with respect to the basic meson-baryon interactions. We shall follow this latter development. At the level of the RPA approximation in a 1p1h (Δ-hole) space, we have first to discuss isobar-hole interactions together with what would be the analogue of the Hartree-Fock potential for isobars.

An important question is the structure of the $N\Delta \rightarrow \Delta N$ exchange interaction. We choose the representation (with $t = \omega^2 - \vec{k}^2$):

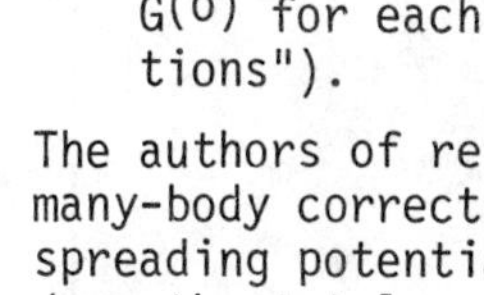

$$W(\vec{k},\omega) = \frac{f^{*2}(t)}{m_\pi^2} \frac{\vec{S}_1\cdot\vec{k}\ \vec{S}_2\cdot\vec{k}}{t-m_\pi^2+i\delta}\ \vec{T}_1\cdot\vec{T}_2^{+} + W'(\vec{k},\omega), \tag{9a}$$

$$W'(\vec{k},\omega) = \frac{f^{*2}}{m_\pi^2}\left[g'(k,\omega)\vec{S}_1\cdot\vec{S}_2^{+} + h'(k,\omega)S_{12}^{*}(\hat{k})\right]\vec{T}_1\cdot\vec{T}_2^{+}, \tag{9b}$$

where the first (long range) term is OPE, while g' and h' summarize all spin-spin and tensor interactions other than OPE (here $S_{12}^{*}(\hat{k}) = 3\ \vec{S}_1\cdot\hat{k}\ \vec{S}_2^{+}\cdot\hat{k} - \vec{S}_1\cdot\vec{S}_2^{+}$). To the extent that these interactions are short ranged, g' is a smooth function of ω and k. In particular, g' = 1/3 corresponds to the standard EELL correction. The small tensor correlations [5,6] represented by h' will not be further developed here.

An extensive discussion in the literature has been concerned with the size of g'(k,ω), once it has been emphasized that a large part of it should come from 2π exchange (i.e. ρ exchange) in combination with additional short range correlations [33]. Fig. 4 shows g'(k,ω=o) as calculated in a model [5] with ρ exchange and a baryon-baryon correlation function with a range determined by the inverse mass of the ω meson, for "weak" and "strong" ρNΔ coupling. (The additional ω dependence increases g' by up to 10 % at $\omega = 2m_\pi$). Clearly, g' is strongly model dependent. On the other hand, recent investigations of the role of g' in the quenching of Gamow-Teller transitions [16] (in the limit ω = o and $k \ll m_\pi$) do indicate that it would be very hard to understand these data unless $g'(k = o, \omega = o) \gtrsim 0.5$. It is important to establish such constraints, since our present knowledge of NΔ interactions is certainly not of predictive quality.

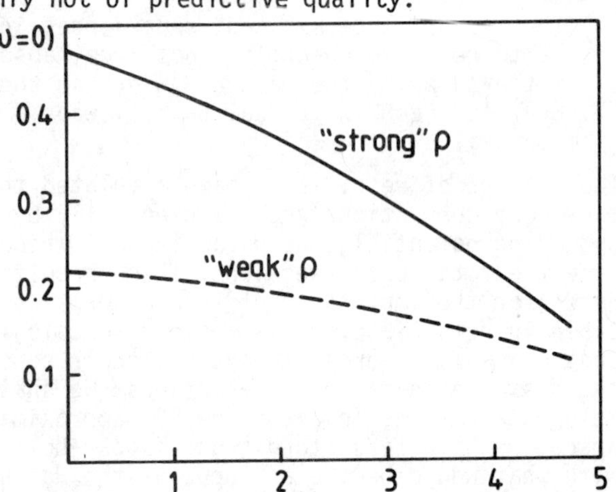

Figure 4: The quantity g' calculated in a model with ρ exchange, π exchange and a baryon-baryon correlation function, according to ref. [5]. A cutoff Λ = 1.2 GeV is used in the πNΔ vertex. "Strong ρ" and "weak ρ" correspond to $f_\rho^{*2}/4\pi$=16.8 and 3.7, respectively.

The Hartree mean field for isobars is assumed to be represented as a local energy-independent Hartree background potential. This Hartree potential gives a binding correction to the isobar mass and is usually taken to be identical to the nucleon single particle potential, for no strong reason except that the role of σ and ω meson exchange in the direct NΔ → NΔ interaction is similar for nucleons and isobars [5]. The Fock term of Fig. 5 is very important since its imaginary part determines the narrowing of the isobar decay width due to the Pauli exclusion principle, while the real part gives a further binding correction which comes from the real part of OPE together with W' of eq. (9).

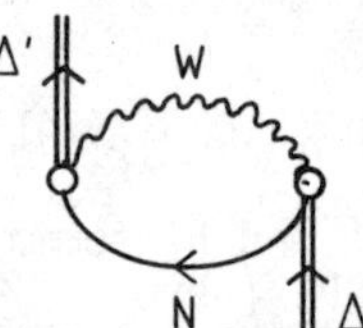

Figure 5: Δ isobar Fock term; the sum over intermediate states covers the filled nucleon Fermi sea. W is the non-static NΔ -interaction of eq. (9).

The essential damping mechanism in the propagation of isobar-hole excitations is introduced by their strong coupling to 2p2h continuum configurations. The most important diagrams of this type are shown in Fig. 6.

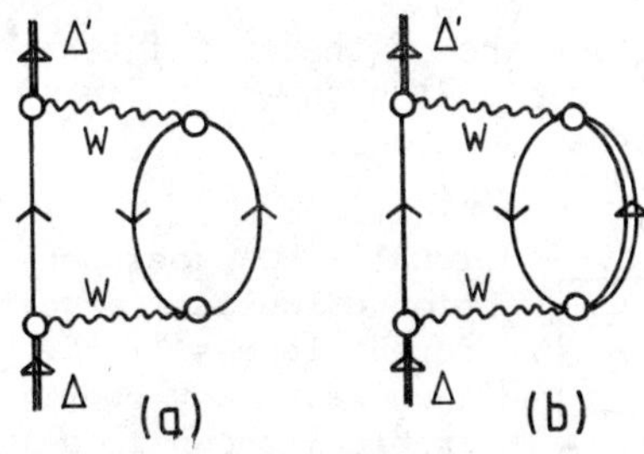

Figure 6:
Coupling of isobar to 2p1h continuum configurations:
(a) is the main contribution from two-nucleon absorption channel.
(b) incorporates resonant rescattering (reflection term) in the quasifree channel.

Microscopic calculations of these processes have been performed in refs. [5,6] for ^{4}He and ^{16}O, where the parameters of the interaction W have been chosen to fulfill the important requirement of consistency with a calculation of πD absorption [22]. For ^{4}He, the absorption width has also investigated in ref. [7].

The pion-nuclear T-matrix with inclusion of these many-body corrections is now given by

$$2\omega\,(\vec{k}\,|T(\omega)|\,\vec{k}) = \sum_{\alpha\beta} (\vec{k}'\,|\delta H_\Delta|\beta)\,(\beta|G(\omega)|\alpha)\,(\alpha|\delta H_\Delta|\vec{k}), \tag{10}$$

where $\qquad G(\omega) = (\omega - \mathcal{H})^{-1}$

is the full isobar-hole Green's function. For simplicity, we write only the direct part of it which defines the Tamm-Dancoff approximation (TDA), while the inclusion of crossed terms (i.e. the step from TDA to RPA) is shown in detail in ref. [5]. The G is obtained by inversion of the complex matrix

$$\mathcal{H}_{\beta\alpha}(\omega) = \bar{\varepsilon}_\alpha(\omega)\delta_{\beta\alpha} + \Sigma_{\beta\alpha}(\omega) + W_{\beta\alpha}(\omega), \tag{11}$$

where $\bar{\varepsilon}_\alpha$ is given by ε_α of eq. (8), except that the free isobar kinetic energy ε_Δ is now replaced by the single particle energy in the isobar Hartree potential. The $W_{\beta\alpha}$ are the isobar-hole matrix elements of the interaction, eq. (9), while $\Sigma = \Sigma\,(\text{Fock}) + \Sigma(\text{2p2h})$ summarizes the self-energy corrections to the masses and widths of isobar-hole states in each partial wave J generated by the Fock terms (Σ(Fock), Fig. 5) and the absorption and reflection contributions (Σ(2p2h), Fig. 6).

We show in Fig. 7 the influence of higher order corrections on the total π^{16}O cross section [6]. While Σ(Fock) (through its positive imaginary part) leads to strong narrowing of the isobar width in all partial waves, additional inclusion of Σ(2p2h) provides the necessary damping (Im $\Sigma(\text{2p2h}) < o$). At low energy, the absorptive width comes exclusively from two-nucleon absorption (Fig. 6 (a)), which is smoothly energy dependent, whereas the reflection term, Fig. 6(b), shows strong energy dependence. The decomposition of σ_{tot} into partial wave cross sections is shown in Fig. 8. In the isobar-doorway picture or refs. [2,8], each one of these is interpreted as an isobar-doorway state of given J^{π}. Also shown in Fig. 9 are selected examples of differential cross sections.

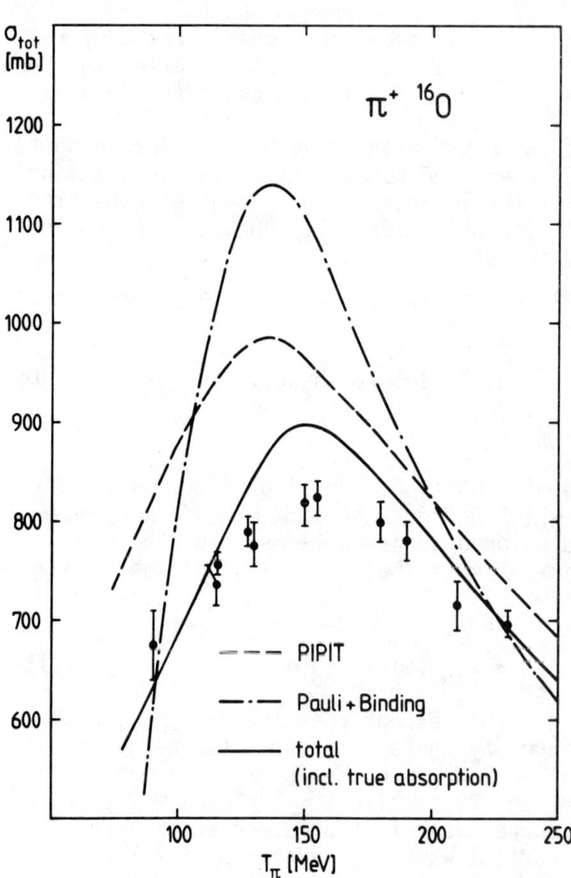

Figure 7:
Total $\pi^{16}O$ cross section calculated at different levels of the π-nuclear many-body problem: shown is a standard first order momentum space calculation (PIPIT) for reference. The other two curves are calculated within the RPA scheme of refs. [5,6] (see eqs. (10-11)):
<u>Pauli + Binding:</u> incorporates isobar Hartree potential, Fock term Σ(Fock) and full interaction W of eq. (9).
<u>Total:</u> includes in addition true absorption and reflection terms Σ(2p2h) (see Fig. 6).
<u>Input:</u> π and ρ exchange plus baryon-baryon correlation function. (cutoff in πNΔ vertex: Λ = 1.2 GeV; ρNΔ coupling: $f^*_{\rho}/\sqrt{4\pi} = 4.1$ ("strong ρ"). Calculation includes all partial waves $J^{\pi}=0^-, 1^+, \ldots$ up to 8^-. For further details see refs. [5,6]. Exp. data: ref. [34]

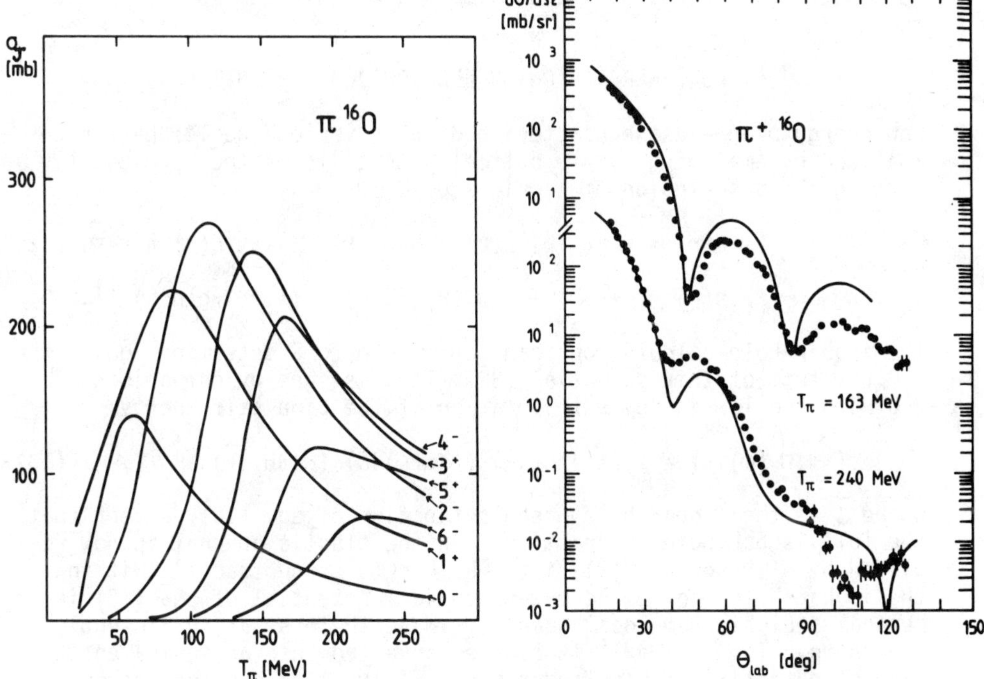

Figure 8:
Decomposition of the $\pi^{16}O$ cross section into partial wave cross sections, once all higher order corrections are included (taken from ref. [6]).

Figure 9:
Examples of differential cross sections for $\pi^{16}O$ including all medium corrections. Experimental data are from ref. [35] (taken from ref. [6]).

The apparent success of these and other isobar-hole calculations should not obscure the basic uncertainties. One of the conclusions from such calculations is that the shifts and widths of isobar-hole states depend strongly on details of the NΔ-interaction which are not very well under control (cutoff factors, ρNΔ coupling constant, short range NΔ correlations etc.) It seems that the quasifree and absorptive widths are reasonably well reproduced once the parameters of the basic baryon-baryon interactions are fixed in accordance with the πD → NN reaction. The shifts in the positions of isobar-hole states are however certainly not determined better than within ~ 30 MeV. There are notorious problems in the 0^- partial wave which, according to ref. [2], require a careful reinvestigation of higher order corrections related to s-wave πN interactions. Furthermore, since the reflection terms (Fig. 6(b)) are large in the 3.3 resonance region, it seems necessary to treat them in higher order. This introduces a problem of self-consistency, which is investigated in ref. [36] within Foldy-Walecka theory, but needs also additional studies for finite nuclei [37].

4. LOW ENERGY PION-NUCLEUS OPTICAL POTENTIAL

Low energy pion elastic scattering data on various nuclei have been analysed by employing simple optical potentials of the form usually used in the description of pionic atoms [10]:

$$2\omega\, U(\vec{r}) = -4\pi\,[b_o\, \rho(r) + B_o \rho^2(r)] + 4\pi\vec{\nabla}\, L(r)\, n(r)\vec{\nabla}, \quad (12)$$

$$n(r) = c_o\rho(r) + C_o\rho^2(r),\; L(r) = [1 + 4\pi g' n(r)]^{-1}.$$

The isobar-hole calculations can contribute to a determination of the $\vec{\nabla}\, C_o\rho^2\vec{\nabla}$ term of this potential [38]. This is done by comparison of $2\omega U$ with the low-energy extrapolation of the pion self-energy

$$(\vec{k}'\,|\Pi_\Delta(\omega)|\,\vec{k}) = \sum_{\beta\alpha} (\vec{k}'\,|\delta H_\Delta|\beta)\;(\beta|\bar{G}(\omega)|\alpha)\;(\alpha|\delta H_\Delta|\vec{k}), \quad (13)$$

where $\bar{G}$ is the isobar-hole Green's function of eq. (10), except that the full isobar-hole interaction W in the elastic channel is now replaced by W', (see eq. (9)) i.e. OPE has to be subtracted. Within the isobar-hole model, the p-wave optical potential of eq. (13) is in general a highly non-local quantity. Nevertheless, at sufficiently low energy ($T_\pi \leq 50$ MeV), it can be converted into an equivalent r-space potential of Kisslinger type $\vec{\nabla}F(r)\vec{\nabla}$, to a good degree of accuracy. At higher energies, in particular in the 3.3 resonance region, such a pseudo-local representation is completely inadequate [39]. For ^{16}O, the extrapolation of eq. (13) down to threshold, if compared with an equivalent ρ^2-potential, yields Im $C_o = 0.08\ m_\pi^{-6}$ [38], which is close to the "experimental" value of 0.07. The discussion of Re C_o includes the consideration of a variety of many-body effects discussed before, like binding and Pauli corrections as well as dispersive real parts from two-nucleon absorption. Calculations have again to face the uncertainties from the off-shell structure of the relevant meson-baryon Lagrangians. The isobar - hole model gives a positive (i.e. attractive) Re C_o of about the same magnitude as Im C_o, the precise value depending on the treatment of binding effects (but Re C_o remains positive even if the isobar Hartree potential is set to zero). We note that a positive Re C_o seems also to be suggested by recent high precision data on shifts and widths of 2p levels in pionic atoms.

The question about the size and sign of Re C_o calls for a reopening. In fact it is clear that there exists a cooperation to order ρ^2 between Re C_o and the EELL effect (determined by the parameter $g'(\omega = m_\pi,\ k = o)$). Increasing repulsion due to short range correlations in g' can be compensated by a corresponding increase of attraction in Re C_o.

The fact that the coupling of isobar-hole states to the two-nucleon continuum alone gives acceptable values for Im C_o at threshold (see also ref. [40]) might appear surprising in the light of recent investigations [41] which point out that the rescattering process, Fig. 3, receives a comparably large additional contribution at low energy from nucleon intermediate states. On the other hand the p-wave scattering volumes for pions coupling to nucleon-hole intermediate states vanish in the limit $k \to o$, regardless of self-energy corrections, so that such terms are expected to be greatly reduced. Nevertheless, this point needs further clarification.

5. CONCLUSIONS

The isobar-hole model is an appropriate scheme to incorporate higher order many-body corrections due to the coupling of the elastic channel to quasifree and absorption modes. These corrections modify the propagation of isobars in the nuclear environment and provide mass shifts and changes of the widths of isobar-hole states in all partial waves. We have emphasized, however, that a reliable microscopic calculation of such higher order term suffers from uncertainties at the level of the basic meson-baryon interactions as well as the short-range correlations between baryons. More precisely, the range of the $\pi N\Delta$ vertex, and the coupling of $N\Delta$ to the $(\pi\pi)$ system (ρ exchange) needs careful investigation. A detailed analysis of $\pi d \leftrightarrow pp$ differential cross sections and polarisation parameters is expected to be a relevant source to provide such information.

For a quantitative comparison to data, it is of course necessary to include the background parts of the π-nucleus optical potential. In particular, the role of second order corrections to the s-wave πN interaction, and there especially the dispersive parts related to absorption, are not sufficiently well understood [44,45].

It is important to establish connections between the π-nuclear T-matrix and the low frequency limit of the response function for pion-like nuclear excitations, where virtual isobar-hole states renormalize the properties of spin-isospin dependent operators. Such considerations are hoped to put constraints on the short range part of the relevant baryon-baryon interactions.

References

[1] M. Hirata, F. Lenz and K. Yazaki, Ann. of Phys. 108 (1977) 16; see also: F. Lenz, Proc. 7th Int. Conf. on High Energy Physics and Nuclear Structure, Zürich 1977, p. 175

[2] M. Hirata, J.H. Koch, F. Lenz and E.J. Moniz, Phys. Lett. 70 B (1977) 281; preprint (1978), to be published in Ann. of Phys.

[3] W. Weise, Nucl. Phys. A 279 (1977) 402)

[4] K. Klingenbeck, M. Dillig and M.G. Huber, Phys. Rev. Lett. 41 (1978) 387

[5] E. Oset and W. Weise, Phys. Lett. 77 B (1978) 159; Nucl. Phys. A (1979), in print

[6] E. Oset and W. Weise, preprint Univ. of Regensburg (1978)
[7] H.M. Hofmann, Z. Physik A 289 (1979) 273
[8] L.S. Kisslinger, in: Theoretical Methods in Medium Energy and Heavy Ion Physics (K.W. McVoy and W.A. Friedman, eds.), Plenum Publ. Co. (1978); L.S. Kisslinger and W.L. Wang, Ann. of Phys. 99 (1976) 374
[9] G.E. Brown, B.K. Jennings and V. Rostokin, preprint Stony Brook (1977)
[10] K. Stricker, H. McManus and J. Carr, preprint MSU (1978), to be published in Phys. Rev. C
[11] R.H. Landau and A.W. Thomas, Nucl. Phys. A 302 (1978) 461
[12] R. Rockmore, E. Kanter and P. Goode, Phys. Lett. 77 B (1978) 149
[13] L.C. Liu and C.M. Shakin, Phys. Rev. C 16 (1977) 333
[14] G.A. Miller, Phys. Rev. C 16 (1977) 2324
[15] A.S. Rinat, preprint Rehovot (1978)
[16] E. Oset and M. Rho, Phys. Rev. Lett. 42 (1979) 47
[17] I.S. Towner and F.C. Khanna, Phys. Rev. Lett. 42 (1979) 51
[18] G. Rowe, M. Salomon and R.H. Landau, Phys. Rev. C 18 (1978) 584
[19] F. Myhrer and A.W. Thomas, NORDITA preprint (1978)
[20] T. DeGrand, R.L. Jaffe, K. Johnson and J. Kiskis, Phys. Rev. 12 (1975) 2060; C. deTar, Phys. Rev. D 17 (1978) 302, 323
[21] G.E. Brown and M. Rho, preprint Stony Brook (1979)
[22] M. Brack, D.O. Riska and W. Weise, Nucl. Phys. A 287 (1977) 425
[23] G. Höhler, F. Kaiser, R. Koch and E. Pietarienen, Physics Data 12-1 (1978)
[24] J.A. Niskanen, Nucl. Phys. A 298 (1978) 417
[25] L.S. Kisslinger, preprint Carnegie-Mellon Univ. (1979)
[26] M.A. Alberg, E.M. Henley, G.A. Miller and J.F. Walker, Nucl. Phys. A 306 (1978) 447
[27] K. Holinde, Proceedings 8th Int. Conf. on Few Body Systems and Nuclear Forces, Lecture Notes in Phys. 87 (1979) 21
[28] O.V. Maxwell, M. Brack and W. Weise, contribution to this conf.
[29] J. Chai and D.O. Riska, contribution to this conference
[30] R. Vinh Mau, in: Mesons in Nuclei (M. Rho and D.H. Wilkinson, eds.), North-Holland Publ. Co. (1978)
[31] G. Jones, Proceedings 8th Int. Conf. on Few Body Systems and Nuclear Forces, Lecture Notes in Phys. 87 (1978) 142
[32] Ch. Weddigen, Nucl. Phys. A 312 (1978) 330
[33] G. Baym and G.E. Brown, Nucl. Phys. A 247 (1975) 345
[34] A.S. Clough et al., Nucl. Phys. B 76 (1974) 15
[35] J.P. Albanese et al., Phys. Lett. 73 B (1978) 119
[36] M.B. Johnson and H.A. Bethe, Nucl. Phys. A 305 (1978) 418; M.B. Johnson and B.D. Keister, Nucl. Phys. A 305 (1978) 461
[37] E. Werner, private communication
[38] E. Oset, W. Weise and R. Brockmann, Phys. Lett. B (1979), in print, and contribution to this conference
[39] E. Oset private communication
[40] C.M. Ko and D.O. Riska, Nucl. Phys. A 312 (1978) 217
[41] K. Shimizu and A. Faessler, preprint
[42] J.W. Durso, A.D. Jackson and B.J. Verwest, Nucl. Phys. A 282 (1977) 404
[43] M. Dillig and M. Brack, J. Phys. G 5 (1979) 223
[44] K.P. Lohs and J. Hüfner, Nucl. Phys. A 296 (1978) 349
[45] F. Hachenberg and H.J. Pirner, Ann. of Phys. 112 (1978) 401

Question: C. Shakin, Brooklyn College
Do you include any density dependence for that Δ width which arises from absorption process?

Response: Weise
This density dependence is automatically included, but not, of course, within the frame of some local density approximation. It comes in through the nuclear single particle wave funcions used to describe the hole states in the 2p2h intermediate excitations. I'd like to emphasize again that the non-locality of the absorption process has to be taken seriously. Any calculation using the local density approximation is not on very strong grounds, at least, in the 3-3 resonance region.

Question: M. Krell, Universite' de Sherbrooke
With respect to your using a positive (attractive) ReC_0: I agree completely with decoupling it from ImC_0 and not using just the negative value of the latter. In my opinion, this treatment, which goes back to Brueckner, was based on using potentials in a perturbative way at that time. In the complete calculations which we do now, the dispersive effect of any absorptive potential appears automatically. But what is specifically your reason for giving ReC_0 the definite positive value?

Response: Weise
The reason is that only a small piece of ReC_0 comes from the dispersive part of 2N absorption alone. A larger part arises from the extrapolation of the "reflection" term down to small energies, which clearly provides attraction (i.e. a positive contribution to ReC_0). In addition, short-range pieces other than one-pion exchange in the isobar Fock terms are attractive. Even if one turns off the (probably attractive) isobar potential, ReC_0 stays positive.

APPROXIMATE TREATMENT OF COULOMB EFFECTS IN π d ELASTIC SCATTERING AT LOW ENERGY

N. Giraud, G. H. Lamot, C. Fayard

Institut de Physique Nucléaire, Université Claude Bernard Lyon-I and IN2P3, 43, Bd du 11 Novembre 1918, 69621 Villeurbanne, France

At the present time, many experimental data for pion-deuteron (π d) elastic scattering are available near and above the (3, 3) resonance region, while only two sets of data have been given in the low energy domain ($T_\pi < 100$ MeV), namely by Axen et al.[1] ($\pi^+ d$ at 47.7 MeV) and by the Saclay team[2] ($\pi^\pm d$ at 47 MeV). Two other $\pi^\pm d$ experiments are considered in Saclay at 25 MeV (analysis in progress) and at 70 MeV (starting in the near future).

The data up to the resonance region have been correctly interpreted by means of more or less refined three-body calculations. In the low energy domain, the Coulomb-nuclear interference effects are crucial, and the new $\pi^\pm d$ data of Saclay encourage theoretical investigations.

Even if the three-body scattering problem with two charged particles is now theoretically solved, it is beyond the present computer facility to apply this formalism to any π d calculation including a large number of two-body channels, and one must therefore use approximate methods. Usually, Coulomb effects are roughly treated[3] by writting the full scattering amplitude $X_{\pi^\pm d}$ as $X_{\pi^\pm d} = X^S_{\pi d} \pm X^C_{\pi d}$, where $X^C_{\pi d}$ is the pure Coulomb amplitude, and $X^S_{\pi d}$ is the pure strong interaction amplitude (Method I). Fig. 1 shows the corresponding $\pi^\pm d$ differential cross-sections at 47.7 MeV compared with the data of Saclay. $X^C_{\pi d}$ is calculated at first order in the form factor approximation. $X^S_{\pi d}$ is obtained by solving exactly the three-body equations, using relativistic kinematics for the pion only, and including the N N tensor force and all S, P π N channels.

In fact, the hadronic scattering amplitude should contain the Coulomb distorsion. An approximate method has been proposed by Streit et al.[4] in the two-body case (Method II). The full scattering amplitude is splitted into strong, Coulomb and mixed parts $X = X_S + X_C + X_R$. In practice, the correction is done on the phase shift which is written as $\delta(k) = \delta_S(k) + \delta_R(k) + \delta_C(k)$, where

ISSN:0094-243X/79/540590-01$1.50

the corrective term $\delta_R(k)$ is obtained at first order in α (the fine structure constant) in terms of the strong phase shift δ_S and its first derivative. Recently, this method has been used successfully in the p-d elastic scattering[5], leading to results very close to the exact ones. It seems thus reasonable to extend this method to the πd case. The results at 47.7 MeV are shown on Fig. 2. The pure strong and pure Coulomb parts are calculated in the same conditions as in Method I. The corrective term is obtained by means of parabolic interpolation from the δ_S phase shifts calculated at 35, 47.7 and 60 MeV.

Comparing Fig. 1 and 2, we see that the differences between methods I and II are small in the region of strong interference between Coulomb and nuclear interactions ($\theta_{Lab} < 90°$). In contrast with the results of Method I, we note that the π^+d and π^-d curves obtained with Method II are crossing at backward angles. Nevertheless, it is difficult to draw any conclusion until the data at all angles are available.

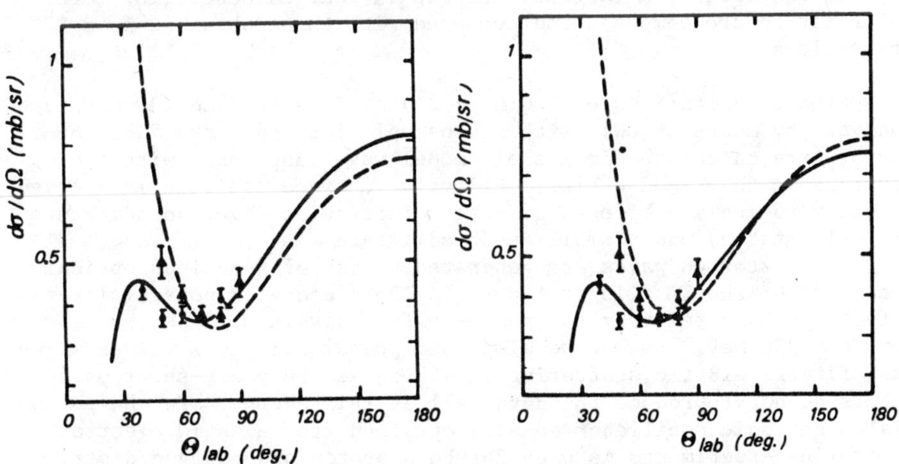

Fig. 1 (Method I) Fig. 2 (Method II)

Theory π^+d (——), π^-d (- - -). Saclay experiment π^+d (ɸ), π^-d (▲)

REFERENCES

1. D. Axen et al., Nucl. Phys. A256, 387 (1976).
2. B. Balestri et al., in Proc. of 8th Int. Conf. on Few Body Systems, Vol. I, p. 227, Graz (1978).
3. A. W. Thomas, Nucl. Phys. A 258, 417 (1976); N. Giraud et al., Phys. Rev. C, to appear.
4. L. Streit et al., in Graz (1978) op. cit., Vol. I, p. 109 and p. 113.
5. H. Ziegelmann, in Graz (1978) op. cit., Vol. I, p. 236.

AN ANALYSIS OF $^{18}O(\pi^{\pm}, \pi^{\pm\prime})$ $^{18}O(2^{+}$, 1.98 MeV) REACTION WITH MICROSCOPIC FORM FACTORS

A. Arima, R. Seki†, K. Yazaki
Department of Physics, Faculty of Science, University of Tokyo
Bunkyo-ku, Tokyo

K. Kume
Department of Physics, Nara Women's University, Nara

and

H. Ohtsubo
Department of Physics, Osaka University, Toyonaka
Osaka

ABSTRACT

Pionic excitation of $^{18}O(0^{+} \rightarrow 2^{+})$ are analyzed by means of DWIA with microscopic form factors. Effective charges describing core polarization are computed and compared with those obtained from E2 transitions.

Pionic excitations of $^{18}O(0^{+} \rightarrow 2^{+})$ at T_{π} = 164 and 230 MeV[1] are analyzed by means of DWIA with microscopic form factors. The form factors are calculated from shell model wave functions[2] with two neutrons in the 1s-0d orbits. Effective charges are then introduced in order to include effects of core polarization (i.e. an admixture of 3p-1h states) and a small 4p-2h admixture.

The distorted waves are generated by use of π-nucleus optical potential of the Kisslinger type. For both energies potential parameters are computed from the phase-shift analysis of Roper et al.[3] For T_{π} = 230 MeV, however we also used phenomenological values which best fit the elastic scattering data because the phase-shift parameters do not reproduce the data well at this energy. In the potential a harmonic oscillator density obtained from elastic electron scattering experiments is used for both proton and neutron distributions.

Fitting the observed $\pi^{\pm}$ excitation cross sections at T_{π} = 164 MeV, we first deduced the effective charges of e_p = 1.32 and e_n = 0.68. We observed that the numerically calculated ratio of π^- and π^+ cross sections, R, is approximately equal to a simple prediction

†Present Address:

Department of Physics and Astronomy, California State University, Northridge, California 91330

W. K. Kellogg Radiation Laboratory, California Institute of Technology, Pasadena, California 91125

ISSN:0094-243X/79/540592-02$1.

based on the 3-3 resonance-dominance, $|(2 + X)/(2 - X)|^2$; where $X = (e_p - e_n)/(e_p + e_n)$, for chosen values of X or e_p/e_n. Note that if there is no core polarization then R = 9 (X = 1), whereas the observed value of R is about 1.9 and is approximately independent of the scattering angle. These effective charges are to be compared with $e_p = 2.05 \pm 0.25$ and $e_n = 0.76 \pm 0.02$ that are computed from the E2 transitions for ^{18}O and ^{18}Ne[4] by use of the present wave functions. We regard that the agreement in e_n's is reasonable, but the difference in e_p's seems to be appreciable. This difference may be due to the fact the proton single-particle wave functions in ^{18}Ne have larger mean-square radii than the neutron wave functions in ^{18}O.

For T_π = 230 MeV we carried a similar calculation by use of the same effective charges. Going from the phase-shift potential parameters to the phenomenological parameters, we observed the agreement with the excitation cross section data improved substantially. This observation indicates a strong correlation between the elastic and inelastic scatterings.

At both energies the calculated excitation cross sections are found to fall at small angles while the observed remain approximately constant. Since the Fourier transforms of the transition form factors vanish at the zero momentum transfer, the decrease in near-forward cross sections is an expected general feature. If the observed trend is indeed true, it is a puzzling feature.

In our analysis we did not include so-called angle-transformation terms both in the optical potential and the transition form factors. The terms are of the order of $E_\pi/2M_N$, which is about 0.15 in the present case. These terms are to be included in a more detailed analysis, but we do not expect any alternation to occur in our discussions and conclusions reported here.

REFERENCES

1. S. Iverson et al., Phys. Rev. Letters 40, 17 (1978); and to be published.
2. T. Inoue et al., Nucl. Phys. 59, 1 (1964); T.T.S. Kuo et al., Nucl. Phys. 85, 40 (1964); and P. J. Ellis et al., Nucl. Phys. A144, 161 (1970) and A181, 368 (1972).
3. L. D. Roper et al., Phys. Rev. 138, B190 (1965).
4. B. A. Brown et al., Nucl. Phys. A277, 77 (1977).

Annihilation Nonlocality Effects in the Pion Optical Potential

A. S. Rosenthal and F. Tabakin
University of Pittsburgh, Pittsburgh, PA 15260

ABSTRACT

The physical annihilation of pions is included in an optical potential by a nonlocal operator, instead of the commonly used $\rho^2(\vec{r})$ form. The dependence of pion-nucleus elastic differential cross sections on the nonlocality in the annihilation operator is determined and found to be strongly linked to the annihilation strength.

INTRODUCTION

Phenomenological optical model analyses of low energy pion-nucleus scattering have employed local, or short ranged terms, of the form $\rho^2(\vec{r})$ to account for meson annihilation or second order multiple scattering effects.[1,2,3] We have treated such effects with a term of the form

$$\Delta U_{s\text{-wave}}(\vec{K},\vec{K}') = B_o \int d\vec{q}\ \rho_n\ (\vec{K}-\vec{q}) S(\vec{q}) \rho_p (\vec{q}-\vec{K}')$$

which is added to the usual optical potential in the relativistic Schrodinger equation[4]

$$\sqrt{K^2+\mu^2}\ \phi(\vec{K}) + \int d\vec{K}'\ U(\vec{K},\vec{K}')\phi(\vec{K}') = E\phi(\vec{K})$$

The nonlocality function was chosen to be of Yukawa form

$$S(q) = \Lambda^2/(\Lambda^2+q^2)$$

and Re B_o was taken to be equal to $-\mathrm{Im}B_o$, corresponding to the usual scalar repulsion. It was found that elastic scattering cross sections alone do not rule out even very large nonlocalities. For example, Table I lists two parameter sets and their associated total cross sections which give rise to virtually identical elastic differential cross sections for ^{208}Pb.

Table I Parameter sets giving the same 50 MeV elastic scattering

$B_o(F^5)$	$\Lambda(F^{-1})$	σ_{el}(mb)	σ_{inel}(mb)	σ_{tot}(mb)
.06-i.06	3.0	823	1297	2121
.10-i.10	1.0	867	1292	2159

ISSN:0094-243X/79/540594-02$1.50

The meson wave functions given by these approximately phase equivalent potentials differ by 10% or more in the nuclear surface. We conclude that reliable estimates of the annihilation nonlocality are necessary if the pion optical potential is to be used with confidence in the analysis of inelastic and reaction processes. Efforts to determine this range theoretically are presently underway.

REFERENCES

1. N. J. DiGiacomo, et. al., Phys. Lett. 66B, 421 (1977)

2. L. C. Liu, Phys. Rev. C17, 1787 (1978)

3. R. H. Landau and A. W. Thomas, Nucl. Phys. A302, 461 (1978)

4. R. A. Eisenstein and F. Tabakin, Comp. Phys. Comm. 12, 237 (1976)

IS THERE A LARGE LOCAL FIELD CORRECTION TO THE π-NUCLEUS OPTICAL POTENTIAL?

S. J. Wallace and M. K. Banerjee†
University of Maryland, College Park, Maryland 20742

The large scattering radius, $(\sigma/\pi)^{1/2} \sim 2.5$ fm, implied by the resonant scattering cross section suggests that pion waves must self-consistently adjust to several nucleons at once in the pion nucleus interaction. These effects arise in multiple scattering theories through triple and higher order terms in the πN interaction whereby the pion scatters at least twice from the same nucleon with intermediate interactions from other nucleons. Such terms are referred to as local field corrections.

Fixed scatterer calculations in a nuclear matter context have indicated that the local field corrections are very large.[1,2] However, the leading order multiple scattering term that contributes, $U_3 \sim \langle\tau_1 G\tilde{\tau}_2 G\tau_1\rangle$, involves triple scattering of the pion by two nucleons as shown in Fig. 1. In the fixed scatterer approximation, the nucleon recoil energy is fixed at a constant value, thereby permitting the pion to resonate with a second nucleon irrespective of the energy lost in the first scattering. When recoil is not neglected, the pion energy shifts away from resonance in the intermediate scattering for recoil momenta $p \gtrsim \sqrt{2M\Gamma} \sim 2\ \mathrm{fm}^{-1}$, where Γ is the resonance width. Additionally, the pion propagation energy is reduced by the recoil causing damping of the propagation for $p \gtrsim \sqrt{2M\omega} \sim 4\ \mathrm{fm}^{-1}$. The momentum cutoff due to pion propagation occurs twice for each intermediate interaction and thus is as important a recoil effect as the shift of the energy away from resonance. Fixed scatterer estimates of local field effects are suspect due to neglect of these effects, and thus the question of the size of the leading local field correction has remained unsettled.

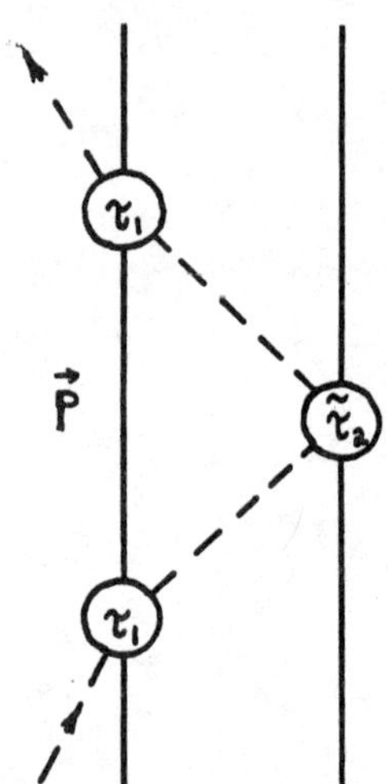

Fig. 1. The pion scatters from nucleon 1 twice, with an intermediate scattering from nucleon 2. The recoil momentum $\vec{p}$ labels the excited states of nucleon 1 between scatterings.

This paper addresses the issue by considering the leading triple scattering process indicated in Fig. 1 using plane waves for the excited states of nucleon

†Supported by the U. S. Department of Energy and the University of Maryland Computer Science Center.

ISSN:0094-243X/79/540596-02$1.50

1. The recoil is included by integrating over recoil momentum, p, with attendant recoil energy $p^2/(2M)$, where M is the nucleon mass. The full pion propagation function is included and a complete treatment of spin and isospin is made for pion scattering from ^{16}O. The calculation of the triple scattering process omits true absorption effects on grounds that the absorption is a logically distinct issue from the local field effect. Thus the intermediate πN amplitude, $\tilde{\tau}_2$, omits the nucleon pole term.

Our result is expressed as the complex ratio, R, of the volume integral of the triple scattering term, U_3, to the first order optical potential term U_1. Figure 2 shows $|R|$ as a function of pion energy. For comparison, the fixed scatterer approximation yields $|R|$ values seven times as large as those shown in Fig. 2. Nonetheless, our value $|R| = 0.7$ at resonance shows that the local field corrections are still very large. The calculation shown here uses a πN off-shell form factor of the form $v(q) = (1+q^2/\alpha^2)^{-1}$, with $\alpha = 940$ MeV/c. If α is reduced to 500 MeV/c, the ratio $|R|$ at resonance change from 0.7 to 0.5, i.e., the local field correction still remains large compared to the first order optical potential. The sign is such that the local field correction reduces the optical potential at resonance. It is important to note that the local field correction typically arises from nucleon separations $\sim$ 1.5 fm, and that the tensor propagation of the pion plays an important role.

In a self-consistent calculation, we expect a reduced local field correction compared with the result reported here for the leading term. Also, the Pauli blocking effect and Δ-recoil effects omitted in this calculation could alter the result. Nonetheless, we must conclude that substantial local field corrections are to be expected for pion scattering near the resonance energy. These corrections are not expected to alter significantly the "black" elastic scattering, however they may provide some insight into the energy dependence of reaction cross sections, such as the single charge exchange cross section.

REFERENCES

1. B. D. Keister, Nucl. Phys. A271, 342 (1976).
2. M. B. Johnson and H. Bethe, Nucl. Phys. A305, 418 (1978); 461 (1978).

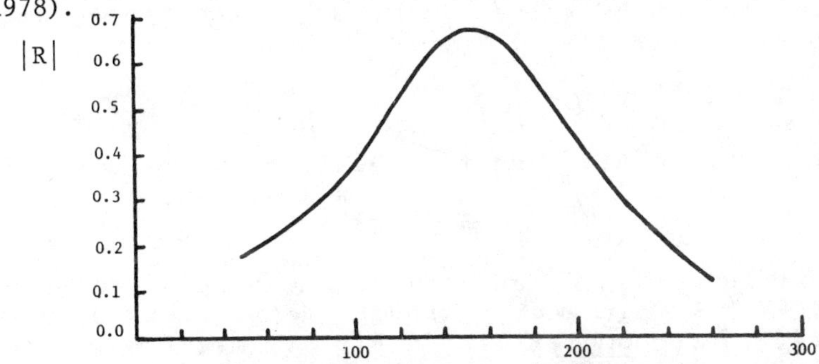

Fig. 2. The magnitude of the complex ratio R, where $R=\int U_3 dV/\int U_1 dV$, expresses the magnitude of the leading local field correction relative to the magnitude of the first order optical potential.

A SOLUBLE KISSLINGER POTENTIAL

M. Bawin
Physique Nucléaire Théorique, Université de Liège, Institut de Physique, Sart Tilman, B-4000 Liège 1, Belgium.

Ericson and Myhrer[1] recently discussed some of the pathological features of the Schroedinger equation associated with a potential of the Kisslinger type :

$$\vec{\nabla}[1 - \alpha(r)]\, \vec{\nabla}\phi(\vec{r}) + [-q(r) + k^2]\, \phi(\vec{r}) = 0 \quad . \tag{1}$$

We wish here to illustrate some of these features in an analytically soluble model, i.e. when $\alpha(r)$ and $q(r)$ have the following form :

$$\alpha(r) = g\, e^{-\mu r} \tag{2}$$

$$q(r) = g\mu\, \frac{e^{-\mu r}}{r} \quad . \tag{3}$$

Specifically, we shall show that for real $\alpha(r)$ the bound states spectrum can arbitrarily be made real or complex and we shall explicitly see that the wavefunction has a logarithmic singularity near the critical radius r_o such that $\alpha(r_o) = 1$. Note that, apart from making eq.(1) analytically soluble, (2) correctly represents the long range part of the nuclear density for suitably chosen values of g and μ.

For s-wave bound states ($E = -\kappa^2/2m$) (1) becomes ($\phi(r) \equiv u(r)/r$) :

$$u'' - \kappa^2 u = \left[\frac{-\kappa^2 g\, e^{-\mu r} - \frac{1}{2}\mu^2 g\, e^{-\mu r}}{(1 - g\, e^{-\mu r})} - \frac{1}{4}\,\frac{\mu^2 g^2\, e^{-2\mu r}}{(1-g\, e^{-\mu r})^2}\right] u \tag{4}$$

Equation (4) can be solved in terms of the hypergeometric function $F(\alpha,\beta,\gamma,z)$:

$$u(r) = e^{-\kappa r}\, F\left(1 + \frac{\kappa}{\mu}\, ,\, \frac{\kappa}{\mu}\, ,\, 1 + \frac{2\kappa}{\mu}\, ;\, g\, e^{-\mu r}\right) \quad , \tag{5}$$

with the bound states spectrum given by :

$$F\left(1 + \frac{\kappa}{\mu}\, ,\, \frac{\kappa}{\mu}\, ,\, 1 + \frac{2\kappa}{\mu}\, ;\, g\right) = 0 \quad . \tag{6}$$

Note that formulas (5) and (6) are meaningful only for $\alpha \equiv g\, e^{-\mu r} < 1$. Indeed, although (5) provides a solution to (1) for <u>complex</u> α of arbitrary magnitude (the solution for $|\alpha| > 1$ can then be obtained through analytic continuation of the solution for $|\alpha| < 1$), the

ISSN:0094-243X/79/540598-02$1.50

hypergeometric function $F(\alpha,\beta,\gamma,z)$ has a cut on the real axis from $z = 1$ to ∞ [2]. Its value on the cut can be prescribed in an arbitrary way, so that the bound states spectrum (6) also becomes arbitrary. In particular, one may choose the value of $u(r)$ on the cut in such a way that the eigenvalues become complex, as shown by Ericson and Myhrer[1] in the framework of their "diffuse plane surface" model. It then becomes mandatory to use complex values of α in the region $|\alpha| > 1$ in order to have a well-defined and unique mathematical solution. Thus, for physical (i.e. complex) values of g, solution (5) is mathematically satisfactory. However, another feature of solution (5) is that, near the critical radius $r = r_o$, ($r_o = \frac{1}{\mu} \log|g|$) it exhibits a logarithmic singularity[3] :

$$u(r) \underset{r\to r_o}{\simeq} \text{constant} \times \log(1 - g\, e^{-\mu r}) \quad . \tag{7}$$

Presumably, as proposed by Ericson and Myhrer[1], such a strong enhancement of $u(r)$ near $r = r_o$ is unphysical and should be replaced with some interpolated wavefunction.

It is a pleasure to thank Professor J. Humblet for helpful discussions.

REFERENCES

1. T.E.O. Ericson and F. Myhrer, Phys. Let. 74B, 163 (1978)
2. I.S. Gradshteyn and I.M. Ryzhik, Tables of Integrals, Series and Products, Academic Press, New York and London 1965, p. 1042.
3. M.A. Abramowitz and I.A. Stegun, Handbook of Mathematical Functions, U.S. Department of Commerce, Applied Mathematics Series, 1964, formula 15.3.10, p. 559.

EXTENSIVE MOMENTUM-SPACE DWIA STUDIES OF PION-NUCLEUS INELASTIC SCATTERING

T.-S. H. Lee, D. Kurath, R. Lawson and S. Chakravarti
Argonne National Laboratory, Argonne, IL 60439*

ABSTRACT

Within the framework of DWIA, we have carried out extensive studies of pion-nucleus inelastic scattering. Our objectives are: (a) to examine critically the extent to which the DWIA method can be valid, (b) to explore nuclear structure from the recent experimental data.

To account for exactly the relativistic nonlocal effects, the inelastic scattering amplitude $\langle \chi^{(-)}_{\vec{k}o'} | U^{(s)}_{fi}(E) | \chi^{(+)}_{\vec{k}o} \rangle$ is calculated directly in momentum space using the numerical method of Lee and Tabakin[1]. The transition potential $U^{(s)}_{fi}(E)$ is determined from the relativistic πN off-shell t matrix and the nuclear transition form factors. The off-shell t matrix is calculated from the finite range πN model using the method[2] of relativistic particle quantum mechanics. The nuclear form factors F_{fi} are calculated either from shell model or from collective model, depending on the nature of the considered excitations. The distorted waves $\chi^{(\pm)}$ are calculated from the optical potential Uo by solving a relativistic scattering equation[1]. Two methods are used to construct the optical potential Uo : (a) the model Hamiltonian of Ref. (2) is used to calculate theoretically $Uo = Uo^{(s)}_o + Uo^{(a)}_o$ with $Uo^{(a)}_o$ describing the effects of pion absorption via the mechanism $\pi NN \rightleftarrows N\Delta \rightleftarrows NN$, (b) improve the model $Uo=Uo^{(s)}_o$ phenomenologically by modifying the radius of $Uo^{(s)}_o$ to fit the elastic scattering cross sections.

Calculations have been performed for the inelastic scattering of pion from (a) the 1p-shell nuclei, (b) ^{18}O, and (c) ^{24}Mg, ^{28}Si, ^{56}Ni and ^{208}Pb. The results of (c) have been presented in Ref. (3). The studies of (a) and (b) are discussed in the other contributed paper (T.-S. H. Lee et. al.).

1. T.-S. H. Lee and F. Tabakin, Nucl. Phys. A226, 253 (1974).
2. T.-S. H. Lee ANL preprint and to be published.
3. S. Chakravarti and T.-S. H.Lee, ANL preprint and to be published.

*Work performed under the auspices of the U.S. Department of Energy.

ISSN:0094-243X/79/540600-01$1.50

Statistical Theory of Pion-Nucleus Reactions Near the (3,3) Region*

D. Agassi and D. S. Koltun
Physics Department, The University of Rochester
Rochester, N.Y. 14627 U.S.A.

ABSTRACT

Pion-nucleus scattering for bombarding energies in the (3,3) region is described in terms of a transport equation, following the propagation, diffusion and true-absorption of the Δ-isobar.

The scattering of a pion by a nuclear target is commonly viewed as a sequence of collisions with individual nucleons, during which the tagged "incoming" pion is emitted and absorbed alternatively. For energies in the (3,3) region this picture is naturally accomodated to the isobar (Δ) model, which is based on two transitions, i.e. $\pi N \rightleftarrows \Delta$ and $\Delta N \rightleftarrows NN$. These can be treated both on and off the energy shell. The fact that the struck nucleons are bound is reflected in two ways. The one is that some of the kinetic energy of the propagating Δ/pion is transferred irreversibly into intrinsic nuclear excitation (heat), thereby leading to a slow-down and diffusion of the Δ/pion. The other effect is the possible importance of off-shell amplitudes, as some crude estimates indicate.

Following previous studies of nuclear-induced reactions and heavy-ion scattering[1], the two above mentioned ingredients have been treated in a transport theory. The nuclear degrees of freedom are treated statistically, in conjunction with the momentum conservation balance pertaining to the transitions $\Delta N \rightleftarrows \pi N$ and $\Delta N \rightleftarrows NN$[2]. It seems advantageous in this context to follow the Δ-isobar (rather than the pion as is the case in multiple-scattering theories) since it is heavier and non relativistic. The nuclear states are described within the exciton model. The central constructs for the statistical considerations are matrix elements of the current operators pertaining to the elementary transitions between nuclear states. Thus for instance $J^{+}_{a,A}(\vec{k}_\pi,\vec{x}_\Delta)$ is related to the annihilation of a Δ at point $\vec{x}_\Delta$ and the creation of a pion with momentum $\vec{k}_\pi$ and nuclear transition $a \rightleftarrows A$, and $j^{+}_{a\alpha}(\vec{x}_\Delta,\vec{k}_1\vec{k}_2)$ to the annihilation of a Δ and the creation of two nucleons with momenta $\vec{k}_1,\vec{k}_2$ and nuclear transition $a \rightleftarrows \alpha$ respectively. We treat $J^{+}_{a,A}\, j^{+}_{a,\alpha}$ as random variables, thereby introducing current-current correlations. The very short life-time of the Δ ($\sim 10^{-23}$sec) implies that the statistical treatment of the nucleus states should be viewed as an idealization.

The resulting transport equation determines $F_{\varepsilon_a(mT_z)}(\vec{x}_\Delta,\vec{k}_\Delta)$

*Supported in part by DOE, contract number Fy-76-5-02-2171

ISSN:0094-243X/79/540601-02$1.50

interpretable as the joint probability distribution for the Δ to be at $\vec{x}_\Delta$ (with respect to the nucleus center), with momentum $\vec{k}_\Delta$, and the nucleus at excitation ε_a, exciton number m and z-isospin projection T_z. The structure of the equation is

$$[\frac{h^2}{m}\vec{k}_\Delta\cdot\vec{\nabla}_{\vec{x}_\Delta} - (\vec{\nabla}_{\vec{x}_\Delta}\cdot U(\vec{x}_\Delta))\cdot\vec{k}_\Delta]F_{\varepsilon_a(mT_z)}(\vec{x}_\Delta,\vec{k}_\Delta) = G - L - In. \quad (1)$$

where G, L and In. represent the "gain", "loss" and inhomogenous terms respectively. They are explicitly given in terms of the correlation functions $\overline{J^+J}$, $\overline{j^+j}$ (2), which are related to the Optical Green-functions for the Δ/pion propagation, and the appropriate density-of-states factors. The ℓ.h.s. of (1) represents a "rate of change". Equation (1) conserves the total probability flux, treats the Δ/pion propagation both on and off-shell and includes true-absorption processes. The off-shell processes, which may play an important role in true-absorption, are not easily included in semiclassical theories such as cascade calculations for meson reactors. The input parameters to the theory are the vertex functions for $\Delta\pi N$, a transition "potential" for $\Delta N \rightleftarrows NN$ and a parametric fit to the energy dependence of the total reactor cross-section. Additional integration of $F(\vec{k}_\Delta,\vec{k}_\Delta)$ yields the following partially inclusive differential cross-sections

$$\frac{d^2\sigma}{d\varepsilon_a d\Omega_\pi}, \quad \frac{d^6\sigma}{d\vec{k}_1 d\vec{k}_2}$$

and their angle-energy integrals, i.e. the total cross-sections for the pion scattering and absorption.

REFERENCES

1. D. Agassi and H. A. Weidenmuller, Phys. Lett. 56B, 305 (1975).
D. Agassi, H. A. Weidenmuller and G. Montzouranis Phys. Rep. 22C, 147 (1975), D. Agassi, C. M. Ko and H. A. Weidenmuller, Ann. of Phys. 107,140 (1977).
2. D. Agassi and D. Koltun, in preparation.

PION-NUCLEUS INELASTIC SCATTERING AND THE NUCLEAR SHELL MODEL

T.-S. H. LEE, D. KURATH and R. LAWSON
Argonne National Laboratory, Argonne, Ill 60439*

The momentum-space DWIA method described in the previous paper by Lee et. al. has been used to carry out calculations of pion inelastic scattering from the lp-shell nuclei, ^{18}O and ^{28}Si, with all nuclear transition form factors calculated from nuclear shell model. In the study of lp-shell nuclei, the validity of DWIA is critically examined using the model of Cohen and Kurath[1]. In the studies of $^{18}O(2^+)$ and $^{28}Si(5^-,6^-)$, sensitivities of pion inelastic scattering to the shell models[2] are investigated. In the following figure, we compare some of our results with recent data[3].

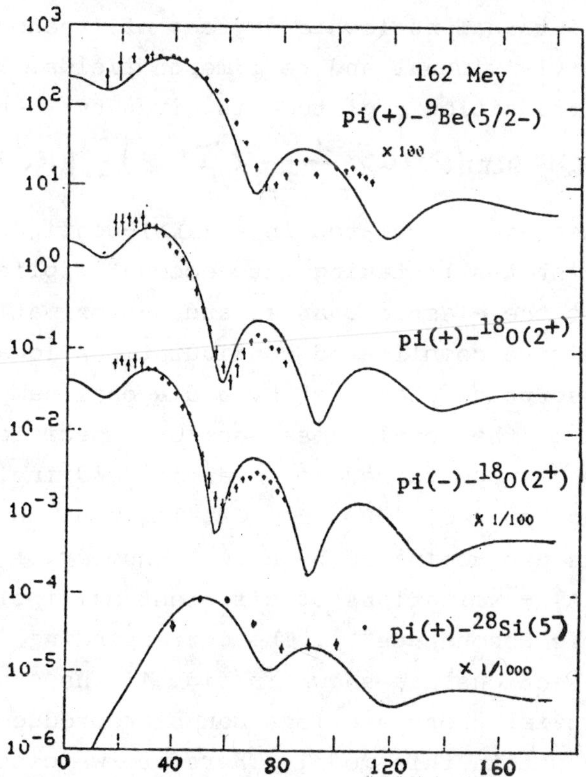

1. S. Cohen and D. Kurath, Nucl. Phys. 73, 1(1965).
2. R. D. Lawson, F. Serduke and H. T. Fortune, Phys. Rev.C14,1245.
3. D. F. Geesaman, C. Olmer and B. Zeidman et. al., Phys. Rev. c18 ,2223(1978),and contribution to this conf. . S. Iversen et. al. Phys. Rev. Lett. 40, 17(1978)

*Work performed under the auspices of the U.S. Department of Energy.

ISSN:0094-243X/79/540603-01$1.50

PIONIC RESONANCE FLUORESCENCE

K. Klingenbeck, M. G. Huber
University of Erlangen-Nürnberg, Erlangen, W-Germany

Pion nucleus scattering in the region of the $\Delta(33)$ resonance is considered as a resonance fluorescence phenomenon where the incoming pion excites resonances $|\mu\rangle$ of the whole many baryon system (A^* resonances[1]). The properties of those A^* resonances are determined by a generalized A-baryon Hamiltonian $\mathcal{H}$, which includes - in addition to the conventional many body degrees of freedom - the internal nucleonic degrees of freedom and their coupling via virtual and real meson fields. The corresponding transition amplitude is given by:

$$T(\underline{k}',\underline{k}) = \sum_{\mu} \langle \underline{k}', A(I)|\mathcal{L}^{+}|\mu\rangle \frac{1}{\omega - \mathcal{E}_{\mu}} \langle \tilde{\mu}|\mathcal{L}|\underline{k}, A(g.s.)\rangle \quad (1)$$

Those eigenmodes are calculated in a $(\Delta\bar{N})$-configuration space (doorway states[2]), taking into account explicitely the coupling to the elastic channel and approximatly the effects of the more complicated configurations (quasi-elastic and absorptive channels) by a diagonal damping width adjusted to the total cross sections. Specific calculations have been performed for elastic (and inelastic) pion scattering for ^{12}C, ^{16}O and ^{40}Ca. It turns out that those reactions are dominated by a small number of collective[3,4] A^* -excitations of different multipolarity ("Giant Isobaric Resonances"[3]). The corresponding partial cross-sections are shown in fig. 1. The experimental differential cross sections can be reproduced satisfactorily within this model. Therefore we conclude that the pattern of fig. 1 is a realistic, although preliminary description of the pionic excitation strength distribution. It remains to be seen to which extend those intermediate A^* resonances dominate also other pion-induced nuclear reactions (such as (π,γ) or (π,p)) in

ISSN:0094-243X/79/540604-02$1.50

this energy region.

1. K. Klingenbeck, M. Dillig, M. G. Huber, Phys. Rev. Lett. 41, 387 (1978)
2. L. S. Kisslinger, W. L. Wang, Phys. Rev. Lett. 30, 1071 (1973)
3. M. Dillig, M. G. Huber, Phys. Lett. 48 B, 417 (1974)
4. M. Hirata, J. H. Koch, F. Lenz, E. J. Moniz, Phys. Lett. 70 B, 281 (1977)

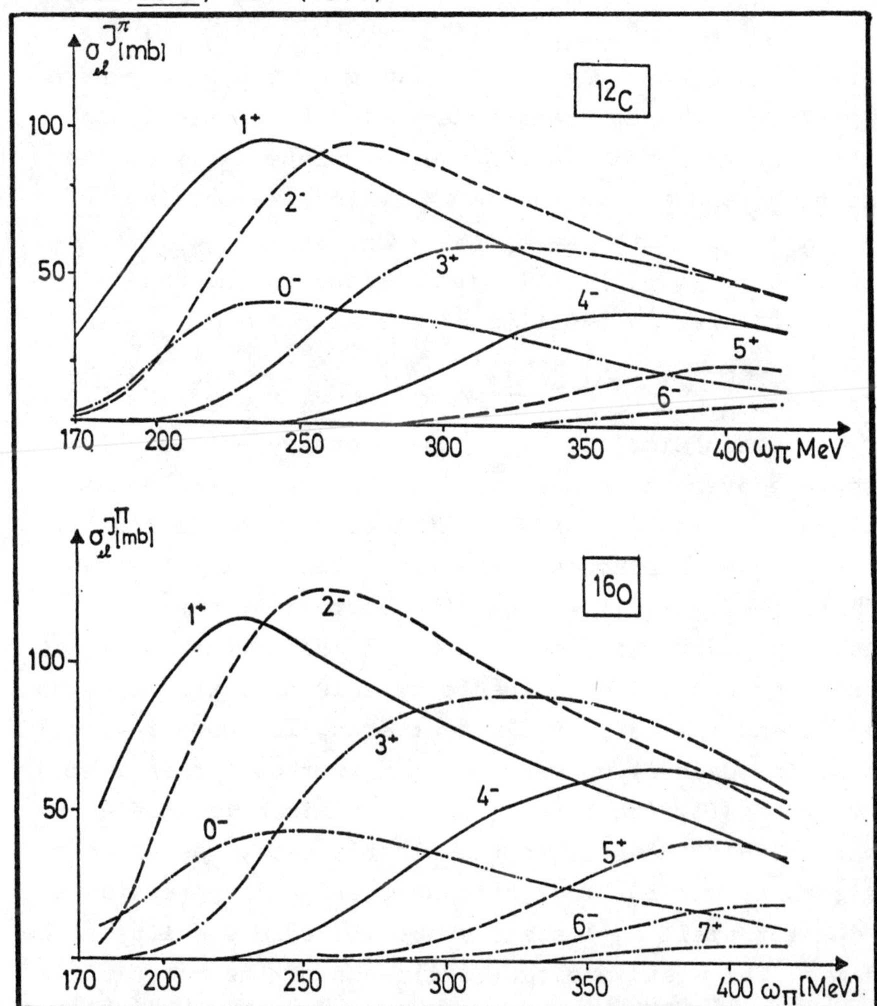

Fig. 1. Pionic excitation strength distributions of ^{12}C and ^{16}O, respectively, for various multipolarities.

RECOIL EFFECTS IN PION NUCLEUS SCATTERING

K. Klingenbeck
University of Erlangen-Nürnberg, Erlangen, W-Germany

If one describes π-nucleus scattering by the isobar excitation of bound nucleons, one of the most important input quantities is the $\pi N\Delta$ transition operator $\mathcal{L}$. The usual form of $\mathcal{L}$ in the πN CM-system is given by:

$$\mathcal{L} = f^*(\kappa^2)/\omega_\pi \; \underline{S}^\dagger \cdot \underline{\kappa} \; \underline{T}^\dagger \cdot \hat{\phi}_\pi \qquad (1)$$

Since $\mathcal{L}$ contains the πN relative momentum $\underline{\kappa}$, we have to perform a proper transformation of kinematical quantities for πA scattering. If we calculate $\underline{\kappa}$ from the momenta $\underline{k}_\pi$ and $\underline{k}_N$ (in the πA CM system) we obtain[1]: $\kappa = \beta(\underline{k}_\pi - \alpha\underline{k}_N)$ with $\beta = M_N/(M_N + \omega_\pi)$ and $\alpha = \omega_\pi/M$. Inserting in eq. (1), this gives raise to the usual static $\mathcal{L}^s$ and in addition a nonstatic term

$$\mathcal{L}^{ns} = -\alpha\beta \frac{f^*(k_\pi^2)}{\omega_\pi} \; \underline{S}^\dagger \cdot \underline{k}_N \; \underline{T}^\dagger \cdot \hat{\phi}_\pi \qquad (2)$$

For the evaluation of this recoil term we explicitely integrate over the momentum distribution of the bound nucleon, according to its wavefunction. Therefore we are able to take into account the Fermi motion in the transformation of eq. (1) without the approximations[2] generally made in the optical model (factorization, frozen nucleus). The influence of this nonstatic term on elastic and inelastic π-^{12}C scattering is shown in fig.1. The solid (dashed) curves are with (without) this recoil term. Typically the first (second) minimum is shifted by about 4 (8-10) degrees in the elastic case. The effects are even larger for inelastic scattering: suppression of $d\sigma/d\Omega$ for small angles and about 15 degrees shift of the minimum. The most important influence of the nonstatic term is seen at large angles, where even the structure of the cross section may be changed. Moreover in this large

ISSN:0094-243X/79/540606-02$1.50

angle region those nonstatic effects show a pronounced energy dependence.

1. M. Hirata, F. Lenz and K. Yazaki, Ann. Phys. (N.Y.) 108, 116 (1977)

2. R. H. Landau, S. C. Phatak and F. Tabakin, Ann. Phys. (N.Y.) 78, 299 (1973)

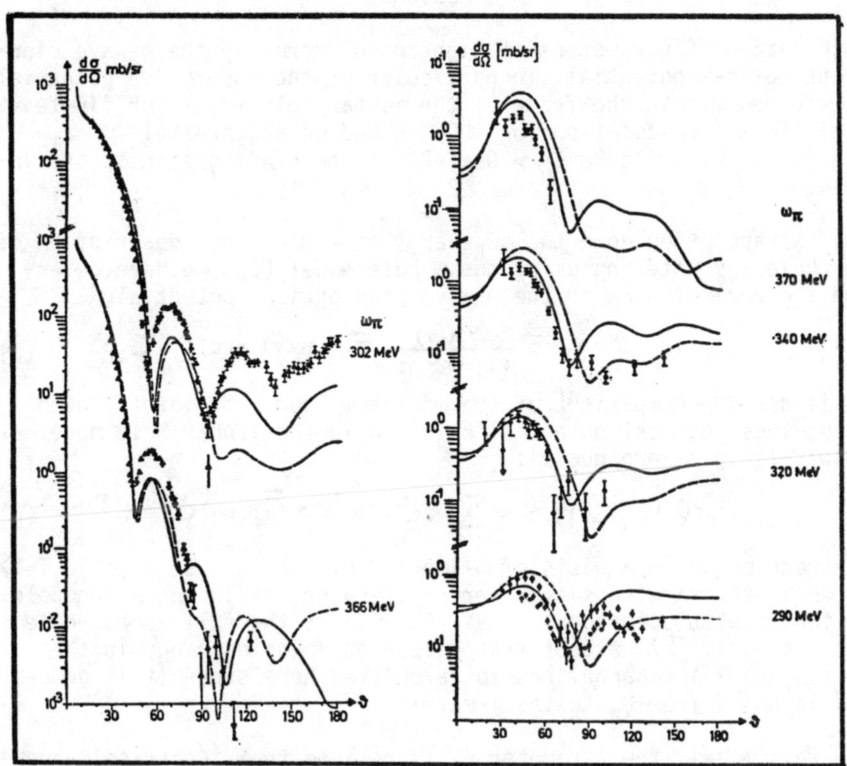

Fig. 1. π-^{12}C scattering for various energies in the resonance region. The solid (dashed) curves are calculated with (without) nonstatic corrections. (a): elastic; data from Piffaretti et al. (1978); (b): inelastic excitation of the 3^- (9.64 MeV) state; data from Binon et al. (1970).

THE ABSORPTIVE PART OF THE LOW-ENERGY PION-NUCLEUS OPTICAL POTENTIAL

E. Oset
Department of Physics, University of Barceolna, Spain
W. Weise
Institute of Theoretical Physics, Univ. of Regensburg, W. Germany

ABSTRACT

A calculation of parameters of higherorder terms in the p-wave pion-nucleus optical potential, in particular of the absorptive potential, is performed within the frame of the isobar-hole model for finite nuclei. From a calculation for ^{16}O, we deduce at threshold Im C_0=0.08 μ^{-6}, while Re $C_0 > 0$ if Pauli and binding effects are included.

By extrapolation down to low energy of a many-body description of pion-nucleus scattering using the Δ-hole model [1], we have determined the parameter C_0 in the p-wave pion optical potential,

$$2\omega\, U_p = 4\pi\, \vec{\nabla} \frac{n(r)}{1+\frac{4\pi}{3}\xi n(r)} \vec{\nabla}\, , \quad n(r) = c_0\rho + C_0\rho^2. \qquad (1)$$

This is done by comparison of the momentum space form of (1) with the non-local optical potential obtained in the isobar-hole model for spin-isospin-zero nuclei:

$$2\omega \langle \vec{k}' | U_p(\omega) | \vec{k} \rangle = \sum_{s's} \langle \vec{k}' | \delta H | s' \rangle \bar{G}_{s's}(\omega) \langle s | \delta H | \vec{k} \rangle . \qquad (2)$$

The expansion is in a basis of Δ-isobar-hole states $|s\rangle = |(\Delta\bar{N})J,T=1\rangle$. Here δH is the $\pi N\Delta$ transition vertex operator, including a monopole form factor with cutoff Λ=1.2 GeV. $\bar{G}_{s's}(\omega)$ is the full isobar-hole Green's function [1] except that direct one-pion exchange in the (elastic) $\Delta\bar{N} \to \Delta\bar{N}$ channel has to be omitted here since it is generated in going from U_p to the T-matrix.

In this model, the parameter C_0 is related to Δ-isobar self-energy corrections of the type shown in Fig. 1, where Im C_0 comes from diagram (a), while all diagrams contribute to Re C_0. The Lorentz-Lorenz correction ξ is related to vertec corrections in the Δ-isobar-hole interaction [1].

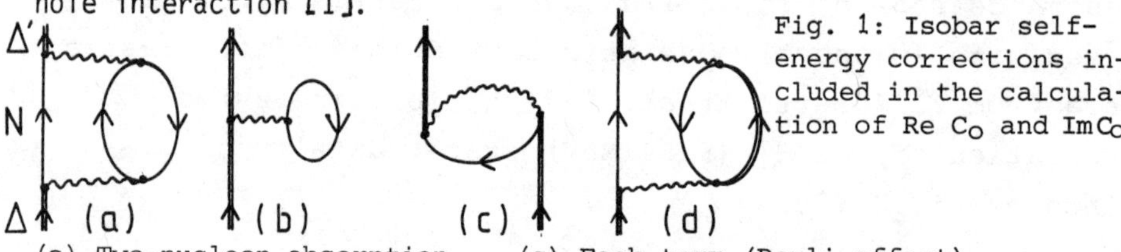

Fig. 1: Isobar self-energy corrections included in the calculation of Re C_0 and ImC_0:

(a) Two-nucleon absorption
(b) Hartree binding) term
(c) Fock term (Pauli effect)
(d) Reflection correction.

ISSN:0094-243X/79/540608-02$1.50

The calculation includes π and ρ exchange plus additional short range correlations as specified in refs. [1,2], where the input parameters are chosen in accordance with an analysis of the $\pi^+ d \rightarrow pp$ reaction. Results for C_0 are shown in Fig. 2. Note that Re $C_0 > 0$; at the same time, the model gives a sizeable Lorentz-Lorenz correction, $\xi = 1.6$.

We have analysed the degree of r-space non-locality of the optical potential, eq. (2). It was found that a representation in terms of a Kisslinger form, eq. (1), is justified at low energy ($T \lesssim 50$ MeV), but becomes meaningless at higher energy [2].

[1] E. Oset and W. Weise, Phys. Lett. 77 B (1978) 159 and Nucl. Phys. A (1979), in print.

[2] E. Oset, W. Weise and R. Brockmann, submitted to Phys. Lett.

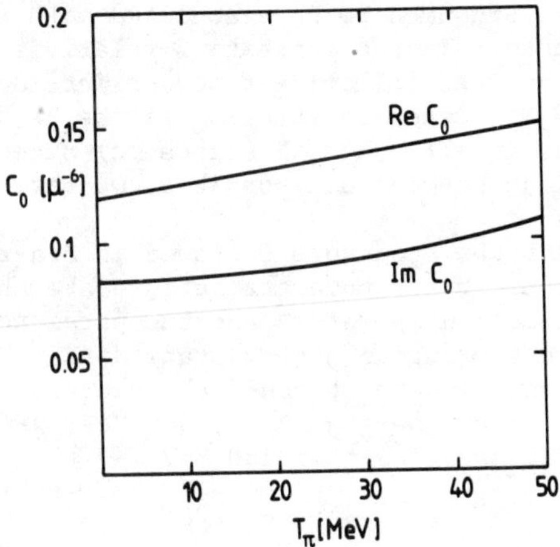

Figure 2:

Values of Re C_0 and Im C_0 deduced from isobar-hole model calculation for ^{16}O, using input parameters as specified in refs. [1,2].

A HAMILTONIAN MODEL FOR NUCLEONS, PIONS AND Δ'S

M. Betz, T.-S. H. Lee and F. Coester
Argonne National Laboratory, Argonne, IL 60439*

ABSTRACT

Within the framework of relativistic particle quantum mechanics, a model is constructed which describes the interactions between nucleons, pions and Δ's at low and medium energy. The parameters of the model are determined phenomenologically by fitting data of simple processes involving these particles. The resulting Hamiltonian should be useful for many-body calculations in which pions may appear explicitly and Δ's are unstable.

OUTLINE OF THE PROGRAM

The degrees of freedom most relevant to low and medium energy nuclear physics are presumed to be associated with three kinds of particles: N, Δ and π. In the context of relativistic particle quantum mechanics[1,2], the following types of interactions are assumed:[3] a) a $\Delta \rightleftarrows N\pi$ vertex interaction in the P_{33} πN states; b) direct interactions in all other πN states; c) direct interactions producing transitions between all possible two-baryon states (NN, NΔ, ΔΔ).

In analogy with the procedure followed in non-relativistic nuclear physics, appropriate parametrizations are chosen for the corresponding interaction operators and the parameters are determined by fitting scattering and bound state data relative to simple systems, such as, in the case at hand: the deuteron bound state; NN elastic scattering at energies up to 800 MeV; π-d elastic scattering and π+d→NN at energies near 150 MeV.

A SIMPLE EXAMPLE

Realization of this program requires further simplification. We expect the following assumptions to represent physically reasonable approximations: i) the Hilbert space may be truncated to three channels:

$$\mathcal{H} \approx \mathcal{H}(NN) \oplus \mathcal{H}(N\Delta) \oplus \mathcal{H}(NN\pi)$$

ii) the πN interaction is small compared to the Δ mass; iii) the baryons may be treated non-relativistically.

The procedure given in ref. (1) is followed to construct a mass operator acting on the three-channel space and satisfying the cluster requirements. Using assumptions ii) and iii) above, one can show that the resulting mass operator differs from that of the free particle system by an interaction term given (up to kinematical

*Work performed under the auspices of the U. S. Department of Energy.

ISSN:0094-243X/79/540610-02$1.50

factors) by the sum of the vertex and direct interactions described above. Standard scattering theory can then be used.

In this communication, we focus our attention on NN scattering in the $^1D_2$ partial wave and on pion scattering and absorption on the deuteron in the resonance region. We neglect all non-resonant πN interactions and assume that the dominant NΔ partial wave is $^5S_2$. Only the $^3S_1$-$^3D_1$ NN interaction is included in πNN states. In order to keep the three-body calculations simple, all baryon-baryon interactions are assumed to be separable. Simple analytic expressions are chosen for all form factors. The parameters of the $\Delta \rightleftarrows \pi N$ vertex and the bare Δ mass are determined by a fit to the P_{33} πN phase-shifts. The NN $^3S_1$-$^3D_1$ interaction is obtained by fitting the phase-shifts and the deuteron properties. These results serve as input to the three-body calculation required to obtain the NΔ T-matrix. The parameters of the baryon-baryon interaction on the space $^1D_2(NN) \oplus {}^5S_2(N\Delta)$ may then be determined by a fit to the data mentioned above.

The influence of the coupling to πNN states on NN phase shifts and the role played by NN interactions in intermediate πNN states are discussed. The importance of treating the scattering and absorption channels consistently in the study of π-d interactions is assessed.

REFERENCES

1. F. Coester, Helvetica Physica Acta 38, 7 (1965).
2. S. N. Sokolov, Sov. Phys. Dokl. 22, 198 (1977).
3. T.-S. H. Lee, to be published.

SUMMATION OF WATSON'S MULTIPLE COLLISION SERIES FOR HADRON-NUCLEUS SCATTERING AT MEDIUM ENERGIES

R. Dymarz* and A. Małecki
Instytut Fizyki Jądrowej, 31-342 Kraków, Poland

ABSTRACT

The leading order corrections to the Glauber model are evaluated. It is stressed that the main contents of the theory of multiple collisions can be reproduced with the aid of the dynamical optical potential which represents an improvement both over the eikonal model and the static Watson potential.

The Glauber model[1] may be considered as a transposition of the Watson theory of multiple collision[2] to the eikonal mechanics. The eikonal approximation to the scattering propagator G=Geik allows, under the usual assumptions of impulse and closure, to effectively sum the infinite Watson series up.

The leading order corrections to the Glauber scattering amplitude F_{G1} can be evaluated following Wallace[3]. They come from three sources: the non-eikonal character of the propagation(N), the presence of the nuclear target Hamiltonian(H) and the transformation between the projectile-nucleon and the projectile-nucleus c.m. systems (W). The example shown in Fig.1 reveals that at a few hundred MeV the non-eikonal correction is very important while the remaining two are smaller and tend to cancel each other.

Disregarding the corrections H and W and calculating the correction N to all orders the eikonal expansion of Watson's series can be obtained:

$$T=T_{G1}+T_{G1}(G-G_{eik})T_{G1}+T_{G1}(G-G_{eik})T_{G1}(G-G_{eik})T_{G1}+\dots \quad (1)$$

where T_{G1} is the scattering matrix of the Glauber model[1].

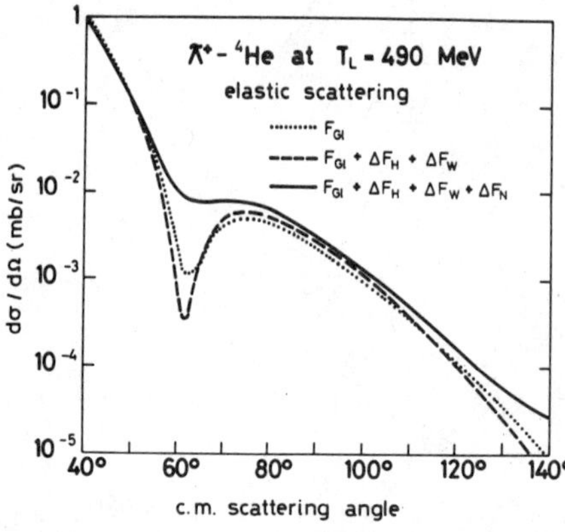

Fig.1. The leading order corrections to the Glauber model. The ^{4}He density and the π^+-N scattering amplitude are approximated as Gaussians with the nuclear radius R=1.37 fm, the total π^+-N cross-section σ=25 mb, the Re/Im ratio α=0,2, and the slope a = 5.3 GeV^{-2}.

* Now at JINR, Dubna, USSR.

ISSN:0094-243X/79/540612-02$1.50

The importance of the non-eikonal effect necessitates a careful treatment of Eq.1 which represents the main contents of the multiple collision theory at medium energies.

An implicit summation of the eikonal expansion (1) can be performed by solving the Schrödinger wave equation with the optical potential[4]:

$$V(r) = -\frac{2iv}{\pi}\int_0^\infty dz\,\frac{d}{dr^2}\,\ln\left[1-\Gamma\left(\sqrt{r^2+z^2}\right)\right] \qquad (2)$$

where the nuclear profile $\Gamma(b)$ is the Fourier-Bessel transform of the Glauber scattering amplitude and v denotes the projectile velocity. The microscopic potential (2) includes the effects of multiple scattering and has a dynamical character[5] since it admits the possibility of virtual excitations of the target during scattering. This should be contrasted with the static potential (referred to as the Watson potential[2]) which arises by taking in Eq.2 only the terms linear in the projectile-nucleon amplitude.

In Fig.2 we compare the results of the Glauber model with those of the partial-wave analysis using the Watson and multiple scattering potentials. The non-eikonal propagation is most important at the minima of angular distribution while the multiple scatterings decisively enhance the cross-section at large momentum transfers. The importance of both the effects indicates that a good understanding of the mechanism of nuclear scattering should be prior to looking for the effects of nuclear structure and for the properties of the elementary interaction.

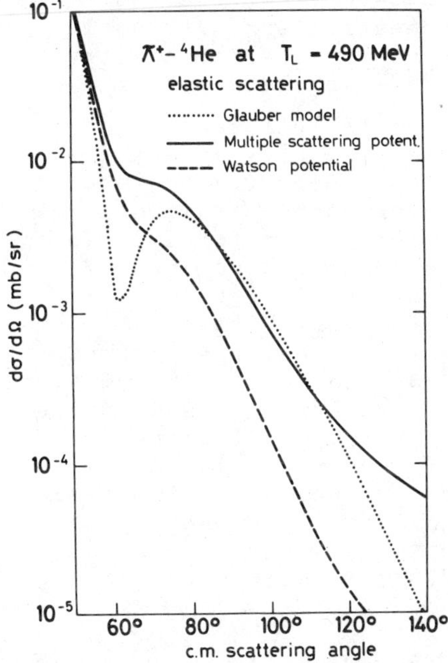

Fig.2. The comparison of the Glauber model with the microscopic optical model using the Watson and multiple scattering potentials. Same parameters as in Fig. 1.

REFERENCES

1. R.J. Glauber, Lectures in Theor.Phys.(Interscience N.Y.,1959) Vol.I., p.315.
2. K.M. Watson, Phys.Rev.89, 575 (1953) ; 105, 1388 (1957).
3. S.J. Wallace, Phys.Rev. C.12, 179 (1975).
4. R. Dymarz and A. Małecki, Phys.Lett.66B, 413 (1977).
5. R. Dymarz and A. Małecki, IFJ Kraków 1021/PH (1978).

P-WAVE ABSORPTIVE AND DISPERSIVE EFFECTS ON PION-NUCLEUS SCATTERING*

G. E. Brown and B. K. Jennings
State University of New York, Stony Brook, N. Y. 11794

R. M. Rockmore
Rutgers University, New Brunswick, N. J. 08903

The approach of Brown and co-workers[1] to low energy pion-nucleus scattering which derives from nuclear matter considerations in which the $\Delta(1230)$ isobar plays a fundamental role is extended to include the effects of P-wave absorption and dispersion. For absorption we make the simplest extension, which, in terms of many-body graphs with two nucleon-hole pair intermediate states, consists mainly in that shown in the inset in Fig. 1. For the exchange of virtual pions and ρ-mesons we use the effective interactions $V_{\pi,\rho}(r,t)g(r)$ which take into account the effects of finite-ranged (π,ρ)NN and $(\pi,\rho)N\Delta$ vertices and short-range correlations. In the case of finite-range effects we use the monopole form factors with cutoffs typically, $\Lambda_\pi = 1.0$ GeV, $\Lambda_\rho = 1.5$ GeV, and with couplings $f^{*2}/4\pi = 0.32$, $g_\rho^{\,2}/4\pi = 0.52$ with $\kappa = 6.6$ and $f_\rho^* = \sqrt{72/25}\, f_\rho$, which provide a reasonable fit to the πD absorption data.[2] Short-range correlations are represented by the two-body correlation function $g(r)$, taken in the schematic form, $g(r) = 1 - j_0(q_c r)$, with $q_c \approx m_\omega$. If we incorrectly replace the resonant denominator $[\omega_\Delta - \omega_q - i\Gamma/2]^2$ in the graph of Fig. 1 by $[(\omega_\Delta - \omega_q)^2 + \Gamma^2/4]$ in order to make contact with the results of Ko and Riska[3] who work in a pair-wave formulation of the Fermi gas model, we find that the results of our simple approach compare favorably with theirs, suggesting a threshold value for $\mathrm{Im}C_0$ in the range $(0.06 - 0.08)m_\pi^{-6}$. From considerations of crossing we associate with the graph of Fig. 1, a like corresponding resonant graph with crossed external pions. The dispersive contribution from this enlarged set of graphs has the correct threshold behavior. Again making the incorrect replacement of the recurring resonant denominator as before we find for purposes of comparison the corresponding curves of $\mathrm{Re}C_0$ shown in Fig. 2. Note that $\mathrm{Re}C_0$ has the same sign as $\mathrm{Im}C_0$ and shows increased cutoff dependence. The threshold value for $\mathrm{Re}C_0$ lies in

*Work supported in part by U.S.D.O.E. contract #EY-76-S-02-3001.

ISSN:0094-243X/79/540614-02$1.50 Copyright 1979 American Institute of Physics

the range $(0.05 - 0.08)m_\pi^{-6}$. At low energies, the Lorentz-Lorenz effect must be included to get a satisfactory fit to cross section data. After the inclusion of our dispersive contribution, we find that at 50 MeV for example, ξ increases from 1.2 to 1.5. A table of relative reaction cross section changes with incident π kinetic energy is included for comparison.

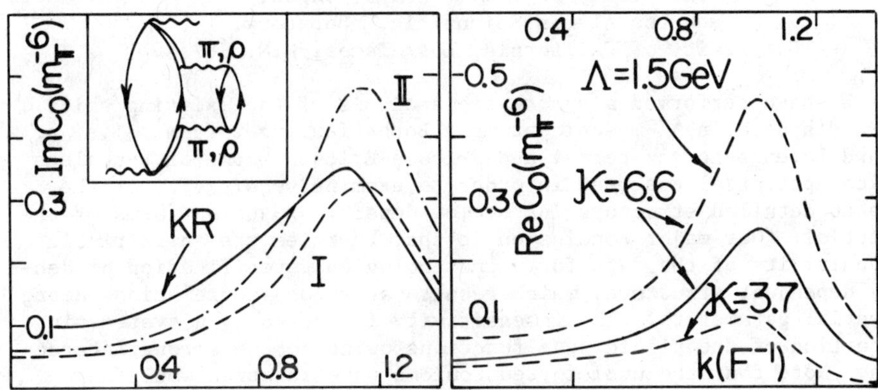

Fig. 1. Graphs of ImC_o for π and ρ rescattering with the principal many-body mechanism in the inset. q_c = 400 MeV (I), 600 MeV (II); Ko and Riska (KR) take Λ = 1.2 GeV.

Fig. 2. Graphs of ReC_o for π and ρ rescattering.

TABLE I. $[(\Delta\sigma_R/\sigma_R(\text{no abs.})] \times 100\%$

Element	T_π = 50 MeV	70 MeV	90 MeV	110 MeV
^{12}C	66	52	28	7
^{16}O	46	36	17	2

REFERENCES

1. G. E. Brown, B. K. Jennings and V. I. Rostokin, Phys. Repts. (in press).
2. M. Brack, D. O. Riska and W. Weise, Nucl. Phys. A287, 425 (1977).
3. C. M. Ko and D. O. Riska, Phys. Rev. (to be published).

HOW MUCH INFORMATION OF π^--NUCLEUS OPTICAL POTENTIAL CAN π^--MESONIC ATOM DATA PROVIDE?

Ryoichi Seki
California State University, Northridge, Calif. 91330†;
W. K. Kellogg Radiation Laboratory,
California Institute of Technology, Pasadena, Calif. 91125†;
University of Tokyo, Tokyo, Japan;
and Los Alamos Scientific Laboratory,
University of California, Los Alamos, N.M. 87545

We have performed a systematic analysis of the existing shift and width data in π^--mesonic atoms. Both Kisslinger form[1] (i.e. no ρ^2 and Lorentz-Lorenz terms) and Ericson-Ericson form[2] of π-nucleus optical potential are used in order to examine sensitivity of the data to detailed structure in the low density expansion forms of the potential. Our major conclusion is that there exists an intrinsic insensitivity of the data for phenomenological specification of density dependent structure, which results as strong correlations among potential parameters. The insensitivity is caused by a systematic distortion of the pionic wave functions owing to the strong interaction. Note that the undistorted Coulomb wave function would have had a differentiating capability because of r^ℓ-type dependence near the nucleus.

The conclusion is perhaps most clearly demonstrated by the fact that two potential forms with six parameters each (i.e. $R_eB_o=R_eC_o=0$ in the Ericson-Ericson form) reproduced equally well with the chi-square per number of freedom (χ^2/N) of about 3. The pionic wave functions for both forms of the potential yield the ratios of expectation values, $R_2 = \langle\rho^2\rangle/\langle\rho\rangle \approx 0.24\mu^3$ and $R_2' = \langle\nabla\rho^2\cdot\nabla\rangle/\nabla\rho\cdot\nabla\rangle \approx 0.38\mu^3$, almost independent of the nuclear mass number. [Incidentally, these values are less than the nuclear matter value of ρ, $0.5\mu^3$, favouring the low-density expansion of the potential.] All of our various fits (by varying up to all 9 parameters) and also more restricted fits previously published[3] are found to satisfy

$$\begin{aligned} &b_o + B_oR_2 && (0.03 - 0.01i)\mu^{-1} \\ &(c_o/\gamma)[1 + C_oR_2'(C_o^{-1}+\gamma^{-1})] \quad \sim && (-0.2 - 0.02i)\mu^{-1} \end{aligned}$$

where $\gamma = 1 + (4\pi/3)\xi c_o\rho$.

A few further comments: 1) The insensitivity of the wave function persists even in the atoms in the p state where the shift changes its sign (i.e. the both forms of the potential yield remarkably close results.) For a controversial ^{110}Pd in the 3p state,[4] all of our fits yield a shift of 11∿13 keV and a width of 30∿32 keV by use of the Fermi distribution with $\rho_p/Z = P_n/N$. 2) R_2 and R_2' do become small in atoms of small shifts and width. Experimental

†Present address.

ISSN:0094-243X/79/540616-02$1.50

efforts ought to be made in more accurate direct and intensity measurements in these atoms. 3) Despite the strong correlation among the parameters we observe indication of $R_e\ B_o > 0$ and $R_e\ C_o < 0$ in our various fits. (Note that our conversion is the imaginary part of a parameter <0.) 4) Because of the insensitivity of data we feel that a purely phenomenological extraction of the neutron distribution from the π^- atom data alone is hopeless at present.

Without changing the method of analysis we also included the recent 2p state data[5] that are an-order-of-magnitude more precise. We have found that our discussions and conclusions presented above remain the same. However, the new set of the data favors somewhat the Ericson-Ericson form, $\chi^2/N = 3.8$ over 4.7. An increase in χ^2/N is contributed appreciably from the 1s state atoms. We have also noticed a large reduction in c_1 in the six-parameter fit for the Ericson-Ericson form, from $-0.25\pm0.05\mu^{-3}$ to $-0.14\pm0.05\mu^{-3}$, whereas other parameters remain about the same. We are refining the present analysis in order to accomodate the accurate data. Clearly a series of new measurements of the similar quality is called for in atoms of 1s and 2p states.

The previous values of the parameter in a fit depend on the nuclear densities used. There is no space here to list our fits in a satisfactory way. For the sake of presenting a clude idea of the present status of our various fits, however, we quote just one for the new data set: For $\xi=1$, $c_o = -0.21\mu^{-3}$, and $c_1 = -0.18\mu^{-3}$ fixed, by use of the Fermi distribution with $\rho_p/Z = \rho_n/N$ we obtain $b_o = -0.003\pm0.008\mu^{-1}$, $b_1 = 0.143\pm0.006\mu^{-1}$, $B_o = 0.15\pm0.04-i(0.046\pm0.003)\mu^{-4}$, $C_o = -0.11\pm0.01 -i(0.09\pm0.01)\mu^{-6}$ with $\chi^2/N=3.5$.

This work was supported in part by NSF through INT 76-82961 and DOE.

REFERENCES

1. L. S. Kisslinger, Phys. Rev. 98, 761 (1955).
2. M. Ericson and T.E.O. Ericson, Ann. Phys. (N.Y.) 36, 323 (1966).
3. M. Krell and T.E.O. Ericson, Nucl. Phys. B11, 521 (1969); D. K. Anderson, D. A. Jenkins and R. J. Powers, Phys. Rev. 188, 9 (1969); L. Tauscher, Proc. Int. Seminar on Pion-Nucleus Interactions, Strasboug, 1971 (CNRS, Strasboug, 1971) p. 45; and L. Tauscher and W. Schneider, Z. Physik 271, 409 (1974).
4. M. Leon et al., Phys. Rev. Lett. 37, 1135 (1976); and J. F. Dubach, E. J. Moniz, and G. D. Nixon, MIT-CTP preprint #749 (1978).
5. C. J. Batty et al., Phys. Rev. Lett. 40, 931 (1978).

MICROSCOPIC THEORY OF THE Δ SPREADING POTENTIAL IN A SIMPLE APPROACH TO PION-NUCLEUS SCATTERING*

R. Brockmann, G. E. Brown, and B. K. Jennings
State University of New York, Stony Brook, N. Y. 11794

R. M. Rockmore
Rutgers University, New Brunswick, N. J. 08903

ABSTRACT

We explore the consequences of a simple optical model approach to pion-nucleus scattering, based on nuclear matter considerations. The introduction of a Δ spreading potential based on a microscopic low-energy model of pair excitation is a novel feature of our treatment.

We explore the consequences of a simple optical model approach to pion-nucleus scattering outlined by Moniz _et al._[1] but not exploited by them in their analysis of scattering data in the resonance region. Our work is based almost entirely on nuclear matter considerations, with shell effects arising as "closure corrections". Thus, for example, we find the size and trend of the Pauli width is well-represented in this approach. In our exploration of this model we do not introduce the Δ spreading potential as in Ref. 1, where its strength is determined phenomenologically by fitting the forward elastic amplitude. Instead, we use the spreading potential which follows from a low-energy model of pair excitation. A comparison of the crude phenomenological (imaginary) isobar-spreading potential with that derived from our microscopic model is shown in Fig. 1; note that our model (imaginary) spreading potential appears to follow the trend of the phenomenological one well up into the resonance region. On the other hand, the phenomenological curve of the real part of the spreading potential of Moniz _et al._[1] is not well-represented by the analogous result in our model; this suggests that other many-body mechanisms may become of importance at low-energies.

*Work supported in part by U.S.D.O.E. contract #EY-76-S-02-3001.

ISSN:0094-243X/79/540618-02$1.50

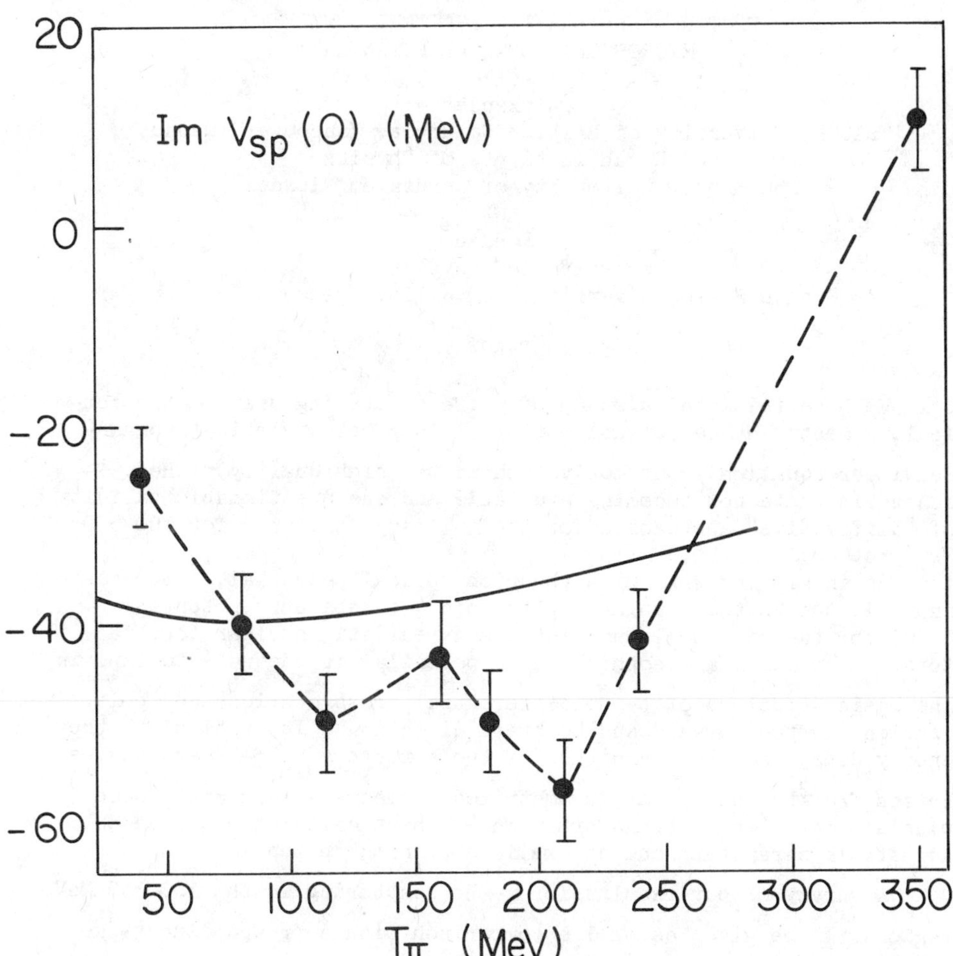

Fig. 1. The imaginary isobar spreading potential fit to forward elastic scattering amplitude (dashed lines) is shown along with that (solid line) predicted by the microscopic model of pair-excitation.

REFERENCES

1. E. J. Moniz, Pion-Nucleus Scattering in the Isobar Formalism (lectures delivered at the NATO Advanced Study Institute, June 1978, Madison, Wis.); F. Lenz, E. J. Moniz and K. Yazaki, to be published; M. Hirata, K. H. Koch, F. Lenz and E. J. Moniz, to be published.

A CALCULATION OF INTERMEDIATE ENERGY PROTON - ^{4}He ELASTIC SCATTERING WITH A MICROSCOPIC OPTICAL POTENTIAL

Y. Alexander
TRIUMF, University of British Columbia, Vancouver, Canada
and Racah Institute of Physics,
The Hebrew University of Jerusalem, Israel*

R. H. Landau+
Department of Physics
Oregon State University, Corvallis, Oregon 97331

ABSTRACT

We have initiated a study of p-^{4}He scattering using a theoretical, momentum space optical potential in a relativistic Lippmann-Schwinger equation. Our motivation is the high quality p-^{4}He scattering data now becoming available and the questionable validity of "diffractive" approaches for the full angular range for 100-500 MeV protons.

As in similar work with the pion optical potential,[1)] our basic input is the NN phase shifts (with spin), a separable potential model for the off-shell functions, and realistic nuclear form factors. Even our least sophisticated potential is adequate to explain the basic structure of p-^{4}He differential cross sections and polarizations in the forward and backward directions. In particular, the energy dependence and magnitude of the structure at backward angles arises from the use of antisymmetrized nucleon-nucleon amplitudes[2)] plus a proper "angle transformation." These calculations have no adjustable parameters and no explicit exchange potential.

A survey of our results for p-^{4}He scattering in the 100-350 MeV region will be given as will a comparison with corresponding π-^{4}He results.

REFERENCES

1. R. H. Landau, Phys. Rev. C15, 2124 (1977).
2. S. A. Gurvitz, Y. Alexander, and A. S. Rinat, Ann. Phys. 98, 346 (1976).

*Present address
+Supported in part by the National Science Foundation PHY76-82659.

ISSN:0094-243X/79/540620-01$1.50 Copyright 1979 American Institute of Physics

OVERVIEW OF K^--NUCLEAR REACTIONS

R.H. Dalitz
Department of Theoretical Physics, Oxford Univeristy, England

INTRODUCTION

The primary capture reactions are $\bar{K}N \to \Lambda\pi$ and $\Sigma\pi$. Measurements to date generally observe an escaping π-meson, whose kinematics reflect the interaction of the Λ or Σ hyperon with the residual nucleus, so that these studies are intimately linked with the properties of the hyperon-nucleon interaction.

1. BARYON RESONANCES IN NUCLEI

The present view is that all known baryonic systems are confined 3-quark states. For example, the octet of spin-1/2 baryons, N, Λ, Σ and Ξ, all have the same internal structure except that there is one strange quark in Λ and Σ, and two strange quarks in Ξ. One interesting question is the effect of their interaction with nuclear matter on their observed properties. The case of the spin-3/2 Δ(1236) state in nuclear matter has already been discussed at this meeting, although regarded more as a πN scattering resonance. We emphasize here that the N(939) and Δ(1236) states are actually closely related objects; they have the same 3-quark configuration, except that the quark spins are aligned to S=3/2 for Δ(1236) rather than to S=1/2, as for N(939). The spin-3/2 decuplet states, Δ, Σ^*, Ξ^* and Ω, are all as elementary as the nucleon. The main difference between N(939) and Δ(1236), for example, is due to their mass difference, large on the nuclear energy scale, which allows the spontaneous decay $\Delta(1236) \to \pi N(939)$, with width Γ=115 MeV in the free state.

These heavier baryonic states can be excited in nuclei by the impact of energetic mesons of appropriate energy. The excitation of Δ(1236) in light nuclei is particularly well known. The use of K^- mesons can, in a similar way, be used to excite Λ^* and Σ^* states within nuclear material. Some interesting examples are as follows:

Λ^*(1405). This state lies below the K^-p threshold, with decay width Γ=40 MeV. However, it can be excited off-mass shell sufficiently to modify the properties of the $K^-p \to \Sigma^\pm\pi^\mp$ reactions on nuclei (cf. talk by Wycech).

Λ^*(1520). This state is directly accessible energetically. This object is particularly interesting because its width is narrow (Γ=16 MeV) in the free state and it has quite a number of decay modes which can be studied.

Σ^*(1785). This has spin-parity $5/2^-$ and is quite strongly excited in the K^-n interaction. In the free state, it has width $\Gamma \simeq 120$ MeV, and many decay modes.

These baryonic states all belong to the lowest 3-quark configuration with negative parity, that with internal orbital angular momentum

ISSN:0094-243X/79/540621-13$1.50

L=1 and mixed permutation symmetry.

The quantities of interest to measure for such a baryonic state, when bound in nuclear matter, are

(i) the mass shift, which will reflect the nuclear well depth experienced by this baryonic object, in first approximation at least,

(ii) the change in its total width Γ. This may be due to (a) the momentum distribution of the nucleons within the nucleus, (b) the change in mass relationships (e.g. the $\Lambda^*(1520)$ width could be quite sensitive to the difference between the nuclear well depths for $\Lambda^*(1520)$ and $\Sigma(1190)$, since the dominant decay channels are D-wave), and (c) to induced decay modes which occur only because of the nuclear environment (e.g. de-excitation processes such as $N\Lambda^*(1520)\rightarrow N\Lambda(1115)$ are now available and will contribute to the total decay rate).

(iii) changes in the branching ratios for its various decay modes, and the relative importance of the non-mesonic de-excitation processes just mentioned.

Some study of these effects has already been made for the state $\Delta(1236)$ in light nuclei. For mass number A=2, de-excitation process $n\Delta^{++}\rightarrow pp$ has been known for several decades from studies of the processes $\pi^+d\rightarrow n\Delta^{++}\rightarrow pp$ and $pp\rightarrow n\Delta^{++}$.

2. K^--NUCLEUS REACTIONS

2.1. <u>Capture from Rest</u> For nuclei Z≥2, the nuclear capture takes place from high-lying atomic orbitals for the K^-meson about the nucleus. After coming to rest, the K^- meson is captured into an atom, usually with large orbital angular momentum ℓ, and then cascades down through its atomic levels, by a sequence of Auger electron emission processes, until its wavefunction overlaps the nucleus sufficiently for the nuclear reaction to compete successfully with further atomic transitions. Since the s-wave interactions $\bar{K}N\rightarrow\Sigma\pi$ and $\Lambda\pi$ are exceedingly strong, the nuclear capture will occur from states with large ℓ, and therefore, as illustrated in Fig. 1, dominantly on the outermost part of the nuclear surface. In this situation, a substantial fraction of the Λ and Σ hyperons formed can escape from the nucleus without further interaction.

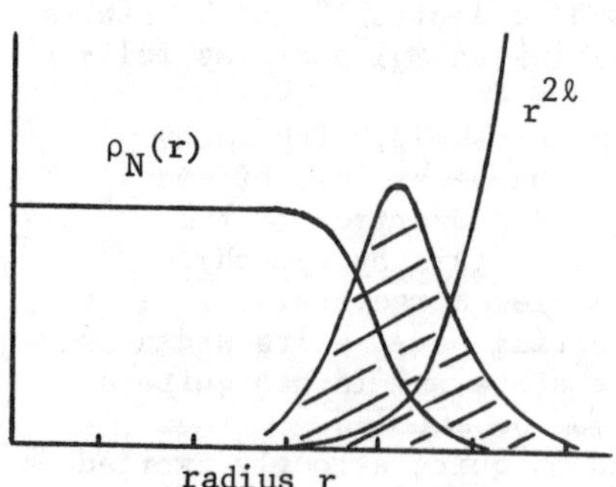

Fig. 1. The nuclear and K^- densities, showing the region where $\bar{K}N$ capture then occurs.

In general, the nuclear capture reaction involves a two-step process. First, the Λ or Σ hyperon is formed by the $\bar{K}N$ interaction on the surface. Second, if directed sufficiently inwards, the hyperon undergoes secondary interactions with the nucleons, especially the reactions $\Sigma N\rightarrow\Lambda N$ and the scattering process $\Lambda N\rightarrow\Lambda N$, often being trapped by the nucleus. Observation of the outgoing $\pi^{\pm}$ energy can tell us only about the first step, at best, and so does not reflect the energy of the final hypernuclear state. This is well illustrated by an old experiment[1], measuring π^- momenta following K^- capture from rest in Br, with the aim of determining D_Λ, the nuclear well depth for a Λ hyperon, from the upper limit of the π^- spectrum. About 2/3 of K^- captures in Br are known to result in 'Λ trapping', leading to the emission of a light hyperfragment or to the formation of a heavy residual hyperfragment (a so-called cryptofragment, which reveals itself through its energetic non-mesonic hypernuclear decay mode), yet not even one π^- meson had momentum corresponding to negative energy for an associated Λ hyperon, in a selected sample of about 300 π^- emissions. The result was that this experiment could give no useful limit on D_Λ.

It is of interest to mention here that non-mesonic two-nucleon capture processes have been observed for the K^- meson, for example,

$$\text{(a) } K^-pn \to p\Sigma^-, \qquad \text{(b) } K^-pp \to n\Sigma^+ \text{ and } p\Lambda. \tag{2.1}$$

They involve a large energy release, so they are characterised by a fast outgoing hyperon (momentum of order 500 MeV/c) and no $\pi^\pm$ emission. Relative to K^- capture leading to π^-emission, the branching ratios which have been measured are as follows:

	deuterium[2]	helium[3]	heavier nuclei (e.g. emulsion)[4]
B.R. (all two-nucleon processes)	1.2±0.1%	16±3%	20-30%

Little is known concerning the detailed spin and isospin characteristics of these capture processes (2.1).

2.2. Reaction in flight. In contrast with the above, in-flight reactions are essentially a volume effect, although the strong damping of the incident K^- wave and the outgoing π^- wave as they pass through nuclear matter does suppress the contributions from the central region of the nucleus and does strongly reduce the total reaction rate. Calculations for these reactions therefore require a good knowledge of K^- and π^- elastic scattering by the nuclear target considered. Although much data is now available for π^-, the first precise data available for K^- is that reported by Marlow et al. (the Carnegie Mellon-Houston-BNL collaboration) at this conference, on ^{12}C at 800 MeV/c. However, for the present purpose, much more data is needed for both K^- and π^- elastic scattering on the various nuclei of interest, and at least for the momentum range 650-1000 MeV/c.

The non-spin-flip amplitude for $K^-n \to \Lambda\pi^-$ effectively transforms the initial nucleon wavefunction $\psi_N(\underline{r})$ to a final Λ wavefunction of the form

$$\psi_\Lambda(\underline{r}) = e^{i\underline{q}\cdot\underline{r}} F(K^- \to \pi^-, \underline{r}) \psi_N(\underline{r}) \tag{2.2}$$

where $\underline{q}$ is the $K^- \to \pi^-$ momentum transfer, and the factor F describes the distortions of the K^- and π^- waves at the interaction point $\underline{r}$, due to their prior and posterior nuclear interactions. The kinematics of the 0° reaction aA→bB are well known[5] (cf. Fig.2). When

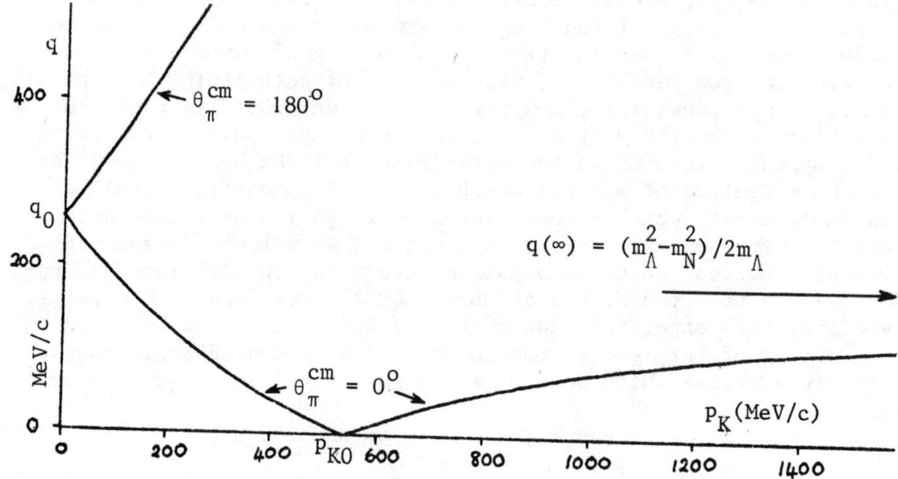

Fig. 2. Momentum transfer q in $\bar{K}N\to\Lambda\pi$ at 0° vs. p_K(lab.)

$(m_a-m_b)>(m_B-m_A)>0$, there exists a special laboratory momentum p_{a0} for which q=0 holds, given by

$$p_{a0} = \{\Pi(m_B-m_A\pm m_a\pm m_b)\}^{\frac{1}{2}}/2(m_B-m_A), \tag{2.3}$$

with value about 530 MeV/c for the case $K^-p\to\Lambda\pi^-$. For a nuclear target, if the nuclear distortions of the K^- and π^- waves could be neglected, so that F=1 held, then, for this special momentum p_{K0}, we would have $\psi_\Lambda(\underline{r}) \equiv \psi_N(\underline{r})$. In other words, the transition $N\to\Lambda$ would then deposit the Λ particle in precisely the same wavefunction as held for the initial neutron. If the nuclear potential acting on the Λ particle were the same as that for the neutron N, then the final state would be a hypernuclear eigenstate and it would exhaust the strength of this transition. As pointed out by Kerman and Lipkin[6], the transitions for each neutron of the initial nucleus would then be coherent and the reaction would be precisely a strangeness analogue transition. In reality, the nuclear well depth D_Λ is much less than D_N, and the detailed properties of the ΛN interaction are quite different from those for NN interaction (e.g. V(NN) has a very strong tensor component; $V(\Lambda N)$ is calculated to have a relatively weak tensor component). Nevertheless, the low-lying nuclear orbitals for the Λ particle are quite similar to those for the neutron, since the nuclear wells they see do have similar size and shape, despite their difference in depth. Even when the strangeness analogue state is split into a number of components by these deviations from an $np\Lambda$ SU(3) Sakata symmetry, the excitation of these separate components will still be strong, since there will still remain a considerable degree of coherence within each of them. With $\theta=0^o$ and $q\simeq 0$, these prominent excitations will all have the same spin-parity as the target nucleus, as pointed out by Feshbach and Kerman[7].

Still with $\theta=0^o$, other states can also be excited when the kinematic conditions are such that $q \neq 0$ and/or $F \neq 1$. However, these are still limited to transitions obeying the selection rule "parity change = $(-1)^{\Delta J}$". For example, we may have, at 0°,

$$K^- + {}^{12}C \to \pi^- + {}^{12}_{\Lambda}C^*(J^P=0^+,1^-,2^+,\ldots) \qquad (2.4)$$

Excitations of this kind will be discussed later in this session by Dr. Kilian and by Dr. May.

Next we may consider non-forward angles, $\theta \neq 0^o$. At present energies, these configurations involve much larger values for q, and the states $J^P=1^-,2^+,..$ are then much more strongly excited. Further, we can then study these excitations as function of q. For example, the excitation of ${}^{12}_{\Lambda}C^*(1^-)$ in (2.4) requires the transition $1p_N \to 1s_\Lambda$, and its rate is essentially proportional to the factor

$$|M_1(q)|^2 = |(\int \psi_{1p} e^{i\underline{q}\cdot\underline{r}} \psi_{1s} d^3\underline{r})|^2, \qquad (2.5)$$

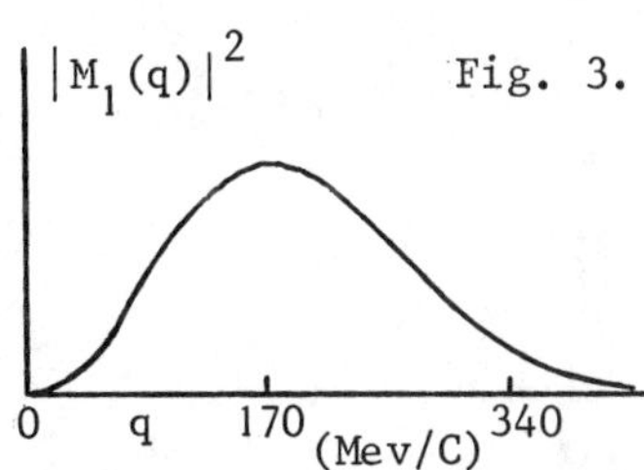

Fig. 3.

whose dependence on q is shown on Fig. 3, its peak value being for $q \simeq 170$ MeV/c. Dr. May will show data on non-forward $K^- \to \pi^-$ excitations, which follow this dependence quite well. Also, for $\theta \neq 0^o$, spin-flip transitions will become effective and these will allow the excitation of other hypernuclear states, with other spin-parity values.

The reaction process $K^-N \to \Sigma\pi^-$ leads to a somewhat similar story. The special momentum p'_{K0} is lower, at about 285 MeV/c. More important, the Σ hyperon reacts strongly with nuclei, through the endothermic reaction $\Sigma N \to \Lambda N$ which releases about 77 MeV, so that it is very doubtful whether any well-defined Σ-hypernuclear states should be expected to exist. Dr. Bertini will show us data on this question later in this session.

2.3 <u>Quasi-free reactions</u>. Quasi-free $K^-n \to \Lambda\pi^-$ reactions will occur when q is sufficiently large. They represent a background to the coherent excitations and the excitations of specific levels emphasized above. They are observed as a broad background hump in the 800 MeV/c experiments, despite the low value for q ($\simeq$ 100 MeV/c). The reason for their persistence to such low q is that there is no Pauli principle restriction on the states available for Λ and Σ hyperons within nuclear matter. They have a particular interest in that the median of their spectrum gives us directly the value for the well depth D_Λ or D_Σ, as the case may be. From the 800 MeV/c data appropriate to the reaction $K^-n \to \Lambda\pi^-$, we can already deduce the value $D_\Lambda = 31 \pm 2$ MeV, which agrees quite well with older estimates based on quite different phenomena, namely measurement of the energy released in the decay process $\Lambda \to p\pi^-$ for Λ particles in the ground state of a large ($A \simeq 100$) hypernucleus.

2.4 <u>Reaction Cross sections</u>. For the 0° reactions, which are those mostly studied to date, the cross section is proportional to $d\sigma(0^o)/d\Omega$ for $\bar{K}N \to \Lambda\pi$ or $\Sigma\pi$, as appropriate. The elementary cross sections for $\Lambda\pi^-$, $\Sigma^o\pi^-$ and $\Sigma^+\pi^-$ vary greatly, and rather independently of one another, over the K^- momentum range 500 to 1500 MeV/c. The $\Lambda\pi^-$ cross section is largest in the K^- momentum range 700 to 900 MeV/c, which is one of the reasons why the experiments to date have concentrated on this momentum range. The net cross section is strongly affected also by the nuclear absorption of the K^- and π^- mesons, as mentioned above. At present, the degree of agreement between the measured and calculated cross sections is not completely satisfactory[9].

3. K^--DEUTERIUM REACTIONS AND THE HYPERON-NUCLEON INTERACTIONS

The reaction $K^-d\rightarrow\Lambda p\pi^-$ for stopping K^- mesons, has been studied in much detail by Tan[9] following the early survey by Dahl et al.[10]. The final $p\pi^-$ state is limited to T=1/2, so that $\Delta(1236)$ can play no direct role in the final configuration. The production and decay of $\Sigma^*(1385)$ plays only a minor role. However, the Λp mass distribution (see Fig.4) shows striking strucure, both in the region $m(\Lambda p)\sim m_N+m_\Lambda$ and in the region $m(\Lambda p)\sim m_N+m_\Sigma$.

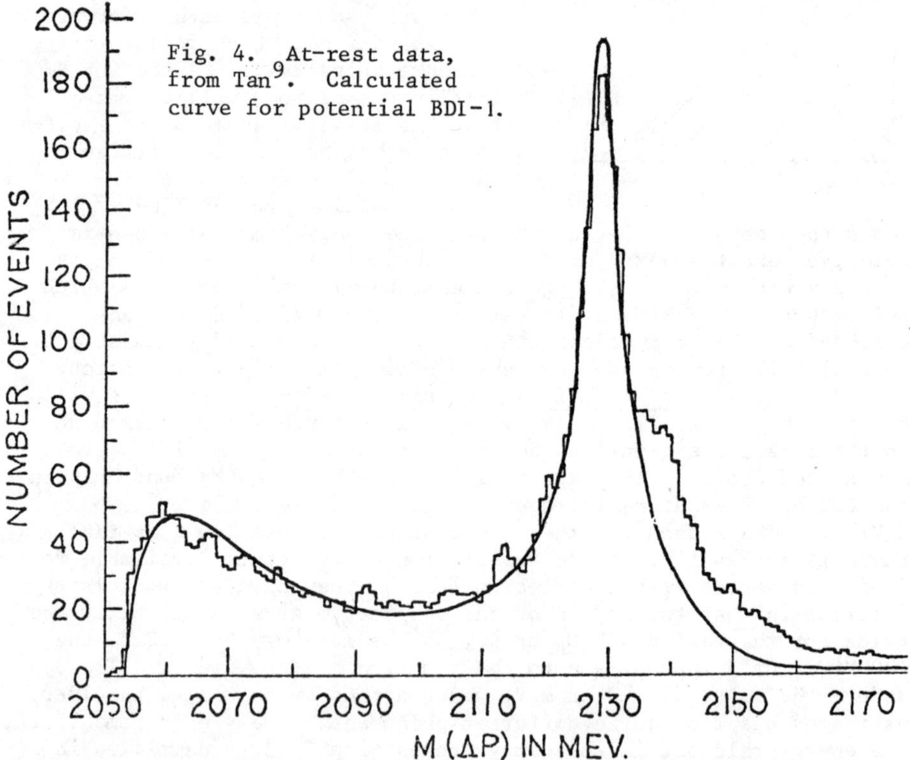

Fig. 4. At-rest data, from Tan[9]. Calculated curve for potential BDI-1.

The in-flight reaction $K^-d\rightarrow\Lambda p\pi^-$ has been studied by a number of groups, over a wide range of momenta. By far the largest body of data is that of Braun et al.[11] at 800 MeV/c. The small angle data ($\cos\theta_\pi>0.9$, thus including θ_π as large as 20°) shows quite similar behaviour, with peaks in the same mass regions; as shown in Fig. 5, the large angle data still show a substantial peak near the limit (m_N+m_Λ), but no strong effect at the mass (m_N+m_Σ).

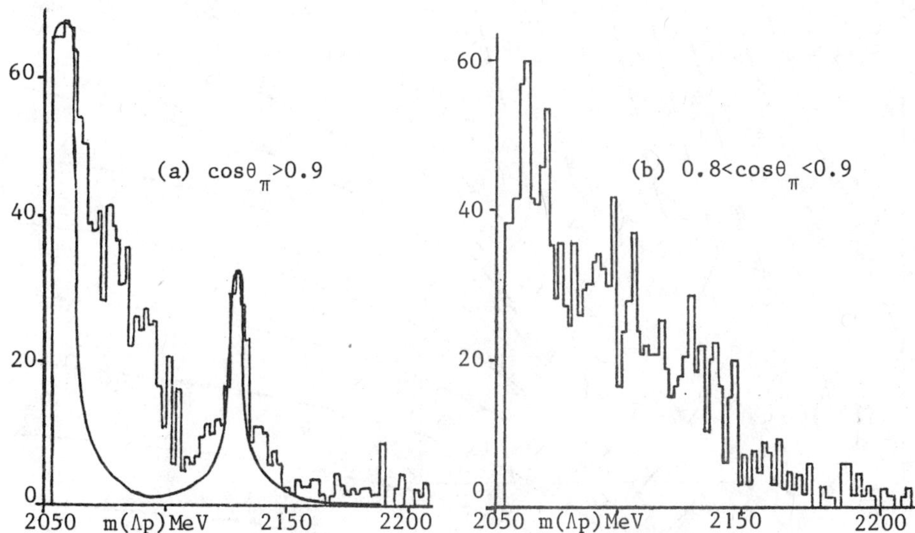

Fig. 5. The m(Λp) distributions given by Braun et al.[10] for $K^-d\to\Lambda p\pi^-$ at 700 MeV/c.

It is interesting to examine the 0° kinematics as function of the mass m(Λp), as is shown on Fig. 6. The special momentum p_{KO} falls with increasing m(Λp) from the value 500 MeV/c at (m_N+m_Λ) until it reaches zero, beyond which it no longer exists. At 700 MeV/c, the momentum transfer at 0° ranges from q=42 MeV/c at (m_N+m_Λ) to 115 MeV/c at (m_N+m_Σ). Since q is small, and the c.m. momenta k_Λ and k_Σ are small, we are concerned only with non-relativistic phenomena and low partial waves in these systems. Since the primary amplitudes have no spin-flip at 0°, the hyperon-nucleon interactions of most importance here are those in the $^3S_1$ state. The most immediate interpretations for these two structures in Fig. 4 are as follows:

(a) ΛN threshold. This region is excited primarily by the $\bar{K}N\to\Lambda\pi$ amplitude and the low-mass peak is due to the nuclear attraction in the $^3S_1$ Λp system. As remarked by Tan, the shape observed is quite well fitted by a $^3S_1$ scattering length a(ΛN) ∿ 3fm. This is quite compatible with the low-energy Λp scattering data.

(b) ΣN threshold. Here both $\bar{K}N\to\Lambda\pi$ and Σπ amplitudes can contribute, in principle. However, the extrapolation from the ΛN threshold indicates that the $\bar{K}N\to\Lambda\pi$ amplitude contributes relatively little to the large peak observed at the ΣN threshold and this view is supported by the detailed calculations made by Morris[12] and by Hemming[13] at Oxford. Hence, the natural interpretation is that the capture interaction $\bar{K}N\to\Sigma\pi$ first excites the ΣN channel and that the nuclear interaction $\Sigma N\to\Lambda N$ then leads to the final state observed,

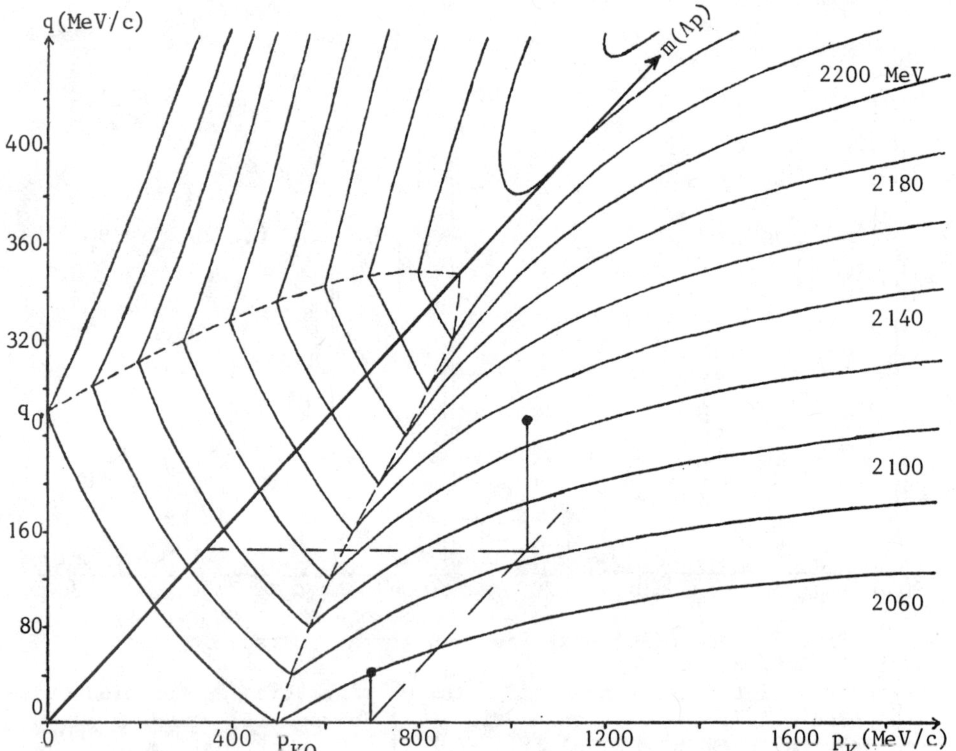

Fig. 6. Momentum Transfer in reaction $K^-d \to \Lambda p \pi^-$ as function of m(Λp) and p_K. The dashed line depicts q_0 and p_{K0} as function of m(Λp).

a two-step process first discussed by Karplus and Rodberg[14] and by Kotani and Ross[15]. Above the ΣN threshold, the ΣN→ΛN transition rate is given by the expression

$$R = \{v_{\Sigma N}\sigma(\Sigma N \to \Lambda N)\} \propto b/|(1-ik_\Sigma(a+ib))|^2 \tag{3.1}$$

where (a+ib) is the $T=\frac{1}{2}$ ΣN scattering length. The capture rate in deuterium is given by multiplying R by a factor $|D|^2$ which measures the mean intensity of the intermediate Σ wave at the other nucleon. In zero-range approximation, and neglecting d-waves we have

$$D(q,k_\Sigma) = \int \psi_d(\underline{r})\exp(i\underline{q}.\underline{r})\{\exp(ik_\Sigma r)/r\}d^3r/4\pi. \tag{3.2}$$

Near the ΣN threshold, this may be expanded as

$$D(q,k_\Sigma) = D(q_t,0)\{1+ia^*k_\Sigma+o(k_\Sigma^2)\} \tag{3.3}$$

where q_t is the threshold value, and the length a^* is characteristic of the deuteron (and of the hyperon-nucleon forces, to a lesser extent), having value about 2.4 f. Just above threshold, the leading term of the rate $R|D|^2$ is

$$R|D|^2 = b|D(q_t,0)|^2\{1-2bk_\Sigma+o(k_\Sigma^2)\}, \tag{3.4}$$

which implies a rapid cusp-like fall in the rate, as function of $m(\Lambda p)$. The expressions (3.1) and (3.2) are valid also below threshold, when k_Σ is replacedby $+i|k_\Sigma|$, so that the leading term there is

$$R|D|^2 = b|D(q_t,0)|^2\{1-(a+a^*)|k_\Sigma|+o(k_\Sigma^2)\}, \tag{3.5}$$

This rate has a cusp-like behaviour below threshold, rising or falling as $m(\Lambda p)$ falls below (m_N+m_Σ) according as $(a+a^*)$ is negative or positive. Thus, the simplest interpretation would be that $(a+a^*)>0$ holds and that the $m(\Lambda p)$ distribution has an upward cusp at the ΣN threshold.

However, this is far from being the whole story. Why is this cusp so prominent? This has to do with the forces acting in the hyperon-nucleon system. We have to consider four coupled channels, ΛN and ΣN for the $^3S_1$ and $^3D_1$ states. If $m_\Sigma = m_\Lambda = m_N$ were to hold, then SU(3) symmetry would tell us that the system would have a bound state, a strange counterpart to the deuteron, which would appear in the hyperon-nucleon S-matrix as a pole on the real energy axis, below the ΛN threshold energy. When $m_\Sigma \neq m_\Lambda \neq m_N$, the SU(3) symmetry is broken, and there does not then necessarily exist any bound state or resonance in the hyperon-nucleon system. Neverthess the interactions are very strong and there will generally be a pole in the S-matrix amplitude on the unphysical sheets of the complex energy plane (but not necessarily adjacent to the physical axis), which is due to the migration of the deuteron pole as the masses m_Λ and m_Σ are varied from m_N to their physical values. The branch cuts and possible pole locations are shown on Fig. 7. Case (a)

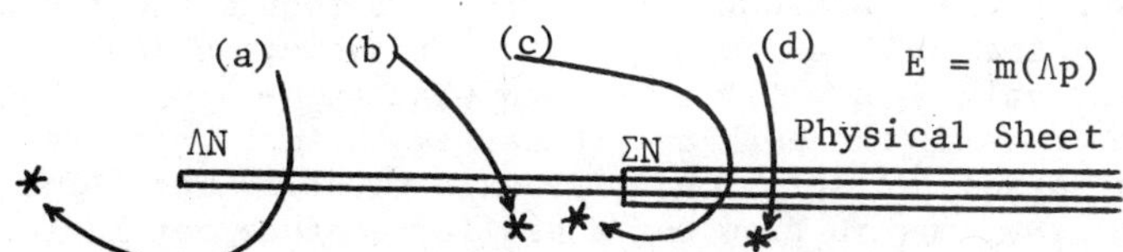

Fig. 7. The branch cuts of the complex energy plane and possible pole locations, assuming C.I.

would not account for the largeness of the cusp since the pole is then far from the ΣN threshold. Cases (b) and (d) correspond to a resonance for a mass $m(\Lambda p)$ below or above the ΣN threshold. Case (c) gives no resonance, but the pole lies close to the ΣN threshold and can generate a large cusp. Which situation applies physically depends on the details of the forces acting.

There is one further complication. The ΣN threshold is split appreciably by electromagnetic mass differences, the Σ^+n threshold being at 2128.9 MeV and the Σ^0p threshold at 2131.4 MeV. The "at rest" data of Tan and the "in-flight" data of Braun et al. both locate the peak at 2129 MeV, as do all of the other experiments, although with poorer statistics. Hence the calculations must be carried out with an explicit treatment of all three channels Λp, $\Sigma^+ n$ and $\Sigma^0 p$.

For these calculations of K^-d capture, there are a number of One Boson Exchange potentials available for test, due primarily to Brown, Downs and Iddings (BDI)[16] and to Nagels, Rijken and DeSwart (NRS)[17]. These potentials fit all the data available on hyperon-nucleon scattering, and include the exchange of all the vector, pseudoscalar and scalar mesons. For the most recent potentials[18], NRS-E and NRS-78, all the available SU(3) relationships between the meson-baryon couplings are used as constraints to relate these potentials with those known for nucleon-nucleon scattering, so that they then fit all of the baryon-baryon scattering data. The various potentials proposed differ qualitatively, as regards the pole location. In the limit of charge-independence (C.I.), BDI-1 and NRS-78 correspond to the case (b) illustrated in Fig. 7, while BDI-2 and NRS-D appear to correspond to case (c). However, whichever sheet the pole is on, it still lies close to the ΣN threshold.

Calculations have been made for the $m(\Lambda p)$ spectrum in the reaction $K^-d \rightarrow \Lambda p \pi^-$ using these potentials.[12,13] For illustration, the results with potential BDI-1 are shown on Figs. 4 and 5. The calculated curves have no free parameters, except for the over-all normalization.

For the at-rest data, the fit to the ΛN threshold region could be improved by increasing the $^3S_1$ ΛN scattering length somewhat, relative to the value used by BDI to determine their potential parameters. The prominent cusp occurs at the $n\Sigma^+$ threshold, in accord with the data; only a weak cusp is predicted for the $p\Sigma^o$ threshold, as shown in Fig. 8. The ratio of the ΣN and ΛN threshold enhancements is also about correct. In the C.I. limit, BDI-1 predicts an S-shaped cusp, as can be deduced from Fig.9, although the calculated peak lies less than 0.01 MeV below the ΣN threshold; we note that the calculated value $a_{\Sigma N} = -3.3$ f is such that $(a_{\Sigma N} + a^*) < 0$. This circumstance broadens the peak and increases the net rate over the ΣN threshold region, in agreement with experiment. The BDI-2 potential predicts an upward cusp and a smaller enhancement in the ΣN threshold region, relative to the ΛN threshold region. When the two ΣN thresholds are separated, the $n\Sigma^+$ cusp predicted by BDI-1 is no longer S-shaped, but the general agreement persists. The interpretation recently favoured by Dosch and Hepp[19] assumes the existence of a peak below the $n\Sigma^+$ threshold. This can certainly be arranged, by modification of BDI-1 for example. Nishimura[20] has recently used s-wave separable potentials to provide such a fit to all of the data, with a peak at 0.5 MeV below this threshold.

These calculations provide no interpretation for the shoulder in Tan's data, on the upper side of the peak. Tan[9] suggested that this might be a resonance effect; others have suggested that it might be a d-wave effect. Morris[12] suggested that it might be specific to the at-rest data, representing p-orbit capture of the K^- meson, although this is not the conventional view of K^- capture in Z=1 nuclei.

For the in-flight data, the BDI-1 calculation predicts the ΛN and ΣN threshold regions well but does not give a good account of the spectrum for intermediate m(Λp). However, these data include all events with out-going pion angle θ_π up to 20°. Since q can then range up to 240 MeV/c, the p- and d-waves for the Λp system will be excited strongly, relative to their excitation for $\theta_\pi=0°$, and this is just the m(Λp) range to which these higher waves are expected to contribute.

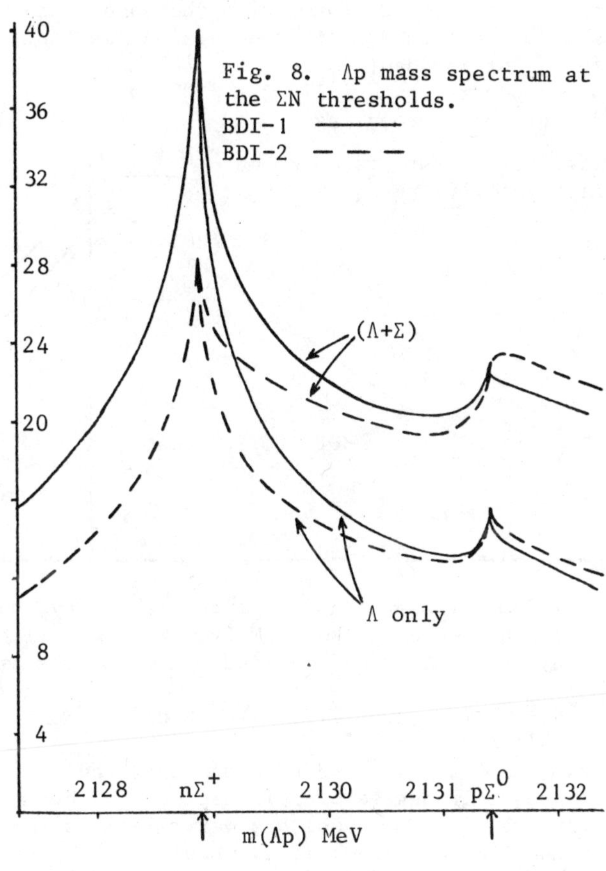

Fig. 8. Λp mass spectrum at the ΣN thresholds.

Finally, we remark that the counter experiments measure only the angle and the momentum of the outgoing π^-. Since the hyperon is not observed, their measurements will include also the final states $n\Sigma^+\pi^-$ and $p\Sigma^0\pi^-$, when the "missing mass" m lies above the appropriate threshold value. The mass distributions calculated for BDI-1 and BDI-2 under these circumstances are shown in Fig. 8. We note that, for BDI-2, the rate rises as m increases above the upper threshold, giving an S-shaped cusp there, a behaviour which is permissible only for a total rate. No data is available yet for counter experiments on deuterium, but Dr. Bertini will present some data of this kind for heavier targets, such as ^{9}Be and ^{12}C.

4. HYPERNUCLEAR SPECTROSCOPY

The strangeness exchange reactions $K^-\rightarrow\pi^-$ are now giving some significant data on excited states of Λ hypernuclei. Dr. Kilian will show spectra for a number of Λ hypernuclei, obtained with the 0° reaction. For example, four states are now known for $^{16}_{\Lambda}$O. The two most prominent states are believed to be strangeness analogue states, with $J^P=0^+$ as holds for the target nucleus ^{16}O;

the others are the ${}^{16}_{\Lambda}O$ ground state, believed to have $J^P=1^-$, and the first excited state with the same spin-parity. Dr. May will show a spectrum for ${}^{12}_{\Lambda}C^*$, determined as function of momentum transfer q for incident K^- momentum 800 MeV/c. Dr. Gal will discuss such spectra, especially for ${}^{16}_{\Lambda}O^*$, with an emphasis on the role of ΛN spin-orbit couplings.

Fig. 9. Capture rate vs. k_Σ near ΣN(C.I.) threshold.

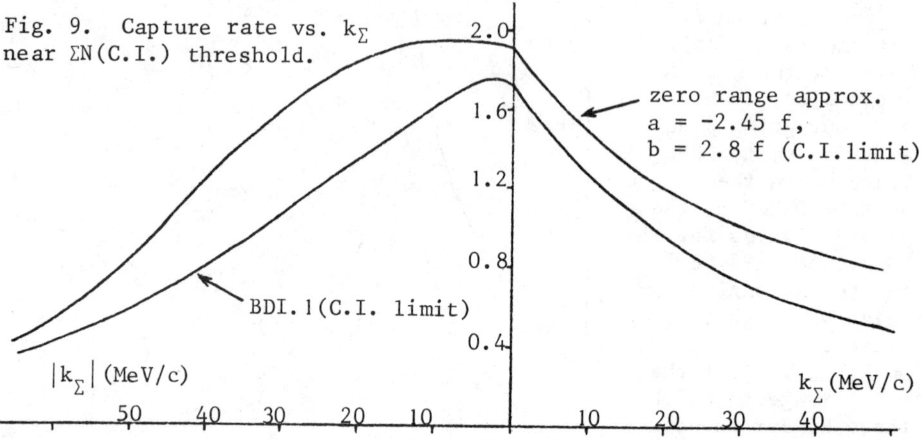

Some evidence on Λ-hypernuclear γ-rays has been reported at this conference. The CERN-Lyon-Warsaw collaboration has observed γ-rays from the $J^P=1^+$ excited states of the A=4 hypernuclei:

$${}^{4}_{\Lambda}H^* \rightarrow {}^{4}_{\Lambda}H+\gamma: \quad 1.04\pm0.04 \text{ MeV}; \qquad {}^{4}_{\Lambda}He^* \rightarrow {}^{4}_{\Lambda}He+\gamma: \quad 1.15\pm0.04 \text{ MeV}$$

From these results, we conclude that the ΛN interaction does have a considerable degree of spin-dependence, but that its charge-symmetry-breaking component has very little spin-dependence, contrary to the simplest theoretical models.

We may view this experiment as the first step toward the observation of γ-ray emission in coincidence with the $K^-\rightarrow\pi^-$ reaction in flight, the important next step in Λ-hypernuclear spectroscopy. One such experiment has been carried out[21], using ^{7}Li target, but its results require confirmation. Such experiments will be of the greatest interest, for determining not only the spectrum of excited states, but also their spins. At present, the only evidence we have on the spins of excited states is that reported here by Dr. May, who shows that the excitation rates for the two lowest states of ${}^{12}_{\Lambda}C$ have the q-dependence (2.2), which implies that $J^P=1^-$ for them both. This area of Λ-hypernuclear spectroscopy deserves to be one of the growth areas of intermediate energy physics in the coming few years.

REFERENCES

1. F. Abraham et al., Nucl. Phys. 81 (1966) 609.
2. V.R. Veirs and R.A. Burnstein, Phys. Rev. D1 (1970) 1883.
3. P.A. Katz et al., Phys. Rev. D1 (1970) 1267.

4. D.A. Evans et al. Proc. Roy. Soc. A262 (1961) 73.
5. R.H. Dalitz and A. Gal, Ann. Phys. (N.Y.) 116 (1978) 167.
6. A.K. Kerman and H.J. Lipkin, Ann.Phys.(N.Y.) 66 (1971) 738.
7. H. Feshbach and A.K. Kerman, in "Preludes in Theoretical Physics" (eds. A. De-Shalit et al., North-Holland, Amsterdam, 1966) p.260
8. G.N. Epstein et al., Phys. Rev. C17 (1978) 1501.
9. T.H. Tan, Phys. Rev. Letters 23 (1969) 395.
10. O. Dahl et al., Phys. Rev. Letters 6 (1961) 142.
11. O. Braun et al., Nucl. Phys. B124 (1977) 45.
12. E. Morris, "Final state interactions in $\bar{K}$-deuterium at rest", D.Phil. thesis (Oxford University, 1974).
13. C. Hemming, "$K^- \to \pi^-$ reactions at 0° on deuterium and other nuclei" D.Phil. thesis (Oxford University, 1978).
14. R. Karplus and L. Rodberg, Phys. Rev. 115 (1959) 1058.
15. T. Kotani and M. Ross, Nuovo Cim. 14 (1959) 1282.
16. J.T. Brown, B.W. Downs and C.K. Iddings, Ann.Phys.(N.Y.) 60 (1970) 148.
17. M.M. Nagels, T.A. Rijken and J.J. DeSwart, Phys. Rev. D15 (1977) 2547.
18. M.M. Nagels, T.A. Rijken and J.J. DeSwart, preprint THEF-NYM-78.4 (Nijmegen University, 1978)
19. H.G. Dosch and V. Hepp, "Analysis of the Λp enhancement at 2.129 GeV in the reaction $K^-d \to \pi^- \Lambda p$", preprint TH.2310-CERN (July, 1977)
20. A. Nishimura, "An analysis of the Λp enhancement", preprint UT-309 (Tokyo, July, 1978).
21. J.G. Herrera et al., Phys. Rev. Letters, 40 (1978) 158.

PROPERTIES OF THE ELEMENTARY KN AND $\bar{K}$N INTERACTIONS*

C. B. Dover**
University of Pennsylvania
Philadelphia, Pennsylvania 19104
and
Princeton University
Princeton, New Jersey 08540

ABSTRACT

Kaon beams offer exciting prospects for the study of both nuclear and hypernuclear physics. Experiments on hypernucleus formation via the (K^-,π^-) reaction, as well as elastic and inelastic scattering of $K^{\pm}$ from nuclei, are already underway. The theoretical analysis of such experiments requires an understanding of the underlying two-body $K^{\pm}N$ interaction. We review here the main features of these elementary amplitudes, as revealed through phase shift analyses and meson exchange models. We indicate how the properties of the two body interaction (isospin dependence, Y* and Z* resonance formation, strangeness exchange, etc.) are reflected in reactions induced by kaons in nuclei.

*Invited talk presented at the Second International Topical Conference on Meson - Nuclear Physics, Houston, March 5 - 9, 1979.

**Permanent Address: Brookhaven National Laboratory, Upton, New York; partially supported by the Department of Energy under Contract No. EY-76-C-02-0016. Also partially supported by the National Science Foundation at the University of Pennsylvania.

ISSN:0094-243X/79/540634-24$1.50

The existence of secondary kaon beams at high energy accelerators (Brookhaven, CERN, KEK) offers the intriguing possibility that such beams may be used for medium energy nuclear physics experiments in addition to their traditional application to the study of elementary particle processes. Some nuclear experiments with kaons have already been performed, in particular, the investigation of hypernuclear spectroscopy via the (K^-,π^-) reaction [1,2]. The potential virtues of both K^- and K^+ as nuclear structure probes, as well as their relation to more familiar medium energy probes such as π, e, p, and α, have been recently delineated[3,4]. This discussion hinges to a large extent on the properties of the elementary $K^\pm$-nucleon interactions. In this talk, we review the present level of experimental and theoretical understanding of the $K^\pm N$ amplitudes.

The average strong interaction amplitudes $(f_{\pi^\pm n} + f_{\pi^\pm p})/2$ for pions on nucleons are the same for π^+ and π^-, assuming charge independence. Both π^+ and π^- interactions with nucleons are dominated by s-channel resonance formation for low energies, $p_{lab} \leq 1 - 2$ GeV/c. For instance, the well-known (3,3) resonance dominates the pion-nucleon interaction below 300 MeV or so. In sharp contrast to the situation for pions, the average interactions of $K^\pm$ with nucleons are very different in character. The low-energy K^--nucleon cross section is resonance-dominance and has an average value of the order of 40 mb, typical of hadronic cross sections. The total cross sections for K^--nucleon scattering in states of pure isospin $I = 0,1$ are shown in Fig. 1 in the low-energy region $p_{lab} \leq 1$ GeV [5]. The presence of a number of $I = 0$ resonances (Y_0^*) and $I = 1$ resonances (Y_1^*) is clear from the data. The energies, widths and quantum numbers of the well-established Y^* resonances[6] are given in Table 1. On the other hand, the low energy K^+-nucleon cross sections are small ($\sim$10 mb) and rather constant in the region $p_{lab} \leq 700$ MeV/c below the onset of strong inelastic processes. This behavior is shown in Fig. 2. At higher energies ($p_{lab} \geq 700$ MeV/c), the K^+-nucleon cross section experiences a sharp rise, and the possibility of exotic Z_0^* and Z_1^* resonances exists. The total and inelastic cross sections in this region are shown in Fig. 3.[6]

The fact that the K^+ and K^- have opposite strangeness $S = \pm 1$ plays a crucial role in understanding these qualitative differences. In the quark model, the K^- is made up of a strange quark λ $(S = -1)$ and a non-strange antiquark $\bar{p}$. In Fig. 4, we illustrate how the $S = -1$ Y^* resonances are formed in a natural way through the annihilation of a quark-antiquark pair $\bar{q}q$. The Y^* resonances are ordinary three quark composites. In contrast, we see that possible <u>exotic</u> K^+-nucleon resonances (called Z^*), having $S = +1$, cannot be composed of three quarks, the minimum complexity being $\bar{\lambda}qqqq$. Such an <u>exotic</u> configuration would correspond to the {10} and {27} representations of SU(3) for I = 0 and 1, respectively. Due to these striking differences in the elementary K^+ and K^- interactions with nucleons, the $K^\pm$ also offers quite distinct possibilities for probing the nucleus.

TABLE 1

Y* Resonance Energies, Widths, and Quantum Numbers*

State	$L_{I,2J}$	Γ(MeV)	Dominant Channel
Σ (1385)	P_{13}	35 - 40	$\Lambda\pi$
Λ (1405)	S_{01}	40	$\bar{K}N$
Λ (1520)	D_{03}	15	$\bar{K}N$, $\Sigma\pi$
Λ (1600)	P_{01}	60	$\Sigma\pi$, $\bar{K}N$
Σ (1660)	P_{11}	50 - 100	$\Sigma\pi$
Λ (1670)	S_{01}	25 - 50	$\Sigma\pi$
Σ (1670)	D_{13}	50 - 60	$\Sigma\pi$
Λ (1690)	D_{03}	30 - 80	$\Sigma\pi$, $\bar{K}N$
Σ (1750)	S_{11}	60 - 100	$\bar{K}N$
Σ (1765)	D_{15}	120	$\bar{K}N$
Λ (1815)	F_{05}	70 - 100	$\bar{K}N$
Λ (1830)	D_{05}	60 - 100	$\Sigma\pi$
Σ (1840)	P_{13}	120	$\bar{K}N$
Λ (1860)	P_{03}	40 - 100	$\bar{K}N$
Σ (1915)	F_{15}	70 - 130	mixed
Σ (1940)	D_{13}	100 - 300	mixed
Σ (2030)	F_{17}	100 - 200	mixed
Λ (2100)	G_{07}	100 - 200	$\bar{K}N$
Λ (2110)	F_{05}	140 - 200	$\Sigma\pi$, $\bar{K}N$

*We list only the well established resonances of Trippe et al.[6], plus Λ (1600) and Σ (1660), which are more firmly established by recent analyses[22], as well as Σ (1840), which is the lowest lying candidate for a P_{13} state[6]. In the last column, we list the dominant decay mode for each resonance; in cases where two modes are prominent, the larger one is listed first.

We first discuss the main features of the K^+-nucleon scattering amplitudes, starting with a look at the data and results of phenomenological phase shift analyses, and then proceeding to a discussion of models with more theoretical input.

A large amount of experimental work has been done on the K^+p and K^+d systems in the energy region below 2 - 3 GeV. This is the region of principal interest here. These measurements include total cross sections, differential cross sections for elastic and charge exchange scattering, and the corresponding polarizations. The K^+n cross sections are obtained from the K^+d results by unfolding the effects of Fermi motion and binding effects in the deuteron.

The elastic scattering angular distributions for K^+p scattering are shown in Fig. 5 for $p_{lab} \leq 2$ GeV/c [7]. The differential cross sections are essentially isotropic up to about 800 MeV/c, consistent with a dominant $S_{1/2}$ amplitude for I = 1. Above this momentum, higher partial waves ($L \geq 1$) enter the problem and a diffraction-like minimum develops. Note that $d\sigma/d\Omega$ exhibits a backward peak; this is characteristic of baryon (for instance, Λ) exchange.

In Fig. 6, we display the differential cross sections for K^+p at very low momenta of 145 and 175 MeV/c [8]. The constructive interference of Coulomb and nuclear amplitudes at forward angles is clearly evident. This interference is in fact constructive in the <u>entire</u> low energy region[8,9] and demonstrates that the dominant $S_{1/2}$ <u>I = 1</u> strong amplitude is <u>repulsive</u>.

For $p_{lab} \leq 800$ MeV, the total K^+p cross section consists almost entirely of elastic scattering. Even though the threshold for single pion production is at $p_{lab} \approx 525$ MeV/c, significant inelasticities do not develop until one is above the K Δ (1236) and K*(890)N thresholds at roughly 870 and 1075 MeV/c, respectively. Recall that K*(890) and Δ(1236) correspond to S-wave πK and p-wave πN resonances. These two channels dominate the total inelastic cross section below 1.5 GeV/c, as shown qualitatively in Fig. 7 [10]. Note that for I = 1, the KΔ process is most important, although interference with K*N is observed[11], while for I = 0, KΔ does not enter.

After the onset of strong inelastic processes around 800 MeV/c, the total elastic cross sections drop continuously. This is shown in Fig. 7 for K^+p. The cross section σ_{ce} for the charge exchange process $K^+n \rightarrow K^0p$ is shown in Fig. 8 [12]. Charge exchange is seen to be relatively unimportant at low momenta relative to elastic scattering. Above 800 MeV/c, charge exchange, like elastic scattering, decreases fairly rapidly. The region around 600 - 800 MeV/c thus seems optimum for charge exchange scattering on nuclei.

The K^+N system is a natural candidate for a phase shift analysis, the scattering being dominated by only a few partial waves (S and P waves up to 800 MeV/c or so). Numerous such analyses have been carried out for the K^+p system[13-15]. Extraction of the K^+n cross sections from K^+d experiments has also led to a number of analyses of the I = 0 amplitudes[16-18], keeping the I = 1 amplitudes fixed. Recently a simultaneous analysis of both I = 0 and 1 partial waves has appeared[19].

The partial wave expansion for the non-spin flip K^+N scattering amplitude assumes the form

$$f^I(s,\theta) = \frac{1}{k} \sum_{\ell=0}^{\infty} [(\ell+1) f^I_{\ell+}(s) + \ell f^I_{\ell-}(s)]\, P_\ell(\cos\theta) \tag{1}$$

where $\ell^{\pm}$ refers to $j = \ell \pm 1/2$, respectively. The inelasticity parameter η and the real phase shift δ are defined for each partial wave by

$$f^I_{\ell\pm}(s, \theta = 0) = (\eta^I_{\ell\pm}\, e^{2i\delta^I_{\ell\pm}} - 1)/2i \tag{2}$$

In phase shift analyses of K^+N data, one strategy is to perform energy independent analyses at a series of energies, and then try to join the solutions in a continuous fashion. An alternative is to perform an energy dependent analysis in which one parametrizes the K^+N amplitudes as continuous functions of energy and determines the free parameters by a χ^2-fit to data extending over a range of energies. We now discuss two representative models, due to B. Martin[19] and the BGRT group [13,16].

The BGRT group parametrizes $\delta_{\ell\pm}$ by a power series expansion

$$\delta_{\ell\pm}(s) = k^{2\ell+1}\,(A_{\ell\pm} + B_{\ell\pm} k^2 + C_{\ell\pm} k^4) \tag{3}$$

and η by a function which decreases smoothly above the inelastic threshold

$$\eta_{\ell\pm}(s) = (1 + y^2_{\ell\pm}(s))^{-1}$$

$$y_{\ell\pm}(s) = \begin{cases} 0 & s \le S_\ell \\ \left(\frac{s - S_\ell}{s}\right)^{3/2} \left[\left(D_{\ell\pm} + E_{\ell\pm}\left(\frac{s - S_\ell}{s}\right)\right)\right] & s > S_\ell \end{cases} \tag{4}$$

where S_ℓ = 2.48, 2.72, and 2.88 GeV^2 for ℓ = 0, 1, 2 . This parametrization was introduced by Lea, Martin and Oades[20], who performed the first extensive phase shift analysis of K^+p data.

Martin[19] has introduced a parametrization of the inverse amplitude $f^{-1}_{\ell\pm}$

$$f^{-1}_{\ell\pm}(s) = D_{\ell\pm}(s) - i\,(1 + \theta\,(k - k_{in})\, R^2_{\ell\pm}(k)) \tag{5}$$

where

$$D_{\ell\pm}(s) = C^{-1}_\ell(k) \sum_{n=0}^{N_\ell} a^{\ell\pm}_n (k/k_{max})^n$$

$$C_\ell(k) = k^{2\ell+1} / (1 + (k/k_{in})^{2\ell+1}) \tag{6}$$

$$R_{\ell\pm}(k) = \sum_{m=1}^{M_\ell} b^{\ell\pm}_m \left(\frac{k - k_{in}}{k_{max} - k_{in}}\right)^m$$

The factor $C_\ell(k)$ contains the correct penetrability factor $k^{2\ell+1}$ for small k, but smoothly approaches a constant for large k. This is an improvement on Eq. (3), which has no cutoff factor for large k. This form of the amplitudes has been used[19] to perform an extensive analysis of all the available K^+p and K^+d data below P_{lab} = 1.5 GeV. This represents the most complete simultaneous analysis of I = 0 and 1 amplitudes to date.

We now discuss the qualitative features of the K^+N phase shift analyses. First consider the region near threshold. Here $\eta = 1$ and the phase shifts δ can be well represented by the effective range expansion

$$k^{2\ell+1} \cot \delta^I_{\ell\pm} = \frac{1}{a^I_{\ell\pm}} + \frac{1}{2} r^I_{\ell\pm} k^2 \tag{7}$$

Values for $a^I_{\ell\pm}$ and $r^I_{\ell\pm}$ have been obtained in a variety of analyses[8,9,13-15,19,20]. For I = 1, we have no ambiguity about the signs of the low energy K^+N I = 1 amplitudes

S1/2	repulsive		
P1/2	repulsive	(I = 1)	(8)
P3/2	attractive		

In addition, the value of the combination $a_{P1/2} + 2a_{P3/2}$, which enters into the total cross section for $\ell = 1$, is very small.

For I = 0, the low energy S1/2 interaction is much weaker that for I = 1, but even its sign is not yet determined. The results of extrapolations of phase shift analyses to zero energy stand in sharp contrast to recent theoretical speculations of Cottingham and collaborators[21], who use a multichannel meson exchange model to predict that the threshold S1/2 I = 0 interaction is <u>strongly attractive</u>. The signs of the I = 0, $\ell = 1$ amplitudes at threshold correspond to

P1/2	attractive	(I = 0)	(9)
P3/2	repulsive		

We now discuss some of the results of phase shift analyses at higher energies, starting with I = 1. In Fig. 9, we show the momentum dependence of the I = 1 phase shifts δ and inelasticities η corresponding to the parametrization of Eqs. (3) and (4) of the BGRT group[13]. The pattern of signs given by Eq. (8) is preserved in the entire region below 1.5 GeV, with a dominant repulsion in the S1/2 partial wave. The different solutions (i) to (iv) are seen from Fig. 9 to be very similar below 800 MeV/c. At higher momentum, they differ primarily in the way in which the inelastic cross section is made up out of the various channels. All solutions have strong inelasticity in the I = 1 P3/2 channel.

In Fig. 10, we display for comparison the results of the most recent analysis of K^+p data by Cutkosky et al.[15] The plot shows the real and imaginary parts of the S1/2, P1/2 and P3/2 amplitudes f, both separately and in the form of Argand diagrams. These amplitudes display the same general features as those in Fig. 9, repulsion in S1/2 and P1/2 and attraction in P3/2, with much more inelasticity in P3/2 than S1/2 as we pass above 800 MeV/c.

From Figs. 9 and 10, we see that the only candidate for an I = 1 exotic K^+N resonance is in the P3/2 channel. There has been a great deal of discussion on whether the attraction which occurs in this channel corresponds to a resonance pole. Several analyses[14,19,22] have led to an extraction of a resonance energy E_R and width Γ for the P_{13} wave. Rough average values are

$$E_R \approx 1800 \text{ MeV}$$
$$\Gamma \approx 200 \text{ MeV} \qquad (10)$$

The elasticity parameter χ is typically in the range 0.1 - 0.2 . Thus the P_{13} corresponds to a very inelastic resonance at best. In meson exchange models, the t-channel exchange of ρ, ϵ, and ω mesons[21] does not produce sufficient attraction in the P_{13} channel to yield a resonance. Substantial additional attraction from coupling to inelastic channels, particularly $K\Delta$, is required. Other analyses[15] do not find any need for a P_{13} resonance, but indicate only sizable attraction in this channel. An analysis of polarization and differential cross section data alone is not able to provide a unique resolution of this question.

We now consider the results of phase shift analyses for the I = 0 K^+N amplitudes[17,19,23]. As an example, we show in Fig. 11 the phase shifts δ and inelasticities η as a function of p_{lab} for the solutions A, C, and D of the BGRT group[16]. The analysis was made using energy dependent parametrizations of Eqs. (3) and (4), using fixed I = 1 amplitudes taken from an earlier work[13]. We show the various solutions to illustrate the range of possibilities; in fact, solution A is very unlikely, at least near 600 MeV/c, since it fails to reproduce the observed large polarization of the non-spectator proton in the process $K^+d \rightarrow K^0pp$ [18]. In addition, only solution D produces a resonable fit to the recent high precision total cross section data from Brookhaven[24]. The description of low momentum ($\leq$ 1.5 GeV/c) I = 0 amplitudes involves more parameters than for I = 1. It was in fact already suggested by early emulsion results with kaons[25] that I = 0 p-wave K^+N interactions are required at 300 MeV/c. The simplest way to produce the large charge-exchange polarization which is observed[18] is via the interference of a large attractive P1/2 I = 0 amplitude with the dominant repulsive S1/2 I = 1 amplitude. The problem is that this interference would also produce a backward peak in $K^+n \rightarrow K^0p$, which is not observed. Thus some amount of D-wave is also needed to compensate for this, an attractive D3/2 amplitude being the most likely possibility (as in solution D of BGRT). In this picture, a large attractive P1/2 wave accounts for the dominantly elastic peak in $\sigma_{TOT}(I = 0)$ around 800 MeV/c and the D3/2 wave contributes importantly to the inelastic cross section and produces the second peak in $\sigma_{TOT}(I = 0)$ around 1.2 GeV/c (see the data in Fig. 3). A large attractive I = 0 P1/2 amplitude is also predicted by dispersion relation sum rules[26]. Such an amplitude is also favored by the observed shallow minimum at 90° in the angular distribution for $K^+n \rightarrow K^+n$ [19]. Thus the evidence for a solution with a strongly attractive I = 0 P1/2 wave looks impressive.

A more recent analysis by B.R. Martin[19] treats I = 0 and 1 amplitudes simultaneously, including some newer data not available to the BGRT group. The Argand diagrams for the Martin analysis[19] are shown in Fig. 12. These results are qualitatively similar to those of the BGRT group[13,16]. In detail, however, there are considerable differences.

The phase shift analyses indicate that only the S1/2, P1/2, and D3/2 I = 0 channels are possible candidates for exotic Z_0^* resonances. Martin[19] found no nearby poles on the second sheet in any of the I = 0 amplitudes; P1/2 has a low-lying pole which lies outside the region where the continuation is valid. The BGRT group[16] has fit a Breit-Wigner form to their I = 0 P1/2 amplitude. This gives a c.m. resonance energy of $E_R \approx 1740$ MeV, $\Gamma \approx$ 300 MeV and $\chi \approx 0.85$, suggesting a very broad but largely elastic structure. As for the I = 1 case, there is considerable disagreement in the literature on whether one or more of the I = 0 amplitudes contain true exotic Z_0^* resonances or just moderate to strong attraction. From the point of view of kaon-nucleus scattering, one is not likely to be able to differentiate models which give a wide Z_0^* resonance from those which have a moderately strong attraction but no resonance in the same channel. The most qualitative differences in the various models occur in the I = 0 S1/2 channel. Some models give strong attraction at low energies[21], while others give weak repulsion[19] for S1/2. Elastic and inelastic scattering of K^+ from nuclei at low energies may shed some light on this question.

We now turn to the theoretical interpretation of the kaon-nucleon phase shifts. The quark model is unable at present to provide estimates of scattering phase shifts, but has been extensively used to predict the Y* spectrum[27] and, more recently, the possible Z* resonances[28]. The three quark sector (Q3) in the SU(6) x O(3) version of the quark model[27] incorporates essentially all the known Y* resonances, and predicts numerous additional states which have not yet been detected experimentally. We refer the reader to refs. (27, 28) for a review of baryon spectroscopy in the quark model. The recent work of Strottman[29] predicts the Z* resonances of negative parity which arise from $Q^4\bar{Q}$ complexes with all quarks in S-states. He used the M.I.T. bag model, the parameters of which are adjusted to the spectroscopy of low-lying $Q\bar{Q}$ mesons and Q^3 baryons. He finds Z* candidates in the following channels:

$$\begin{matrix} S_{01} & & S_{11} \\ D_{03} & & D_{13} \end{matrix} \qquad (11)$$

The S_{01} and D_{03} states are indeed possible candidates for Z_0^* in some analyses, although, as we have argued above, the negative parity P_{01} is more likely. The main problem with (11) arises for I = 1. The only Z_1^* candidate in the data is P_{13}; the S_{11} and D_{13} phase shifts are not only non-resonant, they are in fact repulsive.

Meson exchange models provide an alternative framework for discussing the kaon-nucleon interaction. It is well known that much of the physics of NN scattering at low energies can be explained by meson exchange models[30]. It is very tempting to also adopt such a viewpoint for the K^+N system, since particle exchanges stand out more clearly for such an exotic channel than for the $\bar{K}N$ system, where their effect is somewhat masked by S-channel resonances. This view has been developed in an elegant series of papers by Cottingham et al.[21,31]. The physical content of this approach is seen by the consideration of a simple coordinate space potential model. While this discussion is inadequate for a quantitative treatment of the problem, it does reveal several key aspects of the physics. Such a model can be constructed as in the nucleon-nucleon case. The main differences for K^+N are that a) there is no single π or η exchange term, and b) there are no tensor or quadratic spin-orbit forces (since $S_K = 0$). The dominant contributions come from vector (ρ, ω) and scalar (ϵ) meson exchange, and lead to potentials of the form

$$
\begin{aligned}
V_\rho(r) &= \underset{\sim}{\tau}_K \cdot \underset{\sim}{\tau}_N \left(V_0^\rho(r) - \underset{\sim}{\ell} \cdot \underset{\sim}{s}_N V_{LS}^\rho(r)\right) \\
V_\omega(r) &= V_0^\omega(r) - \underset{\sim}{\ell} \cdot \underset{\sim}{s}_N V_{LS}^\omega(r) \\
V_\epsilon(r) &= -V_0^\epsilon(r) - \underset{\sim}{\ell} \cdot \underset{\sim}{s}_N V_{LS}^\epsilon(r)
\end{aligned}
\qquad (12)
$$

where V_0 and V_{LS} are positive definite Yukawa functions.

The qualitative features of the K^+N phase shifts can be understood by considering the coherences between different components of the meson exchange potentials. For instance, for I = 1, all vector meson and spin-orbit exchange terms are repulsive for $j = \ell - 1/2$, while all scalar exchange and spin-orbit terms are attractive for $j = \ell + 1/2$. This produces a very large splitting of I = 1 P-wave phase shifts, with $\delta_{P13} > 0$. Due to the absence of such overall coherences for I = 0, the difference $\delta_{P01} - \delta_{P03}$ should be considerably smaller than $\delta_{P13} - \delta_{P11}$. This behavior is indeed observed in the empirical phase shifts. The potential model also tells us that the strongest attraction for each L occurs in the P_{01}, D_{03} channels for I = 0 and P_{13}, D_{15} for I = 1. Thus P_{01} and P_{13} are the principal Z* candidates which emerge, in agreement with a number of analyses.

The essence of this naive coordinate space potential treatment is found in more rigorous relativistic formulations of KN scattering involving partial wave dispersion relations and the multichannel N/D method[32]. Here, if one enumerates the main t- and u-channel singularities, a good guide to the systematics of partial wave amplitudes for πN and $K^{\pm}N$ scattering can be obtained by including only the nearby singularities, i.e. the long range forces. This discussion[32] parallels that of the potential models, and predicts a Z_0^* exotic resonance in the P1/2 channel.

We now discuss the work of Cottingham and his collaborators[21,31]. This approach preserves the spirit of a non-relativistic potential model, but is much more quantitative. The method involves first a calculation of the two-pion exchange potential in K^+-nucleon scattering, using the Mandelstam double spectral function[31,33]. In addition to the 2π exchange contributions, 3π exchanges are also included, in the form of elementary exchanges of ω and ϕ mesons, with coupling constants to kaons taken from the quark model. The Blankenbecler-Sugar equation equation is then solved, with the driving term ("potential") taken as a sum of Reggeized meson exchanges. This equation incorporates relativistic kinematics.

The I = 0 and S- and P-wave phase shifts obtained with this approach[31] are shown in Fig. 13. The corresponding I = 1 phase shifts are shown in Fig. 14. Since K*N and $K\Delta$ inelasticities are not included the predicted phase shifts may be sensibly compared with the results of phase shift analyses only below 0.8 GeV/c or so. In this region, there is good agreement, both in sign and magnitude, between predicted and empirical phase shifts, with the exception of the P3/2, I = 1 channel. Note that the I = 1, S1/2 wave seems to require a bit more repulsion, which may indicate the need for a short range repulsive core. For I = 1, P3/2, the calculated phase shift is attractive and of the correct size at low energy, but nowhere near as large as the empirical phase shift near 1.2 GeV, so no exotic resonance is indicated in this channel. The driving mechanism for such a resonance, if it exists, must then be the opening of the $K\Delta$ channel, which would provide additional attraction in an effective K^+N potential.

The work of ref. (31) was extended in ref. (21) to include the coupling of the K*N channel to elastic scattering. The main effect of this coupling is to augment the already sizable attraction in the P_{01} wave, thereby increasing the possibility of an exotic Z_0^* resonance. Whether this wave actually resonates or not depends on details of the model.

The most remarkable feature of this analysis[21] is that the S_{01} wave is <u>attractive</u> at low energies, becoming repulsive before the inelastic threshold. This feature is not present in any existing empirical analysis. This result is somewhat sensitive to a short-ranged repulsive core. However, a core which eliminates the low energy attraction would have to exhibit a rapid momentum dependence in order not to make the higher energy phase shifts too repulsive.

In the $K^{\pm}$-nucleus situation, the presence of a sizable attraction in the S_{01} wave would tend to cancel out some of the repulsive S_{11} contribution to the optical potential. The pattern of Coulomb-nuclear interference would thus be altered at low energy. In addition, there would be a marked change of the selection rules for K^+ inelastic scattering from nuclei. If the S_{11} elementary amplitude dominates, we will excite mostly $\Delta T = 0$ particle-hole states starting from an $N = Z$ target. If $I = 0$ and 1 amplitudes are comparable, on the other hand, $\Delta T = 0,1$ states will be more or less equally populated.

The meson exchange model of Cottingham and collaborators[21,31] provides a clear dynamical mechanism for Z_0^* formation in the P_{01} state. This is at odds with the conclusions of Aaron et al.[17] and the $Q^4\bar{Q}$ calculations[29] where it is argued that Z_0^* resonances are more likely in the S1/2 and D3/2 channels than P1/2. In ref. (17), the coupling of KN to K*N is included, using a three body formalism, but the diagonal $K^+N \rightarrow K^+N$ potential is not treated as realistically as in ref. (21).

We now consider the corresponding situation for the K^-N interaction. As displayed in Fig. 1, the low energy cross sections are resonance dominated. The parameters of the Y* resonances were given in Table 1[6]. There is considerable debate on the energies, widths, and inelasticities of some of the Y* resonances. The most recent extensive analysis is that of Alston-Garnjost et al.[34]

The quantum numbers of the Y* resonances and their classification into multiplets has been frequently discussed in the context of the quark model[27,28,35]. In the usual form of the harmonic oscillator quark model, the decay rates of resonances have been analyzed in terms of an $SU(6)_W$ scheme[36]. The relative signs of decay amplitudes predicted by $SU(6)_W$ enable one to group the observed Y* resonances into multiplets. These multiplets correspond most naturally in the oscillator quark model to $SU(6) \otimes O(3)$ symmetry. The quark model predicts a number of Y* resonances which have not yet been discovered, perhaps because they are too weakly coupled to the $\bar{K}N$ channel. For instance, the $SU(6) \otimes O(3)$ model predicts eight Y* resonances in the P13 channel[37], whereas only one ($\Sigma(1385)$) is well established experimentally. Thus there remain a number of uncertainties in the quark model description of the Y* resonances.

We now discuss the Y* resonances in the context of the simple potential model introduced in Eq. (12) for the K^+N system. The model is less well motivated for the $\bar{K}N$ system, due to the presence of many S-channel resonances, and the intrinsic multichannel nature of the problem at low energy. However, as we shall see, the main features of the level order of Y* resonances can be predicted in this model, which thus serves as a useful mnemonic for understanding the $\bar{K}N$ interaction at low and medium energies. Similar considerations enter in the description of $\bar{K}N$ scattering via partial wave dispersion relations and the N/D method[32]. Here one includes u-channel exchanges as well, which are important for a quantitative description.

The diagonal $\bar{K}N$ potential can be constructed from the K^+N potential of Eq. (12) by inverting the sign of the ω exchange potential.

Several features are evident from the form of the resulting K^-N potential:

i) the central part displays an attractive coherence of ϵ and ω potentials in all channels. For $I = 0$, ρ exchange further supports this coherence. Thus K^-N central potentials are always more attractive than K^+N potentials, for which ϵ and ω tend to cancel. This coherence favors resonance formation for K^-N systems.

ii) unlike K^+N, there are no overall coherences of $\underset{\sim}{\ell} \cdot \underset{\sim}{s}_N$ forces for K^-N. However, ρ spin-orbit terms are the most important, so channels with $\underset{\sim}{\tau}_K \cdot \underset{\sim}{\tau}_N \, \underset{\sim}{\ell} \cdot \underset{\sim}{s}_N > 0$ gain additional spin-orbit attraction. The K^-N potentials can then be arranged in order of decreasing attraction. This leads to the corresponding ordering of Y* resonances which couple primarily to the $\bar{K}N$ channel

$$P_{01} < P_{13} < P_{11} < P_{03}$$
$$D_{03} < D_{15} < D_{13} < D_{05} \qquad (13)$$

and similarly for higher partial waves. A potential of the same form as Eq. (12) can also be written down for the $\pi\Sigma$ channel, except that the attractive ω exchange is now absent. Arguments similar to those outlined above enable one to order the Y* resonances coupling mostly to the $\pi\Sigma$ channel

$$P_{01} < P_{11} < P_{13} < P_{03}$$
$$D_{03} < D_{13} < D_{15} < D_{05} \qquad (14)$$

Note that maximum attraction in <u>both</u> $\bar{K}N$ and $\Sigma\pi$ channels occurs for $I = 0$, $j = \ell - 1/2$. We thus expect a low-lying P_{01}, D_{03}, F_{05}, G_{07},... band for both $\bar{K}N$ and $\Sigma\pi$. Of course, these two bands are coupled, since $\bar{K}N \leftrightarrow \Sigma\pi$ via t-channel K or K*(890) exchange or u-channel baryon exchange. We expect <u>two</u> states in each channel P_{01}, D_{03}, etc., the lowest lying member of the pair coupling more strongly to $\bar{K}N$. This is what is observed experimentally (Table 1). The $I = 0$, $j = \ell-1/2$ $\bar{K}N$ band consists of

Λ (1115)	Λ (1520)	Λ (1815)	Λ (2100)	(15)
P_{01}	D_{03}	F_{05}	G_{07}	

While the corresponding $\Sigma\pi$ band is

Λ (1600)	Λ (1690)	Λ (2110)	(16)
P_{01}	D_{03}	F_{05}	

Note that in this naive "bootstrap" picture, the Λ itself is considered to be a bound state coupling most strongly to $\bar{K}N$. This is consistent with expectations based on SU(3) symmetry.

In summary, many features of the observed Y* spectrum can be understood qualitatively in terms of simple potential concepts. This discussion is of course somewhat old-fashioned compared to the more modern classification scheme of the quark model. However, the potential and quark models may be viewed in some respects as offering alternative descriptions of some of the same phenomena. For instance, the well-known (3,3) resonance may be interpreted as a pion-nucleon composite or as a three quark state, so these two descriptions are complementary. Similar considerations apply to the $\bar{K}N$ system. We emphasize the potential picture here in order to render strange particle physics a bit more understandable for nuclear physicists. Particle physicists are invited to consult other references[27,36] for a discussion in a language more familiar to them.

In the potential language, the spin-orbit force plays a crucial role in determining the level ordering of Y* resonances. For instance, the strong spin-orbit component yields states with $L \neq 0$ which lie near or below the S-states, i.e. Σ(1385) and Λ(1520). Theoretically, we expect the strongest spin-orbit potential to arise from ρ exchange. Rough estimates based on the SU(3) coupling constants of ref. (31) yield

$$V_{LS}^{\rho} : V_{LS}^{\omega} : V_{LS}^{\varepsilon} = 1 : 1/2 : 1/6 . \tag{17}$$

Now let us consider some particular K^--induced reactions of interest for nuclear physics studies. The cross section for $K^-p \to \bar{K}^0 n$ charge exchange[47] is shown in Fig. 15. The Λ(1520) resonance is clearly visible as a 5 mb enhancement at 400 MeV/c; the "background" cross section (in mb) is approximately $\sigma_{TOT}(K^-p \to \bar{K}^0 n) \approx 2.4/p_{lab}$. The dip at 730 MeV/c and the shoulder at 900 MeV/c seen in Fig. 15 are due to the opening of the $K^-p \to \Lambda\eta$ and $\Sigma^0\eta$ channels, respectively. The largest charge exchange cross sections on nuclei (for energies where beams are available) are expected in the region of 1.05 GeV/c, where the elementary cross section has a maximum.

The cross sections for the processes $K^-p \to \pi^0\Sigma^0$, $\pi^0\Lambda$, $\pi^+\Sigma^-$ are shown[37] in Fig. 16. These strangeness exchange reactions are of great interest for the formation of Λ and Σ hypernuclei. Clearly, hypernuclear formation experiments are most favorable below p_{lab} = 1 GeV/c or so. Above this region, the elementary cross sections decrease and the momentum transfer imparted to the nucleus increases, so the Λ or Σ has less probability of sticking in the nucleus.

A variety of theoretical and phenomenological techniques have applied to the analysis of K^--nucleon scattering, including multichannel K-matrix[38,39] and M-matrix[40] fits, energy dependent[41] and energy independent[42] partial wave analyses, single channel treatments[43], studies of quasi-two-body channels[44], etc. We do not discuss these methods in detail here, but instead refer the reader to a variety of review articles[22,45,46].

REFERENCES

1. G.C. Bonazzola et al., Phys. Rev. Lett. 34, 683 (1975); W. Brückner et al., Phys. Lett 62B, 481 (1976); B. Povh, Reports on Progress in Physics, Vol. 39 (1976).
2. M. May, invited talk at this conference.
3. C.B. Dover, Proc. of the Summer Study Meeting on Kaon Physics and Facilities, BNL, June, 1976 (BNL-50579); C.B. Dover and P.J. Moffa, Phys. Rev. C16, 1087 (1977).

4. C.B. Dover and G.E. Walker, Phys. Rev. C, April, 1979.
5. K.K. Li, in Proc. of CMU Conference on Meson-Nuclear Physics, May, 1976; T. Kycia and K.K. Li, private communication.
6. T.G. Trippe et al., Rev. Mod. Phys. 48, S1 (1976).
7. B.J. Charles et al., in Proc. of the Conf. on Baryon Resonances, Purdue, 1973, p. 179.
8. W. Cameron et al., Nucl. Phys. B78, 93 (1974).
9. S. Goldhaber et al., Phys. Rev. Lett 9, 135 (1962).
10. R.W. Bland et al., Nucl. Phys. B13, 595 (1969).
11. R.W. Bland et al., Nucl. Phys. B18, 537 (1970); F. Griffiths et al., Nucl.Phys. B38, 365 (1972); A. Berthon et al., Nucl. Phys. B63, 54 (1973); E. Lesquoy et al., Nucl. Phys. B99, 346 (1975).
12. C.J.S. Damerell et al., Nucl. Phys. B94, 374 (1975).
13. G. Giacomelli et al., Nucl. Phys. B37, 577 (1972); Nucl. Phys. B20, 301 (1970).
14. M.G. Albrow et al., Nucl. Phys. B30, 273 (1971); R.A. Arndt et al., Phys. Rev. Lett. 33, 987 (1974); R. Ayed et al., Phys. Lett. 32B, 404 (1970); S. Kato et al., Phys. Rev. Lett. 24, 615 (1970).
15. P.C. Barber et al., Nucl Phys. B61, 125 (1973); C.J. Adams et al., Nucl. Phys. B66, 36 (1973); C. Lovelace and F. Wagner, Nucl. Phys. B28, 141 (1971); R.E. Cutkosky et al., Nucl. Phys. B102, 139 (1976).
16. G. Giacomelli et al., Nucl. Phys. B71, 138 (1974).
17. W. Slater et al., Phys. Rev. Lett. 7, 378 (1961); V.J. Stenger et al., Phys. Rev. 134B, 1111 (1964); R. Aaron, R.D. Amado and R.R. Silbar, Phys. Rev. Lett. 26, 407 (1971); R.C. Johnson and S. Vlassopoulos, Phys. Lett. 50B, 343 (1974); B.C. Wilson et al. Nucl. Phys. B42, 445 (1972); M. Sakitt et al., Phys. Rev. D15, 1846 (1977); R.G. Glasser et al., Phys. Rev. D15, 1200 (1977); R. Aaron et al., Phys. Rev. D7, 1401 (1973).
18. A.K. Ray et al., Phys. Rev. 183, 1183 (1968).
19. B.R. Martin, Nucl. Phys. B94, 413 (1975).
20. A.T. Lea, B.R. Martin and G.C. Oades, Phys. Rev. 165, 1770 (1968).
21. J.W. Alcock, W.N. Cottingham and A.C. Davis, Univ. of Bristol preprint, Jan. 1977.
22. Proc. of the Topical Conf. on Baryon Resonances, Oxford, July, 1976; see reviews by R.T. Ross, B.R. Martin and P.J. Litchfield on $\bar{K}N$ analyses and B.R. Martin on Z*'s.
23. A.A. Hirata et al., Phys. Rev. Lett. 21, 1485 (1968); Nucl. Phys. B30, 157 (1971).
24. A.S. Carroll et al., BNL-17915 (unpublished).
25. O.R. Price et al., Phys. Rev. Lett. 1, 212 (1958); M.A. Melkanoff, Phys. Rev. Lett. 5, 108 (1960); M. Grilli et al., Nuovo Cim. 10, 205 (1958).
26. C.P. Knudsen and B.R. Martin, Nucl. Phys. B61, 307 (1973).
27. R. Horgan and R.H. Dalitz, Nucl. Phys. B66, 135 (1973); R. Horgan, Nucl. Phys. B71, 514 (1974).

28. P.J. Litchfield, in Proc. of the 17th Intern. Conf. on High Energy Physics, London, 1974; R.H. Dalitz in Proc. of Summer Symposium on New Directions in Hadron Spectroscopy, Argonne 1975.
29. D. Strottman, to appear in Phys. Rev. D.
30. G.E. Brown and A.D. Jackson, The Nucleon-Nucleon Interaction, North-Holland, 1976.
31. J.W. Alcock and W.N. Cottingham, Nucl. Phys. B31, 443 (1971); B56, 301 (1973); B62, 533 (1973); see also Nucl. Phys. B91, 509 (1975); B102, 173 (1976); B111, 233 (1976).
32. G. Gustafson, H. Nielsen and G.C. Oades, Proc. London Conf., 1974, Session C4.
33. S. Mandelstam, Phys. Rev. 112, 1334 (1958).
34. M. Alston-Garnjost et al., Phys. Rev. D (1978).
35. D. Faiman, Nucl. Phys. B115, 478 (1976); B133, 477 (1976).
36. A.J. Hey, P.J. Litchfield and R.J. Cashmore, Nucl. Phys. B95, 516 (1975).
37. B.R. Martin and M.K. Pidcock, Nucl. Phys. B126, 266 and 285 (1977).
38. R.H. Dalitz and S.F. Tuan, Ann. Phys. 3, 307 (1960).
39. B.R. Martin and M. Sakitt, Phys. Rev. 183, 1345 and 1352 (1969); J.K. Kim, Phys. Rev. Lett. 19, 1074 (1967); A.T. Lea et al., Nucl. Phys. B56, 77 (1973).
40. M.L. Gupta, R.A. Arndt and L.D. Roper, Nucl. Phys. B37, 173 (1972).
41. G.P. Gopal et al., Nucl. Phys. B119, 362 (1977).
42. P.N. Hansen, J. Moss and G.C. Oades, Topic Conf. on Baryon Resonances, Oxford, 1976.
43. P. Baillon and P.J. Litchfield, Nucl. Phys. B94, 39 (1975); A.J. Van Horn, Nucl. Phys. B87, 145 and 157 (1975); A. de Bellefon and A. Berthon, Nucl. Phys. B109, 129 (1976).
44. M.J. Corden et al., Nucl. Phys. B92, 365 (1975); Nucl. Phys. B121, 365 (1977).
45. B.H. Bransden, in High Energy Physics, Ed. E.H.S. Burhop, New York, 1969, Vol. 3; B.R. Martin, in Springer Tracts in Modern Physics, Ed. G. Hohler, Springer Verlag, Berlin, 1970, Vol. 55, p. 73 - 141.
46. C.B. Dover and G.E. Walker, Physics Reports, in preparation.
47. T.S. Mast et al., Phys. Rev. D14, 13 (1976); M. Alston-Garnjost et al., Phys. Rev. Lett. 38, 1003 and 1007 (1977).

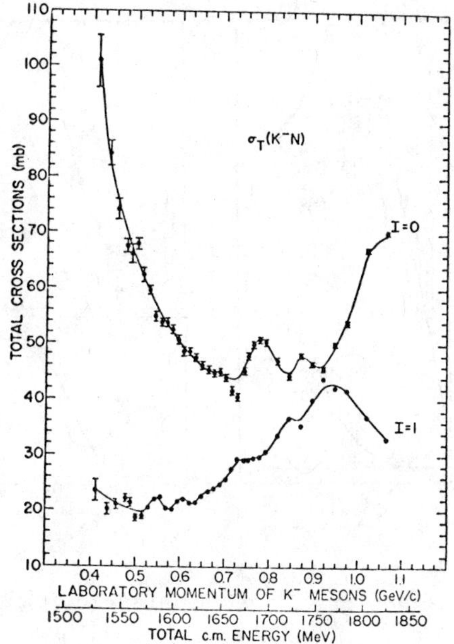

Fig. 1 Total cross sections for K^-- nucleon scattering in isospin I = 0,1 states (5)

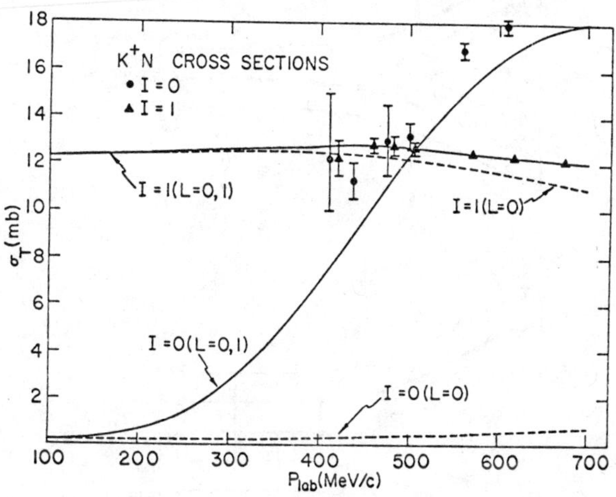

Fig. 2 Total K^+- nucleon I = 0,1 cross sections in the low momentum region $p_{lab} \leq 700$ MeV/c. The solid curves are taken from the analysis of Martin (19); the points with small error bars are from BNL data (5).

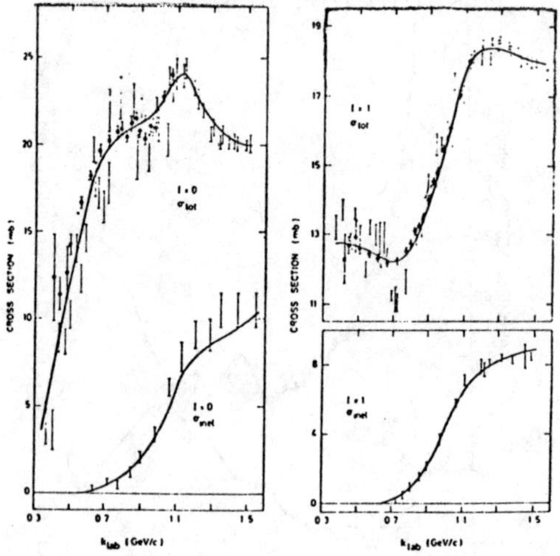

Fig. 3 K^+N cross sections in the inelastic region above 700 MeV/c.

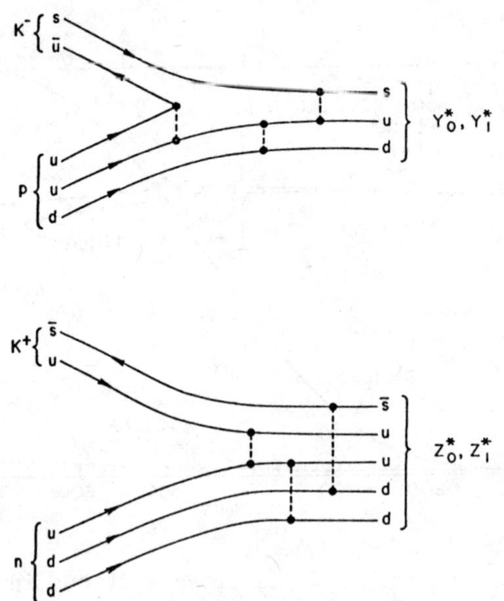

Fig. 4 Quark Model diagrams for the formation of Y* and Z* resonances.

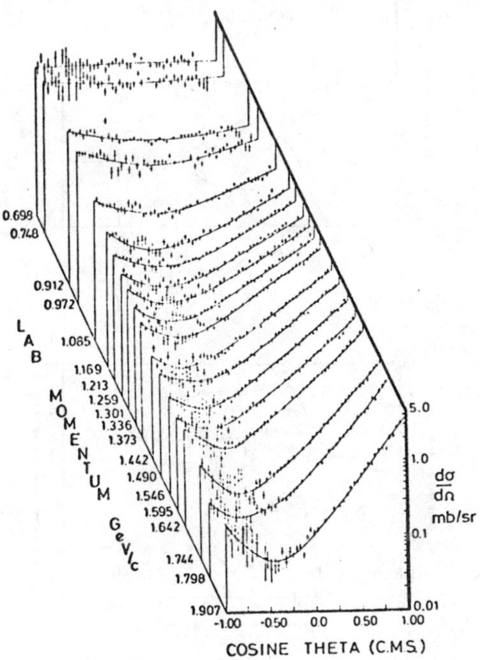

Fig. 5 Elastic scattering K^+p angular distributions for $p_{lab} \leq 2$ GeV/c (7).

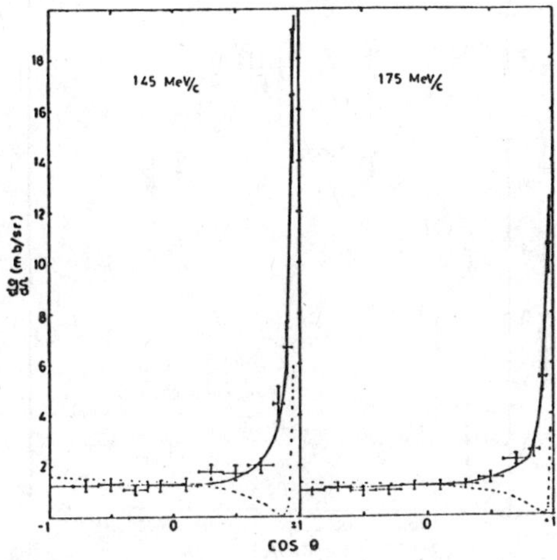

Fig. 6 Differential cross sections for K^+p elastic scattering at very low momenta (8).

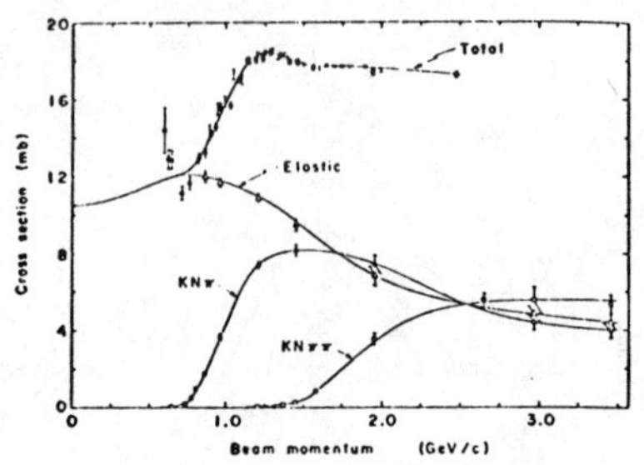

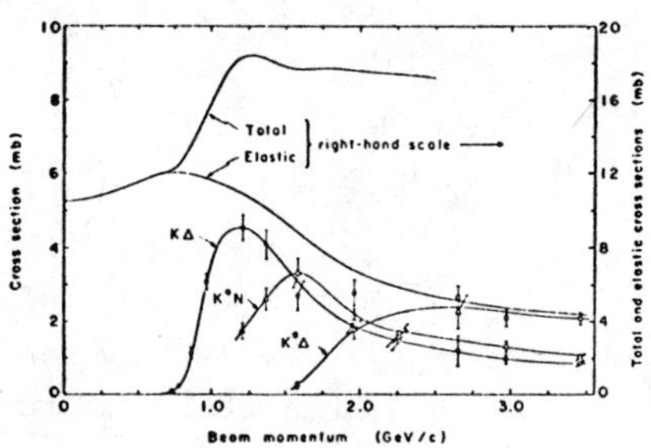

Fig. 7 Quasi-two-body contributions to the total K^+p cross section below 1.5 GeV/c (10).

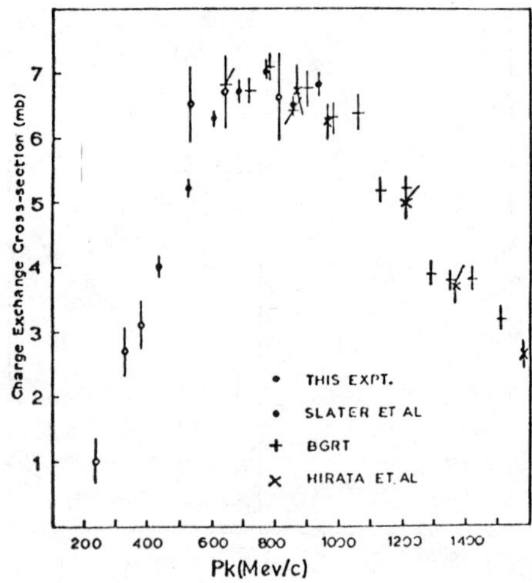

Fig. 8 Total charge exchange cross section $K^+n \rightarrow K^0p$ (12).

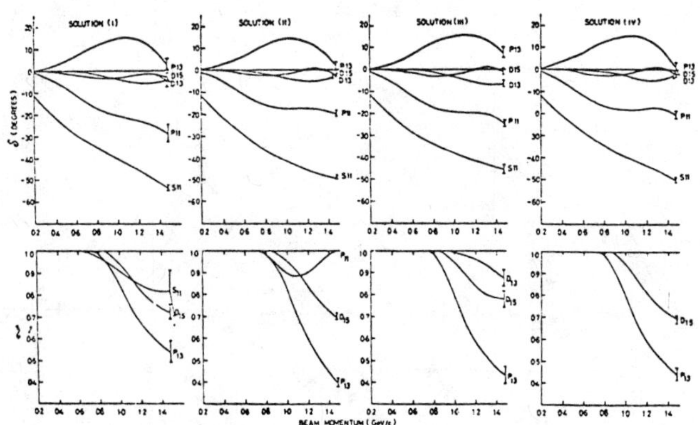

Fig. 9 Phase shifts δ and inelasticities η for I = 1 K^+ - nucleon scattering, according to the BGRT group (13).

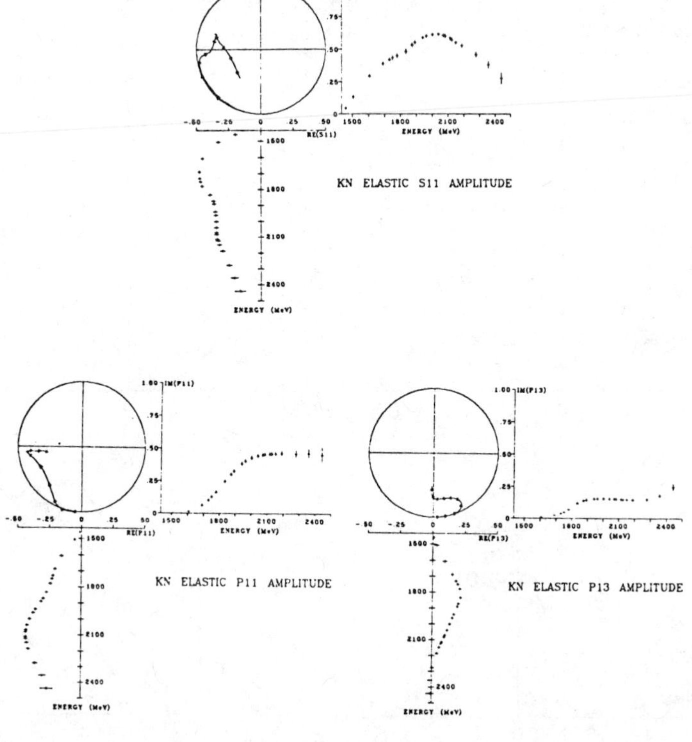

Fig. 10 S and P-wave amplitudes for K^+p scattering, from Cutcosky et al. (15)

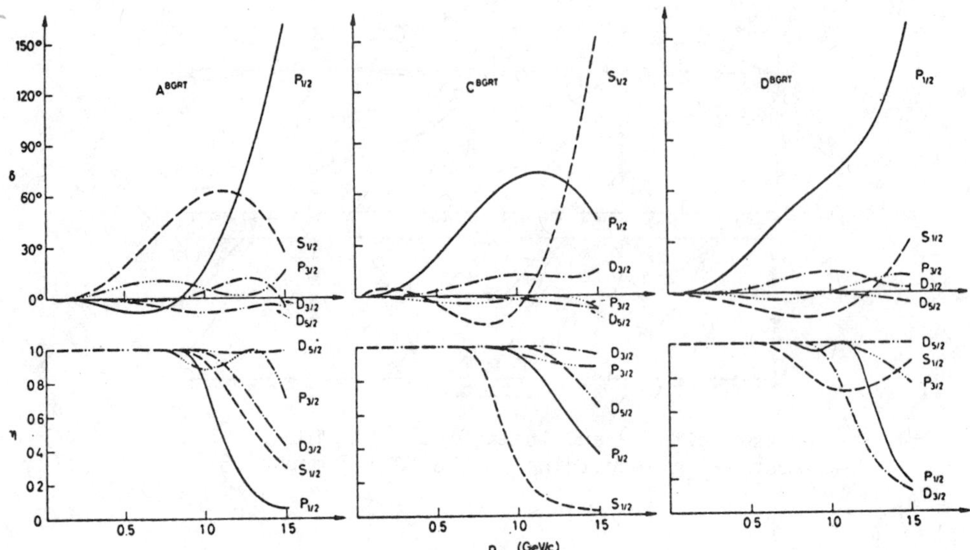

Fig. 11 Phase shifts δ and inelasticities η for I = 0 K^+N scattering, according to the BGRT group (16).

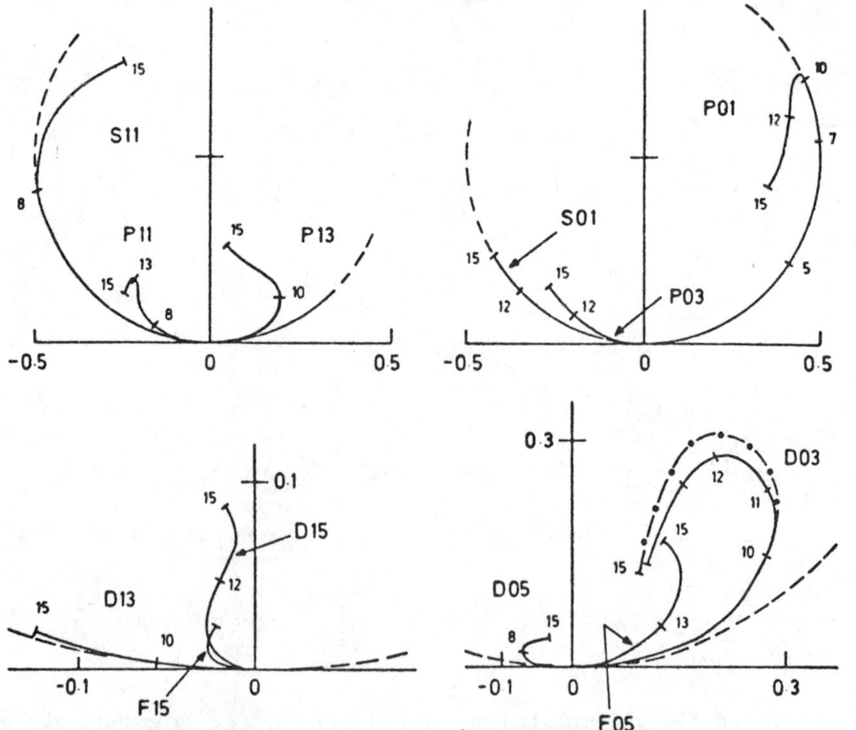

Fig. 12 Argand diagrams for I = 0, 1 K^+N amplitudes from the parametrizatio[n] of Martin (19).

Fig. 13 Phase shifts δ for I = 0 S and P-waves of the K^+N system, as per Cottingham et al. (31).

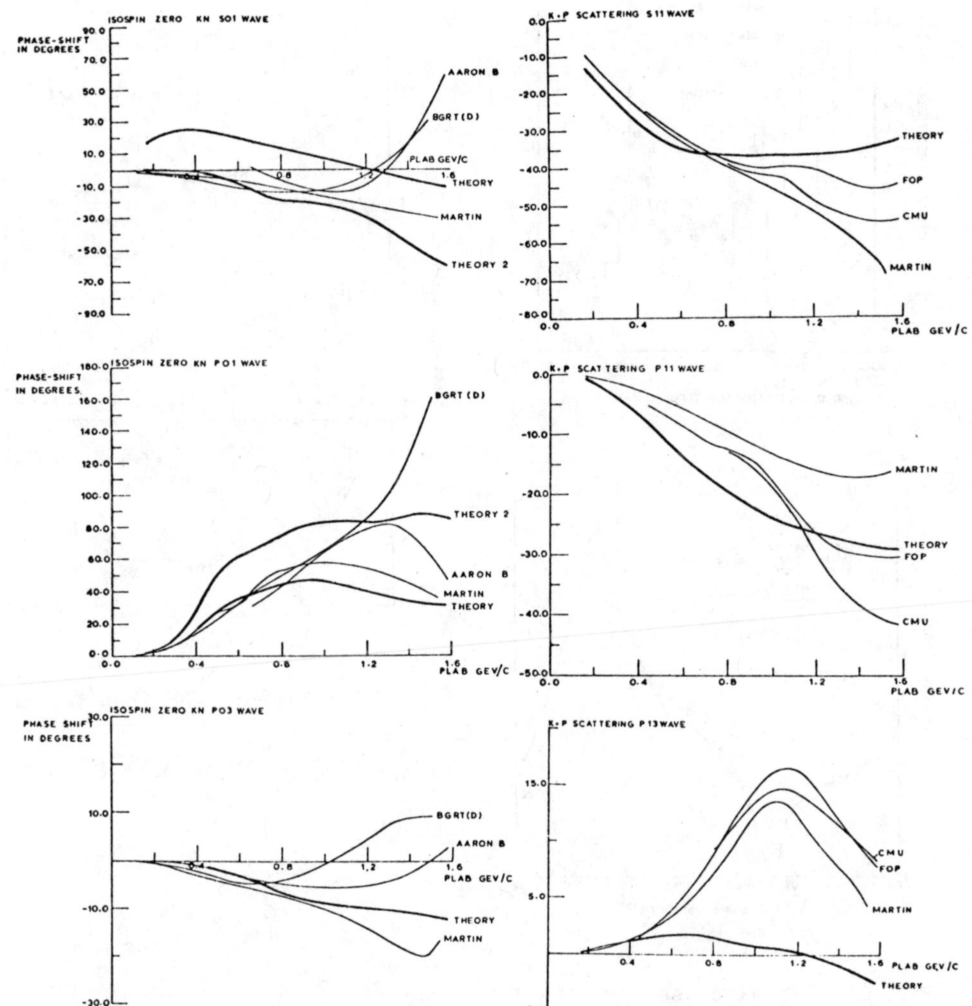

Fig. 14 Same as Fig. 13 for I = 1 (31).

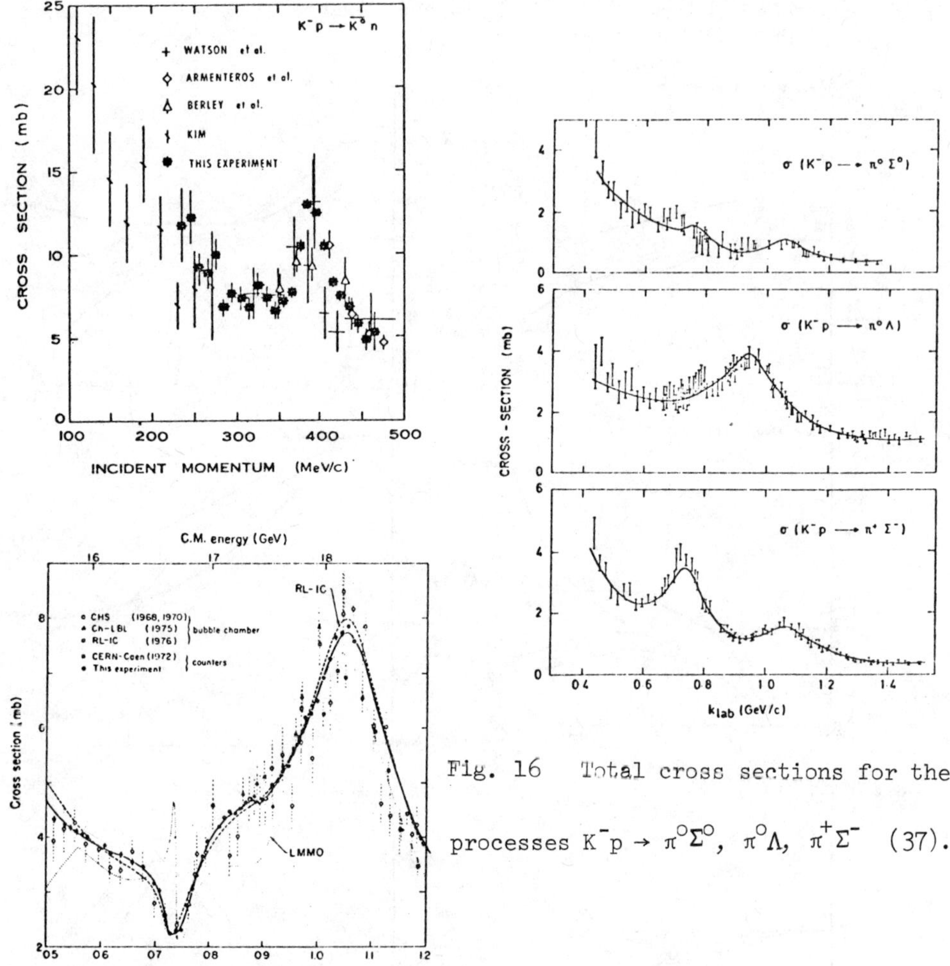

Fig. 16 Total cross sections for the processes $K^-p \to \pi^0\Sigma^0$, $\pi^0\Lambda$, $\pi^+\Sigma^-$ (37).

Fig. 15 Total cross sections for $K^-p \to \bar{K}^0 n$ charge exchange (47).

Question: A. Gal, Hebrew University

Can you generate by t-channel exchanges also πN resonances above the $\Delta(1232)$?

Response: Dover

To some extent, yes, although the u-channel exchanges and other coupled channels such as ρN are also important in a quantitative treatment.

Question: F. Tabakin, University of Pittsburgh

I believe Cutkosky and Deo and also Arndt, *et al* have done phase shift analysis of K^+N and have their results differ concerning the Z^* resonances. Cutkosky has proposed an experiment involving measuring the polarization of the final proton in a K^+d process, as a test of the different ideas concerning the Z^*. Can you tell us about the present status of the BNL experiment on this process?

Response: Dover

The data on the new BNL experiment on charge exchange polarization in $K^+N \rightarrow K^0p$ are currently being analyzed by M. Sakitt, *et al*. Large polarization would favor P_{01} as a candidate for a Z_0^*.

(K,π) Reaction at Brookhaven*

M. May
Brookhaven National Laboratory, Upton, New York 11973

Abstract

The strangeness exchange reaction (K^-,π^-) has been used to study the spectrum of $^{12}_{\Lambda}C$. The angular distributions of the ground state and the peak at 11 MeV excitation have been measured. Evidence is found for the existence of two components of the 11 MeV peak. Results of a search for bound states of the Σ^- neutron system are discussed.

Spectrum of $^{12}_{\Lambda}C$

We[1] report here the results of a study of the hypernucleus $^{12}_{\Lambda}C$ using the reaction

$$K^- \, ^{12}C \rightarrow \pi^- \, ^{12}_{\Lambda}C. \quad (1)$$

The momentum of the incident K^- is 800 MeV/c. The experiment covered a range of angles between K^- and π^- from 0 to 19° corresponding to a variation in momentum transfer to the hypernucleus of 70 to 260 MeV/c. Previous experiments[2,3] using the (K^-,π^-) strangeness exchange reaction to form $^{12}_{\Lambda}C$ have been restricted to a small region near $\theta_{K\pi}=0°$ and consequently to momentum transfer to the nucleus less than 100 MeV/c.

Measurement of the angular distribution of the π^- aids in assigning spins to the hypernuclear states. Large momentum transfer permits the excitation of states not accessible at 0°.

The experiment was performed at the Brookhaven National Laboratory Alternate Gradient Synchrotron in the low energy separated beam (LESB-I). The beam intensity is 2 x 10^5 K^- per pulse with kaons composing 7% of the total flux. A differential Cerenkov counter CK (Fig. 1) with a lucite radiator is used to identify kaons in the beam.

The apparatus (Fig. 1) consists of two magnetic spectrometers to determine the momentum of the K^- and π^-, respectively. The K spectrometer consists of one dipole and four quadrupole magnets, with multi-wire proportional counters (P1, P2, P3) to determine particle trajectories. The K spectrometer is tuned to give point to point focusing and a dispersion of 24 mm/% between P2 and the target. Placing most of the material near foci minimizes the contribution of multiple scattering to the resolution. The momentum of the kaon is determined to first order by its position at P2 and at the target. The angle measurement between P1 and P2 provides higher order corrections.

ISSN:0094-243X/79/540658-08$1.50

The π^- from hypernucleus formation is detected in a second, similarly constructed, spectrometer which is mounted on a platform that can be easily rotated up to an angle of 30° with respect to the beam direction. The optics of the π spectrometer include an intermediate focus in the non-bending plane to enhance acceptance. The spectrometer subtends a solid angle of 12 msr. The acceptance of the π spectrometer was determined both by Monte Carlo calculations and by the measurement of the known[4] cross section for π^- production on hydrogen, using the hydrogen in a scintillator target. The resolution in this experiment was determined primarily by the target thickness and amounts to 4 MeV FWHM for data taken with a 5 g/cm^2 target and 2.5 MeV for data taken with a 2 gm/cm^2 target.

A kaon reaching the target is defined by the coincidence S1·CK·ST (Fig. 1). Typically, the kaon flux on target was 2.5 x 10^4 per pulse. Pions produced in the target are identified by a threshold liquid hydrogen Cerenkov counter CH. The $K\pi$ trigger used for data taking was S1·CK·ST·CH·S2 with the timing between S1 and ST adjusted for kaons and that between ST and S2 adjusted for pions. Data were similtaneously taken using the K beam trigger, S1·CK·ST. Analysis of these beam triggers gave the effective number of kaons on target. The transmission of the K spectrometer depends on the quality of the beam. The efficiency of wire planes and counters in the K spectrometer are sensitive to rate effects. Both these effects were continuously accounted for by the analysis of the K beam events.

Restrictions applied to the data were as follows:

a) The interaction vertex was required to lie within the target.

b) The measured position of the particle in P3 (P4) in the non-bending plane was required to agree with the position calculated by tracing the trajectory through the spectrometer.

c) The pulse height in S1 and the number of hits in P1 and P2 were required to be consistent with a single particle initiating the event. (The event losses due to this restriction were determined by analyzing the K beam events described above.)

d) The pulse height in the target scintillator was required to be large compared to a minimum ionizing event. The decay of the hypernucleus into charged particles produces large pulse heights for real hypernuclear events.

These criteria were necessary to distinguish hypernuclear events from background events which arose primarily from the 2-body decay modes of kaons in the beam.

The spectra obtained show two prominent peaks; one peak is interpreted as the ground state of $^{12}_{\Lambda}C$ and the other corresponds to one or more excited states near 11 MeV excitation. A spectrum taken at 15° is shown in Fig. 2. The broad structure extending to higher excitation energies is interpreted as quasi-free Λ production.

We can consider $^{12}_{\Lambda}C$ as a Λ particle coupled to a ^{11}C core ($3/2^-$ ground state). If the Λ is in an s-orbit a 1^- or 2^- state results, the 1^- being presumably the ground state. With the Λ in a $p_{3/2}$ orbit, the possible excited states have spin-parity 0^+, 1^+, 2^+ and 3^+. According to the calculations of Dover et al.[5] unnatural parity states will be excited with strength 10^{-2} to 10^{-3} of the natural parity components in the kinematic range of this experiment. Thus we expect only the 1^- ground state and the 0^+ and 2^+ excited states. A Λ in a $p_{1/2}$ orbit will give rise to an additional 2^+ state. It is known that $^{12}_{\Lambda}C$ is particle unstable above 10 MeV. Thus we expect no narrow states at higher excitation energy.

The angular distribution of the 11 MeV peak is shown in Fig. 3. There is a 20% uncertainty in the normalization of the data. The sharp peaking at small angles is characteristic of a 0^+ state formed on a 0^+ target, which requires no angular momentum transfer. However calculations of Dover et al.[5] indicate that by 15° the 0^+ contribution to the cross section should have decreased to a negligible fraction of the measured cross section. They attribute the data at large angles to the appearance of the unresolved 2^+ components of the multiplet. Under this assumption we can place limits on the energy splitting between the 0^+ and 2^+ components. This splitting is sensitive to the Λ-nucleon spin-orbit coupling.

The ground state requires a unit of angular momentum transfer for its formation and should reach its maximum around 10°. The angular distribution of the ground state (Fig. 4) is consistent with this interpretation.

The angular distributions measured in this experiment contain information about the spatial extent of the Λ wave functions in nuclear matter, the pion optical potential, and the kaon optical potential. Theoretical efforts to understand this data in a quantitative way will expand our understanding of the nuclear physics of strange particles.

Sigma-Neutron Bound States

Bound states of the Σ^- neutron system, if they exist, are expected to be narrow since in the absence of protons, the Σ^- is stable against strong conversion to a Λ. We have used the apparatus described above to search for bound states of the Σ^-n system. The liquid hydrogen Cerenkov counter was not used, since the presence of hydrogen would have resulted in an undesirable background. The scintillator target was replaced by a 3 g/cm^2 liquid deuterium target and the spectrometer was tuned for positive pions from the reaction

$$K^-d \rightarrow (\Sigma^-n)\ \pi^+. \tag{4}$$

The kaon momentum was 800 MeV/c.

The existence of Σ^- neutron bound states has long been an open question.[7] Several experimental searches have been performed without definitively resolving the issue.[8,9] Since the Σn scattering lengths suggest that the singlet state is more likely to be bound, we have performed the experiment at $\theta_{K\pi}$=20° to enhance the probability of spin flip.

We estimate that if a bound state exists, it should be produced at a level of 6 μb/sr. The factors which enter into this estimate are:

a) The cross section for Σ^- production on a free proton.

b) The probability of spin flip calculated from the KN amplitudes of Gopal.[9]

c) The probability of binding calculated using a Hulthen wave function for the deuteron and for the Σ^-n state.

The data (Fig. 5) shows no evidence for bound states. To calculate an upper limit for the rate of formation of bound states in the region between 0 and 10 MeV binding energy we have used the data with binding greater than 10 MeV for a background subtraction. A precise knowledge of the location of threshold is crucial for this experiment. A measurement of Σ^- production on hydrogen enabled us to correct for any offsets in the energy scale.

Table 1 gives the cross section limits obtained at the 95% confidence level. We can set an upper limit of 1.6 μb/sr for the production of Σ^-n bound states near zero binding energy.

Table 1: Upper limit for production of Σ^-n bound states in the reaction $K^-d \rightarrow (\Sigma^-n)\ \pi^+$ for kaon momentum 800 MeV/c and $\theta_{K\pi}$ 20°.

Binding Energy (MeV)	Upper Limit μb/sr 95% Confidence Level
0	1.6
2	2.3
4	2.0
6	2.5
8	2.8

REFERENCES

* The submitted manuscript has been authored under Contract No. EY-76-C-02-0016, with the U. S. Department of Energy.

1. R. E. Chrien, M. May, H. Palevsky, R. Sutter, R. Cester, M. Deutsch, S. Bart, E. V. Hungerford, B. Mayes, L. Pinsky, T. Williams, R. L. Stearns, P. Barnes, S. Dytman, D. Marlow, and F. Takeutchi, AGS Exp. No. 646/728 collaboration.
2. W. Bruckner *et al*., Phys. Letters 62B (1976) 481.
3. W. Bruckner *et al*., submitted to Physics Letters 1978.
4. R. Armenteros *et al*., Nuclear Physics B21 (1970) 15.
5. C. Dover and A. Gal, C.E. Walker, R.H. Dalitz, sub. to Phys. Lett.
6. R. H. Dalitz and A. Gal, "The Formation of, and the γ-Radiation From, the p-Shell Hypernuclei", University of Oxford, preprint Ref. 8/78.
7. A. Pais and S. B. Treiman, Phys. Rev. 107 (1957) 1396.
8. O. Dahl *et al*., Phys. Rev. Letters 4 (1960) 428.
9. R. A. Burnstein *et al*., Phys. Rev. 177 (1968) 1945.
10. G. P. Gopal *et al*., Nucl. Phys. B119 (1977) 362.

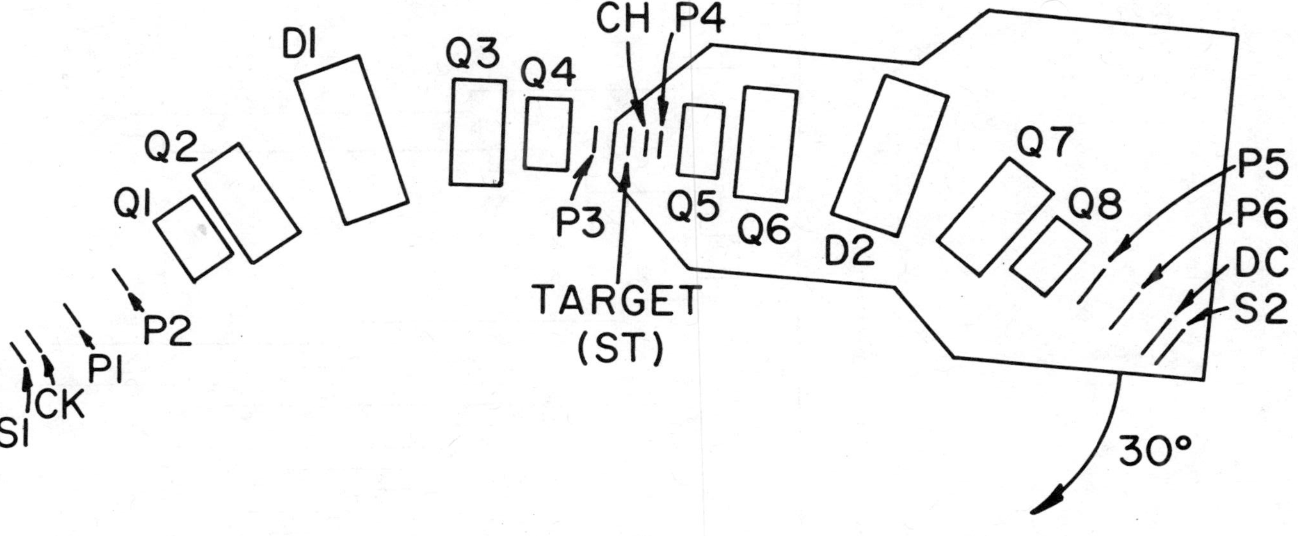

Fig. 1: The Hypernuclear Spectrometer at Brookhaven

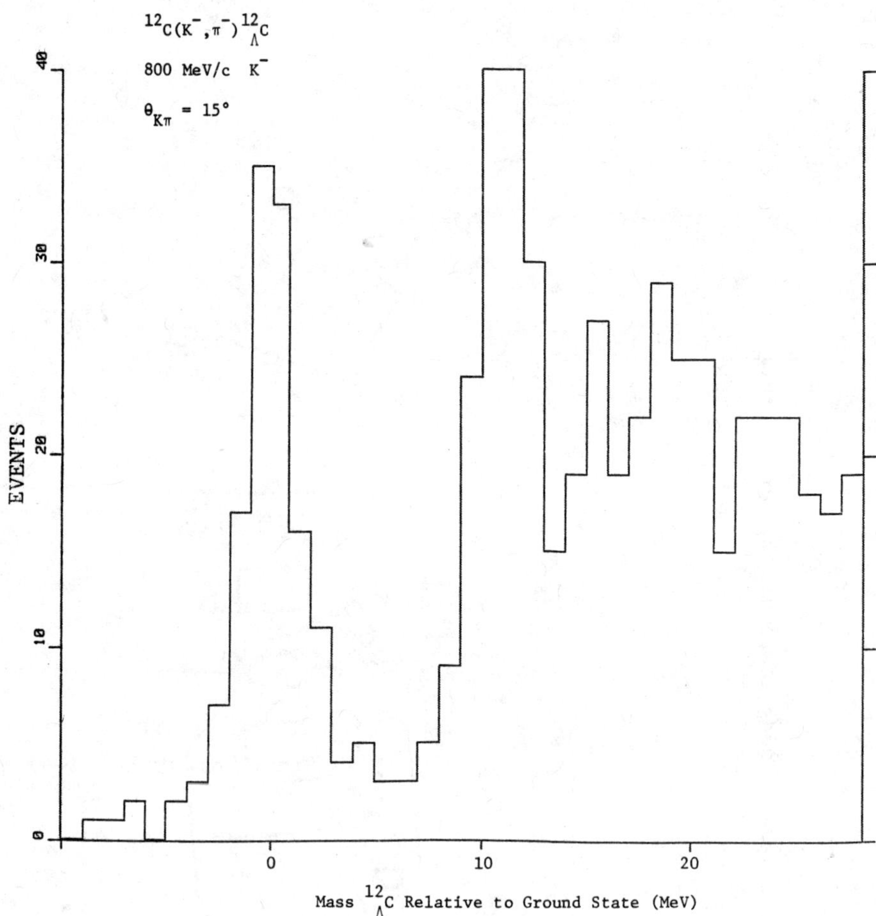
$^{12}C(K^-,\pi^-)^{12}_{\Lambda}C$
800 MeV/c K^-
$\theta_{K\pi}$ = 15°
EVENTS
0
10
20
30
40
0
10
20
Mass $^{12}_{\Lambda}C$ Relative to Ground State (MeV)

Fig. 2

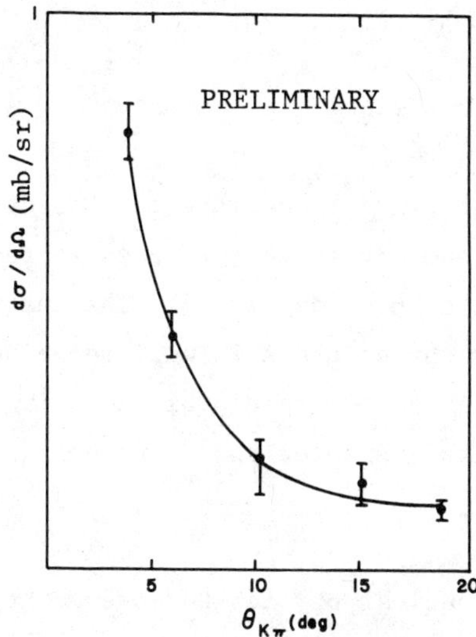

Fig. 3: Differential production cross section for the peak at 11 MeV excitation in ${}^{12}_{\Lambda}C$.

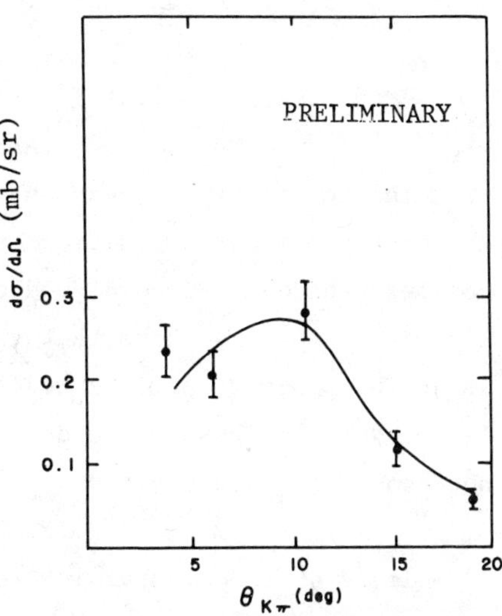

Fig. 4: Differential production cross section for the ground state of ${}^{12}_{\Lambda}C$.

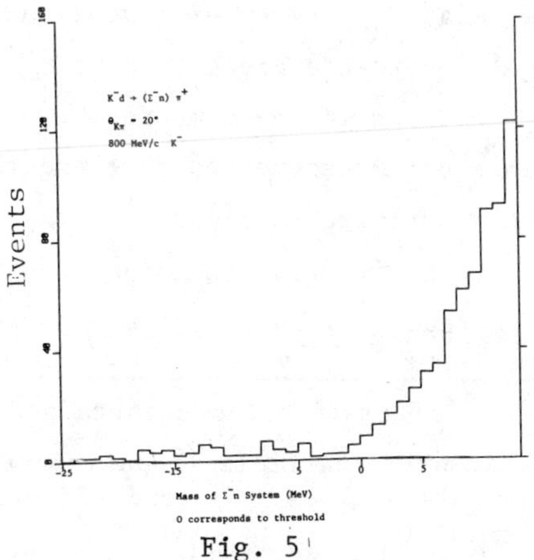

Fig. 5

$\Sigma^{-}n$ Mass Distribution with Sensitivity .18 μb/sr/event

THE ($K^-\pi^-$) REACTION AT CERN

K. Kilian*

Max Planck Institut für Kernphysik, Heidelberg, Germany

ABSTRACT

In the strangeness-exchange reaction ($K^-\pi^-$) on different targets the most prominent transitions to hypernuclear states produced in recoilless and quasi-free Λ production have been identified. The analysis of the spectra shows that the behaviour of the Λ is well described in an independent particle model. The Λ-nucleus spin orbit coupling is found to be about one order of magnitude smaller than for the nucleon-nucleus interaction.

INTRODUCTION

It is of fundamental interest for nuclear physics to observe hyperons in a nucleus[1]. A hyperon can be considered as a probe which allows the testing of models developed for the understanding of the nuclear many-body system. We can get basically new experimental information to check the relation between elementary and effective baryon interaction. Also quasi-particle properties of a nucleon in a nucleus and of hyperons in a nucleus may be compared. Cross checks indicated by arrows in the table below become feasible and they might improve our understanding of the underlying physics.

According to the experimental situation, the following discussions are restricted to hypernuclei with only one hyperon. The role of the

Probe particle	Information can be obtained about the elementary particle interactions from	quasi-particle properties and effective interactions from
nucleon	nucleon-nucleon two-body systems and scattering	nuclear physics ↕
hyperon	hyperon-nucleon two-body systems and scattering ↔	hypernuclear physics

* Visitor at CERN, Geneva, Switzerland.

ISSN:0094-243X/79/540666-14$1.50

Λ particle will be emphasized. But all the qualitative arguments developed will hold for Σ hypernuclei as well. In our experiment we have evidence that several light Σ hypernuclear states have been observed[2].

The best way to produce hypernuclei is the strangeness exchange $K^- + N \rightarrow \Lambda + \pi$ on a nucleon in a nucleus. The produced Λ may then be bound on the residual nucleus or a fragment, thus forming a hypernucleus. An experimental problem was and probably will always be the very low intensity of strange-particle beams. If one wants to extract useful physical information, one has to concentrate the low available reaction rate on some few characteristic hypernuclear states with a very well-known and simple structure.

Hypernuclear ground states have, of course, a simple configuration: the Λ is in the 1S state bound on a nuclear core ground state. Twenty ground-state Λ binding energies up to ${}^{15}_{\Lambda}N$ have been revealed, mainly in emulsion experiments[3]. It could be deduced from them, for example, that the attractive potential depth a Λ feels in nuclear matter is ∿ 28 MeV, only about half the value a nucleon feels.

When in the late sixties stronger K^- beams became available, counter experiments opened up the possibility of investigating also excited hypernuclear states. Especially states which have exactly the simple spin space configuration of the target nucleus ground states are of great interest.

In a first generation of experiments[4] it has been shown that these states are narrow and that they can be selectively produced by a kinematical trick, the recoilless production[5].

RECOILLESS PRODUCTION AND QUASI-FREE SATELLITE STATES

In the strangeness-exchange reaction (1) on neutrons bound on a residual nucleus R

$$K^- + \binom{R}{n} \rightarrow \binom{R}{\Lambda}^* + \pi^- \qquad (1)$$

hypernuclei are produced whose excitation levels show up in the $(K^-\pi^-)$ energy loss spectrum. For K^- momenta between ∿ 300 and ∿ 1000 MeV/c and for π^- detected near 0°, the recoil momentum on the Λ particle is small compared to Fermi momenta of the target nucleons

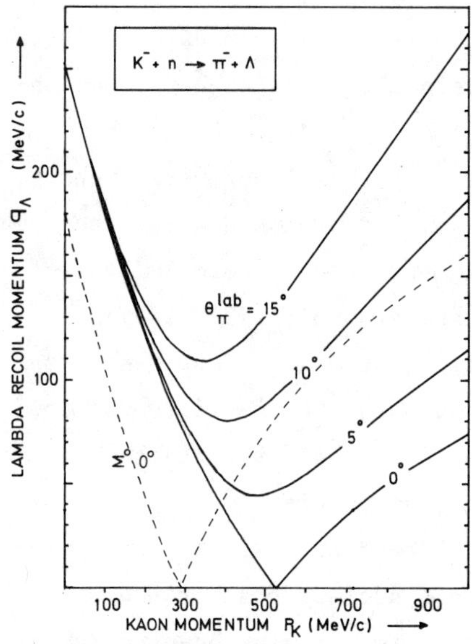

Fig. 1 : Recoil on the Λ particle in the elementary reaction $K^- + n \to \Lambda + \pi^-$ as a function of the kaon laboratory momentum for different $K^-\pi^-$ laboratory angles [dotted line: recoil on a produced Σ^0 at C° ($K^-\pi^-$) angle]

(∿ 250 MeV/c). This situation is shown in Fig. 1. There is a high probability of forming a hypernuclear system where the Λ stays within the well of the residual nucleus. The 0° kinematics and the high K^- and π^- energies favour a direct one-step reaction where single-particle (shell-model) configurations are preferably produced.

The structure of transition amplitude F for hypernucleus production according to reaction (1) is approximated by a product of the elementary ($K^- + n \to \Lambda + \pi^-$) interaction amplitude $f(q_\Lambda)$ with an overlap integral

$$F_{i\to k}(q_\Lambda) \simeq f(q_\Lambda) \int d^3r\, \psi_k^* \exp(i\vec{q}_\Lambda\vec{r})\psi_i \; . \tag{2}$$

$F_{i\to k}$ describes the transfer of a target neutron state ψ_i with, for example, shell-model quantum number i = (nℓj) into a Λ shell-model state ψ_k with quantum numbers k = (n′ℓ′j′).

For small momentum transfer q_Λ the exponential term in Eq. (2) becomes ≈ 1 all over the nuclear volume; the overlap is best if i = k. Only transition amplitudes $F_{i\to i}$ are strong. We expect to see one hypernuclear *recoilless* peak for each neutron hole state i of the target nucleus. Neutron hole and Λ particle couple to the spin parity combination of the target nucleus ground state. In recoilless production no change of angular momentum will occur and at 0° reaction angle also spin flip is excluded. The very pronounced peak in the $^{12}_{\Lambda}C$ spectrum in Fig. 2 is interpreted as being "recoilless" with a neutron hole in the outer $P_{3/2}$ shell of ^{12}C coupled with a Λ in a $P_{3/2}$ configuration to spin parity 0^+ of the ^{12}C target. In our experiment the Λ recoil is only (Fig. 1) q_Λ ≈ 40 to 70 MeV/c. With

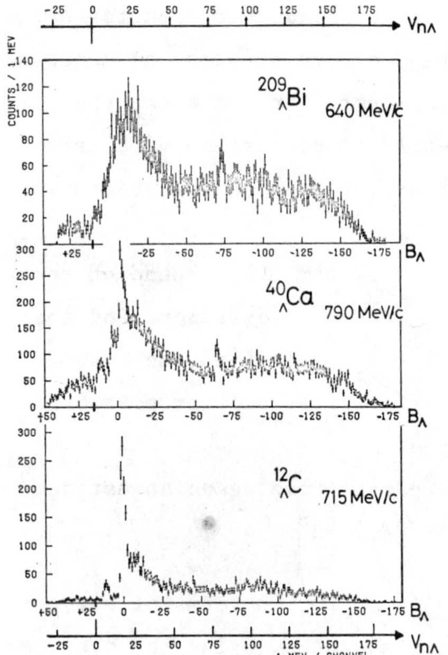

Fig. 2 : Mass spectra of hypernuclei produced in ($K^-\pi^-$) strangeness exchange on ^{12}C, ^{40}C, and ^{209}Bi targets. The quasi-elastic peak dominates in light targets ($^{12}_{\Lambda}C$), while it deteriorates with increasing atomic weight of the target nucleus.

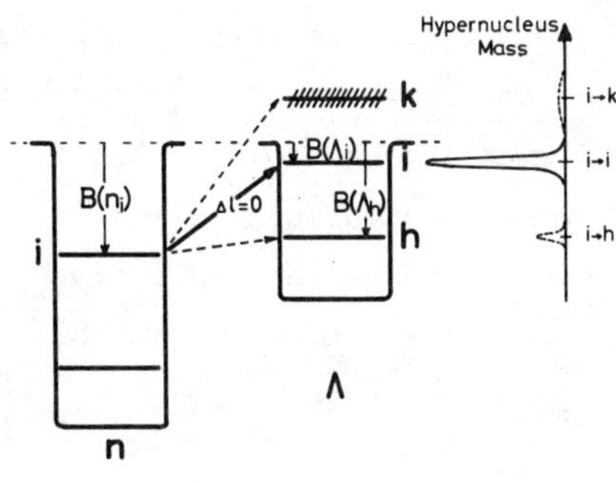

Fig. 3 : Schematic spectrum of hypernuclear states built upon a neutron hole in state i = (nℓj). For small Λ recoil the quasi-elastic transition $n_i \to \Lambda_i$ dominates and quasi-free satellite states, $n_i \to \Lambda_h$ or $n_i \to \Lambda_k$, are weak. The level spacings reflect the different Λ binding energies.

increasing q_Λ (depending on p_K and increasing with the reaction angle) transition amplitudes $F_{i\to k}$ become important. The recoilless peaks shrink and get satellites with Λ particle neutron hole configuration $(i^{-1}k)_{n\Lambda}$, which differ in their configuration from the target nucleus ground state. They are called *quasi-free* states. The single Λ can go freely into configurations already filled with nucleons because there is no Pauli blocking.

The energies of quasi-elastic and quasi-free states can be interpreted according to Fig. 3, in a single-particle picture. They are given by the neutron binding energy $B(n_i)$ [known, for example, from (p,d) experiments] minus the Λ binding energy $B(\Lambda_j)$ (deduced from shell-model calculations for the Λ with a well depth of a ∿ 28 MeV).

For heavier targets hypernuclear states built upon neutron holes in the last shell will dominate the shape of the spectra: a) because of the generally high neutron occupation number in the outer shell, b) because the outer shell is not screened by distortion effects, c) because outer hole states have narrow width, which will be reflected in the hypernuclear states.

Of course the shape of the spectra will strongly depend on the transition probability $P_{i\to i}$ into target-like configurations and the complementary probability $\Sigma_{f\neq i} P_{i\to f} = (1 - P_{i\to i})$ to feed the weaker but more numerous quasi-free satellite states. $P_{i\to i}$ is proportional to the square of the transition amplitude $F_{i\to i}$.

In close analogy to the Mössbauer effect Povh[6] gave an estimate

$$P_{i\to i} \simeq \exp\left[-\left(2n_i + 1\right)\frac{E_\Lambda}{D}\right] . \tag{3}$$

E_Λ/D measures the Λ recoil energy $E_\Lambda = (q_\Lambda)^2/2m_\Lambda$ in terms of a typical level distance D for Λ single-particle states. Using the oscillator approximation, n_i is identified as the oscillator quantum number of the neutron hole state. D can be extrapolated for a target atomic weight A from the known value[4] $D(^{12}_{\Lambda}C) \simeq 11$ MeV by

$$D(A) \simeq D(A_0)\left(\frac{A_0}{A}\right)^{1/3} . \tag{4}$$

$P_{i\to i}$ is the "nuclear Debye-Waller factor". For the cases shown in Fig. 2 from Eqs. (3) and (4) one estimates the recoilless strength $P_{i\to i}$ in the last shell to be about 57% for $^{12}_{\Lambda}C$, 25% for $^{40}_{\Lambda}Ca$, and only 2.8% for $^{209}_{\Lambda}Bi$. This is confirmed in the spectra of Fig. 2. In the $^{12}_{\Lambda}C$ spectrum the recoilless configuration $(1P_{3/2}^{-1}, 1P_{3/2})_{n\Lambda}$ dominates. In $^{40}_{\Lambda}Ca$ still the $(1D_{5/2}^{-1}, 1D_{5/2})_{n\Lambda}$ peak can be seen; for $^{209}_{\Lambda}Bi$, however, no separate recoilless peak appears any more.

The present experiment was designed in order to resolve the fine structure of the recoilless states and to determine the strongest quasi-free transitions. For this purpose, we had to improve the production rates for hypernuclei and the background reduction with respect to earlier experiments.

EXPERIMENT

The $(K^-\pi^-)$ reaction on 11 different target nuclei was studied at K^- momenta around 680 to 790 MeV/c by means of the separated K^-

beam k_{22} at the CERN Proton Synchrotron (PS). The experimental set-up is shown schematically in Fig. 4. Particles produced in the production target are refocused to an achromatic focus of $3.5 \times 0.8\ cm^2$ in the experimental target. Under normal operation with about $1.3 \times \times 10^{12}$ protons on a 3 cm long tungsten production target $2 \times 10^4\ K^-$ reached the experimental target at a kaon momentum of 715 MeV/c. The π^-/K^- ratio at the target was only about 12. The kaons were identified by their time of flight in the kaon spectrometer (Fig. 4), whereas the accompanying background pions were vetoed by a liquid-hydrogen Čerenkov counter in front of the target. The kaon momentum was determined by measuring the particle trajectories in the kaon spectrometer with a hodoscope of plastic counters installed in the first focal plane directly in the mass and momentum slits and with a set of wire chambers in front of the target. Pions from the reaction target, produced in the forward direction, were analysed by the specially designed spectrometer SPES II of Saclay, described in detail elsewhere[7]. The angular acceptance of SPES II is $\gtrsim 20$ msr, the accepted momentum bite ±18%. It could rotate around the target from -5° to 30°. Pions in the spectrometer were identified by their time of flight and by a second liquid-hydrogen Čerenkov counter just behind the target. The pion trajectories in the SPES II were determined with one set of wire chambers at the entrance to the spectrometer and a second set at the exit. The momentum resolution of the SPES II was better than 0.5 MeV/c at 715 MeV/c momentum, whereas the energy loss resolution of the two spectrometers together was about 1.3 MeV. However, the over-all energy resolution in the experiment, of about 2.0 MeV, was determined by the energy-loss difference for kaons and pions in the target and by the straggling in the target and trigger counters. A high rate of hypernucleus production (up to one per minute was obtained by a) 0° kaon production angle, b) the very short high acceptance (~ 9 msr) beam line, c) the large acceptance of SPES II.

A very powerful background reduction already in the on-line trigger was due, besides some other precautions, to the twin Čerenkov counter with liquid-hydrogen radiator. This counter is shown

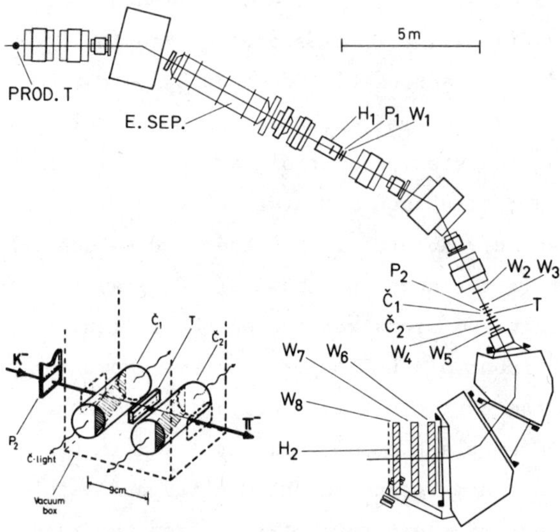

Fig. 4 : Experimental set-up. Kaon and pion momenta are analysed by means of a magnetic system. The particle trajectories are determined by the hodoscopes H_1 and H_2 and the wire chambers W_1-W_8. Kaons and pions are identified: a) by the time-of-flight measurements between plastic scintillators P_1, T, and H_2; b) by the liquid-hydrogen Čerenkov counters $\check{C}_1$ and $\check{C}_2$ before and after the target T. The cylindrical Čerenkov cells contain a thin inclined mirror foil to improve the light collection in the direction of the cylinder axis. This is shown in the schematic view of the target region.

schematically in the inset of Fig. 4. It allowed restriction of the volume where the K^- to π^- reactions occur to ±2.5 cm around the target. This is the distance between the two radiator cells. On an average 8 photoelectrons were collected for a passing pion event in the photomultipliers looking on one radiator cell. The inefficiency for pion identification therefore was less than 5×10^{-4}. For the final off-line evaluation of the $(K^-\pi^-)$ reaction events it was required that kaon trajectories measured in front of the target and pion trajectories measured after the target really met in the target volume. Finally, meson decays in flight in SPES II could be ruled out by comparing the measured track coordinates at the entrance with their values reconstructed from the exit trajectories in SPES II.

For the $(K^-\pi^-)$ events left after the software evaluation we know the measured kaon and pion momentum and the reaction angle θ between kaon and pion track. This is sufficient to determine in the assumed reaction $K^- + T \rightarrow \pi^- + Y$ the mass m_Y of particle Y in its rest frame. In Fig. 5b $(K^-\pi^-)$ events obtained with a ^{6}Li target (95.5% enriched) are plotted according to their calculated hypernuclear mass and their reaction angle θ, assuming T to be ^{6}Li. This

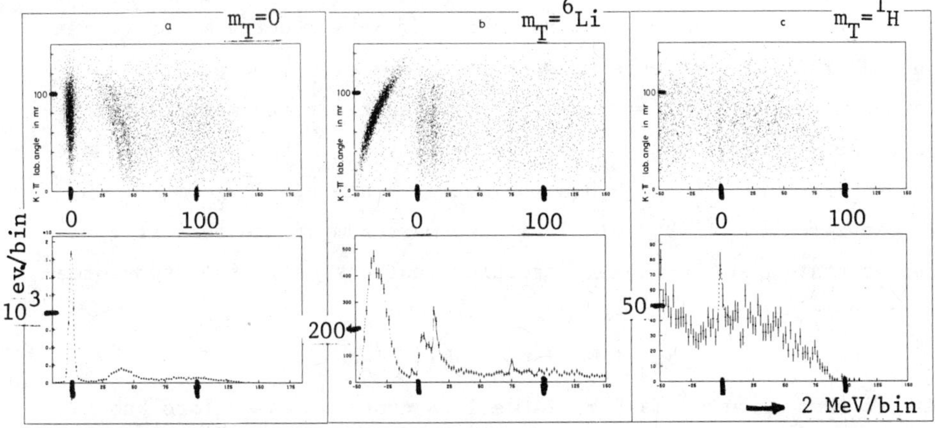

Fig. 5 : Sample of $(K^-\pi^-)$ events obtained with a 790 MeV/c K^- beam on 95.5% enriched ^{6}Li at 0° spectrometer setting. In the reaction $K^- + T \rightarrow \pi^- + Y$ different target masses m_T are assumed to evaluate the mass m_Y of the recoil particle. The density of events depending on m_Y and the $(K^-\pi^-)$ reaction angle is shown and their projection on the m_Y mass scale. a) Assuming $m_T = 0$ there is a group of events reproducing the mass of π^0. These are $K \rightarrow 2\pi$ decays in flight. Plotted is $m_Y - m_{\pi^0}$. b) $m_T = {}^6$Li is assumed. Near $m_Y = m[^5\text{Li(g.s.)}] + m_\Lambda$ clear hypernuclear states appear. In the projection to the mass scale the $K \rightarrow 2\pi$ decay events produce a broad distribution. Plotted is $m_Y - \{m[^5\text{Li(g.s.)}] + m_\Lambda\}$. c) The high mass end of the sample is evaluated for a proton target $m_T = {}^1$H. A peak at $m_Y = m_{\Sigma^+}$ shows reaction events due to $K^- + p \rightarrow \pi^- + \Sigma^+$. Plotted is $m_Y - m_{\Sigma^+}$. Cases (a) and (c) are used for calibration.

plot and its projection to the mass scale clearly reveal hypernuclear states, where the reconstructed mass is independent of θ as expected, and a strong background, which is grouped in a bent distribution at the low mass end. Eventual Σ hypernuclear states appear in this plot shifted to higher mass by the Σ-Λ mass difference (∿ 77 MeV). If one evaluates the same sample of events assuming $m_T = 0$, then as shown in Fig. 5a the bent distribution is regrouped at the mass of the neutral pion. So these events are $K^- \rightarrow \pi^- + \pi^0$ decays in flight (between the two liquid-hydrogen cells). This background can be cut easily in the linearized plot of Fig. 5a and we use it as a check of our spectrometer calibrations. Another calibration reaction exists at the high-energy end of the Λ-hypernuclear spectrum as shown in Fig. 5c. There, we reconstruct a peak at the mass of the Σ^+ hyperon assuming a proton as target. So we observe here the reaction $K^- + p \rightarrow \pi^- + \Sigma^+$ (on protons in the second liquid-hydrogen cell).

FURTHER RESULTS

In order to present the hypernuclear mass spectra m_Y of course one subtracts a reference mass of the order of m_Y. We get the excitation energy scale $E^* = m_Y - m_{g.s.}$ if we subtract the mass of the hypernucleus ground state $m_{g.s.}$, but usually $m_{g.s.}$ is not known. More common is therefore the Λ binding energy scale $B_\Lambda = -[m_Y - m_Y(B_\Lambda = 0)]$. We can calculate $m_Y(B_\Lambda = 0)$. It is the mass sum of the nuclear core ground state and the free Λ particle. Instructive is the transformation energy scale

$$V_{n\Lambda} = m_Y - m_T - (m_\Lambda - m_n) .$$

Target mass m_T and Λ mass m_Λ as well as neutron mass m_n are known. The value $V_{n\Lambda} = 0$ for a hypernuclear state means that the transformation of a bound target neutron into a Λ in the hypernucleus needs the same energy as if n and Λ were free (namely the Λ-n mass difference $m_\Lambda - m_n = 176$ MeV/c^2). The physical meaning of $V_{n\Lambda}$ is (compare Fig. 2)

$$V_{n\Lambda}(ik) = B(n_i) - B(\Lambda_k) . \tag{5}$$

$V_{n\Lambda}$ shows the balance between the energy $B(n_i)$ one has to pay for the separation of a target neutron in a state i and the energy $B(\Lambda_k)$ one gains by the binding of the produced Λ in a state k. If we replace the Λ mass by the Σ hyperon mass we get the scales for Σ hypernuclei.

For neutrons the nuclear potential is deeper than for the Λ. We expect, therefore, recoilless configurations on deep-lying neutrons to have the highest transformation energies. The $(1S_{1/2}^{-1}, 1S_{1/2})_{n\Lambda}$ configuration in the spectrum of ${}^6_\Lambda$Li (Fig. 6) should correspond to the pronounced peak found at highest energy, $V_{n\Lambda}(S_{1/2}^{-1}, S_{1/2}) = 19.4$ MeV.

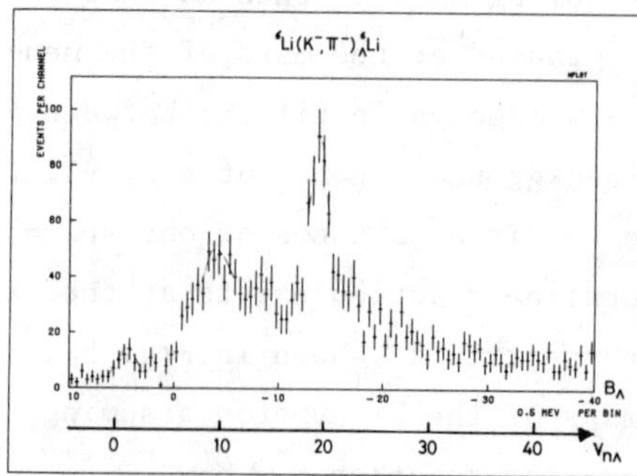

Fig. 6 : Expanded ${}^6_\Lambda$Li mass spectrum of Fig. 5b. The prominent spikes at $V_{n\Lambda} = 19.4$ MeV and 9.4 MeV are attributed to the recoilless particle hole configurations in the 1S and 1P shell, respectively. The weaker state at $V_{n\Lambda} = 1.2$ MeV should be the quasi-free transition to the Li ground state with a $(1S^{-1}, 1P)_{n\Lambda}$ configuration.

The recoilless $(1P_{3/2}^{-1}, 1P_{3/2})_{n\Lambda}$ configuration should show up in the broader and weaker peak at $V_{n\Lambda}(P_{3/2}^{-1}, P_{3/2}) = 9.4$ MeV. The intensity ratio is in rough agreement with the neutron occupation 2:1 for S and P neutrons in ^{6}Li. There is a weak but clear transition to a state at $V_{n\Lambda} = 1.2$ MeV. This could be the $^6_\Lambda$Li ground state with a $(P_{3/2}^{-1}, S_{1/2})_{n\Lambda}$ configuration. This ground state was unknown so far. From the general relation $V_{n\Lambda} = -B_\Lambda + B(n_{\text{last shell}})$ and $B(n_{\text{last shell}}) = 5.66$ MeV [8] we get a lower limit for $B_\Lambda(\text{g.s.}) = 4.5$ MeV. [For $^6_\Lambda$He the mirror hypernucleus of $^6_\Lambda$Li, $B_\Lambda(\text{g.s.}) = 4.18$ MeV is known and it is of similar magnitude.]

For ^{6}Li the S and P state n-separation energies are known from (pd) experiments[9] to be $B(n_{1S}) = 22.3$ MeV and $B(n_{1P}) = 5.66$ MeV, respectively. With the single-particle assignments of the three peaks we can calculate separation energies $B(\Lambda_k)$ for Λ shell-model states on a ^{5}Li core. Using the relation (5) we find:

$$B(\Lambda_{1S}) = B(n_{1S}) - V_{n\Lambda}(S_{1/2}^{-1}, S_{1/2}) = 2.9 \text{ MeV}$$

$$B(\Lambda_{1S}) = B(n_{1P}) - V_{n\Lambda}(P_{3/2}^{-1}, S_{1/2}) = 4.5 \text{ MeV}$$

$$B(\Lambda_{1P}) = B(n_{1P}) - V_{n\Lambda}(P_{3/2}^{-1}, P_{3/2}) = -3.7 \text{ MeV} .$$

The Λ is bound in the 1S state. But it has less binding on the excited (S hole) ^{5}Li core (2.9 MeV) than on the more compressed ^{5}Li ground state (4.5 MeV). The Λ is unbound in the 1P state (-3.7 MeV). The difference of single-particle energies for the Λ bound in $1S_{1/2}$ and $1P_{3/2}$ states on a ^{5}Li ground state core corresponds directly to the 8.2 MeV distance between the quasi-free peak $(1P_{3/2}^{-1}, 1S_{1/2})$ at $V_{n\Lambda} = 1.2$ MeV and the $(1P_{3/2}^{-1}, 1P_{3/2})$ recoilless peak at $V_{n\Lambda} = 9.4$ MeV.

In Fig. 7 we compare results on ^{12}C and ^{16}O targets[10]. The carbon spectrum shows no surprising results. The dominant peak at $V_{n\Lambda} = 18.7$ MeV ($B_\Lambda = 0$) belongs to the recoilless Λ production on the outer $P_{3/2}$ shell. The weaker state at $V_{n\Lambda} = 7.7$ MeV ($B_\Lambda = 11$ MeV) was already observed[4] in a previous experiment using stopped K^- for producing hypernuclei. It belongs to the $(1P_{3/2}^{-1}, 1S_{1/2})_{n\Lambda}$ configuration and has $J^\pi = 1^-$. The energy difference between these two states of 11 MeV discloses the distance between the $1S_{1/2}$ and $1P_{3/2}$ shells of a Λ particle bound on a ^{11}C ground-state core.

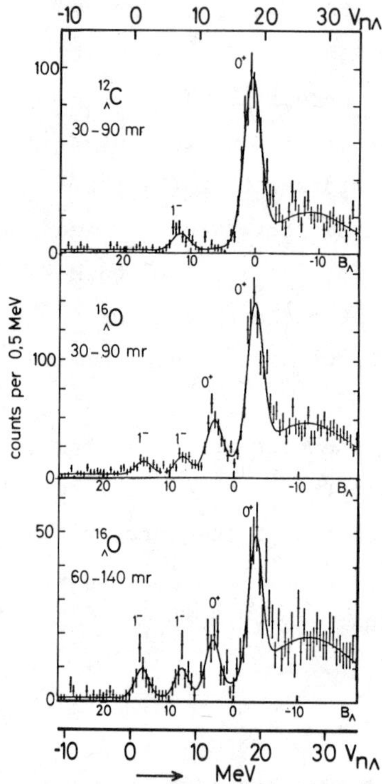

Fig. 7 : Hypernuclear mass spectra ($V_{n\Lambda}$) obtained from the ($K^-\pi^-$) reaction on ^{12}C (top) and ^{16}O (centre) at a kaon momentum of 715 MeV/c and 0° spectrometer setting. In addition, the Λ binding energy B_Λ is plotted for each spectrum. At the bottom a $^{16}_{\Lambda}O$ spectrum taken at 5° spectrometer setting is shown. The curves are fit with 2 and 4 narrow Gaussians and a background for $^{12}_{\Lambda}C$ and $^{16}_{\Lambda}O$, respectively. In the two $^{16}_{\Lambda}O$ spectra the ratio of quasi-free (1^- states) to quasi-elastic (0^+ states) strength increases (by more than a factor of 2) when increasing the average reaction angles.

The transformation energies for the two narrow levels in $^{12}_{\Lambda}C$ $[(1P^{-1}_{3/2}, 1P_{3/2})_{n\Lambda}$ and $(1P^{-1}_{3/2}, 1S_{1/2})_{n\Lambda}$ configuration$]$ coincide well with two peaks observed in the oxygen spectrum. Moreover, the relative intensities are equal in both cases. It is rather obvious that in neighbouring nuclei the transformation energy for the nucleons in the same orbit will be the same in first approximation. We therefore conclude that the two transitions in oxygen at $V_{n\Lambda}$ = 19.2 MeV and $V_{n\Lambda}$ = 8.7 MeV belong to the $(1P^{-1}_{3/2}, 1P_{3/2})_{n\Lambda}$ and $(1P^{-1}_{3/2}, 1S_{1/2})_{n\Lambda}$ configurations, respectively.

In oxygen, two additional transitions are observed, which we ascribe to the additional $P_{1/2}$ neutrons available in ^{16}O. These two states coincide with the carbon spectrum if shifted to a lower excitation $V_{n\Lambda}$ by 6 MeV. A measurement with ^{16}O at higher reaction angles reveals that the states at $V_{n\Lambda}$ = 19.2 MeV and $V_{n\Lambda}$ = 13.2 MeV are produced in recoilless Λ production (Fig. 7); they decrease in intensity with respect to the states at $V_{n\Lambda}$ = = 8.7 MeV and $V_{n\Lambda}$ = 2.7 MeV by more than a factor of 2. The states increasing with angle are therefore assigned to quasi-free Λ production. We relate the $V_{n\Lambda}$ = 13.2 MeV peak to the $J^\pi = 0^+$ state with $(1P^{-1}_{1/2}, 1P_{1/2})_{n\Lambda}$, the one at $V_{n\Lambda}$ = 2.7 MeV to the $J^\pi = 1^-$ state with $(1P^{-1}_{1/2}, 1P_{1/2})$, the one at $V_{n\Lambda} = 2.7$ MeV to the $J^{\pi} = 1^{-}$ state with $(1P^{-1}_{1/2}, 1S_{1/2})_{n\Lambda}$.

The splitting between the hypernuclear states with the $1P^{-1}_{3/2}$ and $1P^{-1}_{1/2}$ nuclear core is 6 MeV for both the recoilless and the quasi-free produced states. This discloses directly that the spin-orbit interaction of the Λ particle in the p-shell nuclei is small. In fact, the energy difference of 6 MeV between the $(1P^{-1}_{3/2}, 1P_{3/2})_{n\Lambda}$ and the $(1P^{-1}_{1/2}, 1P_{1/2})_{n\Lambda}$ configurations agrees so well with the 6.1 MeV energy splitting between the $P_{1/2}$ and $P_{3/2}$ hole states in ^{15}O and the 6 MeV difference between the $(1P^{-1}_{3/2}, 1S_{1/2})_{n\Lambda}$ and the $(1P^{-1}_{1/2}, 1S_{1/2})_{n\Lambda}$ states, that for p-shell hypernuclei we can given an upper limit of 0.3 MeV for the spin-orbit splitting in the Λ-nucleus interaction.

A similar comparative discussion of our new results on ^{32}S and ^{40}Ca is presented in a contribution to this Conference[11]. The results are: i) the level spacings for Λ single-particle states on these medium heavy nuclei are approximately 9 MeV; ii) Also in the $D_{3/2}$, $D_{5/2}$ shells there is no significant spin-orbit splitting found for the Λ particle.

For a quantitative description of the shape of hypernuclear strangeness-exchange spectra, distortion and mixing have to be taken into account. Bouyssy[12] performed a distorted wave impulse approximation (DWIA) calculation using the shell model to describe the structure aspect of hypernuclei. He also considered energetically close-lying non-target-like satellite states. His calculations reproduce excellently the energies as well as the intensities of the transitions observed in our present experiment.

The spin-orbit force plays a very important role in nuclear physics. Nevertheless its origin as well as its strength are still not fully explained. It is puzzling that the strength of the Λ-nucleus spin-orbit force is at least an order of magnitude smaller than that of the nucleon-nucleus one. Using a boson-exchange model for calculating the baryon-baryon interaction, however, Brockmann and Weise[13] have come to the conclusion that the spin-orbit force should be essentially weaker in the Λ-nucleus interaction than in the nucleon-nucleus interaction. It is hoped that with the present results, which supply an additional constraint on the choice of the baryon-nucleus interaction, the understanding of the spin-orbit force in nuclei can be improved.

REFERENCES

1. B. Povh, Ann. Rev. Nucl. Sci. 28, 1 (1978).
   R.H. Dalitz, Invited paper at this Conference.
   A. Gal, Invited paper at this Conference.
   B.F. Gibson, Invited paper at this Conference.
   M. May, Invited paper at this Conference.
   M. Bedjidian et al., Contributed paper at this Conference.
   J.A. Johnstone and J. Law, Contributed paper at this Conference.
   Q.N. Usmani, Contributed paper at this Conference.
2. R. Bertini, Invited paper at this Conference.
3. T. Cantwell et al., Nucl. Phys. A236, 445 (1974).
   M. Jurić et al., Nucl. Phys. B52, 1 (1973).
4. W. Brückner et al., Phys. Lett. 62B, 481 (1976).
5. M.I. Podgoretski, Zh. Eksp. & Teor. Fiz. 44, 695 (1963).
   H. Feshbach and A.K. Kermam, Preludes in theoretical physics (North Holland, Amsterdam, 1966), p. 260.
6. B. Povh, Z. Phys. A279, 159 (1976).
7. E. Aslanides et al., *in* Abstracts of papers submitted to the 7th Internat. Conf. on High-Energy Physics and Nuclear Structure, Zurich, 1977 (SIN, Villigen, 1977), p. 376.
8. F. Ajzenberg-Selove and T. Lauritzen, Nucl. Phys. A227, 1 (1974).
9. B. Fagerström et al., Phys. Scr. 13, 101 (1976).
10. W. Brückner et al., Phys. Lett. 79B, 157 (1979).
11. R. Bertini et al., Contributed paper at this Conference.
12. A. Bouyssy, Nucl. Phys. A290, 324 (1977).
13. R. Brockmann and W. Weise, Phys. Lett. 69B, 167 (1977).

Question: E. Lomon, MIT

Do you get a more accurate value of the Λ central potential depth?

Response: Kilian

This data has not been analyzed in detail yet but is consistent with 20-30 MeV.

Question: E. Auld, UBC

Are the cross sections for the (K^-,π^-) nuclear data consistent with the free K^--N interaction?

Response: Kilian

From our new data we can deduce an effective neutron number on the order of one for Λ hypernucleus production.

Question: A. Gal, Hebrew University

It seems to me perplexing that you assign the narrow excitation observed by you in ^{6}Li to the $(1s)_n^{-1}(1s)_\Lambda$ configuration. The $(1s)_n^{-1}$ state is generally known to be considerably wider. Also, if the broader excitation is assigned by you to $(1p)_n^{-1}(1p)_\Lambda$, then it seems from your data that it takes about 7.5 MeV to excite the Λ from the 1s shell to the 1p shell; however, this excitation is expected to be significantly higher than the 11 MeV observed for this excitation in ^{12}C since ^{6}Li is a considerably lighter nucleus than ^{12}C.

Response: Kilian

The neutron $1s^{-1}$ strength is determined experimentally. In the case of ^{6}Li it has a very narrow component, even more narrow than the $1p^{-1}$ contribution. The discrepancies in the 1s-1p energy splitting for the Λ which you mention and the larger width of the assumed $(1p^{-1}1p)_{n\Lambda}$ configuration might be explained by the fact that the Λ is already unbound in the 1p level.

HYPERNUCLEAR SPECTROSCOPY

A. Gal*
The Racah Institute of Physics
The Hebrew University of Jerusalem, Jerusalem, Israel

ABSTRACT

The recent experiments on (K^-,π^-) reactions in ^{16}O and ^{12}C are analyzed, with special attention paid to the Λ-nuclear spin-orbit coupling and the exchange-mixture of the ΛN effective interaction. The potentiality of $(K^-,\pi^-\gamma)$ reactions for the establishment of the low-lying hypernuclear spectrum in light elements is demonstrated, with specific proposals made to unravel the spin-dependence of the ΛN effective interaction.

INTRODUCTION

Recent data from (K^-,π^-) experiments at BNL and CERN, with kaons of momentum between 700 to 800 MeV/c in the laboratory, were reported in the preceding talks by Drs. M. May and K. Kilian, respectively. The CERN new information, while achieving quite good resolution of about 2 MeV and extending over several nuclei, is yet incomplete because of its restriction to the very forward direction ($\theta \lesssim 5^{\circ}$) and the selectivity of its presentation. For example, the Sulfur and Calcium spectra issued recently for publication [1] show clear evidence for three or four hypernuclear states which probably originate from a neutron hole in the 2s-1d shell; however, more extensive angular distributions are required for the conclusive assignment of spectroscopic configurations to these states so that a satisfactory choice among various forms of Λ-nuclear effective interactions can be readily made. On the other hand, the BNL data,[2] although extending the measurement of angular distributions up to $\theta=19^{\circ}$, suffer currently from rather poor resolution of about 4 MeV and are essentially restricted to ^{12}C. For the latter, and also for ^{16}O, some detailed small-angle distributions have been recently published [3] by the CERN group.

In general, one may divide a typical small angle (K^-,π^-) hypernuclear spectrum into three regions: (i) the high excitation region for which the Λ is formed under quasi-free circumstances and which dominates hypernuclear production for $q>>R$, where q is the momentum transfer and R is the nuclear radius. A typical quasi-free spectral shape relates the shift in the mean energy transfer to the well-depth D of the Λ-nuclear potential and the width to the Λ effective mass. Thus, the earlier 900 MeV/c CERN

*Supported in part by the U.S.-Israel Binational Science Foundation.

ISSN:0094-243X/79/540680-11$1.50

data [4] for ^{12}C, ^{16}O, ^{32}S and ^{40}Ca gave a value [5] of D=(31±2) MeV, consistent with the DWIA fitting of these data by Bouyssy and Hufner,[6] D=(28±3) MeV. The Λ effective mass is so far consistent with a value $m^* \approx m_\Lambda$.

(ii) The excitation region about $B_\Lambda=0$, where "analog" excitations are clearly observed at 0°. Because of the strong nuclear absorption of the incoming kaon and the outgoing pion, these states are tentatively assigned to configurations of the type $(\ell_j)_n^{-1}(\ell_j)_\Lambda$, where ℓ_j is a neutron orbital of the last major shell to be filled in the target nucleus, with spin-parity J^π identical to that of the target ground-state (0^+ for most of the targets analyzed so far). Since, depending mostly on the strength of the Λ-nuclear spin-orbit coupling relative to the empirically determined neutron-nuclear spin-orbit coupling, these unperturbed configurations may lie close to each other, the residual central ΛN interaction may play a decisive role in the proper diagonalization of the energy matrix appropriate to these states. Furthermore, because of the expected near identity of the orbitals of the neutron-hole and of the Λ-particle for such configurations, permutation symmetry of the Λ with respect to the valent neutrons may provide an approximate basis in which the exchange properties of the residual central ΛN interaction play a major role. Thus,[7] the two ingredients of the ΛN interaction most likely to affect the structure of the "analog" region, particularly in light elements, are the Λ-nuclear spin-orbit coupling and the exchange mixture of the ΛN residual interaction. In this respect all past quantitative attempts to determine the strength of the Λ spin-orbit coupling, although a-posteriori might stand out, are incomplete.

(iii) The low-lying excitations, in particular the hypernuclear ground-state configuration, where the Λ is in the 1s shell. Such low-lying states, with tentative spin-parity assignment 1^-, have been observed in $^{12}{}_\Lambda C$ and $^{16}{}_\Lambda O$[2,3] and, with somewhat speculative spin-parity assignments,[1] in $^{32}{}_\Lambda S$ and $^{40}{}_\Lambda Ca$. If these states are particle-stable, γ-spectroscopy [8] becomes feasible for their experimental study. In particular, the possibility arises to determine hypernuclear $(1s_{1/2}\Lambda)$ doublet-splittings which are smaller than the resolution of about 1 MeV attempted in (K^-,π^-) experiments. Some of these doublet-splittings are intimately connected with specific components of the ΛN spin-dependent effective interactions, such as spin-spin, spin-orbit and tensor interactions, and we hope to obtain from these splittings--when observed--useful information on the strength of the spin-dependent components of the ΛN interaction.

In this talk I shall almost exclusively deal with light hypernuclei (A≲16), for which the available data are more suggestive than for heavier elements. The $^{16}{}_\Lambda O$ spectrum is known to provide a good example for the occurrence of "analog" states and, as explained above, this implies some constraints on the Λ-nuclear spin-orbit coupling together with the ΛN exchange mixture. My present remarks on $^{16}{}_\Lambda O^*$ are directed along this line of approach. Less known is the recent suggestion,[10] following the measurements of ref. 2, that the "analog" configuration in $^{12}{}_\Lambda C$ offers a similar information which, however, is to be extracted from well-resolved angular distributions in the range $10^o \leq \theta \leq 20^o$. This suggestion is here elaborated on. Finally, I shall comment on several specific spectra [8] of low-lying hypernuclear levels ($1s_{1/2}\Lambda$) which may be derived in the near future by doing the $(K^-,\pi^-\gamma)$ reaction in light nuclei. As mentioned above, this may unravel information on the spin properties of the ΛN residual interaction.

SPIN-ORBIT FOR Λ IN THE P-SHELL

The most recent experimental evidence [3] on the two 0^+ states in ${}^{16}{}_{\Lambda}O$, which are strongly excited at $\theta=0^o$ and are therefore assumed to belong to the "analog" configuration $(1p)_n^{-1}(1p)_{\Lambda}$, is for a spacing of about $\Delta E=6$ MeV and ratio of intensity of about $R\approx 0.35$ between the lower of these states to the upper one. If zero Λ spin-orbit interactions are assumed, the spacing ΔE is calculated [7,11] to be in the range 8 to 10 MeV, depending on the exchange mixture of the ΛN central effective interaction used in these calculations. To reach $\Delta E\sim 6$ MeV, with the introduction of one-body Λ spin-orbit potential, two solutions for each given value of the exchange mixture are obtained, both with the sign of Λ spin-orbit coupling the same as for nucleons. The first solution, advocated in ref. 6 for zero exchange-mixture, corresponds to Λ one-body spin-orbit coupling about 0.3 of that for nucleons in this mass range. However, the calculated ratio of intensities is found in refs. 7 and 11 to be then about 0.1 or less, in contrast to the value of R quoted above. The other solution, pointed out in ref. 11, corresponds to Λ one-body spin-orbit coupling about twice that for nucleons, almost independently of the exchange-mixture in the range of Wigner to Serber forces, with intensity ratio between R=0.9 to 0.4, respectively. For such strong Λ spin-orbit one-body potential, the underlying two-body ΛN spin-orbit matrix elements may not be neglected, which introduces an additional element of uncertainty into this calculation. One concludes that the $\theta=0^o$ ${}^{16}{}_{\Lambda}O^*$ spectrum does not uniquely determine, at present, the Λ spin-orbit coupling, but that the sign of the Λ spin-orbit splitting is likely to be the same as that for nucleons.

On the theoretical side, OBEP evaluations,[12,13] or those of similar models [14] suggest ΛN spin-orbit interactions which are comparable in sign (for the symmetric interaction) and strength to the NN spin-orbit interaction. The resulting estimate for the one-body Λ spin-orbit potential is about 1/3 to 1/2 of that for the nucleon spin-orbit potential. The phenomenological analysis [9] of hypernuclear ground-states in the p-shell suggests, however, ΛN symmetric spin-orbit interaction of opposite sign to that mentioned above and a strong ΛN antisymmetric spin-orbit interaction which add up to yield a large reversed spin-orbit splitting for the Λ. This embarrassing finding comes about due to the dominance of the term $\vec{s}_{\Lambda}\cdot\vec{\ell}_N$ in the $1s_{1/2}\Lambda$ doublet-splittings, the sign of which is determined, for example, by the assignment [15] of spin J=1 to ${}^{12}{}_{\Lambda}B$ ground state and also, indirectly, by the (4.4±0.7) MeV spacing observed [2] between the two lowest 1^- "states" in $^{12}{}_\Lambda$C. A strong ΛN tensor interaction could in principle change the results of the phenomenological analysis of doublet splittings, but such interaction is not believed to be significant; some of the doublet-splittings which are sensitive to the tensor interaction are mentioned in the next section.

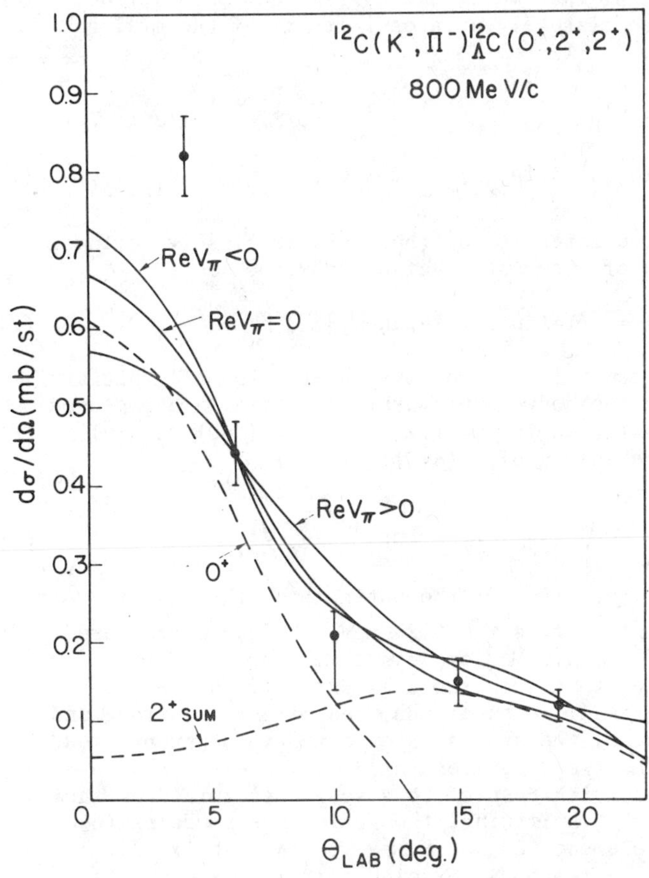

Fig. 1

Spin-orbit effects cannot be directly deduced from the $\theta=0^o$ $^{12}{}_\Lambda$C spectrum, because only one 0^+ "analog" excitation, tentatively assigned as $(1p_{3/2})_n^{-1}(1p_{3/2})_\Lambda$, is observed. However, more states are expected to participate in the formation of the "analog" configuration as θ increases. Fig. 1 shows DWIA calculated [10] angular distributions for the sum of the $0^+,2^+,2^+$ levels of the $(1p_{3/2})_n^{-1}(1p)_\Lambda$ configuration in $^{12}{}_\Lambda$C, normalized to the $\theta=6^o$ data [2] point. The unnatural-parity levels of this configuration are calculated to be only weakly excited and may safely be ignored in the relevant angular range. The effect of the 2^+ states, which differ from each other to zero-order by the Λ spin-orbit coupling, is to produce a shoulder in the summed cross section between 15° and 20°, where the 2^+ states peak. The unresolved data show clear evidence for this shoulder in the angular range characteristic of J=2 excitations, as the 0^+ excitation alone is expected to fall off much more rapidly with angle than the data indicate. In future experiments, with better energy resolution, it may be possible to resolve the 0^+ and 2^+ states which probably include also relatively weakly excited states of approximate structure $^{11}C^*(1/2^-)(1p_{1/2})_\Lambda$ and $^{11}C^*(1/2^-)(1p_{3/2})_\Lambda$, respectively. The inclusion of the latter states requires a departure from the description of ^{12}C in terms of closed jj shells. In the event that the coupling of these states to the main triplet of levels, $0^+,2^+,2^+$, may be neglected the analysis[10] outlined below is useful.

The relative intensity of the 2^+ states will be very sensitive to configuration mixing induced by the ΛN residual interaction. Introducing the appropriate eigenstates in terms of the mixing angle α:

$$\psi_1 = \cos\alpha\ (1p_{3/2})_n^{-1}(1p_{3/2})_\Lambda + \sin\alpha\ (1p_{3/2})_n^{-1}(1p_{1/2})_\Lambda$$

$$\psi_2 = -\sin\alpha\ (1p_{3/2})_n^{-1}(1p_{3/2})_\Lambda + \cos\alpha\ (1p_{3/2})_n^{-1}(1p_{1/2})_\Lambda, \qquad (1)$$

the predicted relative intensity of these states closely follows, in the angular range of interest, that of $\theta=0°$,

$$R \equiv I(\psi_2)/I(\psi_1) = [(1-\tan\alpha)/(1+\tan\alpha)]^2 . \qquad (2)$$

For illustration, a central ΛN effective interaction of arbitrary exchange mixture and one-body Λ spin-orbit potential are employed, leading to the following energy matrix in the $(\psi_1(\alpha=0),\psi_2(\alpha=0))$ basis relative to the energy of $\psi_1(\alpha=0)$:

$$\begin{pmatrix} 0 & 2F_2 \\ 2F_2 , & -3F_2 + \Delta\varepsilon_\Lambda \end{pmatrix} , \qquad (3)$$

where $\Delta\varepsilon_\Lambda=\varepsilon_\Lambda(p_{1/2})-\varepsilon_\Lambda(p_{3/2})$. For the estimate[7] of the Slater integral of the ΛN interaction $F_2\sim-0.24$ MeV, the two states are only weakly mixed if $\Delta\varepsilon_\Lambda\gtrsim0.5$ MeV or $\Delta\varepsilon_\Lambda\lesssim-1$ MeV and both are expected to be excited to a comparable extent. If this is not the case, and $\Delta\varepsilon_\Lambda$ is small in magnitude, the mixing is strong and the excitation of one of the states is expected to be suppressed. Unlike the case of the two 0^+ states in $^{16}_{\Lambda}O$, where the Λ spin-orbit splitting enters with respect to a scale determined by the 6 MeV nuclear spin-orbit splitting, the scale here is determined by the off-diagonal element, $2F_2\sim-0.5$ MeV, of the matrix (3). The introduction of the two-body ΛN spin-orbit matrix elements will not qualitatively change these remarks. Furthermore, the unknown ΛN exchange mixture cancels out in the $^{12}_{\Lambda}C^*$ matrix (3). The set of 2^+ states of $^{12}_{\Lambda}C$ may, therefore, emerge as a fine analyzer for the determination of $\Delta\varepsilon_\Lambda$.

It is clear from the above discussion of $^{16}_{\Lambda}O^*$ and $^{12}_{\Lambda}C^*$ that more and better data are required before a self-consistent analysis of the various main ingredients of the Λ-nuclear effective interaction may be attempted within the configuration $(1p)_n^{-1}(1p)_\Lambda$. At present, the angular distributions published by the CERN group [3] for very small angles ($\theta\lesssim5°$) in ^{12}C and ^{16}O have all cross-sections for the coherent 0^+ states which fall too rapidly with angle. Thus, if the angular distributions are parametrized by $\sim\exp(-b^2q^2)$, then the slope b^2 is measured to be about 5 fm^2, whereas a straightforward calculation yields b^2=2.0 to 2.5 fm^2 in this mass range. It is difficult to think of reasonable admixtures to the coherent excitations which would increase the calculated slope to such extent.

LOW LYING STATES IN THE P-SHELL

A systematic discussion of the γ-yield expected for the electromagnetic decay of hypernuclear low-lying particle-stable states in the 1p shell, following their formation in the (K^-,π^-) reaction in the forward direction, has recently been given [8]. The assignment of spin-parity values to the corresponding hypernuclear states should be facilitated by measuring the angular distributions for γ-rays relative to the forward $\theta_\pi=0°$. Since the primary transition $(1p)_n \to (1p)_\Lambda$ has $\Delta\ell=1$, this angular distribution can only be of the form

$$w(\theta_\gamma) \sim (1+A(\pi\gamma)\cos^2\theta_\gamma) , \qquad (4)$$

with coefficients $A(\pi\gamma)$ calculable for each case.

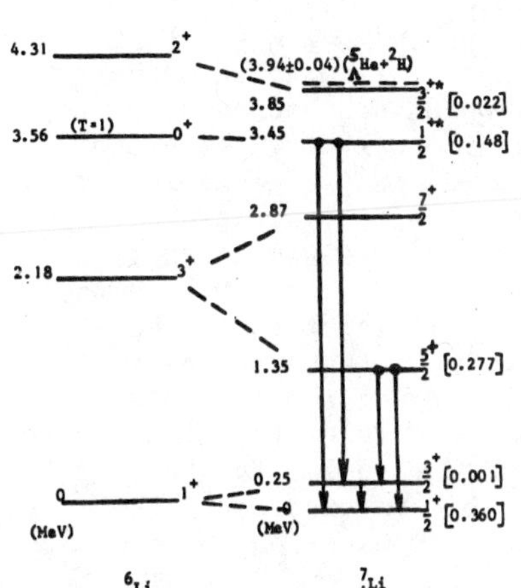

Fig. 2

The case of ^{7}Li is considered fundamental enough to warrant display, since $^7_\Lambda$Li is the lightest p-shell hypernucleus from which γ emission will definitely occur. The particle-stable states expected for $^7_\Lambda$Li are shown in Fig. 2, together with their dominant nuclear parents, for one of the most favorable B_Λ fits [9] performed in the 1p shell. The numbers in the square brackets give the (DWIA) calculated formation rates for the reaction $K^-\ {}^7Li \to \pi^-\ {}^7_\Lambda Li^*$ at 0°, relative to the total $(1p)_n \to (1s)_\Lambda$ transition rate. The strongest γ-ray expected is the E2 $5/2^+ \to 1/2^+$(g.s.). Determination of this γ-ray will indicate the splitting of the $(5/2^+,7/2^+)$ hypernuclear doublet built on the 3^+ first excited ^{6}Li core state. This splitting is predicted to mostly arise from a term $\vec{s}_\Lambda \cdot \vec{\ell}_N$ induced by the ΛN two-body spin-orbit interaction. This is in contrast with the measured [16] doublet splitting for $^4_\Lambda$H which is dominated by the spin-spin term $\vec{s}_\Lambda \cdot \vec{s}_N$. The latter interaction is responsible for the splitting of the g.s. doublet, and $^7_\Lambda$Li is about the only p-shell hypernucleus for which this occurs. Herrera et al. [17] have recently looked for hypernuclear γ-rays in coincidence with the $K^- \to \pi^-$ reaction in flight and reported the observation of an 0.79 MeV γ-ray from a ^{7}Li target. However, the cross-section evaluated by these authors is of the order of magnitude expected of nuclear lines, about one order of magnitude larger than the estimated [18] hypernuclear rate. Furthermore, for $E_\gamma \lesssim 0.8$ MeV the $5/2^+$ level is expected to become isomeric in view of the competing weak decay of hypernuclei; no confirmation of such isomerism has been reported from the B_Λ distribution observed for $^7_\Lambda$Li in emulsion. Finally, it is far from clear that the $1/2^{+*}$, T=1, $^7_\Lambda$Li state is particle-stable so that any observation of fast γ-rays with energies in the range of 3.5 to 4 MeV will be instructive.

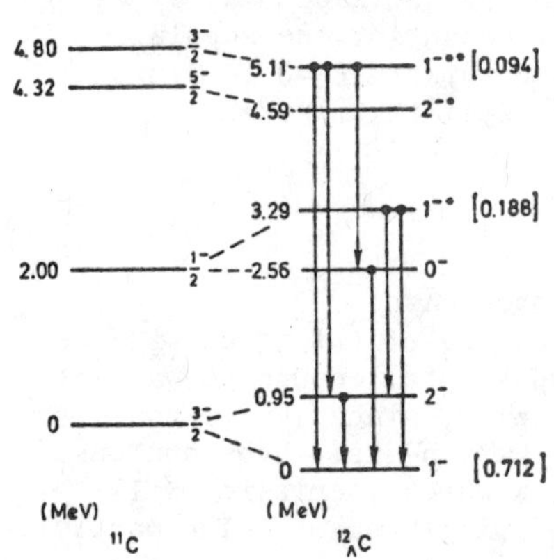

Fig. 3

The low-lying spectrum of $^{12}_\Lambda$C is depicted in Fig. 3. Only 1^- states are expected to be significantly formed and, indeed, the angular distribution of the lowest state observed [2] in $^{12}_\Lambda$C is well fitted by a calculated [10] angular distribution of a 1^- state of the structure depicted in the figure for the g.s. The 1^{-*} state, predicted to lie at 3.3 MeV, lies probably[2] between 3 to 4 MeV, thus confirming that the 1^- state, which had been assumed to be the ground state, is indeed such; a reversal of ordering within any of the first two doublets in $^{12}_\Lambda$C would result in a decrease of about 1 MeV in the calculated excitation energy of the 1^{-*} state relative to the 1^- state, contrary to the recent observation. Of the ΛN spin-dependent interactions, the excitation energy of the 1^{-*} state depends strongly only on the ΛN antisymmetric spin-orbit interaction and this new observation is the only evidence so far for the sign and strength of the antisymmetric spin-orbit interaction. The cascade nature of the γ-spectrum of Fig. 3 is expected to make the $2^- \to 1^-$ spin-flip M1 transition the most intense one. Determination of the ground-state doublet-splitting will be very instructive, since this splitting depends not only on the $\vec{s}_\Lambda \cdot \vec{\ell}_N$ term, but also (probably to a lesser degree) on the ΛN tensor interaction.

The particularly high particle-instability threshold of $^{12}_\Lambda$C against proton emission at (9.8±0.1) MeV deserves special attention, since hypernuclear states of the $(1p_{3/2})$ Λ orbital may turn out to be particle-stable, notably the lowest of the 2^+ states discussed in the previous section. A schematic representation is given in Fig. 4, where the 0^+ "analog" state is depicted at 11 MeV excitation [3]. For simplicity, the mixing between the 2^+ states is ignored and only the one within the configuration shown in the figure is considered. The spacing between the 0^+ and the 2^+ states does not depend on the Λ one-body spin-orbit coupling, and

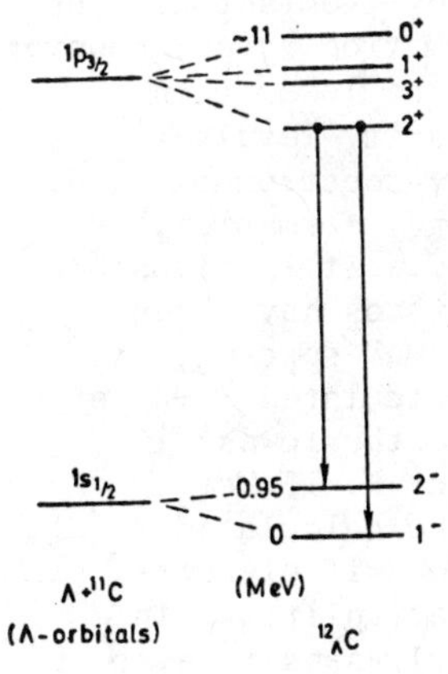

Fig. 4

for a central residual ΛN interaction with exchange-mixture parameter ε' [0 for Wigner interaction and 1/2 for Serber interaction] the 2^+ lies below the 0^+ state, by an amount given [10] by

$$E(0^+)-E(2^+) = -8(1-\varepsilon')F_2-2\varepsilon'(F_0-F_2), \qquad (5)$$

which is estimated in the range of 2 to 4 MeV as ε' is varied between the values 0 and 1/2. Thus, it is very likely that the lowest 2^+ state is particle-stable and will decay by fast E1 emission to the $(1^-,2^-)$ g.s. doublet.

The level diagrams for the charge-doublet states $({}^{14}{}_{\Lambda}N, {}^{14}{}_{\Lambda}C)$ are shown in Figs. 5 and 6. The expected γ-spectra of these mirror hypernuclei are rather different from each other, the formation rates of particle-stable states in ${}^{14}{}_{\Lambda}C$

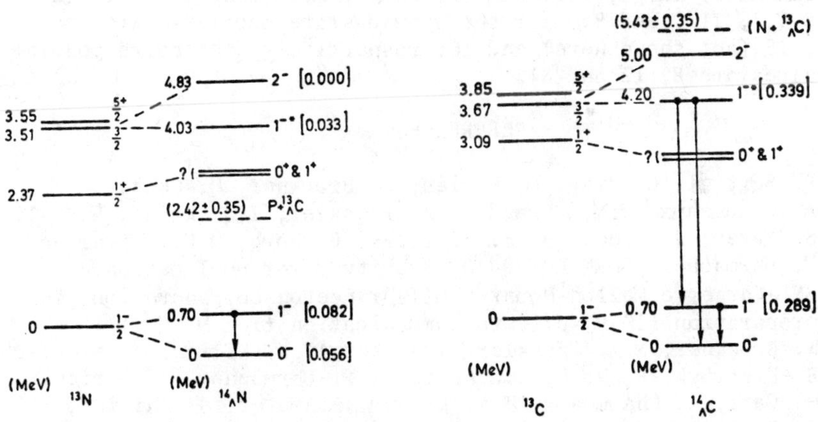

Fig. 5

Fig. 6

being considerably higher. The only advantage of the ^{14}N target is that only one γ-line at most is expected in ${}^{14}{}_{\Lambda}N$, yielding a direct measurement of the $(0^-,1^-)$ doublet-splitting; if the doublet is inverted with respect to the figures, no γ-rays should appear. As expected [9] for the $p_{1/2}$ shell, this splitting is especially sensitive to the ΛN tensor interaction; the calculated spacing of 0.70 MeV in Figs. 5 and 6 ignores this interaction and is based on the $\vec{s}_{\Lambda}\cdot\vec{\ell}_N$ term mentioned above. The ${}^{14}{}_{\Lambda}C$ low-lying spectrum, in addition to the normal transitions depicted, offers the possibility of exciting non-normal parity levels (0^+ and 1^+ in Fig. 6) and their

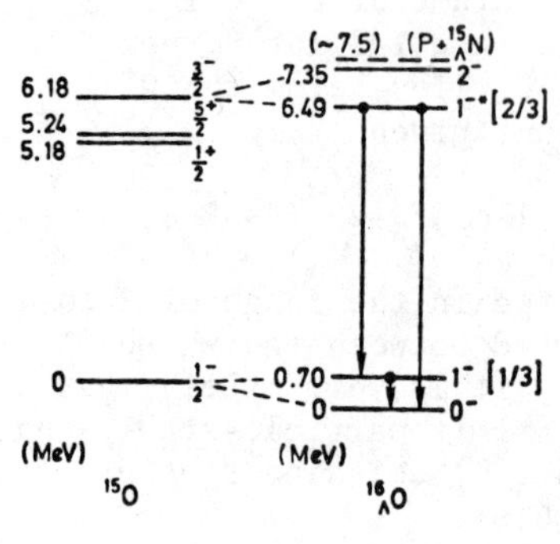

Fig. 7

subsequent decay by γ-emission. The corresponding formation rates, however, are estimated [8] to be lower than those of normal-parity levels.

Finally the γ-spectrum of ${}^{16}{}_{\Lambda}O$, following 0° (K^-,π^-) formation, is depicted on Fig. 7. It is plausible that the two 1^- states have been observed [3] with 6 MeV spacing, compared to the calculated [9] value of 5.8 MeV. Also, the lowest 1^- state is found [3] at B_{Λ}=13 MeV, coinciding with the calculated [9] value. As for the A=14 elements, the $(0^-,1^-)$ doublet splitting in ${}^{16}{}_{\Lambda}O$ is particularly sensitive to the ΛN tensor interaction.

We have at present no evidence on the strength of the ΛN tensor force. Theoretical estimates [12] suggest that the tensor parameter T may have a value of order 0.05 MeV, which would play an appreciable role in the $p_{1/2}$-shell hypernuclei, since the coefficient with which it appears in various doublet-splittings can be remarkably large; for example, this coefficient is evaluated [8] to be 7.3, 11.7 and 8.0 for the ground-state doublet-splittings of A=14, 15 (not shown here) and 16, respectively (the corresponding jj values are 8, 12 and 8).

REFERENCES

1. R. Bertini, O. Bing, P. Birien, W. Brückner, H. Catz, A. Chaumeaux, J.M. Durand, M.A. Faessler, T.J. Ketel, K. Kilian, B. Mayer, J. Niewisch, B. Pietrzyk, B. Povh, H.G. Ritter and M. Uhrmacher, CERN-EP/79-10, Submitted for publication.
2. BNL-Carnegie Mellon-Houston-MIT-Princeton Collaboration, in preparation (1979; private communication from M. May).
3. W. Brückner, M.A. Faessler, T.J. Ketel, K. Kilian, J. Niewisch, B. Pietrzyk, B. Povh, H.G. Ritter, M. Uhrmacher, P. Birien, H. Catz, A. Chaumeaux, J.M. Durand, B. Mayer, J. Thirion, B. Bertini and O. Bing, Phys. Letters 79B, 157 (1978).
4. W. Brückner, B. Granz, D. Ingham, K. Kilian, U. Lynen, J. Niewisch, B. Pietrzyk, B. Povh, H.G. Ritter, and H. Schröder, Phys. Letters 62B, 481 (1976).
5. R.H. Dalitz and A. Gal, Phys. Letters 64B, 154 (1976).
6. A. Bouyssy and J. Hüfner, Phys. Letters 64B, 276 (1976).
7. R.H. Dalitz and A. Gal, Phys. Rev. Letters 36, 362 (1976).
8. R.H. Dalitz and A. Gal, Ann. Phys. (N.Y.) 116, 167 (1978).
9. A. Gal, J.M. Soper and R.H. Dalitz, Ann. Phys. (N.Y.) 113, 79 (1978)
10. C.B. Dover, G.E. Walker, A. Gal and R.H. Dalitz, in preparation (1979).
11. A. Gal, in Proceedings of the Banff Summer School, edited by B. Castel et al. (Plenum, N.Y., 1979).

12. A. Gal, J.M. Soper and R.H. Dalitz, Ann. Phys. (N.Y.) 63, 53 (1971).
13. C.B. Dover, private communication (1978).
14. R. Brockman and W. Weise, Phys. Letters 69B, 167 (1977).
15. D. Zieminska and R.H. Dalitz, Nucl. Phys. A238, 453 (1975).
16. M. Bedjidian, A. Filipkowski, J.W. Grossiord, A. Guichard, M. Gusakov, S. Majewski, H. Piekarz, J. Piekarz and J.R. Pizzi, Phys. Letters 62B, 467 (1976).
17. J.C. Herrera, J.J. Kolata, H. Kraner, C.L. Wang, R. Allen, D. Crockley, A. Hassan, A. Kanofsky and G. Lazo, Phys. Rev. Letters 40, 158 (1978).
18. R.H. Dalitz and A. Gal, J. Phys. G4, 889 (1978).

Question: L. Kisslinger, Carnegie-Mellon University

It is clear why you take a pure configuration for the 0^+ states. Can you use the structure of the 2^+ states (especially in the light of the strong collective effects) to extract information about the Λ-N force.

Response: A. Gal

My simplified presentation assumed a purely closed jj shell description of ^{12}C, which in a more quantitative evaluation will require modification due to core-polarization effects. In this case not only the 2^+ states are likely to acquire a more involved structure, but also the 0^+ state will do so; the simplest admixtures to consider are given by coupling the 1p Λ to $^{11}C^*$(2.0 MeV).

KAON INTERACTIONS WITH VERY LIGHT NUCLEI

B. F. Gibson*
Theoretical Division, Los Alamos Scientific Laboratory
Los Alamos, New Mexico 87545

ABSTRACT

Low energy kaon interactions (both K and $\overline{K}$) with very light nuclei are reviewed. Limitations upon present K-nucleus studies due to uncertainties in the K-N amplitudes are emphasized along with promising uses. A brief review of some of the many interesting aspects of $\overline{K}$-nucleus scattering is given. Comparison of the limited $\overline{K}$-d and $\overline{K}$-^{4}He elastic data with theory is made. The $\overline{K}d \rightarrow \pi\Lambda p$ reaction is discussed including the possible ΣN virtual bound state.

INTRODUCTION

The previous speakers have now thoroughly discussed several topics which might otherwise fall in my purview. Therefore I will forego the usual detailed discussion of K-N and $\overline{K}$-N amplitudes, any consideration of the involved topic of K-mesic atoms, further mention of the stimulating (K^-,π^-) SEX reaction, and serious examination of the related topic of hypernuclear physics. In light of the excellent overview opening this session, I can proceed immediately to the specific points of interest as if you are all experts.

Kaon scattering from light nuclei covers two disparate subjects, since K-N and $\overline{K}$-N amplitudes are not related by crossing symmetry. The K has strangeness +1 and the $\overline{K}$ has strangeness -1. Because historical interest has favored $\overline{K}$ physics, I shall emphasize $\overline{K}$ reactions over K reactions. Also, because of the speculative nature of the field to date, I shall restrict my remarks primarily to those targets for which limited data already exists or appears to be feasible.

Let me once again remind you that it is our hope to use such probes to unravel the mystery of the nucleus. The K and $\overline{K}$ hold promise because of their non-zero strangeness; they do not mediate the N-N force and are therefore not subject to the same overcounting problems as those associated with the pion. In a similar vein, study of the associated Y (or Y*) propagation in the nuclear medium may aid in our effort to understand how to treat "Δ propagation" within the nucleus. Clearly our ideas must be put to the test in the very light nuclei, where we have some hope of treating the theory correctly and therefore resolving such questions as what is the proper off-shell amplitude extrapolation, etc.

K-NUCLEUS INTERACTIONS

The K^+ meson is perhaps the more attractive probe of nuclear

*Work performed under the auspices of the U. S. DOE.

ISSN:0094-243X/79/540691-12$1.50

structure. Because of its strangeness, the low-energy KN amplitudes are not resonant--there are no known S = + 1 baryons. Thus, the K is one of the weakest interacting hadrons, especially when compared to the more usual hadronic probes such as the N, the π, or the α. On the other hand, its interaction is more than electromagnetic. This has led some to propose the K as an ideal probe of the neutron density.[1] (Other possible nuclear structure uses of the K are discussed in Ref. 2-6.) The fact that it is weakly absorbed and therefore "sees" the entire nucleus is certainly an argument that cannot be ignored. However, it is likely that at least the surface features of ρ_n will be determined from proton and pion scattering prior to the existence of extensive K^+-nucleus data.

At the present time our knowledge of the K-N interaction is not broad, as we have heard today. The amplitude is elastic below pion production threshold, if one includes charge exchange in the definition of elastic as we do in pion scattering. We know[7] that $\sigma_T \sim$ 8-10 mb for $p_{Lab} \leqslant 600$ MeV/c so that the mean free path of the K is long, being some ~7 fm for a nuclear density assumption of ~1/7 fm^{-3}. For these reasons we expect that i) single scattering is important, perhaps dominant, ii) DWBA might prove to be a good approximation, and iii) K-nucleus scattering should be a volume phenomena. However, the previously published K-N phases[7,8] are in disagreement, and neither the latest K^+p data[9] (a bubble chamber experiment with differential cross sections at 5 points in the range $178 \leqslant p_{Lab} \leqslant 580$ MeV/c) nor the companion K^+d data[10] are included in those amplitude analyses.

The I = 1 scattering length and effective range from Ref. 9 are -0.314 ± .007 fm and 0.36 ± .07 fm respectively. The constructive Coulomb-nuclear interference shows the amplitude to correspond to a repulsive interaction. From Ref. 7 (see also Ref. 1) we find that $\sigma_{I=1} >> \sigma_{I=0}$ for low energy ($p_{Lab} \leqslant 300$ MeV/c), so that K^+p dominates; we also know that only $\ell = 0$ is significant in that momentum range. Between 300 and 700 MeV/c laboratory momentum, the cross section contains both $\ell = 0$ and 1; $\sigma^{\ell=1}_{I=0} >> \sigma^{\ell=1}_{I=1}$ in that momentum range so that K^+n dominates the $\ell = 1$ amplitude. Dover[11] argues that for $p_{Lab} \leqslant 350$ MeV/c the K will act as a probe similar in nature to the α-particle, except that the α is strongly absorbed. Thus the K may make a better $\Delta S = 0$, $\Delta I = 0$ probe. The changing isospin composition of the K-N amplitude (as a function of p_{Lab}) holds promise of being useful in studying nuclear structure. But as we shall see below, our ability to utilize this probe at present is severely restricted by our inadequate knowledge of the basic KN amplitudes.

The first and only kinematically complete K-d scattering calculation was performed some 15 years ago by Hetherington and Schick. They used a Faddeev type multiple scattering formalism with two-body, S-wave separable potentials of the Yamaguchi form. Coulomb forces as well as $K^+ - K^0$ and other mass differences were neglected. Theoretically they found that in the momentum range of

110-230 MeV/c the impulse approximation was within 25% of the correct answer and that double scattering was good to within 10%. However, for the optical theorem to yield a good result for the total

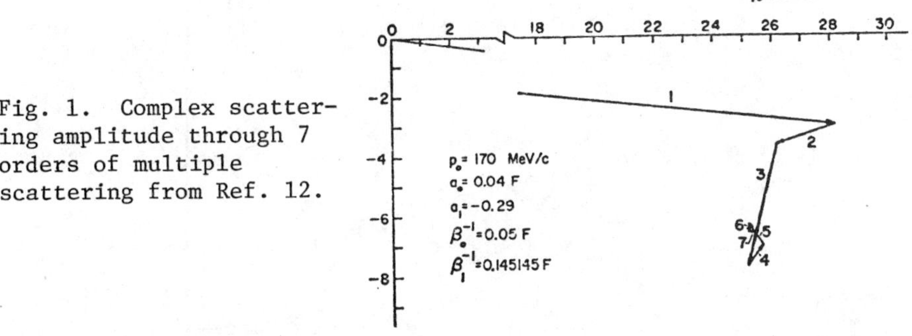

Fig. 1. Complex scattering amplitude through 7 orders of multiple scattering from Ref. 12.

cross section, triple scattering terms had to be included. (See Fig. 1). A more detailed study was not warranted at the time because of the nature of the available K-N input.

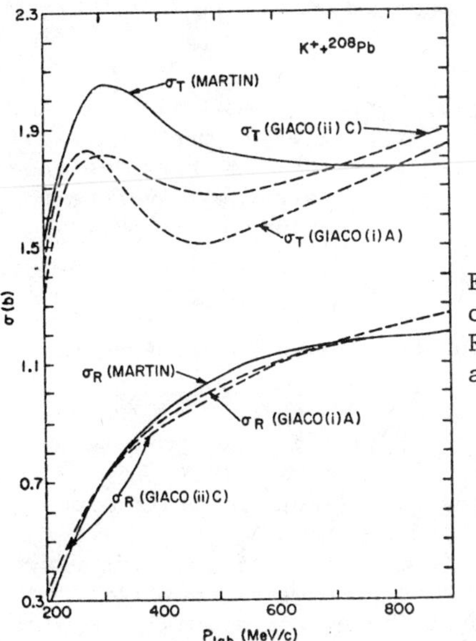

Fig. 2a. Total and reaction cross section predictions from Ref. 1 based upon 3 different KN amplitude sets.

Seeing the elastic scattering studies of Dover and Moffa[1] and the inelastic scattering work of Cotanch,[6,13] it is not clear that the K-N situation has improved. DWBA effects are of the order of 50% in the low momentum region,[13] but more important is that differences between various theoretical predictions with the amplitudes of Ref. 7 and 8 can be just as large or larger,[1,3,13] (See Fig. 2a, b).

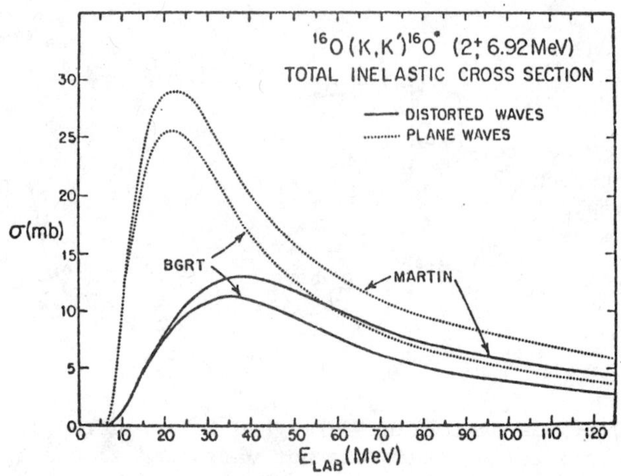

Fig. 2b. Total inelastic cross section predictions from Ref. 13 based upon 2 different KN amplitude sets.

For this reason alone, meaningful K-nucleus analysis will have to await definitive K-N amplitudes. The K^+-d experiments will necessarily be required in combination with K^+-p to determine the K^+-n amplitudes. We have a long row to hoe before we can hope to reap the rewards promised by K-nucleus scattering.

$\overline{K}$-NUCLEUS INTERACTIONS

The $\overline{K}$-nucleus scattering process has proved interesting for several reasons, two of which are:

i) the strength of the $\overline{K}N$ interaction which couples to many resonances, the lowest of which is the $Y^*(1405)$ lying just below threshold.

ii) the strangeness exchange and other reactions which produce hypernuclei of both the Λ and Σ varieties.

The strong nature of the $\overline{K}$-N interaction became evident very early through K-mesic atom studies. The strength was sufficient to make the "scattering lengths" appear to come from a weakly repulsive force if they were incorrectly interpreted in terms of a single-channel potential.[14,15] Along with hypernuclear production one can study coincident γ's;[16] e.g. in the ${}^{4}_{\Lambda}He^* \xrightarrow[M1]{} {}^{4}_{\Lambda}He$ and ${}^{4}_{\Lambda}H^* \xrightarrow[M1]{} {}^{4}_{\Lambda}H$ transitions. Thus, one can extract information about excited states of hypernuclei with (K,γ) just as (π,γ) coincidence measurements yield information about nuclear excitations. The (K^-,γ) reaction has been proposed as a means of studying the Λn interaction[17] as well as the "recoilless" production of hypernuclear states.[18] There exists the possibility of finding Σ-hypernuclear final states which do not convert strongly to Λ-systems; e.g., the reaction $K^{-3}H \rightarrow \pi^{+}(\Sigma^{-}nn)$. Finally, the $(\bar{K},K)$ reaction can take us to possible double Λ hypernuclei or exotic cascade (Ξ) hypernuclei, both with S = - 2. The $K^{-6}Li \rightarrow K^{0}(\Lambda\Lambda nnpp)$ reaction can lead to the bound,[19] triply-closed shell ${}^{6}_{\Lambda\Lambda}He$. The $K^{-4}He \rightarrow K^{+}(\Lambda\Lambda nn)$ reaction is not expected to produce a bound $(\Lambda\Lambda nn)$ system (based upon our present scant knowledge of the $\Lambda\Lambda$ force), although the $K^{-4}He \rightarrow K^{0}(\Lambda\Lambda np)$ reaction might produce a bound ${}^{4}_{\Lambda\Lambda}H$ since ${}^{3}_{\Lambda}H$ is bound. The $K^{-4}He \rightarrow K^{+}(\Xi^{-}pnn)$ reaction could result in an exotic bound state, but it will decay rapidly through $\Xi^{-}p$ conversion; therefore, a more interesting possibility is $K^{-3}H \rightarrow K^{+}(\Xi^{-}nn)$.

In the specific $\bar{K}$ scattering and reaction processes that I shall discuss, I will illustrate primarily with deuterium. It is not that heavier systems are uninteresting. It is that out knowledge of the $\bar{K}N$ amplitudes is so poor that even $\bar{K}d$ discussions remain somewhat speculative.

The low energy K-N interaction has been recently summarized by Martin.[20] Therefore, I will only briefly remind you of the salient features. The open channels are

$$I = 0 \left\{ \begin{array}{l} \bar{K}N \rightarrow \bar{K}N \\ \quad\;\; \rightarrow \pi\Sigma \end{array} \right. \qquad I = 1 \left\{ \begin{array}{l} \bar{K}N \rightarrow \bar{K}N \\ \quad\;\; \rightarrow \pi\Sigma \\ \quad\;\; \rightarrow \pi\Lambda \end{array} \right. .$$

Zero-range K matrix analyses of the low energy data have been published in Ref. 21-23. Effective-range, M matrix analyses[24] have been carried out by Kim[25] and by Berley, et al.[26] Martin[27] has used dispersion relations to constrain the low energy $K^{-}p$ data analysis in a non-diagonal effective range, M matrix analysis. Since there is no direct information about $\pi Y \rightarrow \pi Y$ scattering, the M matrix elements are not uniquely determined. Nevertheless, the $\bar{K}N$ scattering lengths do seem to be well defined:

$$a_o = -1.66 + i0.75 \text{ fm}$$
$$a_1 = \;\; 0.35 + i0.66 \text{ fm}.$$

Hetherington and Schick have again published the only exact calculation of $K^{-}d$ scattering including the breakup reaction.[28,29] From their work we can conclude that, up to laboratory momenta of 300 MeV/c, the S-wave is a significant fraction of the total, that there exists significant multiple scattering as in low energy n-d scattering, and that multiple scattering corrections to the total elastic and reaction cross sections are more significant than those in the total cross section due to cancellations. In particular, it is clear that the double scattering terms are an unreliable guide to the importance of multiple scattering (see Fig. 4); i.e., final state rescattering calculations are to be viewed with suspicion.

This was also demonstrated by Myhrer[30] in a model calculation of elastic scattering at zero energy.

The available data for K^-d scattering and reaction processes exist primarily for $p_{Lab} \geqslant 400$ MeV/c.[31] Additional data in this range were reported by Carroll, et al.[32] and compared with previous work. Recently, the model of Ref. 28 was extended to include the hyperon channels implicitly and applied in estimating K^-d elastic and total cross sections.[33] The momenta reached were not large enough to permit confrontation of the data, although the implicit channel approximation is apparently in reasonable agreement with a more complete calculation[34] for these processes. Nevertheless, it

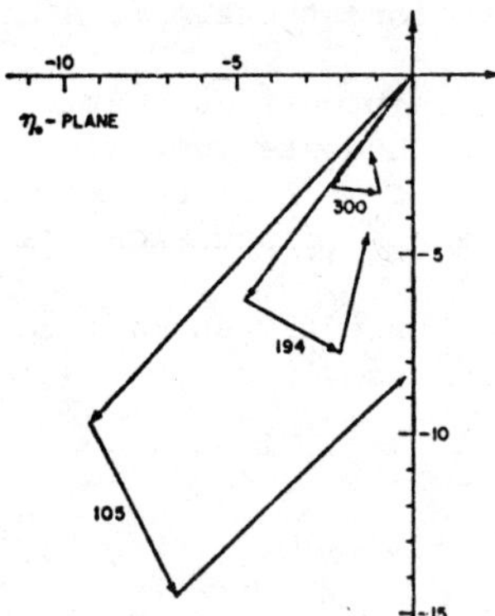

Fig. 4. Contributions to the complex S-wave amplitude for single scattering, double scattering, and full multiple scattering solutions from Ref. 28 for p_{Lab} as indicated.

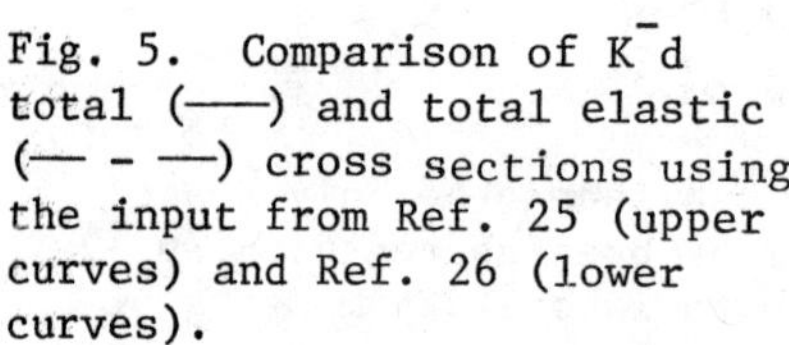
Fig. 5. Comparison of K^-d total (——) and total elastic (— - —) cross sections using the input from Ref. 25 (upper curves) and Ref. 26 (lower curves).

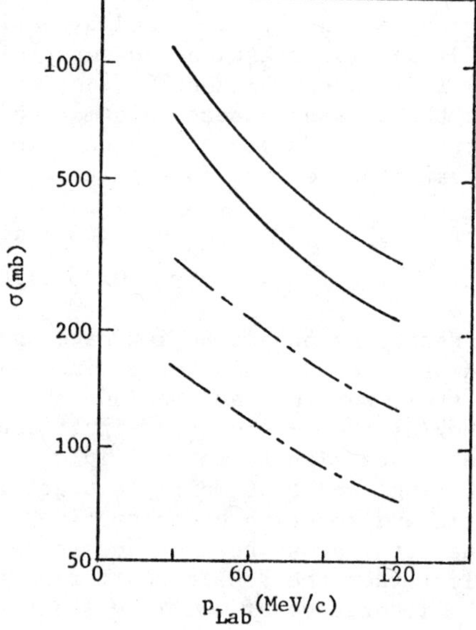

is clear from the numerical results that good $\bar{K}p$ and $\bar{K}d$ experiments at low energy should be a useful aid in differentiating the correct $\bar{K}N$ S-wave amplitudes. (See Fig. 5) The $\bar{K}d$ cross sections predicted using the $\bar{K}N$ input from Ref. 25 and 26 differ by 25-35% throughout the momentum range of 0-120 MeV/c. (One should be cognizant that above 300 MeV/c it is likely more than just S-wave is needed to describe $\bar{K}N$.)

In the Hetherington and Schick charge exchange breakup calculation,[29] results lie a factor of 2 above the sparse data. They did not take into account actual mass differences and charge states, which may explain the overestimate.[35] At least in the elastic scattering calculation,[28] where it is possible to include these effects without an inordinate amount of effort, the difference between the complete and incomplete model calculation is large enough to account for a factor of 2 in the charge exchange reaction, where the cross section is an order of magnitude smaller than the elastic.

Data also exist for the elastic scattering of K^- from ^4He at low momenta.[36,37] In the latter experiment differential cross sections were measured using a bubble chamber at the ZGS by Mazur, et al.[37] in the momentum intervals 100-150 MeV/c and 150-200 MeV/c. They employed a zero-range approximation to extract from the data an S-wave scattering length, a feat not possible in the earlier measurement. Seki claimed some success in fitting the data of Ref. 36 using a multiple scattering formalism and input from Ref. 25. He showed that the real part of the K^-d scattering length in his model agreed with that of Hetherington and Schick, so that one might hope to trust the $K^-{}^4He$ result. However, neglect of the inelastic channels in the intermediate states (meaning that the optical theorem was not satisfied) makes the result difficult to interpret, especially in view of the established importance of multiple scattering through many orders. Deloff and Law[39] then examined the $K^-{}^4He$ elastic scattering problem in some detail obtaining their $\bar{K}$-nucleus potential from folding a $\bar{K}$-N finite range complex potential with the nuclear density distribution. They found a reasonable representation of the data with potentials equivalent to having a 0.4 - 0.5 fm range in a Yukawa model. A comparison of their fit with the data from Ref. 37 and a curve corresponding to the phenomenological potential optical model of Koch and Sternheim[40] is shown in Fig. 6. Reference 39 contains some interesting comments concerning the optical model approach to $\bar{K}$-nucleus low energy scattering, which Deloff continues in Ref. 15. In particular, he points out that the Lorentz-Lorenz effect can give rise to the "change of sign" of the "scattering lengths" from mesic atom data. He also concludes that one must take into account the $\bar{K}N$ finite range--just as in the pion-nucleus optical model, where the zero-range approximation is nonsense.

Let us now turn our attention to the $K^-d \rightarrow \pi^-\Lambda p$ reaction. Considerable interest has been generated by the apparent Λp final state enhancement near the ΣN threshold.[41-43] Braun, et al.[44] reported results from a bubble chamber experiment seeming to confirm

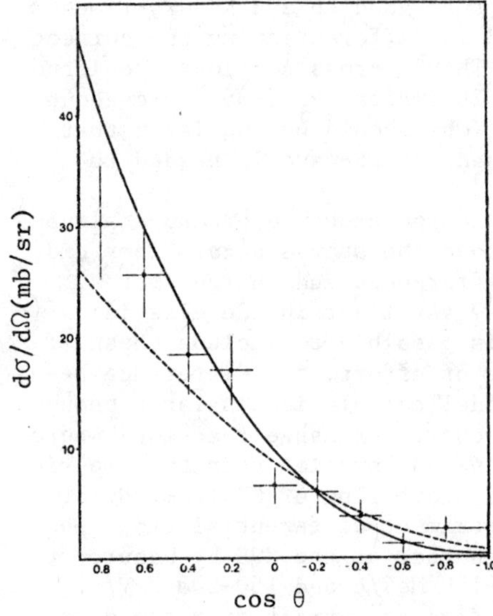

Fig. 6. The $\overline{K}$ - ^{4}He differential cross section (Ref. 37) for p_{Lab} = 100-150 MeV/c. Solid curve corresponds to Ref. 39; the dashed curve to Ref. 40.

an enhancement in the Λp mass distribution which is interpretable[45] as a ΣN virtual bound state (i.e., the ΣN system would possess a bound state were it not coupled to the Λp continuum). In such a picture, the cross section structure is attributed to ΣN → Λp conversion. A corresponding analysis of Λp elastic scattering[44,46] also demonstrated the possible existence of a Σ^-p virtual bound state contribution to the cross section within the limits set by the data.

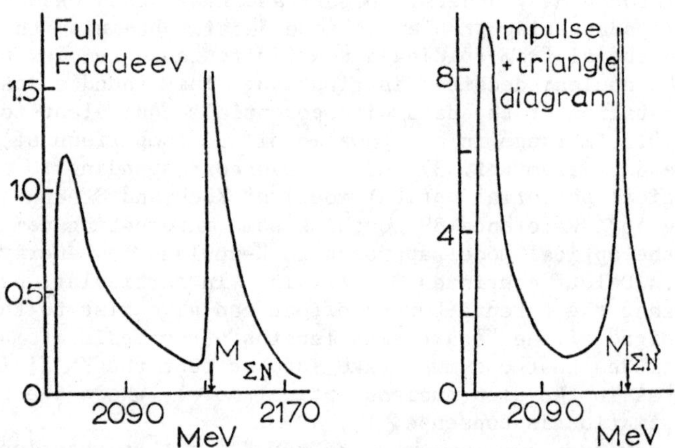

Fig. 7. The model $\overline{K}$d → πΛp spectrum as reported in Ref. 47.

In order to understand the reaction, Toker et al.[34,47] have undertaken the very ambitious task of carrying out a complete Faddeev calculation for the coupled "3-body" system $\{\bar{K}d, \bar{K}NN, \pi YN\}$ including the proper matrix representations of the $(\Lambda N\text{-}\Sigma N)$, $(\bar{K}N\text{-}\pi\Sigma)_{I=0}$, and $(\bar{K}N\text{-}\pi\Sigma\text{-}\pi\Sigma)_{I=1}$ 2-body interactions. The report of their preliminary results at this conference[47] shows that the various rescattering processes in a Faddeev calculation substantially modify the spectrum from the naive impulse-plus-triangle-diagram estimate (see Fig. 7). No comparison with the data is yet warranted, but the authors are to be commended for their efforts.

Let me close this subject with these comments: Kisslinger has looked at applying symmetry arguments to the OBE potentials (as has been done to predict $\bar{N}N$ potentials) and concluded that there is no ΣN virtual bound state.[48] I would suggest that we seek an answer from the YN coupled potentials of Nagels, Rijken, and deSwart;[49] i. e., does the ΣN channel (without coupling to the ΛN channel) support a bound state? If not, is there a strong virtual antibound state as in the case of the np singlet? I remind you that in the $nd \rightarrow nnp$ reaction the $(np)_{S=0}$ final state interaction contributes much more to the peak in the spectrum than does the $(np)_{S=1}$.

THE OUTLOOK

It is our fervent hope that the next few years will bring some clarification of the rather murky picture that we see of both the KN and $\bar{K}N$ low energy amplitudes. Certainly the time is ripe for more precise low energy scattering experiments involving the proton and the deuteron. Without that data, the promise held out for the K as a nuclear structure probe will go unrealized, and our understanding of the interesting $\bar{K}$ reactions will remain speculation.

I have not meant to imply that only the K^+ and K^- data are to be sought. In fact, $\bar{K}_0d$ experiments are to be encouraged due to the absence of Coulomb effects which can be large in low energy elastic scattering. One would also be anxious to see kaon scattering from A = 3 and 4 targets, where we think we understand the nuclei and very useful experiments have been carried out with pions. The question of $\bar{K}$ absorption has not really even been addressed, and yet it must be an important open channel which must be considered.

Our ideas about kaon-nucleus interactions are not far from primative. It is in the very light nuclei that they are apt to be most severely tested. The time has come to begin that testing.

REFERENCES

1. C. B. Dover and P. J. Moffa, Phys. Rev. C 16, 1087 (1977).
2. G. E. Walker, B. A. P. S. 21, 646 (1976).
3. S. R. Cotanch and F. Tabakin, Phys. Rev. C 15, 1379 (1977).
4. R. D. Koshel, P. J. Moffa, and E. F. Redish, Phys. Rev. Lett. 39, 1319 (1977).
5. Y. Alexander and P. J. Moffa, Phys. Rev. C 17, 676 (1978).

6. S. R. Cotanch, Nucl. Phys. A308, 253 (1978).
7. B. R. Martin, Nucl. Phys. B94, 1413 (1975).
8. G. Giacomelli, et al., Nucl. Phys. B20, 301 (1970); B71, 138 (1974). This is commonly referred to as the BGRT group.
9. R. A. Burnstein, et al., Phys. Rev. D 10, 2767 (1974).
10. R. G. Glasser, et al., Phys. Rev. D 15, 1200 (1977).
11. C. B. Dover, BNL 50579, p. 9 (1977).
12. J. H. Hetherington and L. H. Schick, Phys. Rev. 138, B1411 (1965).
13. S. R. Cotanch, Phys. Rev. C 18, 1941 (1978).
14. See, for example, R. Seki and C. E. Wiegand, Ann. Rev. Nucl. Sci. 25, 241 (1975).
15. A Deloff, Nukleonika 22, 875 (1977).
16. See, for example, B. Povh, Ann. Rev. Nucl. Part. Sci. 28, 1 (1978) for a discussion and further references.
17. B. F. Gibson, et al., BNL 18335, p. 296 (1973).
18. H. Feshbach, "Meson-Nuclear Physics-1976," AIP Conf. Proc. 33, p. 521 (1976).
19. D. J. Prowse, Phys. Rev. Lett. 17, 782 (1966).
20. A. D. Martin, Nukleonika 22, 857 (1978).
21. B. R. Martin and M. Sakitt, Phys. Rev. 183, 1345 (1969).
22. A. D. Martin and G. G. Ross, Nucl. Phys. B16, 479 (1970).
23. Y. A. Chao, et al., Nucl. Phys. B56, 46 (1973).
24. M. H. Ross and G. L. Shaw, Ann. of Phys. (NY) 13, 147 (1961).
25. J. K. Kim, Phys. Rev. Lett. 19, 1074 (1964).
26. D. Berley, et al., Phys. Rev. D 1, 1996 (1970); D 3, 2297E (1971).
27. A. D. Martin, Phys. Lett. 65B, 346 (1976).
28. J. H. Hetherington and L. H. Schick, Phys. Rev. 137, B935 (1965).
29. J. H. Hetherington and L. H. Schick, Phys. Rev. 156, 1647 (1967).
30. F. Myhrer, Phys. Lett. 45B, 96 (1973).
31. E. Bracci, et al., CERN/HERA 72-2 (1972).
32. A. S. Carroll, et al., Phys. Rev. Lett. 37, 806 (1976).
33. L. H. Schick and B. F. Gibson, Z. Physik A 288, 307 (1978).
34. G. Toker, A. Gal, and J. M. Eisenberg, Progress Report on "The $K^-d \rightarrow \pi^-\Lambda p$ Reaction at Low Energies in the Faddeev Formalism," January 1979.
35. L. H. Schick (private communication).
36. J. J. Boyd, et al., Phys. Rev. Lett. 19, 1405 (1967).
37. P. O. Mazur, et al., Phys. Rev. D 1, 20 (1970).
38. R. Seki, Phys. Rev. 178, 2316 (1969).
39. A. Deloff and J. Law, Phys. Rev. C 10, 1688 (1974).
40. J. H. Koch and M. M. Sternheim, Phys. Rev. Lett. 28, 1061 (1972).
41. D. Cline, R. Laumann, and J. Mapp, Phys. Rev. Lett. 20, 1452 (1968).
42. G. Alexander, et al., Phys. Rev. Lett. 22, 483 (1969).
43. T. H. Tan, Phys. Rev. Lett. 23, 395 (1969).
44. O. Braun, et al., Nucl. Phys. B124, 45 (1977).
45. H. G. Dosch and V. Hepp, "Analysis of the Λp Enhancement at 2.129 GeV in the Reaction $K^-d \rightarrow \pi^-\Lambda p$," Th. 2310-CERN (1977).

46. Y. Gell, G. Alexander, and I. Stumer, Nucl. Phys. B22, 583 (1970).
47. G. Toker, A. Gal, and J. M. Eisenberg, "The $K^-d \rightarrow \pi^-\Lambda p$ Reaction at Low Energies," contribution to this conference.
48. L. S. Kisslinger (private communication).
49. M. M. Nagels, T. A. Rijken, and J. J. deSwart, Phys. Rev. D 15, 2547 (1977).

Question: C. Dover, University of Pennsylvania

The elementary K^+N amplitudes are much better known in the region of ≥600 MeV/c, where K^+ nucleus scattering can be measured with existing beams. Much of the ambiguities in the K^+ nucleus analysis due to uncertainty in the elementary amplitudes is thus removed in this momentum region, although you are correct in stating that there would be sizeable uncertainties in the region of 100-300 MeV/c.

Response: Gibson

While it is true that the KN amplitudes are better known above 600 MeV/c than below, the differences in K-Pb scattering due to different KN amplitude parameterizations at 600 MeV/c are still too large (15%) to permit useful analysis of kaon-nucleus scattering for nuclear structure purposes. If one wishes to utilize the I=1 dominance of the s-wave KN interaction, then the problems presently associated with experiments in the 300 MeV/c momentum range must be overcome.

Question: S. Hoath, Rutherford Laboratory

Our present experiment on K-p x-rays gives a measure of the K-p scattering length right at threshold, independently of extrapolation and the multichannel problem, that is consistent with zero real part. It disagrees with Martin's analysis by 3-5 standard deviations. The K-p experiment will be running again at CERN in 1979.

Response: Gibson

I should have mentioned that the primary disagreement among the $\overline{K}N$ scattering length predictions is the real part of a_1; the analysis of Kim and of Berley, et al. are certainly consistent with zero. It is pleasing to hear that you have a direct measurement of a_p.

Σ INTERACTIONS IN NUCLEI

R. Bertini
Centre de Recherches Nucléaires, Strasbourg, France
and
CERN, Geneva, Switzerland

An efficient way of producing hypernuclei is to expose the nuclear target to low-momentum negative kaons with strangeness S = -1. In the reactions

$$K^- + N \rightarrow \Lambda + \pi \quad (1)$$

and

$$K^- + N \rightarrow \Sigma + \pi , \quad (2)$$

which transfer the strangeness from kaon to nucleon, a fraction of the K^- interactions with the nucleus leads to the formation of the hypernucleus.

Since the pioneer work of Pniewski[1)], which was the first to show experimental evidence for hypernuclei, two major technical developments have made possible a deeper investigation of this particular nuclear structure where a strange baryon is added to the usual aggregate of neutrons and protons. The first development has been the design of new, short, secondary beams of K^- mesons with a reasonable intensity (typically, at 700 MeV/c, 2×10^4 $K^-/1 \times 10^{12}$ p); the second, the construction of new magnetic spectrometers with high-momentum resolution, large momentum acceptance, and big solid angle. Of this new generation of spectrometers, the one which has the best performance is the SPES II [2)] ($\Delta p/p_0 = 5 \times 10^{-4}$, p_0 = 700 MeV/c, $\Delta\Omega = 2 \times 10^{-2}$ sr, momentum acceptance = $p_0 \pm 18\%$), utilized by us in our experiment.

The experimental set-up near the CERN Proton Synchrotron (PS) kaon beam is shown in Fig. 1. In the first part, the kaon beam is

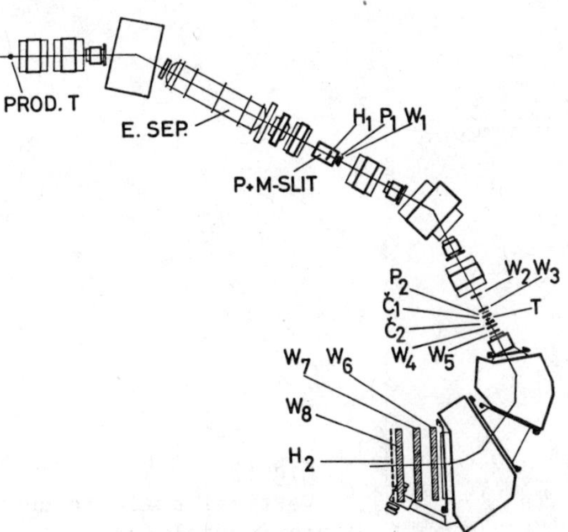

Fig. 1

ISSN:0094-243X/79/540703-11$1.50

separated from the incoming proton beam and cleaned out by the pions; in the second part, its momentum is analysed; and in the third part (SPES II) the momentum of the pions produced in reactions (1) and (2) is measured. Because of the strong contamination of the kaon beam by pions (typically we had 1 K^- for 100 π^- near the reaction target), particular care has been taken with the kaon identification. This was done by measuring the time of flight P_1-P_2 and by rejecting the pions detected by the liquid-hydrogen Čerenkov $\check{C}_1$. The kaon trajectories were determined with the X,Y hodoscope H_1 (3 mm resolution) and the wire chambers W_2, W_3 (1 mm resolution).

Almost 20% of the 700 MeV/c kaons decay in flight in a length of 1 m. The $K^- \to \pi^- + \pi^0$ branch (21%) can obscure the hypernuclei region. Hence pions are identified against decay products and passing kaons by the time of flight P_2-P_3 and the liquid-hydrogen Čerenkov $\check{C}_2$; and their trajectories by the wire chambers (1 mm resolution) W_i (i = 4 to 8). In Fig. 2 is shown a spectrum obtained in the reaction $K^- + {}^7Li \to \pi^- + {}_{hyp}^{7}Li$. As we increase the absolute momentum of the K^- and the π^-, we get directly the transformation energy $M_{hyp} - M_A$ that corresponds to the replacement in the target nucleus A of a neutron by the hyperon. The whole excitation energy measured simultaneously is about 200 MeV. The over-all experimental energy resolution is about 2.5 MeV, due essentially to the straggling in the dE/dx of kaons and pions in the detectors and to the difference in the dE/dx of kaons and pions in the reaction target.

As pointed out by Povh[3)] and shown experimentally by Brückner et al.[4)], two simple mechanisms can be used to explain the strangeness-exchange reaction: the recoilless production and the quasi-free production. In the first mechanism the strange baryon replaces the target nucleon, taking the same space and spin wave function.

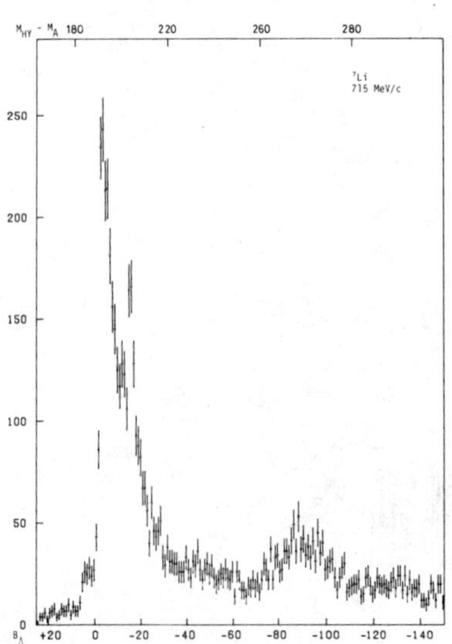

Fig. 2
Vertical scale: counts/1 MeV;
horizontal scales (top and bottom):
1 MeV/channel

The production rate is a function of the number of nucleons in the outer shell that can be transformed into hyperons and depends strongly[3,4] on the transfer momentum q [as $\exp -(aq)^2$], a being of the order of nuclear radius.

The collinear geometry, where the pions are detected near 0°, is hence the most appropriate to feed these states, and the incoming momentum has to be chosen (as shown in Table 1) in such a way that q is minimum.

In the second production mechanism the hyperon can jump into any of the free or already occupied orbits of the nucleus. Here, as pointed out by Dalitz and Gal[5], quasi-free has a more extended meaning than in the normal quasi-free nuclear reaction. The restriction, due to the Pauli principle, for nucleon recoils of momentum $q \leq k_F$, where k_F denotes the Fermi momentum, does not exist, and the hyperon can reach any final state. Therefore this production mechanism is important also for small hyperon recoils q and can easily take place in the collinear geometry. Because of the difference in mass (about 80 MeV) between Λ and Σ hyperons (see Table 1) the optimum K^- momentum for the recoilless Σ production is quite different from the optimum for the recoilless Λ production. The intensity of the present kaon beams decreases very strongly below 700 MeV/c, and the experiments cannot be performed in a reasonable time at the optimum momentum for recoilless Σ production. Moreover, at 700 MeV/c there is a maximum in the cross-section of the elementary reaction $K^- + n \rightarrow \pi^- + \Sigma^0$, whereas the contribution of the reaction $K^- + p \rightarrow \pi^- + \Sigma^+$ is one order of magnitude smaller and hence neglected.

To produce Σ hypernuclei, we have studied reaction (2) in two different experimental set-ups: either detecting a π^- or a π^+. In the first case a Σ^0 hypernucleus can be produced as well as a Λ hypernucleus, as shown in Fig. 2. Reversing the polarity of SPES II we get the second case, where the Σ^- hypernucleus can be produced. The mass (excitation energy) scale is the same. Hence the spectra of Σ^0 and Σ^- hypernuclei, as well as of Λ hypernuclei, can easily be compared, free of relative systematic errors.

TABLE 1 Kinematics of the reaction $K^- + N \rightarrow hyp + \pi$ at 0°

K^- beam momentum	100	200	300	400	500	600	700	800	900	1000
Λ recoil momentum	180	121	74	37	8	16	34	50	63	73
Σ^0 recoil momentum	103	42	7	45	75	100	119	135	148	160
Σ^- recoil momentum	96	35	14	52	82	106	126	142	156	167

In order to emphasize the dependence of the relative contribution of the two production mechanisms (recoilless and quasi-free) on the transfer momentum for Λ hypernuclei and to discuss their implication for the Σ^0 hypernuclei, the results of the present experiment are shown in Fig. 3 and compared with those of Refs. 4 and 6. For all the experiments performed at 715 MeV/c (^{7}Li, ^{9}Be, ^{12}C) there is evidence for the excitation of a large bump in the Σ^0 region. The reaction on ^{6}Li has been studied at 790 MeV/c, and here no statistical evidence appears for this excitation. This result is in agreement with the strong decrease of the elementary cross-section between 715 and 790 MeV/c (a factor of seven), as well with the increase of the transfer momentum. Let us take as an example the (K^-,π^-) reaction on a ^{12}C target that leads to the formation of either a Λ hypernucleus or a Σ^0 hypernucleus. For the first case, we have in the present experiment a strong recoilless peak sitting on a large bump close to quasi-free production and to the recoilless production on the large 1s neutron states. In the old experiment[4)] the recoilless production is strongly reduced and the Σ^0 region is non-existent.

From the centre of gravity ($\bar{\omega}_{exp}$) of the large bump we can try to make an estimate of the Σ^0-nucleus potential. The experimental values of $\bar{\omega}_{exp}$ for the spectra shown in Fig. 3 are exhibited in Table 2, both in the transformation energy and binding energy scale. For the spectra of Ref. 4 a correction is applied in order to have the same energy calibration as we have in the present experiment. As first suggested by Walecka[7)] and subsequently applied by Moniz[8)], the Fermi gas model can successfully reproduce the quasi-free part of the excitation energy spectra. This method was applied to the analysis of the old Λ hypernuclei experiment[4)] by Dalitz et al.[5)]. A more complete study of the different possible contributions to this bump was made by Bouyssy[9)] in the shell-model frame. Their results are in agreement once the same ground-state binding energy has been taken into account.

TABLE 2

Target nucleus	$\bar{\omega}^{\Lambda}_{exp}$		$\bar{\omega}^{\Sigma^0}_{exp}$	
	$M_{HY}-M_A$	B_Λ	$M_{HY}-M_A$	B_{Σ^0}
^{6}Li	195	13	-	-
^{7}Li	196	13	277	17
^{9}Be	195	17	276	21
^{12}C	205	10	285	13

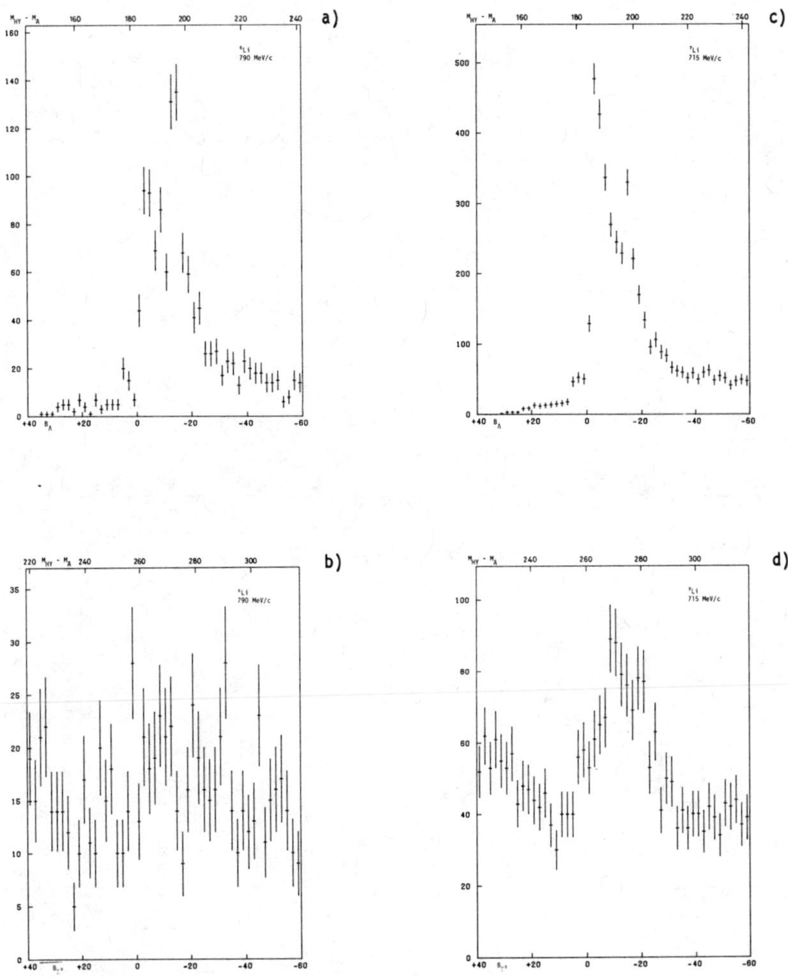

Fig. 3 Vertical scale: counts/2 MeV; horizontal scale (top and bottom): 2 MeV/channel. Note that spectra (f) and (i) come from Ref. 4 and have been corrected to have the same energy calibration as in the present experiment.

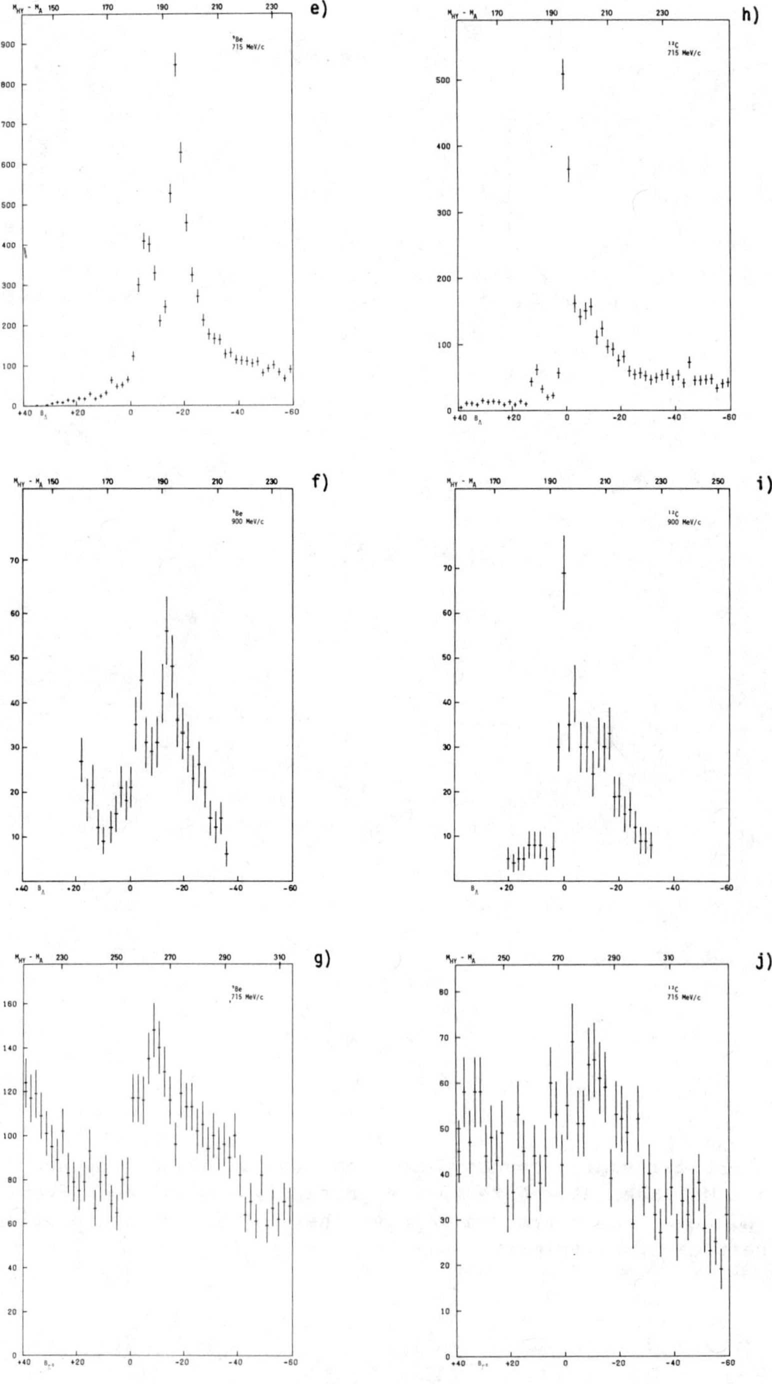
e)
h)
f)
i)
g)
j)
^{9}Be 715 MeV/c
^{12}C 715 MeV/c
^{9}Be 900 MeV/c
^{12}C 900 MeV/c
^{9}Be 715 MeV/c
^{12}C 715 MeV/c

Fig. 3 (cont.)

Let me now compare the $\bar{\omega}_{exp}$ values shown in Table 2 for Λ and Σ^0 hypernuclei (the data are given with a ±2 MeV error in this preliminary analysis). We see a general shift of about 4 MeV. If the above-mentioned analysis is applied for the ^{12}C, this difference in the binding energy between Λ and Σ^0 would imply a Σ^0 potential of about 21 MeV (a neutron potential of 51 MeV has been used in the calculation). To show the effect of such a small potential on the energy spectra of the hypernucleus, let me mention that the difference between the $1s_{1/2}$ and $1p_{3/2}$ shell in ^{12}C measured in the (p,2p) reaction is about 20 MeV. The energy difference between the same single-particle levels has recently been measured[6] to be 11 MeV in $^{12}_{\Lambda}C$, consistent with a constant potential D_{Λ} = 29 MeV.

The same shell-model calculation in an extended configuration frame already applied to the Λ case[9] has been performed by Bouyssy for the $^{12}_{\Sigma^0}C$ hypernucleus[10]. This very preliminary calculation, made with a central potential depth of 20 MeV and a spin orbit potential V_{LS} = 0, is compared with our experimental data in Fig. 4. The quasi-free (QF) and recoiless (R) contributions are also shown separately. The values of N_{eff} defined by the relation

$$\left(\frac{d\sigma}{d\Omega}\right)_{\theta=0^\circ} = N_{eff} \cdot \left(\frac{d\sigma}{d\Omega}\right)_{K^- N \to \Sigma^0 \pi^-}$$

are N^{QF}_{eff} = 1.20 and N^{R}_{eff} = 0.32, in agreement with the predominant quasi-free contribution shown by the data. The total curve does not, however, fit the data nicely, probably because of a too small width for the particle-hole state introduced in the calculation. On the contrary, as far as the quasi-free contribution only is considered, the agreement is quite good, and the centres of gravity of both the theoretical and the experimental curves fit nicely.

In the $^{9}_{\Sigma^0}Be$ spectrum a narrower structure appears above the large quasi-free bump. Its width is about 12 MeV, but a mixing of two different states is possible. If we compare this spectrum with the $^{9}_{\Lambda}Be$ excitation energy spectrum of Ref. 6, we see a similar structure. A recoilless production is possible here on the three neutrons of the $1p_{3/2}$ shell. (There are only two $p_{3/2}$ neutrons in ^{7}Li and a higher nuclear radius in ^{12}C).

There is no reliable estimate of the lifetime of the Σ^0 hyperon in the nucleus. A conservative limit can be obtained by comparing the Σ^0 hypernuclear states that release 80 MeV in their decay with the nuclear states with corresponding excitation energy. These states, essentially hole states in the 1s shell, have about 20 MeV width. This width can be strongly reduced by the restricted possibility of the Σ^0 bound at the nuclear surface to interact with the nuclear matter.

The existing data on the ΣN interaction at low momentum are quite scarce. The Σ hyperon has isospin I = 1 and can couple with a nucleon to I = ½ and 3/2. The $\Sigma^+ p$ interaction is pure in the I = 3/2 state. Its measured total cross-section at an incident Σ^+ momentum

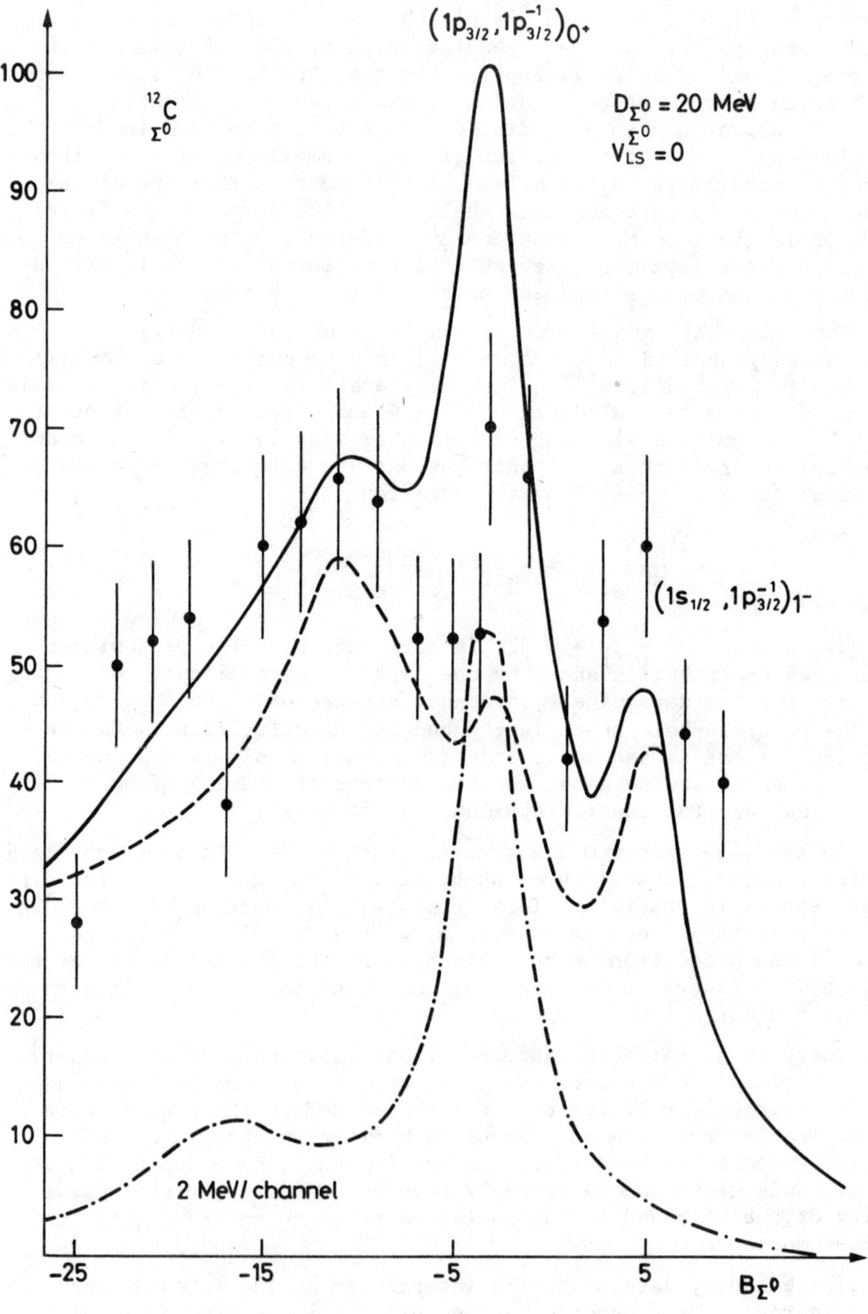
$(1p_{3/2}, 1p_{3/2}^{-1})_{0^+}$
$^{12}_{\Sigma^0}C$
$D_{\Sigma^0} = 20$ MeV
$V_{LS}^{\Sigma^0} = 0$
$(1s_{1/2}, 1p_{3/2}^{-1})_{1^-}$
2 MeV/ channel
B_{Σ^0}
100
90
80
70
60
50
40
30
20
10
-25
-15
-5
5

Fig. 4

of 160 MeV/c is about 110 mb. This low value (10 times less than the unitary limit) suggests that bound states or low-lying resonances in the highest states with $I = 3/2$ are unlikely[11]. The $\Sigma^- + p \rightarrow \Lambda + n$ has been studied with polarized Σ^- hyperons. The Λ particles emerge with non-zero polarization for the $\Sigma^- p$ interaction, which presumably takes place in an s-wave[11,12]. Since only the triplet state can contribute to the polarization, this information suggests that the $\Sigma^- p$ interaction takes place predominantly in the triplet state. In the deuterium bubble chamber a strong enhancement in the $K^- + d \rightarrow \Lambda + p + \pi^-$ cross-section has been observed at the invariant mass close to the ΣN threshold[13]. This narrow experimental peak (about 6 MeV wide) has been interpreted as possibly corresponding to the $I = 1/2$ ΣN system in the triplet state.

The interpretation of our Σ^- nucleus states is more difficult owing to the strong distortion in the high excitation region. A conservative interpretation would suggest a Σ^- nucleus potential even less deep than in the Σ^0 nucleus case. The successful result of the preliminary exploration of this new field of nuclear physics will, we hope, encourage the development of even better kaon beams to enable the study of this kind of hypernuclei in a transfer momentum region more appropriate to the excitation of recoilless states. With the Λ ($I = 0$) and Σ ($I = 1$) hyperons we have indeed two unique probes for studying the nuclear structure even in the more deeply bound states, which cannot be reached in the conventional nuclear reactions.

REFERENCES

1) J. Pniewski and M. Danysz, Phys. Lett. 1, 142 (1962).
2) E. Aslanides, R. Bertini, O. Bing, P. Birien, B. Bricaud, F. Brochard, H. Catz, J.M. Durand, J.C. Faivre, D. Garreta, Ph. Gorodetzky, F. Hibou, J. Pain and J. Thirion *in* Abstracts of papers submitted to the 7th Int. Conf. on High-Energy Physics and Nuclear Structure, Zürich, 1977 (SIN, Villigen, 1977), p. 376.
3) B. Povh, Z. Phys. A279, 159 (1976).
4) W. Brückner, B. Granz, D. Ingham, K. Kilian, U. Lynen, J. Niewisch, B. Pietrzyk, B. Povh, H.G. Ritter and H. Schröder, Phys. Lett. 62B, 481 (1976).
5) R.H. Dalitz and A. Gal, Phys. Lett. 64B, 154 (1976).
6) W. Brückner, M.A. Faessler, T.J. Ketel, K. Kilian, J. Niewisch, B. Pietrzyk, B. Povh, H.G. Ritter, M. Uhrmacher, P. Birien, H. Catz, A. Chaumeaux, J.M. Durand, B. Mayer, J. Thirion, R. Bertini and O. Bing, Phys. Lett. 79B, 157 (1978).
7) J.P. Walecka, High-energy physics and nuclear structure (ed. S. Devons) (Plenum, New York, 1970), p. 1.

8) E.J. Moniz, Phys. Rev. 184, 1154 (1969).
E.J. Moniz, I. Sick, R.R. Whitney, J.R. Ficenec, R.D. Kephart and W.P. Trower, Phys. Rev. Lett. 26, 445 (1971).
9) A. Bouyssy, Nucl. Phys. A290, 324 (1977).
10) A. Bouyssy, private communication.
11) G. Alexander and O. Benary, Proc. Int. Conf. on Elementary Particles, Amsterdam, 1971 (eds. A.G. Tenner and M. Veltman)(North-Holland Publ. Co., Amsterdam, 1972). p. 417, and references therein.
12) M.M. Nagels, T.A. Rijken and J.J. De Swart, THEF-NYM-78-4 report, and references therein.
13) H.G. Dosch, V.F. Filthuth, V. Hepp and E. Kluge, Phys. Lett. 21, 236 (1966).

Question: A. Gal, Hebrew University

The fact that Σ^0 turns out to be attractive in the nucleus (D≃21 MeV in your presentation), whereas Σ^- comes out unbound may have nothing to do with the charge state of the Σ baryon, but with the isospin value of the Σ-hypernuclear system. Take for example ^{12}C. The Σ^0 hypernuclear states excited in (K^-,π^-) are probably dominated by I=1/2. The (K^-,π^+) reaction, however, necessarily leads to I=3/2 (and higher values) for the Σ^- hypernuclear states. Hence, it could as well be that we have here a measure of the symmetry energy of the Σ baryon in the nucleus. This point certainly deserves further work.

Comment: K. Kilian, CERN

In my opinion in order to determine differences in the optical potential well depth for Σ^0 and Σ^- one promising way is to compare the measured nucleon-hypernucleon transformation energies $V_{NY} \sim B(N_i) - B(Y_k)$. If one can identify states with identical spin space configurations in the cases of neutron hole - Σ^0 particle and proton hole - Σ^- particle, then one can estimate comparable single particle energies $B(Y_k)$ for Σ^0 and Σ^-. In this way the influence of different neutron and proton binding energies $B(N_i)$ can be taken into account. These Σ^0 and Σ^- single particle energies give an indication of the corresponding potential well depth.

K^- - ^{12}C ELASTIC SCATTERING AT 800 MeV/C

D. Marlow, P. Barnes, S. Dytman,
R. Eisenstein, W. Wharton, F. Takeutchi
Carnegie-Mellon University, Pittsburgh, PA 15213

S. Bart, D. Hancock, R. Hackenburg,
E. Hungerford, L. Pinsky, T. Williams
University of Houston, Houston, TX 77004

R. Chrien, H. Palevsky, R. Sutter
Brookhaven National Laboratory, Upton, NY 11971

We have measured the differential cross-section for elastic K^- scattering from ^{12}C at 800 MeV/C ($T_K \sim 440$ MeV), for angles $5° \lesssim \theta_{Lab} \lesssim 30°$. The data were collected at the BNL AGS using a recently constructed beam handling system and energy-loss spectrometer.[1] The overall system resolution is ~1.5 MeV, the spectrometer solid angle is ~15 msr, the angular acceptance ~8°, and $\Delta p/_p$ is typically ±1-1/2%. With 4 x 10^{12} protons on the production target, the K^- rate was 20,000/sec.

The energy spectra obtained show good separation for all states below 10 MeV. The preliminary elastic angular distribution with a large normalization uncertainty is shown below, compared to a calculation by Dover and Walker,[2] who used the K-N amplitudes of Gopal et al[3] to construct the first order optical potential. The calculations are a reasonable representation of the data.

1. BNL proposals 646 and 692.
2. C. Dover and G. Walker, preprint (1978).
3. G. P. Gopal, et al., Nuc. Phys. B119 (1977) 362.

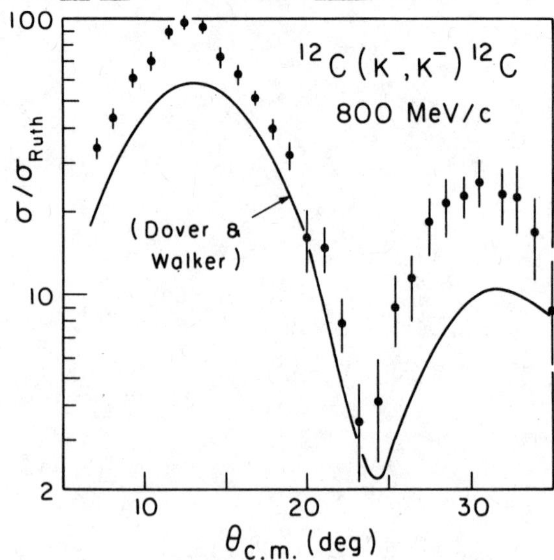

ISSN:0094-243X/79/540714-01$1.50

OBSERVATION OF KAONIC HYDROGEN ATOM X-RAYS

J D Davies*, G J Pyle, G T A Squier
University of Birmingham, Birmingham, UK

C J Batty, S F Biagi, S D Hoath, P Sharman
Rutherford Laboratory, Chilton, Didcot, Oxon, UK

A S Clough
University of Surrey, Guildford, UK

INTRODUCTION

The study of kaonic hydrogen atom X-rays, which should be shifted in energy and have a broadened line shape because of the K^--p strong interaction, should provide a method of determining the S-wave K^--p scattering length at threshold quite independently of the more conventional scattering experiments[1]. However these X-rays have a low yield in a liquid hydrogen target because of strong absorption of the kaon from high n atomic s-states populated by stark effects during collisions between the electrically neutral K^--p atom and other protons[2]. Previous attempts to detect the 6.5keV 2p - 1s X-rays from kaonic hydrogen, which have used both liquid and high pressure gas targets, have been unsuccessful[3]. Using a specially constructed low Z cryogenic target fitted with a high resolution Si(Li) detector, and the K17 beamline[4] at the Rutherford Laboratory, we have obtained some data on K-He, π-He, K-H and π-H which indicate the observation of kaonic hydrogen X-rays.

TARGET AND DETECTION SYSTEM

The target consisted of a $\sim$ 1mm pure Aℓ "spinning" welded to a thick Aℓ top plate enclosing a total volume of $\sim$ 12 litres. Thin Be windows permitted X-rays to pass from the target volume to a 300mm^2 x 5mm thick Si(Li) crystal mounted above the top plate. A retractable ^{57}Co source was used to test the detector in situ. The detector O/P pulse height was digitized by a 16K ADC and routed in 32ns blocks of relative time between the conventional particle stop trigger and the detector O/P pulse, in order to cover a 500ns wide range of the instrumental timing. The system had a timing resolution of 30ns FWHM at 11keV and an energy resolution of 260eV FWHM at 6.5keV. Instrumental timing effects originating in the detector were found to vary with incident X-ray energy in a consistent manner, and the K-H data was optimised by taking these effects into account. Contribution of Compton events induced by high energy photons was reduced relative to true low energy photons by such optimization, which improved the observation of the low yield K-H X-rays.

*Present address: EP Division, CERN

ISSN:0094-243X/79/540715-02$1.50

RESULTS

All the spectra obtained were clear of contaminant X-ray lines, except for π-Be 3 - 2 X-rays at 8.1keV. The K^- - He data improved upon the published spectrum of Wiegand and Pehl[5], and demonstrated the detection of high yield 6.4keV X-rays from the target volume. The π-He data showed no X-rays in the 6.5keV region of interest for K-H, and detailed fits including computed compton scattering distributions were used to check the ^{57}Co calibration against the very precise data of Backenstoss et al[6]. Limited data on K-H and π-H showed a peak at 6.5keV in the K-H spectrum which had the timing characteristics of a true low energy X-ray. Detailed analysis[8] showed the peak had an energy of 6.52 ± 0.06keV, a poorly determined width consistent with the instrumental resolution and an area of 45 ± 23 counts equivalent to a yield of about 0.1% per K^- stop. Searches for the 3p - 1s K^--p X-rays at 7.7keV, were limited by the 32ns timing block width to an upper limit of about 0.1% per K^- stop.

REMARKS

The observed peak was consistent with an essentially unmodified 1s - state of the K^--p atom, implying a rather small S-wave K^--p scattering length at threshold. The difference between the result and the previous extrapolation methods is about (0.7 ± 0.2) fm. The experiment is being continued at CERN[9] this year.

REFERENCES

1. J Wrzecionko and S Wycech, Czech. J. Phys. B24, 1293 (1974).
2. M Leon and H A Bethe, Physical Review 127, 636 (1962).
3. A Roberts et al, Rutherford Laboratory Proposal 156 (1974); L Tauscher, private communication (1978).
4. M K Craddock, C J Reason and R A J Riddle, RL-76-091.
5. C E Wiegand and R H Pehl, Phys. Rev. Lett 27, 410 (1971).
6. G Backenstoss et al, Nucl. Phys. A232, 519 (1974).
7. C J Batty et al, to be published by Nucl. Phys. A(1979).
8. J D Davies et al, to be published by Phys. Lett. B (1979).
9. J D Davies et al, CERN PS-165.

KAON-NUCLEUS SCATTERING AND KAON-NUCLEAR COUPLING CONSTANTS

Olgierd Dumbrajs*)
Division de Physique Théorique
Institute de Physique Nucléaire, 91406 Orsay, France

Possibilities are examined of obtaining information on kaon-nuclear coupling constants without referring to any specific models. The basis for such a type of analysis is the use of analytic properties of the scattering amplitudes. Particularly useful is the forward dispersion relation.

It is known [1] that the most suitable dispersion relation for determination of meson-nuclear coupling constants is the one for the antisymmetric amplitude. In the case of the kaon-nuclei scattering the use of the antisymmetric amplitude is even more justified because both the real and the imaginary parts are directly measured in regeneration experiments.

As a practical example we consider the $K^{\pm\,12}C$ elastic scattering. We write down the disprsion relation in the form:

$$\mathrm{Re}\, f^{-}(\omega) = \frac{2\omega\gamma}{\omega^2 - \omega_p^2} + \frac{2\omega}{\pi} P \int_{\omega_0}^{\infty} \frac{\mathrm{Im} f^{-}(\omega')}{\omega'^2 - \omega^2} d\omega' \quad (1)$$

Here $f^{-}(\omega)$ is the antisymmetric $K^{\pm\,12}C$ scattering amplitude in the lab. system, ω is kaon lab. energy, γ is the residue of the effective pole. We place it at the position which corresponds to the pole due to the ground state of the ${}^{12}_{\Lambda}B$-hypernucleus. The integration starts from the beginning of the unphysical region due to the ${}^{11}B\Lambda$-cut. The amplitude $f^{-}(\omega)$ can be either calculated on the basis of experimental data on $K^{\pm\,12}C$ elsatic scattering or, via isotopic invariance, connected to the measured regeneration amplitude.

We have used the following input in eq.(1). The integral in the energy range from ω_0 to m_K was disregarded. At the physical threshold we put $\mathrm{Im} f^{-}(m_K)=1.18$ fm in accordance with the value of the K^{-}-carbon scattering length [2]. In the energy range between the threshold and 0.83 GeV we assumed that $\mathrm{Im} f^{-}$ is linearly rising with energy. Experimental data on $K^{\pm\,12}C$ total cross section are available in the energy range between 0.83 GeV and 3.3 GeV [3,4]. Above 3.3. GeV $\mathrm{Im} f^{-}$ was evaluated on the basis of regeneration data parametrised in the form

*) Address after 1st February 1979: Institute of Physics, University of Aarhus, DK-8000 Aarhus C, Denmark.

ISSN:0094-243X/79/540717-02$1.50

given in [5].

Values of Ref^- are obtained by combining the measured values of Re$f_{K^{\pm}\,^{12}C}$ at 1.75, 2.07 and 2.31 GeV [6] and on the basis of regeneration data [5]. The residue of the effective pole was determined as γ=12.3±3.0.

It is known [7] that the effective pion-nuclei residues are systematically smaller than the cohrent sum of the elementary πNN residues. This reflects the occurrence of excahnge or shadowing effects in nuclei [8]. What is the situation in our case?

We define the coherent sum of the elementary KNΛ and the KNΣ residues

$$\gamma_{coh}=\left[6g_{\Lambda}^{2}\,\frac{(m_{\Lambda}-m_{p})^{2}-m_{K}^{2}}{4m_{p}^{2}}+18g_{\Sigma}^{2}\frac{(m_{\Sigma}-m_{p})^{2}-m_{K}^{2}}{4m_{p}^{2}}\right]$$

Inserting the values of elementary coupling constants g_{Λ}^{2}=13.7±1.9, g_{Σ}^{2}=3.7±1.3 [9]we obtain γ_{coh}=8.4±1.9. Consequently, the effective residue determined from the forward dispersion relation is larger than the coherent sum of elementary residue. This finding coincides with the claim of the authors of ref. [10]. However, this coincidence is fortuitous, a result of two cancelling errors in ref. [10]: the superflouous factor 2 in the pole term makes γ two times smaller, the unrealistic high energy model for Imf^- makes γ about two times larger.

Obviously the K^--nucleon and K^--nucleus interactions near the threshold are dominated by different physical mechanisms.

REFERENCES

1. T.E.O. Ericson and M.P.Locher, Nucl.Phys.A148 (1970).
2. R. Seki, Phys.Rev.Lett. 29, 240 (1972).
3. D.V.Bugg et al.,Phys.Rev. 168, 1466 (1968).
4. R.J.Abrams et al., Phys.Rev. D4, 3235 (1971).
5. B.Diu and A. Ferraz de Camargo filho, Nuovo Cimento 47A, 495 (1978).
6. B. Gobbi et al., Phys.Rev.Lett. 29, 1278 (1972).
7. M.P.Locher and T.Mizutani, Phys.Reports 46, 43 (1978).
8. M.Ericson, Ann.Phys. 63, 562 (1971).
9. A.D.Martin, Phys.Letters 65B , 346 (1976).
10. K.Arai, I.Eudo and M.Kikugawa, Progr.Theor.Phys. 56, 1345 (1976).

OBSERVATION OF A SET OF NUCLEAR SHELL ORBITALS FOR THE Λ PARTICLE IN $^{32}_{\Lambda}S$ AND $^{40}_{\Lambda}Ca$

R. Bertini[1**], O. Bing[1], P. Birien[2], W. Brückner[3], H. Catz[2*], A. Chaumeaux[2**], J.M. Durand[2], M.A. Faessler[3], T.J. Ketel[3], K. Kilian[3], B. Mayer[2], J. Niewisch[3], B. Pietrzyk[3], B. Povh[3], H.G. Ritter[3], M. Uhrmacher[3]

ABSTRACT

Energy excitation spectra have been obtained for hypernuclei in the s-d shell from the (K^-,π^-) reaction in flight on $^{32}_{\Lambda}S$ and $^{40}_{\Lambda}Ca$. The analysis is carried out in a simple particle hole model and leads to the results that the Λ behaves in the nucleus like an independant particle in a harmonic oscillator potential well with no significant spin orbit term.

This paper reports on the study of $^{32}_{\Lambda}S$ and $^{40}_{\Lambda}Ca$ hypernuclei via the (K^-,π^-) reaction at 720 and 790 MeV/c close to 0°. The momentum transfer is thus very low and the production is said either recoilles (if the Λ remains in the same shell as the neutron hole) or quasi-free in the Λ jumps in another shell.

The experiment has been performed on the K22 secondary beam at the CERN Proton Synchrotron (P.S.), the outgoing pions being analysed through the magnetic spectrometer SPES II from Saclay. Fig. 1 shows the measured spectra. These are best compared if plotted as a function of $E_K-E_\pi \simeq M_{HY}-M_A$ which is the energy necessary to transform a neutron into a Λ. As the transormation energy in one specific shell configuration is expected to vary slowly from a nucleus to a neighboring one, the peaks appearing at the same position for calcium and sulphur will be assigned to the same configuration. The most prominent recoilless transitions are expected to stem from the (K^-,π^-) reaction on the neutrons of the last shells. In the case of calcium these are the $1d_{5/2}$, $2s_{1/2}$ and $1d_{3/2}$ whereas for sulphur the $1d_{3/2}$ is excluded. As the transition strength is roughly proportional to the number of neutrons in an orbital the transition at the binding energy $B_\Lambda = -5$ MeV observed in calcium and sulphur is assigned to the $(1d_{5/2}, 1d_{5/2}{}^{-1})_{\Lambda n}$ configuration and the one at $B_\Lambda = 0$ observed in calcium but not in sulphur in assigned to $(1d_{3/2}, 1d_{3/2}{}^{-1})_{\Lambda n}$. The recoiless peaks having a strong angular dependence[2] the comparison of the spectra at 0° and 5° in calcium gives support to this interpretation. These two peaks

[1] Centre de Recherches Nucléaires and Université Louis Pasteur, Strasbourg, France.

[2] Département de Physique Nucléaire, CEN Saclay, Gif-sur-Yvette, France.

[3] Max Planck-Institut, Heidelberg, Germany and Physikalishes Institut der Universität, Heidelberg, Germany (supported in part by the Bundesministerium für Forshung und Technologie).

* CERN Fellow.

** CERN Associate.

ISSN:0094-243X/79/540719-02$1.50

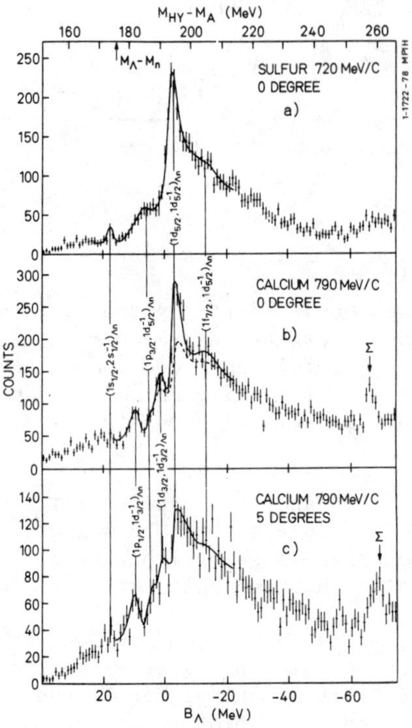

Fig. 1 - Spectra obtained for the (K^-,π^-) reaction on ^{32}S and ^{40}Ca. The 5° spectrum on ^{40}Ca has been inserted in Fig. 1b as a dotted line and normalised on the $(1p_{1/2}, 1d_{3/2}^{-1})$ state. The peak indicated as Σ^+ is due to free Σ production on liquid hydrogen Cerenkov counters. The lines are only given to guide the eye.

are 5 MeV apart, as much as the splitting between the $d_{5/2}$ and $d_{3/2}$ neutron holes in ^{40}Ca[3] so that we do not observe a significant spin-orbit splitting for the Λ, in agreement with our results on the p-shell nuclei[1]. Regarding the quasi-free peaks $\Delta\ell = 2$ or spin-flip transitions are forbidden so that the states we can observe are $(1p_{3/2}, 1d_{5/2}^{-1})_{\Lambda n}$ in calcium and sulphur and $(1p_{1/2}, 1d_{3/2}^{-1})_{\Lambda n}$ in calcium only. The shoulder at $B_\Lambda = 7$ MeV in $^{40}_{\Lambda}Ca$ and $^{32}_{\Lambda}S$ is assigned to the first of these configuration and the state at $B_\Lambda = 11$ MeV in $^{40}_{\Lambda}Ca$ to the second. One peak appears at $B_\Lambda = 17.5$ MeV in $^{32}_{\Lambda}S$ which is the expected value for the ground state[4] with the configuration $(1s_{1/2}, 2s_{1/2}^{-1})_{\Lambda n}$. In $^{40}_{\Lambda}Ca$ this part of the spectrum is obscured by the $K^- \to \pi^\circ + \pi^-$ decay.

Taking into account the known energy of the neutron holes we come to the following conclusions : i) there is no (or little) spin-orbit splitting in the p and d shells for the Λ nucleus interaction ; ii) the shells are approximately equidistantly spaced with a separation of about 9 MeV. In a very crude manner we can say that the Λ particle behaves like a spinless particle in a harmonic oscillator potential well.

REFERENCES

1. W. Brückner et al., Phys. Lett. 79B (1978) 157.
2. B. Povh, Ann. Rev. Nucl. Phys. 28 (1978).
3. G. Jacob and A.J. Marris, Rev. Mod. Phys. 45 (1973) 6.
   P. Doll et al., Nucl. Phys. A263 (1976) 210.
4. A. Bouyssy, Nucl. Phys. A290 (1977) 324.

ON THE STRUCTURE OF STRANGENESS ANALOGUE RESONANCE IN $^{6}_{\Lambda}$Li HYPERNUCLEUS

L. Majling, M. Sotona, J. Žofka
Institute of Nuclear Physics, Řež, Czechoslovakia

V. N. Fetisov
P. N. Lebedev Physical Institute, Moscow, USSR

R. A. Eramzhyan
Joint Institute of Nuclear Research, Dubna, USSR

ABSTRACT

The excitation of $^{6}_{\Lambda}$Li in the reaction ^{6}Li(K^-, π^-)$^{6}_{\Lambda}$Li*, formation of the strangeness analogue resonance (SAR) and its decay have been studied. It is suggested, how the predictions on the structure of $^{6}_{\Lambda}$Li and their implications for the Λ-nucleons coexistence may be verified in the experiment.

INTRODUCTION

Shell model proved to be a useful tool to treat the low lying hypernuclear states [1]. There were attempts to apply it to the nuclear resonance states, which appear in (K^-, π^-) reactions, as well. In this contribution, the excited states of positive parity in $^{6}_{\Lambda}$Li have been treated in translationally invariant shell model. In $^{6}_{\Lambda}$Li, the model space is not large and an additional nuclear wave function symmetry simplifies the identification of SAR components.

RESULTS AND DISCUSSION

The extensive results on structure and cross sections obtained [2] may be explained in a simplified qualitative form as follows: In the recoilless reaction ^{6}Li(K^-, π^-)$^{6}_{\Lambda}$Li*, predominantly 1^+ strangeness analogue states (SAS) are excited. The ground state of ^{6}Li is well described by $|s^4p^2\ {}^{13}S_1\rangle$ [3] and thus the "doorway states" with explicit nuclear symmetry [2] in the spurious-free representation (requiring $\omega_\Lambda = \omega_N$) are:

ISSN:0094-243X/79/540721-02$1.50

$$\left|SAS,\ {}^{6}_{\Lambda}Li\right\rangle = \sqrt{\frac{20}{45}}\left|[32]\ {}^{24}S_{3/2},\ s_{\Lambda}\right\rangle + \sqrt{\frac{2}{45}}\left|[41]\ {}^{22}S_{1/2}, s_{\Lambda}\right\rangle$$

$$-\sqrt{\frac{5}{45}}\left|[32]\ {}^{22}S_{1/2}, s_{\Lambda}\right\rangle - \sqrt{\frac{18}{45}}\left|[41],\ ({}^{22}P\ p_{\Lambda} - {}^{22}S_{sp} s_{\Lambda}), {}^{3}S_{1}\right\rangle$$

Two strong components of SAS decay differently: $\left|[32]\ {}^{24}S_{3/2},\ s_{\Lambda}\right\rangle$ via ${}^{4}_{\Lambda}He + d$ (or ${}^{4}_{\Lambda}He^{*} + d$, if energetically possible), while $\left||41|,\ ({}^{22}P\ p_{\Lambda} - S_{sp} s_{\Lambda}),\ {}^{3}S_{1}\right\rangle$ via ${}^{5}Li + \Lambda$ or ${}^{5}_{\Lambda}He + p$. This is seen in fig. 1., where positions of doorway states and decay spectroscopic factors are visualized for one interaction. In the π^{-} spectrum, four resonances at approx. 12, 18, 19, 25 MeV should be detected (further fragmentation due to tensor forces is possible), that at 18 MeV being clustered and narrow. Higher lying ones may decay via ${}^{6}_{\Lambda}Li^{*} \rightarrow {}^{4}_{\Lambda}He^{*} + d$, giving an explanation for the known hypernuclear 1.1 MeV γ-quanta. Observation of deuterons in the decay of ${}^{6}_{\Lambda}Li^{*}$ would thus give an evidence for the hypernuclear clusterization. Splitting of SAR give an information on Λ single particle energies and Λ-N interaction[2].

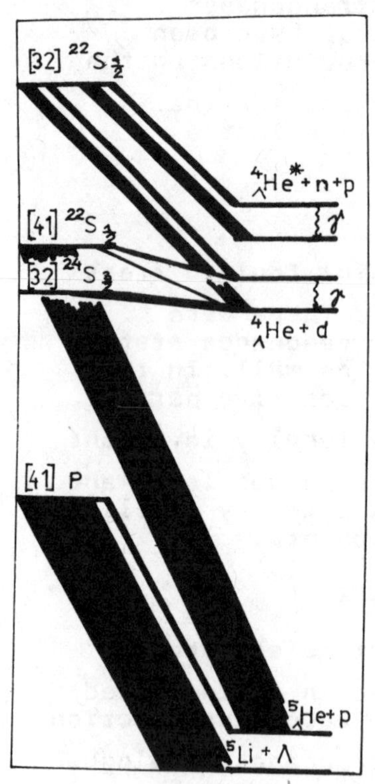

REFERENCES

1. R. H. Dalitz, A. Gal, Prepr. Oxford 8/78.
2. L. Majling et al, to be published.
3. T. W. Donelly, J. D. Walecka, Phys. Lett. 44B (73).

FURTHER EVIDENCE FOR THE EXISTENCE OF THE γ DECAYING STATES OF ${}^4_\Lambda$H AND ${}^4_\Lambda$He

CERN-LYON-WARSAW Collaboration

M. Bedjidian, E. Descroix, J. Y. Grossiord, A. Guichard, M. Gusakow, M. Jacquin, J. R. Pizzi
Institut de Physique Nucléaire (and IN2P3)
Université Claude Bernard Lyon-I, Villeurbanne (France)

M. J. Kudla, H. Piekarz, J. Piekarz, J. Pniewski
Institute of Experimental Physics, University of Warsaw and Institute of Nuclear Research, Warsaw (Poland)

The γ spectra induced by stopped K^- mesons in ^{6}Li and ^{7}Li targets were obtained in coincidence with the accompanying charged and neutral pions.

The detection of pions from two-body decays of the ${}^4_\Lambda$H and ${}^4_\Lambda$He hypernuclei

$$ {}^4_\Lambda H \rightarrow \alpha + \pi^- \quad (E_{\pi^-} = 53 \text{ MeV}) $$

$$ {}^4_\Lambda He \rightarrow \alpha + \pi^o \quad (E_{\pi^o} = 57 \text{ MeV}) $$

allowed to observe and identify the γ transitions in these hypernuclei.

The γ line at (1.04 ± 0.04) MeV (Fig. 1a) observed with charged pions was identified as a γ transition in ${}^4_\Lambda$H. This result confirms our previous observation for this line [1]. A second line at (1.15 ± 0.04) MeV was observed. As it was detected in coincidence with π^o mesons, it was ascribed to a γ transition in ${}^4_\Lambda$He (Fig. 1b).

Using these new values for the excitation energies of the 1^+ levels of ${}^4_\Lambda$H and ${}^4_\Lambda$He, the parameters of the phenomenological Λ-N potential for the S- state interaction were recalculated. It turned out that the potential having an intrinsic range of 1.5 fm and a hard core radius of 0.45 to 0.60 fm gave the best fit simultaneously to the Λ-p elastic scattering cross-section data and to the binding energy of ${}^3_\Lambda$H. Assuming a charge symmetry breaking potential of the form :

$$ U_{CSB} = - \tau_3^i \, W_o(r) \, (\alpha + \vec{\sigma}_\Lambda \cdot \vec{\sigma}_N) $$

it was also found that the spin dependant term was weak.

REFERENCE

1) M. Bedjidian et al., Physics Letters, 62B (1976) 467.

ISSN:0094-243X/79/540723-02$1.50

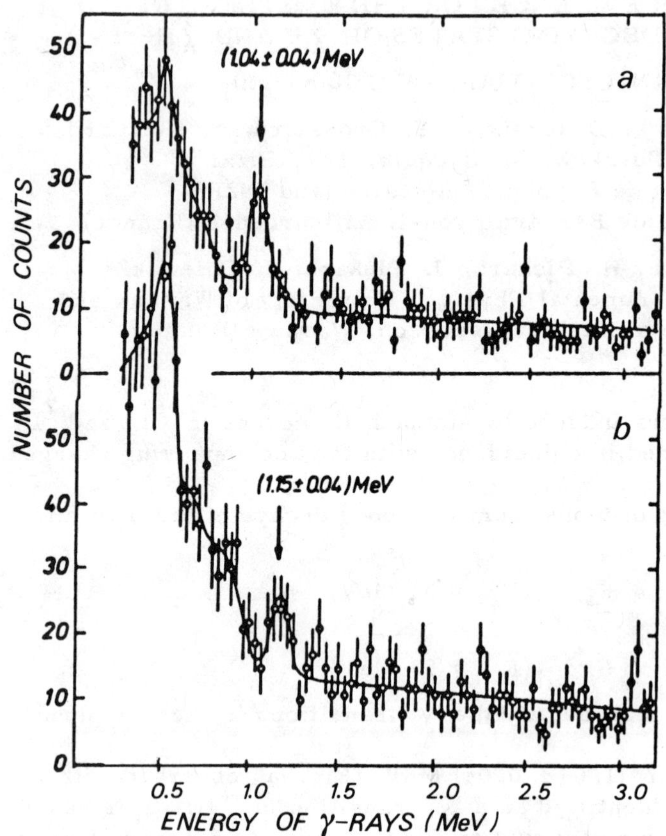

Fig. 1. Added γ- spectra from ^{6}Li and ^{7}Li targets

a- in coincidence with charged pions
b- in coincidence with neutral pions.

THE $K^-d \to \pi^-\Lambda p$ REACTION AT LOW ENERGIES*

G. Toker and A. Gal
Racah Institute of Physics, The Hebrew University
Jerusalem, Israel

J.M. Eisenberg
Department of Physics and Astronomy, Tel-Aviv University
Tel-Aviv, Israel

We report here on progress in performing a Faddeev calculation for the process $K^-d \to \pi^-\Lambda p$ for kaons at rest. The main motivation for this study is the enhancement in the Λp mass distribution observed near $\Sigma^+ n$ threshold, both for capture at rest[1] and in flight[2]. This enhancement might represent a ΣN bound state embedded in the Λp continuum. The two-body input takes the interactions NN, πN, YN and $\bar{K}N \to \bar{K}N, \pi Y$ in s-wave separable forms, except in the πN case for which the important p-wave 3,3 channel is also included. At a later stage, the ^3S-^3D baryon-baryon coupling may be incorporated. Nonrelativistic kinematics are used and the isospin formalism is exploited to reduce the number of coupled integral equations.

The lowest-order diagrams for this reaction are shown in Fig. 1 and include impulse approximation (a), $\Sigma N \to \Lambda N$ conversion and ΛN rescattering (b), and initial (final) state interaction (c). Of these, the Σ-Λ conversion provides the dominant effect at ΣN threshold, leading to a cusp there; see Fig. 2a for the mass spectrum as calculated from the lowest-order graphs, as yet without πN p-wave effects. The Faddeev result of Fig. 2b, which iterates all of

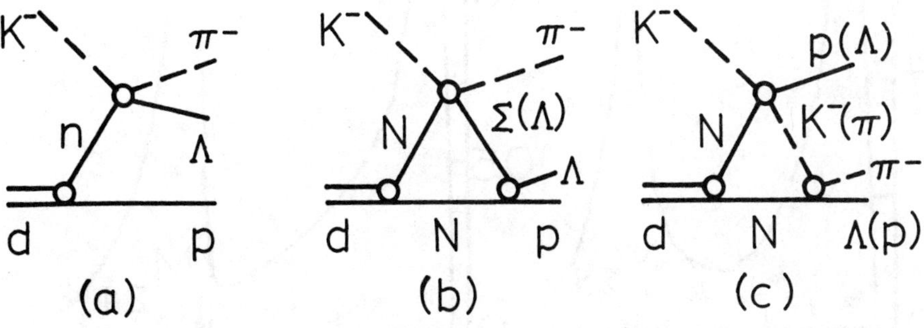

Fig. 1. Lowest-order diagrams contributing to $K^-d \to \pi^-\Lambda p$: (a) impulse approximation, (b) $\Sigma N \to \Lambda N$ conversion or ΛN rescattering, (c) initial or final (K or π) meson interaction.

*Work supported in part by the U.S.-Israel Binational Science Foundation.

ISSN:0094-243X/79/540725-02$1.50

the various rescattering processes, shows a substantially modified spectrum. In particular, the detailed properties (height and width) of the narrow enhancement at the ΣN threshold are considerably modified by the full Faddeev calculation. To date, our treatment of the YN interaction channel has been somewhat restricted, and no additional structure above ΣN threshold has been found in the calculation. Thus we have, as yet, no explanation for a structure of this nature which may be indicated in the experiment [1,2].

REFERENCES

1. T.H. Tan, Phys. Rev. Letters 23, 395 (1969).
2. O. Braun et al., Nucl. Phys. B124, 45 (1977) and references cited there.

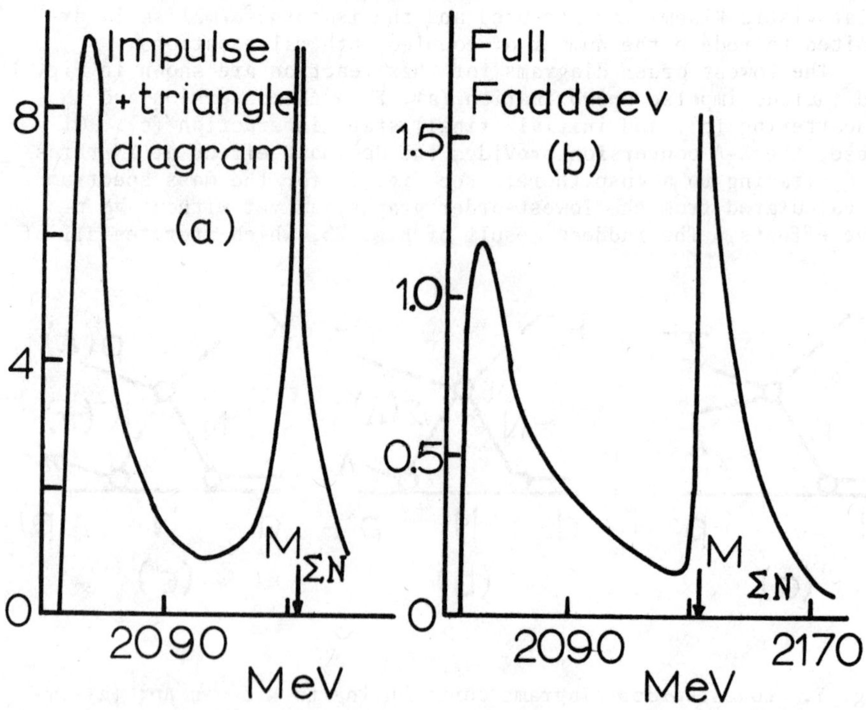

Fig. 2. (a) Mass spectrum for Λp as obtained from summing contributions of diagrams in Fig. 1. (b) Mass spectrum calculated from full Faddeev formalism with all rescatterings.

EFFECTIVE LAMBDA-NUCLEON POTENTIALS*

F. Mujib, M. Shoeb, Q.N. Usmani, and M.Z.R. Khan
Dept. of Physics, A.M.U., Aligarh-202001, India

ABSTRACT

The lambda binding energy data of p-shell hypernuclei have been analysed within the shell model frame work using central, charge-independent, spin- and state-dependent effective Λ-nucleon potentials of Gaussian form and Skyrme type. The c.m. energy in each case has been accounted for approximately and also due consideration has been given to the variation of core size and the Λ-particle wavefunction. The fits to the data in the present analysis indicate that in earlier analyses the role of non-central and three-body forces has been somewhat over emphasized.

INTRODUCTION

To explore the character of lambda-nucleon (ΛN) interaction extensive analyses[1,2] of the Λ-binding energy (B_Λ) of p-shell hypernuclei were carried out. Gal et al[1] and Lee et al[2] find that simple assumption of a central two-body ΛN force is unable to produce the B_Λ data unless noncentral and three-body forces are included. In these analyses reduced matrix elements of ΛN interaction were treated as parameters thus three important effects have been neglected. The first is the sensitivity of B_Λ to the oscillator size parameter of nucleon. The second is the variation of the Λ-wavefunction as the core nucleus varies from ^{4}He to ^{14}N. The use of a mean Λ and nucleon wavefunction means that kinetic energy and the ΛN interaction matrix elements are both in error. The third effect which is left out in the above analyses is that of center of mass (c.m.) motion. We find that after taking these effects into account the simple assumption of a central, charge-independent, spin- and state- dependent two-body ΛN potential is adequate to explain the binding energy of p-shell hypernuclei.

CHOICE OF THE ΛN POTENTIAL, LAMBDA AND NUCLEON WAVEFUNCTION

We have taken ΛN potentials which are central, charge-independent, spin- and state-dependent. For the

*A detail analysis is to appear in Journal of Physics G (1979).

ISSN:0094-243X/79/540727-02$1.50

shape function of the potential we have taken Gaussian form and Skyrme type without three-body part.

For Λ-particle we use single Gaussian wavefunction which is a fairly good approximation[1] for the p-shell hypernuclei. The size parameter b_Λ of Λ-wavefunction is determined for each hypernucleus by minimizing the corresponding Λ-particle energy, thereby enabling the Λ-wavefunction to be different for different nuclei. The core nuclei are described by the intermediate coupled wavefunctions of Soper and the nucleons are assumed to move in a harmonic oscillator well [$b=(\hbar/m\omega)$ is obtained from r.m.s. radii].

RESULT AND DISCUSSION

In the analysis we could include B_Λ data[3] of those hypernuclei for which r.m.s radii of core are known from electron scattering experiments[4] namely: ${}^5_\Lambda He$, ${}^7_\Lambda Li$, ${}^8_\Lambda Li$, ${}^{10}_\Lambda Be$, ${}^{10}_\Lambda B$, ${}^{11}_\Lambda B$, ${}^{12}_\Lambda B$, ${}^{13}_\Lambda C$, ${}^{14}_\Lambda C$, and ${}^{15}_\Lambda N$. The potential parameters are searched by minimizing the conventional χ^2. The size parameters (b) are also varied within their experimental uncertainty which in some cases is very small. Finally the values of the χ^2 for fairly good fit to the data are 3.4 and 5.0 for Skyrme type and Gaussian potential respectively. The Λ-binding in nuclear matter has also been estimated for $k_F=1.35 fm^{-1}$. A value of 27.5MeV and 32.7MeV for Skyrme and Gaussian potential respectively are obtained which are consistent with the empirical estimates which is about 27 MeV.

Our ΛN potentials have appreciable spin-dependence and are qualitatively different in the p-state. The need to incorporate noncentral and three-body forces in our analysis does not seem to arise on the basis of these data.

REFERENCES

1. A. Gal, J.M. Soper, and R.H. Dalitz Ann. of Phys. (N.Y.) 63, 53 (1971) and 72, 445 (1972).
2. T. Lee, S. Hsieh, and C. Chen-Tsai Phys. Rev. C2, 366 (1970).
3. M. Juric et al. Nucl. Phys. B52, 1 (1973); B_Λ of ${}^{13}_\Lambda C$ and ${}^{14}_\Lambda C$ were taken from T. Cantwell et al. Nucl. Phys. A236, 445 (1974).
4. F. Ajzenberge-Selove Nucl. Phys. A268, 1 (1976) and references therein.

WEAKLY SPIN-DEPENDENT EFFECTIVE ΛN POTENTIALS

M. Shoeb, F. Mujib, Q.N. Usmani and M.Z.R Khan,
Dept. of Physics, A.M.U. Aligarh-202001, U.P.India.

ABSTRACT

An analysis of p-shell hypernuclei has been carried out within the shell model framework with central, charged-independent, state-dependent and weakly spin-dependent ΛN forces of Skyrme type. The calculations reported here are preliminary.

INTRODUCTION

In our earlier analyses[1], being presented to this conference, we have analysed Λ-binding energy data of p-shell hypernuclei within the shell model framework assuming central, charge-independent, spin- and state-dependent effective ΛN potentials. These potentials gave a good account of the data. The effective potentials reported in ref. 1 were strongly spin and state dependent. However, there are some indications that the ΛN force is not as strongly spin dependent[2,3]. We, therefore, carried out a search to see if effective potentials with relatively weaker spin dependence could fit the data satisfactorily. We were successful in finding such sets.

CALCULATIONS, RESULTS, AND DISCUSSION

The formulation is the same as described in ref. 1. Only Skyrme type ΛN potentials have been employed in the present calculations. The potential is of the following form

$$V_{i\Lambda}=V^{\ell}_{s,t}\left[\delta(\vec{r}) - a\hbar^{-2}\left\{p^2\delta(\vec{r})+\delta(\vec{r})\ p^2-2\vec{p}.\delta(\vec{r})\vec{p}\right\}\right],$$

where $V^{\ell}_{s,t}$ are the volume integrals of the potential in the singlet and triplet states of relative angular momentum ℓ and a is a parameter.

It is seen from table I that we could fit the data satisfactorily with relatively weak spin-dependent ΛN potentials. We have at least one set in which the spin dependence is compareable to that indicated in ref. 2. So far, we have not estimated well depth of Λ particle in nuclear matter with these sets of potential parameters. It seems that to decide among various potentials of this analysis and those of ref. 1 further calculations have to be carried out for the hypernuclear excited states and nuclear matter. Such a study is in progress.

ISSN:0094-243X/79/540729-02$1.50

Table 1: Potential parameters and fits to B_Λ data. $\beta_0 = V_t^0/V_s^0$, $\beta_1 = V_t^1/V_s^1$, $V_{s,t}$ are in units of $MeV fm^3$ and a is in units of fm^2.

${}^A_\Lambda Z$	Experimental B_Λ (MeV)	$\beta_0=\beta_1=0.75$, $V_s^0=-1227.01$, $V_s^1=5117.20$, $a = 0.948$; $\chi^2=19.7$ — Theoretical B_Λ (MeV)	$\beta_0+0.5=\beta_1=1.25$, $V_s^0=-848.01$, $V_s^1=1752.7$, $a = 0.71$; $\chi^2=16.1$ — Theoretical B_Λ (MeV)
${}^5_\Lambda He$	3.12±0.02	3.107	3.123
${}^7_\Lambda Li$	5.58.0.03	5.526	5.569
${}^8_\Lambda Li$	6.80±0.03	6.792	6.794
${}^{10}_\Lambda Be$	9.30±0.26	9.023	9.081
${}^{10}_\Lambda B$	8.89±0.12	8.795	8.791
${}^{11}_\Lambda B$	10.24±0.05	10.218	10.174
${}^{12}_\Lambda B$	11.37±0.06	11.346	11.370
${}^{13}_\Lambda C$	11.69±0.12	11.922	11.920
${}^{14}_\Lambda C$	12.17±0.33	13.016	13.005
${}^{15}_\Lambda N$	13.59±0.15	13.876	13.839

It may be interesting to remark that our Chi-square values are much lower than those of Gal et al[3].

REFERENCES

1. F. Mujib, M. Shoeb, Q.N. Usmani, and M.Z.R. Khan submitted in this conference and a detailed analysis is to appear in J.Phys.G.Nucl.Phys.(1979).
2. A. Bamberger et al Nucl. Phys. B60, 1(1973).
3. A. Gal, J.M. Soper, and R.H. Dalitz Ann. Physics (N.Y.) 113,79(1978) and R.H. Dalitz invited paper presented at the Seminar "Kaon-Nucleus Interaction and Hypernuclei" of the Labedev Institute, Moscow, held at Zvenigorod on 12-15 September,1977.

Λ BINDING TO NUCLEAR MATTER IN FERMI HYPER-NETTED CHAIN APPROXIMATION

Q.N. Usmani*
Department of Physics, A.M.U., Aligarh, India

ABSTRACT

The Fermi Hyper-Netted-Chain (FHNC) technique has been extended to the problem of calculating the Λ binding to nuclear matter. By making explicit calculations it is shown that in FHNC approximation the overbinding problem[1,2] is alleviated to a considerable extent and also this technique seems to provide a resolution of the 'disturbing' discrepancy between the g matrix[2] and 'constrained' variational[1] (CV) calculations.

INTRODUCTION

It has been a persistent feature of the calculation of Λ binding to nuclear matter (to be denoted as D) that one obtains binding energy values much larger than the accepted empirical estimates[2] which is about 27 MeV. Moreover, the disagreement between the g matrix[2] and CV calculations[1] seems disturbing. Therefore, to look into the causes of overbinding problem and the discrepancy between the above mentioned two approaches, we extend the FHNC technique to the calculation of D.

FORMULAE AND THEORY

To a very good approximation[3] we may obtain for D in FHNC

$$D=n\int g_\Lambda(r_{\Lambda 1})\left(V_{N\Lambda}-\frac{\hbar^2}{4}\frac{m_\Lambda+m}{mm_\Lambda}(\nabla^2 \ln f_\Lambda(r_{\Lambda 1}))\right)d\vec{r}_{\Lambda 1} \quad - (1)$$

where $V_{N\Lambda}$ is spin averaged ΛN potential and n is nuclear matter density. g_Λ is the ΛN pair distribution function obtained by solving hypernetted chain equations[3]. For the ΛN correlation factor, $f_\Lambda(r_{\Lambda 1})$, we have assumed the form

$$f_\Lambda(r)=\begin{cases}0, & r<c\\ \left[1-\exp(-a(r-c))\right]\cdot\left[1+q\exp(-a(r-c))\right], & r\geq c\end{cases} \quad - (2)$$

where c is the hard core radius and a and q are variational parameters. For the NN correlation factor we

*This work was carried out at Service de Physique Theorique, Centre d'Etudes Nucleares de Saclay (France).

ISSN:0094-243X/79/540731-02$1.50

have found it sufficicient to take the same form as (2) with q=0. The hard core ΛN potentials of Herndon and Tang were taken from table I of ref. 2.

RESULTS, DISCUSSION AND CONCLUSIONS

The results are given in table I for k_f=1.366fm^{-1}. It is seen that in all the cases the D in FHNC approximation is significantly lower compared to the g matrix and CV results. In view of the success of FHNC approximation[4] we arrive at the conclusion that the CV of Mueller and Clark[1] do not give an upper bound to D. Thus, there is nothing to worry as far as the disagreement between the two approaches is concerned. However, the differences between the reaction matrix and the FHNC calculations are larger than the expected 10% difference that may arise from the hole-rearrangement (r-h) energy diagrams considered by Dabrowski and Kohler[5]. This may be due to the fact that the (r-h) diagram is only one of an infinite subset of ring diagrams of RPA type. The FHNC approximation takes care of all such contributions. This in fact is a happy circumstance. In the present study all the calculations were carried out by assuming the ΛN potential same in relative s and p states. If the reaction matrix calculations are any guide, one may hope to reduce D by taking a weaker potential in p state along with the supression of ΣN-ΛN coupling which should also be helpful in reducing D. Hopefully, this may bring D into accord with the empirical estimates.

Table I

Pot.	g matrix D(MeV)	CV calculations D(MeV)	FHNC D(MeV)	FHNC a	FHNC q
C'	40.5	-	26.4	6.55	-0.1
E'	56.0	62.1	49.7	4.45	0.0
E	61.7	68.6	55.2	4.80	-0.1
H	56.8	72.2	51.1	4.55	-0.1

REFERENCES

1. G. Mueller and J.W. Clark; Nucl. Phys., B7, 227 (1968).
2. D.M. Rote and A.R. Bodmer; Nucl. Phys., A148, 97 (1970).
3. Q.N. Usmani, Submitted in Nucl. Phys. A.
4. S. Fantoni and S. Rosati; Lett. Nuovo Cimento, 16, 531 (1976), and references therein.
5. J. Dabrowski and H.S. Kohler; Phys. Rev. 136, B162 (1964).

LOW ENERGY Σ^--NUCLEUS INTERACTION

J. A. Johnstone
McMaster University, Hamilton, Ontario, Canada

J. Law
University of Guelph, Guelph, Ontario, Canada

ABSTRACT

We have fitted consistently the Σ^- atomic and the ΣN scattering data within a finite range potential model.

The analysis of the ΣN scattering data by G. Alexander et. al.[1], produced two sets of complex scattering lengths. Together with the data on atomic level widths and shifts of Σ^- atoms of C.J. Batty et.al.[2] it is possible to analyse the combined data to see if they are consistent and also to see if the atomic data can distinguish between the two sets of scattering lengths.

The calculations were done by first fitting Yukawa shape complex potentials

$$V_{\Sigma N}(r) = V_{JI}\, e^{-\mu r}/\mu r$$

to reproduce the complex scattering lengths as a function of the potential range by varying μ from 1.5 fm^{-1} to 3.0 fm^{-1}. These potentials were then folded[3] into the nuclear distribution to produce the Σ^- nucleus effective potential, and the Schrodinger equation solved for the level widths and shifts. The inverse range μ was fixed at the value which minimized the deviation of the calculated level widths and shifts from the measured data. As a further step, the depths of the potentials were then varied within the uncertainty in the fit to the scattering lengths, to minimize the overall χ^2 further. The resultant fits are shown in the tables.

Table I. Scattering lengths and well-depths for best fit of data with μ = 2.6476 and 2.7268 fm^{-1} for sets 1 and 2 respectively.

	Experimental				Calculated				Potential Depth (Mev)			
	Set(1)		Set(2)		Set(1)		Set(2)		Set(1)		Set(2)	
Scattering Length (fm)	Re(a)	Im(a)	Re(a)	Im(a)	Re(a)	Im(a)	Re(a)	Im(a)	Re(v)	Im(v)	Re(v)	Im(v)
$\bar{\Sigma}n(S=0)$	1.4±1.3	-	-0.8±2.0	-	.666	-	-.798	-	-834.85	-	-1463.00	-
$\bar{\Sigma}n(S=1)$	-0.7±0.4	-	-0.8±0.7	-	-.785	-	-	-	-1373.06	-	-	-
$\bar{\Sigma}p(S=0)$	-0.4±0.5	0.5±0.6	0.6±1.0	1.0±1.0	-.390	.545	.593	.929	-1332.76	133.32	-1672.73	691.69
$\bar{\Sigma}p(S=1)$	0.5±0.3	1.1±0.5	0.6±0.3	1.0±1.0	.605	.875	-	-	-1520.91	754.00	-	-

ISSN:0094-243X/79/540733-02$1.50

Table II. Best fits to the level widths and shifts and the corresponding χ^2

Element and Transition		Experimental ε (eV)	Experimental Γ (eV)	Calculated Set(1) ε	Calculated Set(1) Γ	Calculated Set(2) ε	Calculated Set(2) Γ
O	4→3	-320±230	-	-162.71	216.82	-150.64	200.61
Mg	5→4	- 25±40	<70	- 35.27	34.35	- 35.97	40.95
Al	5→4	- 68±28	43±75	- 77.43	64.95	- 76.99	65.83
Si	5→4	-159±36	217±110	-162.57	217.26	-158.37	225.54
S	5 4	-360±220	870±700	-566.62	759.98	-552.49	821.81
Reduced χ^2				0.17		0.19	

The results indicate that the Σ^- atomic data can be fitted consistently with both sets of scattering lengths. However, the fit to set 1, reproduces a Σ^- neutron potential which binds this system with a binding energy of 2.96 MeV. The experimental verification of this bound state or otherwise could be the basis for determining the correct set of parameters. Calculations are in progress on whether Σ^-nn is also a bound system.

REFERENCES

1. G. Alexander et.al. Phys. Rev. D6, 2405 (1972).
2. C.J. Batty et.al. Measurement of strong interaction effects in Σ atoms. Rutherford preprint (1978).
3. A. Deloff and J. Law, Phys. Rev. C10, 1688 (1974).

DECAY OF $\Sigma^{-}n$ DIBARYONS

B. Budick
New York University, New York, N.Y. 10003

ABSTRACT

We have calculated the momenta of neutrons and pion from the decay of a hypothetical $\Sigma^{-}n$ hyperfragment, including final-state interaction between the two neutrons.

INTRODUCTION

The possible existence of a bound $\Sigma^{-}n$ dibaryon has been the subject of several theoretical [1-4] and experimental [5-8] papers. A recent search yielding a negative result was reported at the conference[9].

NEUTRON SPECTRA FROM $\Sigma^{-}n$ DECAY

This problem was first considered by Common[3)]. He neglected final-state interactions and showed that the Pauli principle led one to expect a two peak spectrum. A low momentum peak arises from the momentum of the spectator neutron as the Σ^{-} decays in the dibaryon. A high momentum peak at 193 Mev/c is just the fast neutron from the Σ^{-} two-body decay.

In our approach we use a properly symmetrized final state wave-function including final state interaction similar to that employed by Chen[10].

We find neutron distributions shown in figure 1, (A)-(D), corresponding to binding energies of 0.05, 0.1, 0.25 and 0.5 MeV respectively. The effect of final state interactions is evident in the large number of events in which the two neutrons recoil together with momenta of roughly 108 MeV/c. The three groups of neutrons have kinetic energies of 0.4, 6.3 and 20 MeV, respectively.

We have also examined the cascade of exotic atoms that incorporate $\Sigma^{-}n$ as a satellite. We find that a large fraction of dibaryons that stop in insulators of low atomic number will decay before being absorbed.

The author wishes to acknowledge helpful discussions with K. Kilian, A. Gal, M. Mandelkern and L. Rosenberg. J. Kast and R. Anigstein assisted with some of the final-state calculations.

ISSN:0094-243X/79/540735-02$1.50

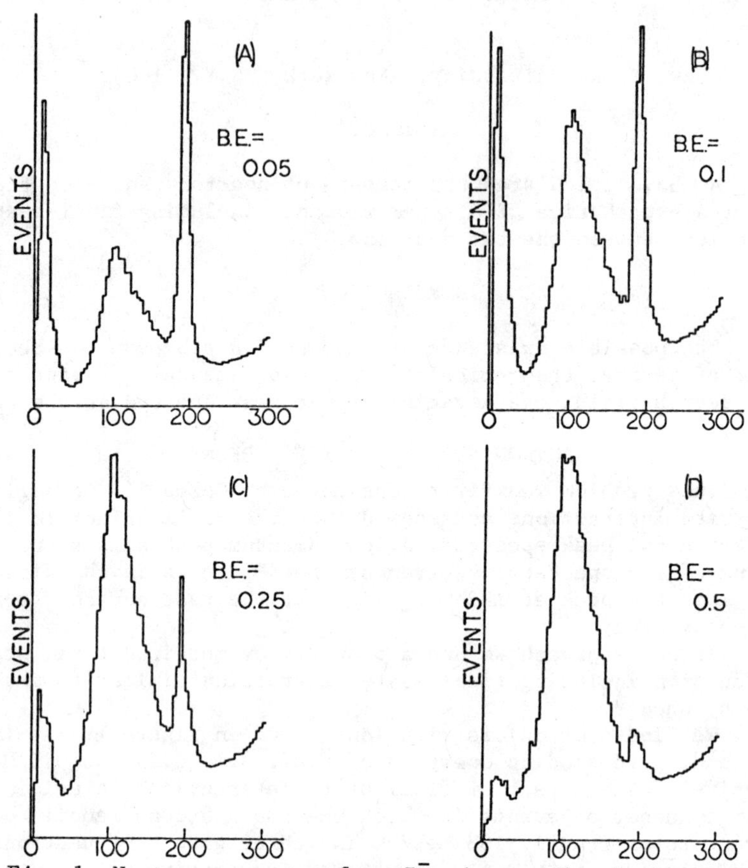

Fig. 1. Neutron momentum from $\Sigma^{-}n$ decay. The event rate in arbitrary units is plotted vs. momentum in MeV/c. Assumed dibaryon binding energies in MeV appear on each figure.

REFERENCES

1. A. Pais and S.B. Treiman, Phys. Rev. 107, 1396 (1957).
2. G.A. Snow, Phys. Rev. 110, 1192 (1958).
3. A.K. Common, Nuovo Cimento 20, 1133 (1961).
4. M.M. Nagels, T.A. Rijken and J.J. de Swart, Univ. of Nijmegen preprint
5. O. Dahl et al., Phys. Rev. Letters 4, 428 (1960).
6. T.B. Day, J. Sucher and G.A. Snow, Phys. Rev. Letters 3, 61 (1959).
7. R.G. Ammar et al., Phys. Rev. 126, 1914 (1060).
8. R.A. Burnstein et al., Phys. Rev. 177, 1945 (1969).
9. M. May, this conference.
10. Y.Y. Chen, Nuovo Cimento 19, 36 (1961).

New Facilities at SIN

J.Domingo

This morning I have been asked to talk to you about new facilities at SIN. In the short time available I will only be able to give you a very brief outline of the planned facilities. I will first present the facilities which are already under construction and end by mentioning some of our ideas for future facilities.

The layout of the SIN experimantal hall is shown in Fig. 1. As most of you probably know we have two accelerators. The injector cyclotron (a modified commercial cyclotron) which accelarates the protons to 72 MeV, and the ring which accelerates them from 72 to 590 MeV. The present current and reliability limits are dictated by the injector cyclotron and the major new facility being constructed at SIN is the new injector. This is also shown in Fig. 1. During the last months before the present shutdown routine operation was obtained with currents of 90 - 110 μa, but although the extraction efficiency of the injector has been raised to around 90 %, the activation limits on it will probably never allow us to exceed 150 μa. Since the extraction efficiency from the ring is very high (99.9 %) and the present target stations can be modified to handle currents of several hundred microamperes, we have been working on the design of a new injector for the past several years. About one year ago we received governmental approval to proceed with the project and at present the building is under construction and many of the major components have been ordered.

The new injector will consist of a 860 keV pre-injector Cockroft-Walton feeding a 4 sector separated magnet cyclotron. The injector ring will be quite similar in principle to the main ring and the physical size will be almost as large. The sector magnets will be approximately wedge shape with a wedge angle of 28°, a gap width of 35 mm, and a individual weight of 180 metric tons. The r.f. system differs from that of the main ring and consists of two λ/2 delta resonators for the main acceleration systems, and two 3rd harmonic cavities similar to those used in the ring. The delta cavities are utilized in order to be able to meet the beam optics requirement of a peak voltage which rises from 100 kV at injection

ISSN:0094-243X/79/540737-13$1.50

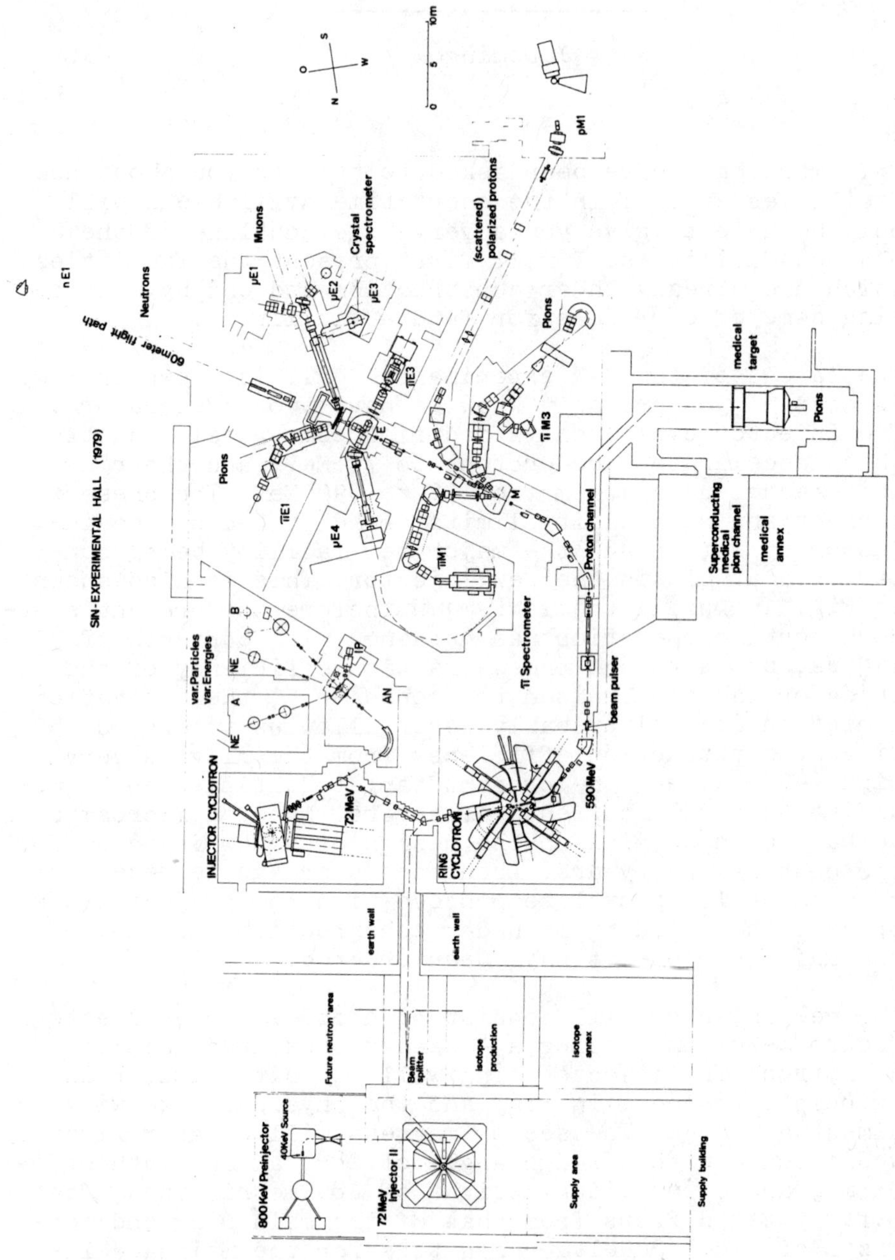

Fig. 1 Schematic view of SIN experimental hall showing the location of Injector II presently under construction

to 250 kV at extraction. The estimated power consumption per cavity is approximately 140 KW. A full scale model of these cavities will be constructed this year so that final tests of the design can be carried out. One of the most critical parts of the new injector design is the high current ion source, and a 300 keV ion source - accelerating column test facility has been constructed to study the proposed 15 ma source. The Swiss government gave us permission to construct the new injector, but no additional funds so its completion date depends upon how much money we can make available from our normal annual budget. At present the planning is to bring the machine on the air in mid 82. Once the present bottle neck caused by injector I has been removed it is interesting to investigate just what the current limitations of SIN facilities will be. Fig. 2 shows the estimated limits due to various sources for both the new injector and ring. The slopes of the horizontal lines indicate the uncertainity in these limits. The first limits we will hit will be due to the rf power presently available to the ring and the heat dissipation limit of the present beam dump and thick target station. Both of these limits lie somewhere arond 4 - 500 μa. The r.f. power limitation is relatively easy to remove and it is quite likely that we will raise the available power to 2 MW very soon after the new injector comes into operation. Since the 3rd harmonic cavity has already been installed in the ring and two are forseen for the new injector this should allow us to accelerate beams of around 2 ma. The question of how we will upgrade our production targets and beam dump to be able to handle this current is presently undergoing discussion, and I will try to present some of our ideas later. The question of the ultimate current limit of the accelerator system is rather uncertain and depends upon how much one can increase the voltage gain per turn in the two machines. As a rough guess this limit is somewhere in the range 3 - 5 ma.

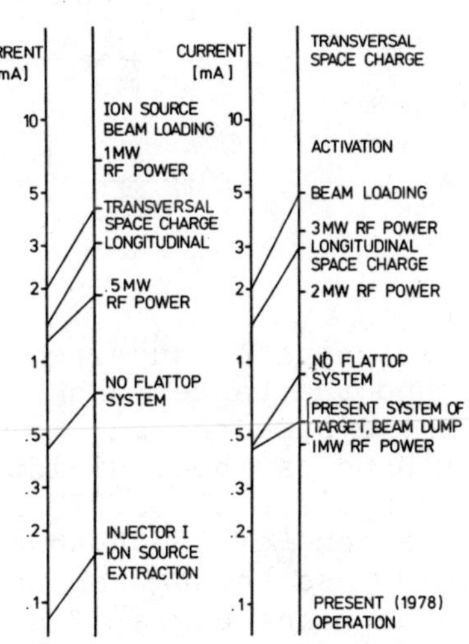

Fig. 2 Current limits of injector II and the 600 MeV ring cyclotron.

The second major project presently nearing completion is the new medical facility. Fig. 3 shows a schematic view of the treatment apparatus. The low energy pions (200 Mev/c) are transported from the production target to the patient

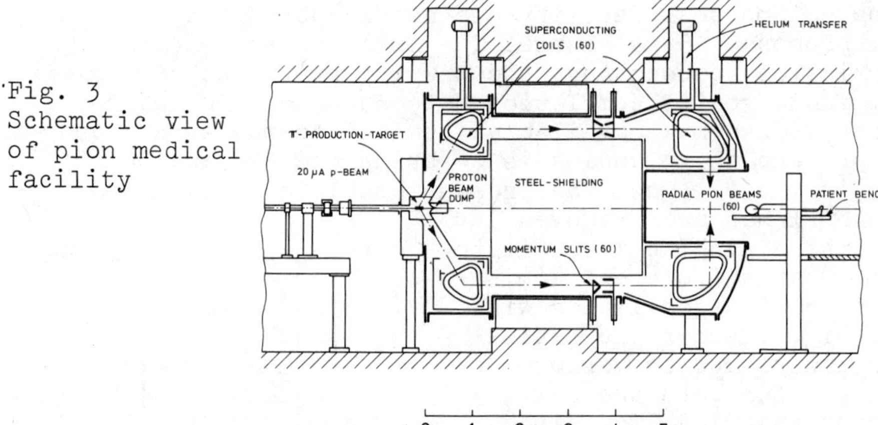

Fig. 3
Schematic view of pion medical facility

by 60 pairs of superconducting coils arranged axially around the target and patient. Each of these 60 channels has the same central momentum, but the transported momentum band can be individually controlled by means of 60 intermediate slits. These slits can be used to completely close some of the channels in case the beams in question would pass through an undesired region of the patient. The entrance angle of a channel is centered about 60° and the exit angle about 90°. The coils have curved boundaries in order to correct for various higher order aberrations. The superconducting coils are cooled by a 500 Watt gas turbine refrigerator system similar to that used on the superconducting μ channels. The facility is designed to operate with a nominal proton beam of 20 μa; this will produce a dose of between 30 and 50 Rad/min. in a one liter volume. The proton beam for the medical area will be derived from the main beam by means of a high voltage septum; this will allow an essentially continuous variation in the splitting ratio with no degradation in the duty cycle. The present long shut down is due to the installation of this transport system. At present the superconducting coils are just being delivered and it is hoped to have the facility in operation by mid summer. The first clinical tests are scheduled for early 80.

Since it is expected that the doctors will only use the medical facility during normal working hours, a second facility is being developed to use the 20 μa beam during nights and weekends. This system is being developed by a collaboration between Gatchina, E.T.H.Z. and SIN and will use the direct production technique pioneered at Gatchina to study pionic atoms via a bent crystal spectrometer. The basic idea of the Gatchina scheme is to place the target to be investigated in the immediate vicinity of the pion production target and look at those events where the pions are captured in the secondary target. Although most of the pions produced in the production target are not captured in the secondary target, the effective solid angle is so large that one gains roughly 2 orders of magnitude over a conventional system using a pion channel. Of course this technique can only be utilized where one is essentially insensitive to the large background from the production target. As an example of the increased rates expected for this system, the group has estimated that for a 20 μa beam incident on a target sandwich consisting of 25 mm thick Be primary target disks surrounding a 6 mm thick Mg secondary target they will have a stopping rate in the Mg of greater than 10^9 per sec and a counting rate in the main diffraction peak of a 8X10 cm crystal of about 20 counts per second.

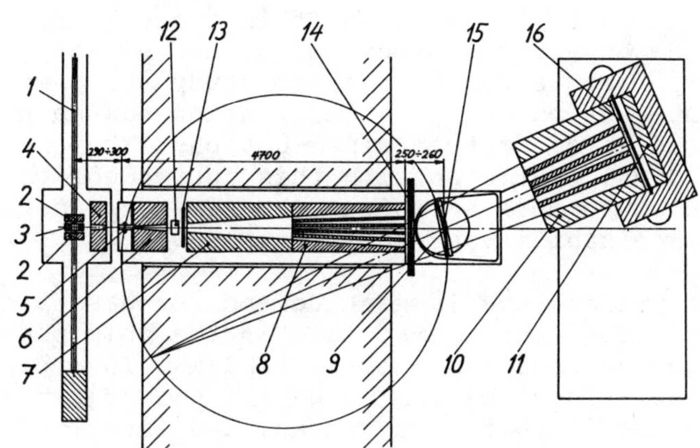

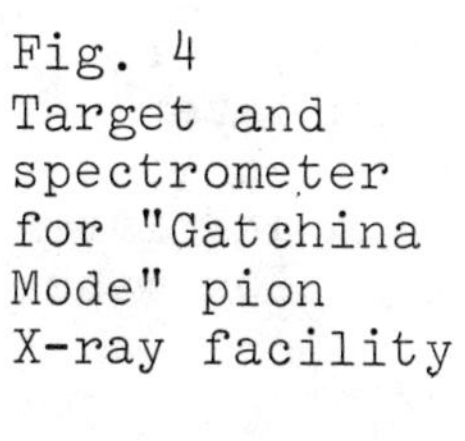
Fig. 4
Target and spectrometer for "Gatchina Mode" pion X-ray facility

Fig. 4 shows the layout of the spectrometer facility which will be located in the ПМ3 beam area. The proton beam obtained from the medical beam transport system by simply turning off the next to last bending magnet will be incident on the two Be pion production targets (2) placed on either side of the secondary target to be investigated (3). This secondary target will have a ring form so that the main proton beam will pass through it without interaction. Several secondary targets will be mounted on a support system such that they can be remotely moved in to position; thus rapid comparisions can be made between different secondary materials. The main entrance slit (5) consists of 2 tungsten cylinders mounted rigidly in a parallelogram block the opening of which can be varied by means of a lead screw which distorts the parallelogram. Objects (4, 6, 7 and 8) are various collimators; the bent crystal (15) diffracts those X-rays satisfying the Bragg condition through a fifth collimator (10) onto the detector (11). The Bragg angle will be measured in the early phase by a sine screw mechanism this will later be replaced by a laser interferometer. The crystal and spectrometer system will mainly be provided by Gatchina and the target and control system by SIN. The beam dump and shielding are being installed during the present shut down, and it is planned that the target system and spectrometer will be installed this autumn so that the first test runs can be undertaken later this year. The first experiments which the group wishes to undertake with the system are measurements of shifts and widths of the 2p states in $^{16,18}O$, ^{19}F, ^{23}Na, $^{24,25,26}Mg$, $^{28,29,30}Si$ and $^{32,34}S$ with an accuracy of about 1 %. Since the dominant part of the 2p shift in these atoms is essentially independant of nuclear structure effects, these first experiments can be used to study the p wave Π-nuclear interaction and this information can be used in later studies of neutron distributions. They hope to achieve a final precision with this instrument of 1 ppm and later go on to more ambitious experiments - negative pion mass, polarizability of the pion, etc.

At present the largest demand for beam time at SIN comes from the muon users; this is the only place where we have serious scheduling problems. In order to try to ease this situation we have modified two existing beams to provide better μ fluxes. The first modification is to the μE4 or "Hahn" beam derived from the short superconducting solenoid. The floor area available has been considerably enlarged and the beam transport has been improved by the addition of a QQDQQ analysing system followed by a slit and a quadrupole triplett. Since this beam is derived at a 90° production angle it has optimum characteristics for low momentum muons (50 to 80 Mev/c). It offers a much larger choice of range (100 mg/cm^2 to 2 g/

cm^2) due to the possibility of operating the analysis system in either a chromatic or achromatic mode. Due to the 90° production angle it has a lower neutron background than that of the main μ channel, and it also does not have the disadvantage of having its momentum setting coupled to another channel. The maximum μ flux is however approximately an order of magnitude lower than for the main μ channel. The characteristics of this channel are shown in Fig. 5.

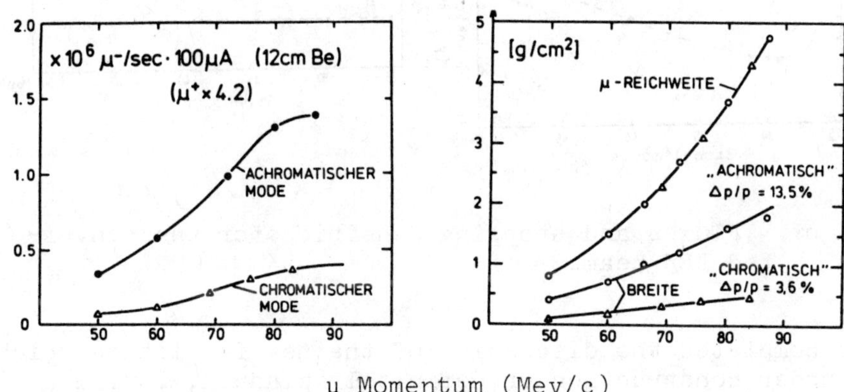

Fig. 5 μ fluxes, ranges and range widths for the modified μE4 beam operating in the "achromatic" and "chromatic" modes.

In addition to the modification of the μE4 channel, the old biomedical channel ΠE3 will be converted for the Arizona mode of operation. Since the biological experiments will move to the new medical area this channel will be fully available for physics users. The main modification to this channel will be to remove the two 1 mm thick A windows presently separating the channel vacuum system from the machine vacuum system. The window on the channel has been removed in the present shut down and, providing no problem develops with the conical vacuum seal to the target chamber, the proton channel window will be removed in early summer. In addition to removing the vacuum windows an appendix has been constructed to bend the beam back into a horizontal direction (the original medical beam was vertical). A degrader will also be inserted into a dispersive focus to remove most of the electron contamination. The expected low momentum fluxes and stopping densities as shown in Fig. 6.

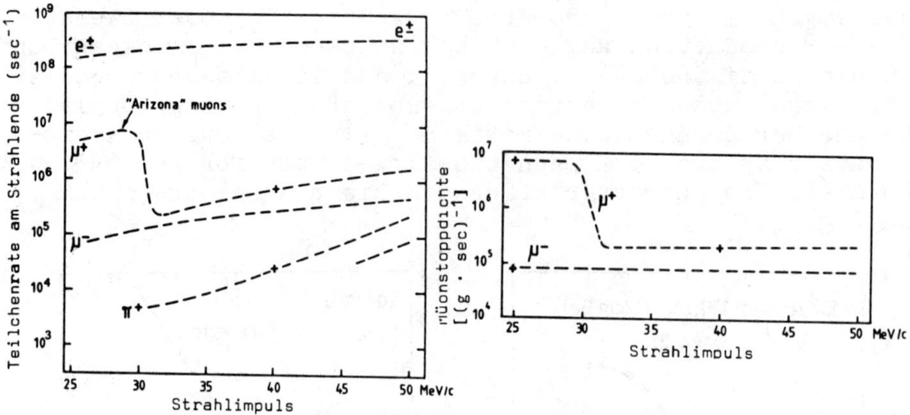

Fig. 6 μ Fluxes and stopping densities for the converted ΠE3 beam.

This completes the discussion of the new facilities which are under construction or definitely planned; I will now move to the more speculative projects for the future.

The major new project being seriously studied at the present time is a spallation neutron source. Since the new injector will allow us to produce beams exceeding the thermal limit of the present beam dump, the idea of using a new beam dump as a spallation neutron source seems quite attractive. This proposal also gains support from the fact that the only reasonable research reactor in Switzerland has been closed down, and in general there is a much larger demand for research reactor time than can presently be met. In particular Germany has been seriously ly considering the construction of a medium flux reactor because of the severe overdemand for time on the high flux reactor in Grenoble. A study programm started last spring in collaboration with various Europian neutron physicists has concluded that the most useful type of source for SIN to consider is a continuous source employing a liquid PbBi target and a liquid deuterium cold moderator. With such a cold source and a proton beam of 3 ma we would have the best cold neutron facility presently available. For thermal neutrons a proton beam of 3 ma would make us roughly equivalent to the proposed German medium flux reactor. Fig. 7 shows a schematic layout of the spallation target system and Fig. 8 gives the expected distribution of the thermal neutron flux for a proton beam of 1 ma. The location of this source is strongly linked to the construction of new secondary beam lines

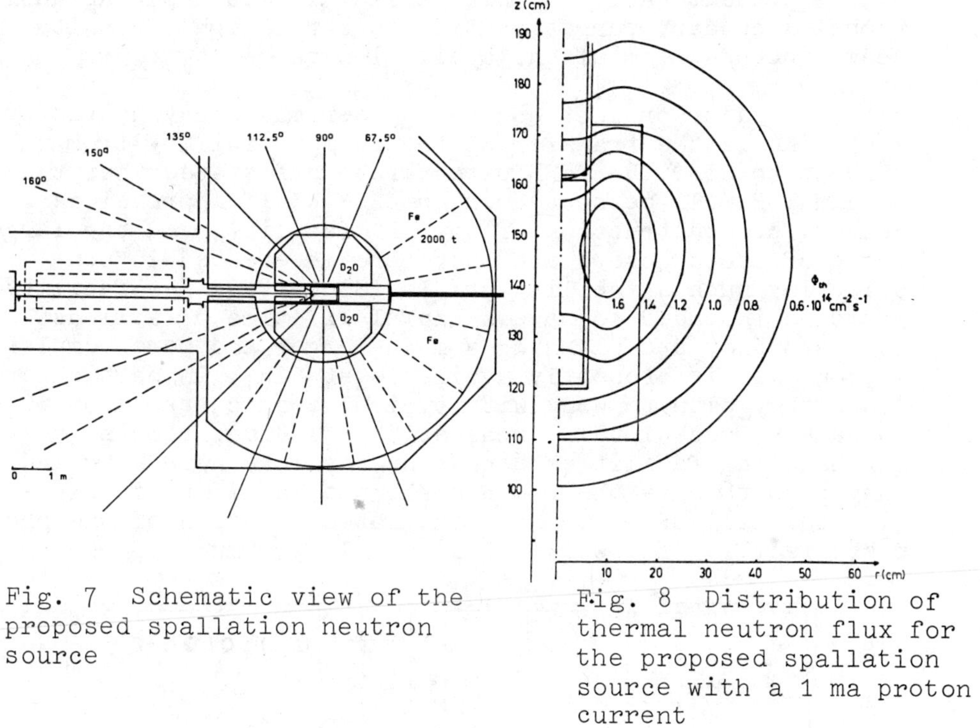

Fig. 7 Schematic view of the proposed spallation neutron source

Fig. 8 Distribution of thermal neutron flux for the proposed spallation source with a 1 ma proton current

and extension of the present experimental hall; this in turn is a largely financial question coupled with an eventual participation by other European countries in the construction of the neutron source. At present we are still very much in the "sand box" stage of planning, but we have two rough classes of layouts: the poor man's and optimum solutions. In the poor man's solution we would reconstitute the beam after the thick target and replace the present beam dump by the spallation target. In this case none of the present secondary beam lines would be changed and modifications to the existing proton beam line and production targets could be kept relatively simple. In the optimal solution we would construct a new high intensity beam line and use the spallation target for the beam dump of this new beam line. The present thick target and beam dump set up would be retained but the current on this line would be limited to 4 - 500 μamps. This lower intensity beam would be split off the main beam by means of a second splitter similar to that used to split off the medical beam. The mostly likely position of this splitter is between the present thick and thin targets so that the high intensity proton beam would pass along the central axis of the experimental hall; we would then simply extend the length of the present hall. In this version we could put an additional secondary target station into the high intensity beam line upstream of the spallation target.

A second major project undergoing serious study at the present time is the transfer of the Isolde facility from the CERN-SC to SIN. The CERN directorate has stated that the SC and ISR will be shut down in 82 - 83 if approval is received for construction of the LEP facility, and the large group of Isolde users are therefore interested in constructing such a facility at SIN. Since this new facility would be intended to have a useful lifetime of at least 10 years they are planning a much larger and more complex system than is presently available at CERN. In particular since they plan to work with typical beam currents of 10 - 20 µa with occasionally runs at 50 - 100 µa, remote target handling facilities become an absolute necessity and they plan to have their own small hot lab in their experimental hall. Fig. 9 shows the latest version of the proposed facility and while it is still very much in the pre-

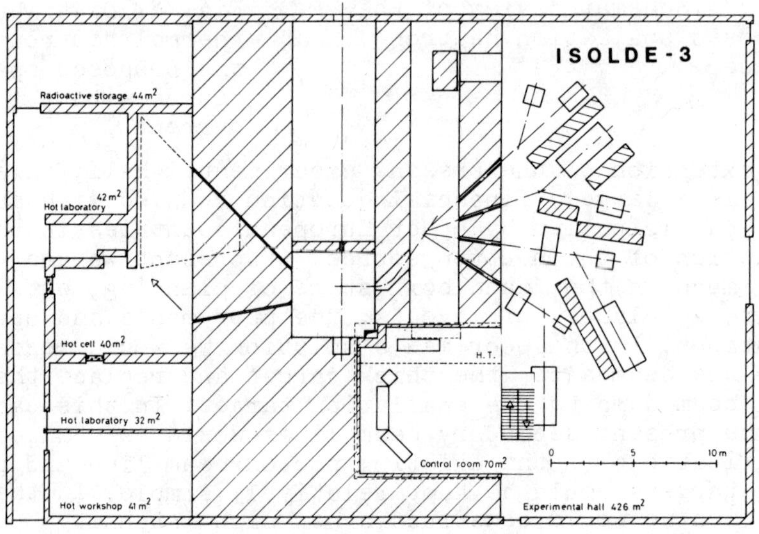

Fig. 9 Proposed Isolde facility

liminary planning stage one can already see that the area considered is quite a large one. The beam for this facility would be derived from a second beam splitter similar to the one for the medical channel, but again the final system adopted is strongly coupled to the layout of new high intensity beam line. The status of this project is very uncertain at present since its construction will be quite expensive and cannot be undertaken without some political-financial arrangement between CERN and the Swiss government.

Finally let me mention a possible major new piece of experimental apparatus. It seems that the experimentalists working at SIN are so busy taking data that the majority have no time for thinking about major new pieces of apparatus such as PCT systems or a wall of NaI, luckily, I might add from the point of view of our very tight budget situation for the next couple of years. Nevertheless we all realize that sometime soon we have to sit down and try to decide what we will do in 3 or 4 years when the new injector comes into operation. We are organizing various study groups to try to examine this question, and we will later try to bring them together for a workshop at something like a Zuoz school.

One piece of apparatus that is currently undergoing extensive discussion is the large solid angle magnetic detector system Ares currently being constructed at Dubna. Since the Dubna-SC will definitely (I have been told) be shut down this summer for rebuilding, Dubna has proposed that SIN and Dubna collaborate on the final construction of the device and that it then be brought to SIN. Fig. 10 shows a schematic view of Ares which consists of a nest

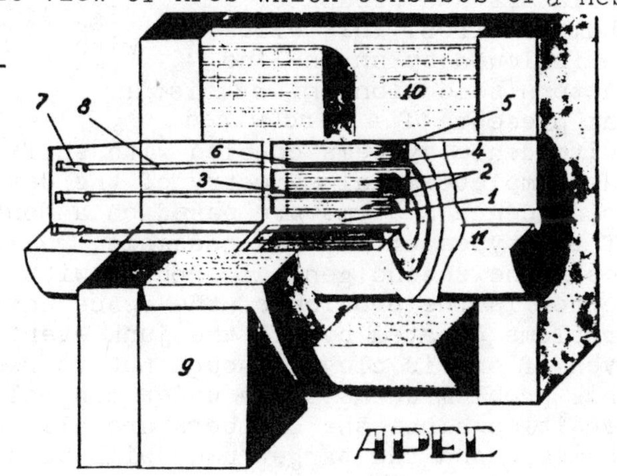

Fig. 10 ARES spectrometer

1. Target
2.5. Wire chambers
3.4. Scintillators
6. γ Converter
7. P.M.'s
8. Light guides
9. Yoke
10. Coils
11. Pole

of cylindrical wire chambers and scintillators placed in an axial magnetc field. The diameter and length of this chamber assembly is approximately 1 meter. Most of the chambers are simple 1-dimensional read out but some of them use the charge induced on spiral cathode strips to provide an axial coordinate. For detection of γ's a converter foil is placed at a fairly large radius and the last large cylindrical chambers then operate as a pair spectrometer. Fig. 11 shows a possible μ → e γ set up and such an event although I hasten to add that the present plans are to use this system first to look for μ → 3e not μ → eγ. The problem with all such large acceptance devices when looking for rare events is that if one doesn't have an extremely good fast on-line event selection system he is soon buried in tons of tape containing nothing but uninteresting events. The Ares will have a very fast rough selection using ordinary nanosecond electronics operating on segmented scintillators and groups of chamber wires. This will be followed by an on-line pattern selection unit consisting (at present) of a hard wired coincidence systems coupled with shift registers to treat the complete axial symmetry of the device. The hard wired coincidence systems are based on a Monte-Carlo analysis of the types of signals possible from the events of interest. The second general problem with such a rare event detector is the problem of the space charge and dead time problems created by all the junk events; they create charge even if one is clever enough not to read them out. In Ares this problem seems to be under control for present beam intensities since the chambers are all small gap proportional chambers and the target used will be a low density foam so

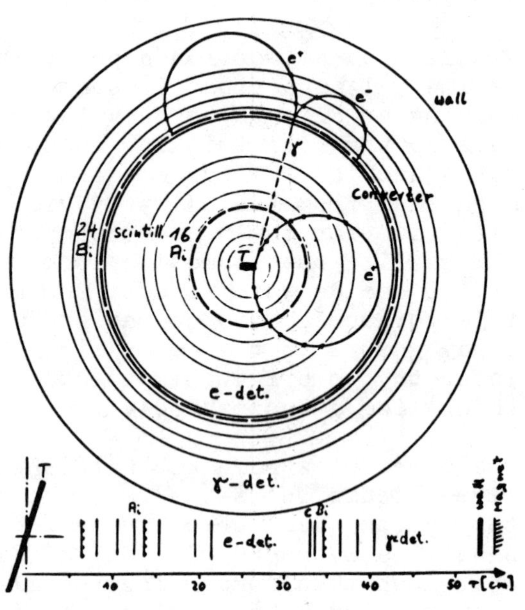

Fig. 11 ARES detector system with a μ → eγ event.

that the axial space charge load is well distributed. On paper one is well below Charpak's limit of 10^4 sec^{-1} per mm of wire, but what will happen when one actually puts the thing in a magnetic field may be another story.

At present a prototype system is about to be tested at the Dubna SC; this system will be placed in an old magnet and used to look for u → 3e events before the SC shuts down in June. The full scale magnet should be finished late this year so in principle the system could be shipped to SIN sometime in 1980, but at present negotiations on the collaboration are still underway so it is too early to make any definite statements.

Finally let me mention the new development by Charpak's group at CERN of a triggerable wire chamber. This device seems to me to be of vital importance for the future of high intensity machines, and we are beginning to study some prototype chambers. Basically they have succeeded in dividing the usual 10^6 chamber gain into 2 separate segments each having roughly 10^3 gain with a delay and trigger stage between the two so that only those few events events of interest receive the full 10^6 gain. This plus the fact that the gain region is not localized about the wires should allow one to utilize the device in beams two to three orders of magnitude more intense than with the present type of wire chamber.

1. G. Charpak and F. Sauli: The Multistep Avalanche Chamber, Submitted to physics Letters B

NEW FACILITIES AND KAON FACTORY DESIGNS FOR TRIUMF

M.K. Craddock
University of British Columbia and TRIUMF,
Vancouver, B.C., Canada V6T 1W5

ABSTRACT

The TRIUMF 520 MeV H^- cyclotron came into operation in December 1974 and first delivered a 100 μA beam in July 1977. Facilities recently commissioned include a second proton line for polarized and high resolution experiments (single turn extraction has been demonstrated at 200 MeV), a pion spectrometer, a 70/90 MeV proton beam for isotope production and a new slow (≤130 MeV/c) pion/muon channel. Major projects under way include a high-resolution fast pion channel, dc and rf separators to give purer stopping muon beams, and a time projection chamber.

Further in the future are the possibilities of building a third (simultaneous) 400-500 MeV proton beam and high-flux muon and biomedical π^- channels. More fascinating still is the possibility of elevating TRIUMF from a pion to a kaon factory, producing currents 100-1000 times more intense than those at present available in the GeV range. Two post-accelerator designs are being considered—a 30 μA rapid cycling 8-10 GeV proton synchrotron (which could in turn inject into a higher-energy machine)—and a 400 μA isochronous ring cyclotron built in two stages, the first to 3 GeV and the second to 8.5 GeV.

INTRODUCTION

Since the cyclotron was commissioned in December 1974 the regularly operated beam current has been gradually increased from a few nA to 20 μA. A 100 μA beam was first run in July 1977 and since the commissioning of the permanent beam stop and thermal neutron facility (TNF) with a 120 μA beam in February 1978 100 μA shifts have been scheduled with low but increasing frequency. (The status of the cyclotron has most recently been reviewed by K.L. Erdman.[1]) During 1979 just over 10% of the high-current running is being scheduled at 100 μA. The integrated charge delivered will be ~75,000 μA-h, a factor 3 greater than that delivered in 1978 (27,000 μA-h—itself a factor 3 greater than the figure for 1977).

The reason for not scheduling a greater percentage of 100 μA running is to keep the radiation levels in the cyclotron vacuum tank low while significant manual work remains to be done there. (This work includes installation of additional extraction systems and various improvements to the RF accelerating system, including full remote handling capability and the addition of third harmonic flat-topping.) Although the use of peripheral graphite absorbers, lead shields and other techniques has reduced the radiation level by a large factor,[2] we are now at a stage where the skilled workforce receives close to its permitted quarterly dose during a one-month

ISSN:0094-243X/79/540750-10$1.50

shutdown.

Besides its role as a meson factory, TRIUMF has a second important role to play as a source of variable energy proton beams between 180 and 520 MeV; indeed it is at present the only accelerator producing protons directly at these energies, without degradation. Experiments requiring unpolarized beam and regular energy spread (1 MeV FWHM) are of course mounted on lines 4A or 4B (see Fig. 1) and run simultaneously with pion or muon experiments fed from targets in line 1A. The fraction of beam delivered to line 4 can be varied between 1/1 and 1/10,000 by suitable choice of the shape and height of the extraction foil.

However, a polarized source is also available and slit-selected beams of higher energy resolution are being developed. These beams are too weak to be useful for pion production, so the cyclotron is dedicated to their use for 20-30% of the time, depending on demand. The Lamb shift polarized source[3] provides a 200 nA extracted proton beam with >70% polarization at all energies. Furthermore a beam of highly polarized monoenergetic neutrons can be obtained via the $d(\vec{p},\vec{n})2p$ reaction with a liquid deuterium target.[4] The high value of the Wolfenstein R_t parameter results in neutron polarizations rising from ~50% near 500 MeV to 65% near 200 MeV; the intensity is 2×10^6 neutrons/sec through a 5 cm aperture.

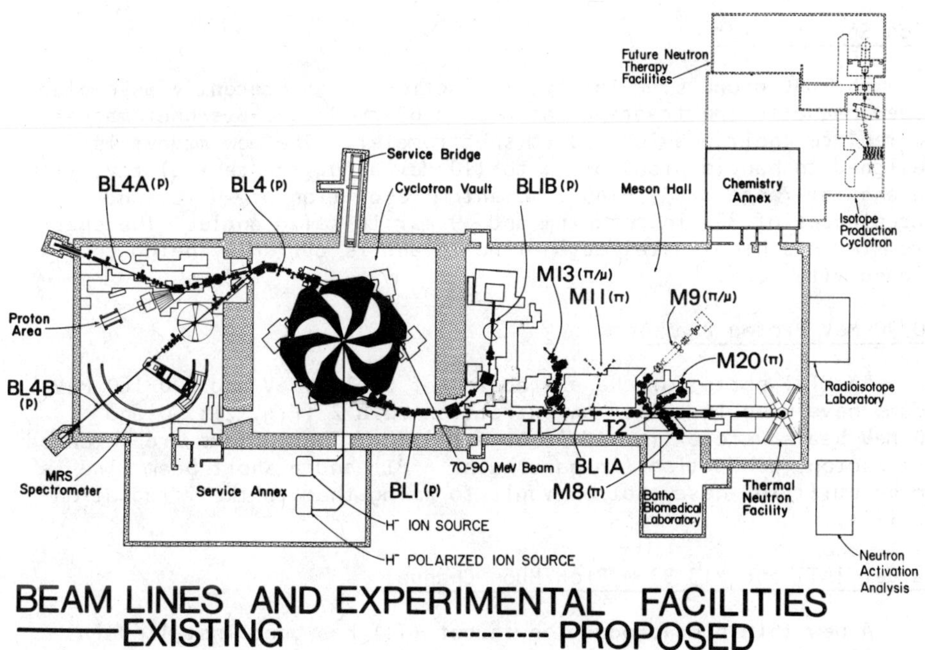

Fig. 1. Beam lines and experimental areas.

RECENTLY COMMISSIONED FACILITIES

Internal Cyclotron Slits (for reducing the proton beam energy spread)

Four pairs of internal slits, located in the 3 and 30 MeV regions, were commissioned during 1978. With these it has been possible to select beams of small emittance, narrow phase acceptance and small coherent betatron oscillations so that the turns remain separated out to 200 MeV. (Improvements in cyclotron stability have also been necessary.) That a single turn could be extracted was confirmed by observing the time-of-flight spectrum of the external beam.[5] In this case the energy spread should be $\leq$320 keV; work is in hand to measure this directly.

Proton Line 1B

This "peanuts" line (so-called because of its supposed cost philosophy) was completed and commissioned in January 1979. It is intended for polarized or high resolution beams, allowing them to be used simultaneously on lines 1B and 4; beam intensity is limited to 10 nA. Beams of 200, 250, 300 and 350 MeV have so far been commissioned; the spot sizes are $<5 \times 10$ mm.

Pion Spectrometer

The UBC group studying (p,π) reactions[6] have recently assembled a new magnetic spectrograph based on a 65 cm Browne-Buechner magnet, to replace their previous 50 cm spectrometer. The new magnet is designed to handle pions of up to 110 MeV along the central ray, with an angular range of 35°-145°, momentum resolution of $\sim$0.1%, and acceptances of 30% in momentum and $\sim$9 msr in solid angle. The spectrometer has been installed on line 1B and is currently being commissioned with beam.

70/90 MeV Proton Beams

Special ports for the extraction of 65-100 MeV and 100-180 MeV beams have been installed on extraction horn 2 (Fig. 2); 70 and 90 MeV beams were extracted in August 1978. These beams are intended for isotope production (in particular ^{123}I) and a short beam line is to be built in the cyclotron vault to bring them to the irradiation facility.

Target 1AT1 and M13 Slow Pion/Muon Channel

A new thin pion production target (T1) has been installed in beam line 1A. Like T2 there are five targets mounted, but all $\leq$4 g/cm^2. T1 is designed to feed the M11 fast pion channel (see below) and the M13 slow pion/muon channel. M13 takes off at 135° in imitation of M9, which has very low electron and fast neutron contamination. The channel can operate up to 130 MeV/c (50 MeV pions) and is 9.5 m long with two 60° bends in opposite directions. Experiments

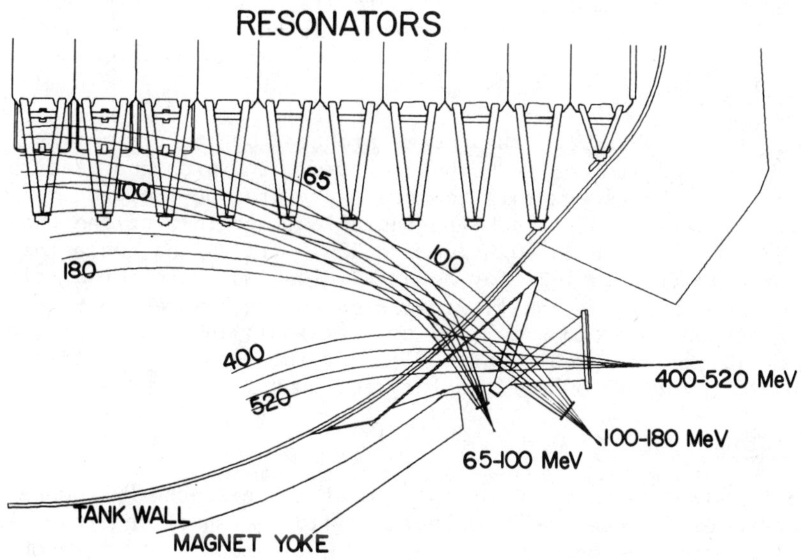

Fig. 2. Beams leaving exit horn 2.

with an α-source have shown the channel properties to be very much as predicted (solid angle acceptance 30 msr, momentum resolution 0.8% with slits closed and a small source). T1 is being commissioned with a proton beam during the week of this conference.

FACILITIES UNDER CONSTRUCTION OR DEVELOPMENT

Separated Turn Operation

To extend separated turn operation of the cyclotron from 200 MeV to 500 MeV it appears that the stability of the RF voltage is close to the required tolerance, but that the stability of the magnetic field has to be improved by about a factor 3. Third harmonic flat-topping also has to be applied to the RF accelerating system to widen the phase acceptance and increase the beam intensity to ~1 μA. Work on the RF system is proceeding satisfactorily. Separated turn operation should allow beams to be extracted with energy spreads as low as 100 keV.

M11 High Resolution Fast Pion Channel

This channel is designed to accept 30-350 MeV pions leaving the T1 target at 2.5°, and give ±0.1% momentum resolution with 6% momentum acceptance. Because of brazing problems with the septum magnet (which separates the pions from the proton beam) installation of the channel has been delayed and its design has been reconsidered. The first (60°) bending magnet is being modified to ensure a more constant field shape at different excitations. The second bending magnet will now bend 60° left for the achromatic and "normal" dispersed modes, and 30° right for the "reversed" dispersed mode. Five sextupole magnets will be used to correct second-order aberrations.

M9 Channel Extension

The cloud and surface muon beams produced by the present M9 channel have good luminosity but suffer from high pion and electron contamination. To remedy this temporarily a crossed-field dc velocity-filter separator is being installed, lengthening the channel from 8.5 m to 20 m. With 40 kV/cm and 226 G fields the contamination is expected to drop to 1-2% for negative pions and zero for electrons. For 77 MeV/c muons (μ^-) the flux expected in a 10% momentum bite is 5×10^5/sec/100 μA protons on a 10 cm Be target; the small vertical aperture (10 cm) and great length (3 m) of the separator plates unfortunately cut out 60% of the incident muons.

RF Separator

To improve the vertical acceptance of the extended M9 channel it is proposed to replace the dc velocity filter by an rf separator in 1980. This would be only 70 cm long and would consist of a resonant cavity excited at 23 MHz, the cyclotron rf and pulse repetition frequency. The separator would be located about 12 m downstream of the target so that 77 MeV/c pion and muon pulses would arrive about half a cycle apart and be deflected in opposite directions; a compensating magnetic field (~110 G) steers the muons back on target. With a 24 kV/cm fundamental field and 20% third harmonic flat-topping virtually all the muons (~10^6 μ^-/sec/100 μA) should be transmitted to the target,with <1% pion contamination.

Time Projection Chamber

To observe the rare decay modes of the pion or muon a detector is required having large solid angle, high resolution and good background rejection. The time projection chamber[7] satisfies these requirements and one is being built for installation on the extended M9 channel. The chamber itself is complete and is undergoing initial tests. It measures 80 cm long × 100 cm diameter and runs on an 80:20 argon-methane mixture at 1 atmosphere in a 150 V/cm electric field. There are 12 endcap wires for each of the six 60° sectors. The cube magnet (on loan from the University of Chicago) provides a 9.2 kG field homogeneous to 0.5% over the chamber volume. The TPC will initially be used to search for the reactions $\mu^-Z \rightarrow e^-Z$ and $\mu^-Z \rightarrow e^+(Z+2)$.

Polarized Hydrogen Target

The principal components of a 4 cm^3 polarized proton target have been purchased from the University of Liverpool. The target consists of frozen butanol and the components include the superconducting magnet, cryostat, and ^{3}He gas handling system. The target is now being reassembled for use in p-p scattering studies.

FUTURE POSSIBILITIES

A number of additional projects are under consideration. They may be classified broadly as proton beams, secondary channels, spectrometers and post-accelerators. The latter we leave for the next section. For spectrometers there are proposals for a more powerful (645 MeV/c) pion spectrometer, and for upgrading the MRS (medium resolution spectrometer). Interest in new channels has centred on proposals for high flux channels, both of muons and of negative pions for cancer therapy. Besides designs similar to those at other laboratories, Lobb[8] has proposed an axi-symmetric channel with an annular solid angle acceptance of 1 sr; pions or muons are selected with the help of an annular degrader. Any high flux channel would probably have to be fed from a new proton line. Work will start this year on installing the extraction system and combination magnet for an eventual line 2A (400-500 MeV, see Fig. 2) before the radiation levels around the cyclotron become too high. Finally there remains the possibility of extracting 100-180 MeV beams from the same exit horn; TRIUMF would then be providing beams of continuously variable energy from 70 to 520 MeV.

KAON FACTORIES

The present meson factories accelerate currents of the order of 100 μA (6×10^{14} protons/sec) to energies of 500-800 MeV. These currents are several hundred times larger than those being accelerated into the GeV range by the present generation of proton synchrotrons, whose intensities are limited by their injectors and the problems of injection. It is therefore tempting to consider the possibility of using the meson factories themselves as injectors to higher energy machines which would accelerate 10 or 100 μA to many GeV with the aim of producing kaon or even antiproton beams orders of magnitude more intense than available at present.

The uses to which intense kaon beams might be put (and some methods by which they might be produced) were discussed in 1976 at the Brookhaven Summer Study Meeting on "Kaon Physics and Facilities".[9] Intense non-strange beams would also be available—nucleons, pions, muons and neutrinos—kaon decay providing a possibly unique source of electron-neutrinos of moderate energy. Some experimental topics of particular interest are:

1) K^--N interaction - the phase shift analysis is "still very confused" and unconfirmed resonances abound.[10]
2) K^+-nucleus scattering - the K^+ is the only strongly interacting probe weakly absorbed by the nucleus.
3) Hypernuclei - intense beams permit counter experiments and spectroscopy of hypernuclear excited states.
4) (K^-,K^+) reactions - could give stable baryonic systems with strangeness -2 such as $(\Lambda\Sigma^-)$ and $^6He_{\Lambda\Lambda}$.
5) Kaonic and hyperonic atom X-rays - give information on K^- and Σ^--nucleus interactions.
6) K, Σ, Λ-decays - several possible channels involving neutral

currents have yet to be observed.
7) Neutrino-induced reactions at intermediate energies.
8) K^0 regeneration on nuclei.
9) p-p scattering - polarized beam would allow spin-dependent effects to be explored.
10) π-N and π-nuclear scattering and reactions.
11) Electromagnetic breeding of nuclear fuel by spallation neutrons - a possible practical application.

To produce intense and clean kaon beams energies well above threshold are needed. Berley[10] has analysed various kaon beams, showing that their intensities rise very strongly with incident proton energy at first, but flatten off at $\sim$7 GeV for K^+ and $\sim$9 GeV for K^-. At TRIUMF, two options are being considered to accelerate intense proton beams to 8-10 GeV—a fast-cycling synchrotron and a two-stage isochronous ring cyclotron.

PROTON SYNCHROTRON

J.R. Richardson[11] has proposed building a fast-cycling synchrotron to accelerate a 32 μA proton beam from TRIUMF to 8-10 GeV. In this case the chief design problem is in matching the time structure of the two machines, the synchrotron being pulsed at 20 Hz, while the 500 MeV cyclotron operates cw at 23 MHz. Three extraction schemes from TRIUMF have been considered:

- conventional proton extraction by stripping foil (giving minimal interference with cyclotron operation)
- resonant extraction of H^- ions at the $\nu_r = 3/2$ resonance (430 MeV); this would circumvent Liouvillean restrictions on filling the synchrotron phase space
- pulsed extraction of "100 turn stacks".

At the moment the last option appears to be the most promising, since it would provide an intermediate pulse frequency. The beam would be allowed to drift 90° out of phase at 450 MeV, after which it would begin decelerating back towards the centre (this behaviour has already been observed experimentally). The outermost 50 accelerating and 50 decelerating turns, which would be located in a 25 mm wide radial interval, could then be extracted in one bunch 0.44 μsec long by pulsing an axial electric field. The repetition period would be 22 μsec, corresponding to 14 synchrotron turns. Fast bumper magnets and rf acceleration would be used to move the equilibrium orbit away from the injection system. 200 macropulses per synchrotron cycle would be adequate to achieve an intensity of 2×10^{14} p sec^{-1} (32 μA), assuming 400 μA in the present cyclotron. To accommodate the long injection time (9% duty factor) the magnet cycle would be flat-bottomed using 60 Hz third harmonic. This machine in turn could act as injector to a high intensity 40 GeV synchrotron.

Table I. TRIUMF → fast cycling synchrotron.

Number of stacked turns - TRIUMF	~100
Energy spread of extracted beam	450 ± 2.5 MeV
Injected beam microstructure	10 each pulses of 14 nsec
macrostructure	0.4 μsec every 21 μsec
Beam loss - stacking in TRIUMF	<2%
Phase space ratio	$\sim 10^5$
Synchrotron frequency	20 Hz
Space charge limit	8×10^{14} p sec^{-1}
Phase space limit	10^{15} p sec^{-1}
Synchrotron orbit time	1.5-1.14 μsec
Macropulses per synchrotron cycle	200
Injection time	4.4 msec
Mean synchrotron radius	50 m
Average energy gain per turn	0.45 MeV
Radio frequency	0.67-0.93 MHz
Beam loss - extraction, injection and capture	<5%
Final energy	8-10 GeV

ISOCHRONOUS RING CYCLOTRONS

The alternative proposal[12] is for a two-stage isochronous ring cyclotron to accelerate protons to 8.5 GeV. (Isochronous cyclotron designs for several GeV have previously been considered by Sarkisyan[13] and Gordon.[14]) There being no incompatibility in time structure the full beam currrent from the present cyclotron (eventually 400 μA) could in principle be injected into the ring machines. The first stage of 15 sectors and 10 m radius would take a 450 MeV beam from TRIUMF to 3 GeV, the acceleration being completed by a second stage of 30 sectors and 20 m radius (Fig. 4). Superconducting magnets would

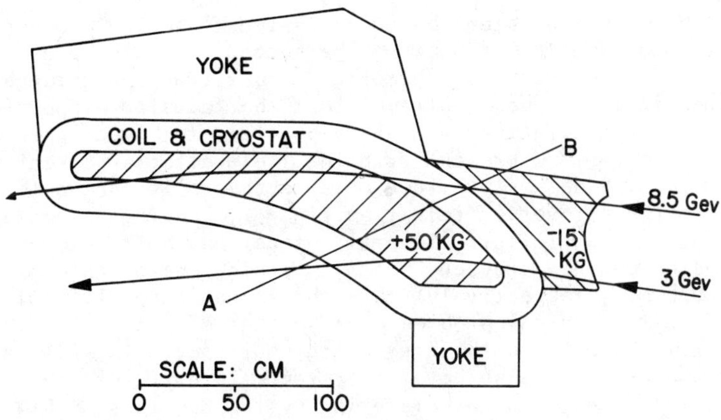

Fig. 3. Superconducting magnet design for second stage 3 to 8.5 GeV 30 sector cyclotron.

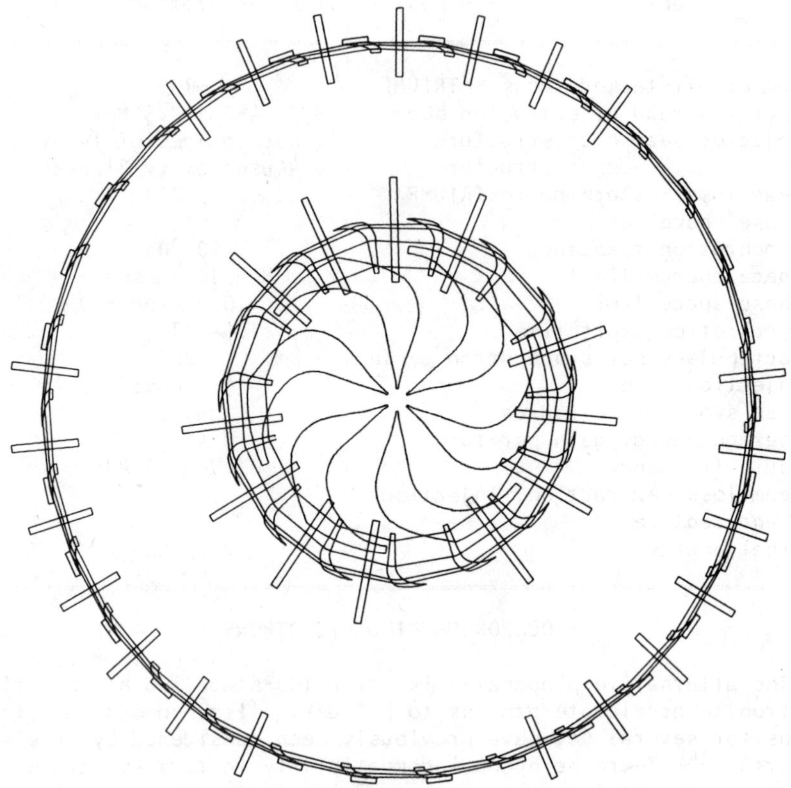

Fig. 4. A possible kaon factory design for TRIUMF, utilizing 3 GeV and 8 GeV (20.6 m radius) isochronous ring cyclotrons with superconducting magnets and SIN-style cavities (the 520 MeV cyclotron magnet poles are shown to scale for comparison).

be used, the weight of steel being estimated to be 2000 tonnes for the first stage and 1800 tonnes for the second, less in total than in the present 500 MeV machine. Numerical orbit tracking through simulated magnetic fields has confirmed that the focusing properties of the design are satisfactory and has emphasized the importance of using small pole-gaps to prevent fringing field effects weakening the edge focusing. Steel is provided outside the coils on the focusing edge to provide a reverse field (see Fig. 3), keeping the edge hard and increasing the flutter. Several integer and half-integer radial resonances have to be crossed, but with a high energy gain per turn (3 MeV and 8 MeV, respectively) this should cause no difficulty. The most difficult technical problem will be that of extracting the beam efficiently. Here again the high energy gain per turn will be important; resonances will also be of assistance in exciting coherent radial oscillations. The accelerating system consists of SIN-style cavities, with flat-topping provided by operating some at the second harmonic (first stage) or third harmonic (second stage). The phase compression effect is also utilized to allow higher fundamental frequencies (and hence smaller cavities) to be used on successive stages.

Table II. Ring cyclotron kaon factory.

	First Stage	Second Stage
Injection energy (MeV)	450	3000
Extraction energy (MeV)	3000	8500
Cyclotron radius c/ω_p (m)	10.3	20.6
Number of sectors	15	30
Primary cavities	8 at 46 MHz	15 at 69 MHz
Harmonic cavities	4 at 92 MHz	6 at 207 MHz
Cavity dimensions: primary (m^2)	5.9 × 3.6	4 × 2.6
secondary (m^2)	5.9 × 1.6	4 × 1.5
Total rf power (MW)	2.0	1.7
Energy gain/turn: (MeV) injection	1.2	7.9
extraction	3.6	7.9
$\Delta E/\Delta r$ (MeV/mm): injection	0.23	1.9
extraction	5.6	30
Radius gain/turn (mm): injection	5.3	4.2
extraction	0.64	0.26
Magnet weight (tonnes)	2000	1800
Approx. number of turns	900	700

Both designs are very much in their preliminary stages; during the coming year they will have to be looked at in more depth and the feasibility assessed of submitting a formal proposal for one or the other.

The author thanks all his colleagues at TRIUMF and the participating universities whose work he has had the privilege of reporting.

REFERENCES

1. K.L. Erdman, IEEE Trans. NS-26(2), B-2 (1979).
2. E.W. Blackmore et al., IEEE Trans. NS-26(2), F-4 (1979).
3. J.L. Beveridge et al., IEEE Trans. NS-26(3), G-19 (1979).
4. C. Amsler et al., Nucl. Instr. & Meth. 144, 401 (1977); C. Amsler et al., Nucl. Instr. & Meth. 157, 203 (1978).
5. E.W. Blackmore et al., IEEE Trans. NS-26(3), G-20 (1979).
6. G. Jones, invited paper at this conference.
7. D.R. Nygren, LBL internal report, February 1974.
8. D.E. Lobb, IEEE Trans. NS-26(2), B-17 (1979).
9. Proc. Summer Study Meeting on Kaon Physics & Facilities, Brookhaven, June 1976, (ed. H. Palevsky) BNL-50579.
10. Particle Data Group, Review of Particle Properties, 1978, LBL-100.
11. J.R. Richardson, IEEE Trans. NS-26(2), G-6 (1979).
12. M.K. Craddock, C.J. Kost, J.R. Richardson, ibid, C-6 (1979).
13. L.A. Sarkisyan, Proc. 2nd All-Union Conf. on Particle Accelerators (Moscow, 1972) I, 33; Nucl. Instr. & Meth. 142, 393 (1977).
14. See H. Blosser, "Cyclotrons-1972", AIP Conf. Proc. 9, 16 (1972).

PHYSICS USES OF INTENSE PROTON BEAMS AT 10-20 GeV

D. E. Nagle
LASL; Los Alamos, New Mexico 87545

The so-called "Meson Factories," i.e., LAMPF SIN, TRIUMF, are accelerators designed to make beams of protons, pions, and muons with orders of magnitude more intensity than did the earlier generation of synchrocyclotrons which they replaced. These facilities by and large are fulfilling the promises made for them when they were proposed: namely, they are providing intense primary and secondary beams, are attracting a broad clientele among physicists interested in nuclei and in basic interactions of particles, and are producing an impressive flow of new experimental and theoretical results and insights, as evinced by published data, workshops, and conferences, such as this one.

Looking to the future one can ask what would be the impact of a "Kaon factory" ten years hence. Would the scientific benefits justify investing scarce resources of money and of able people in this way rather in some other way? Is such a project technically feasible? How are the energy and current to be specified?

Many studies have been done to outline interesting new physics in the 10-20 GeV region. Examples are the 1976 BNL Summer Study on Kaon Physics, the 1977 Cern Workshop on Intermediate Energy Physics, conferences on polarized beams and targets at the Argonne Laboratory in 1975 and 1978, the 1978 BNL workshop on the AGS Fixed Target programs. A series of working seminars on physics in this energy region is being held at LAMPF, spring 1979. Some of the thoughts expressed in this paper come out of that seminar.

Some specific research fields are listed in Table I. These are active today, and one must ask where we might be after the next decade of work with existing facilities. The remarks below obviously are not intended as a scholarly assessment of the status of these fields, but only to indicate where emphasis might be placed in developing a detailed justification for a kaon factory. More detailed information is in the references.

TABLE I

PHYSICS AT 10-20 GeV

1. Physics with Polarized Beams and Polarized Targets
2. Pion Scattering
3. Exotic Hadronic States
4. Hypernuclear Physics
5. Kaonic Atomic X-rays
6. Nuclear Scattering with Kaons
7. Nuclear Spectroscopy with Kaons
8. Neutral Kaon Physics
9. Rare Decays of the Kaon
10. Antiproton Physics
11. Neutrino-Electron Scattering
12. Hadronic Atoms

ISSN:0094-243X/79/540760-05$1.50

At present we are seeing a very rapid development of fundamental theories of the strong interactions, the unified weak and electromagnetic interactions, and various attempts at more comprehensive unified theories, including gravity. We may expect that in ten years time this development will have made very substantial progress. Some compare this epoch to that of the 1920's when quantum mechanics was developed. For the strong interactions we now have quantum chromodynamics (QCD), still inchoate but with successes in the description of deep inelastic lepton-proton scattering in terms of moments of distributions of the constituents of baryon (quarks and gluons). At lower energies we have quark bag models, of the baryons and perhaps of the nucleus. We may hope for a better understanding of the quark masses, bag radius, perhaps bag thickness. The further study of known resonances in meson-baryon scattering, search for exotic states, etc., can be pursued in this connection.

Concerning the unified weak and electromagnetic interactions, any future theory must preserve the remarkable successes of the Weinberg-Salam model (W-S); i.e., it might imbed SU(2) x U(1) in an expanded group. Within the present "standard" W-S model, although θ_W is well determined, one needs to know the mass of the W, and Z, and of the Higgs bosons. Presumably the next generation of superaccelerators will try to address the question of M_W and M_Z, but maybe via studies of kaon rare decays the kaon factories will say something more about the mass of any Higgs which carries lepton flavor violation. Indeed, muon rare decays are being vigorously pursued at the pion factories with this objective in mind.

Although very elaborate and beautiful experiments have been done on K_S^0 - K_L^0 physics, it seems likely that in ten years, with improved instrumentation and intensity, one should further pursue this most important field, the only one to date where CP violation is manifested. An experiment is now underway at Brookhaven to test CP invariance in $k\mu3$ decays to a level of parts per thousand. Weak decays of polarized Σ^+ have also been discussed by R. K. Adair, as a place to search for a CP violating amplitude. Much better limits can obviously be placed on such amplitudes with better beams.

Generally speaking, the rapid progress in our theoretical understanding of fundamental physics will increase the interest in much better fundamental experiments in this energy region.

Nuclear structure as revealed by scattering of medium energy particles from nuclei was one of the major justifications for meson factories. Indeed, from the meson factories there is coming a wealth of information about nuclear scattering by pions, protons, and neutrons. The qualitative change from prefactory times is that we are producing <u>routinely</u> quantities of high resolution spectra, thanks to HRS, the high resolution proton spectrometer, and to EPICS, the high resolution pion spectrometer, and the other spectrometers at LAMPF, and at TRIUMF, and at SIN (such as SUSI). Pion physicists are now acquiring routinely elastic and inelastic spectra for π^+ and π^- at a wide variety of angles, energies, and nuclei, thus delineating new features of these nuclei. The π^0 spectra from charge exchange scattering of pions on nuclei are now being developed, as are nucleon charge exchange reactions. This whole field of so-called medium energy physics, now being so brilliantly exploited, is being detailed in numerous conferences, including this one.

K^+ elastic scattering will be an interesting probe of the nucleus. More massive than the pion, the kaon has a much shorter de Broglie wavelength, but it is more weakly absorbed than is the pion near resonance. The mean free path of the K^+ in nuclear matter is $\sim$ 6 fm. So it better probes the interior of the nucleus than the pion. K^+ scattering resembles electron scattering in this respect, but differs in being sensitive to the neutron distribution, and in the K having spin zero.

The K^-N system forms the so-called Y^* resonances which may propagate in nuclei in ways analogous to the Δ resonance in π^- nucleus scattering. The isobar-doorway model of π^- nucleus scattering may have an extension to K^--nucleus scattering. (See for example, M. Dillig, Proceedings of this conference.)

The $(K^-\pi^-)$ reaction in nuclei can lead to Λ or Σ hypernuclei via strangeness exchange to a bound nucleon. The hypernuclei are a new form of matter whose study will pay big intellectual dividends to our discipline. This is indeed nuclear physics, but of a new kind. Hypernuclear physics has been discussed in detail at yesterday afternoon's session of this conference. Very beautiful results were presented, but it was clear that large increases in beam intensities would be especially valuable in this sector.

Antiproton-nucleus and antiproton-nucleon interactions are extremely interesting. C. Dover[1] and D. Wolf[2] give recent reports. In the quark model the $p\bar{p}$ potential should be strongly attractive, leading to low lying bound states (baryonium). Experiments are underway at CERN and BNL to look for such low lying states of $p\bar{p}$.

The $\bar{p}$ will be very strongly absorbed in nuclei, making the scattering sensitive only to the tail of the nuclear density distribution. The protonium atom, analogous to positronium, antiprotonic atoms, low energy $\bar{p}$-nucleus scattering, and modes of nuclear breakup following $\bar{p}$ absorption are some fields to be studied.

On the technical side, the recent successes at Novosibirsk and at CERN with electron cooling of p beams allows us to anticipate getting zero energy, cooled $\bar{p}$ beams of higher stopping density. Such beams will greatly improve $p\bar{p}$ and $\bar{p}$ nucleus experiments.

The answer to the accelerator feasibility question is yes. There are doubtless several viable solutions. There was a proposal in 1956 to build a high-current (25μA), 2 GeV, two-stage cyclotron at LASL.[3] The Mura project proposed a DC magnet with alternating gradient to reach 10 GeV with high intensity.[4] At the Bloomington 1978 conference on cyclotrons and their applications, Richardson,[6] Craddock,[7] and others presented a number of possible schemes, viz., a two-stage isochronous ring cyclotron for protons, final energy 8.5 GeV, or TRIUMF injecting into a fast-cycling synchrotron.

Another possibility would be to use the 800 MeV linac at LASL to inject into a fast-cycling synchrotron similar to the booster synchrotron at Fermilab.[5] The Fermilab booster has recently run at 0.45×10^{14}/sec, rather close to the design value. It was built 5 years ago at a cost of about $17 M. Table II lists the properties of the LAMPF linac and Table III the properties of a fast cycling ring.

TABLE II

PROPERTIES OF LAMPF

		Present	Planned
Energy:	800 MeV		
Current:	H^+	500 μA	1 mA
	H^-	∿30 μA	400 mA
		Current	rms
Emittance:	H^+	300 μA	.035 π cm mr
	H^-	"	.06 π cm mr
Rep Rate:	120 pulse/sec		
Pulse Length:	600 μs		
Polarized Beam Available:	Yes		

TABLE III

PROPERTIES OF THE SYNCHROTRON

Final Energy	16 GeV
Final Average Current	∿50 μA
Rep Rate	15 Hz
Diameter of Orbit	200 m
Magnets Laminated Iron:	
Peak Field	6.7 kG
Magnet Power	2.6 MW
RF Frequency Swing f_f/f_i	1.19

The principal advantages of LAMPF as an injector are:

1. High current: ∿100 μA av is easily supplied by the linac. The linac could continue operating as a pion factory.
2. High injection energy means small frequency swing, ∿19%.
3. H^- injection and stripping would be used. Polarized beam would be available.
4. The linac has a high brightness beam.
5. Space charge limit at injection into the synchrotron is high.

Table II gives the properties of the LAMPF linac.

The choice of 16 GeV in Table III is rather arbitrary. For kaon production of say 750 MeV/c there is a pronounced knee at about 12 GeV after which the curve flattens.[8] For antiproton production the knee is less pronounced, but 16 GeV would appear to be a reasonable compromise between adequate cross section and cost. The maximum diameter ring at LAMPF would be ∿300 m.

Very important to the success of kaon beam physics and of kaon-nuclear spectroscopy would be advances in the design of secondary beam lines. These lines should have high f number; i.e., big aperture and short length. The use of superconducting magnets is certainly indicated. New designs of separators must be developed. The experience at LAMPF with high-power targets and secondary beam lines underlines the importance of designing a target, main beam line, secondary beam line, shield and a maintenance and servicing complex, as an integrated system to achieve reliability, good access, and bright beams.

At Vancouver in August 1979, during the "Eighth International Conference on High Energy Physics and Nuclear Structure," a concurrent "Workshop on Kaon Factories" will be scheduled. Panel sessions on kaon factories will be included in a workshop on "Future Program Options in Medium Energy Physics" to be held at Los Alamos August 20-31, 1979.

We can look forward to lively discussions this year of the prospects for such facilities.

1. C. Dover, workshop on the A65 Fixed Target Program BNL50947.
2. D. Wolfe, ibid.
3. F. Ribe and D. Nagle, "Proposal for a 2 GeV High Current Cyclotron," (unpublished).
4. MURA Proposal (unpublished).
5. Y. G. Basargin, E. G. Komar, V. M. Lobashov, and I. A. Sheikedo, Dokl. Akad. Nauk. SSR, 209, 4, 819, (1973).
6. J. R. Richardson, TRI-pp-78-18.
7. M. K. Craddock et al. TRI-pp-78-20. See also Craddock, Proceedings of this conference.
8. Bunce, BNL 50874.

THE CONCEPT OF AN EASTERN MESON FACILITY

Robert T. Siegel
College of William and Mary, Williamsburg, Va. 23185

ABSTRACT

The design concept and scientific justification for an Eastern Meson Facility is described. The proposed facility would provide high intensity pion and muon beams up to 300 MeV for physical research, and would place an equal emphasis on pion therapy and physics research. The main accelerator would be a 580 MeV CW separated-sector cyclotron of the design at the Swiss Institute for Nuclear Research.

INTRODUCTION

We will describe briefly our current thinking about a pion facility which would complement LAMPF by providing high intensity secondary beams on a CW duty cycle. While the initial cost would be substantial, the sharing of capital outlay between at least two agencies and the projected low operating cost will hopefully generate interest in the physics community.

DESCRIPTION

In late 1977 and early 1978 a group of about two dozen physicists met in Williamsburg and divided among themselves the task of studying the scientific justification for an eastern meson facility, since nicknamed EMF. Their work was summarized in a preliminary EMF proposal which presented the justification for EMF from the point of view of physical science research. With that report those of us interested in building EMF became convinced that it could be justified for physical research to an extent, but at about the same time the first cost estimates showed that it would be unrealistic to expect construction funding to come entirely from the physics and chemistry budgets of the granting agencies. We were not surprised at this, because we had always felt that another pion factory would only be built if it were to be largely engaged in biomedical research, to be more specific, pion therapy. So we undertook to study the interest in pion therapy among physicians who would be accessible to EMF, and also to secure advice in preparing a proposal to obtain a large part (about half) of the construction funds specifically for pion therapy. That task has now been completed, and the finished (approx. $50M) proposal to build EMF will appear in a few days from now.

What do we propose, and why? The properties of EMF are as follows:

ISSN:0094-243X/79/540765-10$1.50

1. Proton energy - 590 MeV
2. Proton Current - 100 μA initially, 1-2 ma within 2-3 years of initial operation.
3. Physics Production Target Stations - one, with three secondary beams from this target.
4. Pion Therapy Facilities - Initially two treatment rooms, to operate alternatively, and eventually four areas. Area of pion therapy facility - approx. 20,000 sq. feet.
5. Location of EMF - old SREL building and site, Newport News,Va.

We comment on each of these items below. First, the proton energy and current are fixed by our choice of duplicating the SIN separated - sector cyclotron and in constructing their planned injector II as the preceding stage. The first stage (ahead of Injector II) is an 800 keV Cockcraft-Walton. But why do we choose for EMF 600 MeV protons and the duplication of an existing machine? Well, we think that the traditional structure in the U.S. of nuclear, intermediate, and high-energy physics has its virtues. Let me remind you that that structure had a pyramidal form, with many low-energy machines, and a decreasing number of machines of higher energy until there were only one or two of the highest energy proton machines in operation at any given time. For example, in the 1950's and 1960's there were five synchrocyclotrons simultaneously in operation, and this led to a competitive situation which was healthy for the science. And already there were multi-GeV machines in operation at this time.

The low energy nuclear physics community has been able to maintain its vigor with a fairly large number of accelerators still in operation. Even in the age of new heavy-ion machines there are serious plans for several accelerators, and apparently at least two of them will be built with ≳ 200 MeV/nucleon available over a large variety of projectiles. There will also no doubt be many existing machines at low energy continuing in operation.

In medium energy proton energy there is at present only one pion factory in operation in this country, with an operating budget over five times that for all the synchrocyclotrons operating before 1970. We think one of the important arguments for EMF is that its operating budget will be under $5M per year, and that this will result in a highly competitive rate of production of physics per dollar of operating funds invested.

But of course pion factories don't come at bargain prices these days. Indeed, after we decided that a high duty-factor machine was desirable for EMF, and had come to the conclusion that the proven reliability of the SIN machine made it the choice for EMF, we called on an independent firm (Brobeck Associates) to prepare a cost estimate for us. That was because we didn't trust ourselves to make a cost estimate because of the well-known tendency of those promoting a project to make it seem as attractive as possible by minimizing its apparent cost. (Sometimes it's hard to find out how much something cost even after it is finished!) The result is that a year ago Brobeck figured that it would cost about $35M to install a duplicate of the SIN building in the existing SREL building. We have modified his estimate to take account of the cost of the bio-medical annex, inclusion of the new SIN injector instead of the AVF cyclotron originally built, and inflation, to come up with a 1979 estimate for EMF which is $49 M, of which $26 M is being requested from HEW and $23 M from NSF.

The first figure shows how the SIN cyclotron fits into the existing SREL building. The main ring machine fits neatly into the existing cyclotron room, and the injector out on the floor, and there is one target station. (SIN has two target stations and experimental areas for the injector with perhaps 15% more building space.)

We see that the external proton beam from the main ring is split and a portion directed through a part in the South (upper) wall of the building to a pion therapy (biomedical) facility to be built outside. The conceptual design for the pion clinical facility is shown in figures 3 & 4. The idea is basically to use so called "pion concentrators" as first built at Stanford. These are superconducting "orange section" spectrometers which gather pions produced at 60° to the incident proton beam and have an effective solid angle of about one steradian. This is about 25 times greater than a conventional lens system subtends at the target, and makes it practical to do therapy with 20 μA of primary protons. There may be disadvantages to these concentrators, but within a couple of years one should be able to determine the most effective way to use them. They do illustrate one thing, namely that ingenuity (Stanford's) can lead to devices which minimize the primary beam intensity needed to provide a given flux of secondary pions or tertiary muons, and thus minimize maintenance cost and also residual activity. With the current level of concern over radiation exposure to personnel, one of the main emphasis at EMF will certainly be to develop more devices of this kind.

I have dwelt on the hardware proposed for EMF before talking about the physics which could be done there. That physics is described in the proposal, and we can point out first that since EMF is designed to complement LAMPF, there will not be large spectrometers like HRS and EPICS. One expects to specialize in low energy secondary beams of pions and muons, some of them designed for maximum fluxes for stopping particles. So there would be a superconducting muon channel, a surface muon beam, and a low energy pion beam to begin with. These would serve the kinds of experiments in which there was the greatest interest expressed by the study group, including:

1. Pion-nucleus interactions, including absorption studied by coincidence experiments (π, NN) (π,πX) for example.
2. Pion-nucleon scattering at low energies.
3. Pion decays.
4. Muon-decay parameters.
5. Gauge theory tests in rare muon decays.
6. Muon capture in hydrogen.
7. Radiative muon capture to specific final states.
8. Lepton number conservation tests.
9. Neutrino transformations and new neutrino-like particles.
10. QED studies involving the muon.
11. Muon spin rotation - μSR.
12. Nuclear chemistry with muons, pions, and protons, etc.

One sees from this list a heavy emphasis on weak interaction physics and on muons in particular. Such experiments characteristically require special set-ups which are unique to the experiment, so there is no application for large spectrometer facilities at EMF, as we have mentioned above. We do think that there will be plenty of demand for beam time at EMF, particularly from physicists on the east coast.

Of course we are thinking too of the need for a pion therapy facility on the east coast. I discuss this even at a physics meeting because in the future there will be an increasing need to justify physics research facilities by their applied research potential as well as their basic science research potential. It is natural enough for society to want to see some practical outcome from the investment in basic research. We who do research because we like it are a privileged group, and the argument that eventually someone will put our discoveries and measurements to use has not really satisfied either the Congress or the public. So we should feel more secure in the future of our work if there is some applied research going on at the same place where we pursue our basic studies.

As far as pion therapy is concerned, it will be necessary to expand the statistics being gathered at LAMPF if pion therapy is to become a standard modality of treatment. While the early results at LAMPF are truly spectacular, many more patients must be treated before the results convince the medical community. There is a large patient population on the east coast which can be served by EMF, and there is reason to believe that at least 15-20% of all malignancies now treated by conventional radiations would be more effectively treated by pions. That represents a strong incentive to provide superior therapy facilities at EMF, and the preliminary design has attempted to do this.

Since this talk is between two from representatives of operating pion factories, and since the subject of kaon-antiproton factories is already under consideration at those two labs, I must make the obligatory remarks to the effect that a next step for high intensity machine design could well be for 100 μA of protons at 20 GeV, and that EMF would produce a high-quality 600 MeV proton beam for injection into the $k\bar{p}$ machine. We have done nothing serious yet on such $k\bar{p}$ design, however.

We think that the idea of an Eastern Meson Facility deserves serious consideration by the physics community.

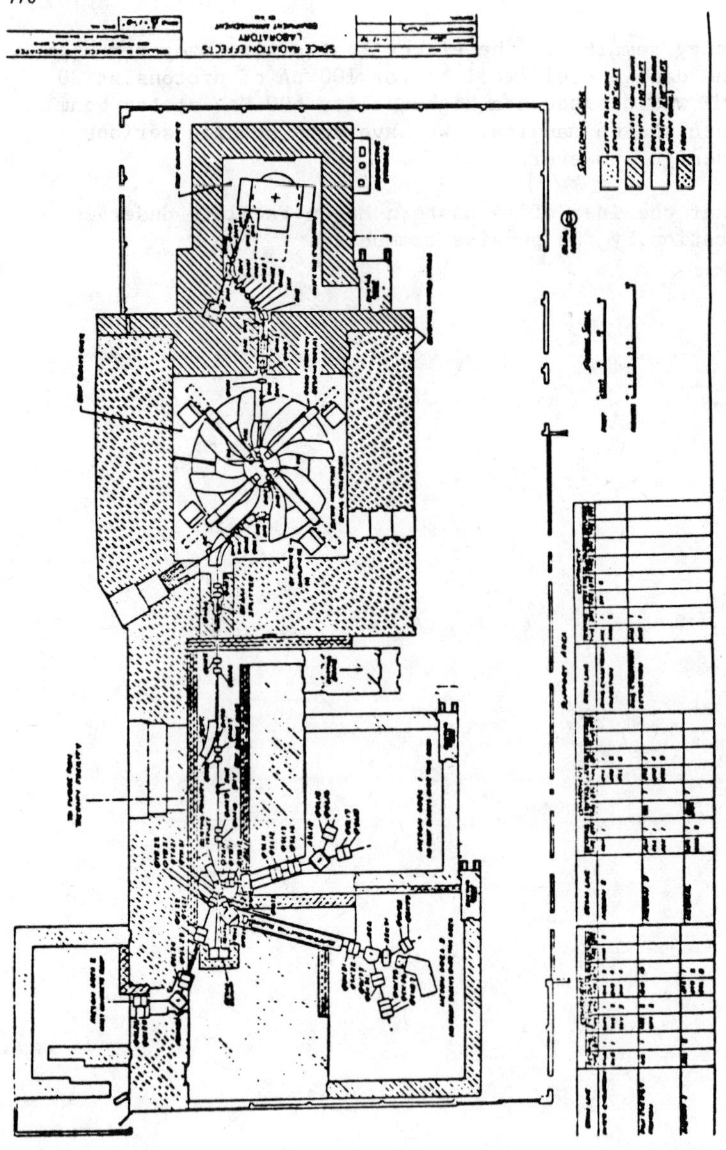

Fig. 1. Proposed Layout of Accelerator and Physics Research Area at EMF.

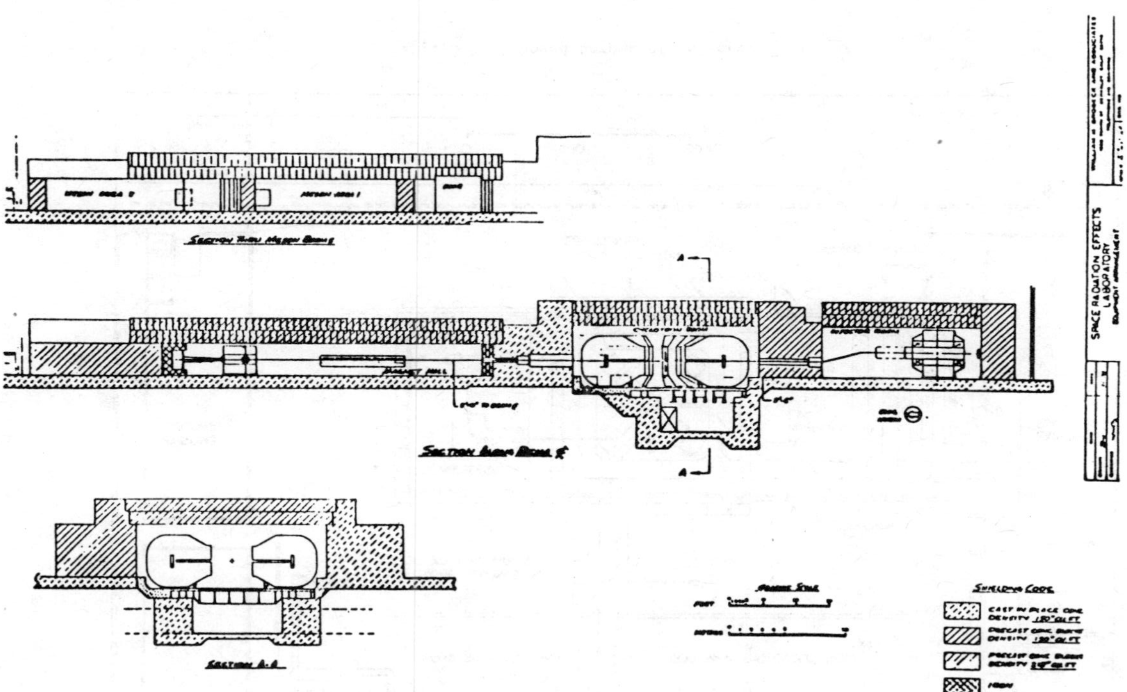

Fig. 2. Elevations (Sections) of the Main EMF Building.

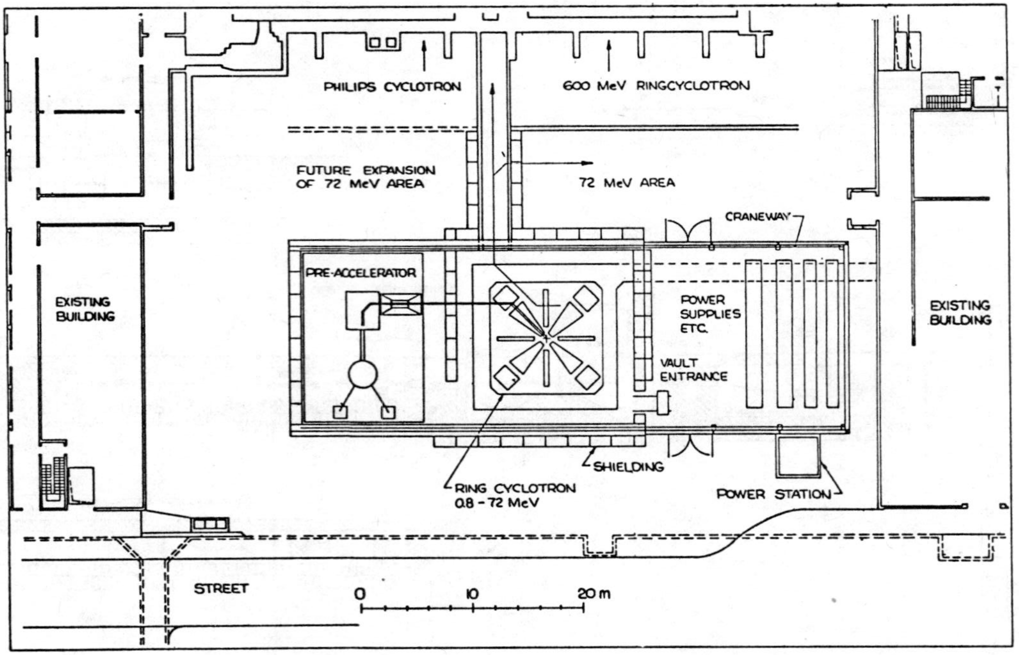

Fig. 3. Proposed Layout of Injector II at SIN.

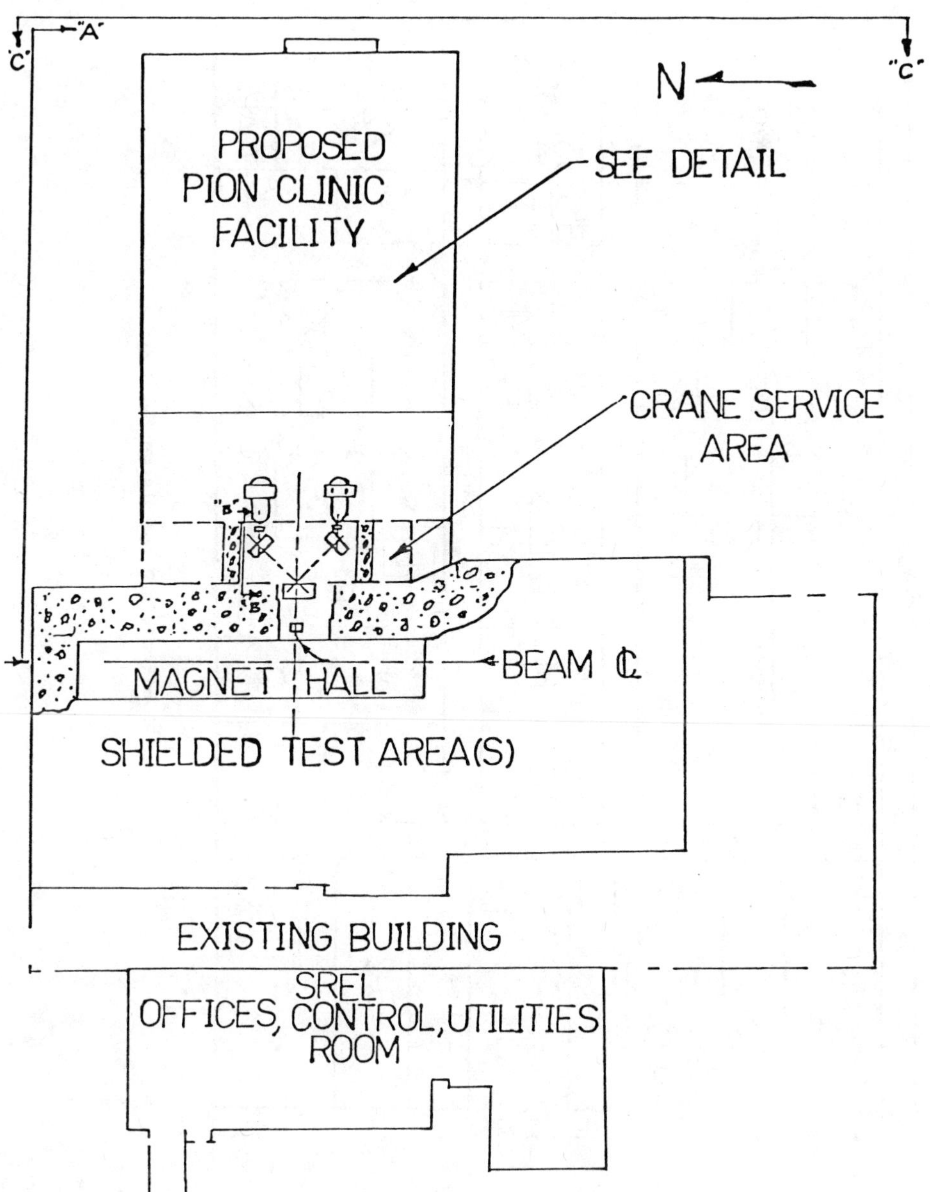

Fig. 4. Proposed Layout of Pior Therapy Facility at EMF.

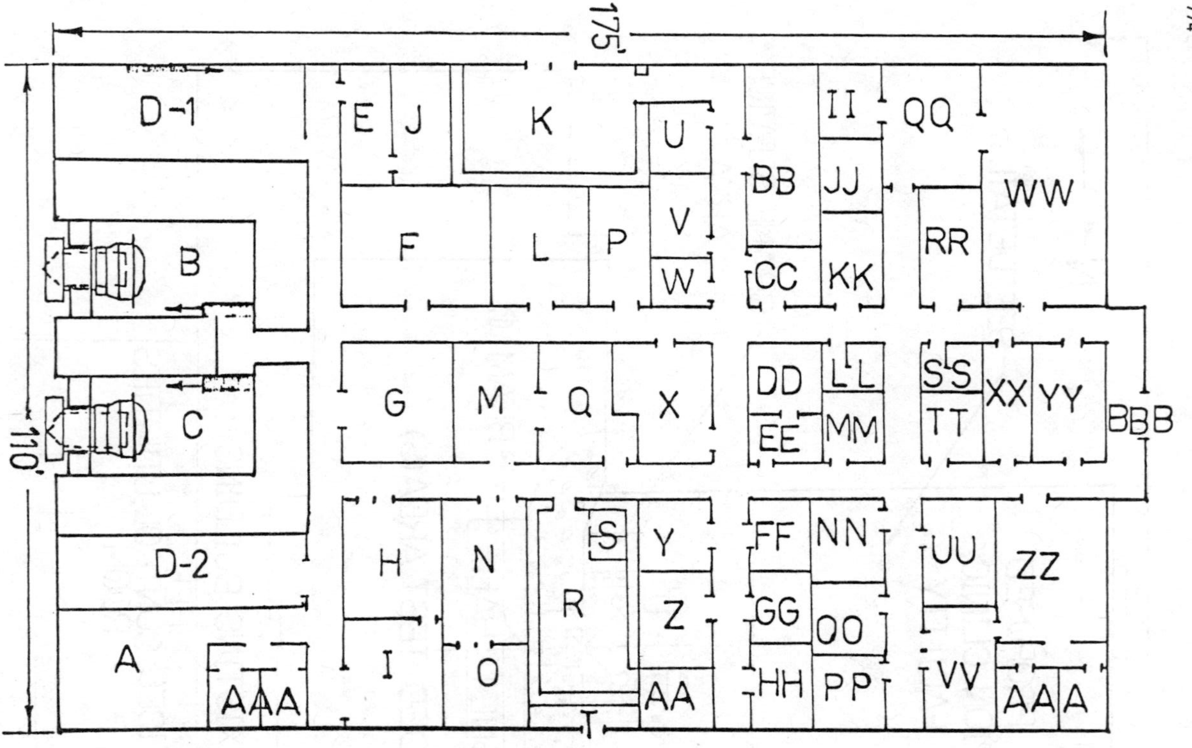

Fig. 5. Floor Plan of EMF Pion Therapy Facility.

AUTHOR INDEX

CONFERENCE PARTICIPANTS

D. AGASSI, Univ. of Rochester, Rochester, NY, USA
J. ALLRED, Los Alamos Scientific Lab., Los Alamos, NM, USA
J. ALSTER, TRIUMF, Univ. of British Columbia, Vancouver, Canada
K. AMOS, Univ. of Melbourne, Parkville, Victoria, Australia
R. ANDERSON, Los Alamos Scientific Lab., Los Alamos, NM, USA
D. ASHERY, Univ. of Tel Aviv, Tel Aviv, Israel
E. AUERBACH, Brookhaven National Lab., Upton, NY, USA
E. AULD, Univ. of British Columbia, Vancouver, Canada
S. BAKER, Rice University, Houston, TX, USA
B. BAKKER, Nat. Lab. V.U., Amsterdam, Netherlands
M. BANERJEE, Univ. of Maryland, College Park, MA, USA
B. BASSALLECK, Univ. of British Columbia, Vancouver, Canada
M. BAWIN, Univ. de Liege, Liege, Belgium
M. BEDJIDIAN, IPN de Lyon, Villeurbanne, France
G. BEER, Univ. of British Columbia, Vancouver, Canada
R. BENT, Indiana University, Bloomington, IN, USA
P. BERTIN, Lab. Physique Corpusculaire, Aubiere, France
R. BERTINI, CERN, Geneve, Switzerland
F. BERTRAND, Oak Ridge National Lab., Oak Ridge, TN, USA
M. BETZ, Argonne Nat. Lab., Argonne, IL, USA
B. BHAKAR, Univ. of Manitoba, Winnipeg, Canada
R. BHALERAD, Brooklyn College of CUNY, Brooklyn, NY, USA
M. BLECHER, VPI & SU, Blacksburg, VA, USA
J. BOLGER, Los Alamos Scientific Lab., Los Alamos, NM, USA
B. BONNER, Los Alamos Scientific Lab., Los Alamos, NM, USA
Y. LE BORNEC, IPN, Orsay, France
N. DE BOTTEN, CEN Saclay, Gif-sur-Yvette, France
R. BOUDRIE, Los Alamos Scientific Lab., Los Alamos, NM, USA
W. BRAITHWAITE, Univ. of Texas, Austin, TX, USA
T. BRESSANI, Istituti di Fisica, Cagliari, Italy
W. BRISCOE, Univ. of California, Los Angeles, CA, USA
R. BUCHANAN, Los Alamos Scientific Lab., Los Alamos, NM, USA
B. BUDICK, New York University, New York, NY, USA
G. BURLESON, New Mexico State Univ., Las Cruces, NM, USA
J. CAMMARATA, VPI & SU, Blacksburg, VA, USA
A. CASS, Univ. of Melbourne, Parkville, Victoria, Australia
H. CATZ, CEN Saclay, Gif-sur-Yvette, France
C. CERNIGOI, Istutu di Fisica, Univ. Trieste, Italy
J. CHAI, Michigan State Univ., EAst Lansgin, MI, USA
M. COOPER, Los Alamos Scientific Lab., Los Alamos, NM, USA
M. CRADDOCK, Univ. of British Columbia, Vancouver, Canada
B. CRAFT, Massachusetts Inst. of Tech., Cambridge, MA, USA
H. CRANNELL, Catholic Univ., Washington, D.C., USA
K. CROWE, Lawrence Berkeley Lab., Berkeley, CA, USA
F. CVERNA, Los Alamos Scientific Lab., Los Alamos, NM, USA
R. DALITZ, Univ. of Oxford, Oxford, U.K.
A. DELOFF, Univ. of Guelph, Guelph, Ontario, Canada

P. DEUTCHMAN, Univ. of Idaho, Moscow, ID, USA
J. DEUTSCH, Univ. de Louvain, Louvain-la-Neuve, Belgium
M. DILLIG, Univ. of Erlangen-Nurnberg, Erlangen, Germany
J. DOMINGO, SIN, Villigen, Switzerland
W. DONNELY, Stanford University, Stanford, CA, USA
C. DOVER, Univ. of Pennsylvania, Philadelphia, PA, USA
J. DUBACH, Los Alamos Scientific Lab., Los Alamos, NM, USA
I. DUCK, Rice University, Houston, TX, USA
O. DUMBRAJS, IPN, Orsay, France
S. DYTMAN, Carnegie-Mellon Univ., Pittsburgh, PA, USA
J. EGGER, Institute de Physique, Neuchatel, Switzerland
J. EISENBERG, Univ. of Tel Aviv, Tel Aviv, Israel
R. EISENSTEIN, Carnegie-Mellon Univ., Pittsburgh, PA, USA
R. ELLIS, Los Alamos Scientific Lab., Los Alamos, NM, USA
G. EPSTEIN, Massachusetts Inst. of Tech., Cambridge, MA, USA
D. ERNST, Texas A & M Univ., College Station, TX, USA
H. FEARING, TRIUMF, Univ. of Brisith Columbia, Vancouver, Canada
D. FITZGERALD, Los Alamos Scientific Lab., Los Alamos, NM, USA
J. FRANK, Los Alamos Scientific Lab., Los Alamos, NM, USA
E. FRIEDMAN, Hebrew University, Jerusalem, Israel
M. FURIC, Rice University, Houston, TX, USA
A. GAL, Hebrew University, Jerusalem, Israel
M. GALLIO, INFN Serione di Torino, Torino, Italy
R. GARFAGNINI, Istututo di Fisica Universita, Torino, Italy
D. GESSAMAN, Argonne National Lab., Argonne, IL, USA
J. GERMOND, Institut de Physique, Neuchatel, Switzerland
B. GIBSON, Los Alamos Scientific Lab., Los Alamos, NM, USA
D. GIEBINK, Los Alamos Scientific Lab., Los Alamos, NM, USA
P. GLODIS, Univ. of California, Los Angeles, CA, USA
K. GOTOW, VPI & SU, Blacksburg, VA, USA
J. GREBEN, TRIUMF, Univ. of British Columbia, Vancouver, Canada
P. GRETILLAT, Institut de Physique, Neuchatel, Switzerland
E. GROSS, Oak Ridge National Lab., Oak Ridge, TN, USA
C. GUARALDO, Lab., Naziouali di Frascati, Frascati, Italy
M. GUSAKOV, IPN, de Lyon, Villeurbanne, France
M. GYULASSY, Lawrence Berkeley Lab., Berkeley, CA, USA
R. HACKMAN, VPI & SU, Blacksburg, VA, USA
M. HAMM, Los Alamos Scientific Lab., Los Alamos, NM, USA
E. HAMMEL, Inst. f. Radiumforschung & Kernphysik, Vienna, Austria
M. HASINOFF, Univ. of British Columbia, Vancouver Canada
T. HENNINO, IPN, Orsay, France
W. HERMANS, IKO, Amsterdam, Netherlands
S. HOATH, Rutherford Lab., Chilton Didcot Oxon, England
H. HOFMANN, Univ. of Erlangen, Erlangen, Germany
J. HOFTIEZER, Rice University, Houston, TX, USA
B. HOISTAD, Los Alamos Scientific Lab., Los Alamos, NM, USA
E. HUNGERFORD, Univ. of Houston, Houston, TX, USA
C. INGRAM, SIN, Villigen, Switzerland
G. IGO, Univ. of Calif., Los Angeles, CA, USA
H. JACKSON, Argonne National Lab., Argonne, IL, USA

R. JEPPESEN, Univ. of Montana, Missoula, MT, USA
R. JOHNSON, Univ. of British Columbia, Vancouver, Canada
G. JONES, Univ. of British Columbia, Vancouver, Canada
D. JUDD, Rice University, Houston, TX, USA
J. KALLNE, Univ. of Virginia, Charlottesville, VA, USA
A. KANOFSKY, Leigh Univ. Bethlehem, PA, USA
K. KILIAN, CERN Geneve, Switzerland
L. KISSLINGER, Carnegie-Mellon Univ., Pittsburgh, PA, USA
D. KOLTUN, Univ. of Rochester, Rochester, NY, USA
R. KOSHEL, Univ. of Maryland, College Park, MD, USA
K. KOWALSKI, Case Western Reserve Univ., Cleveland, OH, USA
M. KRELL, Univ. de Sherbrooke, Sherbrooke, Canada
R. LANDAU, Oregon State Univ., Corvallis, OR, USA
E. LOMON, Massachusetts Inst. of Tech., Cambridge, MA, USA
V. LIND, Utah State Univ., Logan, UT, USA
L. LIU, Brooklyn College of CUNY, Brooklyn, NY, USA
C. LUNKE, Inst. de Physique, Neuchatel, Switzerland
W. MADIGAN, Rice University, Houston, TX, USA
E. MATHIE, Univ. of British Columbia, Vancouver Canada
J. MATTHEWS, Massachusetts Inst. of Tech., Cambridge, MA, USA
R. MARRS, TRIUMF, Univ. of British Columbia, Vancouver, Canada
M. MAY, Brookhaven National Laboratory, Upton, NY, USA
B. MAYES, Univ. of Houston, Houston, TX, USA
R. MC ADAMS, Utah State Univ., Logan, UT, USA
R. MC LEOD, Texas A & M, College Station, TX, USA
H. MC MANUS, Michigan State Univ., East Lansing, MI, USA
G. MILLER, Univ. of Washington, Seattle, WA, USA
J. MILLER, CEN Saclay, Gif-sur-Yvette, France
E. MONIZ, Massachusetts Inst. of Tech., Cambridge, MA, USA
C. MOORE, Univ. of Texas, Austin, TX, USA
C. MORRIS, Univ. of Texas, Austin, TX, USA
G. MUTCHLER, Rice University, Houston, TX, USA
D. NAGLE, Los Alamos Scientific Lab., Los Alamos, NM, USA
G. NIXON, Massachusetts Inst. of Tech., Cambridge, MA, USA
F. OBENSHAIN, Oak Ridge National Lab., Oak Ridge, TN, USA
J. O'CONNELL, National Bureau of Standards, Washington, D.C., USA
E. OSET, Univ. of Barcelona, Barcelona, Spain
O. OTTESON, Utah State Univ., Logan, UT, USA
M. PAUL, Argonne National Lab. Argonne, IL, USA
G. PEPIN, Rice University, Houston, TX, USA
C. PERDRISAT, College of William & Mary, Williamsburg, VA, USA
R. PETERSON, National Science Foundation, Washington, D.C., USA
G. PHILLIPS, Rice University, Houston, TX, USA
L. PINSKY, Univ. of Houston, Houston, TX, USA
G. PIRAGINO, Ist. di Fisica Universita, Torino, Italy
B. PREEDOM, Univ. of South Carolina, Columbia, SC, USA
K. RAO, Rensselaer Polytechnic Inst., Troy, NY, USA
R. REDWINE, Massachusetts Inst. of Tech., Cambridge, MA, USA
M. REINER, University of Rochester, Rochester, NY, USA
A. REJ, Fysisk Inst, Univ. Trondheim, Trondheim, Norway

A. RINAT, Weizmann Inst. of Science, Rehovet, Israel
R. ROCKMORE, Rutgers Univ., New Brunswick, NJ, USA
L. ROSEN, Los Alamos Scientific Lab., Los Alamos, NM, USA
A. ROSENTHAL, Univ. of Pittsburgh, Pittsburgh, PA, USA
M. SADLER, Los Alamos Scientific Lab., Los Alamos, NM, USA
A. SAHARIA, TRIUMF, Univ. of British Columbia, Vancouver, Canada
M. SOLOMON, Univ. of British Columbia, Vancouver, Canada
B. SAGHAI, CEN Saclay, Gif-sur-Yvette, France
W. SAPP, Massachusetts Inst. of Technology, Cambridge, MA, USA
F. SCHLEPUETZ, Univ. Zuerich, Zuerich, Switzerland
D. SCHNEIDER, Univ. of Rochester, Rochester, NY, USA
E. SCHWARZ, Institut de Physique, Neuchatel, Switzerland
R. SEKI, California State University, Northridge, CA, USA
K. SETH, Northwestern Univ., Evanston, IL, USA
C. SHAKIN, Brooklyn College, Brooklyn, NY, USA
R. SIEGEL, College of William & Mary, Williamsburg, VA, USA
D. SOBER, Catholic University, Washington, D.C., USA
A. STANOVNIK, Inst. Jozef Stefan, Ljubljana, Yugoslavia
G. STEPHENSON, Los Alamos Scientific Lab., Los Alamos, NM, USA
M. STERNHEIM, Univ. of Massachusetts, Amherst, MA, USA
F. TABAKIN, Univ. of Pittsburgh, Pittsburgh, PA, USA
F. TAKEUTCHI, Carnegie-Mellon Univ., Pittsburgh, PA, USA
H. THIESSEN, Los Alamos Scientific Laboratory, Los Alamos, NM, USA
A. THOMAS, TRIUMF, Univ. of British Columbia, Vancouver, Canada
M. TSANGARIDES, Indiana Univ., Bloomington, IN, USA
S. TURPIN, Rice University, Houston, TX, USA
H. ULLRICH, Univ. Karlsruhe, Karlsruhe, Germany
Q. USMANI, Aligarh Muslim Univ., Aligarh, India
P. VERNIN, CEN Saclay, Gif-sur-Yvette, France
B. VER WEST, Texas A & M Univ., College Station, TX, USA
G. WALKER, Indiana Univ., Bloomington, IN, USA
S. WALLACE, University of Maryland, College Park, MD, USA
W. WEISE, Univ. of Regensburg, Regensburg, Germany
D. WEISS, Los Alamos Scientific Lab., Los Alamos, NM, USA
W. WHARTON, Carnegie-Mellon Univ., Pittsburgh, PA, USA
C. WILKIN, University College, London, England
T. WILLIAMS, Univ. of Houston. Houston, TX, USA
S. WYCECH, Inst. for Nuclear Research, Warsaw, Poland
B. ZEIDMAN, Argonne National Lab., Argonne, IL USA

	AIP Conference Proceedings	L.C. Number	ISBN
No.27	Topics in Statistical Mechanics and Biophysics: A Memorial to Julius L. Jackson (Wayne State University, 1975)	75-36309	0-88318-126-6
No.28	Physics and Our World: A Symposium in Honor of Victor F. Weisskopf (M.I.T., 1974)	76-7207	0-88318-127-4
No.29	Magnetism and Magnetic Materials - 1975 (21st Annual Conference, Philadelphia)	76-10931	0-88318-128-2
No.30	Particle Searches and Discoveries - 1976 (Vanderbilt Conference)	76-19949	0-88318-129-0
No.31	Structure and Excitations of Amorphous Solids (Williamsburg, Va., 1976)	76-22279	0-88318-130-4
No.32	Materials Technology - 1975 (APS New York Meeting)	76-27967	0-88318-131-2
No.33	Meson-Nuclear Physics - 1976 (Carnegie-Mellon Conference)	76-26811	0-88318-132-0
No.34	Magnetism and Magnetic Materials - 1976 (Joint MMM-Intermag Conference, Pittsburgh)	76-47106	0-88318-133-9
No.35	High Energy Physics with Polarized Beams and Targets (Argonne, 1976)	76-50181	0-88318-134-7
No.36	Momentum Wave Functions - 1976 (Indiana University)	77-82145	0-88318-135-5
No.37	Weak Interaction Physics - 1977 (Indiana University)	77-83344	0-88318-136-3
No.38	Workshop on New Directions in Mössbauer Spectroscopy (Argonne, 1977)	77-90635	0-88318-137-1
No.39	Physics Careers, Employment and Education (Penn State, 1977)	77-94053	0-88318-138-X
No.40	Electrical Transport and Optical Properties of Inhomogeneous Media (Ohio State University, 1977)	78-54319	0-88318-139-8
No.41	Nucleon-Nucleon Interactions - 1977 (Vancouver)	78-54249	0-88318-140-1
No.42	Higher Energy Polarized Proton Beams (Ann Arbor, 1977)	78-55682	0-88318-141-X
No.43	Particles and Fields - 1977 (APS/DPF, Argonne)	78-55683	0-88318-142-8
No.44	Future Trends in Superconductive Electronics (Charlottesville, 1978)	77-9240	0-88318-143-6
No.45	New Results in High Energy Physics - 1978 (Vanderbilt Conference)	78-67196	0-88318-144-4
No.46	Topics in Nonlinear Dynamics (La Jolla Institute)	78-057870	0-88318-145-2
No.47	Clustering Aspects of Nuclear Structure and Nuclear Reactions (Winnepeg, 1978)	78-64942	0-88318-146-0
No.48	Current Trends in the Theory of Fields (Tallahassee, 1978)	78-72948	0-88318-147-9
No.49	Cosmic Rays and Particle Physics - 1978 (Bartol Conference)	79-50489	0-88318-148-7
No.50	Laser-Solid Interactions and Laser Processing - 1978 (Boston)	79-51564	0-88318-149-5
No.51	High Energy Physics with Polarized Beams and Polarized Targets (Argonne, 1978)	79-64565	0-88318-150-9
No. 52	Long-Distance Neutrino Detection - 1978 (C.L. Cowan Memorial Symposium)	79-52078	0-88318-151-7
No. 53	Modulated Structures - 1979 (Kailua Kona, Hawaii)	79-53846	0-88318-152-5
No. 54	Meson-Nuclear Physics - 1979 (Houston)	79-53978	0-88318-153-3